D0723401

REVIEWS IN MINERALOGY
AND GEOCHEMISTRY

VOLUME 55 2004

RECEIVED
MAY 18 2004
UTSA Library Serials

GEOCHEMISTRY OF NON-TRADITIONAL STABLE ISOTOPES

EDITORS:

Clark M. Johnson *University of Wisconsin-Madison*
Madison, WI, U.S.A.

Brian L. Beard *University of Wisconsin-Madison*
Madison, WI, U.S.A.

Francis Albarède *Ecole Normale Supérieure de Lyon*
Lyon, France

FRONT COVER: Periodic Table of the Elements (lanthanides and actinides omitted) illustrating elements that have two or more stable isotopes (light blue background), including those that have very long half-lives. Geochemistry of the stable isotopes of H, B, C, O, and S have been discussed in RiMG volumes 16, 33, and 43.

Series Editor: ***Jodi J. Rosso***

MINERALOGICAL SOCIETY OF AMERICA
GEOCHEMICAL SOCIETY

COPYRIGHT 2004

MINERALOGICAL SOCIETY OF AMERICA

The appearance of the code at the bottom of the first page of each chapter in this volume indicates the copyright owner's consent that copies of the article can be made for personal use or internal use or for the personal use or internal use of specific clients, provided the original publication is cited. The consent is given on the condition, however, that the copier pay the stated per-copy fee through the Copyright Clearance Center, Inc. for copying beyond that permitted by Sections 107 or 108 of the U.S. Copyright Law. This consent does not extend to other types of copying for general distribution, for advertising or promotional purposes, for creating new collective works, or for resale. For permission to reprint entire articles in these cases and the like, consult the Administrator of the Mineralogical Society of America as to the royalty due to the Society.

REVIEWS IN MINERALOGY AND GEOCHEMISTRY

(Formerly: REVIEWS IN MINERALOGY)

ISSN 1529-6466

Volume 55

Geochemistry of Non-Traditional Stable Isotopes

ISBN 093995067-7

Additional copies of this volume as well as others in this series may be obtained at moderate cost from:

THE MINERALOGICAL SOCIETY OF AMERICA
1015 EIGHTEENTH STREET, NW, SUITE 601
WASHINGTON, DC 20036 U.S.A.
WWW.MINSOCAM.ORG

Library
University of Texas
at San Antonio

DEDICATION

Dr. William C. Luth has had a long and distinguished career in research, education and in the government. He was a leader in experimental petrology and in training graduate students at Stanford University. His efforts at Sandia National Laboratory and at the Department of Energy's headquarters resulted in the initiation and long-term support of many of the cutting edge research projects whose results form the foundations of these short courses. Bill's broad interest in understanding fundamental geochemical processes and their applications to national problems is a continuous thread through both his university and government career. He retired in 1996, but his efforts to foster excellent basic research, and to promote the development of advanced analytical capabilities gave a unique focus to the basic research portfolio in Geosciences at the Department of Energy. He has been, and continues to be, a friend and mentor to many of us. It is appropriate to celebrate his career in education and government service with this series of courses in cutting-edge geochemistry that have particular focus on Department of Energy-related science, at a time when he can still enjoy the recognition of his contributions.

WITHDRAWN
UTSA Libraries

GEOCHEMISTRY OF NON-TRADITIONAL STABLE ISOTOPES

55 *Reviews in Mineralogy and Geochemistry* 55

FROM THE SERIES EDITOR

The review chapters in this volume were the basis for a two-day short course on non-traditional stable isotopes held prior (May 15–16, 2004) to the spring AGU/CGU Meeting in Montreal, Canada. The editors (and conveners of the short course) Clark Johnson, Brian Beard and Francis Albarède and the other chapter authors/presenters have done an exceptional job of familiarizing us with the cutting edge of this exciting field of study.

Errata (if any) can be found at the MSA website *www.minsocam.org*.

Jodi J. Rosso, Series Editor
West Richland, Washington
March 2004

FOREWORD

Until only a few years ago, I would never have imagined that a volume on the stable isotope geochemistry of elements like Mg, Fe or Cu would be written. In fact, a comic book of blank pages entitled "The Stable Isotope Geochemistry of Fluorine" would have been a more likely prospect. In volume 16 of this series, published in 1986, I wrote: *Isotopic variations have been looked for but not found for heavy elements like Cu, Sn, and Fe. … Natural variations in isotopic ratios of terrestrial materials have been reported for other light elements like Mg and K, but such variations usually turn out to be laboratory artifacts.* I am about ready to eat those words.

We have known for many years that large isotopic fractionations of heavy elements like Pb develop in the source regions of TIMS machines. Nonetheless, most of us held fast to the conventional wisdom that no significant mass-dependent isotopic fractionations were likely to occur in natural or laboratory systems for elements that are either heavy or engaged in bonds with a dominant ionic character. With the relatively recent appearance of new instrumentation like MC-ICP-MS and heroic methods development in TIMS analyses, it became possible to make very precise measurements of the isotopic ratios of some of these non-traditional elements, particularly if they comprise three or more isotopes. It was eminently reasonable to reexamine these systems in this new light. Perhaps atomic weights could be refined, or maybe there were some unexpected isotopic variations to discover. There were.

Around the turn of the present century, reports began appearing of biological fractionations of about 2–3 per mil for heavy elements like Fe and Cr and attempts were made to determine the magnitude of equilibrium isotope effects in these systems, both by experiment and semi-empirical calculations. Interest emerged in applying these effects to the study of environmental problems. Even the most recalcitrant skeptic now accepts the fact that measurable and meaningful variations in the isotopic ratios of heavy elements occur as a result of chemical, biological and physical processes.

Most of the work discussed in this volume was published after the year 2000 and thus the chapters are more like progress reports rather than reviews. Skepticism now focuses on whether isotopic variations as small as 0.1 per mil are indeed as meaningful as some think, and the fact that measured isotopic fractionations of these non-traditional elements are frequently much smaller than predicted from theoretical considerations. In fact the large fractionations suggested by the calculations provide much of the stimulus for working in this discipline. Clearly some carefully designed experiments could shed light on some of the ambiguity. My optimism for the future of this burgeoning new field remains high because it is in very good hands indeed.

James R. O'Neil

Dept. of Geological Sciences
University of Michigan
Ann Arbor, MI

PREFACE & ACKNOWLEDGMENTS

Approximately three-quarters of the elements in the Periodic Table have two or more isotopes. RiMG volumes 16 and 43 were devoted to H, C, O, and S isotope variations, and B isotope variations were discussed in RiMG volume 33. The importance of these elements to geochemistry may be illustrated by a *GeoRef* search of O isotope publications, which yields over 25,000 papers, theses, and abstracts spanning over five decades. Isotopic variations of the remaining 56 elements that have two or more isotopes, however, remains relatively little explored, but is gaining rapid attention, in part driven by advances in analytical instrumentation in the last 5–10 years. Our goal for this volume was to bring together a summary of the isotope geochemistry of *non-traditional* stable isotope systems as is known through 2003 for those elements that have been studied in some detail, and which have a variety of geochemical properties. In addition, recognizing that many of these elements are of interest to workers who are outside the traditional stable isotope fields, we felt it was important to include discussions on the broad isotopic variations that occur in the solar system, theoretical approaches to calculating isotopic fractionations, and the variety of analytical methods that are in use. We hope, therefore, that this volume proves to be useful to not only the isotope specialist, but to others who are interested in the contributions that these *non-traditional* stable isotopes may make toward understanding geochemical and biological cycles.

Many people have contributed to bringing this volume to fruition. The editors thank the authors of the individual chapters for their contributions. In addition, many scientists provided reviews of the chapters, including Francis Albarède, Ariel Anbar, Brian Beard, Tom Bullen, Don Canfield, Rick Carlson, Tom Chacko, Lui Chan, Robert Clayton, Max Coleman, Robert Criss, Susan Glasauer, Gideon Henderson, James Hogan, Clark Johnson, Mark Kurz, Peter Larson, Jean-Marc Luck, Brian Marshall, Peter Michael, Eta Mullane, Thomas Nägler, James O'Neil, Olivier Rouxel, Sara Russell, Edwin Schauble, John Valley, Ed Young, and six anonymous reviewers. Alex Speer and others at the MSA business office helped us greatly with the logistics of both the volume and the accompanying Short Course. We gratefully acknowledge the financial support provided by the U.S. Department of Energy. And finally, we are indebted to Jodi Rosso, series editor, for all of her work - she was a joy to work with, and it is through her dedication that the RiMG volumes continue to flourish.

Clark M. Johnson
Madison, Wisconsin

Brian L. Beard
Madison, Wisconsin

Francis Albarède
Lyon, France

TABLE OF CONTENTS

1 Overview and General Concepts

Clark M. Johnson, Brian L. Beard and Francis Albarède

2 An Overview of Isotopic Anomalies in Extraterrestrial Materials and Their Nucleosynthetic Heritage

Jean Louis Birck

3 Applying Stable Isotope Fractionation Theory to New Systems

Edwin A. Schauble

4 Analytical Methods for Non-Traditional Isotopes

Francis Albarède and Brian L. Beard

5 Developments in the Understanding and Application of Lithium Isotopes in the Earth and Planetary Sciences

Paul B. Tomascak

6 The Isotope Geochemistry and Cosmochemistry of Magnesium

Edward D. Young and Albert Galy

7 The Stable-Chlorine Isotope Compositions of Natural and Anthropogenic Materials

Michael A. Stewart and Arthur J. Spivack

8 Calcium Isotopic Variations Produced by Biological, Kinetic, Radiogenic and Nucleosynthetic Processes

Donald J. DePaolo

9 Mass-Dependent Fractionation of Selenium and Chromium Isotopes in Low-Temperature Environments

Thomas M. Johnson and Thomas D. Bullen

10A Fe Isotope Variations in the Modern and Ancient Earth and Other Planetary Bodies

Brian L. Beard and Clark M. Johnson

10B Isotopic Constraints on Biogeochemical Cycling of Fe

Clark M. Johnson, Brian L. Beard, Eric E. Roden
Dianne K. Newman and Kenneth H. Nealson

11 The Stable Isotope Geochemistry of Copper and Zinc

Francis Albarède

12

Molybdenum Stable Isotopes: Observations, Interpretations and Directions

Ariel D. Anbar

Reviews in Mineralogy & Geochemistry
Vol. 55, pp. 1-24, 2004
Copyright © Mineralogical Society of America

Overview and General Concepts

Clark M. Johnson, Brian L. Beard

Department of Geology and Geophysics
University of Wisconsin-Madison
1215 West Dayton Street
Madison, Wisconsin 53706, U.S.A.

Francis Albarède

Ecole Normale Supérieure de Lyon
46 Allée d'Italie
69007 Lyon, France

INTRODUCTION

Of the eighty-three naturally occurring elements that are not radioactive or have half lives long enough to be considered stable ($\geq 10^9$ yrs), nearly three-quarters have two or more isotopes. Variations in the isotopic ratios of a number of these elements, including H, C, N, O, and S, provide the foundation for the field of *stable isotope geochemistry*. Investigations of variations in the isotopic compositions of these *traditional* elements have provided important constraints on their sources in natural rocks, minerals, and fluids. These studies have focused on a range of problems including planetary geology, the origin and evolution of life, crust and mantle evolution, climate change, and the genesis of natural resources. Much less attention, however, has been paid to stable isotope variations of other elements that are also geochemically important such as certain metals and halogens. In part this has been due to analytical challenges, although first-order variations for several systems have been constrained using long-standing analytical methods such as gas- and solid-source mass spectrometry. With the advent of analytical instrumentation such as multi-collector, inductively-coupled plasma mass spectrometry (MC-ICP-MS), large portions of the Periodic Table are now accessible to stable isotope studies.

In this volume, the geochemistry of a number of *non-traditional* stable isotopes is reviewed for those elements which have been studied in some detail: Li, Mg, Cl, Ca, Cr, Fe, Cu, Zn, Se, and Mo. This volume is intended for the non-specialist and specialist alike. The volume touches on the multiple approaches that are required in developing new isotopic systems, including development of a theoretical framework for predicting possible isotopic fractionations, perfecting analytical methods, studies of natural samples, and establishment of a database of experimentally-determined isotope fractionation factors to confirm those predicted from theory. In addition to the systems discussed in this volume, we expect that more elements will be subject to isotopic studies in the next several years, significantly broadening the field that is known as *stable isotope geochemistry*.

Chapter 1 is intended to provide an overview of basic concepts of stable isotope geochemistry that are applicable to the chapters that follow on specific topics and isotopic systems. There are a number of excellent reviews of stable isotope geochemistry that have tended to focus on H, C, O, and S, including two prior volumes of Reviews in Mineralogy and

1529-6466/04/0055-0001$05.00

Geochemistry (Valley et al. 1986; Valley and Cole 2001), and a few texts (e.g., Criss 1999; Hoefs 2004). The concepts discussed in these works are entirely applicable to the isotopic systems discussed in the present volume. Because our discussion here is restricted to essential concepts, we refer the reader who is interested in more depth to the works above.

ISOTOPIC ABUNDANCES AND NOMENCLATURE

There are many sources for information on the distribution of the elements and isotopes (e.g., Lide 2003), as well as discussions that are pertinent to stable isotope geochemistry (e.g., Criss 1999). In Chapter 2 Birck (2004) reviews in detail the isotopic distribution and nucleosynthetic origin of many elements that are of geochemical interest. He highlights the fact that isotopic variations for many elements are greatest for extraterrestrial samples, where evidence for a variety of fractionation mechanisms and processes are recorded, including mass-dependent and mass-independent fractionation, radioactive decay, and incomplete mixing of presolar material. Below we briefly review a few general aspects of isotope distribution that are pertinent to isotopic studies that bear on nomenclature, expected ranges in isotopic fractionation, and analytical methodologies.

The number of stable isotopes for the naturally occurring elements tends to increase with increasing atomic number, to a maximum of 10 for Sn (Fig. 1). Elements with low atomic numbers tend to have the lowest number of stable isotopes, limiting the possible ways in which isotopic compositions can be reported. Both H and C have only two stable isotopes (Fig. 1), and therefore isotopic compositions are reported using one ratio, D/H and $^{13}C/^{12}C$, respectively. Single ratios can only be used for B and N, and data are reported as $^{11}B/^{10}B$ and $^{15}N/^{14}N$, respectively. Of the *non-traditional* stable isotope systems discussed in this volume, only three have just two stable isotopes (Li, Cl, and Cu; Fig. 1).

The choice of isotopic ratios for reporting data increases, of course, for elements with three or more isotopes. Although O isotope compositions are almost exclusively reported in terms

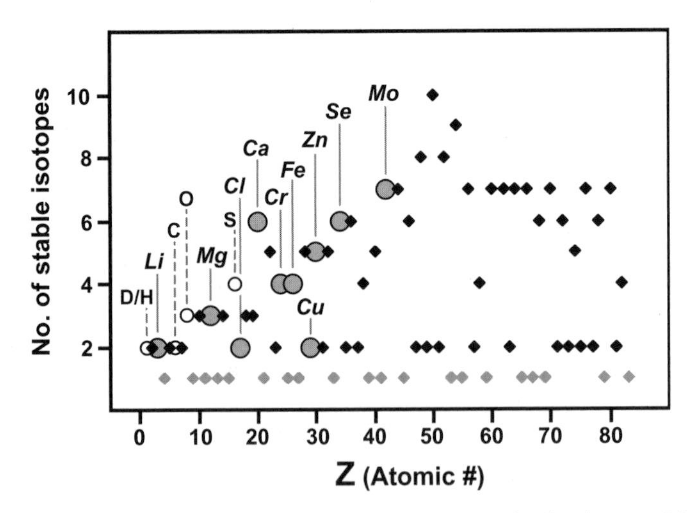

Figure 1. Number of stable isotopes relative to atomic number (Z) for the elements. Mono-isotopic elements shown in gray diamonds. Elements discussed in this volume are shown as large gray circles. Other elements that have been the major focus of prior isotopic studies are shown in small white circles, and include H, C, O, and S. Nuclides that are radioactive but have very long half-lives are also shown in the diagram.

of $^{18}O/^{16}O$ ratios, data for $^{17}O/^{16}O$ variations are reported, and are valuable in distinguishing mass-dependent and mass-independent isotopic variations (see Chapter 2; Birck 2004). The majority of S isotope data are reported using the ratio of the two most abundant isotopes, $^{34}S/^{32}S$, although studies of mass-independent isotope variations report data for three or all four stable isotopes of S (e.g., Farquhar et al. 2000). Magnesium, Ca, Cr, Fe, Zn, Se, and Mo all have three or more stable isotopes (Fig. 1), and therefore the isotopic compositions for these elements may be reported using a number of ratios. In some cases, isotopic abundances limit the isotopic ratios that can be measured with reasonable precision. For many elements, however, isotopic abundances are relatively high for a number of isotopes and, in these cases, data are reported using a number of ratios.

The vast majority of isotopic data are reported in "delta notation", where the isotopic composition is cast as the deviation of an isotopic ratio relative to the same ratio in a standard (e.g., O'Neil 1986a):

$$\delta^i E_X = \left(\frac{R_X^{i/j} - R_{STD}^{i/j}}{R_{STD}^{i/j}} \right) 10^3 \tag{1}$$

where i and j are the specific isotopes used in ratio R of element E, X is the sample of interest, and STD is a standard reference material or reservoir. It is traditional to use isotope i in the δ value, and it is important to note the specific ratio $R^{i/j}$ that is used. The units for the $\delta^i E_X$ value are in parts per thousand or "per mil", which is commonly noted using the per mil sign (‰). In the case of the $\delta^{18}O$ value, $R = {}^{18}O/{}^{16}O$, following the traditional protocol of expressing $R^{i/j}$ as the abundance ratio of the rare isotope to the major isotope (e.g., O'Neil 1986a). Similarly, δD, $\delta^{13}C$, $\delta^{15}N$ and $\delta^{34}S$ values are defined using the isotopic ratio of rare isotope over major isotope, that is, D/H, $^{13}C/^{12}C$, $^{15}N/^{14}N$, and $^{34}S/^{32}S$, respectively. The conventional definitions of $R^{i/j}$ for H, C, N, and S correspond to *heavy over light* masses, leading to a consistent nomenclature where a positive $\delta^i E_X$ value refers to a sample that is relatively enriched in the heavy isotope (a high $R^{i/j}$ ratio relative to the standard). Although some groups have reported isotopic compositions in parts per 10,000 using the ε unit (e.g., Rehkämper and Halliday 1999; Zhu et al. 2000), this nomenclature is not very common and should probably be discontinued to avoid confusion.

The standard reference used for reporting isotopic compositions of H and O is Standard Mean Ocean Water (SMOW), although the fossil marine carbonate PDB standard is also used for oxygen (O'Neil 1986a). For the new isotopic systems discussed in this volume, the choice of standard reference material or reservoirs has not been settled in many cases (Table 1), requiring careful attention when comparing $\delta^i E$ values. Some groups have chosen a major geologic reservoir such as ocean water or bulk earth, following the convention used for some light stable isotope systems, as well as several radiogenic isotope systems such as Nd, Hf, and Os. Reporting $\delta^i E$ values relative to a major geologic reservoir may be advantageous in interpretations, where relative enrichment or depletion in heavy isotopes are referenced to a geologic component. The disadvantage of using major geologic reservoirs as a reference is that they are not distinct substances that can be measured in the laboratory. In other cases, an international or other widely-available purified standard is used. For some isotopic systems, data are reported relative to in-house standards (Table 1), which may not allow easy cross-calibration with other laboratories. Data should be reported for several internationally-available standards whenever possible so that δ values may be compared using equivalent reference materials or reservoirs.

The motivation for defining the $\delta^i E_X$ value as *rare isotope over major isotope* lies in the fact that the mathematical forms of mixing relations and other physical processes are greatly simplified in cases where the rare isotope i is very low in abundance, which leads to the simplification that the abundance isotope j may be treated as invariant, particularly when the

Table 1. Nomenclature and standards for isotope systems.

Element	Ratio	Typical Precision	Standard Reference Material or Reservoir	Other Notation	Ratio	Other Standards	Isotopic Composition
δ^7Li (‰)	^{7}Li/^{6}Li	0.5-1.0‰	NIST SRM-8545	δ^6Li (‰)	^{6}Li/^{7}Li	IRMM-016	δ^7Li = 0
δ^{26}Mg (‰)	^{26}Mg/^{24}Mg	0.1-0.2‰	DSM 3	δ^{25}Mg (‰)	^{25}Mg/^{24}Mg	SRM 980_O	δ^{26}Mg = −3.41
				Δ^{26}Mg (non-mass-dependent variations)			
δ^{37}Cl (‰)	^{37}Cl/^{35}Cl	0.1-0.3‰	Std. Mean Ocean Chloride (SMOC)				
δ^{44}Ca (‰)	^{44}Ca/^{40}Ca	0.1-0.2‰	UCB Ca std.			Seawater Ca	δ^{44}Ca = +0.9
						Fluorite Ca	δ^{44}Ca = +0.5
				ε_{Ca} (^{40}Ca/^{42}Ca; radiogenic ^{40}Ca enrichment; parts per 10,000)			
δ^{53}Cr (‰)	^{53}Cr/^{52}Cr	0.1-0.2‰	NIST SRM-979				
δ^{56}Fe (‰)	^{56}Fe/^{54}Fe	0.1-0.5‰	Bulk Earth (BE)	δ^{57}Fe (‰)	^{57}Fe/^{54}Fe	IRMM-14	δ^{56}Fe = −0.09
				$\delta^{57/56}$Fe (‰)	^{57}Fe/^{56}Fe		
				ε^{57}Fe	^{57}Fe/^{54}Fe (parts per 10,000)		
				ε^{56}Fe	^{56}Fe/^{54}Fe (parts per 10,000)		
δ^{65}Cu (‰)	^{65}Cu/^{63}Cu	0.1-0.2‰	NIST SRM-976				
δ^{66}Zn (‰)	^{66}Zn/^{64}Zn	~0.1‰	NIST SRM-682	δ^{67}Zn (‰)	^{67}Zn/^{64}Zn	NIST SRM-683	δ^{66}Zn = +2.4
				δ^{68}Zn (‰)	^{68}Zn/^{64}Zn		
				ε^{66}Zn	^{66}Zn/^{64}Zn (parts per 10,000)		
				ε^{67}Zn	^{67}Zn/^{64}Zn (parts per 10,000)		
				ε^{68}Zn	^{68}Zn/^{64}Zn (parts per 10,000)		
δ^{80}Se (‰)	^{80}Se/^{76}Se	0.2-0.3‰	CDT	δ^{82}Se (‰)	^{82}Se/^{76}Se	NIST SRM-3149	δ^{80}Se = 0
δ^{97}Mo (‰)	^{97}Mo/^{95}Mo	0.1-0.2‰	Rochester JMC Mo	δ^{98}Mo (‰)	^{98}Mo/^{95}Mo	Seawater Mo	δ^{97}Mo = +1.5

Note: *Standard Reference Material or Reservoir* refers to R^{ij}_{STD} in equation 1 in the text.

range in isotopic compositions is relatively restricted. For example, the exact mixing relation for a single element between components A and B which differ in their isotopic compositions is given by:

$$\frac{M_A}{M_B} = -\left(\frac{C_B}{C_A}\right)\left(\frac{R^*_{MIX} - R^*_B}{R^*_{MIX} - R^*_A}\right) \tag{2}$$

where M_A and M_B are the masses of components A and B, respectively, C_A and C_B are the concentrations of the element in components A and B, respectively, and R^* is defined as the ratio of the mass of the rare isotope over the total mass of the element (Eqn. 1.14, p. 22-23, Criss 1999). In the case of O, where the abundances of ^{18}O and ^{16}O are 0.20% and 99.76%, respectively, R^* is very nearly equal to the $^{18}O/^{16}O$ ratio, allowing us to relate the exact mixing equation directly to the measured isotopic compositions. We may further simplify Equation (2) using δ notation as:

$$\delta^i E_{MIX} \approx \delta^i E_A f + \delta^i E_B (1 - f) \tag{3}$$

where f is the fraction of component A in the two-component mixture.

Although the simplicity of Equation (3) makes it quite useful, it is important to note that it is valid only for cases when the abundance of the isotope in the numerator of R is very low or when the difference between $\delta^i E_A$ and $\delta^i E_B$ is small. For example, use of Equation (3) for mixing two components that differ in their $^{18}O/^{16}O$ ratios by 50‰ produces an error of 0.0016‰ at 50:50 mixing (assuming equal oxygen abundances) relative to use of the exact mixing Equation (2). However, in the case of Fe isotopes, for example, where ^{56}Fe and ^{54}Fe have abundances of 91.76% and 5.84%, respectively, the assumption that ^{54}Fe is a rare isotope is not valid. Assuming that two components differ in their $^{56}Fe/^{54}Fe$ ratios by 50‰, use of the approximate Equation (3) produces an error of 0.61‰ at 50:50 mixing (assuming equal iron abundances) relative to use of the exact mixing Equation (2). If mixing relations are calculated using $^{57}Fe/^{54}Fe$ ratios, as is reported by a number of workers, isotope i is even closer in abundance to isotope j, and the error introduced using Equation (3) in the above example becomes larger, at 0.90‰. Fortunately, we will see that except for Li, isotopic variations in nature for the elements discussed in this volume are generally less than 10‰, and in most cases Equation (3) may be used without significant loss in accuracy. This may not be the case, however, for experiments that involve enriched isotope tracers.

It is not always possible to follow the convention of *rare isotope over major isotope* in the definition of $\delta^i E_X$ *and* maintain the convention *heavy over light* for elements across the Periodic Table. If we restrict R to be the ratio of the major isotope to the next-most-abundant isotope, it will be *heavy over light* for only half of the elements under consideration (Fig. 2). The combined definition of *rare over major* and *heavy over light* is satisfied for Mg, Cl, Ca, Cr, Cu, and Zn, but is *not* satisfied for Li, Fe, Se, and Mo (Fig. 2). If the *rare over major* definition of R is maintained, as has been the norm for over five decades of work in the *traditional* stable isotopes, then there will be inconsistencies in the sign of the $\delta^i E_X$ value, where, in some cases, a positive value will reflect a relative enrichment in the heavy isotope, and in others, a depletion. Given the fact that some nucleosynthetic processes produce values of R (as defined in Fig. 2) that approach unity with increasing atomic number (Fig. 2), we suggest that the isotope community should abandon the practice of *rare over major* in defining R for new isotopic systems at the intermediate- to heavy-mass range, because as R approaches unity, the simplifications of mixing and other equations become less accurate, regardless of defining isotopic ratios as *rare over major* or *major over rare*. Instead, defining R for new isotopic systems as *heavy over light* is probably preferred, because this will maintain the same convention used in *traditional* stable isotope systems, where a positive $\delta^i E_X$ value

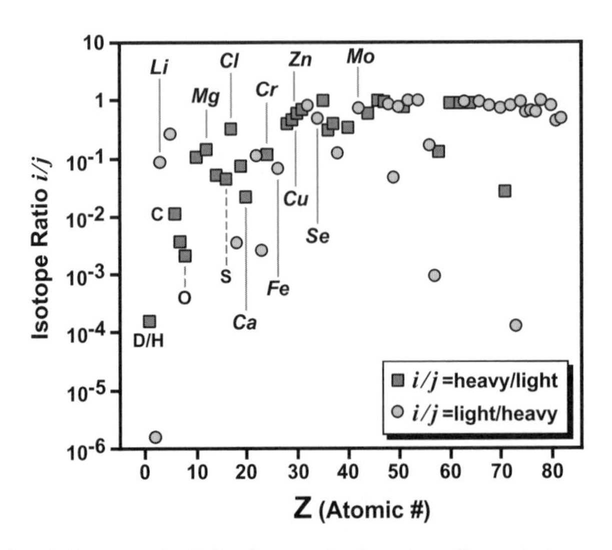

Figure 2. Variations in isotope ratio *i/j* for elements that have two of more isotopes relative to atomic number (Z), where *j* is the most abundant mass, and *i* is the next-most abundant mass. Elements where *i/j*, as defined here, reflects the ratio of a heavy mass *i* relative to that of *j* are shown in gray squares, whereas elements where *i/j* reflects a lighter mass *i* over heavy mass *j* are shown in gray circles. The lighter elements tend to have low *i/j* ratios, from ~10^{-6} to ~10^{-1}, which simplify mixing relations in terms of δ plots. In the case of elements that have been extensively studied for their isotopic variations, such as H, C, O, and S, *i/j* is heavy over light, leading to the traditional convention that a high δ value reflects a relative enrichment of the heavy isotope. The convention *i/j* does not remain consistently *heavy over light* for a number of non-traditional stable isotopes, including those discussed here.

indicates a relative enrichment in the heavy isotope relative to a standard (or, more precisely, an enrichment in the heavy/light isotopic *ratio*, as discussed above), and a negative $\delta^i E_X$ value indicates a relative depletion in the heavy isotope. The *heavy over light* convention is followed in this volume.

ISOTOPIC FRACTIONATIONS

As discussed by O'Neil (1986b), and also in Chapter 3 of this volume (Schauble 2004), isotopic fractionations between species or phases depend on a number factors, including relative mass difference, the nature of the bonding environment, and redox state. Ignoring issues related to bonding and redox state, we generally expect that the range in isotopic variations will decrease with increasing atomic number (Z) because the relative mass difference also decreases (Fig. 3), generally following the relation of $1/Z$. The relatively large mass differences for the *traditional* stable isotopes such as H, C, O, and S, ranging from 66.7% for H to 3.0% for S (calculated as $\Delta m/\bar{m}$; Fig. 3), are in part responsible for the large ranges in isotopic ratios that have been measured in natural samples for these elements, from 10's to 100's of per mil (‰). As reviewed in Chapter 5 (Tomascak 2004), of the elements discussed in this volume, Li shows the greatest range in isotopic composition, commensurate with its low mass, high relative mass difference, and bonding environment (Fig. 3). Despite relatively small mass differences on the order of ~1%, however, significant isotopic variations are seen up through Mo (Chapter 12), and, in fact, are seen for heavier elements such as Hg ($Z = 80$; Jackson 2001) and Tl ($Z = 81$; Rehkämper et al. 2002), indicating that new isotopic systems are likely to be developed across the Periodic Table as analytical precision improves, even for

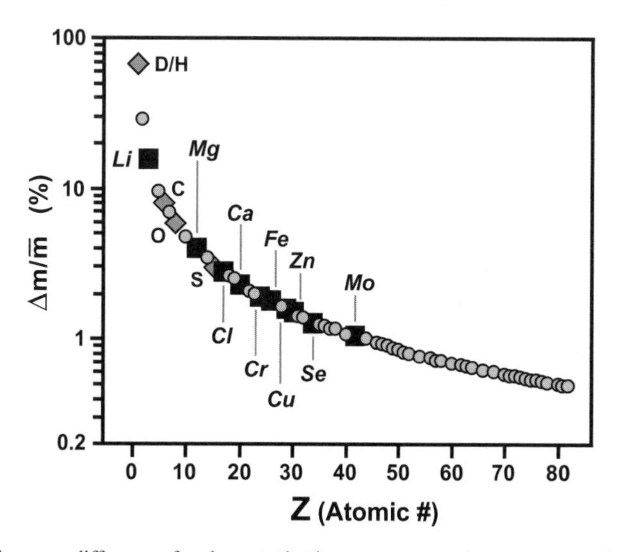

Figure 3. Relative mass differences for elements that have two or more isotopes, cast as $\Delta m/\bar{m}$, where Δm is the unit mass difference ($\Delta m = 1$), and $\bar{m}$ is the average mass of the isotopes of that element, as a function of atomic number (Z). Note that $\Delta m/\bar{m}$ is reported in percent, and is plotted on a log scale. Elements that are discussed in this volume shown in large black squares. Other elements that have been the major focus of isotopic studies shown in gray diamonds, and include H, C, O, and S. The relatively large mass differences for the light elements generally produce the largest isotopic fractionations, whereas the magnitude of isotopic fractionation is expected to markedly decrease with increasing mass.

heavy elements where the relative mass differences are only a few tenths of a percent.

The relatively small mass differences for most of the elements discussed in this volume requires very high-precision analytical methods, and these are reviewed in Chapter 4 by Albarède and Beard (2004), where it is shown that precisions of 0.05 to 0.2 per mil (‰) are attainable for many isotopic systems. Isotopic analysis may be done using a variety of mass spectrometers, including so-called *gas source* and *solid source* mass spectrometers (also referred to as *isotope ratio* and *thermal ionization* mass spectrometers, respectively), and, importantly, MC-ICP-MS. Future advancements in instrumentation will include improvement in *in situ* isotopic analyses using ion microprobes (secondary ion mass spectrometry). Even a small increase in precision is likely to be critical for isotopic analysis of the intermediate- to high-mass elements where, for example, an increase in precision from 0.2 to 0.05‰ could result in an increase in signal to noise ratio from 10 to 40.

The isotope fractionation factor and mass-dependency of fractionation

Following standard practice (e.g., O'Neil 1986a), the isotopic fractionation factor between two substances A and B is defined as:

$$\alpha_{A\text{-}B} = \frac{R_A^{i/j}}{R_B^{i/j}} \tag{4}$$

which may be cast in terms of $\delta^i E$ values as:

$$\alpha_{A\text{-}B} = \frac{1000 + \delta^i E_A}{1000 + \delta^i E_B} \tag{5}$$

Note that $\alpha_{A\text{-}B}$ simply reflects the contrast in isotopic compositions between two substances and, in terms of physical processes, could reflect equilibrium or non-equilibrium partitioning of isotopes. For an isotope exchange reaction in which one atom is exchanged, $\alpha_{A\text{-}B}$ is equal to the equilibrium constant. Because $\alpha_{A\text{-}B}$ is very close to unity, generally on the order of 1.00X, we may take advantage of the relation that $10^3\ln(1.00X) \approx X$, which provides the useful relation:

$$10^3 \ln \alpha_{A\text{-}B} \approx \delta^i E_A - \delta^i E_B \equiv \Delta_{A\text{-}B} \qquad (6)$$

This allows us to describe isotopic fractionations by simply subtracting the $\delta^i E$ values of substances A and B (carefully keeping the order of subtraction consistent). Assuming a fractionation of 10‰ ($\alpha_{A\text{-}B} = 1.010$), an error of only 0.05‰ is introduced if the fractionation is described using $\Delta_{A\text{-}B}$ as compared to $\alpha_{A\text{-}B}$, which will be immaterial for most of the elements discussed in this volume, unless the fractionations are very large (several tens of per mil or greater).

Isotopic fractionations are sometimes discussed using the ε notation, and this has been most commonly used to describe kinetic isotope effects. This usage is common in the S isotope literature (e.g., Canfield 2001), and is also used in Chapter 9 (Johnson and Bullen 2004). The fractionation between reactant A and product B is defined as:

$$\varepsilon_{A\text{-}B} = \left(\alpha_{A\text{-}B} - 1\right) 10^3 \qquad (7)$$

$\varepsilon_{A\text{-}B}$ therefore has units of per mil.

For elements that have three or more isotopes, isotopic fractionations may be defined using two or more isotopic ratios. Assuming that isotopic fractionation occurs through a mass-dependent process, the extent of fractionation will be a function of the relative mass differences of the two isotope ratios. For example, assuming a simple harmonic oscillator for molecular motion, the isotopic fractionation of $R^{i/j}$ may be related to $R^{k/j}$ as:

$$\alpha_{k/j} = \left(\alpha_{i/j}\right)^Z \qquad (8)$$

where $Z = (m_i/m_k)\{(m_k - m_j)/(m_i - m_j)\}$, and m refers to the masses of the individual isotopes i, j, and k (e.g., Criss 1999). We illustrate this mass-dependent relation for Mg in Figure 4, where Equation (8) becomes:

$$\alpha_{25/24} = (\alpha_{26/24})^{0.521} \qquad (9)$$

Over small ranges in isotopic composition, Equation (8) may be approximated by the linear form:

$$\delta^k E \approx \left(\frac{k-j}{i-j}\right) \delta^i E \qquad (10)$$

where i, j, and k are integer masses. In the case of Mg, this relation would be:

$$\delta^{25}\text{Mg} \approx 0.5\ \delta^{26}\text{Mg} \qquad (11)$$

Equations (8) and (10) are applicable to stable isotope systems where isotopic fractionation occurs through mass-dependent processes which comprise the majority of cases described in this volume. These relations may also be used to identify mass-independent fractionation processes, as discussed in Chapter 2 (Birck 2004). Mass-dependent fractionation laws other than those given above distinguish equilibrium from kinetic fractionation effects, and these are discussed in detail in Chapters 3 and 6 (Schauble 2004; Young and Galy 2004). Note that distinction between different mass-dependent fractionation laws will generally require very

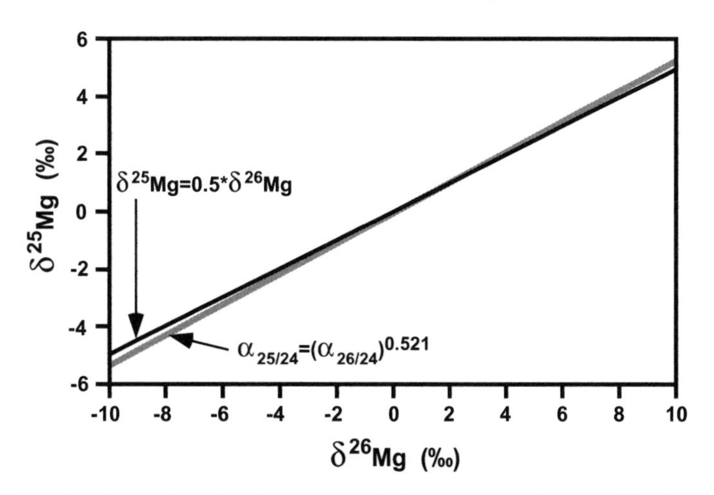

Figure 4. Illustration of mass-dependent fractionation of Mg isotopes, cast in terms of δ values. δ^{26}Mg and δ^{25}Mg values based on ^{26}Mg/^{24}Mg and ^{25}Mg/^{24}Mg ratios, respectively. A common equilibrium fractionation model, as defined by exponential relations between α values (fractionation factors) for different isotope ratios, is shown in the gray line. A simple linear relation, where the slope is proportional to the mass difference of the isotope pair, is shown in the black line. Additional mass-dependent fractionation laws may be defined, and all are closely convergent over small ranges (a few per mil) in isotope compositions at δ values that are close to zero.

high-precision isotopic analyses, depending on the range in isotopic compositions that are produced (Chapters 4 and 6).

The choices for defining $\delta^{i}E$ values for new isotope systems that contain multiple isotopes will be largely determined by consideration of the analytical precision. For example, Ca isotope ratios might be reported using the extreme end members ^{40}Ca and ^{48}Ca, representing an 8 mass-unit spread. But, as discussed in Chapter 8, the very low abundance of ^{48}Ca (0.19%) makes this a poor choice. Although ^{44}Ca/^{40}Ca fractionations are one-half those of ^{48}Ca/^{40}Ca fractionations, the much higher precision with which the ^{44}Ca/^{40}Ca ratio may be determined makes this ratio a superior choice. In the case of Mg, where the two minor isotopes ^{25}Mg and ^{26}Mg have nearly equal abundances of 10.00% and 11.01%, respectively, ^{25}Mg/^{24}Mg and ^{26}Mg/^{24}Mg ratios should be determined with equal precision, making the ^{26}Mg/^{24}Mg ratio the clear choice for providing the largest signal to noise ratio because ^{26}Mg/^{24}Mg fractionations will be approximately twice those of ^{25}Mg/^{24}Mg fractionations for mass-dependent processes. In practice, however, both ratios are commonly measured to provide insight into different mass-dependent processes, as well as monitor possible anomalies in ^{26}Mg abundances due to decay of ^{26}Al, which may be an issue for extraterrestrial samples; these issues are discussed in detail in Chapters 2 and 6 (Birck 2004; Young and Galy 2004).

Predicted and measured isotopic variations

The general "rules of thumb" for isotopic fractionations discussed in Chapter 3 (Schauble 2004), including bonding, redox state, and relative mass differences, lay out broad expectations for the extent of isotopic fractionations that may be observed in new isotopic systems. Several examples of isotopic fractionations for various elements are illustrated in Figure 5. Commensurate with its large relative mass difference, ^{18}O/^{16}O fractionations may be quite large at low temperatures, exceeding, for example 20‰ for $CaCO_3$–H_2O fractionations at 50°C (Fig. 5). Although isotopic fractionations are relatively large at low temperatures

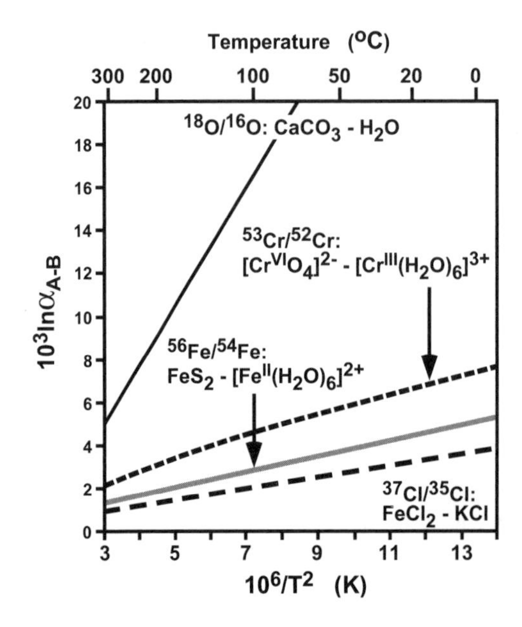

Figure 5. Examples of predicted and measured isotopic fractionations for O, Cr, Fe, and Cl, as cast in the traditional $10^3\ln\alpha_{A\text{-}B}$ - $10^6/T^2$ diagram. The quantity $10^3\ln\alpha_{A\text{-}B}$ places the isotope fractionation factor $\alpha_{A\text{-}B}$ in units of per mil (‰). Isotopic fractionations for relatively light elements, such as O, are generally higher than those of higher-mass elements, as expected based on changes in their relative mass differences (Fig. 3). $CaCO_3$ - H_2O curve for $^{18}O/^{16}O$ fractionations based on experiments from O'Neil et al. (1969). $[Cr^{VI}O_4]^{2-}$ - $[Cr^{III}(H_2O)_6]^{3+}$ curve for $^{53}Cr/^{52}Cr$ fractionations based on calculations from Schauble et al. (2004). FeS_2 - $[Fe^{II}(H_2O)_6]^{2+}$ curve for $^{56}Fe/^{54}Fe$ fractionations based on calculations from Polyakov and Mineev (2000) and Schauble et al. (2001). $FeCl_2$ - KCl curve for $^{37}Cl/^{35}Cl$ fractionations based on calculations by Schauble et al. (2003).

they are generally much smaller for higher-mass elements, where they are rarely expected to exceed 10‰ at temperatures of 100°C or less. For example, calculations of $^{37}Cl/^{35}Cl$, $^{53}Cr/^{52}Cr$, and $^{56}Fe/^{54}Fe$ fractionations for the systems $FeCl_2$–KCl, $Cr(VI)_{aq}$–$Cr(III)_{aq}$, and FeS_2–$Fe(II)_{aq}$, respectively, range from 1 to 5‰ at 100°C (Fig. 5; Schauble 2004). Although these isotopic fractionations are significantly smaller than those found for H, C, O, and S, they are still significant relative to current state-of-the-art analytical precisions (Chapter 4; Albarède and Beard 2004).

In Figure 6 we summarize the range in isotopic compositions measured for natural terrestrial samples for the elements discussed in this volume (Li, Cl, Ca, Cr, Fe, Cu, Zn, Se, and Mo). As discussed in Chapter 2, as well as in several subsequent chapters, the range in isotopic compositions for many of these elements is extended considerably if extraterrestrial samples are included. The very large relative mass difference of 15% for ^6Li and ^7Li has produced a very large range in isotopic compositions in nature of ~75‰ (Fig. 6). As reviewed in Chapter 6 (Tomascak 2004) the largest Li isotope fractionation is produced in relatively low-temperature environments, including seafloor weathering and marine hydrothermal systems. Most igneous rocks have relatively homogenous δ^7Li values, unlike other light isotopes such as those of oxygen, suggesting that larger fractionations are associated with covalently bonded O in silicates as compared to the ionic bonds that characterize Li. The isotopes of Mg have one-third of the relative mass difference of Li, but, so far, the range in $\delta^{26}Mg$ values are proportionally much smaller than suggested by the contrast in relative mass differences of the two elements (Fig. 6). $\delta^{26}Mg$ values for Mg-bearing carbonates appear to vary by ~2‰, but little is yet known about Mg isotope fractionations at low temperatures. Because Mg is a light element which has two of its three isotopes that are similar in abundance, it is ideally suited to investigation of different mass-dependent fractionation mechanisms that may operate in nature, and Young and Galy (2004) discuss this in detail in Chapter 6.

Chlorine isotope compositions vary by up to 15‰ (Chapter 7; Stewart and Spivack 2004). These large variations in Cl isotope compositions are found in marine environments, including mid-ocean ridge basalts, seafloor and hydrothermal alteration products, and sedimentary pore

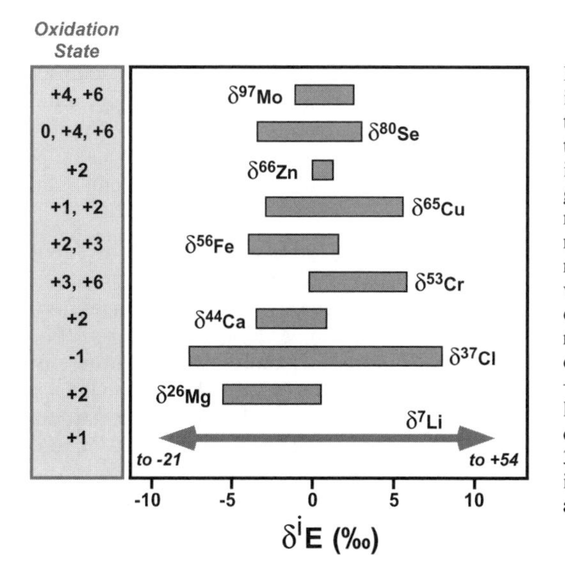

Figure 6. Summary of ranges in isotopic compositions for natural terrestrial samples as discussed in this volume. Isotopic variability in extraterrestrial samples is often greater. Isotopic compositions reported as δ values in units of per mil (‰), based on isotopic ratios and reference standards as used in this volume (Table 1). Note that the range of isotopic compositions for Li is much greater than the scale used in the diagram, where δ^7Li values vary from −21 to +54. In many cases, relatively large isotopic fractionations occur during redox reactions (see Chapter 3), and the common oxidation states in near-surface natural environments are listed on the left.

fluids. δ^{37}Cl values of fumarolic gases are also quite variable, suggesting a potential genetic tracer. Stewart and Spivack (2004) also note that significant Cl isotope variations are found in chlorinated organic compounds produced by industrial processing, similar to those found for industrially processed Li. These observations highlight the potential environmental applications of Li and Cl isotope variations.

Mantle-derived igneous rocks appear to be homogenous with respect to mass-dependent Ca isotope variations, although some K-rich rocks may have anomalously low δ^{44}Ca values reflecting ^{40}Ca enrichments from ^{40}K decay (Chapter 8; DePaolo 2004). One of the more important contrasts in Ca isotope compositions in the Earth is that measured between seawater Ca and igneous Ca, which appears to reflect Ca isotope fractionation during weathering and transport to the oceans. In contrast to the relatively high δ^{44}Ca values for seawater, biologically-processed Ca has relatively low ^{44}Ca/^{40}Ca ratios, and this appears to systematically decrease up the food chain. These observations highlight the potential for using Ca isotopes to trace biological processing of Ca in higher-order animals.

The isotope geochemistry of two redox-sensitive elements, Cr and Se, are compared in Chapter 9 by Johnson and Bullen (2004). Because Cr(VI) is quite soluble in natural waters, and is highly toxic, the significant fractionation in ^{53}Cr/^{52}Cr ratios that occurs during redox cycling of Cr is likely to find wide application in environmental geochemistry. Production of high δ^{53}Cr values for aqueous Cr seems likely to be a fingerprint for chromium reduction in natural systems. Selenium commonly occurs in three oxidation states in near-surface natural environments, Se(0), Se(IV), and Se(VI), and thus it is not surprising that significant isotopic fractionations are observed during redox cycling. The largest Se isotope fractionations appear to occur during reduction of Se(VI) and Se(IV) oxyanions, and this may occur through microbial activity, or through abiotic reduction.

Iron isotope geochemistry is reviewed in Chapter 10A (Beard and Johnson 2004) where summaries are given of the variations in ^{56}Fe/^{54}Fe ratios in inorganic experimental systems and natural igneous, metamorphic, and sedimentary rocks. Large fractionations of ^{56}Fe/^{54}Fe ratios occur during redox cycling of Fe, where low δ^{56}Fe values appear to be characteristic of reducing near-surface environments. In contrast, terrestrial surface weathering does not appear to fractionate Fe isotopes for bulk sedimentary materials, although isotopic variations during hydrothermal alteration of oceanic crust is well documented. Significant Fe isotope variations are also found in some high-temperature samples. In Chapter 10B, Johnson et al. (2004) discuss experimental evidence for biological fractionation of Fe isotopes, during microbial reduction and oxidation, as well as supporting studies in mineral dissolution and sorption. Microbial reduction of ferric oxides produces aqueous Fe(II) that has low δ^{56}Fe values whose origin lies in a number of pathways, whereas microbial oxidation produces ferric Fe precipitates that have relatively high δ^{56}Fe values.

Review of the isotope geochemistry of the transition metals is continued by Albarède (2004) in Chapter 11, where isotopic variations in Cu and Zn are discussed. The significant changes in bonding environments of Cu(I) and Cu(II) produce significant differences in δ^{65}Cu values for oxidized and reduced Cu compounds, and isotopic variations of up to 9‰ are observed in nature. Isotopic variations of Zn are significantly more restricted, where δ^{66}Zn values vary by less than 2‰, but systematic variations are recorded in Fe–Mn nodules from the ocean floor. Measurable isotopic variations are found for Cu and Zn in sedimentary rocks, as well as ore deposits, and this remains a promising aspect of future Cu and Zn isotope studies.

Molybdenum isotope variations appear to be on the order of 3.5‰ in ^{97}Mo/^{95}Mo ratios, where the largest fractionation is seen between aqueous Mo in seawater and that incorporated in Fe–Mn crusts and nodules on the seafloor (Chapter 12; Anbar 2004). This isotopic contrast is interpreted to reflect fractionation by Mo sorption to Mn oxide-rich sediments relative to aqueous Mo. The δ^{97}Mo values for euxinic sediments in turn are distinct from those of Fe–Mn crusts, highlighting the isotopic contrasts between major repositories of Mo in surface and near-surface environments. As discussed by Anbar (2004) in Chapter 12, a major focus of research on Mo isotopes has been the potential use as a paleoredox indicator in marine systems.

Processes that may produce isotopically distinct reservoirs

We briefly review processes in which isotopic fractionations may be recorded in isotopically distinct reservoirs that are preserved in nature. These concepts have been extensively covered in the H, C, O, and S isotope literature, and we illustrate several examples for the *non-traditional* stable isotope systems discussed in this volume. One of the simplest processes that produces isotopically distinct reservoirs would be slow reaction of substance A to B, where A and B remain open to complete isotopic exchange during the process. This is commonly referred to as *closed system equilibrium*, and the changes in isotopic compositions that occur may be defined by the exact relation:

$$\delta^i E_B = \frac{\alpha_{B\text{-}A}\,\delta^i E_{SYS} + 1000 f(\alpha_{B\text{-}A} - 1)}{\alpha_{B\text{-}A} - \alpha_{B\text{-}A} f + f} \tag{12}$$

where $\alpha_{B\text{-}A}$ is the *B-A* fractionation factor, $\delta^i E_{SYS}$ is the $\delta^i E$ value for the total system, and f is the fraction of A remaining ($f = 1$ when the system is entirely A) (Eqn. 3.19, p. 105, Criss 1999). Note that Equation (12) is simpler if cast in terms of $\alpha_{B\text{-}A}$ rather than $\alpha_{A\text{-}B}$, as was defined in Equation (4) above. If $\alpha_{B\text{-}A}$ is close to unity, Equation (12) may be simplified to:

$$\delta^i E_B \approx \delta^i E_{SYS} + f(\alpha_{B\text{-}A} - 1)\,10^3 \tag{13}$$

(Criss 1999), which may be further simplified to:

$$\delta^i E_B \approx \delta^i E_{SYS} + f\Delta_{B\text{-}A} \tag{14}$$

using the approximation $\Delta_{B\text{-}A} \approx (\alpha_{B\text{-}A} - 1) \times 10^3$ (note that this is equivalent to $\varepsilon_{B\text{-}A}$, following Eqn. 7 above). Following these simplifications, the corresponding $\delta^i E_A$ value, at a given f, is:

$$\delta^i E_A \approx \delta^i E_B - \Delta_{B\text{-}A} \tag{15}$$

(Criss 1999). Equations (14) and (15) describe a straight line in terms of $\delta^i E_A$ or $\delta^i E_B$ as a function of f or the fraction of B produced (Fig. 7). For example, in a system that is initially composed only of A and has an initial $\delta^i E$ value of zero, where $10^3\ln\alpha_{B\text{-}A} \approx \Delta_{B\text{-}A} = -1.5‰$, the first fraction of B to form will have $\delta^i E_B = -1.5‰$ (Fig. 7). As the reaction proceeds, the shifting mass balance of phases A and B will require shifts in their $\delta^i E$ values while maintaining a constant isotopic fractionation between A and B (Fig. 7). When the reaction is

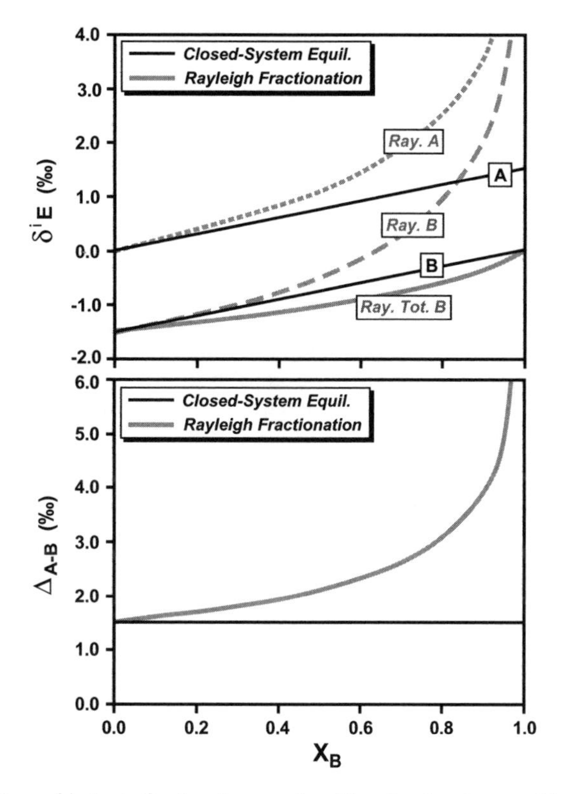

Figure 7. Comparison of isotopic fractionations produced by closed-system equilibrium and Rayleigh fractionation processes where phase A is reacted to phase B, as a function of the proportion of B (X_B) that is produced. Isotopic compositions are reported as $\delta^i E$, in units of per mil, and the fractionation factor $\alpha_{B\text{-}A}$ is 0.9985 ($\alpha_{A\text{-}B}$=1.0015). In the upper diagram, solid black lines mark the $\delta^i E$ values of reactant A and product B that reflect the isotopic mass-balance constraints imposed by a closed system where A and B continuously maintain isotopic equilibrium. During Rayleigh fractionation (gray lines), the product B is isolated from isotopic exchange with A immediately after formation, producing more extreme isotopic variations than in closed-system equilibrium fractionation. The $\delta^i E$ values of remaining phase A are shown in the short-dashed line in the upper diagram, and the $\delta^i E$ values of the instantaneous product B is shown in the long-dashed lines in the upper diagram. Variations in the $\delta^i E$ values for the *total product B* shown in solid gray line. In the lower diagram, the measured isotopic contrast between A and B, which may be conveniently defined as $\Delta_{A\text{-}B}$ (see text), is constant for closed-system equilibrium fractionation (solid horizontal line), but changes during Rayleigh fractionation (curved gray line), even for a constant fractionation factor $\alpha_{A\text{-}B}$ that may reflect equilibrium conditions.

complete and the system is entirely B, mass balance requires that δ^iE_B is the same as the initial δ^iE_A value (Fig. 7). The importance of this illustration is that in cases where reactions go to completion, if there is no addition or loss of element E from the system, there will be no net change in isotopic composition, even if $\alpha_{B\text{-}A}$ is significantly different than unity. Although data that fall along linear $\delta^iE - f$ trends are likely to be indicative of isotopic equilibrium, there are cases where incremental (Rayleigh) fractionation may occur under kinetic conditions that may be mistaken for closed-system equilibrium (e.g., p. 154-157, Criss 1999), requiring other approaches to establishing isotopic equilibrium such as use of enriched isotope tracers.

In the case of reactions where the products do not continue to exchange with other phases in the system, as might be the case during precipitation of a mineral from solution, Rayleigh fractionation may best describe the changes in δ^iE values for the individual components. The well-known Rayleigh equation (Rayleigh 1902) is:

$$\frac{\left[R^{i/j}\right]}{\left[R^{i/j}\right]_{\mathrm{i}}} = f^{(\alpha_{B\text{-}A}-1)} \tag{16}$$

where $[R^{i/j}]_{\mathrm{i}}$ is the initial ratio $R^{i/j}$ (which may be defined for either A or B), and f is the fraction of A remaining. Cast in terms of δ^iE values, Equation (16) becomes:

$$(1000 + \delta^iE)/(1000 + \delta^iE_{\mathrm{i}}) = f^{(\alpha_{B\text{-}A}-1)} \tag{17}$$

where δ^iE may be defined for either A or B, and the subscript i refers to the initial δ^iE value.

The products of Rayleigh fractionation are effectively isolated from isotopic exchange with the rest of the system immediately upon formation. If the process occurs slowly, such that each increment of product B forms in isotopic equilibrium with the reactant A prior to isolation of B from the system, then $\alpha_{B\text{-}A}$ would be an equilibrium isotope fractionation factor. However, if the process of formation of B is rapid, incremental formation of B may be out of isotopic equilibrium with A. In this case, $\alpha_{B\text{-}A}$ would be a kinetic isotope fractionation factor, which may be a function of reaction rates or other system-specific conditions.

Because the product in Rayleigh fractionation is progressively isolated, large changes in δ^iE values in the remaining components may occur (Fig. 7). A common process by which this occurs would be condensation or precipitation. Assuming that $\alpha_{B\text{-}A}$ is constant, the differences in the instantaneous δ^iE values for A and incremental formation of B will be constant (Fig. 7). However, in practice, the isotopic composition of the *total* condensed or solid phase after a given extent of reaction is of interest, and the changes in δ^iE values for the total phase B are more modest than for the incremental portions of B (Fig. 7). Because the δ^iE values of the remaining reactant A change dramatically toward the end of the reaction, the apparent value of $\Delta_{B\text{-}A}$ between A and the *total* condensed or solid phase deviates strongly from the fractionation factor as the reaction proceeds, and this is a key feature of Rayleigh fractionation (Fig. 7). As can be seen in Figure 7, the difference in the measured δ^iE values for A and total B most closely match that of the true fractionation factor at the beginning of the reaction, where the measured isotopic contrast is insensitive to the specific mechanism (e.g., closed-system equilibrium or Rayleigh) by which A and B are physically separated (Fig. 7).

Adsorption of Mo to Mn oxyhydroxides produces an isotopic fractionation that appears to follow that of a closed-system equilibrium model as a function of the fraction of Mo adsorbed (Fig. 8). Barling and Anbar (2004) observed that the δ^{97}Mo values for aqueous Mo (largely the $[MoO_4]^{2-}$ species) were linearly correlated with the fraction (f) of Mo adsorbed (Fig. 8), following the form of Equation (14) above. The δ^{97}Mo–f relations are best explained by a Mo_{aq}–Mn oxyhydroxide fractionation of +1.8‰ for ^{97}Mo/^{95}Mo, and this was confirmed through isotopic analysis of three solution-solid pairs (Fig. 8). The data clearly do not lie

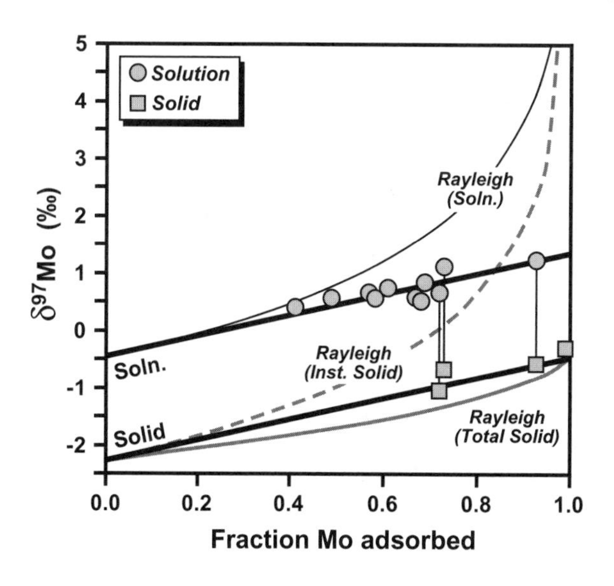

Figure 8. Example of apparent closed-system equilibrium fractionation, where Mo in solution is sorbed to Mn oxides (Barling and Anbar 2004). The δ^{97}Mo values of the Mo remaining in solution during sorption follow the linear trends that are consistent with closed-system equilibrium fractionation where isotopic equilibrium is continuously maintained between Mo in solution and that sorbed to the Mn oxides. Three aqueous-solid pairs (shown with tie lines) are consistent with this interpretation. The isotopic data cannot be explained through a Rayleigh process, where the product of the reaction (sorbed Mo) is isolated from isotopic exchange with aqueous Mo.

along a Rayleigh fractionation trend (Fig. 8), as would be expected if the adsorbed Mo did not remain in isotopic communication with Mo_{aq} at different degrees of total adsorption.

In contrast, reduction of Cr(VI) in solution, followed by precipitation of hydrated $Cr^{III}_2O_3$, produces fractionation in $^{53}Cr/^{52}Cr$ that follows a Rayleigh process (Fig. 9). When plotted as a function of the fraction of $Cr(VI)_{aq}$ reduced (f), the δ^{53}Cr values increase non-linearly, lying about a Rayleigh fractionation trend calculated for a $[Cr^{VI}O_4]^{2-}$–hydrous $Cr^{III}_2O_3$ fractionation of +3.4‰ (Fig. 9). The δ^{53}Cr–f trends are not consistent with the linear trends expected for a closed-system equilibrium model. In the Cr(VI) reduction experiments of Ellis et al. (2002), the fractionation factor, $\alpha_{Cr(VI)-Cr(III)}$, derived from the data is interpreted to reflect kinetic isotope fractionation, although it is important to note that not all Rayleigh processes are associated with kinetic isotope fractionation.

DETERMINATION OF ISOTOPIC FRACTIONATION FACTORS

Isotopic fractionation factors may be calculated based on theory (e.g., Urey 1947; Bigeleisen and Mayer 1947), and the various approaches used for such calculations are discussed in detail in Chapter 3 (Schauble 2004). Calculated fractionation factors are extremely important for new isotope systems because they constrain the isotopic fractionations that may be anticipated in nature, and, in some cases, provide the only means for constraining fractionation factors for systems that are inaccessible to experiments. Isotopic fractionation factors may also be inferred from natural assemblages (e.g., Bottinga and Javoy 1975), although issues of attainment of isotopic equilibrium or differential exchange upon cooling remain in many cases (e.g., Giletti 1986; Eiler et al. 1992; Kohn and Valley 1998). Ultimately, however, experimental verification of calculated fractionation factors is desirable. In their compilation of experimentally determined isotope fractionation factors for H, C, and O isotopes, Chacko et al. (2001) summarize measurements for 306 systems, many of which involved several independent studies. We are far from reaching this level of experimental calibration of isotope fractionation factors for the new isotopic systems discussed in this volume, and expanding the database of isotope fractionation factors remains a high priority for the isotope community.

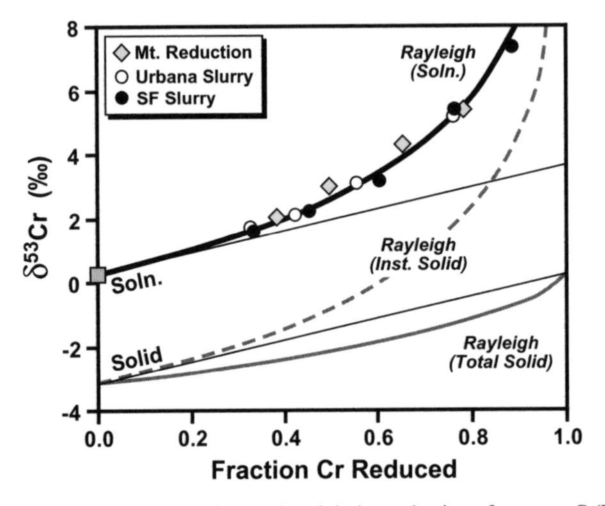

Figure 9. Example of Rayleigh fractionation produced during reduction of aqueous Cr(VI) to Cr(III), as reported in three reduction experiments by Ellis et al. (2002). Because the reduced product is a Cr(III)-hydroxide that does not continue to isotopically exchange with the aqueous Cr(VI), the δ^{53}Cr values of the remaining Cr(VI) increase along a Rayleigh trend. Although each of the three experiments of Ellis et al. (2002) produced slightly different Cr(III)-Cr(VI) fractionation factors, the curves illustrated are calculated for a single $\alpha_{Cr(III)\text{-}Cr(VI)}$=0.9966. Ellis et al. (2002) interpret the inferred $\alpha_{Cr(III)\text{-}Cr(VI)}$ as reflecting kinetic isotope fractionation during reduction and precipitation.

The majority of experimental studies of isotope fractionation have involved fluid-mineral pairs, reflecting the common means by which minerals form, particularly in low-temperature environments. In fluid-mineral systems, isotopic exchange is limited by the exchange properties of the solid phase, where the mechanisms of exchange may include replacement, recrystallization, Oswald ripening, surface diffusion, volume diffusion, grain boundary diffusion, and dislocation/lattice diffusion (e.g., Cole and Chakraborty 2001). Attainment of isotopic equilibrium in fluid-mineral systems is often difficult in experimental systems, especially at low temperatures, where the rates of diffusion are very low. Determination of equilibrium isotopic partitioning is of particular interest because equilibrium fractionation factors provide insights into bonding environments and other physico-chemical properties of the species of interest and equilibrium isotope fractionation factors are independent of kinetic issues or pathways of formation.

Fluid-mineral fractionation factors

In addition to providing the means for calculating the isotopic compositions of ancient fluids based on analysis of minerals, mineral-fluid isotope fractionation factors provide an opportunity to combine fractionation factors when there is a common substance such as water. A fundamental strategy for compiling databases for isotopic fractionation factors is to reference such factors to a common substance (e.g., Friedman and O'Neil 1977). For example, the quartz-water fractionation factor may be combined with the calcite-water fractionation factor to obtain the quartz-calcite fractionation factor at some temperature. It is now recognized, however, that the isotopic activity ratio of water in a number of experimental determinations of $^{18}O/^{16}O$ mineral-fluid fractionation factors has been variable, in part due to dissolution of mineral components at elevated temperature and pressure. The influence of solute composition on mineral-fluid fractionation factors may be quite significant, and this effect has been termed the *isotope salt effect* (e.g., O'Neil and Truesdell 1991; Horita et al. 1993a,b; Hu and Clayton 2003). Using substances other than water as the exchange medium is one approach to addressing this issue. Carbonate (Clayton et al. 1989), CO_2 (O'Neil and Epstein 1966; Mattey et al. 1990; Matthews et al. 1994; Fayek and Kyser 2002), and H_2 (Vennemann and O'Neil 1996) have been used successfully. In the case of $^{18}O/^{16}O$ fractionation factors for minerals of wide geologic interest, it is generally accepted that use of carbonate as the exchange medium offers a superior reference compared to water (e.g., Clayton et al. 1989; Chacko et al. 1996; Hu and Clayton 2003). Alternatively, mineral-mineral fractionation factors may be obtained in three-phase systems that contain two minerals and water, where the isotopic effect of dissolved components in the fluid cancels out (Hu and Clayton 2003). As new isotopic systems are developed, it seems likely that a variety of reference exchange media will be required, depending on the element, as databases of fractionation factors are developed.

The significant effects on $^{18}O/^{16}O$ fractionations due to dissolved components in mineral-water systems, up to 2‰ in high-temperature experiments, highlight the importance of speciation and activity in experimental systems in terms of their effects on isotopic fractionation factors. Thermodynamic data bases (e.g., Johnson et al. 1992; Shock et al. 1997; Sverjenski et al. 1997; Holland and Powell 1998) demonstrate that for all of the elements discussed in this volume, as well as many others, speciation in aqueous solutions may be highly variable, suggesting that the *isotope salt effects* that have been observed for oxygen in mineral-water systems are likely to be much greater for metals and halogens in mineral-fluid systems; indeed, calculated isotopic fractionation factors are, in general, highly variable among various aqueous species for metals and halogens, as reviewed in Chapter 3 (Schauble 2004). As the experimental databases for isotopic fractionation factors expand for the *non-traditional* stable isotope systems, the choices of exchange media for determining isotopic fractionation factors will be critical, as will consideration of the exact species involved.

Evaluating approach toward isotopic equilibrium in experiments

In this last section, we briefly review some of the methods that have been used to evaluate isotopic exchange in experiments, highlighting approaches used in well-studied systems such as oxygen, and note some new applications in the *non-traditional* stable isotopes. One classic method for determining equilibrium fractionation factors is to approach isotopic equilibrium from both sides (O'Neil 1986b). Alternatively, assessment of isotopic equilibrium may be done using the *three isotope method* (Matsuhisa et al. 1978, 1979; Matthews et al. 1983a,b). This approach involves starting materials that lie on two separate, but parallel, mass-dependent fractionation lines, where movement toward isotopic equilibrium will move the phases onto a new mass-dependent fractionation line at an intermediate position in a ratio-ratio diagram (Fig. 10). In its application to oxygen isotopes, the $^{18}O/^{16}O$ ratios of the starting materials have been chosen to lie close to those expected under equilibrium, whereas the $^{17}O/^{16}O$ ratios may be chosen to be far from equilibrium, providing a sensitive measure of the approach to isotopic equilibrium (Fig. 10). The *three isotope method* remains one of the most elegant and rigorous means for constraining the approach toward isotopic equilibrium, and is ideally suited for isotope systems where two isotope ratios may be determined independently.

In principle, the three isotope method may be widely applied to new isotope systems such as Mg, Ca, Cr, Fe, Zn, Se, and Mo. Unlike isotopic analysis of purified oxygen, however, isotopic analysis of metals that have been separated from complex matrices commonly involves measurement of several isotopic ratios to monitor potential isobars, evaluate the internal consistency of the data through comparison with mass-dependent fractionation relations (e.g., Eqn. 8 above), or use in double-spike corrections for instrumental mass bias (Chapter 4; Albarède and Beard 2004). For experimental data that reflect partial isotopic exchange, their isotopic compositions will not lie along a mass-dependent fractionation line, but will instead lie along a line at high angle to a mass-dependent relation (Fig. 10), which will limit the use of multiple isotopic ratios for isobar corrections, data quality checks, and double-spike corrections.

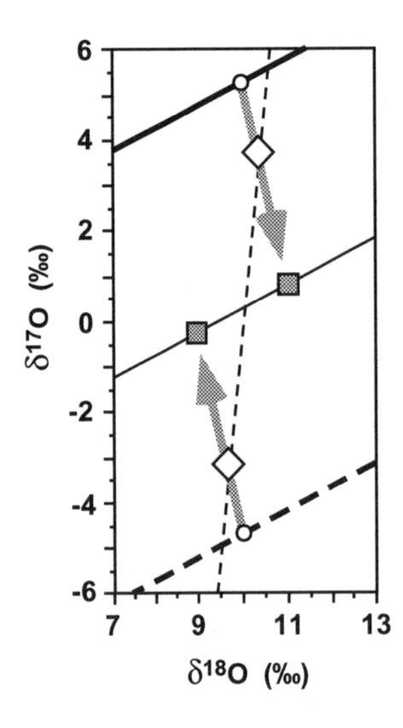

Figure 10. Three isotope method for assessing attainment of isotopic equilibrium for O isotope exchange experiments, based on the approach taken by Matthews et al. (1983a,b). Heavy solid black line (top line) reflects the terrestrial fractionation line for $^{17}O/^{16}O$-$^{18}O/^{16}O$ variations, and the small open circle on this line represents a natural mineral sample. The lower heavy dashed line represents a fluid that lies below the terrestrial fractionation line that was obtained through mixing of two highly disparate isotope compositions (MEOW II water developed at the University of Chicago). Movement of the two phases toward isotopic equilibrium will be occur in the directions of the arrows; assuming equal proportions of oxygen in the fluid and mineral phases, isotopic equilibrium will be marked by data that lie in the middle (thin solid line), where the final isotopic compositions of fluid and mineral are separated by the equilibrium fractionation factor. Partial exchange will produce isotopic compositions (diamonds) that do not lie along a mass-dependent fractionation line, where mineral-fluid pairs will define a line (dashed thin line) that lies at a high angle to a mass-dependent line.

An alternative to the three isotope method is to use an enriched isotope tracer in experiments that are run in parallel to those involving "normal" isotope compositions under identical run conditions. Use of enriched isotope tracers to evaluate the kinetics of isotopic exchange has a long history (e.g., Mills and Urey 1940). The extent of exchange toward isotopic equilibrium may be defined as:

$$F = (\delta^i E - \delta^i E_I)/(\delta^i E_E - \delta^i E_I) \tag{18}$$

where $\delta^i E_I$ and $\delta^i E_E$ are the initial and equilibrium isotopic compositions, respectively, and $\delta^i E$ is the isotopic composition observed at any time of interest (e.g., Criss et al. 1987). The fraction of isotopic exchange (F) varies from 0 to 1 as isotopic equilibrium is approached. Inspection of Equation (18) shows that F may be calculated with great sensitivity if enriched isotope tracers are used because the differences in the δ values will be large. Moreover, calculation of F using enriched isotope tracers will be relatively insensitive to the final equilibrium fractionation factor, which may be unknown at the start of an experiment. Calculation of F using Equation (18) is valid for ranges in $\delta^i E$ values up to a few hundred per mil. For more extreme isotopic enrichments, F must be calculated using the atomic abundances and molecular weights of the two components:

$$F = \frac{W_N\left(X_S^i - R_{MEAS}^{i/j} X_S^j\right)}{W_S\left(R_{MEAS}^{i/j} X_N^j - X_N^i\right) + W_N\left(X_S^i - R_{MEAS}^{i/j} X_S^j\right)} \tag{19}$$

where W_N and W_S are the molecular weights of the normal and isotopically enriched element, respectively, $X^i{}_S$ and $X^j{}_S$ are the atomic abundances of isotopes i and j in the isotopically enriched "spike", respectively, and $X^i{}_N$, and $X^j{}_N$ are the atomic abundances of isotopes i and j for the isotopically "normal" component, respectively. Equation (19) may be derived from standard two-component mixing relations (e.g., Albarède 1995).

Substituting F into a general rate equation produces:

$$\frac{-d(1-F)}{dt} = k_n(1-F)^n \tag{20}$$

where k is the rate constant, and n is the order of the reaction, generally an integer from 0 to 3. Although most isotope exchange reactions appear to follow a first-order rate law ($n = 1$) (e.g., Criss et al. 1987; Chacko et al. 2001), isotopic exchange may also follow a second-order rate law (e.g., Graham 1981; Johnson et al. 2002). Integration of equation 20 for first- and second-order rate laws ($n = 1$ and 2) yields the following linear forms:

$$\ln(1-F) = -k_1 t \quad \text{for } n = 1 \tag{21}$$

$$\frac{F}{(1-F)} = k_2 t \quad \text{for } n = 2 \tag{22}$$

An example using enriched isotope tracers to study the kinetics of isotopic exchange is shown in Figure 11, where an enriched ^{57}Fe tracer for the ferric Fe phase was used to determine the kinetics of isotopic exchange between aqueous ferric and ferrous Fe. In the system $[Fe^{III}(H_2O)_6]^{3+}$–$[Fe^{II}(H_2O)_6]^{2+}$, Fe isotope equilibrium is rapidly attained, where 95% isotopic equilibrium is established within ~60 seconds at 22°C (Fig. 11), or within ~5 minutes at 0°C, where isotopic exchange occurs via a second-order rate law. Despite the rapid exchange kinetics, the ferric Fe species was separated essentially instantaneously, effectively "freezing in" the isotopic composition of hexaquo Fe(III) and Fe(II), with minimal isotopic re-equilibration (~10–20%) during species separation (Johnson et al. 2002; Welch et al. 2003). If the timescale of species separation was on the order of that required for isotopic equilibrium between aqueous species, the measured isotopic compositions of the separated components would *not* preserve the instantaneous compositions in solution. For the range in isotopic compositions shown in Figure 11, calculation of F for the rate equations using Equation (18) produces an error of less than 0.005 relative to that calculated using Equation (19), which is insignificant.

Although closed-system isotopic equilibrium often may be attained among aqueous species at low temperatures on reasonable timescales (Fig. 11), it is much more difficult to reach isotopic equilibrium in fluid-mineral systems at low temperatures (a few hundred degrees or less). Solid-state diffusion rates at low temperatures are far too slow to allow isotopic equilibrium to be established through diffusion alone, and instead can only occur in low-temperature experiments through recrystallization (e.g., O'Neil 1986b; Chacko et al. 2001; Cole and Chakraborty 2001). Alternatively, estimates of mineral-fluid isotope fractionation factors via direct mineral synthesis through slow precipitation is an approach many workers have taken, particularly for carbonates and oxides (e.g., O'Neil et al. 1969; Tarutani et al. 1969; Yapp 1987, 1990; Carothers et al. 1988; Kim and O'Neil 1997; Bao and Koch 1999). It is, however, recognized that the high activation energies that are associated with

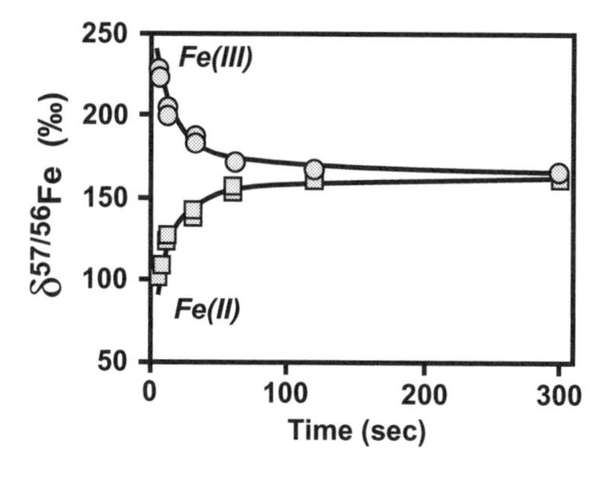

Figure 11. Determination of ferrous-ferric isotope exchange kinetics in dilute aqueous solutions using ^{57}Fe-enriched tracer solutions. Measured $\delta^{57/56}Fe$ values for ferrous (squares) and ferric (circles) Fe in solution versus time. Initial $\delta^{57/56}Fe$ values for $Fe(II)_{aq}$ ~0‰ and $Fe(III)_{aq}$ ~331‰. The rapid convergence in $^{57}Fe/^{56}Fe$ ratios for the ferric and ferrous species indicates that isotopic equilibrium is attained within minutes. Adapted from Welch et al. (2003).

nucleation and crystal growth at low temperatures may be problematic in terms of estimating equilibrium isotope fractionation factors (Chacko et al. 2001), and these factors are a likely reason for the very wide range in isotope fractionation factors that have been measured in low-temperature experiments. For example, at room temperature, there is a 16‰ range in the experimentally determined $^{18}O/^{16}O$ fractionation factor for the magnetite-water pair (O'Neil and Clayton 1964; Becker and Clayton 1976; Blattner et al. 1983; Rowe et al. 1994; Zhang et al. 1997; Mandernack et al. 1999) and a 11‰ range for the hematite-water and goethite-water pairs (Clayton and Epstein 1961; Clayton 1963; Yapp 1987, 1990; Müller 1995; Bao and Koch 1999). The rate of precipitation of minerals from solution has long been known to influence isotopic fractionations for the light stable isotopes (e.g., Turner 1982; McConnaughey 1989; Romanek et al. 1992; Kim and O'Neil 1997), and point to the importance of kinetic effects during mineral synthesis or precipitation approaches to determining isotopic fractionation factors.

Recrystallization of minerals or precursor phases may promote isotopic exchange with a fluid (e.g., O'Neil and Taylor 1967; Matthews et al. 1983a,b; Bao and Koch 1999), and will often move minerals closer to isotopic equilibrium. Qualitative evidence for dissolution/re-precipitation in experimental runs is often found in SEM or TEM images, but such images do not quantify the amount of an element that has passed through the aqueous and solid phases during dissolution/re-precipitation. Ion-microprobe analysis may identify dissolution and re-precipitation and may allow determination of isotope fractionation factors for systems that have only partially exchanged (e.g., Fortier et al. 1995). Another approach to quantifying the extent of dissolution and re-precipitation is through use of enriched isotopic tracers that are added following initial mineral formation in synthesis experiments. An example of this approach is shown in Figure 12, where an ^{57}Fe-enriched tracer was used to calculate the degree of dissolution and re-precipitation in the system $[Fe^{III}(H_2O)_6]^{3+}$–Fe_2O_3. This system was in approximate chemical equilibrium based on the observation that little change in aqueous Fe(III) contents occurred, demonstrating that the temporal changes in Fe isotope compositions reflect isotopic exchange during dissolution and re-precipitation where their relative rates were in balance; if such rates were slow, they should approach equilibrium conditions in terms of isotopic fractionation.

CONCLUSIONS

The principles of isotopic fractionation apply to all elements, and the methodologies that have been developed for the *traditional* isotopic systems such as H, C, O, and S are completely applicable to "new" isotopic systems, such as those discussed in this volume. Issues of nomenclature and standards will continue to be important as additional isotopic systems are explored, requiring the worker interested in these systems to be aware of differences among various studies that have yet to sort themselves out. Advances in theory and analytical methods promise to add many more isotopic systems to the field of *stable isotope geochemistry* than those reviewed in this volume. In addition to determining the ranges in isotopic compositions that may exist in natural systems, equal effort must be applied to experimental determination of isotope fractionation factors in carefully designed and implemented experimental studies, drawing upon the lessons that have been learned from several decades of study in the *traditional* stable isotope systems.

ACKNOWLEDGMENTS

Robert Criss, James O'Neil, and John Valley kindly reviewed the chapter. In addition, we thank the authors of the other chapters in this volume for their suggestions and comments.

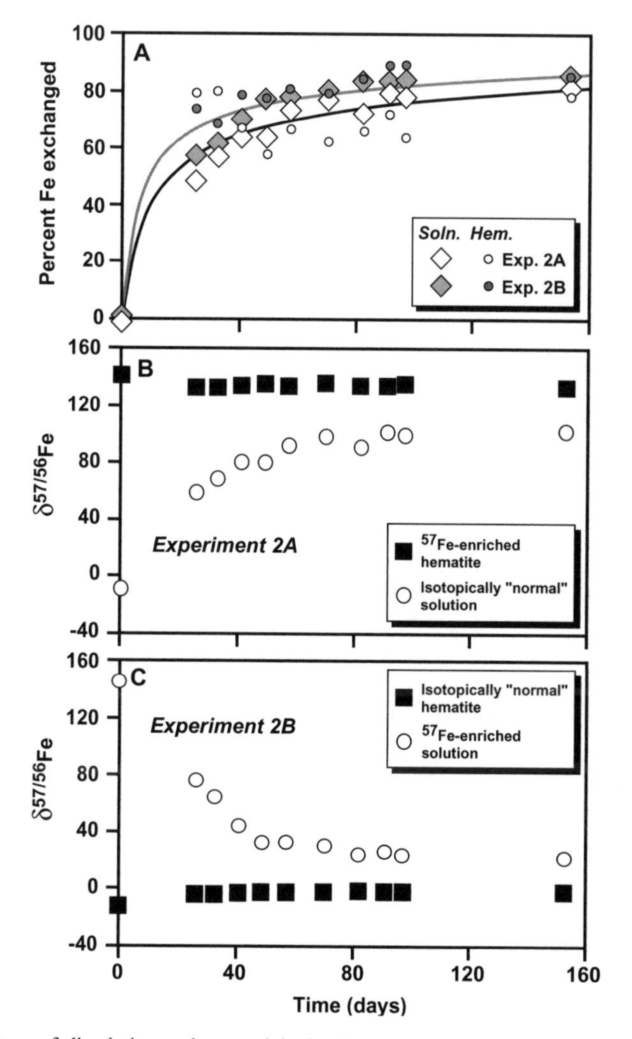

Figure 12. Extent of dissolution and re-precipitation between aqueous Fe(III) and hematite at 98°C calculated using ^{57}Fe-enriched tracers. **A.** Percent Fe exchanged (*F* values) as calculated for the two enriched-^{57}Fe tracer experiments in parts B and C. Large diamonds reflect *F* values calculated from isotopic compositions of the solution. Small circles reflect *F* values calculated from isotopic compositions of hematite, which have larger errors due to the relatively small shifts in isotopic composition of the solid (see parts B and C). Curves show third-order rate laws that are fit to the data from the solutions. **B.** Tracer experiment using ^{57}Fe-enriched hematite, and "isotopically normal" Fe(III). **C.** Identical experiment as in part B, except that solution Fe(III) is enriched in ^{57}Fe, and initial hematite had "normal" isotope compositions. Data from Skulan et al. (2002).

REFERENCES

Albarède F (1995) Introduction to Geochemical Modeling. Cambridge University Press, Cambridge, UK

Albarède F (2004) The stable isotope geochemistry of copper and zinc. Rev Mineral Geochem 55:409-427

Albarède F, Beard BL (2004) Analytical methods for non-traditional isotopes. Rev Mineral Geochem 55: 113-151

Anbar AD (2004) Molybdenum stable isotopes: observations, interpretations and directions. Rev Mineral Geochem 55:429-454

Bao H, Koch PL (1999) Oxygen isotope fractionation in ferric oxide-water systems: low-temperature synthesis. Geochim Cosmochim Acta 63:599-613

Barling J, Anbar AD (2004) Molybdenum isotope fractionation during adsorption by manganese oxides. Earth Planet Sci Lett 217:315-329

Becker RH, Clayton RN (1976) Oxygen isotope study of a Precambrian banded iron-formation. Hamersley Range, Western Australia. Geochim Cosmochim Acta 40:1153-1165

Beard BL, Johnson CM (2004) Fe isotope variations in the modern and ancient earth and other planetary bodies. Rev Mineral Geochem 55:319-357

Bigeleisen J, Mayer MG (1947) Calculation of equilibrium constants for isotopic exchange reactions. J Phys Chem 13:261-267

Birck JL (2004) An overview of isotopic anomalies in extraterrestrial materials and their nucleosynthetic heritage. Rev Mineral Geochem 55:25-64

Blattner P, Braithwaite WR, Glover RB (1983) New evidence on magnetite oxygen isotope geothermometers at 175°C and 112°C in Wairakei steam pipelines (New Zealand). Isotope Geosci 1:195-204

Bottinga Y, Javoy M (1975) Oxygen isotope partitioning among the minerals and triplets in igneous and metamorphic rocks. Rev Geophys Space Phys 13:401-418

Canfield DE (2001) Biogeochemistry of sulfur isotopes. Rev Mineral Geochem 43:607-636

Carothers WW, Adami LH, Rosenbauer RJ (1988) Experimental oxygen isotope fractionation between siderite-water and phosphoric acid liberated CO_2-siderite. Geochim Cosmochim Acta 52:2445-2450

Chacko T, Hu X, Mayeda TK, Clayton RN, Goldsmith JR (1996) Oxygen isotope fractionations in muscovite, phlogopite, and rutile. Geochim Cosmochim Acta 60:2595-2608

Chacko T, Cole DR, Horita J (2001) Equilibrium oxygen, hydrogen and carbon isotope fractionation factors applicable to geologic systems. Rev Mineral Geochem 43:1-81

Clayton RN (1963) High-temperature isotopic thermometry. *In:* Nuclear Geology on Geothermal Areas. Cons Naz Ric Lab Geol Nucl, p 222-229

Clayton RN, Epstein S (1961) The use of oxygen isotopes in high-temperature geological thermometry. J Geol 69:447-452

Clayton RN, Goldsmith JR, Mayeda TK (1989) Oxygen isotope fractionation in quartz, albite, anorthite, and calcite. Geochim Cosmochim Acta 53:725-733

Cole DR, Chakraborty S (2001) Rates and mechanisms of isotopic exchange. Rev Mineral Geochem 43:83-223

Criss RE (1999) Principles of Stable Isotope Distribution. Oxford Univ Press, New York

Criss RE, Gregory RT, Taylor HP (1987) Kinetic theory of oxygen isotope exchange between minerals and water. Geochim Cosmochim Acta 51:1099-1108

DePaolo DJ (2004) Calcium isotopic variations produced by biological, kinetic, radiogenic and nucleosynthetic processes. Rev Mineral Geochem 55:255-288

Eiler JM, Baumgartner LP, Valley JW (1992) Intercrystalline stable isotope diffusion: a fast grain boundary model. Contrib Mineral Petrol 112:543-557

Ellis AS, Johnson TM, Bullen TD (2002) Cr isotopes and the fate of hexavalent chromium in the environment. Science 295:2060-2062

Farquhar J, Bao H, Thiemens M (2000) Atmospheric influence of Earth's earliest sulfur cycle. Science 289: 756-758

Fayek M, Kyser TK (2000) Low temperature oxygen isotopic fractionation in the uraninite-UO_3-CO_2-H_2O system. Geochim Cosmochim Acta 64:2185-2197

Fortier SM, Cole DR, Wesolowski DJ, Riciputi LR, Paterson BA, Valley JW, Horita J (1995) Determination of the magnetite-water equilibrium oxygen isotope fractionation factor at 350°C: a comparison of ion microprobe and laser fluorination techniques. Geochim Cosmochim Acta 59:3871-3875

Friedman I, O'Neil JR (1977) Compilation of Stable Isotope Fractionation Factors of Geochemical Interest. US Geol Surv Prof Paper 440-KK

Giletti BJ (1986) Diffusion effects on oxygen isotope temperatures of slowly cooled igneous and metamorphic rocks. Earth Planet Sci Lett 77:218-228

Graham CM (1981) Experimental hydrogen isotope studies. Diffusion of hydrogen in hydrous minerals, and stable isotope exchange in metamorphic rocks. Contrib Mineral Petrol 76:216-228

Hoefs J (2004) Stable isotope geochemistry. 5th Edition. Springer, Berlin

Holland TJB, Powell R (1998) An internally consistent thermodynamic data set for phases of petrological interest. J Metamorphic Geology 16:309-343

Horita J, Wesolowski DJ, Cole DR (1993a) The activity-composition relationship of oxygen and hydrogen isotopes in aqueous salt solution: I. Vapor-liquid water equilibration of single salt solutions for 50 to 100°C. Geochim Cosmochim Acta 57:2797-2817

Horita J, Cole DR, Wesolowski DJ (1993b) The activity-composition relationship of oxygen and hydrogen isotopes in aqueous salt solution: II. Vapor-liquid water equilibration of mixed salt solutions from 50 to 100°C and geochemical implications. Geochim Cosmochim Acta 57:4703-4711

Hu G, Clayton RN (2003) Oxygen isotope salt effects at high pressure and high temperature and the calibration of oxygen isotope geothermometers. Geochim Cosmochim Acta 67:3227-3246

Jackson TA (2001) Variations in the isotope composition of mercury in a freshwater sediment sequence and food web. Canadian J Fish Aquat Sci 58:185-196

Johnson CM, Skulan JL, Beard BL, Sun H, Nealson KH, Braterman PS (2002) Isotopic fractionation between Fe(III) and Fe(II) in aqueous solutions. Earth Planet Sci Lett 195:141-153

Johnson CM, Beard BL, Roden EE, Newman DK, Nealson KH (2004) Isotopic constraints on biogeochemical cycling of Fe. Rev Mineral Geochem 55:59-408

Johnson JM, Oelkers EH, Helgeson HC (1992) SUPRT92: A software package for calculating the standard molal thermodynamic properties of minerals, gases, aqueous species and reactions from 1 to 5000 bars and 0 to 1000°C. Comput Geosci 18:899-947

Johnson TM, Bullen TD (2004) Mass-dependent fractionation of selenium and chromim isotopes in low-temperature environments. Rev Mineral Geochem 55:289-317

Kim S-T, O'Neil JR (1997) Equilibrium and nonequilibrium oxygen isotope effects in synthetic carbonates. Geochim Cosmochim Acta 61:3461-3475

Kohn MJ, Valley JW (1998) Obtaining equilibrium oxygen isotope fractionations from rocks: theory and example. Contrib Mineral Petrol 132:209-224

Lide DR (2003) CRC handbook of chemistry and physics (84th ed). CRC press, Boca Raton

Mandernack KW, Bazylinski DA, Shanks WC III, Bullen TD (1999) Oxygen and iron isotope studies of magnetite produced by magnetotactic bacteria. Science 285:1892-1896

Matsuhisa Y, Goldsmith JR, Clayton RN (1978) Mechanisms of hydrothermal crystallization of quartz at 250°C and 15 Kbar. Geochim Cosmochim Acta 42:173-182

Matsuhisa Y, Goldsmith JR, Clayton RN (1979) Oxygen isotopic fractionation in the system quartz-albite-anorthite-water. Geochim Cosmochim Acta 43:1131-1140

Mattey DP, Taylor WR, Green DH, Pillinger CT (1990) Carbon isotopic fractionation between CO_2 vapor, silicate and carbonate melts: An experimental study to 30 kbar. Contrib Mineral Petrol 104:492-505

Matthews A, Goldsmith JR, Clayton RN (1983a) Oxygen isotope fractionations involving pyroxenes: The calibration of mineral-pair geothermometers. Geochim Cosmochim Acta 47:631-644

Matthews A, Goldsmith JR, Clayton RN (1983b) On the mechanisms and kinetics of oxygen isotope exchange in quartz and feldspars at elevated temperatures and pressures. Geol Soc Am Bull 94:396-412

Matthews A, Palin JM, Epstein S, Stolper EM (1994) Experimental study of $^{18}O/^{16}O$ partitioning between crystalline albite, albitic glass, and CO_2 gas. Geochim Cosmochim Acta 58:5255-5266

McConnaughey T (1989) ^{13}C and ^{18}O isotopic disequilibrium in biological carbonates: I. Patterns. Geochim Cosmochim Acta 53:151-162

Mills GA, Urey HC (1940) The kinetics of isotopic exchange between carbon dioxide, bicarbonate iron, carbonate ion and water. J Am Chem Soc 62:1019-1026

Müller J (1995) Oxygen isotopes in iron (III) oxides: A new preparation line; mineral-water fractionation factors and paleo-environmental considerations. Isotopes Environ Health Stud 31:301-302

O'Neil JR (1986a) Appendix: Terminology and Standards. Rev Mineral 16:561-570

O'Neil JR (1986b) Theoretical and experimental aspects of isotopic fractionation. Rev Mineral 16:1-40

O'Neil JR, Clayton RN (1964) Oxygen isotope geothermometry. *In:* Isotope and Cosmic Chemistry. Craig H, Miller SL, Wasserburg GW (eds) North-Holland, Amsterdam, p 157-168

O'Neil JR, Epstein S (1966) Oxygen isotope fractionation in the system dolomite-calcite-carbon dioxide. Science 152:198-201

O'Neil JR, Taylor Jr HP (1967) The oxygen isotope and cation exchange chemistry of feldspar. J Geophys Res 74:6012-6022

O'Neil JR, Truesdell AH (1991) Oxygen isotope fractionation studies of solute-water interactions. *In:* Stable Isotope Geochemistry: A Tribute to Samuel Epstein. Taylor Jr. HP, O'Neil JR, Kaplan IR (eds), Geochem Soc Spec Pub 3:17-25

O'Neil JR, Clayton RN, Mayeda TK (1969) Oxygen isotope fractionation in divalent metal carbonates. J Chem Phys 51:5547-5558

Polyakov VB, Mineev SD (2000) The use of Mössbauer spectroscopy in stable isotope geochemistry. Geochim Cosmochim Acta 64:849-865

Rayleigh L (1902) On the distillation of binary mixtures. Phil Mag S.6 4:521-537

Rehkämper M, Halliday AN (1999) The precise measurement of Tl isotopic compositions by MC-ICP-MS: application to the analysis of geological materials and meteorites. Geochim Cosmochim Acta 63:935-944

Rehkämper M, Frank M, Hein JR, Porcelli D, Halliday A, Ingri J, Liebetrau V (2002) Thallium isotope variations in seawater and hydrogenetic, diagenetic, and hydrothermal ferromanganese deposits. Earth Planet Sci Lett 197:65-81

Romanek CS, Grossman EL, Morse JW (1992) Carbon isotopic fractionation in synthetic aragonite and calcite: Effects of temperature and precipitation rate. Geochim Cosmochim Acta 56:419-430

Rowe MW, Clayton RN, Mayeda TK (1994) Oxygen isotopes in separated components of CI and CM meteorites. Geochim Cosmochim Acta 58:5341-5347

Schauble EA (2004) Applying stable isotope fractionation theory to new systems. Rev Mineral Geochem 55: 65-111

Schauble EA, Rossman GR, Taylor Jr. HP (2001) Theoretical estimates of equilibrium Fe-isotope fractionations from vibrational spectroscopy. Geochim Cosmochim Acta 65:2487-2497

Schauble EA, Rossman GR, Taylor Jr. HP (2003) Theoretical estimates of equilibrium chlorine-isotope fractionations. Geochim Cosmochim Acta 67:3267-3281

Schauble EA, Rossman GR, Taylor Jr. HP (2004) Theoretical estimates of equilibrium chromium-isotope fractionations. Chem Geol, in press.

Shock EL, Sassani DC, Willis M, Sverjensky DA (1997) Inorganic species in geologic fluids: Correlations among standard molal thermodynamic properties of aqueous cations and hydroxide complexes. Geochim Cosmochim Acta 61:907-950

Skulan JL, Beard BL, Johnson CM (2002) Kinetic and equilibrium Fe isotope fractionation between aqueous Fe(III) and hematite. Geochim Cosmochim Acta 66:2995-3015

Stewart MA, Spivack AJ (2004) The stable-chlorine isotope compositions of natural and anthropogenic materials. Rev Mineral Geochem 55:231-254

Sverjensky DA, Shock EL, Helgeson HC (1997) Prediction of the thermodynamic properties of aqueous metal complexes to 1000°C and 5 kb. Geochim Cosmochim Acta 61:1359-1412

Tarutani T, Clayton RN, Mayeda TK (1969) The effect of polymorphism and magnesium substitution on oxygen isotope fractionation between calcium carbonate and water. Geochim Cosmochim Acta 33: 987-996

Tomascak PB (2004) Developments in the understanding and application of lithium isotopes in the earth and planetary sciences. Rev Mineral Geochem 55:153-195

Turner JV (1982) Kinetic fractionation of carbon-13 during calcium carbonate precipitation. Geochim Cosmochim Acta 46:1183-1191

Urey HC (1947) The thermodynamic properties of isotopic substances. J Chem Soc (London), p 562-581

Valley JW, Cole DR (eds) (2001) Stable Isotope Geochemistry. Rev Mineral Geochem, Vol 43

Valley JW, Taylor HP Jr, O'Neil JR (eds) (1986) Stable Isotopes in High Temperature Geological Processes. Rev Mineral, Vol 16

Vennemann TW, O'Neil JR (1996) Hydrogen isotope exchange between hydrous minerals and molecular hydrogen: I. A new approach for the determination of hydrogen isotope fractionation at moderate temperature. Geochim Cosmochim Acta 60:2437-2451

Welch SA, Beard BL, Johnson CM, Braterman PS (2003) Kinetic and equilibrium Fe isotope fractionation between aqueous Fe(II) and Fe(III). Geochim Cosmochim Acta 67:4231-4250

Yapp CJ (1987) Oxygen and hydrogen isotope variations among goethites (FeOOH) and the determination of paleotemperatures. Geochim Cosmochim Acta 51:355-364.

Yapp CJ (1990) Oxygen Isotopes in iron (III) oxides 1. Mineral-water factors. Chem Geol 85:329-335

Young ED, Galy A (2004) The isotope geochemistry and cosmochemistry of magnesium. Rev Mineral Geochem 55:197-230

Zhu XK, O'Nions RK, Guo Y, Reynolds BC (2000) Secular variation of iron isotopes in north Atlantic Deep Water. Science 287:2000-2002.

Zhang C, Liu S, Phelps TJ, Cole DR, Horita J, Fortier SM, Elless M, Valley JW (1997) Physiochemical, mineralogical, and isotopic characterization of magnetite-rich iron oxides formed by thermophilic iron-reducing bacteria. Geochim Cosmochim Acta 61:4621-4632

Reviews in Mineralogy & Geochemistry
Vol. 55, pp. 25-64, 2004
Copyright © Mineralogical Society of America

2

An Overview of Isotopic Anomalies in Extraterrestrial Materials and Their Nucleosynthetic Heritage

Jean Louis Birck

Laboratoire de Géochimie et Cosmochimie
Institut de Physique du Globe
4, Place Jussieu
75252 Paris Cedex 05, France

INTRODUCTION

Isotopic anomalies are expected in primitive meteorites since astronomical observation and astrophysical modeling of stars predict a great variety of stellar processes. Protostellar clouds should partially preserve the memory of this diversity in solid grains. Since 1970, high precision mass spectrometry and high resolution ion probes have led to the discovery of numerous isotopic anomalies, which were rapidly associated with nucleosynthetic processes. A general rule is that small isotopic effects (parts in 10^3–10^4) are observed in centimeter size samples, whereas order of magnitude variations are observed at the micron scale in circumstellar grains.

Refractory materials in primitive meteorites were investigated first as they have the best chance of escaping homogenization in the early solar system. Inclusions in C3 carbonaceous chondrites exhibit widespread anomalies for oxygen and the iron group elements. Only a few members, dubbed "FUN" (for "Fractionated and Unknown Nuclear" effects), also display anomalous compositions for the heavy elements. Anomalies in inclusions have generally been connected with explosive or supernova nucleosynthesis.

Several types of presolar circumstellar grains have been separated from the matrix of chondrites: diamonds, silicon carbide, graphite, oxides. The isotopic ratios of the light elements (C-N-O) vary over several orders of magnitude in these grains. Only a few measurement have been performed for heavier elements with generally s-process signatures. AGB stars at different stages of their evolution are thought to be the sources of most circumstellar grains. Nevertheless grains with supernova signatures have also been found. For Cr and Mo in bulk primitive carbonaceous chondrites (C1, C2), large isotopic differences exist between the different major mineral phases of the bulk rock.

A number of now extinct radioactive isotopes have existed in the early solar system. This is shown by the variations that they induce in the abundances in their daughter nuclides. Their main use is in establishing a chronology between their parental presolar stellar sources, and the formation of the solar system and the planets. An active debate is presently going on whether some of these short-lived nuclides could have been made within the early solar system by an intense flux of energetic protons from the young sun.

This chapter presents an updated report of the data with necessarily some limitations, but I will try to keep the most striking features and highlight the clearest relation with nucleosynthesis in stars.

1529-6466/04/0055-0002$05.00

Historical background

In 1970 only three kinds of isotopic variations were known from measurements on solar system samples. Isotopic ratios involving the decay product of long-lived radioactive nuclei (^{40}K-^{87}Rb-^{147}Sm-^{176}Lu-^{187}Re-U-Th) constitute the first family. The second relates to physico-chemical fractionation of the light elements (H-C-N-O) which is mass dependent (hereafter referred to as "linear effects"). The third comprises the products of cosmic-ray spallation which produce very minor amounts of some nuclei (see Reedy et al. (1983) and Honda (1988) for an overview). A number of investigations for extinct radioactivities and isotopic heterogeneities had been carried out earlier but failed to demonstrate resolved isotopic variations (Wasserburg and Hayden 1955; Anders and Stevens 1960; Murthy and Sandoval 1965; Eugster et al. 1969; Schramm et al. 1970; Huey and Kohman 1972). The searches for isotopic variations due to irradiation by an early active Sun failed as well (Fowler et al. 1962; Burnett et al. 1966). Hence until the early 1970s, it was generally accepted that the solar system had undergone a high temperature stage volatilizing all, or almost all solids, and homogenizing at least the inner solar system where the samples were coming from (Cameron 1969). Nucleosynthetic theory in the same time was directed toward reproducing the solar system abundances (Burbidge et al. 1957).

One striking exception was the very early discovery of ^{129}I decay to ^{129}Xe (Jeffery and Reynolds 1961). This discovery reflects the particular properties of rare gases which are nearly absent in telluric planetary bodies. Because they are not diluted by high abundances of isotopically 'normal' noble gases, anomalies in rare noble gas components were the first to be detected. This is also the reason for the Xe record of the fission of ^{244}Pu (Rowe and Kuroda 1965). From the available data on short-lived nuclides at that time, it was concluded that the last nucleosynthetic input into the protosolar cloud predated the formation of the planets by 100-200 Ma.

The first evidence for non-mass dependent isotopic variations came from oxygen in the refractory inclusions of the carbonaceous chondrite Allende (Clayton et al. 1973). The large effects found for oxygen, a major constituent of rocks, acted as a trigger for numerous high precision isotopic studies for as many elements as were possible to measure, and creating a new field in planetology. The outcome of this profusion of successful studies has been summarized in several past reviews, with special emphasis on groups of elements or objects (Clayton et al. 1988; Lee 1988; Anders and Zinner 1993; Zinner 1997; Goswami and Vanhala 2000). The major advances in the field roughly correspond to two waves of new data: the high precision isotopic measurement in the cm size inclusions of the C3 chondrites and the discovery of presolar grains which thereafter were investigated with microprobe techniques. This is not fortuitous and connects to two significant instrumental evolutions. The lunar exploration program led to the development of a new generation of mass spectrometers, able to resolve 10^{-4} (100 ppm) relative isotopic differences (Papanastassiou and Wasserburg 1969; Wasserburg et al. 1969). The sample sizes of the inclusions is the same order of magnitude as some precious lunar samples. The development of high resolution SIMS/ion probe (Lepareur 1980) was also under way during the 1970s and was shown to be suitable for natural micro sample isotopic analysis (Hutcheon 1982).

The definition of isotopic anomalies

In a strict sense "isotopic anomalies" are defined here as isotopic variations that are not understood to have been generated from a once uniform reservoir by processes acting within the solar system. They may result from incomplete homogenization of isotopically very diverse nucleosynthetic components. They potentially possess two types of information: the nature of the nucleosynthetic sources of the material, and the processes acting during their transfer from within stars to their preservation in meteorites. The definition of the "anomalous" composition reflect the status of knowledge around 1970 and by extension has been extended to all variations which were not identified in 1970. Effects related to extinct radioactivities, are also usually referred to as "Anomalies" as well as non-linear physico-chemical effects (Thiemens 1999; Clayton 2002) which will be briefly discussed at the end of this review. Extinct radioactivities can induce isotopic variations within the solar system starting with an isotopically homogeneous reservoir. They then do not strictly correspond to the first definition. Nevertheless they enter into the extended definition and are obviously related to nucleosynthesis occurring shortly before the formation of our planetary system; it is therefore useful to include them in this chapter.

THE MEASUREMENT OF ISOTOPIC ANOMALIES

Many of the isotopic anomalies would not be accessible without the ability to measure isotopic ratios with precisions down to 20-100 ppm. This is well below the level of instrumental fractionation for traditional thermal ionization mass spectrometers. From the experimental point of view, the data are often not interpretable in a straightforward manner and an isolated set of data may lead to ambiguities. These ambiguities will be greatly reduced or eliminated if more than one isotope ratio for the element under study can be measured, an appropriate normalization scheme is used to correct for instrumental fractionation, the presence of anomalous isotopic patterns in neighboring elements, and other astrophysical considerations are also taken into account. In addition, natural linear (or Rayleigh) mass fractionation is often present in primitive meteoritic samples as a result of partial evaporation or condensation. Nevertheless, the main observations, which will be presented in the following review, can be divided in four categories.

Large isotopic variations

Isotopic variations are large enough to exceed terrestrial variations by more than about an order of magnitude. As an example, this is the case for ion probe work on the CNO isotopes of circumstellar grains (see Section "Presolar stardust grains..."). These effects can be attributed unambiguously to nucleosynthetic reactions.

Two isotope elements and small effects

For some elements the variations are commensurate with, or smaller than the instrumental reproducibility or terrestrial variations, and it is difficult to distinguish between linear fractionation due to physico-chemical effects and nuclear effects. All the interpretation depends on the connections between the variation in the different elements and the phase they are in. This is also relevant to a number of radiochronometric systems in which the decay product belongs to an element having only two isotopes (e.g. ^{107}Pd-^{107}Ag, ^{205}Pb-^{205}Tl). Modern ICPMS instruments are able to resolve differences much smaller than the possible natural mass dependent effects (e.g., Carlson and Hauri 2001a, b). A correlation with the parent-daughter elemental ratio usually indicates that radioactive decay has occurred. Nevertheless, this interpretation must be used with caution when the overall isotopic ratio variations in the daughter element are small. It may be possible that in some cases processes such as volatilization, producing the spread in the parent-daughter ratio, could also produce mass dependent fractionation in the daughter element, resulting in a pseudo-isochron. Cross-checking with similar mineralogical situations without radioactive decay from the parent, alleviates this problem.

Three isotope elements and small effects

Non linear variations can be smaller (a few ε units) than instrumental fractionation in TIMS instruments (a few ‰ per mass unit). It is then the choice of the isotopic pair used for normalization which determines that the third isotope displays non-linear effects. ICPMS is able to measure natural fractionation with a precision close to that obtained for non-linear effects (Carlson and Hauri 2001a; Kehm et al. 2003) but, if we take the instrumental fractionation observed in TIMS instruments as the possible extent of fractionation in natural processes, ICPMS does not lift this particular ambiguity. The choice of the normalizing pair is not unambiguous and has to be sustained by interelement comparison and theoretical considerations based on nucleosynthetic models. As an example when an extinct radioactivity is examined, the variations are expected on the daughter isotope in correlation with the elemental parent daughter ratio.

Elements having 4 and more isotopes

They are used as above, but in general they lead to more straightforward interpretations. The correction for instrumental fractionation involves an isotope pair for which the measurements are in agreement with the terrestrial ratio, whether this choice results from the measurement itself or from model considerations. In cases where all the ratios are different from the terrestrial ratios, model considerations are used to interpret the data. In the most common cases, one isotope displays wider variations than the others and constitutes a guideline for modeling the origin of the anomalies.

Double spike techniques have been used occasionally to quantify the fractionation produced in natural processes, but this does not change the magnitude of non-linear effects as the natural fractionation follows closely the same laws as the instrumental fractionation. For the following, the instrumental fractionation correction is not discussed as this is described in other chapters of this volume. I suggest that the reader consult the original papers reporting the data for the details. In general the exponential law is used (Russell et al. 1978). In its outcome, it is very close to the Rayleigh law which has a better grounding in the physics of evaporation, but is simpler to use in on line ratio calculations.

Units

Units which are used in isotopic work depend on the precision of the measurements. Generally δ units are used for stable isotopes and correspond to permil relative deviation. It is used occasionally also for non linear effects and then they are permil (‰) deviations without reference to mass differences between the isotopes. Since the beginning of the 70s (e.g., Papanastassiou and Wasserburg 1969) thermal ionization data are often given in ε units which are fractional deviation from the normal in 0.01%. With the new generation of more precise instruments, results are sometimes given in ppm (parts per million) relative to a terrestrial standard sample.

For the following text, isotopic "anomalies" always stands for non-linear or non-mass dependent variations; "linear" or "mass dependent" have the same meaning although mass dependent fractionation may not be strictly linear (Rayleigh). Usually, in the first approach the difference is not essential for description

STELLAR EVOLUTION AND NUCLEOSYNTHESIS

Almost all of the elements heavier than He are synthesized in the interiors of stars. The work of Burbidge et al. (1957) gives the theoretical framework for the synthesis of the elements. The experimental evidence of active nucleosynthesis came from the discovery of the unstable nuclei of technetium in the spectra of red giants (Merrill 1952). The solar elemental and isotopic abundances which are taken from the primitive carbonaceous chondrites constitute the guidelines for testing such models (Anders and Grevesse 1989). A minimum of eight basic processes are required to reproduce the observed compositions. Nucleosynthetic theory has expanded since the initial work, but this frame is essentially still valid today. It was also realized that materials expelled into the interstellar medium by stellar winds or explosions contribute to the chemical evolution of the galaxy (Timmes et al. 1995) and become part of the raw material for the next generations of stars. As a result the materials constituting the solar system are a mixture of materials from a vast number of stellar sources (see Clayton 1988) for the primary nucleosynthetic sources of the elements up to Ni). The understanding of isotopic anomalies requires some basic concepts of stellar evolution and nucleosynthesis. The purpose here is only to give a broad outline. For further details, many textbooks (Clayton 1983; Rolfs and Rodney 1988; Bernatowicz and Zinner 1997) and review articles are available if the reader is interested in more details (e.g., Woosley 1986; Woosley and Weaver 1995; Arlandini et al. 1999; Busso et al. 1999; Rauscher et al. 2002 and references therein).

The energy powering stars results from nuclear fusion reactions transforming light nuclei into heavier ones. During their different evolutionary stages, a number of different nucleosynthetic process occur in different types of stars. Stars spend most of their lives in the core H-burning phase producing helium. When H is exhausted the star evolves into a red giant where part of the core material mixes convectively with the envelope and changes the surface composition. The next step is the onset of He combustion into ^{12}C, some of which gets converted into ^{16}O. When He is exhausted the future of the star depends on its mass. In the lighter stars (M < 8 Mo, where Mo = 1 solar mass), the core pressure and temperature are too low to proceed with further fusion reactions and becomes inert. The outer layers expand into an Asymptotic Giant Branch star (AGB) while H and He fusion continues just outside the core in thin shells producing large amounts of ^{12}C and heavy elements by slow neutron capture (s-process). Periodic convection episodes bring freshly synthesized materials to the surface. Large stellar winds expel the outer layers of the stars into a planetary nebula leaving a white dwarf star.

Massive stars

Cores of stars heavier than about 8 Mo continue to produce heavier nuclei within their cores, first the silicon mass region and ultimately the iron group nuclei. The resulting structure is an onion shell structure in which each layer has experienced more extensive fusion history than the next overlaying layer. When the core is composed mostly of iron, nuclear fusion can produce no more energy, and the structure collapses and rebounds on the core in an energetic event called supernova. The shock wave expels most of the mass of the star e.g. more than 90% for a 25 Mo star in a type II supernova (Woosley and Weaver 1995). This shock wave also heats the material in the different layers that it crosses and causes extensive nuclear reactions known as explosive nucleosynthesis. Taking together the products of hydrostatic fusion combustion and explosive nucleosynthesis, type II Supernovae are considered to be major producers of elements heavier than H in the galaxy. They are also thought to be the major site of the r- and p- nucleosynthetic processes. In the last stages of the evolution of massive stars, the core reaches very high temperature, above about 4×10^9 K, at which point the nuclear reaction network attains a thermal equilibrium called "nuclear statistical equilibrium" (NSE) or e-process. Variants having high neutron densities of this process have also been described to account for the production of neutron-rich isotopes (Hainebach et al. 1974; Cameron 1979; Hartmann et al. 1985).

Massive stars from 25 to 100 Mo already lose a substantial fraction of their mass in strong stellar winds ranging from 2×10^{-6} Mo/y during the H-Burning phase up to 5×10^{-5} Mo/y in the He-burning phase also known as Wolf-Rayet phase. As a large convective core develops in these stars, fresh nucleosynthetic product are soon exposed on the surface and ejected with the stellar winds (Prantzos et al. 1986). Wolf-Rayet stars may have been the principal source of ^{26}Al in meteorites (Arnould et al. 2000).

The fate of stars also depends on the presence of a companion. Other violent episodes known as Novae or type Ia supernovae result from the accretion of materials from a partner star. Supernovae of type I are thought to be responsible for most of the production of the neutron-rich isotopes of the iron group (Woosley et al. 1995; Höflich et al. 1998). One of the key parameters in stellar evolution, and consequently the nucleosynthetic outcome, is the "metallicity," defined as the proportion of elements heavier than He.

p-r-s-processes

Elements heavier than iron are not produced by the stellar fusion reactions. They are endothermic products and, to reproduce observed isotopic abundances, three production mechanisms are required (Burbidge et al. 1957). Two of them are neutron addition reactions starting with nuclei from the iron group. They are divided according to their slow (s-process) or rapid (r-process) time scale, as compared with the ß decay from unstable nuclei encountered along the neutron capture path. In the s-process, the orders of magnitude of the exposure to neutrons is: neutron densities of 3×10^8 n/cm^3 at temperatures close to 3×10^8 K for pulses of a few years occurring at intervals in the scale of 10^4 y. The r-process is at the origin of the elements beyond bismuth, and especially U-Th. r-process nuclei are produced in a high temperature (>10^9 K) stellar environment having neutron densities in the range 10^{20}-10^{25} n/cm^{-3} as highly unstable precursors which then return to the stability valley by ß-decay (Matthews and Cowan 1990; Meyer 1994). The neutron-rich isotopes of the elements from Zn to U result at least partially, if not totally, from this process. The neutron-deficient nuclei also called the p-nuclei cannot be produced by the two former processes. They represent no more than 1% of the bulk elemental abundances. It is generally considered that radiative proton capture and (γ, n) photodisintegration on preexisting neutron-rich nuclei are playing a key role in the p-process (Rayet et al. 1990; Lambert 1992; Meyer 1994).

In considering the mass balance of the solar system, the main production of s-process nuclei is attributed to the AGB phase. For the r- and p-process the relation with a particular astrophysical site is less straightforward but most models referred in the above reviews relate them to supernova events (Matthews and Cowan 1990; Rayet 1995).

As a concluding remark of this section, the theoretical models of nucleosynthesis within stars show that the isotopic compositions of the elements are highly variable depending on star size, metallicity, companion's presence. From the isotopic data obtained in diverse solar system materials it turns out that most of this material was highly homogenized in the interstellar medium or by the formation of the solar system. The presence of isotopic anomalies preserved in some primitive materials are the last witnesses of the initial diversity of the materials constituting our planetary system.

THE ISOTOPIC HETEROGENEITY IN THE REFRACTORY COMPONENTS OF THE EARLY SOLAR-SYSTEM

The contraction of the interstellar cloud at the origin of the solar system was a strongly exothermic process and the inner solar system went through a temperature peak which caused the loss of a number of volatile elements. This episode also promoted the isotopic homogenization of the elements. The completeness of the later process depends strongly on physical factors, such as the temperature and duration, but also on the mobility of the individual elements. In addition, parent body thermal metamorphism tends to erase isotopic heterogeneities and to reset radioactive chronometers (Göpel et al. 1994). Refractory mineral phases or phase assemblages are expected to better preserve any initial isotopic heterogeneity, and have therefore been the prime focus of isotopic measurements.

Carriers of anomalies were first investigated in carbonaceous chondrites, which are the most primitive meteorites available. For reasons that are not well understood, there is considerable variation in the size, morphology and mineralogy of refractory components between primitive meteorite families. Millimeter to several cm sized inclusions are found in the CV3 carbonaceous chondrites (Christophe 1968). Their compositions are close to those predicted for early condensates in a high temperature gaseous nebula with solar composition (Grossman 1972; Grossman and Larimer 1974). They were first to be investigated for practical reasons, such as, ease of separation.

Other meteorite classes like C2, CO and ordinary chondrites contain much smaller inclusions less than 1mm (MacPherson et al. 1988) and require ion microprobe techniques to evaluate the isotopic compositions. On the least metamorphosed side, C1 have very few inclusions or oxide grains, but are the carrier of the greatest amounts of stellar nanodiamond and other carbides (Anders and Zinner 1993). As will be shown for Cr anomalies in carbonaceous chondrites, the survival of the mineral carriers of the anomalies also depends on the metamorphic grade of the meteorites. Nevertheless, isotopic anomalies have also been found in higher metamorphic grade from other classes, especially in the reduced enstatite chondrites.

Refractory inclusions of the CV3 carbonaceous chondrites (CAIs)

The initial data on isotopic anomalies resulted from the abundant Allende meteorite, of which about 3 tons were recovered in 1969. Inclusions often found in CV3 meteorites like Allende, are straightforward to separate from the surrounding matrix with traditional mineral separations. Their sizes were within the same order of magnitude as the small lunar samples, for which a large experimental development effort had just been made (Gast et al. 1970; Tera et al. 1970). Refractory inclusions are often easily identified in hand-specimen as whitish objects in the dark matrix. They make up about 5% of the bulk mass of the meteorite. Other CV3s contain similar inclusions with some petrographic differences, but their isotopic data fit within the range of observations of Allende inclusions and will not be discussed here. The isotopic compositions for inclusions clearly divide them into two groups.

The first group called "common" or "normal" inclusions, represents the overwhelming majority of the samples. These display isotopic anomalies up to 4% in O and in the iron group elements. For the other elements investigated to date, they have normal isotopic compositions within instrumental resolution. The second group is composed of a very few inclusions showing mass fractionated light elements, much larger and variable effects in the iron group elements, and anomalies in all other elements investigated so far, especially the heavy elements Sr, Ba, Nd, Sm. They have been dubbed "FUN" inclusions for "Fractionated and Unknown Nuclear" effects. "FUN" signatures are indistinguishable from the main group on petrographic or mineralogical grounds, except for an enhanced occurrence in purple spinel-rich inclusions (Papanastassiou and Brigham 1989). They clearly show that in some cases, specific nucleosynthetic products were able to escape complete rehomogenization between their ejection from stars and their incorporation into meteorites.

Common inclusions.

Oxygen. The first isotopic measurements were for oxygen (Clayton et al. 1973). In the 3 isotope diagram (Fig. 1) all terrestrial samples plot on a mass dependent fractionation line of slope 0.52. Refractory inclusions and their minerals plot on a slope 1 line, which deviates by up to 2% from the terrestrial line for the most anomalous samples. This indicates a mixing line between an ^{16}O enriched reservoir that is distinct from the Earth-Moon with a composition of $\delta^{17}O = -42‰$ and $\delta^{18}O = -40‰$, and a terrestrial-like reservoir (Young and Russell 1998). Detailed analysis of mineral separates from a number of inclusions indicate that the most refractory minerals, like spinel and pyroxene, possess the highest ^{16}O enrichments, whereas the smallest enrichments are found in melilite and feldspathoids. This suggests that the carrier of the anomaly is a solid that underwent exchange with a nebular gas close to terrestrial

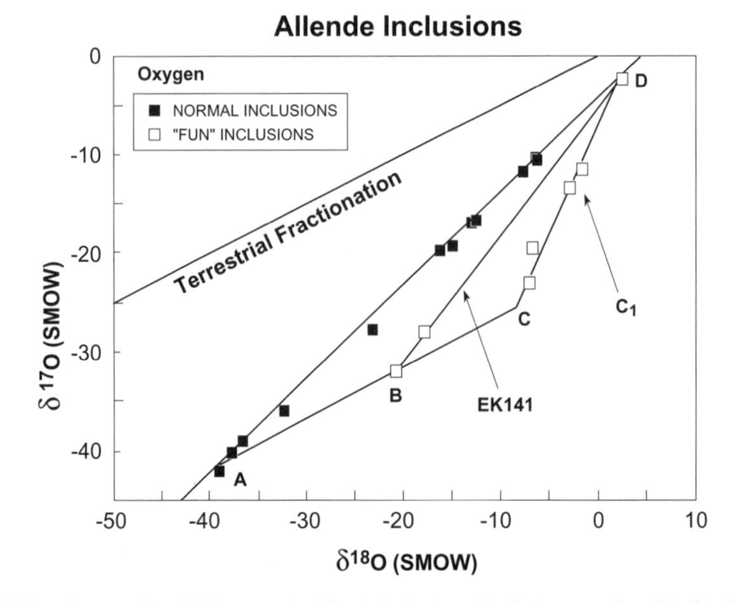

Figure 1. Three isotope plot of O isotopes in Allende inclusions. Deviations are plotted in δ units which are ‰ deviations relative to the terrestrial SMOW standard. In a two stage model, "normal" inclusions had initially a composition close to point A and exchanged with a reservoir poorer in ^{16}O in the region of point D (Clayton et al. 1973). "FUN" inclusions underwent an intermediate step along a fractionation line between point A and point C. Then each inclusion exchanged with the same ^{16}O poor reservoir D (Clayton and Mayeda 1977).

composition. In Figure 1, the most resistant mineral phases are assumed to retain their original isotopic composition. That the refractory inclusions have undergone secondary exchanges is also suggested by the observation of mineralogical alteration by a volatile rich and oxidizing gas (Wark 1986).

The ^{16}O excess was first suggested to have a nuclear origin in stars. Almost pure ^{16}O is produced in He-burning shells in massive stars, and in supernovae. On the other hand it has been shown that non-mass dependent fractionation can be produced in the laboratory by non-nuclear processes (Thiemens and Heidenreich 1983; Thiemens 1988). Similar non-linear effects have been found for O isotopes in atmospheric gases (Schueler et al. 1990; Thiemens et al. 1995). Although stellar nucleosynthesis is indeed at the origin of the O observed in the universe, the link between O isotopic anomalies in inclusions and nucleosynthesis is still under debate (Thiemens 1999; Clayton 2002).

Magnesium. The next element investigated was Mg as it is close to oxygen in several aspects: it is a major element, and has a relatively low atomic mass number. It possesses only three isotopes, and ^{26}Mg variations could possibly be induced by radioactive decay of now extinct ^{26}Al, so other non linear effects cannot be resolved unambiguously. If $^{24}Mg/^{25}Mg$ is assumed to be un-altered by nucleosynthetic inputs, then variations of a few ‰/amu around the average terrestrial value may reflect condensation-evaporation processes in the solar nebula (Wasserburg et al. 1977) and the alteration process already mentioned for oxygen (Young et al. 2002).

The iron group elements (from calcium to zinc). Some typical results on this group of elements are displayed in Figure 2. This group of elements have the most stable nuclei of the

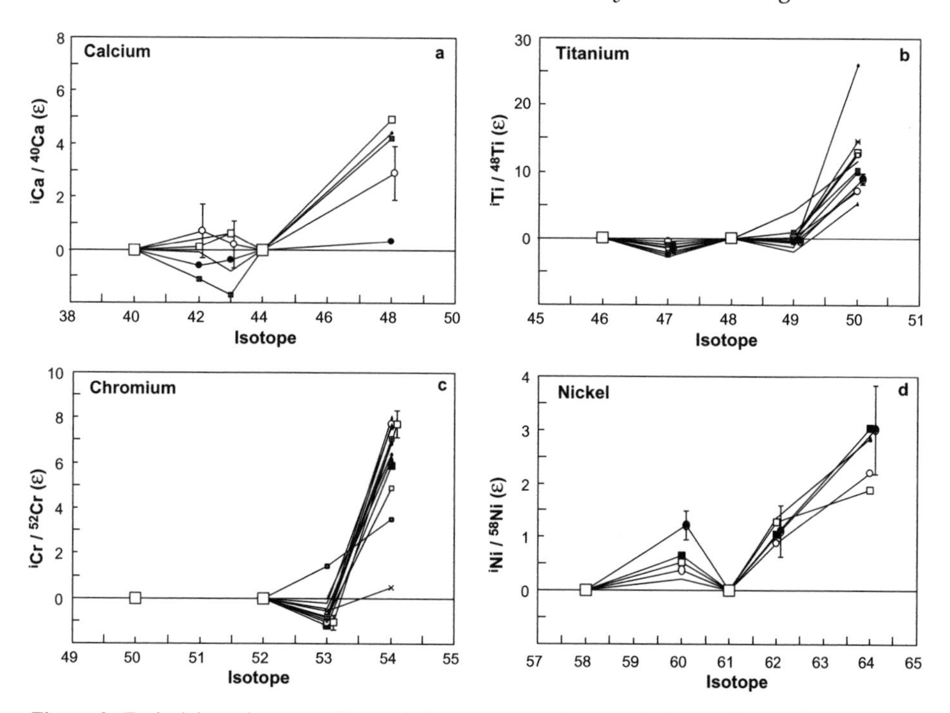

Figure 2. Typical isotopic compositions of the iron group elements in "normal" Allende inclusions (Niederer et al. 1981; Niemeyer and Lugmair 1981; Birck and Allegre 1984; Jungck et al. 1984; Birck and Allegre 1988; Birck and Lugmair 1988; Bogdanovski et al. 2002). The data are plotted as deviations in ε units (fractional deviation in 0.01%) from the terrestrial composition after normalization relative to the two isotopes represented with large open squares. The errors on the individual data points are not represented to not confuse the Figure. Typical experimental errors are represented only on one point in each sub-figure. Variations are clearly resolved for the most neutron-rich isotopes of this group of elements. Inclusions have an excess of a component produced in a neutron-rich nuclear statistical equilibrium.

periodic table. They may constitute the ultimate ashes of quiescent nuclear fusion in massive stars (see Section "Massive stars"). Ca and Ti belong to the refractory group of elements which appear in the first condensates above 1400K, whereas Zn is volatile (Larimer 1967). Cr and Ni are not considered to be refractory elements in the condensation sequence, and are close to the common major elements: Fe, Si, Mg. They are expected to condense with the bulk mass of a nebula of solar composition.

- Calcium. Ca is abundant in refractory inclusions. It has seven isotopes that span a 20% relative mass range. The nucleosynthetic pathway leading to the formation of Ca is expected to be varied because of its location between the silicon group and the iron group (Niederer and Papanastassiou 1984; Clayton 1988). ^{40}Ca which constitutes 96% of the total Ca, is also one of the decay products from ^{40}K , and can vary within the solar system with time, depending on the K/Ca ratio. Fortunately K is a volatile element and is strongly depleted in inclusions relative to Ca. Thus radioactive ingrowth of ^{40}Ca from ^{40}K over the lifetime of the solar system can be neglected in inclusions. This allows the use of the $^{40}Ca/^{44}Ca$ ratio to precisely determine the instrumental mass fractionation. In "Normal" or "Common" inclusions only the most neutron-rich isotope ^{48}Ca shows clearly resolved (more than 5σ) excesses (Fig. 2a). These excesses relative to average solar system composition are small, up to 6 ε with an average of about 3 ε (Lee et al. 1978; Jungck et al. 1984; Niederer and Papanastassiou 1984). No real correlation exists between O and Ca besides the presence of isotopic effects in the same objects.

^{48}Ca unlike the other isotopes of Ca, can be produced in significant amounts only in high neutron density regions of stars. Several processes can produce ^{48}Ca enrichments: neutron-rich Si burning (Cameron 1979), nβ–process (Sandler et al. 1982), s-process (Peters et al. 1972) and neutron-rich statistical equilibrium (Hainebach et al. 1974; Cameron 1979; Hartmann et al. 1985). All these processes take place in massive stars (>8 M_0). Addition of material from a heterogeneous supernova remnant to the solar system could produce regions with ^{48}Ca excess. In fact, Ca isotopic measurements eliminate the r-process, because it would overproduce ^{46}Ca well beyond the detection level despite its small abundance. No such excess is observed. The Ca isotopes alone do not permit one to distinguish between the other possible processes, and the other iron group elements are needed.

Unlike O, mass dependent fractionation is widespread for Ca in inclusions; it ranges from −3.8 to 6.7 ‰ (Niederer and Papanastassiou 1984) which is about four times the terrestrial range (Schmitt et al. 2003). However, 80% of samples fall within an interval of 2‰. The mass fractionation is the result of complex sequences of condensation and vaporization. The connection to Mg isotopic fractionation is not obvious for these samples as the resolution of Mg measurements is much smaller.

- Titanium. Ti has 5 contiguous isotopes from mass 46 to 50. As with Ca, Ti shows clearly resolved effects only on the most neutron-rich isotope: ^{50}Ti (Heydegger et al. 1979; Niederer et al. 1981; Niemeyer and Lugmair 1981). As for ^{48}Ca, a narrow range of excesses is found but the magnitude is about twice that of ^{48}Ca from 7 to 25 ε with an average of about 8 ε (Fig. 2b). The choice of the pair of isotopes to correct for instrumental *fractionation* is not of much importance. The high precision of the measurements gives hints of variations in other isotopes like ^{47}Ti or ^{49}Ti, but their size is an order of magnitude less than for ^{50}Ti. At least 3 different nucleosynthetic components are suspected, but this is better demonstrated in FUN inclusions to be discussed later. The main feature in “common” inclusions is the clear excess in ^{50}Ti.

  The *nucleosynthetic* sources for Ti isotopes are very similar to those of the isotopes of Ca, and ^{50}Ti requires a neutron-rich zone to be produced in significant amounts. In addition to the nonlinear effects, absolute isotopic compositions have been measured in a number of samples using double spike techniques (Niederer et al. 1985). Mass dependent fractionation effects are rarely resolved and are small, below 1 ‰/amu except in one sample, where it reaches 1.3 ‰/amu. In general the fractionation is in favor of the heavy isotopes; partial condensation or evaporation may explain of this observation.

- Chromium. Cr has four isotopes from mass 50 to 54. The choice of the normalizing isotope pair can be more ambiguous for this element (Birck and Allegre 1984; Papanastassiou 1986). ^{50}Cr and ^{52}Cr were used for this purpose because: 1) their nucleosynthetic properties are close to the non-anomalous isotopes of Ca-Ti, 2) excesses are expected on the neutron-rich isotope ^{54}Cr and 3) ^{53}Cr may be subject to variations related to the possible extinct radioactivity of ^{53}Mn. Indeed systematic ^{54}Cr excesses from 3 ε to 8 ε are found (Fig. 2c; Birck and Allegre 1984; Papanastassiou 1986; Birck and Allegre 1988; Bogdanovski et al. 2002). Variations of ^{53}Cr between −1.5 ε to +1 ε are related to ^{53}Mn decay and are discussed in section “53-Manganese –53-Chromium.” When comparing with ^{48}Ca and ^{50}Ti excesses, the ^{54}Cr variations are of similar magnitude and suggest a nucleosynthetic component resulting from neutron-rich statistical equilibrium in a presupernova massive star (5 to 20 solar masses), near the cut-off between the ejected layers and the neutron star (Hainebach et al. 1974; Hartmann et al. 1985).

- Iron. Fe has 4 isotopes of which the heaviest ^{58}Fe has a very small abundance of about 0.3%. The precision of thermal ionization mass spectrometers is around 10 ε on this isotope and there is only a hint in some "normal inclusions" for an excess in ^{58}Fe (Völkening and Papanastassiou 1989). Recent ICPMS measurements at the 2 ε precision level display normal isotopic compositions for Fe in planetary materials but no Allende inclusion was reported in this study (Kehm et al. 2003). If excesses of similar magnitude to ^{48}Ca, ^{50}Ti, ^{54}Cr were present they would not be clearly resolved in agreement with the observations. When ^{54}Fe and ^{56}Fe are used to correct for instrumental *mass* fractionation, ^{57}Fe exhibits normal abundances, suggesting all three isotopes are present in solar relative abundances.

- Nickel. Ni has 5 isotopes, of which ^{60}Ni can be the product of the extinct radioactivity of ^{60}Fe. When considering the possible nucleosynthetic processes at the origin of Ca,Ti,Cr anomalies, several heavy isotopes of Ni may be affected by anomalies leaving only ^{58}Ni the major isotope and ^{61}Ni a minor isotope to correct for instrumental fractionation (Birck and Lugmair 1988). There are only a few data available indicating a systematic excess of ^{62}Ni and ^{64}Ni averaging at about 1ε and 3ε respectively (Fig. 2d). Again considering the other iron group elements, the anomalies suggest a particular nucleosynthetic source: neutron-rich nuclear statistical equilibrium with multiple zone mixing (Hartmann et al. 1985).

- Zinc. Zn is the upper end of the iron group. Although a strong hint exists for excesses of about 0.6 ε in ^{66}Zn in some samples, the results are not as clear as for the other elements of the group (Loss and Lugmair 1990; Völkening and Papanastassiou 1990). The *measurements* show that the effects are about a factor ten smaller than predicted by the models of nucleosynthesis. This does not constitute a problem for the model, but is most probably due to the properties of Zn. Zn is among the volatile elements and has been shown to be orders of magnitude more mobile than the other members of the iron group in circumstellar envelopes (Van Winckel et al. 1992). The longer residence time in the gas phase probably results in a more thorough homogenization of Zn isotopes between the various reservoirs constituting the solar system.

Elements beyond zinc. Despite extensive investigations carried out in parallel with the "FUN" inclusions, clear-cut mass-independent isotopic heterogeneities (e.g., more than 5 σ of a single measurement) have not been identified yet in "normal" inclusions for the rest of the elements besides extinct radioactivities. Nevertheless, there are consistent hints for excesses of about 2 ε in ^{96}Zr (Harper et al. 1991; Schönbächler et al. 2003) which is an r-process isotope for which some coupling may exist with neutron-rich statistical equilibrium (Meyer 1994).

As a partial summary, in "normal" C3V inclusions material from the neutron-rich nuclear statistical equilibrium nucleosynthetic process is in excess relative to the average solar system composition, as well as an ^{16}O-rich component. Nevertheless, the exact composition of this material is somewhat blurred by secondary processes (nebular or interstellar) as the observations show no strict interelement correlation (Jungck et al. 1984; Birck and Lugmair 1988).

"FUN" inclusions.

This group is made up of a very few individual inclusions found among Allende studies. In fact it can be reduced to the two most striking samples: C1 and EK141. EK141 is unique in its isotopic properties, whereas a number of inclusions are closely related to C1 in their isotopic compositions (Papanastassiou and Brigham 1989; Loss et al. 1994). The discussion will therefore be limited to these two particular objects. C1 and EK141 have very similar isotopic patterns for O and Mg but differ radically in the other isotopes (Clayton and Mayeda 1977; Wasserburg et al. 1977).

Oxygen and magnesium. FUN inclusion samples do not plot on the O mixing-line between a solar nebular component and the exotic ^{16}O rich component (Fig. 1). They plot on the ^{18}O-rich side of this line and are thought to have been produced in several successive steps (Clayton and Mayeda 1977). They started as normal inclusions with a 4% ^{16}O enrichment. Then there is an intermediate stage where O mass fractionation occurs along a slope 0.52. The last stage is the same as for normal inclusions: isotopic exchange with a reservoir having a more nearly normal composition, which reduces the anomalies in the most sensitive minerals like melilite, and leaves spinel almost unaffected.

Magnesium is mass fractionated by 2 to 3 %/amu in favor of the heavy isotopes (Wasserburg et al. 1977). This is about 2 times the fractionation of O. When canceling out this mass fractionation by normalizing to the terrestrial $^{25}Mg/^{24}Mg$ ratio, a deficit of ^{26}Mg of about 3 ‰ is found in both inclusions. This is generally not believed to be a deficit of radiogenic ^{26}Mg (from ^{26}Al) but to some unidentified nucleosynthetic effect on one of the 2 other isotopes of Mg: ^{24}Mg, ^{25}Mg.

Silicon. Normal inclusions are spread on a mass dependent fractionation line over a few ‰/amu around the solar system average. FUN inclusions display a heavy isotope enrichment from 5 to 15 ‰/amu in a similar way to Mg. Non-linear effects are small and indicate an excess of ^{29}Si smaller than 0.5 ‰ (Clayton et al. 1984).

The iron group elements. Figure 3 displays the isotope ratios for this group. As a general result, the most neutron-rich isotope nuclei of this group display the largest variations relative to the other isotopes. Deficits are seen for C1, and excesses for EK141.

- Calcium. ^{48}Ca has a 20 ε deficit in C1 and a 141 ε excess in EK141. Clear differences from terrestrial compositions are also apparent for all other isotopes, with magnitudes ranging from 5 to 20 ε. Whatever the choice is for the pair of normalizing isotopes (^{46}Ca is not taken because of its very small abundance), all the others are non-solar. Also, a large mass dependent fractionation of similar amplitude to the non-linear features is apparent between the two samples (Fig. 3a; Lee et al. 1978; Niederer and Papanastassiou 1984) . All this together does not allow one to determine which isotope pair should be used for the normalization. Nevertheless, besides the excess in EK141 and the deficit in C1 of the component originating in neutron-rich statistical equilibrium, there is at least one other nucleosynthetic component present in variable amounts. This component may originate in explosive He or O burning shells, which are the main producers of the lighter Ca isotopes (Clayton 1988).
- Titanium. The isotopic effects are very similar to those of Ca, with the largest effects on the most neutron-rich isotope (Fig. 3b; Niederer et al. 1981). A deficit and an excess on ^{50}Ti for C1 (−51 ε) and EK141 (+37 ε) are observed respectively and agree with Ca data, suggesting a mixing model with a component produced in a neutron-rich nuclear statistical equilibrium. As for Ca the presence of isotopic anomalies up to 15 ε for the other isotopes of Ti and of mass dependent fractionation of up to 0.8 ‰/amu leaves an ambiguity in attributing the anomalies to a component that was produced by a specific nucleosynthetic process (Niederer et al. 1985). Nucleosynthetic processes which can contribute to the production of Ti isotopes are numerous and have been discussed extensively in conjunction with the Ca isotopic data (Niederer et al. 1981; Clayton 1988). Nevertheless in C1, the pattern of Ti isotopes may indicate a deficit of ^{46}Ti, the second most anomalous isotope after ^{50}Ti. In that case one can speculate that the signature of explosive O burning in a supernova shell (Woosley et al. 1973) is indicated.

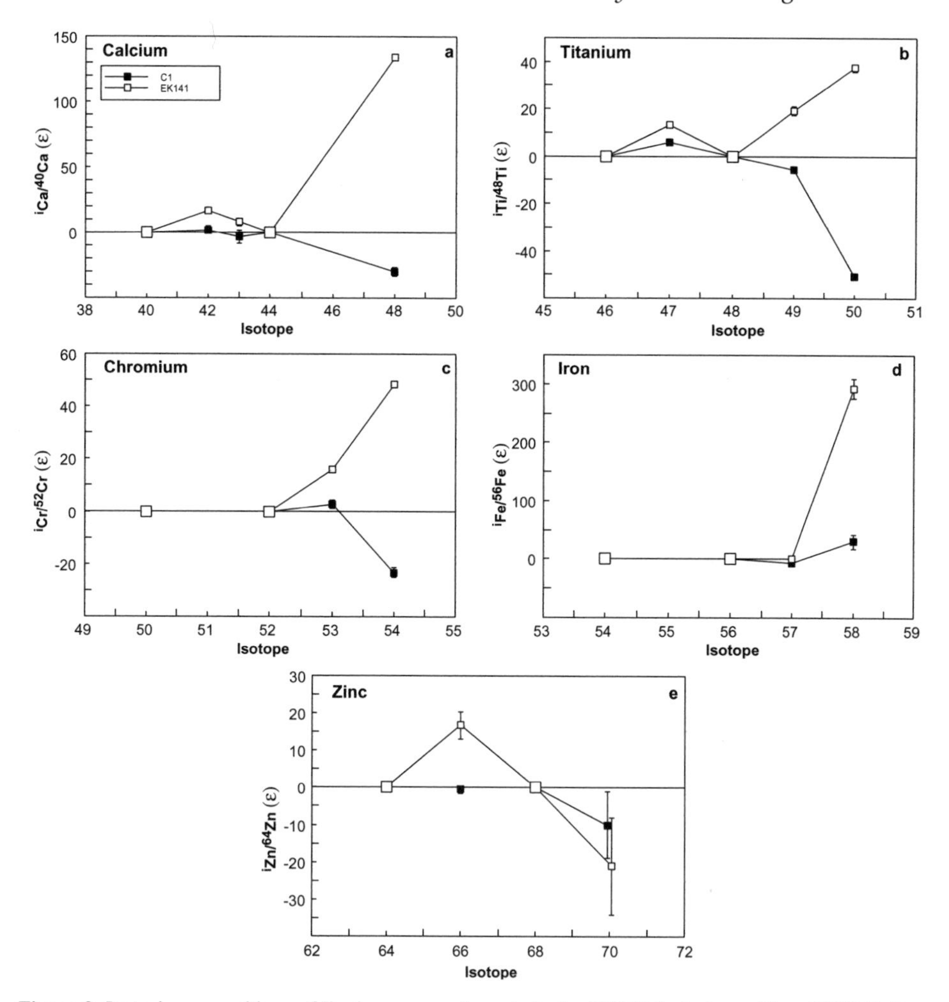

Figure 3. Isotopic compositions of the iron group elements in the "FUN" inclusions: C1 and EK141 (Lee et al. 1978; Niederer et al. 1985; Papanastassiou 1986; Völkening and Papanastassiou 1989; Völkening and Papanastassiou 1990). Scales are as in Figure 2. ■ and □ are used for C1 and EK141 respectively. The individual errors are represented but are often smaller than the dots representing the data. Isotopic variations in "FUN" inclusions are more than an order of magnitude larger than in "normal" inclusions. Both excesses and deficits in the neutron-rich statistical equilibrium produced component are present. EK141 is unique among inclusions, in showing large isotopic variation of similar amplitude for several isotopes in most elements.

- Chromium. The pattern is again very similar to Ti and Ca, with large excesses of the most neutron-rich isotope ^{54}Cr (48 ε) in EK141 and a large deficit in C1 (−23 ε) (Fig. 3c; Papanastassiou 1986). Unlike normal inclusions, ^{53}Cr exhibits a large excess in both inclusions. In nucleosynthetic models, this isotope is mostly produced as ^{53}Mn which decays to ^{53}Cr (Hainebach et al. 1974; Hartmann et al. 1985). Whether this happened in the source of EK141 or that some unidentified process produced ^{53}Cr directly, is not settled yet. The presence of mass dependent fractionation has not been investigated in details for Cr. This is also the case for the 3 remaining iron group elements: Fe, Ni and Zn.

- Iron, zinc (Fig. 3d,e). Nickel has not been investigated in FUN inclusions. For Fe (Völkening and Papanastassiou 1989) and Zn (Völkening and Papanastassiou 1990) only EK141 displays clear anomalies, with excesses in ^{58}Fe (292 ε) and in ^{66}Zn (15.5 ε). C1 has only a hint of an excess in ^{58}Fe (30 ± 15 ε). In an inclusion similar to C1, but extracted from the Vigarano meteorite, a deficit in ^{66}Zn has been evidenced by Loss et al. (1994). This is in agreement with the deficit in neutron-rich isotopes of the other iron-group elements in the same inclusion. The data of EK141 are in agreement with the model deduced from the other iron group elements

Heavy elements (Fig. 4). This is the domain where the FUN inclusions differ the most from normal inclusions. With anomalies in the range of a few ε to a few ‰, and also possible mass dependent fractionation, there can be an ambiguity in using a given isotopic pair to normalize for instrumental fractionation. The corrections may result in attribution of some isotopic anomalies which are actually normal. This can be avoided with a minimum number of assumptions. The heavy elements beyond Zn in the solar system are produced through 3 nucleosynthetic processes: p, r, and s (Burbidge et al. 1957). Heavy elements often have between 5 and 10 isotopes, and the three processes are required to produce all the various isotopes of a single element. To interpret the observed anomalies, the normalization of the measured isotopic ratios can be tried in using a pair of isotopes produced in only one of these processes. This is not always possible and inter-element comparison is also required (Lugmair et al. 1978; McCulloch and Wasserburg 1978a; Lee 1988).

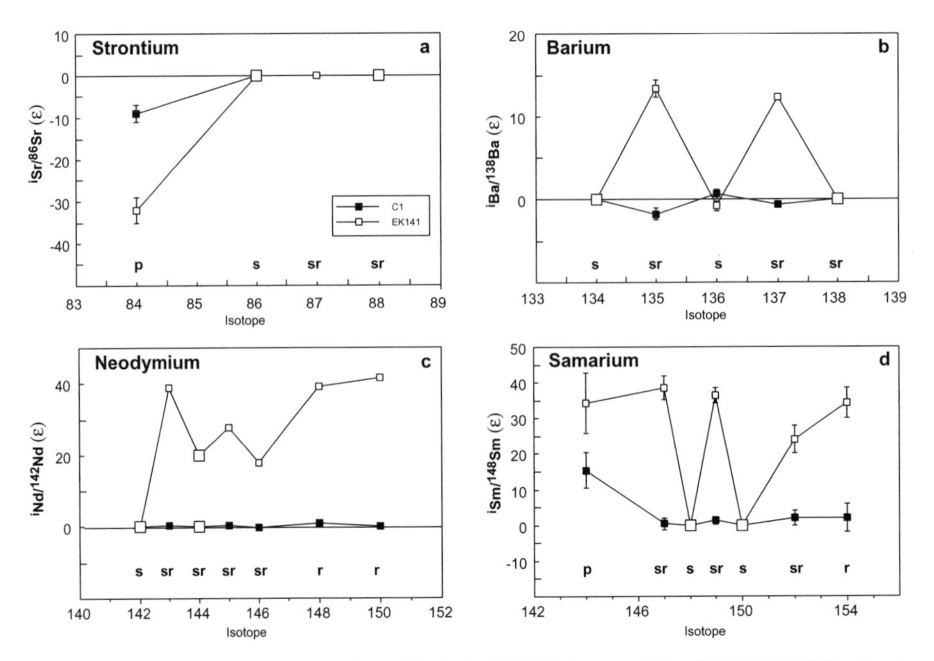

Figure 4. Non-linear effects for Sr, Ba, Nd and Sm in the "FUN" inclusions: C1 and EK141 (McCulloch and Wasserburg 1978a,b; Papanastassiou and Wasserburg 1978). Relative deviations from terrestrial standard ratios are plotted after normalization with the isotope pair represented with large open squares. Each isotope is labeled with its primary nucleosynthetic source. In using s-process isotopes for normalization, clear excesses in r-process nuclei are seen for Ba and Sm in EK141. Sr is normal in both inclusions except for a deficit in the p-process ^{84}Sr. As Nd has only one pure s-process isotope at mass 142, the data in EK141 have been further corrected to yield an excess in ^{150}Nd identical to that of ^{154}Sm as these two isotope are pure r-process nuclei expected to be produced in comparable abundances.

- Strontium. C1 and EK141 display deficits of 8 ε and 32 ε in the least abundant isotope ^{84}Sr which is a p-process nucleus (Fig. 4a; Papanastassiou and Wasserburg 1978). The other three isotopes ^{86}Sr-^{87}Sr-^{88}Sr, which result mainly from s-process nucleosynthesis, are in normal proportions. The isotopic pattern could be interpreted as an extra addition to average matter of a s-process enriched component. This seems implausible because this s-process component would have to have precisely the solar composition for the 3 heavy Sr isotopes, and it is expected that the many stars which contributed to the solar system mix did not produce all 3 isotopes in exactly the same pattern. In C1 and EK141, the contrasting p deficits in the Sr mass region with p-r excesses in the Ba-REE region require that the nucleosynthesis in these two mass regions occurred in distinct nucleosynthetic sites. Two components can be sought: one with the p deficit contributes to the mass region $A < 90$ and the second to the mass region >130.

- Barium, neodymium, samarium (Fig. 4b,c,d). For these 3 elements, C1 displays normal isotopic compositions, except for an excess of the p-produced isotope ^{144}Sm. In EK141, and using s-only isotopes for normalization, clearly resolved excesses of 10 to 40 ε are found for r-produced isotopes of Ba, Nd and Sm (McCulloch and Wasserburg 1978a; McCulloch and Wasserburg 1978b). The remarkable feature is that the particular r-process component found in the FUN inclusion EK141 is like that of the average pattern of r-process isotopes in the solar system.

Inclusions of the CV3 led to the search for isotopic signatures of individual nucleosynthetic processes, or at least for components closer to the original signature than average solar compositions. They have also begun to demonstrate the isotopic variability of matter emerging from these processes in agreement with astrophysical and astronomical expectations. The principal features of inclusions are: an up to 4% ^{16}O enriched reservoir in the early solar system, variations in a component produced in a nuclear neutron-rich statistical equilibrium, and variations in the contribution of p- and r-process products to the heavy elements.

Hibonite rich inclusions of the CM2 and oxide grains in carbonaceous chondrites

Hibonites $CaAl(Ti,Mg)_{12}O_{19}$ is one of the most refractory minerals found in primitive meteorites (Grossman 1972; Lattimer et al. 1978) and is expected to condense at high temperatures just after corundum from a cooling gas of solar composition. Hibonite-bearing inclusions and spherules are common in CM2 meteorites, of which Murchison is the most studied (MacPherson et al. 1983). These inclusions are similar to Allende inclusions in several aspects: they carry isotopic anomalies and they have undergone secondary alteration on their exterior. Hibonite inclusions are too small to be extracted by traditional means (Ireland et al. 1985; Hinton et al. 1987). In general, data are reported on single grains of 10-100 μm in size which have been classified according to their morphology, which correlates reasonably well with the isotopic characteristics (Ireland 1990). Almost all isotopic data available are ion probe measurements, and are often accompanied by measurements of a number of trace element abundances, especially REEs. REEs are refractory elements, and have the signature of earlier volatilization-condensation processes. One group of hibonite grains has a condensation pattern in which the most refractory REEs are depleted (Ireland 1990). Ion probe measurements have a reduced precision by about an order of magnitude (‰ level) relative to thermal ionization measurements. This is due to the small sample size and sometimes to the necessity to correct for isobaric interferences (e.g., ^{48}Ca-^{48}Ti). For minor or trace elements, the precision is even lower and the data base is not very large. However, large isotopic effects are present in these grains.

Magnesium. Corundum-hibonite associations are what could be the first condensates from a solar composition gas. Mg is not a refractory element and is strongly depleted in these inclusions. For an object with such a primitive composition, one would also expect the presence of ^{26}Al in significant amounts. ^{26}Mg effects related to ^{26}Al are rare, but correlated excess is present in others (Fahey et al. 1987; Ireland 1988). The uniform ^{26}Mg composition must be characteristic of the reservoir from which the corundum-hibonite formed. Formation of these primitive objects after ^{26}Al decay is difficult to reconcile with solar system formation models. ^{26}Mg heterogeneity in presolar reservoirs is preferred by some of the authors. Fahey et al. (1987) argue for heterogeneous distribution of ^{26}Al in the solar nebula. Mass-dependent fractionation favoring the heavy isotopes is present in some samples up to 1.8 %/amu resulting from distillation processes (Ireland 1988; Ireland 1990). Effects resulting from ^{26}Al decay (see below) range from 0 to the canonical value found in Allende Inclusions ($^{26}Al/^{27}Al = 5.1 \times 10^{-5}$ in Fig. 9b) and are mostly restricted to certain morphological types of hibonites (Ireland 1988); others classes show little or no evidence for past ^{26}Al effects.

Titanium-calcium. The first evidence for isotopic anomalies in the iron-group was found in Ti showing up to 10% excesses of ^{50}Ti in hibonites from the Murray CM2 meteorite (Hutcheon et al. 1983; Fahey et al. 1985; Ireland et al. 1985; Hinton et al. 1987). Further studies in Murchison showed that ^{50}Ti extended from −7% to +27% associated with ^{48}Ca variation from −6% to +10% (Ireland 1988; Ireland 1990). Except for the magnitude of the variations, this is similar to the results from Allende inclusions. Only a few samples display mass-dependent fractionation for which it ranges up to 1.3 %/amu. In the majority of the samples, it is absent or very low (less than 1 ‰/amu) for Ca-Ti. There is no correlation between the presence of linear fractionation and the magnitude of ^{50}Ti effects. ^{49}Ti variations are also present, but about an order of magnitude smaller than ^{50}Ti. Variations affecting these two isotopes are related but not strictly correlated (Ireland 1988).

As for Allende's inclusions, variable contributions of a component produced in neutron-rich nuclear statistical equilibrium best explains the ^{50}Ti-^{48}Ca data. Some parts of the solar nebula were depleted in these isotopes as deficits are also seen. There are several possibilities for explaining the variations in ^{49}Ti. 1) The neutron-rich component itself may be heterogeneous and incorporate locally less neutron-rich statistical equilibrium products (Hartmann et al. 1985). 2) ^{49}Ti may result from another process like explosive Si or He burning (Clayton 1988). This component would be associated with the neutron-rich component but not completely homogenized. In all cases, carriers are solid grains which may have behaved differently than the gaseous nebula during the formation of the solar system. A minimum number of components may be calculated to account for the Ca and Ti isotopic data, which number up to 3–4 (Ireland 1990) but to be conservative at the 5σ level, clearly resolved effects are present only on 3 isotopes (^{48}Ca,^{49}Ti, ^{50}Ti).

Despite anomalies up to 27%, titanium is still 98.4% of average solar isotopic composition which is a strong argument in favor of a solar system origin for the hibonite grains or aggregates. Little information is available on the isotopic ratios from the other elements. Oxygen isotopic compositions are within the range of Allende's inclusions based on the few data available (Hashimoto et al. 1986; Olsen et al. 1988).

"Presolar" stardust grains: diamond, graphite, carbides and oxides

The subject of this section is "acid-resistant" phases obtained from the matrix by dissolving away the major matrix minerals. They are understood as refractory phases, which is true in general but not strictly equivalent. Diamonds represent at most a few hundred ppm of the total mass of the meteorite, and silicon carbide a few ppm only, but nevertheless, thousands of grains have been analyzed and the presolar origin of the grains is indicated by enormous isotopic ratio differences relative to solar system average. Isotopes of the CNO elements are normally not part of this volume, but in these grains variations are so large that there is no doubt of their nucleosynthetic origin. The word "stardust" is a generic term for the grains.

The first indications of carriers of fresh nucleosynthetic products came from the work on the isotopes of Ne which indicated a component enriched in ^{22}Ne (Black and Pepin 1969; Black 1972). Further work eventually led to the isolation of a carrier of more than 99% pure in ^{22}Ne (Jungck and Eberhardt 1979; Eberhardt et al. 1981). In the mean time, Xe also revealed several components located primarily in trace phases. It turned out that Xe-HL showing the most striking isotopic differences from the planetary average noble gas patterns resides in acid-resistant phases (Lewis et al. 1975). The preparation procedures for isolating presolar grains have been improved over time and involve dissolution of the matrix of the rock and further physico-chemical separations by grain size and density (Tang et al. 1988; Amari et al. 1994). There is now a data base of isotopic analyses on thousands of individuals grains and this chapter can only be a rapid glimpse to the isotopic variety in these objects. For more details see Bernatowicz and Zinner (1997).

Nanodiamonds. Nanodiamonds were first isolated by Lewis et al. (1987). They are very small, with an average size of 2nm and are a stable phase relative to graphite in H-rich environments (Nuth 1987; Badziag et al. 1990). Their small size precludes the analysis of single grains, and isotopic data are averages of many grains. C isotope ratios are close to the average solar value and N which constitutes up to 0.9% of the diamond, is depleted by as much as 35% in ^{15}N (Virag et al. 1989; Russell et al. 1991). Three distinct noble gas components are present in diamonds: two with close to normal isotopic compositions and one with anomalous isotopic compositions, in particular Xe-HL and Kr-H (Fig. 5; Huss and Lewis 1994). The favored explanation of the Xe-HL pattern is that it was produced in supernovae by a combination of p-and r-process (Clayton 1989). According to the small size of the diamonds, only 1 grain in 10^6 has a single Xe atom. Thus it is possible that most diamonds come from other sources (Alexander 1997) and that most of them could in fact be produced in the solar system (Dai et al. 2002). Thermal processing before accretion is also indicated by the data (Huss et al. 2003). Very few other isotopic data exist on these objects. A study on Sr and Ba isotopes showed only marginal effects occurring at the r-process isotopes, which is in agreement with a supernova origin (Lewis et al. 1983, 1991). Te isotopic variations point to a component with virtually no light isotopes and requires a separation of the r-process isotopes of Te-Xe from their precursors within a few hours (Ott 1996; Richter et al. 1998). The supernova component found in nanodiamond has to go through series of complex processes to account for the observation.

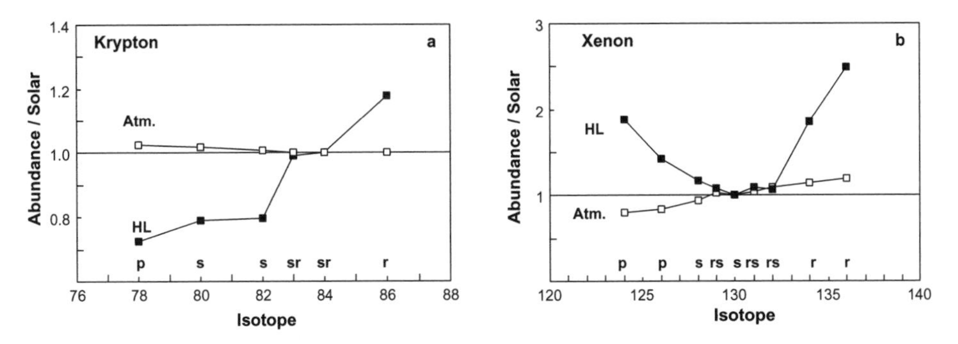

Figure 5. Relative abundances of the Kr, Xe isotopes (Huss and Lewis 1994) in presolar diamonds have been measured in bulk samples (= many grains) and are plotted relative to solar wind abundances. The terrestrial atmosphere is shown for comparison and displays a pattern close to mass dependent fractionation relative to the solar wind. The primary nucleosynthetic processes at the origin of the different nuclei are also listed. Both Kr and Xe are elevated in the r-process isotopes, whereas only Xe is also enriched in the p-isotopes. These patterns are a strong argument in favor of a supernova origin for the diamonds. Ne isotopes in presolar diamond is within the field of bulk meteorite data.

Silicon carbide and graphite grains. An overview of the thousands of data available is displayed in Figures 6 and 7. This fraction resulting from the separation procedure has been shown to be the carrier of s-process Xe (Xe-s). Graphitic carbon and silicon carbide have been shown to be the carrier of Ne E (Fig. 7a). The nucleosynthetic source of Neon E (pure ^{22}Ne) has long been thought to be ^{22}Na ($T_{1/2}$ = 2.6 yr) (Black 1972; Brown and Clayton 1992), but can also be produced in He-burning shells of AGB stars, like the CNO anomalies (Ott 1993). This question is not settled yet.

Grains larger than 0.5 μm are common and can be individually analyzed. Tremendous variations, over 4 orders of magnitude, are observed in the isotopic ratios of CNO (Fig. 6) and over 1.5 order of magnitude for Si. If the isotope ratios reflect the nucleosynthetic processes without too much intercomponent mixing and fractionation, then the variations in most grains reflect various degrees of He shell and H shell burning, and mixing in AGB stars. The data base is now numerous and reviews exist in which references can be found (Anders and Zinner 1993; Busso et al. 1999; Nittler 2003).

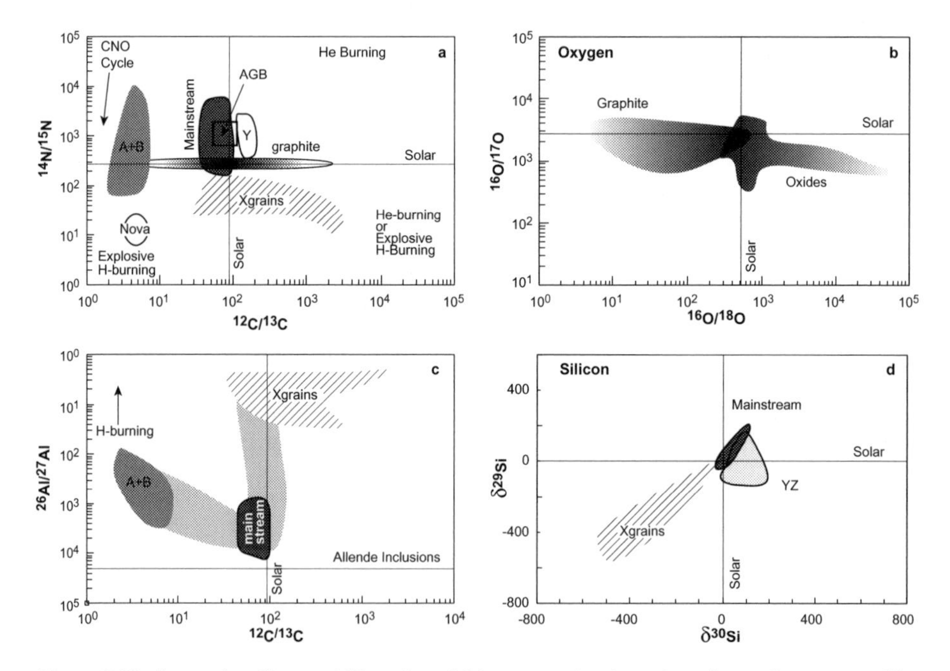

Figure 6. Single presolar silicon carbide grains exhibit a tremendous isotopic variety and are separated in several populations on the basis of grain size and C-N-O-Si isotopic properties. Graphite grains are also represented in (a) for comparison. Figure after Hoppe and Ott (1997), Zinner (1997), Amari et al. (2001) and Nittler (2003). The data are plotted as absolute values in (a), (b), (c), and the solar average value is also shown. Si is plotted as ‰ deviations relative to the terrestrial values in (d). ^{26}Al/^{27}Al is calculated assuming that all excess ^{26}Mg found in the samples results from ^{26}Al decay. All data displayed have been obtained on individual grains ranging from 0.3 to 5 μm in size by ion probe measurements (Amari et al. 1994, 2001). Model predictions in AGB stars are also shown. X grains are thought to be produced in supernovae (Nittler et al. 1996; Besmehn and Hoppe 2003). The data represented here represent thousands of individual grains. The main stream of grains (more than 90% by number) are produced at different stages of evolution of AGB stars. Oxygen isotopic data are only available in graphite and oxide grains. In (a) and (b) the gradation of the shading reflects roughly the data density of population in the given areas of the diagram. There are a few grains located in the space between the defined populations; they are represented by shaded areas. There are also a very few grains with extraordinary isotopic signatures located outside the fields displayed here and which are not represented here for clarity.

Large excesses in ^{26}Mg have been interpreted as resulting from the decay of ^{26}Al (Fig. 6c; Hoppe et al. 1994; Huss et al. 1997). Initial values as high as 0.5 have been found in $^{26}Al/^{27}Al$. High ^{26}Al is produced in the H-burning shell, but, if it can match such high ratios in agreement with other elements, is not known yet. Calcium and titanium display excesses in ^{42}Ca, ^{43}Ca, ^{44}Ca and a V shaped pattern for Ti (Fig. 7b; Amari et al. 2001). The results are qualitatively consistent with a s-process during He burning.

The heavy elements carry clear excesses in a s-process component as can be seen from Figure 7. This has been demonstrated for Kr, Sr, Xe, Ba, Nd, Sm (Ott and Begemann 1990; Prombo et al. 1993; Lewis et al. 1994; Richter 1995; Hoppe and Ott 1997). The models can be made to fit very precisely the measured data (Lattanzio and Boothroyd 1997; Busso et al. 1999). Mo and Zr can occur as microcrystals of Mo-Zr-C within graphite grains. Typical s-process patterns are observed, with isotopic variations of about a factor of more than 5 (Nicolussi et al. 1997; Nicolussi et al. 1998a).

Some X-SiC grains possess r-process excesses (see Nittler, 2003, and references therein) as well as a fraction of the graphite grains (Nicolussi et al. 1998c). The variability in the grains is such that subgroups (the X SiC, low density graphite and Si_3N_4 grains) carry the signature of supernova nucleosynthesis in Si (Amari et al. 1999), as well as for excess ^{44}Ca interpreted as decay product from ^{44}Ti (Nittler et al. 1996; Amari and Zinner 1997; Besmehn and Hoppe 2003).

Oxide grains. Corundum, hibonite and spinel represent a very minor proportion of the acid resistant residue (Nittler et al. 1994; Zinner et al. 2003) in the range of 0.05 ppm of the bulk meteorite. This may also be due to their having a smaller size distribution or the grains are more susceptible to destruction in the meteorite and in sample preparation. Only a small number of such grains has been analyzed and among these less than 10 have proven to possess very exotic isotopic signatures (Huss et al. 1994). A few other grains with similar signatures have been found in Murchison and Bishunpur which is a low metamorphic grade ordinary chondrite of type LL3 (Choi et al. 1998).

In these investigations, due to practical limitations, measurements have been limited to grains larger than about 1 μm in size, but progress in the instrumentation will lead to more subtle phases and intra-grain analyses. Most probably new effects will be seen, and will increase the observed variability in isotopic components. The modeling of nucleosynthesis and circumstellar processes is also at stake. The puzzle at the moment is quite complicated as different nucleosynthetic sources (AGB stars, supernovae) are represented in the major families of grains, except for the first one containing the supernova diamonds. It turns out that some stellar zone are very specific isotopic "spike" factories, which thereafter mix with other components depending on stellar dynamics.

Solar system processes are also among the targets, because of the disappearance of this heterogeneity with the metamorphic grade of the meteorites. Studies of temperature and duration of metamorphism are still in their early stages (El Goresy et al. 1995; Huss 1997)

THE ISOTOPIC HETEROGENEITY IN NON-REFRACTORY EARLY SOLAR SYSTEM MATERIALS

Presolar stardust discovered to date in meteorites constitute no more than 0.5‰ of the total mass of the samples, and one common chemical property is that they are acid resistant. Isotopic heterogeneity could also be present in less refractory phases like silicates, provided parent body metamorphism did not erase the differences. Noble gases are not discussed here because they are depleted by many orders of magnitude relative to the Sun and can be dominated by trace exotic minerals. Nitrogen is not discussed for the same reason. The

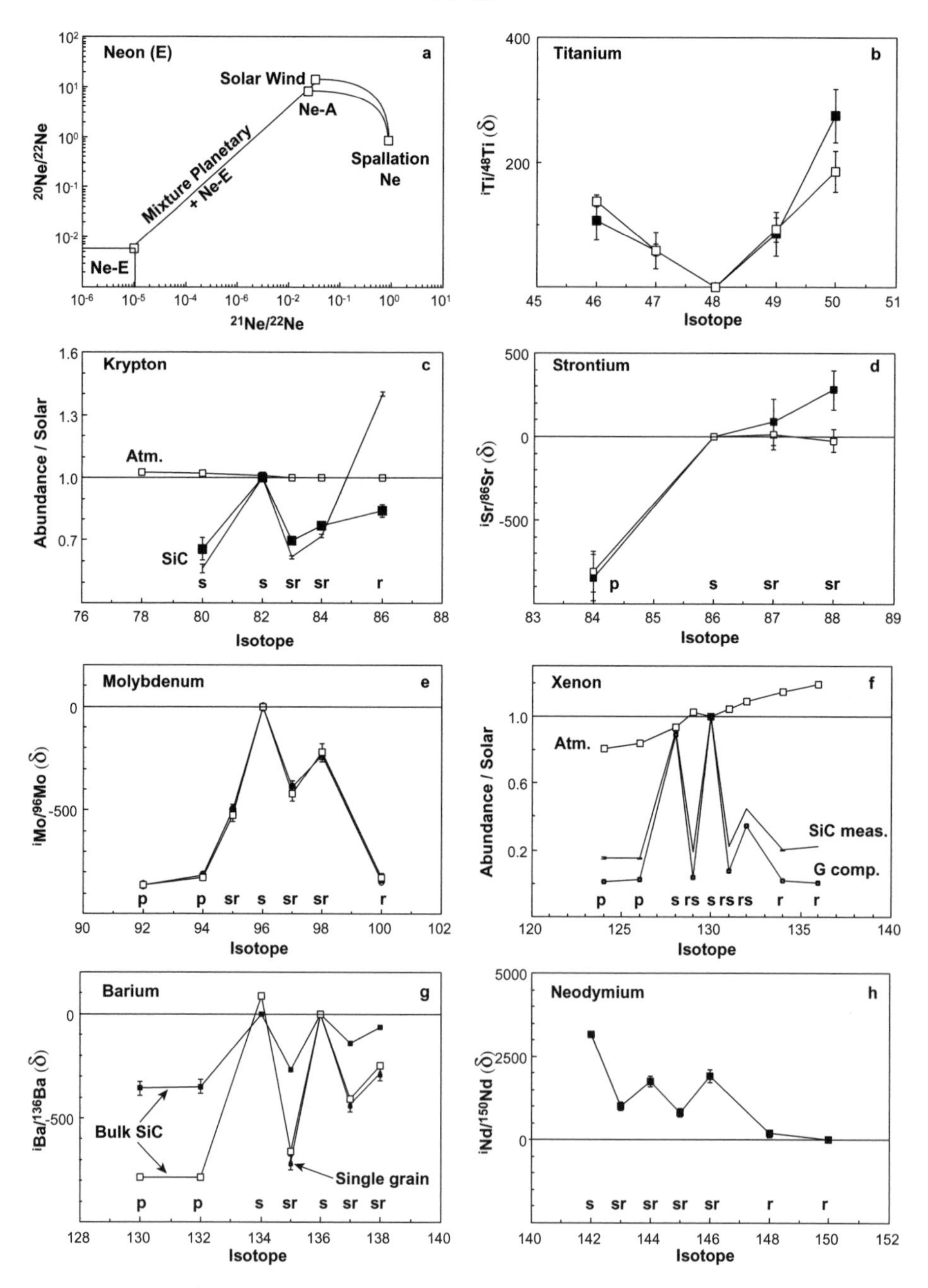

Figure 7. A few examples of isotopic patterns of Ne, Ti and heavy elements in SiC and graphite grains are displayed. Absolute ratios are plotted for Ne (a) whereas abundance ratios relative to solar wind composition are plotted for Kr (c) and Xe (f). The remaining elements are plotted as ‰ deviations from laboratory standards. The data have been obtained on bulk SiC separates by traditional mass spectrometry for Ne (Jungck and Eberhardt 1979), Kr (Ott et al. 1988; Lewis et al. 1994), Sr, Ba (Ott and Begemann 1990; Prombo et al. 1993) and Nd (Richter 1995). SIMS techniques *(caption continued on facing page)*

purpose here is to look to elements which are close to bulk solar proportion when compared to Si as a reference. This field has not received as much attention as the refractory grains but a few striking examples of isotopic diversity exist in separates from bulk samples.

Oxygen

The O isotopes show significant heterogeneity between the different meteorite classes (Fig. 8a; Clayton et al. 1976, 1977). Differences are small, but, each chondrite group has a distinct bulk O isotopic composition. O isotopes also indicate the close ties between the Earth and the Moon. O therefore can be used to identify members of a family that formed from a common reservoir, which is the definition of a tracer. Such differences are also found between chondrules within the same meteorites related to their size (Gooding et al. 1983). This is a survival of the initial isotopic heterogeneity in already high temperature processed materials like chondrules.

Chromium

C1 chondrites have very fine-grained predominantly phyllosilicate mineralogy. Mineral separates of phyllosilicates are not practically achievable. In using increasing strength acids, the stepwise dissolution/leaching procedure exhibits large isotopic heterogeneity for Cr isotopes (Fig. 8b) which represents about 0.3% of the mass. The size of the non-linear variations ranges up to 2.2% for the most neutron-rich isotope ^{54}Cr (Rotaru et al. 1992; Podosek et al. 1997).Variation on ^{53}Cr are smaller and have been interpreted as resulting from ^{53}Mn decay.

54-Chromium. The isotopic heterogeneity is limited to this isotope which can be compared with the "normal" refractory inclusions of Allende. Both ^{54}Cr deficits and excesses are found ranging from $-7.6\ \varepsilon$ to $+210\ \varepsilon$ (Fig. 8b). The fractions showing the highest enrichment in ^{54}Cr with no correlated effects in ^{50}Cr, ^{52}Cr, ^{53}Cr points towards a nucleosynthetic component, which is 99% pure in ^{54}Cr. This component is probably the same as the component found in the CV3 inclusions, and which is produced in a neutron-rich nuclear statistical equilibrium in presupernova massive stars.

None of the dissolution fractions has normal isotopic compositions. This implies that the solar system results from mixing of major Cr-bearing components, none having the solar (terrestrial) isotopic compositions. Data obtained from higher metamorphic grade carbonaceous chondrites show that the isotopic heterogeneity is erased by parent body metamorphism with an amplitude decreasing from 220 ε in the C1 chondrites, through 40 ε in

---

Figure 7 *(caption continued from facing page).*

have been used for single grain analysis of Ti (Amari et al. 2001) and by laser-ablation-RIMS techniques for Sr (Nicolussi et al. 1998b), Mo (Nicolussi et al. 1998a) and Ba (Savina et al. 2003a). Bulk analyses mostly reflect the isotopic ratios of the main stream population. The magnitude of the effects does not require one to correct for mass fractionation. As displayed on one sample for Ba, individual grains are often close to the bulk samples. With the exception of Nd for which a r-process isotope is taken for reference, abundant s-process isotopes are usually taken as reference to compare the compositions. In the resulting patterns, the isotopes produced by p-, r-process are highly depleted or absent within the resolution of the experiments. These results point to strongly s-process dominated components originating in AGB stars. The spectrum of Kr is somewhat more complex with, in some sets of data, an over-abundance of r-process ^{86}Kr. Adjustments in the neutron exposure within the star can resolve this apparent discrepancy. It is also noticeable that ^{86}Sr, ^{87}Sr, ^{88}Sr are produced in proportion close to the average solar system composition despite the variability in the neighboring elements. In general for an element, numerous individual grain data are available and the measured composition can be interpreted as a mixture of a component close to normal solar system composition and a component representing the signature of the nucleosynthetic process (G-component). This G-component is represented for Xe. It can be seen that with carefully prepared samples the G-component is dominant and precisely defined. For the other elements, grains very close to the original nucleosynthetic isotopic signature are also found.

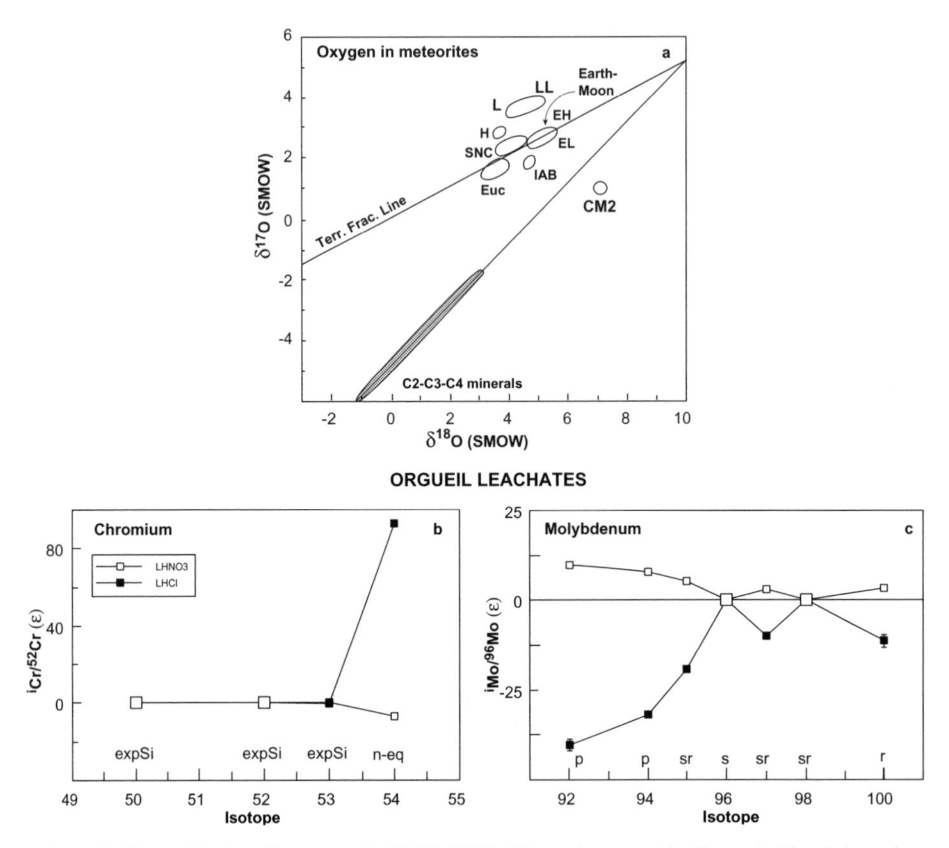

Figure 8. Figure (a) after Clayton et al. (1976, 1977). The scales are as in Figure 1. The O isotopic compositions of the different meteorite classes are represented: ordinary chondrites (H, L, LL), enstatite chondrites (EH, EL), differentiated meteorites (eucrites, IAB irons, SNCs) and some components of the carbonaceous chondrites. As the different areas do not overlap, a classification of the meteorites can be drawn based on O isotopes. Cr (b) and Mo (c) isotope compositions obtained by stepwise dissolution of the C1 carbonaceous chondrite Orgueil (Rotaru et al. 1992; Dauphas et al. 2002), are plotted as deviations relative to the terrestrial composition in ε units. Isotopes are labeled according to their primary nucleosynthetic sources. "ExpSi" is for explosive Si burning and "n-eq" is for neutron-rich nuclear statistical equilibrium. The open squares represent a HNO_3 4 N leachate at room temperature. The filled square correspond to the dissolution of the main silicate phase in a HCl-HF mix. The M pattern for Mo in the silicates is similar to the s-process component found in micron-size SiC presolar grains as shown in Figure 7.

the C2 chondrites, and to below 5 ε in the C3 chondrites. Ti and Fe isotopic analysis at the 2 ε precision level have shown no evidence for isotopic anomalies in the same fractions of the C1 Orgueil. From the CAI observations of related effects between ^{54}Cr, ^{48}Ca, ^{50}Ti, ^{58}Fe effects of a few tens of ε are expected for Ti and Fe. This is not observed and is most probably due to the mineralogy of the matrix (Rotaru et al. 1992). Cr is present in at least 2 different carriers behaving differently in the dissolution procedure: one with depleted ^{54}Cr and one with the neutron-rich component carrying pure ^{54}Cr. Ti and Fe are obviously not in the same carriers. The two nucleosynthetic components of Ti and Fe are probably also present in the samples but they are located in similar mineralogical sites which so far could not be resolved.

Bulk rock analyses of carbonaceous chondrites exhibit a ^{54}Cr excess from 1 to 2 ε (Rotaru et al. 1992; Shukolyukov et al. 2003). The carbonaceous chondrites are not exactly solar in their Cr isotopic bulk composition, but taking into account that the components are more than 220 ε apart, the match is very close and the idea that C chondrites are a fair representation of the solar system average is still reasonable.

53-Chromium. Much smaller variations are evident in this isotope: usually a few ε. They are not correlated with ^{54}Cr variations but with the Mn/Cr elemental ratios. They are also present in the higher metamorphic grades with even larger spreads. ^{53}Cr is not produced directly in stars but through ^{53}Mn which then decays to ^{53}Cr. These arguments favor the interpretation of ^{53}Cr as solely due to in situ ^{53}Mn decay (Birck et al. 1999).

Molybdenum

The same stepwise dissolution procedure revealed Mo isotopic variations of up to 50 ε (Dauphas et al. 2002). Mo belongs to the heavy elements beyond the iron peak. It has r-, s-process isotopes but differs from most other elements in having two abundant p-isotopes ^{92}Mo (14.8%) and ^{94}Mo (9.2%). ^{94}Mo may also be s-produced. The largest differences from the terrestrial or average solar composition appear in the same steps as the most striking ^{54}Cr isotopic effects (Fig. 8c). As for Cr, very little Mo has average solar system composition. The largest variation corresponds to an excess of s-process Mo as was observed in SiC presolar grains (Nicolussi et al. 1998a) and the host phase of this particular component is thought to be SiC (Dauphas et al. 2002).The complementary pattern is not expected to be in a specific host phase because r and s components have no reason to be coupled. Variable Mo isotopic compositions have also been found in iron meteorite with differences between classes of up to 4 ε on ^{92}Mo but with no differences within a given class . The Cr and Mo data demonstrate that the signature of nucleosynthesis is also present in major mineral phases of primitive meteorites and parent body metamorphism is an efficient homogenization process (Rotaru et al. 1992).

There are also indications that Ru displays systematic deficits in ^{100}Ru in iron meteorites and in Allende relative to terrestrial composition (Chen et al. 2003). A deficit in s-process isotopes of Ru is the favored interpretation.

EXTINCT RADIOACTIVITIES

Isotopic variations can be generated from a once uniform reservoir by decay of radioactive nuclides. Extinct radioactivities are the radioactive nuclides which were present in the early system, but have now completely decayed. Table 1 shows the radioactive nuclei that have been positively or tentatively identified. The separation from long period chronometers is between ^{40}K ($T_{1/2}$ = 1.2 b.y.), which is still present, and ^{146}Sm ($T_{1/2}$ = 104 m.y.) of which a few atoms could theoretically still be detected, but not with standard instrumentation. Half-lives from a few m.y. to 100 m.y. cover the time scale which is expected for solar system and planet formation. Occasionally nuclides of a few years half-life have been sought to explain isotopic variations as for NeE in presolar silicon carbide. A number of reviews have been published on extinct radioactivities and their possible nucleosynthetic sources (Wasserburg and Papanastassiou 1982; Arnould et al. 2000; Goswami and Vanhala 2000; Russell et al. 2001). I present here only the broad outlines of this field. Extinct radioactivities are demonstrated by the existence, in coexisting phases, of a positive correlation between the elemental parent-daughter ratio and isotopic variation involving the daughter isotope. Due to their short half lives, the nuclides have small abundances compared to stable isotopes accumulated during 10 b.y. of galactic evolution. Moreover the protostellar cloud could have stayed closed to nucleosynthetic inputs before contraction into the solar system (Wasserburg and Papanastassiou 1982). This would further reduce the abundances of the short lived

Table 1. Radionuclides of half-life from 0.01 m.y. to 150 m.y.

Parent	*Daughter*	*Half-Life (m.y.)*	*Detected (§)*	*Initial solar system Abundance*	*Reference*
^{10}Be	^{10}B	1.5	+	$^{10}Be/^{9}Be = 9\times10^{-4}$	McKeegan et al. 2000
^{26}Al	^{26}Mg	0.73	+	$^{26}Al/^{27}Al = 5.1\times10^{-5}$	Lee et al. 1977
^{36}Cl	^{36}Ar	0.28	–		
^{41}Ca	^{41}K	0.1	+	$^{41}Ca/^{40}Ca = 1.4\times10^{-8}$	Sahijpal et al. 1998
^{53}Mn	^{53}Cr	3.7	+	$^{53}Mn/^{55}Mn = 4.4\times10^{-5}$	Birck and Allegre 1985
^{60}Fe	^{60}Ni	1.5	+	$^{60}Fe/^{56}Fe \approx 1.5\times10^{-6}$	Birck and Lugmair 1988
^{79}Se	^{79}Br	0.06	–		
^{81}Kr	^{81}Br	0.21	–		
^{92}Nb	^{92}Zr	36	+	$^{92}Nb/^{93}Nb \approx 1\times10^{-5}$	Schönbächler et al. 2002
^{97}Tc	^{97}Mo	2.7	?		
^{98}Tc	^{98}Ru	4.3	?		
^{99}Tc	^{99}Ru	0.21	?		
^{107}Pd	^{107}Ag	6.5	+	$^{107}Pd/^{108}Pd = 2.4\times10^{-5}$	Chen and Wasserburg 1990
^{129}I	^{129}Xe	16	+	$^{129}I/^{127}I \approx 1\times10^{-4}$	Reynolds 1963
^{135}Cs	^{135}Ba	2.3	?		
^{137}La	^{137}Ba	0.06	–		
^{146}Sm	^{142}Nd	103	+	$^{146}Sm/^{144}Sm \approx 0.005–0.015$	Lugmair et al. 1983; Prinzhofer et al. 1989
^{150}Gd	^{146}Sm	1.8	–		
^{182}Hf	^{182}W	9.4	+	$^{182}Hf/^{180}Hf \approx 1\times10^{-4}$	Yin et al. 2002
^{202}Pb	^{202}Hg	0.6	–		
^{205}Pb	^{205}Tl	14.1	?		
^{239}Pu	α,SF	0.02	–		
^{242}Pu	α,SF	0.35	–		
^{244}Pu	α,SF	81	+	$^{244}Pu/^{238}U \approx 0.004–0.007$	Rowe and Kuroda 1965
^{247}Cm	^{235}U	16	?		
^{248}Cm	^{244}Pu	0.35	–		

§ (+) positive detection (?) hint (–) undetected or uninvestigated

nuclides. The detectability of their decay products is always a difficult task, requiring state of the art precision and sensitivity in mass spectrometry. Due to their rapid decay, they have a potential of being high resolution chronometers, but this assumes that these nuclides were homogeneously distributed in the early solar system. Since they are now extinct, a reference object with a well known age determined with a long-lived nuclide (U-Pb generally), is used to convert relative ages determined with short-lived radionuclides to absolute ages. The ideal reference object would be a common well-equilibrated meteorite which was not disturbed thereafter. Half-lives and early solar system contents of the short-lived nuclides are given in Table 1. Examples of the correlations resulting from a number of extinct radioactivities are given in Figure 9.

10-Beryllium - 10-Boron ($T_{1/2}$ = 1.5 m.y.)

B has only 2 isotopes and natural fractionation may be present as observed on the Earth. Only highly radiogenic phases avoid this ambiguity. B is a mobile and volatile element and alteration by a gas phase has partially re-equilibrated oxygen in inclusions. Nevertheless, the presence of live ^{10}Be has been established by ion probe work (Fig. 9a; McKeegan et al. 2000). The interest of this extinct radioactivity is that ^{10}Be is not produced significantly in stellar nucleosynthesis and an irradiation process is required; the question being if it was by the early sun or somewhere else close to a presolar star. As B isotopes are measured by ion-probe, the Al-Mg system can be readily investigated on the same analytical spots. There is a good connection between the two chronometers ^{10}Be-^{10}B and ^{26}Al-^{26}Mg (MacPherson et al. 2003) but differential behavior relative to temperature and to mineralogy also occurs (Sugiura et al. 2001) and contributes to establishment of the thermal history of the inclusions.

26-Aluminum - 26-Magnesium ($T_{1/2}$ = 0.7 m.y.)

This nuclide is of particular interest for two reasons: its very short half-life and its potential as a heat source for early planetary objects (Urey 1955; Lee et al. 1977). Detailed reviews with emphasis on this nuclide are available (Wasserburg and Papanastassiou 1982; MacPherson et al. 1995). An early study in feldspars which have high Al/Mg ratios, extracted from a variety of chondrites and eucrites did not detect evidence for past ^{26}Al (Schramm et al. 1970). The first positive evidence came from the refractory inclusions of Allende with ^{26}Mg excesses correlated with Al/Mg (Gray and Compston 1974; Lee and Papanastassiou 1974) but the clear evidence that this was not a fossil correlation, was demonstrated by an isochron from inclusions which were initially homogenized by melting (Lee et al. 1976; Clayton 1986) (Fig. 9b). Numerous results were also thereafter obtained with ion probe in situ measurements. A number of well-defined isochrons from refractory inclusions give the so-called "canonical" value of ^{26}Al/^{27}Al = 5.1×10^{-5} (see MacPherson et al. (1995) for a review) Many inclusions display lower values which can be interpreted in 3 different ways: 1) ^{26}Al was heterogeneously distributed in the early solar system, 2) secondary processes have partially re-equilibrated the samples with normal Mg, 3) delayed formation after partial decay of ^{26}Al. An illustration of protracted formation is given by the detailed work of Hsu et al. (2000) on different locations within a single CAI. Several stages of melt addition within a few 10^5 years are suggested.

In ordinary chondrites, Ca-Al rich inclusions are also present and give the canonical value (Russell et al. 1996), but other object like chondrules or mineral grains give reduced values by a factor of 5 to 100 (Hinton and Bischoff 1984; Hutcheon and Hutchison 1989). Delayed formation relative to CAIs is a probable cause. High precision ICPMS measurements of Mg have been used to address the timing of chondrule formation and show the importance of gas during the formation process (Galy et al. 2000).

Parent body metamorphism also resets the ^{26}Al-^{26}Mg chronometer as it does for the traditional long-period chronometers. Plagioclases separated from a few of the oldest chondrites display a good agreement between ^{26}Al-^{26}Mg and U-Pb chronometers (Göpel et al.

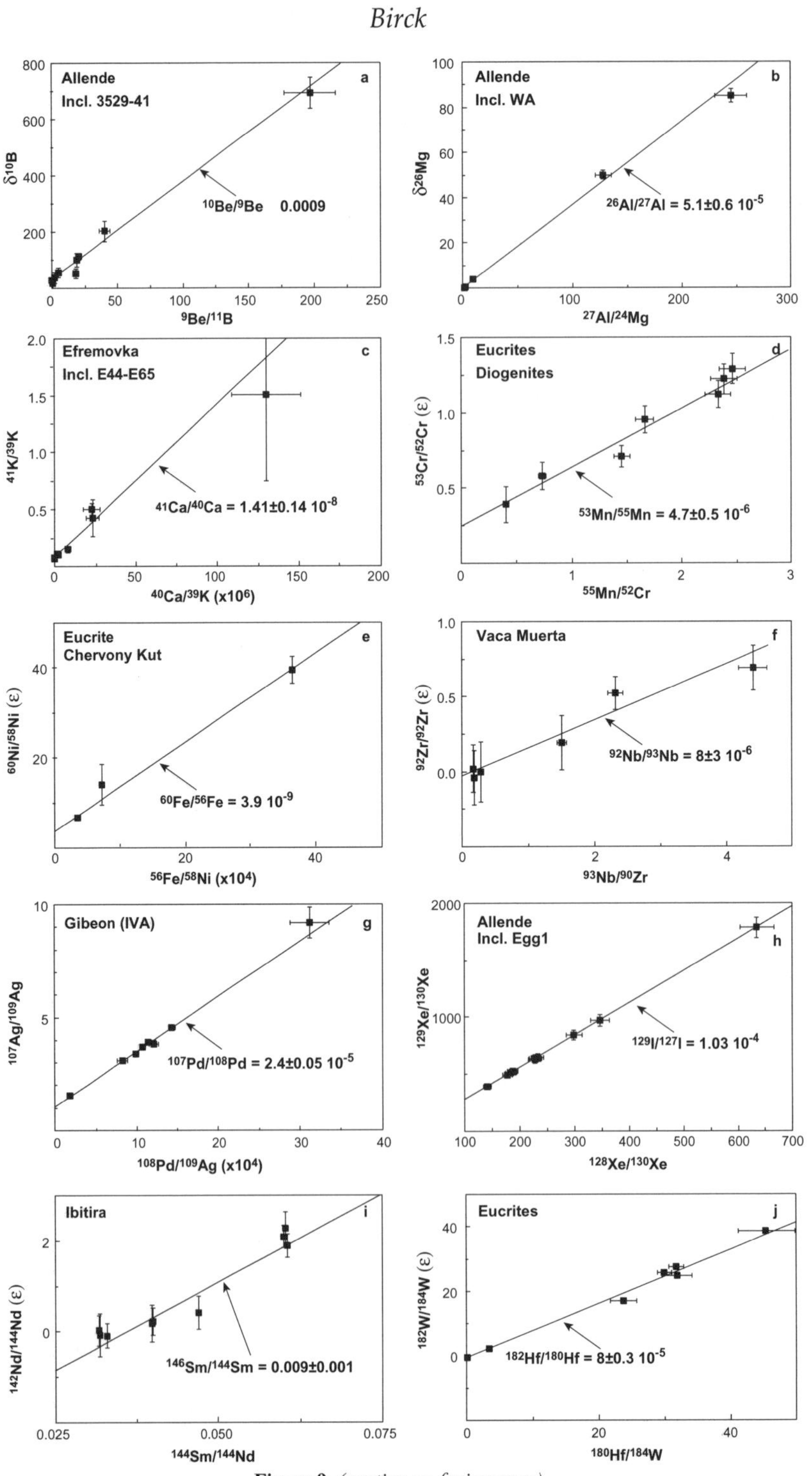

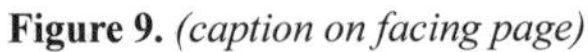
Figure 9. *(caption on facing page)*

1994; Zinner and Göpel 2002). This work is an argument in favor of ^{26}Al being homogeneously distributed in the solar system at least in chondrites (Huss et al. 2001). ^{26}Al has also been found at very low levels in basaltic achondrites (Srinivasan et al. 1999; Wadhwa et al. 2003) and the data show that secondary effects (such as shock?) have often affected the Al-Mg system. It is not known yet if the chronometry stemming from ^{26}Al in eucrites is in agreement with other chronometers or not. The connection so far looks complex. If ^{26}Al was homogeneously distributed in the early solar system, which is not proven yet, it might have melted the interior of bodies of a few km radius and may constitute an efficient trigger for early magmatism.

^{26}Al is produced efficiently in a number of nucleosynthetic processes like hydrostatic burning or explosive H burning in novae (Woosley 1986) or Wolf-Rayet stars (Arnould et al. 2000) or in AGB stars (Wasserburg et al. 1994). Its short half-life requires that this has happened no more than 3 Ma before meteorite formation (Wasserburg and Papanastassiou 1982). It has been argued that a late addition of about 10^{-3} to 10^{-4} solar masses of freshly nucleosynthesized material has been added to the solar protostellar cloud just before the formation of the sun (Birck and Allegre 1988), but other hypotheses have not been completely eliminated, such as local irradiation by energetic protons from the young sun (Heymann and Dziczkaniec 1976; Lee 1978; Gounelle et al. 2001). However irradiation should also produce effects on other elements (Clayton et al. 1977a) which have not been observed yet.

41-Calcium - 41-Potassium ($T_{1/2}$ = 0.1 m.y.)

^{41}Ca is very important because its very short half life provide a strong time constraint between a supernova event and the formation of the solar system. It was identified in very low K grains in C2 and C3 chondrites with an average ^{41}Ca/^{40}Ca value of 1.4×10^{-8} (Fig. 9c; Srinivasan et al. 1996). The presence of live ^{41}Ca in the early solar system constrains the time scale for the collapse of the protosolar cloud to be less than a million years (Cameron 1995; Wasserburg et al. 1996). As for ^{10}Be there is a good correlation of the ^{41}Ca-^{41}Ca system with the ^{26}Al-^{26}Mg data in the primitive samples (Sahijpal et al. 1998). The close connection of ^{41}Ca with ^{26}Al and ^{10}Be has been taken as an argument in favor of a production of these isotopes within the solar system by proton irradiation in an active X-wind phase of the young sun (Shu et al. 2000; Gounelle et al. 2001). This would release the time constraint between presolar stellar nucleosynthesis and planet formation, but would complicate the use of short-lived nuclide as chronometers.

53-Manganese - 53-Chromium ($T_{1/2}$ = 3.7 m.y.)

Most of the ^{53}Cr found in the solar system is produced in presupernova neutron-poor nuclear statistical equilibrium as ^{53}Mn which then decays to ^{53}Cr (Hainebach et al. 1974; Hartmann et al. 1985). The first evidence for the presence of ^{53}Mn in the early solar system

Figure 9 ***(on facing page).*** Examples of isotopic variations resulting from extinct radioactivities. Isotopic ratios are plotted as absolute values for K (Srinivasan et al. 1996), Ag (Chen and Wasserburg 1990) and Xe (Swindle et al. 1988). They are plotted as deviation from a terrestrial standard in δ units (‰ fractional deviations) for B (McKeegan et al. 2000) and Mg (Lee et al. 1977), and in ε units (0.01% fractional deviations) for Cr (Lugmair and Shukolyukov 1998), Ni (Shukolyukov and Lugmair 1993), Zr (Schönbächler et al. 2002), Nd (Prinzhofer et al. 1989) and Hf (Quitté et al. 2000). Some systems do not exhibit sufficient isotopic dispersion in Allende inclusions to define a clear isochron. The examples displayed here have been chosen for the clarity of correlation in the isochron diagram as more favorable parent daughter ratios are often found in differentiated meteorites: irons for ^{107}Pd-^{107}Ag, eucrites for ^{182}Hf-^{182}W. Solar system initial isotopic ratios of the parent element can be found in Table 1. The meteorites in which the measurements have been carried out, are given in each sub-figure. For I-Xe the parent daughter ratio is not measured as such. The sample is irradiated with neutrons which convert ^{127}I into ^{128}Xe as ^{128}I decays to ^{128}Xe within a few hours (Swindle et al. 1988). The pattern shown here represent the stepwise degassing from 1400–1950°C, starting from upper right of the diagram. The ^{129}I/^{127}I ages are usually given as relative ages to the L4 chondrite Bjurböle (^{129}I/^{127}I = 1.095 10^{-4}).

was found in Allende CAIs with a $^{53}Mn/^{55}Mn$ ratio of 4.4×10^{-5} (Birck and Allegre 1985). Such a value allows this potential chronometer to be of some practical use over about 30 m.y. after CAI formation. Inclusion data show that about 0.02% of present normal ^{53}Cr was ^{53}Mn at the time of inclusion formation.

^{53}Mn was found in chondrites, pallasites and eucrites (Birck and Allegre 1988; Lugmair and Shukolyukov 1998). Its presence in basaltic eucrites unambiguously established the in-situ nature of ^{53}Mn decay (Fig. 9d). In general, these measurements require very high precision mass spectrometry, to 20 ppm precision and better. Ion probe measurements on very high Mn/Cr phases reveal very radiogenic ratios: sulfides in enstatite chondrites (Wadhwa et al. 1997), phosphates of iron meteorites (Davis and Olsen 1990), olivine in chondrites (Hutcheon et al. 1998). They exhibit $^{53}Mn/^{55}Mn$ ratios ranging from 10^{-8} to 10^{-5} which represent a delayed formation by more than 7 m.y. relative to inclusion formation or re-equilibration in secondary processes. Eucrites together with pallasites and diogenites (the HED family) plot along a well defined bulk rock isochron, but individual samples do not display a coherent picture of internal mineral isochrons which are distributed from the $^{53}Mn/^{55}Mn$ value of 4.7×10^{-6} down to zero slope (Lugmair and Shukolyukov 1998).

As for other extinct radioactivities, to provide absolute ages, the $^{53}Mn/^{53}Cr$ chronometer has to be anchored to a sample for which the absolute age is known from a traditional chronometer, usually U-Pb because of its high precision. There is debate going on how to do this with ^{53}Mn and how to compare the ^{53}Mn-^{53}Cr data to other short-lived and long-lived radiometric systems (Lugmair and Shukolyukov 1998; Birck et al. 1999; Quitté 2001). Further high precision work is still required to settle the debate.

60-Iron - 60-Nickel ($T_{1/2}$ = 1.5 m.y.)

^{60}Fe is a neutron-rich nucleus and is expected to be produced in similar processes to those producing the excess of neutron-rich nuclei of the iron group elements in Allende's CAIs. The first hint for the presence of ^{60}Fe was found in the Ni isotopic compositions in CAIs (Birck and Lugmair 1988). The presence of ^{60}Fe has been clearly identified in eucrites which have high Fe/Ni ratios (Fig. 9e; Shukolyukov and Lugmair 1993). In using ion probes to spot phases with high Fe/Ni ratios, the presence of ^{60}Fe was also demonstrated in chondrites (Mostefaoui et al. 2003). The initial solar system value for $^{60}Fe/^{56}Fe$ is estimated around 1.5×10^{-6} with some assumptions. If homogeneously distributed, ^{60}Fe can constitute a rapid heat source to melt small planetary bodies in addition to the possible presence of ^{26}Al.

92-Niobium - 92-Zirconium ($T_{1/2}$ = 36 m.y.)

As ^{146}Sm, ^{92}Nb can be used to place constraints on the site of p-process nucleosynthesis. After a first hint (Harper 1996; Sanloup et al. 2000), its presence was established in an ordinary chondrite and a mesosiderite (Fig. 9f; Schönbächler et al. 2002). The solar system initial $^{92}Nb/^{93}Nb$ ratio was between 10^{-5} and 3×10^{-5} a value in the same order of magnitude as most of the other extinct radioactivities.

107-Palladium - 107-Silver ($T_{1/2}$ = 6.5 m.y.)

Pd is a siderophile element and is concentrated in iron meteorites. Ag also has some siderophilic character, but is predominantly chalcophile. It also belongs to the volatile elements which are depleted in a number of iron meteorite classes. The chemical properties of this system result in high Pd/Ag ratios in the metal phase of iron meteorites and low ratios in the associated sulfide nodules, if they are present. The existence of ^{107}Ag excess correlated with the Pd/Ag ratio in coexisting phases and between bulk meteorites show that ^{107}Pd was widespread in the early solar system (Fig. 9g; Kelly and Wasserburg 1978; Chen and Wasserburg 1983; Chen and Wasserburg 1990). The narrow range around 2.1×10^{-5} for the ^{107}Pd /^{108}Pd ratio in all meteorites not disturbed by secondary processes shows that all classes of iron meteorites formed in a narrow time interval of less than 10 m.y. This is in agreement with W isotopic data, although the constraint is not very strong. Silver has only two isotopes and can only be measured down to a precision of c.a. 1‰ in TIMS instruments. This system could only be investigated in radiogenic samples. The introduction of ICPMS allows an order of magnitude improvement in precision and hence the investigation of silicate meteoritic materials (Carlson and Hauri 2001a). The results are in agreement with former Pd-Ag data and other chronometric systems.

^{107}Pd results from the same type of r-process as ^{129}I, but it is not well characterized (Wasserburg and Papanastassiou 1982) and can result from a late addition of a component incorporating a number of other short-lived nuclide as well.

129-Iodine - 129-Xenon ($T_{1/2}$ = 16 m.y.)

This was the first extinct radioactivity detected (Jeffrey and Reynolds 1961) and was made possible by the early high sensitivity of rare gas measurements and the low abundance of Xe in rocks. I has only one stable isotope at mass 127. Its abundance is measured as ^{128}Xe after exposing a sample to an adequate neutron flux. The correlation between ^{128}Xe and ^{129}Xe observed in a stepwise degassing of a sample demonstrates that the excess ^{129}Xe results from ^{129}I decay (Fig. 9h). Results in primitive meteorites and inclusions show that ^{129}I/^{127}I is close to 10^{-4}. Chronometry with ^{129}I-^{129}Xe has been widely used in meteorite work (Reynolds 1963; Hohenberg 1967) but occasionally has some difficulties to agree with the other chronometers due to the sensitivity of I to secondary processes and water alteration (Pravdivtseva et al. 2003; Busfield et al. 2004; see also Swindle and Podosek (1988) for an extensive review) .

146-Samarium - 142-Neodymium ($T_{1/2}$ = 103 m.y.)

^{146}Sm has the longest half life for the extinct radioactivities. Theoretically there should be a few atoms left in present day samples, although their detectability is not obvious and could be confused in meteorites with spallation-produced ^{146}Sm. ^{146}Sm is a p-process nuclide. It was detected first in a magmatic meteorite: Angra Dos Reis but at the limit of resolution of the method (Lugmair and Marti 1977; Jacobsen and Wasserburg 1980). Effects were well resolved in Allende acid-resistant residues (Lugmair et al. 1983) giving an initial ^{146}Sm/^{144}Sm ratio of 0.005. Further studies in eucrites showed that it was widespread (Fig. 9i; Prinzhofer et al. 1989) and give a higher initial ^{146}Sm/^{144}Sm ratio of 0.015, 4.56 Gy ago. Several nucleosynthetic models have been proposed to produce ^{146}Sm, some of which are compatible with a constant production in the galaxy without the need for a late spike to introduce the extinct radionuclide in the protosolar cloud (Prinzhofer et al. 1992). The long half life of ^{146}Sm allows for an investigation of crustal fractionation processes in the early earth, but this requires even higher precision mass spectrometry (Sharma et al. 1996) down to 4-5 ppm to resolve differences (Caro et al. 2003)

182-Hafnium - 182-Tungsten ($T_{1/2}$ = 9.4 m.y.)

^{182}Hf is an r-process nuclide similar to ^{107}Pd, ^{129}I and ^{244}Pu. The main interest of this system resides in the chemical properties of the parent-daughter elements. Hf is highly lithophile and partitions into planetary crusts whereas W is strongly siderophile and concentrates into planetary cores. When metal in equilibrium with silicates separates from these silicates, it traps W without Hf and then freezes the isotopic composition of W at the moment of the separation. On the other hand, the silicate part (which still contains very little W) gets very radiogenic and has a rapidly increasing ^{182}W/^{184}W ratio with time, which yields a high resolution chronometer. A deficit in ^{182}W has been detected in iron meteorites by Harper and Jacobsen (1994) and then by Lee and Halliday (1995, 1996). These data along with ^{107}Pd, show that the segregation of iron meteorites was an early and short episode. The last authors applied these measurements to a broad range of samples to study planetary core formation as well as the crystallization of the Moon's magma ocean (Lee et al. 1997). A large data base exists now for this system for iron meteorites as well as for chondrites and achondrites (Fig. 9i; Quitté et al. 2000). Cosmogenic W may constitute a problem for planetary surface samples having large galactic-cosmic-ray exposure ages like the lunar rocks (Leya et al. 2000). W isotopic data have shown that the silicate part of the Earth is more radiogenic by 2 ε than the bulk solar system (Kleine et al. 2002; Yin et al. 2002). It follows that the Earth's core formed rapidly, within 30 m.y. This system is presently under intense investigation because its half-life gives access to the processes forming the larger planets, like the Earth, the Moon and Mars.

244-Plutonium spontaneous fission ($T_{1/2}$ = 81 m.y.)

Excess fission Xe due to the decay of ^{244}Pu, was found in U-Th rich samples (Rowe and Kuroda 1965; Lewis 1975). Its abundance relative to U is about 7×10^{-3}. This extinct radionuclide is a pure r-process product and its abundance can be used to estimate the timing of the r-process contribution to the presolar interstellar cloud (Wasserburg et al. 1996)

Other nuclides

Other nuclides have also been sought for, like ^{135}Cs (Shen et al. 1994; Hidaka et al. 2001), ^{205}Pb (Chen and Wasserburg 1987; Rehkämper and Halliday 1999), ^{247}Cm (Chen and Wasserburg 1981) but so far have not been detected.

In summary, the extinct radioactivities which have a limited time of existence in the solar system, constrain the time interval between the late stages of stellar nucleosynthesis and the formation of the solar system. Some production may also occur within the solar system during active periods of the young Sun. There have been numerous studies about how this matter was added into the solar system as a late spike of about 10^{-3} solar masses of freshly stellar processed material or from constant production in the galaxy (Wasserburg et al. 1996; Goswami and Vanhala 2000; Russell et al. 2001). These models are refined constantly with the input of new data and will probably continue to evolve in the future.

OPEN QUESTIONS AND FUTURE DIRECTIONS

The measurements of isotopic anomalies in meteorites has contributed greatly to the understanding of mixing processes and time scales in the formation of the solar system as well as strong constraints on presolar stellar evolution but it also left unanswered questions and revealed new complexities which are discussed here.

Mass independent fractionation in physico chemical-processes

The case of oxygen is puzzling and still unsolved. When excess ^{16}O was discovered in the inclusions of Allende (Clayton et al. 1973), the most straightforward interpretation was an addition of supernova oxygen in these particular objects, but this is still debated today whereas many elements are unambiguously thought to display anomalies with stellar nucleosynthetic origins. Laboratory experiments showed that it was possible to create non-mass-dependent fractionating O by ultraviolet light photolysis, RF discharge or electrical discharge (see Thiemens (1988) for a review). Effects mimicking a ^{16}O addition up to 70‰ have been observed (Thiemens and Heidenreich 1983), and the implication of such effects in the solar system formation has been discussed and included in the interpretation of meteoritic data (Thiemens 1999; Clayton 2002). A number of questions include: how many such reservoirs can be created and to what extent are other elements involved?

Early proton irradiation within the solar system

The presence of short-lived nuclides like ^{41}Ca and ^{26}Al imply a very short time scale between stellar nucleosynthesis and the formation of planetary bodies. To accommodate this time constraint, it is tempting to try to include some of the short-lived nuclide production within the early solar system (X-wind model of Shu et al. (2000)). Such models have been developed earlier (Fowler et al. 1962; Heymann and Dziczkaniec 1976; Lee 1978) but their past action failed to be proven unambiguously (Burnett et al. 1966). On one hand, stellar nucleosynthesis satisfactorily explains the observed anomalies as discussed above. On the other hand, an irradiation process is required to produce ^{10}Be. This alternative is presently under debate. The model of Shu et al. (2000) is the latest evolution of such irradiation scenarios based on stellar evolution models. Relevant questions are: irradiation should also produce isotopic heterogeneity for some sensitive stable isotopes, and how homogeneously are the materials exposed to protons ? On the other side, can there be some unsuspected stellar site able to produce sufficient amounts of ^{10}Be ?

Presolar silicates

The huge isotopic variations in presolar grains have so far been found in diamonds, carbides and oxides. This is due to the separation procedures using HF to remove the major minerals from the chondrites. It is expected that presolar silicate mineral grains also existed and future work directed toward their separation is anticipated. What nucleosynthetic signatures they could carry is also an open question. The extent of these large isotopic variations within the bulk of the presolar material is also unknown. Are all the grains constituting the protosolar cloud so isotopically dispersed as the presolar grains found to date or is it some peppering process in which matter synthesized in the last few hundred m.y., before solar birth, is injected in a roughly homogeneous medium. The cosmic chemical memory (Clayton 1982; Jones et al. 1997) is still somewhat model dependent on this aspect.

Instrumentation

The instrumentation is constantly progressing in two directions: increased spatial resolution with the evolution of ion probes (Zinner et al. 2003), and increased precision with the evolution of multicollector TIMS and ICPMS instruments. There is no doubt that the possibility of precisely addressing the mass-dependent fractionation effects, new visions of early solar system history will emerge. Smaller presolar grains can now be investigated and a strong push exists to develop specific instruments to analyze as many elements as possible within a single sub-micron grain, and automated isotopic analysis (Savina et al. 2003b; Nittler and Alexander 2003).

There are many other questions underlying the measurement of isotopic anomalies and future planetary explorations will bring new stones to the construction.

CONCLUSION

Despite the numerous and sometimes huge isotopic anomalies found during the past 30 years, the solar system, at a first glance at the meter scale and at the percent level of precision, still looks remarkably homogeneous. The formation of the sun and the planets was a very efficient homogenizing process. The discovery of the isotopic anomalies in the 1970s has opened a fascinating door to the presolar history of the solar system materials. They allowed characterization of a number of nucleosynthetic processes expected to be active in the galaxy and which are required to produce the isotope ratios of elements found in the solar system. A number of processes have been clearly identified and connected to astronomical observations and astrophysical models. By using individual grains ejected by stars, a somewhat unexpected large diversity of effects has been discovered even within the same process. The interpretation is complex and lies between two extremes. One possibility is that measured ratios are straight production ratios within stars and the very large isotopic variability reflects the variability of stars, and mixing processes within stars, having contributed to the solar system. On the other hand, there can be a limited number of stellar processes producing extreme isotopic compositions, and the observations may reflect various mixing ratios between these relatively few components during the long evolution between stellar interiors and planetary bodies in the solar system. Such mixtures are necessary to explain some observations, such as Xe-HL as one example. That secondary processes, like gas phase alteration operated during the evolution of the solar nebula is demonstrated in Allende inclusions, and parent body metamorphism in chondrites. The complexity of the image that we now have from the forming solar system results from the variety of the isotopic measurements. Nevertheless the accumulation of new data and the improvements in the models of planetary formation, will help constrain more strictly the unknowns.

ACKNOWLEDGMENTS

I wish to thank the anonymous reviewer, M. Kurz and R.N. Clayton for the comments and the improvements of the text of this overview. Many thanks to J. Dyon for his help for the Figures. I am also grateful to C. Göpel, G. Manhès, A. Trinquier, M. Moreira and C.J. Allègre for the numerous discussions about radioactive isotopes and isotopic heterogeneity. This is IPGP contribution no 1961.

REFERENCES

Alexander CMOD (1997) Dust production in the galaxy: the meteorite perspective. *In:* Astrophysical Implications of the Laboratory Study of Presolar Materials. Bernatowicz TJ, Zinner E (eds) AIP, New York, p 567-593

Amari S, Lewis RS, Anders E (1994) Interstellar grains in meteorites: I. Isolation of SiC, graphite, and diamond; size distributions of SiC and graphite. Geochim Cosmochim Acta 58:459-470

Amari S, Zinner E (1997) Supernova grains from meteorites. *In:* Astrophysical Implications of the Laboratory Study of Presolar Materials. Bernatowicz TJ, Zinner E (eds) AIP, New York, p 287-306

Amari S, Zinner E, Lewis RS (1999) A singular presolar SiC grain with extreme ^{29}Si and ^{30}Si excesses. Astrophys J 517:L59-L92

Amari S, Nittler LR, Zinner E, Lodders K, Lewis RS (2001) Presolar SiC grains of type A and B: their isotopic compositions and stellar origins. Astrophys J 559:463-483

Anders E, Stevens CM (1960) Search for extinct lead 205 in meteorites. J Geophys Res 65:3043-3047

Anders E, Grevesse N (1989) Abundances of the elements: meteoritic and solar. Geochim Cosmochim Acta 53:197-214

Anders E, Zinner E (1993) Interstellar grains in primitive meteorites: diamond, silicon carbide, and graphite. Meteoritics 28:490-514

Arlandini C, Käppeler F, Wisshak K, Gallino R, Lugaro M, Busso M, Staniero O (1999) Neutron capture in low-mass asymptotic giant branch statrs: cross sections and abundance signatures. Astrophys J 525: 886-900

Arnould M, Meynet G, Mowlavi N (2000) Some selected comments on cosmic radioactivities. Chem Geol 169:83-105

Badziag P, Verwoerd WS, Ellis WP, Greiner NR (1990) Nanometre-sized diamonds are more stable than graphite. Nature 343:244-245

Bernatowicz TJ, Zinner E (1997) Astrophysical Implications of the Laboratory Study of Presolar Materials. AIP, New York

Besmehn A, Hoppe P (2003) A NanoSIMS study of Si- and Ca-Ti-isoptopic compositions of presolar silicon carbide grains from supernovae. Geochim Cosmochim Acta 67:4693-4703

Birck JL, Allegre CJ (1984) Chromium isotopic anomalies in Allende refractory inclusions. Geophys Res Lett 11:943-946

Birck JL, Allegre CJ (1985) Evidence for the presence of ^{53}Mn in the early solar system. Geophys Res Lett 12:745-748

Birck JL, Allegre CJ (1988) Manganese chromium isotope systematics and development of the early solar system. Nature 331:579-584

Birck JL, Lugmair GW (1988) Ni and Cr isotopes in Allende inclusions. Earth Planet Sci Lett 90:131-143

Birck JL, Rotaru M, Allegre CJ (1999) ^{53}Mn-^{53}Cr evolution of the early solar system. Geochim Cosmochim Acta 63:4111-4117

Black DC (1972) On the origins of trapped helium, neon and argon isotopic variation in meteorites - II. Carbonaceous meteorites. Geochim Cosmochim Acta 36:377-394

Black DC, Pepin RO (1969) Trapped neon in meteorites. II. Earth Planet Sci Lett 6:395-405

Bogdanovski O, Papanastassiou DA, Wasserburg GJ (2002) Cr isotopes in Allende Ca-Al-rich inclusions. Lunar Planet Sci XXXIII:1802

Brown LE, Clayton DD (1992) Silicon isotopic composition in large meteoritic SiC particles and ^{22}Na origin of ^{22}Ne. Science 258:970-972

Burbidge EM, Burbidge GR, Fowler WA, Hoyle F (1957) Synthesis of the elements in stars. Rev Mod Phys 29:547-630

Burnett DS, Lippolt HJ, Wasserburg GJ (1966) The relative isotopic abundance of ^{40}K in terrestrial and meteoritic samples. J Geophys Res 71:1249-1259

Busfield A, Gilmour JD, Whitby JA, Turner G (2004) Iodine-xenon analysis of ordinary chondrite halide: implications for early solar system water. Geochim Cosmochim Acta 68:195-202

Busso M, Gallino R, Wasserburg GJ (1999) Nucleosynthesis in asymptotic giant branch stars: relevance for galactic enrichment and solar system formation. Annu Rev Astronom Astrophys 37:239-309

Cameron AGW (1969) Physical conditions in the primitive solar nebula. *In:* Meteorite Research. Millman PM (ed) Reidel, Dordrecht, p 7-12

Cameron AGW (1979) The neutron-rich silicon burning and equilibrium processes of nucleosynthesis. Astrophys J 230:53-57

Cameron AGW (1995) The first ten million years in the solar nebula. Meteoritics 30:133-161

Carlson RW, Hauri EH (2001a) Extending the ^{107}Pd-^{107}Ag chronometer to low Pd/Ag meteorites with muticollector plasma-ionization mass spectrometry. Geochim Cosmochim Acta 65:1839-1848

Carlson RW, Hauri EH (2001b) Silver isotope variations in the Earth as measured by multi-collector ICP-MS. Goldschmidt Conf Hot Springs, Virginia: Abstr 3661

Caro G, Bourdon B, Birck JL, Moorbath S (2003) ^{146}Sm-^{142}Nd evidence from Isua metamorphosed sediments for early differentiation of the Earth's mantle. Nature 423:428-432

Chen JH, Wasserburg GJ (1981) The isotopic composition of uranium and lead in Allende inclusions and meteoritic phosphates. Earth Planet Sci Lett 52:1-15

Chen JH, Wasserburg GJ (1983) The isotopic composition of silver and lead in two iron meteorites: Cape York and Grant. Geochim Cosmochim Acta 47:1725-1737

Chen JH, Wasserburg GJ (1987) A search for evidence of extinct lead 205 in iron meteorites. Lunar Planet Sci 18:165-166

Chen JH, Wasserburg GJ (1990) The isotopic composition of Ag in meteorites and the presence of ^{107}Pd in protoplanets. Geochim Cosmochim Acta 54:1729-1743

Chen JH, Papanastassiou DA, Wasserburg GJ (2003) Endemic Ru isotopic anomalies in iron meteorites and in Allende. Lunar Planet Sci XXXIV:1789

Choi BG, Huss GR, Wasserburg GJ, Gallino R (1998) Presolar corundum and spinel in ordinary chondrites: origins from AGB stars and a supernova. Science 282:1284-1289

Christophe M (1968) Un chondre exceptionnel dans la météorite de Vigarano. Bul Soc Fr Mineral Cristallogr 91:212-214

Clayton DD (1982) Cosmic chemical memory: a new astronomy. Quart J Roy Astron Soc 23:174-212

Clayton DD (1983) Principles of stellar evolution and nucleosynthesis. University of Chicago Press, Chicago

Clayton DD (1986) Interstellar fossil ^{26}Mg and its possible relationship to excess meteoritic ^{26}Mg. Astrophys J 310:490-498

Clayton DD (1988) Stellar nucleosynthesis and chemical evolution of the solar neighborhood. *In*: Meteorites and the Early Solar System. Kerridge JF, Matthews MS (eds) University of Arizona Press, Tucson, p 1021-1062

Clayton DD (1989) Origin of heavy xenon in meteoritic diamonds. Astrophys J 340:613-619

Clayton DD, Dwek E, Woosley SE (1977a) Isotopic anomalies and proton irradiation in the early solar system. Astrophys J 214:300-315

Clayton RN (2002) Self-shielding in the solar nebula. Nature 415:860-861

Clayton RN, Grossman L, Mayeda TK (1973) A component of primitive nuclear composition in carbonaceous meteorites. Science 182:485-488

Clayton RN, Onuma N, Mayeda TK (1976) A classification of meteorites based on oxygen isotopes. Earth Planet Sci Lett 30:10-18

Clayton RN, Mayeda TK (1977) Correlated O and Mg isotopic anomalies in Allende inclusions. I Oxygen. Geophys Res Lett 4:295-298

Clayton RN, Onuma N, Grossman L, Mayeda TK (1977) Distribution of the pre-solar component in Allende and other carbonaceous chondrites. Earth Planet Sci Lett 34:209-224

Clayton RN, MacPherson GJ, Hutcheon ID, Davis AM, Grossman L, Mayeda TK, Molini-Velsko CA, Allen JM, El Goresy A (1984) Two forsterite-bearing FUN inclusions in the Allende meteorite. Geochim Cosmochim Acta 48:535-548

Clayton RN, Hinton RW, Davis AM (1988) Isotopic variations in the rock-forming elements in meteorites. Phil Trans R Soc Lond A 325:483-501

Dai ZR, Bradley JP, Joswiak DJ, Brownlee DE, Hill HGM, Genge MJ (2002) Possible *in-situ* formation of meteoritic nanodiamonds in the early Solar System. Nature 418:157-159

Dauphas N, Marty B, Reisberg L (2002) Molybdenum nucleosynthetic dichotomy revealed in primitive meteorites. Astrophys J 569:L139-L142

Davis AM, Olsen EJ (1990) Phosphates in the El Sampal IIIA iron meteorite have excess ^{53}Cr and primordial lead. Lunar Planet Sci 21:258-259

Eberhardt P, Jungck MHA, Meier FO, Niederer FR (1981) A neon-E rich phase in Orgueil; results obtained on density separates. Geochim Cosmochim Acta 45:1515-1528

El Goresy A, Zinner E, Marti K (1995) Survival of isotopically heterogeneous graphite in a differentiated meteorite. Nature 373:496-499

Eugster O, Tera F, Wasserburg GJ (1969) Isotopic analyses of barium in meteorites and in terrestrial samples. J Geophys Res 74:3897-3908

Fahey AJ, Goswami JN, McKeegan KD, Zinner EK (1985) Evidence for extreme ^{50}Ti enrichments in primitive meteorites. Astrophys J 296:L17-L20

Fahey AJ, Goswami JN, McKeegan KD, Zinner EK (1987) ^{26}Al, ^{244}Pu, ^{50}Ti, REE, and trace element abundances in hibonite grains from CM and CV meteorites. Geochim Cosmochim Acta 51:329-350

Fowler WA, Greenstein JL, Hoyle F (1962) Nucleosynthesis during the early history of the solar system. Geophys J 6:148-220

Galy A, Young ED, Ash RD, O'Nions RK (2000) The formation of chondrules at high gas pressures in the solar nebula. Science 290:1751-1753

Gast PW, Hubbard NJ, Wiesmann H (1970) Chemical composition and petrogenesis of basalts from Tranquility Base. Geochim Cosmochim Acta Suppl 1:1143-1163

Gooding JL, Mayeda TK, Clayton RN, Fukuoka T (1983) Oxygen isotopic heterogeneities, their petrological correlations, and implications for melt origins of chondrules in unequilibrated ordinary chondrites. Earth Planet Sci Lett 65:209-224

Göpel C, Manhès G, Allègre CJ (1994) U-Pb systematics of phosphates from equilibrated ordinary chondrites. Earth Planet Sci Lett 121:153-171

Goswami JN, Vanhala HAT (2000) Extinct radionuclides and the origin of the solar system. *In:* Protostars & Planets IV. Mannings V, Boss AP, Russel SS (eds) University of Arizona Press, Tucson, p 963-994

Gounelle M, Shu FH, Shang H, Glassgold AE, Rehm EK, Lee T (2001) Extinct radioactivities and protosolar cosmic-rays: self-shielding and light elements. Astrophys J 548:1051-1070

Gray CM, Compston W (1974) Excess ^{26}Mg in the Allende meteorite. Nature 251:495-497

Grossman L (1972) Condensation in the primitive solar nebula. Geochim Cosmochim Acta 36:597-619

Grossman L, Larimer JW (1974) Early chemical history of the solar system. Rev Geophys Space Phys 12: 71-101

Hainebach KL, Clayton DD, Arnett WD, Woosley SE (1974) On the e-process. Its components and their neutron excesses. Astrophys J 193:157-168

Harper CL (1996) Evidence for ^{92}Nb in the early solar system and evaluation of a new p-process cosmochronometer from $^{92}Nb/^{92}Mo$. Astrophys J 466:437-456

Harper CL, Wiesmann H, Nyquist LE, Hartmann D, Meyer B, Howard WM (1991) Interpretation of the ^{50}Ti-^{96}Zr anomaly correlation in CAI: NNSE Zr production limits and S/ R/ P decomposition of the bulk solar system abundances. Lunar Planet Sci XXII:517-518

Harper CL, Jacobsen SB (1994) Investigations of the ^{182}Hf-^{182}W systematics. Lunar Planet Sci XXV:509-510

Hartmann D, Woosley SE, El Eid MF (1985) Nucleosynthesis in neutron-rich supernova ejecta. Astrophys J 297:837-845

Hashimoto A, Hinton RW, Davis AM, Grossman L, Mayeda TK, Clayton RN (1986) A hibonite-rich Murchison inclusion with anomalous oxygen isotopic composition. Lunar Planet Sci XVII:317-318

Heydegger HR, Foster JJ, Compston W (1979) Evidence of a new isotopic anomaly from titanium isotopic ratios in meteoritic material. Nature 278:704-707

Heymann D, Dziczkaniec M (1976) Early irradiation of matter in the solar system: magnesium (proton, neutron) scheme. Science 191:79-81

Hidaka H, Ohta Y, Yoneda S, DeLaeter JR (2001) Isotopic search for live ^{135}Cs in the early solar system and possibility of ^{135}Cs-^{135}Ba chronometer. Earth Planet Sci Lett 193:459-466

Hinton RW, Bischoff A (1984) Ion microprobe magnesium isotope analysis of plagioclase and hibonite from ordinary chondrites. Nature 308:169-172

Hinton RW, Davis AM, Scatena-Wachel DE (1987) Large negative ^{50}Ti anomalies in refractory inclusions from the Murchison carbonaceous chondrite: evidence for incomplete mixing of neutron-rich supernova ejecta into the solar system. Astrophys J 313:420-428

Höflich P, Wheeler JC, Thielemann FK (1998) Type Ia supernovae: influence of the initial composition on the nucleosynthesis, light curve, and spectra and consequences for the determination of Ω_M and Λ. Astrophys J 495:617-629

Hohenberg CM, Podosek, FA, Reynolds, JH (1967) Xenon-iodine dating: sharp isochronism in chondrites. Science 156:202-206

Honda M (1988) Statistical estimation of the production of cosmic-ray induced nuclides in meteorites. Meteoritics 23:3-12

Hoppe P, Amari S, Zinner E, Ireland T, Lewis RS (1994) Carbon, nitrogen, magnesium, silicon and titanium isotopic compositions of single interstellar silicon carbide grains from the Murchison carbonaceous chondrite. Astrophys J 430:870-890

Hoppe P, Ott U (1997) Mainstream silicon carbide grains from meteorites. *In:* Astrophysical Implications of the Laboratory Study of Presolar Materials. Bernatowicz TJ, Zinner E (eds) AIP, New York, p 27-59

Hsu W, Wasserburg GJ, Huss GR (2000) High time resolution by use of the ^{26}Al chronometer in the multistage formation of CAI. Earth Planet Sci Lett 182:15-29

Huey JM, Kohman TP (1972) Search for extinct natural radioactivity of ^{205}Pb via thallium-isotope anomalies in chondrites and lunar soil. Earth Planet Sci Lett 16:401-410

Huss GR (1997) The survival of presolar grains in solar system bodies. *In:* Astrophysical Implications of the Laboratory Study of Presolar Materials. Bernatowicz TJ and Zinner E (eds) AIP, New York, p 721-748

Huss GR, Fahey AJ, Gallino R, Wasserburg GJ (1994) Oxygen isotopes in circumstellar Al2O3 grains from meteorites and stellar nucleosynthesis. Astrophys J 430:L81-L84

Huss GR, Lewis RS (1994) Noble gases in presolar diamonds I: Three distinct components and their implications for diamond origin. Meteoritics 29:791-810

Huss GR, Hutcheon ID, Wasserburg GJ (1997) Isotopic systematics of presolar silicon carbide from the Orgueil (C1) carbonaceous chondrite: implications for solar system formation and stellar nucleosynthesis. Geochim Cosmochim Acta 61:5117-5148

Huss GR, MacPherson GJ, Wasserburg GJ, Russell SS, Srinivasan G (2001) Aluminum-26 in calcium-aluminum-rich inclusions and chondrules from unequilibrated ordinary chondrites. Meteorit Planet Sci 36:975-997

Huss GR, Meshik AP, Smith JB, Hohenberg CM (2003) Presolar diamond, silicon carbide, and graphite in carbonaceous chondrites: implications for thermal processing in the solar nebula. Geochim Cosmochim Acta 67:4823-4848

Hutcheon ID (1982) Ion probe magnesium isotopic measurements of Allende inclusions. Amer Chem Soc Symp Series 176:95-128

Hutcheon ID, Steele IM, Wachel DES, MacDougall JD, Phinney D (1983) Extreme Mg fractionation and evidence of Ti isotopic variations in Murchison refractory inclusions. Lunar Planet Sci XIV:339-340

Hutcheon ID, Hutchison R (1989) Evidence from the Semarkona ordinary chondrite for ^{26}Al heating of small planets. Nature 337:238-241

Hutcheon ID, Krot AN, Keil K, Phinney DL, Scott ERD (1998) ^{53}Mn-^{53}Cr dating of fayalite formation in the CV3 chondrite Mokoia: evidence for asteroidal alteration. Science 282:1865-1867

Ireland TR (1988) Correlated morphological, chemical, and isotopic characteristics of hibonites from the Murchison carbonaceous chondrite. Geochim Cosmochim Acta 52:2827-2839

Ireland TR (1990) Presolar isotopic and chemical signatures in hibonite-bearing refractory inclusions fron the Murchison carbonaceous chondrite. Geochim Cosmochim Acta 54:3219-3237

Ireland TR, Compston W, Heydegger HR (1985) Titanium isotopic anomalies in hibonites from the Murchison carbonaceous chondrites. Geochim Cosmochim Acta 49:1989-1993

Jacobsen SB, Wasserburg GJ (1980) Sm-Nd isotopic evolution of chondrites. Earth Planet Sci Lett 50:139-155

Jeffery PM, Reynolds JH (1961) Origin of excess ^{129}Xe in stone meteorites. J Geophys Res 66:3582-3583

Jones AP, Tielens AGGM, Hollenbach DJ, McKee CF (1997) The propagation and survival of interstellar grains. *In:* Astrophysical Implications of the Laboratory Study of Presolar Materials. Bernatowicz TJ, Zinner E (eds) AIP, New York, p 595-613

Jungck MHA, Eberhardt P (1979) Neon-E in Orgueil density separates. Meteoritics 14:439-440

Jungck MHA, Shimamura T, Lugmair GW (1984) Ca isotope variations in Allende. Geochim Cosmochim Acta 48:2651-2658

Kehm K, Hauri EH, Alexander CMOD, Carlson RW (2003) High precision iron isotope measurements of meteoritic material by cold plasma ICP-MS. Geochim Cosmochim Acta 67:2879-2891

Kelly WR, Wasserburg GJ (1978) Evidence for the existence of ^{107}Pd in the early solar system. Geophys Res Lett 5:1079-1082
Kleine T, Münker C, Mezger K, Palme H (2002) Rapid accretion and early core formation. Nature 418:952-955
Lambert DL (1992) The p-nuclei. Abundances and origins. Astron Astrophys Rev 3:201-256
Larimer JW (1967) Chemical fractionation in meteorites. I Condensation of the elements. Geochim Cosmochim Acta 31:1215-1238
Lattanzio JC, Boothroyd AI (1997) Nucleosynthesis of elements in low to intermediate mass stars through the AGB phase. *In:* Astrophysical Implications of the Laboratory Study of Presolar Materials. Bernatowicz TJ and Zinner E (eds) AIP, New York, p 85-114
Lattimer JM, Schramm DN, Grossman L (1978) Condensation in supernova ejecta and isotopic anomalies in meteorites. Astrophys J 219:230-249
Lee DC, Halliday AN (1995) Hafnium-tungsten chronometry and the timing of terrestrial core formation. Nature 378:771-774
Lee DC, Halliday AN (1996) Hf-W Isotopic evidence for rapid accretion and differentiation in the early solar system. Science 274:1876-1879
Lee DC, Halliday AN, Snyder GA, Taylor LA (1997) Age and origin of the moon. Science 278:1098-1103
Lee T (1978) A local proton irradiation model for the isotopic anomalies in the solar system. Astrophys J 224: 217-226
Lee T (1988) Implications of isotopic anomalies for nucleosynthesis. *In*: Meteorites and the Early Solar System. Kerridge JF, Matthews MS (eds) University of Arizona Press, Tucson, p 1063-1088
Lee T, Papanastassiou DA (1974) Mg isotopic anomalies in the Allende meteorite and correlation with O and Sr effects. Geophys Res Lett 1:225-228
Lee T, Papanastassiou DA, Wasserburg GJ (1976) Demonstration of ^{26}Mg excess in Allende and evidence for ^{26}Al. Geophys Res Lett 3:109-112
Lee T, Papanastassiou DA, Wasserburg GJ (1977) ^{26}Al in the early solar system: Fossil or fuel? Astrophys J 211:L107-L110
Lee T, Papanastassiou DA, Wasserburg GJ (1978) Calcium isotopic anomalies in the Allende meteorite. Astrophys J 220:L21-L25
Lepareur M (1980) Le microanalyseur ionique de seconde génération Cameca, modèle 3f. Rev Tech Thomson 12:225-265
Lewis RS (1975) Rare gases in separated whitlockite from the St Séverin chondrite; xenon and krypton from fission of extinct ^{244}Pu. Geochim Cosmochim Acta 39:417-432
Lewis RS, Srinivasan B, Anders E (1975) Host phase of a strange xenon component in Allende. Science 190: 1251-1262
Lewis RS, Tang M, Wacker JF, Anders E, Steel E (1987) Interstellar diamonds in meteorites. Nature 326: 160-162
Lewis RS, Huss GR, Lugmair GW (1991) Finally, Ba & Sr accompanying Xe-HL in diamonds from Allende. Lunar Planet Sci 24:807-808
Lewis RS, Amari S, Anders E (1994) Interstellar grains in meteorites: II. SiC and its noble gases. Geochim Cosmochim Acta 58:471-494
Leya I, Wieler R, Halliday AN (2000) Cosmic-ray production of tungsten isotopes in lunar samples and meteorites and its implications for Hf-W cosmochemistry. Earth Planet Sci Lett 175:1-12
Loss RD, Lugmair GW (1990) Zinc isotope anomalies in Allende meteorite inclusions. Astrophys J 360:L59-L62
Loss RD, Lugmair GW, Davis AM, MacPherson GJ (1994) Isotopically distinct reservoirs in the solar nebula: isotope anomalies in Vigarano meteorite inclusions. Astrophys J 436:L193-L196
Lugmair GW, Marti K (1977) Sm-Nd-Pu timepieces in the Angra dos Reis meteorite. Earth Planet Sci Lett 35: 273-284
Lugmair GW, Marti K, Scheinin NB (1978) Incomplete mixing of products from r-, p-, and s-process nucleosynthesis: Sm-Nd systematics in Allende inclusion EK141. Lunar Planet Sci IX:672-673
Lugmair GW, Shimamura T, Lewis RS, Anders E (1983) Samarium-146 in the early solar system: evidence from neodymium in the Allende Meteorite. Science 222:1015-1018
Lugmair GW, Shukolyukov A (1998) Early solar system timescales according to ^{53}Mn-^{53}Cr systematics. Geochim Cosmochim Acta 62:2863-2886
MacPherson GJ, Bar-Matthews M, Tanaka T, Olsen E, Grossman L (1983) Refractory inclusions in the Murchison meteorite. Geochim Cosmochim Acta 47:823-839
MacPherson GJ, Wark DA, Armstrong JT (1988) Primitive material surviving in chondrites: refractory inclusions. *In*: Meteorites and the Early Solar System. Kerridge JF, Matthews MS (eds) University of Arizona Press, Tucson, p 746-807

MacPherson GJ, Davis AM, Zinner EK (1995) The distribution of aluminum-26 in the early solar system; a reappraisal. Meteoritics 30:365-386
MacPherson GJ, Huss GR, Davis AM (2003) Extinct ^{10}Be in type A calcium-aluminum-rich inclusions from CV chondrites. Geochim Cosmochim Acta 67:3615-3179
Matthews GJ, Cowan JJ (1990) New insights into the astrophysical r-process. Nature 345:491-494
McCulloch MT, Wasserburg GJ (1978a) Barium and neodymium isotopic anomalies in the Allende meteorite. Astrophys J 220:L15-L19
McCulloch MT, Wasserburg GJ (1978b) More anomalies from the Allende meteorite: Samarium. Geophys Res Lett 5:599-602
McKeegan KD, Chaussidon M, Robert F (2000) Incorporation of short-lived ^{10}Be in a calcium-aluminium-rich inclusion from the Allende Meteorite. Science 289:1334-1337
Merrill PW (1952) Technetium in the stars. Science 115:484-486
Meyer BS (1994) The r-, s-,and p-processes in nucleosynthesis. Annu Rev Astronom Astrophys 32:153-190
Mostefaoui S, Lugmair GW, Hoppe P, El Goresy A (2003) Evidence for live iron-60 in Semarkona and Chervony Kut: a nanosims study. Lunar Planet Sci XXXIV:1585
Murthy VR, Sandoval P (1965) Chromium isotopes in meteorites. J Geophys Res 70:4379-4382
Nicolussi GK, Davis AM, Pellin MJ, Lewis RS, Clayton RN, Amari S (1997) s-Process zirconium in presolar silicon carbide grains. Science 277:1281-1283
Nicolussi GK, Pellin MJ, Lewis RS, Davis AM, Amari S, Clayton RN (1998a) Molybdenum isotopic composition of individual presolar silicon carbide grains from the Murchison meteorite. Geochim Cosmochim Acta 62:1093-1104
Nicolussi GK, Pellin MJ, Lewis RS, Davis AM, Clayton RN, Amari S (1998b) Strontium isotopic composition in individual silicon carbide grains: a record of s-process nucleosynthesis. Phys Rev Lett 81:3583-3586
Nicolussi GK, Pellin MJ, Lewis RS, Davis AM, Clayton RN, Amari S (1998c) Zirconium and molybdenum in individual circumstellar graphite grains: new data on the nucleosynthesis of the heavy elements. Astrophys J 504:492-499
Niederer FR, Papanastassiou DA, Wasserburg GJ (1981) The isotopic composition of titanium in the Allende and Leoville meteorites. Geochim Cosmochim Acta 45:1017-1031
Niederer FR, Papanastassiou DA (1984) Ca isotopes in refractory inclusions. Geochim Cosmochim Acta 48: 1279-1293
Niederer FR, Papanastassiou DA, Wasserburg GJ (1985) Absolute isotopic abundances of Ti in meteorites. Geochim Cosmochim Acta 49:835-851
Niemeyer S, Lugmair GW (1981) Ubiquitous isotopic anomalies in Ti from normal Allende inclusions. Earth Planet Sci Lett 53:211-225
Nittler LR (2003) Presolar stardust in meteorites: recent advances and scientific frontiers. Earth Planet Sci Lett 209:259-273
Nittler LR, Alexander CMOD, Walker RM, Gao X, Zinner EK (1994) Meteoritic oxide grain from supernova found. Nature 370:443-446
Nittler LR, Amari S, Zinner E, Woosley SE, Lewis RS (1996) Extinct ^{44}Ti in presolar graphite and SiC: proof of a supernova origin. Astrophys J 462:L31-L34
Nittler LR, Alexander CMOD (2003) Automated isotopic measurements of micron-sized dust: application to meteoritic presolar silicon carbide. Geochim Cosmochim Acta 67:4961-4980
Nuth JA (1987) Small-particle physics and interstellar diamond. Nature 329:589
Olsen EJ, Davis AM, Hutcheon ID, Clayton RN, Mayeda TK, Grossman L (1988) Murchison xenoliths. Geochim Cosmochim Acta 52:1615-1626
Ott U (1993) Physical and isotopic properties of surviving interstellar carbon phases. *In:* Protostars & Planets III. Levy Hand Lunine JI (eds) University of Arizona Press, Tucson, p 883-902
Ott U (1996) Interstellar diamond xenon and timescales of supernova ejecta. Astrophys J 463:344-348
Ott U, Begemann F, Yang J, Epstein S (1988) S-process krypton of variable isotopic composition in the Murchison meteorite. Nature 332:700-702
Ott U, Begemann F (1990) Discovery of s-process barium in the Murchison meteorite. Astrophys J 353:L57-L60
Papanastassiou DA (1986) Chromium isotopic anomalies in the Allende meteorite. Astrophys J 308:L27-L30
Papanastassiou DA, Wasserburg GJ (1969) Initial strontium isotopic abundances and the resolution of small time differences in the formation of planetary objects. Earth Planet Sci Lett 5:361-376
Papanastassiou DA, Wasserburg GJ (1978) Strontium isotopic anomalies in the Allende meteorite. Geophys Res Lett 5:595-598
Papanastassiou DA, Brigham CA (1989) The identification of meteorite inclusions with isotope anomalies. Astrophys J 338:L37-L40
Peters, JG, Fowler WA, Clayton DD (1972) Weak s-process irradiations. Astrophys J 173:637-648

Podosek FA, Ott U, Brannon JC, Neal CR, Bernatowicz TJ, Swan P, Mahan SE (1997) Thoroughly anomalous chromium in Orgueil. Meteorit Planet Sci 32:617-627

Prantzos N, Doom C, Arnould M, De Loore C (1986) Nucleosynthesis and evolution of massive stars with mass loss and overshooting. Astrophys J 304:695-712

Pravdivtseva OV, Krot AN, Hohenberg CM, Meshik AP, Weisberg MK, Keil K (2003) The I-Xe record of alteration in the Allende CV chondrite. Geochim Cosmochim Acta 67:5011-5026

Prinzhofer A, Papanastassiou DA, Wasserburg GJ (1989) The presence of ^{146}Sm in the early solar system and implications for its nucleosynthesis. Astrophys J 344:L81-L84

Prinzhofer A, Papanastassiou DA, Wasserburg GJ (1992) Samarium-neodymium evolution of meteorites. Geochim Cosmochim Acta 56:797-815

Prombo CA, Podosek FA, Amari S, Lewis RS (1993) s-Process Ba isotopic compositions in presolar SiC from the Murchison meteorite. Astrophys J 410:393-399

Quitté G (2001) Etude des météorites à l'aide du système Hf-W: Contraintes sur les événements du système solaire primitif. Thèse Université Denis Diderot (Paris VII)

Quitté G, Birck JL, Allègre CJ (2000) ^{182}Hf-^{182}W systematics in eucrites: the puzzle of iron segregation in the early solar system. Earth Planet Sci Lett 184:63-74

Rauscher T, Heger A, Hoffman RD, Woosley SE (2002) Nucleosynthesis in massive stars with improved nuclear and stellar physics. Astrophys J 576:323-348

Rayet M (1995) The p-process in type II supernovae. Astron Astrophys 298:517-532

Rayet M, Prantzos N, Arnould M (1990) The p-process revisited. Astron Astrophys 227:271-281

Reedy RC, Arnold JR, Lal D (1983) Cosmic-ray record in solar system matter. Science 219:127-135

Rehkämper M, Halliday AN (1999) The precise measurement of Tl isotopic compositions by MC-ICPMS: application to the analysis of geological materials and meteorites. Geochim Cosmochim Acta 63:935-944

Reynolds JH (1963) Xenology. J Geophys Res 68:2939-2956

Richter S (1995) Massenspektrometrische Untersuchungen von interstellarer Materie und Bedingungen im s-Prosess der Nukleosynthese. Ph.D. dissertation. University of Mainz

Richter S, Ott U, Begemann F (1998) Tellurium in pre-solar diamonds as an indicator for rapid separation of supernova ejecta. Nature 391:261-263

Rolfs CE, Rodney WS (1988) Cauldrons in the cosmos. University of Chicago Press, Chicago

Rotaru M, Birck JL, Allegre CJ (1992) Clues to early solar system history from chromium isotopes in carbonaceous chondrites. Nature 358:465-470

Rowe MW, Kuroda PK (1965) Fissiogenic Xe from the Pasamonte meteorite. J Geophys Res 70:709-714

Russell SS, Arden JW, Pillinger CT (1991) Evidence for multiple sources of diamond from primitive chondrites. Science 254:1188-1191

Russell SS, Srinivasan G, Huss GR, Wasserburg GJ, MacPherson GJ (1996) Evidence for widespread ^{26}Al in the solar nebula and constraints for nebula time scales. Science 273:757-762

Russell SS, Gounelle M, Hutchison R (2001) Origin of short-lived radionuclides. Phil Trans R Soc Lond A 359:1991-2004

Russell WA, Papanastassiou DA, Tombrello TA (1978) Ca isotope fractionation on the earth and other solar system materials. Geochim Cosmochim Acta 42:1075-1090

Sahijpal S, Goswami JN, Davis AM, Grossman L, Lewis RS (1998) A stellar origin for the short-lived nuclides in the early Solar System. Nature 391:559-561

Sandler GG, Koonin SE, Fowler WA (1982) Ca-Ti-Cr anomalies in an Allende inclusion and the n-ß process. Astrophys J 259:908-919

Sanloup C, Blichert-Toft J, Télouk P, Gillet P, Albarède F (2000) Zr isotope anomalies in chondrites and the presence of ^{92}Nb in the early solar system. Earth Planet Sci Lett 184:75-81

Savina MR, Davis AM, Tripa CE, Pellin MJ, Clayton RN, Lewis RS, Amari S, Gallino R, Lugaro M (2003a) Barium isotopes in individual presolar silicon carbide grains from the Murchison meteorite. Geochim Cosmochim Acta 67:3201-3214

Savina MR, Pellin MJ, Tripa CE, Veryovkin IV, Calaway WF, Davis AM (2003b) Analysing individual presolar grains with CHARISMA. Geochim Cosmochim Acta 67:3215-3225

Schmitt AD, Stille P, Vennemann T (2003) Variations of the ^{44}Ca/^{40}Ca ratio in seawater during the past 24 million years: evidence from δ^{44}Ca and δ^{18}O values of Miocene phosphates. Geochim Cosmochim Acta 67:2607-2614

Schönbächler M, Rehkämper M, Halliday AN, Lee D, Bourot-Denise M, Zanda B, Hattendorf B, Günther D (2002) Niobium-zirconium chronometry and early solar system development. Science 295:1705-1708

Schönbächler M, Lee DC, Rehkämper M, Halliday A, Fehr AM, Hattendorf B, Günther D (2003) Zirconium isotope evidence for incomplete admixing of r-process components in the solar nebula. Earth Planet Sci Lett accepted

Schramm DN, Tera F, Wasserburg GJ (1970) The isotopic abundance of ^{26}Mg and limits on ^{26}Al in the early solar system. Earth Planet Sci Lett 10:44-59

Schueler B, Morton J, Mauersberger K (1990) Measurement of isotopic abundances in collected stratospheric ozone samples. Geophys Res Lett 17:1295-1298

Sharma M, Papanastassiou DA, Wasserburg GJ, Dymek RF (1996) The issue of the terrestrial record of ^{146}Sm. Geochim Cosmochim Acta 60:2037-2047

Shen JJS, Lee T, Chang CT (1994) Lanthanum isotopic composition of meteoritic and terrestrial matter. Geochim Cosmochim Acta 58:1499-1506

Shu FH, Najita J, Shang H, Li ZY (2000) X-winds: theory and observations. *In:* Protostars & Planets IV. Mannings V, Boss AP and Russel SS (eds) University of Arizona Press, Tucson, p 789-814

Shukolyukov A, Lugmair GW (1993) Live iron-60 in the early solar system. Science 259:1138-1142

Shukolyukov A, Lugmair GW, Bogdanovski O (2003) Manganese-chromium isotope systematics of Ivuna, Kainsaz and other carbonaceous chondrites. Lunar Planet SciXXXIV:1279

Srinivasan G, Sahijpal S, Ulyanov AA, Goswami JN (1996) Ion microprobe studies of Efremovka CAIs: II. Potassium isotope composition and ^{41}Ca in the early solar system. Geochim Cosmochim Acta 60:1823-1835

Srinivasan G, Goswami JN, Bhandari N (1999) ^{26}Al in eucrite Piplia Kalan: plausible heat source and formation chronology. Science 284:1348-1350

Sugiura N, Shuzou Y, Ulyanov A (2001) Beryllium-boron and aluminum-magnesium chronology of calcium-aluminum-rich inclusions in CV chondrites. Meteorit Planet Sci 36:1397-1408

Swindle TD, Caffee MW, Hohenberg CM (1988) Iodine-xenon studies of Allende inclusions: Eggs and the Pink Angel. Geochim Cosmochim Acta 52:2215-2227

Swindle TD, Podosek FA (1988) Iodine-Xenon dating. *In*: Meteorites and the Early Solar System. Kerridge JF and Matthews MS (eds) University of Arizona Press, Tucson, p 1114-1146

Tang M, Lewis RS, Anders E (1988) Isotopic anomalies of Ne, Xe, and C in meteorites. I. Separation of carriers by density and chemical resistance. Geochim Cosmochim Acta 52:1221-1234

Tera F, Eugster O, Burnett DS, Wasserburg GJ (1970) Comparative study of Li, Na, K, Rb, Cs, Ca, Sr and Ba abundances in achondrites and in Apollo 11 lunar samples. Geochim Cosmochim Acta Suppl 1:1637-1657

Thiemens MH (1988) Heterogeneity in the nebula: evidence from stable isotopes. *In*: Meteorites and the Early Solar System. Kerridge JF and Matthews MS (eds) University of Arizona Press, Tucson, p 899-923

Thiemens MH (1999) Mass independent isotope effects in planetary atmospheres and the early solar system. Science 283:341-345

Thiemens MH, Heidenreich JE (1983) The mass independent fractionation of oxygen: a novel isotope effect and its possible cosmochemical implications. Science 219:1073-1075

Thiemens MH, Jackson TL, Brenninkmeijer CAM (1995) Observation of a mass-independent oxygen isotopic composition in terrestrial stratospheric CO_2, the link to ozone chemistry, and the possible occurrence in the Martian atmosphere. Geophys Res Lett 22:255-257

Timmes FX, Woosley SE, Weaver TA (1995) Galactic chemical evolution: hydrogen through zinc. Astrophys J Suppl 98:617-658

Urey HC (1955) The cosmic abundances of potassium, uranium and thorium and the heat balances of the Earth, the Moon, and Mars. Proc Nat Acad Sci 41:127-144

Van Winckel H, Mathis JS, Waelkens C (1992) Evidence from zinc abundances for dust fractionation in chemically peculiar stars. Nature 356:500-501

Virag A, Zinner E, Lewis RS, Tang M (1989) Isotopic compositions of H, C, and N in Cδ diamonds from the Allende and Murray carbonaceous chondrites. Lunar Planet Sci XX:1158-1159

Völkening J, Papanastassiou DA (1989) Iron isotope anomalies. Astrophys J 347:L43-L46

Völkening J, Papanastassiou DA (1990) Zinc isotope anomalies. Astrophys J 358:L29-L32

Wadhwa M, Zinner EK, Crozaz G (1997) Manganese-chromium systematics in sulfides of unequilibrated enstatite chondrites. Meteorit Planet Sci 32:281-292

Wadhwa M, Foley CN, Janney PE (2003) High precision Mg isotopic analyses of achondrites: is the ^{26}Al-^{26}Mg chronometer concordant with other high resolution chronometers? Geochim Cosmochim Acta Suppl 67: A517

Wark DA (1986) Evidence for successive episodes of condensation at high temperature in a part of the solar nebula. Earth Planet Sci Lett 77:129-148

Wasserburg GJ, Hayden RJ (1955) Time interval between nucleogenesis and the formation of meteorites. Nature 176:130-131

Wasserburg GJ, Papanastassiou DA, Nenow EV, Bauman CA (1969) A programmable magnetic field mass spectrometer with on-line data processing. Rev Sci Instr 40:288-295

Wasserburg GJ, Lee T, Papanastassiou DA (1977) Correlated O and Mg isotopic anomalies in Allende inclusions: II. Magnesium. Geophys Res Lett 4:299-302

Wasserburg GJ, Papanastassiou DA (1982) Some short-lived nuclides in the early solar system- a connection with the placental ISM. *In:* Essays in Nuclear Astrophysics. Barnes CA, Clayton DD and Schramm DN (eds) Cambridge University Press, Cambridge, p 77-140

Wasserburg GJ, Busso M, Gallino R, Raiteri CM (1994) Nucleosynthesis in asymptotic giant branch stars: relevance for galactic enrichment and solar system formation. Astrophys J 424:412-420

Wasserburg GJ, Busso M, Gallino R (1996) Abundances of actinides and short-lived non-actinides in the interstellar medium: diverse supernova sources for the r-processes. Astrophys J 466:109-113

Woosley SE (1986) Nucleosynthesis and stellar evolution. *In*: Nucleosynthesis and Chemical Evolution. Hauck B, Maeder A and Meynet G (eds) University of Texas Press, Austin, p 113-154

Woosley SE, Arnett WD, Clayton DD (1973) The explosive burning of oxygen and silicon. Astrophys J Suppl 26:231-312

Woosley SE, Langer N, Weaver TA (1995) The presupernova evolution and explosion of helium stars that experience mass loss. Fresenius Astrophys J 448:315-338

Woosley SE, Weaver TA (1995) The evolution and explosion of massive stars. II. Explosive hydrodynamics and nucleosynthesis. Astrophys J Suppl 101:181-235

Yin Q, Jacobsen SB, Yamashita K, Blichert-Toft J, Télouk P, Albarède F (2002) A short time scale for terrestrial planet formation from the Hf-W chronometry of meteorites. Nature 418:949-951

Young ED, Russell SS (1998) Oxygen reservoirs in the early solar nebula inferred from an Allende CAI. Science 282:452-455

Young ED, Ash RD, Galy A, Belshaw NS (2002) Mg isotope heterogeneity in the Allende meteorite measured by UV laser ablation-MC-ICPMS and comparisons with O isotopes. Geochim Cosmochim Acta 66:683-698

Zinner E (1997) Presolar material in meteorites: an overview. *In:* Astrophysical Implications of the Laboratory Study of Presolar Materials. Bernatowicz TJ and Zinner E (eds) AIP, New York, p 3-26

Zinner EK, Göpel C (2002) Aluminium-26 in H4 chondrites: implications for its production and its usefulness as a fine-scale chronometer for early solar system events. Meteorit Planet Sci 37:1001-1013

Zinner E, Amari S, Guinness R, Nguyen A, Stadermann FJ, Walker RM, Lewis RS (2003) Presolar spinel grains from the Murray and Murchison carbonaceous chondrites. Geochim Cosmochim Acta 67:5083-5095

Reviews in Mineralogy & Geochemistry
Vol. 55, pp. 65-111, 2004
Copyright © Mineralogical Society of America

3

Applying Stable Isotope Fractionation Theory to New Systems

Edwin A. Schauble

Department of Earth and Space Sciences
University of California, Los Angeles
Los Angeles, California 90095-1567, U.S.A.

INTRODUCTION

A basic theoretical understanding of stable isotope fractionations can help researchers plan and interpret both laboratory experiments and measurements on natural samples. The goal of this chapter is to provide an introduction to stable isotope fractionation theory, particularly as it applies to mass-dependent fractionations of non-traditional elements and materials. Concepts are illustrated using a number of worked examples. For most elements, and typical terrestrial temperature and pressure conditions, equilibrium isotopic fractionations are caused by the sensitivities of molecular and condensed-phase vibrational frequencies to isotopic substitution. This is explained using the concepts of vibrational zero-point energy and the partition function, leading to Urey's (1947) simplified equation for calculating isotopic partition function ratios for molecules, and Kieffer's (1982) extension to condensed phases. Discussion will focus on methods of obtaining the necessary input data (vibrational frequencies) for partition function calculations. Vibrational spectra have not been measured or are incomplete for most of the substances that Earth scientists are interested in studying, making it necessary to estimate unknown frequencies, or to measure them directly. Techniques for estimating unknown frequencies range from simple analogies to well-studied materials to more complex empirical force-field calculations and *ab initio* quantum chemistry. Mössbauer spectroscopy has also been used to obtain the vibrational properties of some elements, particularly iron, in a variety of compounds. Some kinetic isotopic fractionations are controlled by molecular or atomic translational velocities; this class includes many diffusive and evaporative fractionations. These fractionations can be modeled using classical statistical mechanics. Other kinetic fractionations may result from the isotopic sensitivity of the activation energy required to achieve a transition state, a process that (in its simplest form) can be modeled using a modification of Urey's equation (Bigeleisen 1949).

Theoretical estimates of isotopic fractionations are particularly powerful in systems that are difficult to characterize experimentally, or when empirical data are scarce. The accuracy of a theoretical calculation is limited by uncertainty in input data, and by errors resulting from simplified thermodynamic treatment of molecular motion. Typical uncertainties are larger than the nominal precision of a careful isotope ratio measurement, although ongoing improvements in the quality of spectroscopic data and molecular modeling methods are helping to close this gap. Nonetheless, the accuracy of relatively simple theoretical models is sufficient to provide a quantitative framework for interpreting the results of a set of measurements. Theoretical calculations are easily extended to temperature conditions that are not easily accessed by experiments, which is especially relevant for low-temperature mineral-solution fractionations where isotopic exchange equilibrium often cannot be achieved on a reasonable laboratory

1529-6466/04/0055-0003$05.00

time scale. Qualitative rules of thumb based on theoretical concepts can be applied to systems that have not been explicitly studied. Complex fractionations, involving a combination of mechanisms, are common, and theoretical techniques can provide a unique perspective to help pick apart the underlying causes of an observed fractionation, and help understand and demonstrate its geochemical significance.

Overview

The purpose of this chapter is to provide a brief, practical guide to the theory of stable isotope fractionations. This subject is particularly apt for inclusion in a volume dedicated to less studied and novel stable isotope systems, because these systems often lack an extensive record of measurements and empirical intuitions to guide the planning and interpretation of analytical campaigns. As technological advances like multiple collector inductively coupled plasma mass spectrometry (MC-ICP-MS) push stable isotope geochemistry into uncharted territory, it becomes increasingly necessary for analysts—as well as interested researchers outside of the field of stable isotope geochemistry—to have an informed perspective on the basic mechanisms driving variations in isotopic abundances in natural samples.

This review will introduce basic techniques for calculating equilibrium and kinetic stable isotope fractionations in molecules, aqueous complexes, and solid phases, with a particular focus on the thermodynamic approach that has been most commonly applied to studies of equilibrium fractionations of well-studied elements (H, C, N, O, and S) (Urey 1947). Less direct methods for calculating equilibrium fractionations will be discussed briefly, including techniques based on Mössbauer spectroscopy (Polyakov 1997; Polyakov and Mineev 2000).

History

The theory of stable isotope fractionation precedes the development of modern mass spectrometry, and includes a few very early studies (Lindemann and Aston 1919; Lindemann 1919; Urey and Greiff 1935). The modern theoretical formulation for calculating equilibrium isotope fractionations can be credited to seminal work by Urey (1947) and Bigeleisen and Mayer (1947), which forms the basis for much of this chapter. Theoretical calculations in these papers successfully predicted the directions, magnitudes, and temperature sensitivities of isotopic fractionations. These calculations suggested the possibility of paleothermometry using $^{18}O/^{16}O$ fractionation in the calcite-water system (Urey 1947), foreshadowing the creation of a major field of geochemical research. Later the Urey formulation was extended to encompass crystals and some amorphous solids (e.g., Bottinga 1968; Kieffer 1982). A key point is that all of these theoretical treatments result from a simplified thermodynamic model of the quantum mechanics of molecular vibration and rotation—making theoretical calculations feasible for many important substances—while retaining enough accuracy to be quantitatively useful. As we will see, limited or imprecise data on molecular or crystal vibrations are a major issue to be overcome in calculating accurate theoretical fractionations. Several excellent reviews of stable isotope fractionation theory applied to the commonly studied elements—H, C, N, O, and S—are available in the literature (Richet et al. 1977; O'Neil 1986; Criss 1991, 1999; Chacko et al. 2001).

There have been relatively few theoretical studies of fractionations involving other elements. Lindemann (1919) probably performed the first theoretical calculation of a stable isotope fractionation, estimating the vapor pressures of the isotopes of lead. Urey (1947) and Urey and Greiff (1935) modeled equilibrium isotope fractionations for several geochemically important molecules containing chlorine, bromine, and iodine. Bigeleisen and Mayer (1947) briefly discuss possible fractionations involving the silicon and tin halides, focusing specifically on the effect of coordination number on fractionations. Later theoretical work was concentrated in the specialized literature of isotope separation. Of particular interest during this period was the development of force-field and quantum-mechanical techniques for estimating unknown vibrational frequencies of molecules containing rare isotopes (Kotaka and Kakihana 1977; Kotaka et al. 1978; Hanschmann 1984). In recent years, as the set of elements with detectable natural isotopic variations has expanded, theoretical estimates of isotopic fractionations for boron (Oi 2000; Oi and Yanase 2001), lithium (Yamaji et al. 2001), chlorine (Paneth 2003; Schauble et al. 2003), chromium (Schauble et al. in press) and iron (Polyakov 1997; Polyakov and Mineev 2000; Schauble et al. 2001, Anbar et al. in press) have been published.

General rules governing equilibrium stable isotope fractionations

It is clear, from an examination of both observed and theoretically calculated isotope fractionations, that there are a number of qualitative chemical rules that can be used to estimate which substances will tend to be enriched in heavy isotopes in a given geochemical system. O'Neil (1986) tabulated five characteristics that are shared by the elements that show large variations in isotopic composition in nature. These include (i) low atomic mass, (ii) large relative mass differences between stable isotopes, (iii) tendency to form highly covalent bonds, (iv) multiple oxidation states or other chemical variability, and (v) availability of multiple isotopes with sufficient abundance to make measurements feasible. The elements covered in the present volume, in general, fail to meet one or more of these criteria. They are heavy (i.e., Fe, Mo, and Cd), do not have large mass differences between the measured stable isotopes (Cr), form bonds that are predominantly ionic (Li, Mg, Ca) rather than covalent, or have homogenous chemistry and a single predominant oxidation state in nature (Li, Mg, Si, Ca). Nonetheless, these rules are strongly supported by theoretical considerations originally derived by Bigeleisen and Mayer (1947), and form the basis for a qualitative guide to stable isotope fractionations in all elements:

Qualitative Rules governing equilibrium stable isotope fractionations:

1. Equilibrium isotopic fractionations usually decrease as temperature increases, roughly in proportion to $1/T^2$ for most substances. Note that exceptions may occur, particularly if the fractionation is very small, or if the element of interest is bonded directly to hydrogen in one phase.
2. All else being equal, isotopic fractionations are largest for light elements and for isotopes with very different masses, scaling roughly as $(m_{heavy} - m_{light})/(m_{light}m_{heavy})$ (often simplified to $\Delta m/m^2$) where m_{light} and m_{heavy} are the masses of two isotopes of an element, Δm is the difference between their masses, and m is the average atomic mass of the element.
3. At equilibrium, the heavy isotopes of an element will tend to be concentrated in substances where that element forms the stiffest bonds (i.e., bonds with high spring constants). The magnitude of the isotopic fractionation will be roughly proportional to the difference in bond stiffness between the equilibrated substances. Bond stiffness is greatest for *short*, *strong* chemical bonds; these properties correlate with:
   a. high oxidation state in the element of interest
   b. for anions like Cl^- and Se^{2-} (and O^{2-}), high oxidation state in the atoms to which the element of interest is bonded
   c. bonds involving elements near the top of the periodic table
   d. the presence of highly covalent bonds between atoms with similar electronegativities
   e. for transition elements, low-spin electronic configurations, also d^3 or d^8 electronic structure for octahedrally-coordinated atoms
   f. low coordination number

4. Substances where the element of interest is directly bonded to hydrogen, or is part of a low-mass molecule, may not be as enriched in heavy isotopes as would be expected from rule (3). This phenomenon, usually of 2nd order importance, is most pronounced for substances with stiff bonds and at low temperatures.

Rules (2) and (3) imply that large equilibrium fractionations are most likely to occur at low temperatures between substances with markedly different oxidation states, bond partners, electronic configurations, or coordination numbers.

It is important to point out that these rules (particularly 3 and 4) are largely untested with respect to fractionations in the less-well studied elements (those other than H, C, N, O, and S) at the present time. Furthermore, within rule (3), it is not known which chemical properties are the most important determinants of bond stiffness. The order of listing reflects a rough estimate of relative importance, based on experiments and theoretical studies in a variety of isotopic systems. However, this order is likely to vary somewhat from one system to another even if it is generally correct. What is known is that measurable isotopic fractionations are typically larger and more common in samples from low-temperature environments than in high-temperature (particularly igneous) environments for Ca (Russell et al. 1978; Skulan et al. 1997), Fe (Beard et al. 2003), Mo (Barling et al. 2001), and Mg (Galy et al. 2002). $^{56}Fe/^{54}Fe$ ratios are higher in aqueous Fe^{3+} than in coexisting aqueous Fe^{2+} (rule 3a) (Johnson et al. 2002), and Fe^{3+}-Cl^{-} complexes retained on ion-exchange resins appear to have lower $^{56}Fe/^{54}Fe$ ratios than Fe^{3+}-OH_2 complexes in solution (rules 3b and 3c) (Roe et al. 2003). Borate solutions with low pH, favoring the 3-coordinate $B(OH)_3$ have a greater affinity for heavy ^{11}B than high pH solutions dominated by 4-coordinated $[B(OH)_4]^-$ (rule 3f) (Kakihana et al. 1977). Clearly, more experiments are needed before a robust assessment of these rules can be made.

These rules suggest that stable isotope measurements of redox-active transition elements (e.g., Cr, Fe, Cu and Mo), and main group elements (Cl, Br, and Te) are likely to provide records of modern and ancient oxidation conditions. For chalcophile and siderophile elements like Fe, Cu, Zn, and Mo, low and moderate temperature partitioning between oxide, sulfide, and/or metal phases is also likely to cause diagnostic isotopic fractionation. In contrast, isotopic fractionation of main group lithophile metals like Li, Mg, Si and Ca are more likely to reflect changes in coordination number. Equilibrium fractionations in this last group (particularly for the heavier elements) are likely to be rather small, and may commonly be overwhelmed by kinetic fractionations in low-temperature and biological systems.

Kinetic isotope fractionations also show systematic behavior, although they are more difficult to describe with a list of widely applicable rules. Here the term kinetic is being used loosely to describe a host of basically one-directional processes occurring under conditions of incomplete isotopic exchange. This definition is useful in its simplicity, although it encompasses a number of distinct fractionation mechanisms, and in practice there is some ambiguity in dealing with "equilibrium-like" fractionations (like oxygen-isotope fractionation in biologically precipitated calcite). One common feature in many kinetic fractionations is that light isotopes, being more reactive, are concentrated in reaction products. This behavior is observed, for instance, in rapid precipitations of Fe^{3+} oxyhydroxide and oxide minerals from Fe^{3+}-bearing solutions (Bullen et al. 2001; Skulan et al. 2002), in evaporation of many substances including silicate melts (Davis et al. 1990), and in numerous biological reactions. Diffusive and evaporative kinetic fractionations do not, in general, decrease in magnitude with increasing temperature in the same way that equilibrium fractionations do (although they may be sensitive to temperature). Kinetic fractionations are usually sensitive to a host of factors (such as reaction rates and the presence of exchange catalysts) in addition to temperature.

Fractionation factors

Before introducing the quantitative theory of isotope fractionations, it will be useful to review terms relevant to calculating fractionation factors.

The isotope fractionation factor for isotopes ${}^{light}X$ (light) and ${}^{heavy}X$ (heavy) between two substances *XA* and *XB* is usually expressed in terms of "α":

$$\alpha_{XA\text{-}XB} = \frac{\left({}^{heavy}X / {}^{light}X \right)_{XA}}{\left({}^{heavy}X / {}^{light}X \right)_{XB}}$$

Fractionations are typically very small, on the order of parts per thousand or parts per ten thousand, so it is common to see expressions like 1000•ln(α) or 1000•(α–1) that magnify the difference between α and 1. α =1.001(1000•[α–1] = 1) is equivalent to a 1 per mil (‰) fractionation. Readers of the primary theoretical literature on stable isotope fractionations will frequently encounter results tabulated in terms of β-factors or equilibrium constants. For present purposes, we can think of β_{XR} as simply a theoretical fractionation calculated between some substance *XR* containing the element *X*, and dissociated, non-interacting atoms of *X*. In the present review the synonymous term α_{XR-X} is used. This type of fractionation factor is a convenient way to tabulate theoretical fractionations relative to a common exchange partner (dissociated, isolated atoms), and can easily be converted into fractionation factors for any exchange reaction:

$$\alpha_{XA\text{-}XB} = \frac{\alpha_{XA\text{-}X}}{\alpha_{XB\text{-}X}} = \frac{\beta_{XA}}{\beta_{XB}}$$

Conversions between equilibrium constants and fractionation factors are more complicated, as it is often necessary to account for molecular stoichiometry and symmetry. For a generic balanced isotopic exchange reaction,

$$\mathrm{n}\bullet {}^{light}X_mA + \mathrm{m}\bullet {}^{heavy}X_nB \leftrightarrow \mathrm{n}\bullet {}^{heavy}X_mA + \mathrm{m}\bullet {}^{light}X_nB$$

where m and n are stoichiometric coefficients, the equilibrium constant K_{eq} is related to the fractionation factor $\alpha_{XA\text{-}XB}$ by the expression,

$$\alpha_{XA\text{-}XB} = \left(\frac{S_{heavy\ X_mA}}{S_{light\ X_mA}} \right)^{\frac{1}{\mathrm{n}}} \left(\frac{S_{light\ X_nB}}{S_{heavy\ X_nB}} \right)^{\frac{1}{\mathrm{m}}} \left(K_{\mathrm{eq}} \right)^{\frac{1}{\mathrm{mn}}}$$

where $S_{heavy\ X_mA}$, $S_{light\ X_mA}$, $S_{light\ X_nB}$, and $S_{heavy\ X_nB}$ are the molecular symmetry numbers for each reactant and product molecule. Unlike equilibrium constants, isotopic fractionation factors for the elements of interest here can be determined accurately without worrying about symmetry numbers, and are simply related to isotopic ratios, and are therefore better suited to an introductory discussion. To avoid confusion, the examples discussed below are deliberately chosen so that calculated fractionation factors and equilibrium constants are equivalent.

EQUILIBRIUM FRACTIONATION THEORY

Fractionation in molecules

Equilibrium stable isotope fractionation is a quantum-mechanical phenomenon, driven mainly by differences in the vibrational energies of molecules and crystals containing atoms of differing masses (Urey 1947). In fact, a list of vibrational frequencies for two isotopic forms of each substance of interest—along with a few fundamental constants—is sufficient to calculate an equilibrium isotope fractionation with reasonable accuracy. A succinct derivation of Urey's formulation follows. This theory has been reviewed many times in the geochemical literature, (e.g., Urey 1947; O'Neil 1986; Criss 1991, 1999; Chacko et al. 2001), and it is whole-heartedly suggested that an interested reader look at several versions—digestion of thermodynamic concepts often requires more than one attempt. A thorough discussion of partition functions can be found in any statistical mechanics textbook, (e.g., Mayer and Mayer 1940). In the present introduction, we will focus on one simple isotope exchange reaction in the Cl-isotope system.

Let us consider a simple isotope exchange reaction, where ^{35}Cl and ^{37}Cl swap between the diatomic gas ClO and an isolated Cl atom:

$$^{35}\mathrm{Cl}^{16}\mathrm{O} + {}^{37}\mathrm{Cl} \leftrightarrow {}^{37}\mathrm{Cl}^{16}\mathrm{O} + {}^{35}\mathrm{Cl}$$

This type of exchange (with an isolated atom as one exchange partner) is the basis of Urey's treatment, because it facilitates a particularly convenient set of simplifications. The equilibrium constant for this exchange reaction,

$$K_{\mathrm{eq}} = \frac{\left({}^{37}\mathrm{Cl}^{16}\mathrm{O}\right)\left({}^{35}\mathrm{Cl}\right)}{\left({}^{35}\mathrm{Cl}^{16}\mathrm{O}\right)\left({}^{37}\mathrm{Cl}\right)}$$

is equivalent to the equilibrium isotopic fractionation factor "α":

$$\alpha_{\mathrm{ClO\text{-}Cl}} = \frac{\left({}^{37}\mathrm{Cl}/{}^{35}\mathrm{Cl}\right)_{\mathrm{ClO}}}{\left({}^{37}\mathrm{Cl}/{}^{35}\mathrm{Cl}\right)_{\mathrm{Cl}}} = \frac{\left({}^{37}\mathrm{Cl}^{16}\mathrm{O}\right)/\left({}^{35}\mathrm{Cl}^{16}\mathrm{O}\right)}{\left({}^{37}\mathrm{Cl}\right)/\left({}^{35}\mathrm{Cl}\right)}$$

This equivalence is not universal—molecules with more than one atom of the element being exchanged may require a somewhat more complicated treatment. However, these complications have a negligible effect on the final result, so we have chosen an example that simplifies the mathematics as much as possible. As with any chemical reaction, the equilibrium constant can be determined from the free energies of the reactants and products, using the familiar expressions,

$$\Delta G^0_{\mathrm{Rxn}} = -\mathrm{R}T\ln\left(K_{\mathrm{eq}}\right)$$

$$K_{\mathrm{eq}} = \exp\left(\frac{-\Delta G^0_{\mathrm{Rxn}}}{\mathrm{R}T}\right)$$

where $\Delta G^0_{\mathrm{Rxn}}$ is the Gibbs free energy of the reaction, R is the molar gas constant (approximately 8.314 J•mol/K), T is the absolute temperature, and K_{eq} is the equilibrium constant. Isotope-exchange reactions are particularly simple because the bond structure and thus the potential energy of each molecule are essentially unchanged by isotopic substitution. For this reason, calculations only need to consider the contributions of dynamic energy associated with atomic motion. Isotopic exchange reactions also do not, in general, involve significant pressure-volume work because the number of molecules on both sides of the reaction is the same, and because isotope substitution has a negligible effect on the molar volumes of condensed and non-ideal phases under normal conditions (see Gillet et al. 1996; Driesner and Seward 2000; Horita et al. 2002 for more thorough discussions of the influence of pressure on isotope fractionations of light elements). Under these conditions, the Gibbs free energy of the exchange reaction $\Delta G^0_{\mathrm{Rxn}}$ is equivalent to the Helmholtz free energy (ΔF_{Rxn})

$$G = F + PV \quad \text{and} \quad \Delta G \approx \Delta F$$

So that the basic energetic expression to be evaluated is:

$$K_{\text{eq}} = \exp\left(\frac{-\Delta F_{\text{Motion}}}{\text{R}T}\right)$$

The energy associated with atomic motion is quantized into discrete levels. These motions can be divided into translations, rotations, and vibrations (Fig. 1). An isolated atom like Cl only has translational degrees of freedom. In a molecule there are three translational degrees of freedom, corresponding to motion of the molecule's center of mass in three dimensional space. Most molecules have three rotational degrees of freedom as well, corresponding to the three orthogonal moments of inertia, but linear molecules (including ClO) have only two rotational degrees of freedom. Because the instantaneous motion of each atom in a molecule can be described with three parameters, corresponding to velocities in each of the Cartesian directions, there are three degrees of freedom of motion for each atom. So a molecule with N atoms must have a total of $3N - 3 - 3$ vibrational degrees of freedom (or $3N - 3 - 2$ for linear molecules). Only vibrational quanta are sufficiently large relative to the ambient thermal energy ($\sim 1.5 \times \text{R}T/\text{N}_0$; N_0 = Avogadro's number, 6.022×10^{23}) to drive chemical reactions at normal temperatures ($T \geq 100$ K) (Fig. 2). In the simple case of a harmonic vibration, the energy levels are evenly spaced, with a half-quantum of energy present even in the lowest state,

$$E(\text{vib})_i = (n_i + \tfrac{1}{2})\text{h}\nu_i \tag{1}$$

where n_i (= 0, 1, 2, …) is the quantum number for the ith vibrational degree of freedom, h is Planck's constant (6.626×10^{-34} J•sec), and ν_i is the oscillation frequency of the vibration. The half-quantum of vibrational energy present when $n_i = 0$ is called the zero-point energy. The frequency of a vibration is determined by the masses of atoms that are in motion and the forces that oppose motion, and is therefore sensitive to isotopic substitution. For a diatomic molecule like ClO (with 1 vibration), the frequency can be expressed simply as:

$$\nu = \frac{1}{2\pi}\sqrt{\frac{\text{k}_\text{s}}{\mu}} = \frac{1}{2\pi}\sqrt{\text{k}_\text{s}\left(\frac{1}{m_{\text{Cl}}} + \frac{1}{m_{\text{O}}}\right)}$$

where k_s is the effective spring constant of the Cl-O bond and μ is the reduced mass of ClO. Substituting heavy ^{37}Cl for the more common ^{35}Cl isotope increases the reduced mass of the molecule while leaving the spring constant unchanged, thus lowering the vibrational frequency from 853.72 cm^{-1} (Burkholder et al. 1987) to 846.45 cm^{-1} and the zero-point energy from 5,105 J/mole to 5,062 J/mole (cm^{-1} is a spectroscopic unit for frequency called a wavenumber, 1 cm^{-1} = 2.9979×10^{10} sec^{-1}). Vibrational frequencies typically range from ~100 cm^{-1} all the way up to 4000 cm^{-1}, corresponding to effective force constants on the order of 50–2000 Newtons/m. One could construct a crude quantitative model of stable isotope fractionations by simply adding up the zero-point energies of the molecules on the right side of the exchange equation, and subtracting the zero-point energies of the left side:

$$\Delta F_{\text{Motion}} \approx \sum \frac{1}{2}\text{h}\nu_{\text{Products}} - \sum \frac{1}{2}\text{h}\nu_{\text{Reactants}} \tag{2}$$

which would correctly predict higher ^{37}Cl/^{35}Cl in ClO equilibrated with atomic Cl (Schauble et al. 2003). This comparison illustrates the basic principle that substances with large zero-point energy shifts on isotopic substitution tend to be enriched in heavy isotopes. Of course, molecules are not always in their ground vibrational states, and it is necessary to include terms to account for the energies of those excited molecules, but it is still generally true that zero-point energy shifts control equilibrium fractionations. The total energy of motion of a system of molecules is determined by means of partition functions. A partition function can be thought

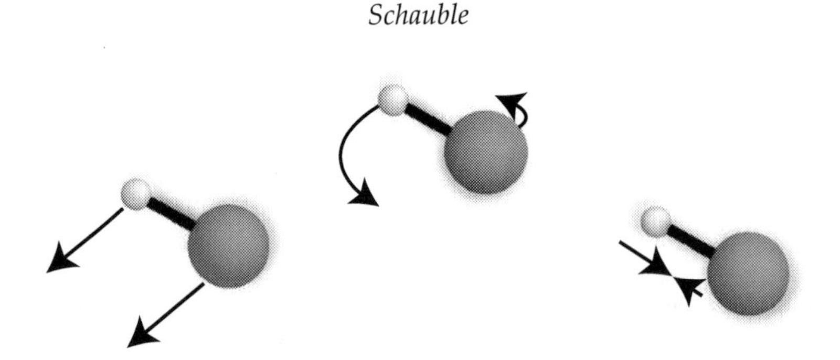

Figure 1. Translation, rotation, and vibration of a diatomic molecule. Every molecule has three translational degrees of freedom corresponding to motion of the center of mass of the molecule in the three Cartesian directions (left side). Diatomic and linear molecules also have two rotational degrees of freedom, about rotational axes perpendicular to the bond (center). Non-linear molecules have three rotational degrees of freedom. Vibrations involve no net momentum or angular momentum, instead corresponding to distortions of the internal structure of the molecule (right side). Diatomic molecules have one vibration, polyatomic linear molecules have $3N$-5 vibrations, and nonlinear molecules have $3N$-6 vibrations. Equilibrium stable isotope fractionations are driven mainly by the effects of isotopic substitution on vibrational frequencies.

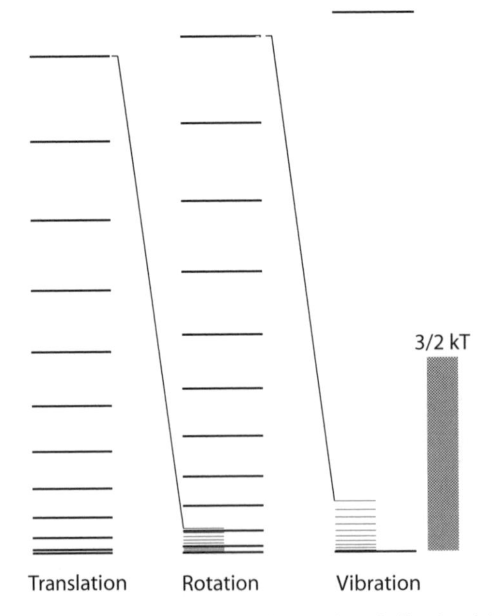

Figure 2. Relative sizes of translational (left), rotational (center), and vibrational (right) energy quanta for a typical diatomic molecule (ClO). The quantum energies of allowed states of motion are calculated using constants tabulated by Burkholder (1987). Presented at the same scale (lower right), rotational quanta are so small that low-lying rotational states are as closely spaced as the thickness of a line. Likewise, the lowest energy translational quanta (assuming a confining volume of 10^{-24} liters) are too small to see on the same scale as rotational energy levels (lower center). The classical thermal energy (3/2 kT) of a particle at 298 K is shown as a gray bar at the same scale as the vibrational energy levels. Because the spacing between rotational and translational energy levels is so much smaller than the ambient thermal energy, it is usually not necessary to include a full quantum-mechanical treatment of these types of motion when calculating equilibrium stable isotope fractionation factors. Rotational and translational quanta become even smaller for larger, more massive molecules.

of as a sum, over all of the energy states of a molecule, of the probabilities that the molecule will occupy a particular state. For instance, the vibrational partition function, Q, of a molecule is defined as a sum over all of its vibrational energies E_n,

$$Q=\sum_n \exp(-E_n/\mathrm{k}T)$$

where k is Boltzmann's constant (1.381×10^{-23} J/K or 0.6951 $\mathrm{cm^{-1}/K}$). Partition functions are closely related to the Helmholtz free energy,

$$F=-\mathrm{R}T\ln(Q)$$

For harmonic vibrations, we can insert Equation (1) above,

$$\begin{aligned}Q_{\text{Vib}}&=\sum_n \exp(-E_n/\mathrm{k}T)\\&=\sum_i\sum_n \exp\left[-\left(n_i+\frac{1}{2}\right)\mathrm{h}\nu_i\Big/\mathrm{k}T\right]\end{aligned}$$

This rather awkward expression can be simplified by recognizing the presence of a geometric series:

$$\begin{aligned}&=\sum_i \exp\left[-\left(\tfrac{1}{2}\right)\mathrm{h}\nu_i\Big/\mathrm{k}T\right]\times\left(\sum_{\mathrm{n}}\exp\left[-(n_i)\mathrm{h}\nu_i/\mathrm{k}T\right]\right)\\&=\sum_i \exp\left[-\frac{\mathrm{h}\nu_i}{2\mathrm{k}T}\right]\times\frac{1}{1-\exp\left[-\mathrm{h}\nu_i/\mathrm{k}T\right]}\end{aligned}$$

leaving a finite sum over the harmonic vibrational frequencies of the molecule.

Energy quanta associated with molecular rotation and translation are so small that they can be treated approximately without an explicit sum over the quantum energies.

$$Q_{\text{Rot}}=\frac{8\pi^2 I\mathrm{k}T}{\mathrm{h}^2}\qquad\qquad Q_{\text{Trans}}=V\left(\frac{2\pi m_{\text{molecule}}\mathrm{k}T}{\mathrm{h}^2}\right)^{3/2}$$

where I is the moment of inertia of the molecule (which can be determined from the molecular structure), m_{molecule} is its mass, and V is the volume of the system. While these expressions look complicated, bear in mind that they will almost completely cancel out by the time we're finished. The translational and rotational free energies, added to the vibrational energy, give the total energy of atomic motion,

$$\begin{aligned}F_{\text{Motion}}&=-\mathrm{R}T\left[\ln(Q_{\text{Trans}})+\ln(Q_{\text{Rot}})+\ln(Q_{\text{Vib}})\right]\\&=-\mathrm{R}T\ln(Q_{\text{Trans}}\times Q_{\text{Rot}}\times Q_{\text{Vib}})\end{aligned}$$

and thus the equilibrium constant for the exchange reaction,

$$\begin{aligned}K_{\text{eq}}&=\exp\left(\frac{-\Delta F_{\text{Motion}}}{\mathrm{R}T}\right)=\exp\left(\frac{-\left[F_{\text{Motion}}\{\text{Products}\}-F_{\text{Motion}}\{\text{Reactants}\}\right]}{\mathrm{R}T}\right)\\&=\exp\left(\sum_{\text{Products}}\ln[Q_{\text{Trans}}\times Q_{\text{Rot}}\times Q_{\text{Vib}}]-\sum_{\text{Reactants}}\ln[Q_{\text{Trans}}\times Q_{\text{Rot}}\times Q_{\text{Vib}}]\right)\\&=\frac{\prod_{\text{Products}}(Q_{\text{Trans}}\times Q_{\text{Rot}}\times Q_{\text{Vib}})}{\prod_{\text{Reactants}}(Q_{\text{Trans}}\times Q_{\text{Rot}}\times Q_{\text{Vib}})}\end{aligned}$$

For our simple exchange equation, this expression becomes

$$\alpha_{\text{ClO-Cl}} = K_{\text{eq}} = \frac{Q_{\text{Trans}}(^{35}\text{Cl})}{Q_{\text{Trans}}(^{37}\text{Cl})} \times \frac{Q_{\text{Trans}} \times Q_{\text{Rot}} \times Q_{\text{Vib}}(^{37}\text{Cl}^{16}\text{O})}{Q_{\text{Trans}} \times Q_{\text{Rot}} \times Q_{\text{Vib}}(^{35}\text{Cl}^{16}\text{O})}$$

$$= \left(\frac{m_{^{35}\text{Cl}}}{m_{^{37}\text{Cl}}}\right)^{3/2} \times \left(\frac{m_{^{37}\text{Cl}^{16}\text{O}}}{m_{^{35}\text{Cl}^{16}\text{O}}}\right)^{3/2} \times \frac{I_{^{37}\text{Cl}^{16}\text{O}}}{I_{^{35}\text{Cl}^{16}\text{O}}} \times \left(\frac{\exp\left[-\frac{\text{h}\nu_{^{37}\text{Cl}^{16}\text{O}}}{2\text{k}T}\right]}{1-\exp\left[-\frac{\text{h}\nu_{^{37}\text{Cl}^{16}\text{O}}}{\text{k}T}\right]}\right) \times \left(\frac{1-\exp\left[-\frac{\text{h}\nu_{^{35}\text{Cl}^{16}\text{O}}}{\text{k}T}\right]}{\exp\left[-\frac{\text{h}\nu_{^{35}\text{Cl}^{16}\text{O}}}{2\text{k}T}\right]}\right)$$

after plugging in the expressions for the different partition functions and simplifying.

The last step in Urey's derivation is the application of the Redlich-Teller product rule (e.g., Angus et al. 1936; Wilson et al. 1955), which relates the vibrational frequencies, moments of inertia, and molecular masses of isotopically substituted molecules. For ClO,

$$\left(\frac{m_{^{35}\text{Cl}}}{m_{^{37}\text{Cl}}}\right)^{3/2} = \left(\frac{m_{^{35}\text{Cl}^{16}\text{O}}}{m_{^{37}\text{Cl}^{16}\text{O}}}\right)^{3/2} \times \frac{I_{^{35}\text{Cl}^{16}\text{O}}}{I_{^{37}\text{Cl}^{16}\text{O}}} \times \frac{\nu_{^{35}\text{Cl}^{16}\text{O}}}{\nu_{^{37}\text{Cl}^{16}\text{O}}} \qquad \text{(Redlich-Teller)}$$

Inserting this into the partition function ratio yields the final result:

$$\alpha_{\text{ClO-Cl}} = \frac{\nu_{^{35}\text{Cl}^{16}\text{O}}}{\nu_{^{37}\text{Cl}^{16}\text{O}}} \times \left(\frac{\exp\left[-\frac{\text{h}\nu_{^{37}\text{Cl}^{16}\text{O}}}{2\text{k}T}\right]}{1-\exp\left[-\frac{\text{h}\nu_{^{37}\text{Cl}^{16}\text{O}}}{\text{k}T}\right]}\right) \times \left(\frac{1-\exp\left[-\frac{\text{h}\nu_{^{35}\text{Cl}^{16}\text{O}}}{\text{k}T}\right]}{\exp\left[-\frac{\text{h}\nu_{^{35}\text{Cl}^{16}\text{O}}}{2\text{k}T}\right]}\right)$$

$$= 1.0096 \ @ \ 298K$$

A more general expression for the equilibrium fractionation of isotopes ${}^{light}X$ and ${}^{heavy}X$, applicable to all diatomic and larger molecules is:

$$\alpha_{XR\text{-}X} = \left[\prod_i \frac{\nu_{light\,XR}}{\nu_{heavy\,XR}} \times \left(\frac{\exp\left[-\frac{\text{h}\nu_{heavy\,XR}}{2\text{k}T}\right]}{1-\exp\left[-\frac{\text{h}\nu_{heavy\,XR}}{\text{k}T}\right]}\right) \times \left(\frac{1-\exp\left[-\frac{\text{h}\nu_{light\,XR}}{\text{k}T}\right]}{\exp\left[-\frac{\text{h}\nu_{light\,XR}}{2\text{k}T}\right]}\right)\right]^{1/\text{n}} \qquad (3)$$

where XR, containing n atoms of element X and N total atoms, is the molecule of interest, ${}^{light}XR$ is XR containing only the light isotope ${}^{light}X$, ${}^{heavy}XR$ contains only ${}^{heavy}X$, and the product is over all $3N-6$ or $3N-5$ vibrational frequencies of XR. The exponent 1/n is a normalizing factor that accounts for multiple substitutions in molecules containing more than one atom of X. Note that α is only affected by vibrational frequencies that change upon isotopic substitution, the vibrational partition function ratio for other frequencies will cancel to unity. The reader is cautioned that α is generally not equivalent to K_{eq} if n > 1. The theoretical expression for α is often called the reduced partition function ratio because of the cancellations of the rotational and translational energy terms. In addition to its role in simplifying the final expression for α, the Redlich-Teller product rule is an important criterion for evaluating the accuracy of a set of vibrational frequencies for a molecule—substantial deviations from the product rule indicate that measurement errors, unusually large anharmonic effects, and/or typographical mistakes are likely to degrade the accuracy of a calculated fractionation.

Equilibrium isotopic fractionations calculated using Urey's approach assume harmonic vibrations and rigid-body rotation, and use a simplified treatment of rotational energies. They also average intra-molecular isotopic fractionations over the entire molecule. These assumptions greatly simplify calculations, because the only inputs needed are vibrational frequencies for isotopically light and heavy molecules. The assumptions are generally reasonable over the typical temperature range of interest to geochemists (~1000 K > T > 100 K) for isotope systems other than H, C, N, O, and S. Fractionations at higher temperatures are usually so small that the progressive breakdown of the first two assumptions has little practical significance. Fractionations calculated at cryogenic temperatures (<100 K) should include a full quantum-mechanical treatment of molecular rotation. Vibrational anharmonicity and vibrational-rotational interactions can be included in theoretical calculations, but are usually practical only for molecules that have been the subjects of extensive, highly precise spectroscopic studies (Richet et al. 1977). For many substances (including essentially all condensed phases), uncertainties associated with measured and/or modeled vibrational frequencies are likely to be larger than any anharmonic and vibrational-rotational effects, so in practice there is probably little reason to use the more complex models.

Fractionation in crystalline materials and solutions

Gas-phase molecules play a relatively minor role in the geochemistry of most elements other than H, C, N, O, and S, so it is important to consider extensions of the theory outlined in the preceding section to other types of materials, particularly aqueous solutions and crystals. In general, the same energetic concepts (especially zero-point energy) apply, but it is necessary to make additional assumptions to deal with the complexities and uncertainties that arise in dealing with condensed phases.

For solutions or crystals, it is apparent that the simple picture of an isolated molecule with a completely known, finite number of vibrational frequencies is impractical. A microscopic crystal 1 μm across might contain 10^{11} or more atoms, implying 3×10^{11} or more vibrational modes! Similarly, an atom, ion or molecule dissolved in water will interact with water molecules and other dissolved species in a continuously changing arrangement of hydrogen bonds and ion pairs. However, these complications can be treated in a simplified way or even ignored while still allowing calculations that are accurate enough to be useful.

When dealing with dissolved molecules and molecular ions that remain more or less intact in solution (e.g., $[ClO_4]^-$, $B(OH)_3$ and CCl_4), or with aqueous complexes where intra-complex bonds are probably much stronger than interactions with the bulk solvent (e.g., $[Cr(H_2O)_6]^{3+}$, $[FeCl_4]^-$, and Mg^{2+} in chlorophyll), a reasonable if crude approximation is to treat each as though it actually *were* a gas-phase molecule. The only inputs needed in such a model are the intra-molecular vibrational frequencies (preferably measured in solution). This approach has been followed in numerous studies (Urey 1947; Kotaka and Kakihana 1977; Schauble et al. 2001), and has proven to be reliable under favorable conditions, at least as a semi-quantitative guide to real fractionations. This gas-in-solution approximation probably works best for molecules and complexes that are substantially heavier than a single atom of the element of interest (i.e., selenium in $[SeO_4]^{2-}$ rather than H_2Se), when those atoms are isolated from direct contact with the solvent (chlorine in $[ClO_4]^-$ rather than $[ClO]^-$), when solvent interactions are weak, and when intra-molecular bonds are strong. It is to be expected, however, that such calculations will not be as accurate as would be possible for a true gas—for instance, most gases exhibit measurable isotopic fractionations between the vapor phase and solution that is only crudely accounted for in this approximation. It is likely that theoretical calculations could be improved by taking the solvent into account more explicitly, but in practice the spectroscopic data are not known, so it is necessary to create a vibrational model of the solution. These models will be introduced in a later section.

Crystals lack some of the dynamic complexity of solutions, but are still a challenging subject for theoretical modeling. Long-range order and forces in crystals cause their spectrum of vibrational frequencies to appear more like a continuum than a series of discrete modes. Reduced partition function ratios for a continuous vibrational spectrum can be calculated using an integral, rather than the finite product used in Equation (3) (Kieffer 1982),

$$\alpha_{XR(\text{crystal})-X} = \left(\frac{m_{light\,X}}{m_{heavy\,X}}\right)^{3/2} \times \exp\left(\frac{1}{n}\,\frac{\int_0^{\nu_{max}} \ln\left(\frac{\exp\left[-\frac{h\omega}{2kT}\right]}{1-\exp\left[-\frac{h\omega}{kT}\right]}\right) g_{heavy\,X}(\omega)\,d\omega}{\int_0^{\nu_{max}} \ln\left(\frac{\exp\left[-\frac{h\omega}{2kT}\right]}{1-\exp\left[-\frac{h\omega}{kT}\right]}\right) g_{light\,X}(\omega)\,d\omega}\right)$$

where $g_{heavy_X}(\omega)$ and $g_{light_X}(\omega)$ are the vibrational density of states of isotopically heavy and light forms of the crystal, respectively, n is the number of atoms of X in each unit cell of XR(crystal), and the integrals extend to the highest vibrational frequencies of the crystal. $g(\omega)$ is proportional to the number of vibrational degrees of freedom in the infinitesimal frequency interval $\omega \rightarrow \omega + \delta\omega$ for a molecule $g(\omega)$ takes the form of a series of delta functions centered at the discrete vibrational frequencies. Alternatively, the continuous vibrational spectrum can be approximated using a representative set of discrete frequencies—even a relatively small number of frequencies (~10 to 20 for a small unit cell) can be sufficient (Elcombe and Pryor 1970). A small subset of the vibrational frequencies in the continuum—those with a phase wavelength much longer than a unit cell—can be measured using conventional infrared and Raman spectroscopy. Unfortunately, these vibrational modes are particularly unrepresentative of the total vibrational spectrum, but they may be sufficient for crystals with ~6 or more atoms per unit cell. In crystals the rotational and translational terms in the Redlich-Teller product rule disappear, yielding the high-T product rule of Kieffer (1982). Crystal vibrational frequencies must obey the high-T product rule, or calculated fractionations will fail to converge to $\alpha = 1$ at high temperatures (see Chacko et al. 1991 and Bottinga 1968 for an example of the use of the high-T product rule in evaluating calculated C and O isotope fractionations in calcite).

For crystals with molecule-like constituents, like the BO_3^{3-} and BO_4^{5-} groups in some borates, semi-quantitative models of the molecular component as a gas-phase entity have been proposed (Oi et al. 1989). This is conceptually similar to the approximation made for species in solution, although in practice most studies of crystals consider additional frequencies that reflect inter-molecular vibrations. The spectroscopic data on these vibrations (which typically have lower frequencies than the intra-molecular vibrations) are often available, at least approximately, from infrared and Raman spectroscopy and elastic properties. This type of hybrid molecule-in-crystal model has been applied to many minerals in theoretical studies of carbon and oxygen isotope fractionation, the most noteworthy being studies of calcite (Bottinga 1968; Chacko et al. 1991) and silicates (Kieffer 1982). Because spectroscopic data are always incomplete (especially for substances substituted with rare isotopes), some amount of vibrational modeling is necessary.

The chemistry of stable isotope fractionation

Bigeleisen and Mayer (1947) simplified the reduced partition function by observing that vibrational frequency shifts caused by isotope substitution are relatively small (except when deuterium is substituted for normal hydrogen). When the dimensionless quantity hν/kT is of the order 5 or less (corresponding to a typical 1000 cm^{-1} vibration at 288 K)—a condition applicable to most geochemical situations,

$$\alpha_{XR-X} \approx 1 + \frac{h^2 \sum_i \nu_{light\ XR}{}^2 - \nu_{heavy\ XR}{}^2}{24 n \mathrm{k}^2 T^2} \tag{4}$$

This relation correctly predicts that most equilibrium stable isotope fractionations are inversely proportional to the square of absolute temperature, and is the basis of equilibrium fractionation rule (1). A detailed derivation of the Bigeleisen and Mayer model has been presented in an earlier review (Criss 1991).

Using a sum-of-squares rule from theoretical vibrational spectroscopy, Bigeleisen and Mayer (1947) then showed that, under the conditions relevant to Equation (4),

$$\sum_i \nu_{light\ XR}{}^2 - \nu_{heavy\ XR}{}^2 \approx \frac{m_{heavy\ X} - m_{light\ X}}{m_{light\ X} m_{heavy\ X}} \frac{1}{4\pi^2} A \tag{5}$$

where A is the sum of all force constants acting on an atom of X, averaged over the molecule. This relation is the origin of rules (2) and (3).

Note that when hν/kT is of order 10 or more, either at low temperatures or for molecules with high-frequency vibrations, essentially all molecules are in the vibrational ground state. Under these circumstances the free energy of the exchange reaction is only sparingly sensitive to temperature, and fractionations more closely approximate Arrhenius-like $1/T$ behavior. Bonds between a heavier atom and hydrogen usually lead to high-frequency bond-stretching vibrations that are somewhat sensitive to isotope substitution of the heavy atom. For this reason, the Bigeleisen and Mayer (1947) analysis tends to break down in molecules where the element of interest in bonded directly to hydrogen. Under these circumstances, heavy isotopes will be less concentrated in the substance than the force-constant analysis might suggest, leading to rule (4) and a special exception to rule (1). The chlorine isotope system (^{37}Cl-^{35}Cl) provides a good example of this behavior (Fig. 3), with $\alpha_{HCl\text{-}Cl}$ displaying marked concave-down curvature when plotted against $1/T^2$. This phenomenon also contributes to the high temperature-sensitivity of oxygen isotope fractionations between anhydrous minerals (lacking O-H bonds) and water.

Other causes of equilibrium isotopic fractionation

A handful of non-vibrational mechanisms responsible for equilibrium fractionations have been proposed, including effects of nuclear spin or shape on electronic energies (Bigeleisen 1998). These non-vibrational phenomena are distinguishable from conventional fractionations, at least in principle, because they are not expected to be mass dependent, and geochemists should be careful not to automatically dismiss data that do not conform to mass-dependent behavior. These effects may be restricted to very heavy elements (i.e., uranium). Claims of non-vibrational fractionations (Fujii et al. 2002) in lighter elements have not been substantiated using modern analytical techniques, to my knowledge, and it is not yet clear how important these unconventional mechanisms are in natural systems. For this reason, the present review focuses on the well established vibrational model (Bigeleisen and Mayer 1947; Urey 1947).

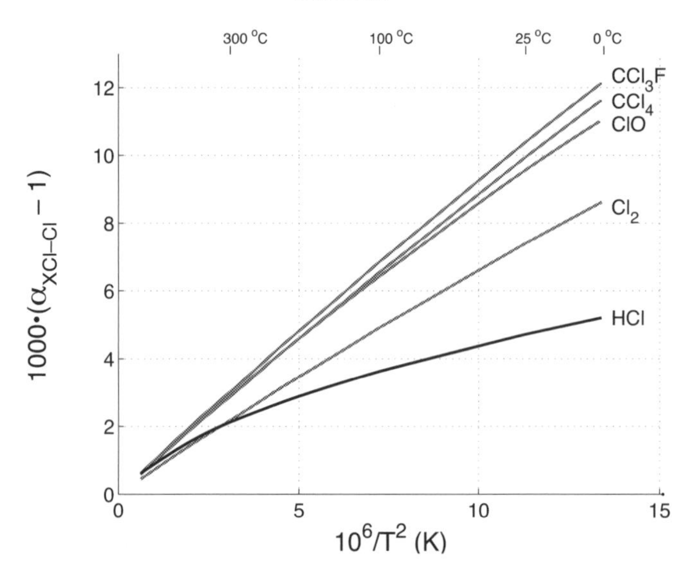

Figure 3. Theoretical ^{37}Cl-^{35}Cl fractionation factors for Cl-bearing molecules, adapted from (Schauble et al. 2003). Typically, stable isotope fractionation factors are nearly linear when plotted against $1/T^2$, as can be seen for CCl_3F, CCl_4, ClO, and Cl_2 in this example (gray lines). This behavior is predicted by eq. 4 (Bigeleisen and Mayer 1947). However, molecules in which the atom of interest is bonded directly to hydrogen (such as HCl, dark line) often show significant concave-down curvature on the same axes, because $h\nu_{HCl}/kT$ is too high ($\approx$ 14.4 at room temperature) for the Bigeleisen and Mayer (1947) simplification to apply. At room temperature, HCl is predicted to have lower $^{37}Cl/^{35}Cl$ than Cl_2, even though the spring constant for the H-Cl bond is considerably higher (530 N/m) than the Cl-Cl spring constant (330 N/m). This anomalous affinity of H-X compounds for light isotopes is the origin of rule 4 and the special exception to rule 1.

APPLYING STABLE ISOTOPE FRACTIONATION THEORY

Estimating unknown vibrational frequencies

The theory of stable isotope fractionation described by Urey (1947) and Bigeleisen and Mayer (1947) has been repeatedly confirmed by comparison of theoretical predictions with laboratory experiments and measurements on carefully chosen suites of natural samples. However, there are a number of hurdles that can make it difficult to apply the theory to novel substances and isotopic systems. Paramount among these difficulties is the limited availability of highly accurate and complete vibrational frequencies in the spectroscopic literature, a problem that is particularly acute for materials containing rare isotopes and for most condensed phases. To get a sense of the difficulty, let us consider a 1 cm^{-1} error in the isotopic shift of one 500 cm^{-1} vibrational frequency. The error is chosen arbitrarily here, but is reasonable for condensed-phase materials and all but the smallest molecules. Using Equation (4), this propagates to a 1.0 per mil error in α at 298 K. This error is much larger than the typical uncertainty of a modern mass-spectrometric measurement, and will tend to increase as the size of a molecule (and thus the number of vibrations) increases. Accurate and complete spectroscopic data *are* available for many small gas-phase molecules that are likely to be of interest in geochemical studies, both in the primary research literature and in spectroscopy reviews and databases. For more complex materials, however, it is usually necessary to approximate unknown vibrational frequencies using some type of model. A number of excellent introductory vibrational analysis texts have been published. The present introduction borrows heavily from two recommended texts—Wilson et al. (1955) and Nakamoto (1997)—the former text contains detailed derivations of basic principles and mathematical techniques, while Nakamoto (1997) provides a concise introduction along with copious data tables, appendices, and references to relevant primary literature.

The simplest type of vibrational model consists of one or more simple rules used to estimate frequencies of isotopically substituted substances from known vibrational frequencies for the common isotopic form. An example is the model developed by Kieffer (1982) to calculate vibrational frequencies in ^{18}O-bearing silicates. This approach is most likely to be effective in systems where spectroscopic data on the dominant isotopic form of a set of related substances is known—as is the case with natural ^{16}O-dominated silicate minerals—and when data bearing on isotopic substitution effects in similar materials is also known. For instance, Kieffer (1982) used results from ^{18}O-substituted silica glass to estimate frequencies in ^{18}O-quartz. Furthermore, the technique is mostly limited to instances where a more-or-less discrete molecule (such as SiO_4^{4-}) with distinctive vibrational properties can be recognized. In fact, Kieffer's results for other silicates were extrapolated from a force-field model of the SiO_4^{4-} "molecule." In heavy element stable isotope systems, the rules-based approach is unlikely to be widely applicable because these fractionations are much more dependent on the low-frequency vibrations that are most sensitive to heavy-element isotopic substitution. These frequencies are more difficult to assign to a particular type of atomic motion, and generally have not been measured with as much accuracy.

A potentially much more adaptable technique is force-field vibrational modeling. In this method, the effective force constants related to distortions of a molecule (such as bond stretching) are used to estimate unknown vibrational frequencies. The great advantage of this approach is that it can be applied to any material, provided a suitable set of force constants is known. For small molecules and complexes, approximate force constants can often be determined using known (if incomplete) vibrational spectra. These empirical force-field models, in effect, represent a more sophisticated way of extrapolating known frequencies than the rule-based method. A simple type of empirical molecular force field, the modified Urey-Bradley force field (MUBFF), is introduced below.

Another way to determine force constants is through *ab initio* quantum mechanical calculations. In this approach, the electronic structure of a molecule is determined through an approximate solution of the Schrödinger equation, and force constants are determined from the 2nd derivatives of the electronic energy of the molecule with respect to atomic displacements. Although quantum mechanical calculations are computationally intensive, advances in processor speed and memory size and performance have made relatively accurate calculations feasible on many desktop computers using well-documented, freely available software (i.e., GAMESS; Schmidt 1993). The independence of *ab initio* force fields from empirical vibrational frequency data means that partial and imprecise spectra can be used to independently verify the accuracy of each force field model. The downside is that it can be tricky to accommodate complicating factors like molecule-solvent interactions (see Anbar et al., in press, for an example calculation that attempts to overcome this difficulty), and to model materials with complex electronic structures. In principle, crystals can also be modeled with both empirical and *ab initio* force fields, although the range of suitable materials is more restricted.

Vibrational force-field modeling

In any force-field model, the molecule to be analyzed is treated as a set of masses connected by springs. Calculating vibrational frequencies for a particular set of coupled masses and springs is essentially a problem of matrix algebra, and the summary presented below is more mathematically intense than preceding sections. The equations may appear daunting at first, but remember that calculations for real molecules can be largely automated using computers, and greatly simplified by taking advantage of molecular symmetries. The geometry of a molecule may be determined from structural measurements (typically X-ray or neutron diffractometry and/or rotational spectroscopy), or from *ab initio* structural relaxation. Given a set of force-constant parameters, the vibrational spectrum of a molecule can then be determined by solving a set of differential equations,

$$\frac{d}{dt}\frac{\partial KE}{\partial\left(\frac{dq_j}{dt}\right)}+\frac{\partial PE}{\partial q_j}=0$$

where KE and PE are the potential and kinetic energies of the molecule, respectively, and q_j are small, mass-weighted displacements of each atom from the equilibrium position in each of the three Cartesian directions:

$$q_1=\sqrt{m_1}\Delta x_1,\ q_2=\sqrt{m_1}\Delta y_1,\ q_3=\sqrt{m_1}\Delta z_1,\ q_4=\sqrt{m_2}\Delta x_2,\ etc.$$

If we define a set of mass-weighted Cartesian force constants f_{ij},

$$f_{ij}=f_{ji}=\frac{\partial^2 PE}{\partial q_i \partial q_j}$$

then all vibrational frequencies ν satisfy the secular equation:

$$\begin{vmatrix} f_{11}-4\pi^2\nu^2 & f_{12} & \cdots & f_{1,3N} \\ f_{21} & f_{22}-4\pi^2\nu^2 & \cdots & f_{2,3N} \\ \cdots & \cdots & \cdots & \cdots \\ f_{3N,1} & f_{3N,2} & \cdots & f_{3N,3N}-4\pi^2\nu^2 \end{vmatrix}=0$$

Once a set of force constants is established, isotopic substitution is modeled by simply changing the relevant mass terms while leaving all other parameters unchanged. In *ab initio* models the force constants can all be determined directly by calculating energies and forces acting on atoms in systematically distorted molecules—in some *ab initio* methods it is even possible to calculate force constants analytically in the minimum-energy configuration of the molecule. In empirical force-field calculations, however, the force constants are solved using a limited set of constraints. The most common constraints are known vibrational frequencies. The number of Cartesian force constants increases roughly as the square of the number of atoms, while the number of vibrations increases linearly. Because of this scaling, it is not generally possible to solve for all of the Cartesian force constants of large molecules (i.e., molecules containing more than three atoms) independently unless vibrational frequencies for two or more isotopic forms of the molecule are known.

When constructing a vibrational model it is important to take note of possible simplifying procedures, particularly for complex phases. Many of the substances that are most interesting from the point of view of isotope geochemistry are complex solutions, crystals, large molecules, and amorphous phases that may be impossible to model in detail. Practical application of theoretical techniques often requires compromising detail and a certain amount of accuracy in order to find a tractable path to useful results. Choosing an appropriate compromise is important, as is seeking out corroborating data that can be used to justify a particular simplification. In studying aqueous molecules and complexes, for instance, the common procedure of ignoring solvent effects seems to yield reasonable, if only semi-quantitative estimates of isotopic fractionations (e.g., Kakihana and Kotaka 1977; Schauble et al. 2001; Johnson et al. 2002). The validity of the simplified treatment can be justified by noticing that vibrational frequencies in many strongly-bonded molecule-like species are not very sensitive to the details of the phase in which they occur. Schauble et al (2001) observed, for instance, that measured vibrational frequencies for many aqueous iron chloride and iron-aquo complexes changed by ~5% or less in solutions of varying chemical composition, and even in crystals held together by weak ionic- or hydrogen-bonding networks.

Empirical force fields

Empirical force field models have long been used to estimate unknown vibrational frequencies in theoretical stable isotope studies (Urey and Greiff 1935; Urey 1947). For small, symmetric molecules and molecule-like aqueous and crystalline species force-field calculations can be easily implemented in typical spreadsheet software or with scientific computation packages. The major difficulty in implementing an empirical force field model is obtaining accurate, well-constrained force constants from known vibrational frequencies – the number of independent force constants to be constrained must be smaller than the number of known frequencies. Numerous schemes have been developed that require a minimal number of force constants, including the valence force field (VFF) (e.g, Wilson et al. 1955), orbital valence force field (OVFF) (Heath and Linnett 1948), and modified Urey-Bradley force field (MUBFF) (Simanouti (Shimanouchi) 1949). These force field methods share more similarities than differences, and we will focus on the MUBFF, which has been applied in numerous theoretical studies (Kotaka and Kakihana 1977; Kotaka et al. 1978; Schauble et al. 2001). The MUBFF accounts for only three types of molecular distortions: bond stretching, bond-angle bending, and repulsion between adjacent, non-bonded atoms (Fig. 4), with the consequence that most of the Cartesian force constants cease to be independent. The choice of force-constant types follows from chemical intuition, as these types of distortion affect the basic structural properties of a molecule. The effectiveness of the MUBFF in calculating frequencies has been reviewed for tetrahedral (XY_4) and octahedral (XY_6) complexes (Basile et al. 1973; Krynauw 1990). More complex force fields, including additional inter-atomic interactions and/or anharmonic potential parameters, can be applied to well-studied molecules, but not in the most common situation where frequencies have only been measured in one isotopic form (or a natural mixture of isotopes that is dominated by one isotope of each element in the molecule).

As is common with empirical force fields, MUBFF calculations are carried out using internal molecular coordinates rather than Cartesian coordinates. Internal coordinates describe the structure of a molecule in terms of bond lengths and angles between bonds. As an example, for a bent tri-atomic molecule *ABC* the three internal coordinates include the lengths of bonds *AB* (r_{AB}) and *BC* (r_{BC}), as well as the angle between them (α_{ABC}). Larger molecules may also

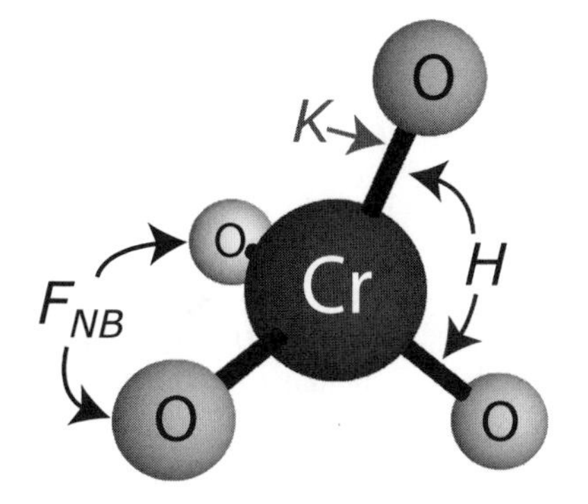

Figure 4. Schematic illustration of force-constant parameters used in Modified Urey-Bradley Force-Field (MUBFF) vibrational modeling (Simanouti (Shimanouchi) 1949). The MUBFF is a simplified empirical force field that has been used to estimate unknown vibrational frequencies of molecules and molecule-like aqueous and crystalline substances. Here, three force constants (K, H, and F_{NB}) describe distortions of a tetrahedral XY_4 molecule, $[CrO_4]^{2-}$ due to bond stretching (Cr-O), bond-angle bending ($\angle$ O-Cr-O), and repulsion between adjacent non-bonded atoms (O..O). Less symmetric molecules with more than one type of bond or unequal bond angles require more parameters, but they will belong to the same basic types.

have additional types of internal coordinates, such as torsional or dihedral angles. Any small perturbation of these coordinates can be described in terms of a vector R, which for the example molecule *ABC* looks like this:

$$R = \begin{bmatrix} \Delta r_{AB} \\ \Delta r_{BC} \\ \Delta \alpha_{ABC} \end{bmatrix}$$

The potential energy of a molecule can then be described in terms of the internal coordinates, i.e.,

$$2PE = R^t FR$$

where R^t is the transpose of R and F is a matrix of mass-independent internal-coordinate force constants. The kinetic energy can also be written in terms of R, albeit with a few intermediate steps. The matrix describing perturbations of the Cartesian coordinates of each atom (with the center of mass as the origin) is defined,

$$X = \begin{bmatrix} \Delta x_1 \\ \Delta y_1 \\ \Delta z_1 \\ \Delta x_2 \\ \cdots \\ \Delta z_N \end{bmatrix}$$

along with an isotope-sensitive diagonal matrix M^{-1} consisting of the reciprocal masses of each atom repeated three times,

$$M^{-1} = \begin{bmatrix} \frac{1}{m_1} & & & & 0 \\ & \frac{1}{m_1} & & & \\ & & \frac{1}{m_1} & & \\ & & & \ddots & \\ 0 & & & & \frac{1}{m_N} \end{bmatrix}$$

Then a third matrix B relating the internal and Cartesian coordinates is determined,

$$R = BX$$

Finally we can write out a kinetic energy matrix G,

$$G = BM^{-1}B^t$$

such that

$$2KE = \frac{dR^t}{dt} G^{-1} \frac{dR}{dt}$$

where G^{-1} is the reciprocal of the special matrix G. Each eigenvalue of the product matrix GF is equal to $4\pi^2\nu^2$, where ν is a vibrational frequency of the molecule. This procedure for calculating vibrational frequencies in terms of internal molecular coordinates is often referred to as the *GF* method.

The eigenvector corresponding to each eigenvalue determines the characteristic motion of each atom in the vibration. The characteristic atomic motions associated with each frequency (called normal modes) are closely related to the problem of determining how vibrational frequencies change when one isotope of an element is substituted for another. Specifically, isotopic sensitivity scales with the intensity of atomic motion: if an atom remains stationary in a particular normal mode, the frequency of that mode will not be affected by isotopic substitution of that atom; if the atom exhibits high-amplitude motion, the frequency of the mode will be highly sensitive to isotopic substitution.

Of course, before the atomic motions and unknown frequencies can be determined it is necessary to constrain the force constants from known frequencies. From the derivation shown above it is not clear how to distinguish one mode from another—how to determine which calculated frequency corresponds to a known measured frequency. This is done by classifying each normal mode according to its symmetry properties. A detailed introduction to the use of molecular symmetry in vibrational analysis is beyond the scope of the present review, several comprehensive reviews have been published elsewhere (Herzberg 1945; Wilson et al. 1955; Cotton 1971; McMillan and Hess 1988; Nakamoto 1997). Of immediate relevance is the fact that molecular symmetries can be used to block-diagonalize the G and F matrices, so that eigenvalues can be determined from the products of submatrices corresponding to each irreducible representation of the point group of the molecule. Since the Raman and infrared selection rules that determine whether a particular vibrational mode can be observed are also determined by the irreducible representation of each mode, it is usually possible to match calculated and observed frequencies in small groups if not strictly on a one-to-one basis.

Optimized MUBFF models typically reproduce the input frequencies within ~10% or better. In order to reap the maximum possible benefit from the known frequencies, the model ratio of frequencies for isotopically substituted molecules can be multiplied by the measured frequency to give a normalized isotopic frequency, i.e.,

$$\left(\nu_{rare\ isotope}\right)_{calc.} = \left(\frac{\nu_{rare\ isotope}}{\nu_{major\ isotope}}\right)_{MUBFF} \times \left(\nu_{major\ isotope}\right)_{measured} \tag{6}$$

GF method calculations are simplified by the systematic behavior of the G matrix elements (Decius 1948). MUBFF calculations, however, are somewhat complicated by the force constants representing interactions between non-bonded atoms—these can be tedious to express in terms of internal coordinates. Computer programs have been written to partially automate calculations, thereby reducing the necessary effort and minimizing opportunities for errors (e.g., Schachtschneider 1964; Gale and Rohl 2003).

In addition, G and F matrix elements have been tabulated (see Appendix VII in Nakamoto 1997) for many simple molecular structure types (including bent triatomic, pyramidal and planar tetratomic, tetrahedral and square-planar 5-atom, and octahedral 7-atom molecules) in block-diagonalized form. MUBFF G and F matrices for tetrahedral XY_4 and octahedral XY_6 molecules are reproduced in Table 1. Tabulated matrices greatly facilitate calculations, and can easily be applied to vibrational modeling of isotopically substituted molecules. Matrix elements change, however, if the symmetry of the substituted molecule is lowered by isotopic substitution, and the tabulated matrices will not work in these circumstances. For instance, $^{12}C^{35}Cl_4$, $^{13}C^{35}Cl_4$, and $^{12}C^{37}Cl_4$ all share full XY_4 tetrahedral symmetry (point group T_d), but $^{12}C^{35}Cl^{37}Cl_3$ and other partially ^{37}Cl-substituted forms do not, and cannot be modeled with the tabulated matrices. The practical significance of this restriction is negligible, however, because one corollary of the rule the mean for isotopic substitution (Bigeleisen 1955) is that calculated fractionation factors are almost completely insensitive to the number of substituted atoms.

Example calculation: $^{53}Cr \leftrightarrow {}^{50}Cr$ substitution in the chromate anion. An example calculation on the chromate anion $[CrO_4]^{2-}$ makes use of tabulated G and F matrix elements.

Table 1. Block diagonalized G and F matrices for tetrahedral XY_4 and octahedral XY_6 molecules, using the modified Urey-Bradley force field. Adapted from Nakmoto (1997). m_X and m_Y are the masses of atoms of X and Y, and r is the length of the X-Y bond. K, H, and F_{NB} are force constants for bond stretching, bond-angle bending, and non-bonded repulsion, respectively.

XY_4 molecules, point group T_d

A_1 (ν_1) – Raman active

$$G(A_1) = 1/m_Y \qquad F(A_1) = (K + 4F_{NB})$$

E (ν_2) – doubly degenerate, Raman active

$$G(E) = 3/(r^2 m_Y) \qquad F(E) = (H + 0.4F_{NB})$$

F_2 (ν_3, ν_4) – triply degenerate, Raman and infrared active

$$G(F_2) = \begin{bmatrix} \frac{1}{m_O} + \frac{4}{3m_{Cr}} & \frac{-8}{3m_{Cr}r} \\ \frac{-8}{3m_{Cr}r} & \frac{1}{r^2}\left(\frac{16}{3m_{Cr}} + \frac{2}{m_O}\right) \end{bmatrix} \qquad F(F_2) = \begin{bmatrix} K + \frac{6}{5}F_{NB} & \frac{3}{5}rF_{NB} \\ \frac{3}{5}rF_{NB} & r^2(H + 0.4F_{NB}) \end{bmatrix}$$

XY_6 molecules, point group O_h

A_{1g} (ν_1) – Raman active

$$G(A_1) = 1/m_Y \qquad F(A_1) = (K + 4F_{NB})$$

E_g (ν_2) – doubly degenerate, Raman active

$$G(E_g) = 1/m_Y \qquad F(E_g) = (K + 0.7F_{NB})$$

F_{1u} (ν_3, ν_4) – triply degenerate, infrared active

$$G(F_{1¼}) = \begin{bmatrix} \frac{1}{m_O} + \frac{2}{m_{Cr}} & \frac{-4}{m_{Cr}r} \\ \frac{-4}{m_{Cr}r} & \frac{2}{r^2}\left(\frac{4}{m_{Cr}} + \frac{1}{m_O}\right) \end{bmatrix} \qquad F(F_2) = \begin{bmatrix} K + 1.8F_{NB} & 0.9rF_{NB} \\ 0.9rF_{NB} & r^2(H + 0.55F_{NB}) \end{bmatrix}$$

F_{2g} (ν_5) – triply degenerate, Raman active

$$G(F_{2g}) = 4/(r^2 m_Y) \qquad F(F_{2g}) = (H + 0.55F_{NB})$$

F_{2u} (ν_6) – triply degenerate, neither Raman nor infrared active

$$G(F_{2u}) = 2/(r^2 m_Y) \qquad F(F_{2u}) = (H + 0.55F_{NB})$$

The chromate anion is a highly soluble, toxic tetrahedral complex (point group T_d) that occurs in oxidized, neutral-basic solutions. It is also one of a small number of aqueous complexes that have been thoroughly characterized by spectroscopic measurements on numerous isotopic compositions (Müller and Königer 1974), so it will be possible to check the vibrational model against real data. Here the MUBFF is applied under the assumption that aqueous chromate can be approximately modeled as a gas-phase molecule.

Due to numerous symmetries in the T_d point group, most of the nine vibrational modes of tetrahedral XY_4 molecules and complexes like the chromate anion are degenerate (meaning that two or more modes have the same frequency), and there are only four distinct frequencies. These frequencies are all observable with Raman spectroscopy, which is particularly suited to studies of dissolved substances. There is one frequency belonging to the A_1 irreducible representation, one belonging to the doubly degenerate E irreducible representation, and two to the triply-degenerate F_2 irreducible representation. F_2 frequencies are also observable with infrared spectroscopy, although water is such a strong infrared absorber that this can be difficult in solution. For the natural isotopic mixture in chromate (dominantly $[^{52}Cr^{16}O_4]^{2-}$) the four distinct frequencies are (Müller and Königer 1974):

$$\nu_1(A_1), 846\ cm^{-1} \quad \nu_2(E), 347\ cm^{-1} \quad \nu_3(F_2), 891\ cm^{-1} \quad \nu_4(F_2), 368\ cm^{-1}$$

The symmetrized G and F matrices for $[CrO_4]^{2-}$ molecules are:

$$G=\begin{bmatrix} \frac{1}{m_O} & 0 & 0 & 0 \\ 0 & \frac{3}{m_O r^2} & 0 & 0 \\ 0 & 0 & \frac{1}{m_O}+\frac{4}{3m_{Cr}} & \frac{-8}{3m_{Cr}r} \\ 0 & 0 & \frac{-8}{3m_{Cr}r} & \frac{1}{r^2}\left(\frac{16}{3m_{Cr}}+\frac{2}{m_O}\right) \end{bmatrix}$$

$$F=\begin{bmatrix} K+4F_{NB} & 0 & 0 & 0 \\ 0 & r^2(H+0.4F_{NB}) & 0 & 0 \\ 0 & 0 & K+\frac{6}{5}F_{NB} & \frac{3}{5}rF_{NB} \\ 0 & 0 & \frac{3}{5}rF_{NB} & r^2(H+0.4F_{NB}) \end{bmatrix}$$

where K, H, and F_{NB} are force constants for Cr-O bond stretching, O-Cr-O bond-angle bending, and O-O repulsion, respectively, and r is the equilibrium Cr-O bond length. The first two terms on the diagonals correspond to the A_1 and E vibrations, while the 2×2 submatrix at lower right corresponds to the F_2 vibrations—each can be solved separately:

$$4\pi^2\nu_1^2=\frac{1}{m_O}(K+4F_{NB})$$

$$4\pi^2\nu_2^2=\frac{3}{m_O}(H+0.4F_{NB})$$

$$4\pi^2\nu_{3,4}{}^2=eigenvalue\left(\begin{bmatrix} \frac{1}{m_O}+\frac{4}{3m_{Cr}} & \frac{-8}{3m_{Cr}r} \\ \frac{-8}{3m_{Cr}r} & \frac{1}{r^2}\left(\frac{16}{3m_{Cr}}+\frac{2}{m_O}\right) \end{bmatrix}\begin{bmatrix} K+\frac{6}{5}F_{NB} & \frac{3}{5}rF_{NB} \\ \frac{3}{5}rF_{NB} & r^2(H+0.4F_{NB}) \end{bmatrix}\right)$$

Here there are four measured frequencies with which to constrain three independent force constants, so the best-fitting force constants can be determined through an iterative least squares fit, minimizing $\Sigma(\nu_{meas}-\nu_{calc})^2$. Assuming average atomic masses of 51.996 and 15.9994 for chromium and oxygen, respectively, the best-fit force constants are K = 495.2 Newtons/m, H = 21.3 Newtons/m, and F_{NB} = 44.7 Newtons/m. These force constants show the typical relationship $K \gg H, F_{NB}$. Calculated frequencies are:

$$\nu_1(A_1),\ 845.6\ \text{cm}^{-1}\quad \nu_2(E),\ 353.1\ \text{cm}^{-1}\quad \nu_3(F_2),\ 891.3\ \text{cm}^{-1}\quad \nu_4(F_2),\ 362.5\ \text{cm}^{-1}$$

with a total squared misfit of 68 cm^{-2}. If the masses of ^{50}Cr and ^{53}Cr are inserted for m_{Cr}, it is found that only the F_2 frequencies ν_3 and ν_4 change, falling by 7.4 cm^{-1} and 2.8 cm^{-1}, respectively, upon substitution of the heavier isotope. The chromium-isotope sensitivity of the F_2 vibrations occurs because these are the only ones in which the chromium atom moves (Fig. 5). A_1 and E modes, by contrast, consist only of motion of oxygen atoms. Infrared measurements on sulfate salts doped with $[^{50}CrO_4]^{2-}$ and $[^{53}Cr^{16}O_4]^{2-}$ show that the F_2 frequencies actually change by 8.5±0.3 cm^{-1} and 2.2±0.3 cm^{-1}, respectively. We can use both calculated and observed isotopic frequencies to calculate reduced partition function ratios for ^{53}Cr-^{50}Cr exchange in chromates. $\alpha_{Chromate-Cr}$ = 1.0365 and 1.0393 at 298 K, respectively, a difference of 2.8 per mil. Published measurements have reported variations in $^{53}Cr/^{52}Cr$ ratios rather than $^{53}Cr/^{50}Cr$ (Ellis et al. 2002); so it is useful calculate analogous fractionation factors for the ^{53}Cr-^{52}Cr exchange reaction. These are 1.0120 and 1.0129, respectively, suggesting that the MUBFF model introduces an error of ~1 per mil in a room temperature calculation of $^{53}Cr/^{52}Cr$ fractionation.

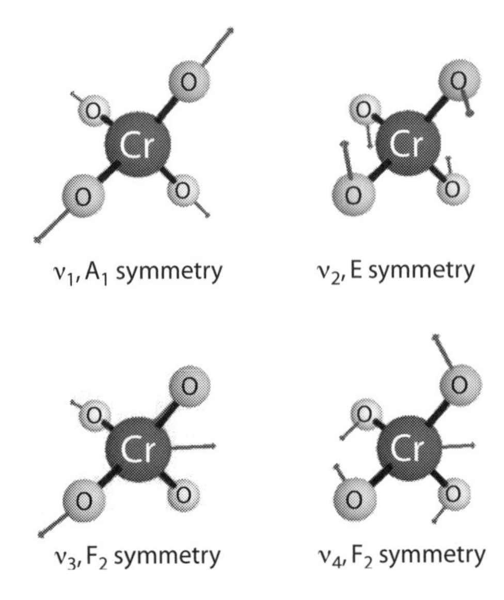

Figure 5. Normal modes for vibration of tetrahedral $[CrO_4]^{2-}$ (chromate). There are four distinct vibrational frequencies, including one doubly-degenerate vibration (E symmetry) and two triply-degenerate vibrations (F_2 symmetry), for a total of nine vibrational modes. Arrows show the characteristic motions of each atom during vibration, and the length of each arrow is proportional to the magnitude of atomic motion. Only F_2 modes involve motion of the central chromium atom, and as a result their vibrational frequencies are affected by Cr-isotope substitution. The normal modes shown here were calculated with an *ab initio* quantum mechanical model, using hybrid Hartree-Fock/Density Functional Theory (B3LYP) and the 6-31G(d) basis set—other *ab initio* and empirical force-field models give very similar results.

Transferable empirical force fields (interatomic potentials)

Other types of models have occasionally been used to calculate unknown vibrational frequencies. One promising approach has been the use of transferable empirical force fields. Transferable force fields are designed to be applicable to a range of compounds, rather than a single molecule, and consequently include parameters that account explicitly for properties like equilibrium bond lengths, ionic charges, and coordination number that change from one substance to another. This flexibility means that transferable force fields (often called interatomic potentials) can also be used to calculate binding, surface and defect energies, and to estimate unknown crystal structures and elastic properties. Interatomic potentials have been developed for numerous oxide, silicate, and carbonate materials (Catlow et al. 1988; Dove et al. 1992; Le Roux and Glasser 1997; Demiralp et al. 1999; Cygan et al. 2002). Much like MUBFF force constants, transferable potentials are determined using experimental measurements of molecular structures, mineral lattice constants and elastic properties, and known vibrational frequencies; they may also be constrained by *ab initio* calculations (Harrison and Leslie 1992). Free and commercial software packages such as GULP (Gale and Rohl 2003) and Cerius 2 (Accelrys) can be used to model vibrational spectra of minerals, molecules, and solutions based on these transferable potentials. The author is not aware of any published theoretical stable isotope studies of non-traditional elements that have taken advantage of transferable force fields, but they have been used in calculations of oxygen and carbon isotope fractionations in silicates (Patel et al. 1991) and carbonate minerals (Dove et al. 1992). An introduction to the construction and use of interatomic potentials was presented in a previous Reviews in Mineralogy and Geochemistry volume (Gale 2001).

Ab initio force-field modeling

Over the last decade, *ab initio* quantum-mechanical force fields have begun to be applied in theoretical stable isotope studies of molecules and dissolved species (Bochkarev et al. 2003; Driesner et al. 2000; Oi 2000; Oi and Yanase 2001). This method shows great promise for future studies, because *ab initio* calculations accurately describe chemical properties such as force constants without the necessity of assuming allowed force-constant types (which may not be universally applicable). *Ab initio* calculations are also ideally suited to molecules with relatively few known vibrational frequencies because there are no empirical parameters that require fitting. Because the minimum-energy molecular structure is typically also calculated, measured structures can be used to verify the appropriateness of a particular model. Numerous reviews of quantum mechanical calculations (e.g., Foresman and Frisch 1996; Thijssen 1999) and their applications in Earth science (Cygan and Kubicki 2001) have been published.

The central problem in any quantum mechanical model is finding accurate solutions to the time independent form of the Schrödinger equation,

$$\boldsymbol{H}\Psi = E\Psi$$

where Ψ is the wavefunction describing the electronic structure of the molecule, H is the Hamiltonian describing the potential field and kinetic energy of electrons and atomic nuclei (nuclei are generally assumed to be fixed in space), and E are the allowed electronic energies of the molecule. For all but the simplest systems, it is not possible to solve the Schrödinger equation exactly. However, reasonably accurate approximate solutions are possible using simplified descriptions of the electronic potentials and wavefunctions. Regardless of the method chosen, determination of the ground state electronic structure of a molecule involves the optimization of a large number of variables, which are usually determined via an iterative procedure to yield a minimum-energy, self-consistent set of electronic wavefunctions. Numerous free and commercial software packages such as Jaguar (Schrödinger, Inc.), Gaussian (Gaussian, Inc.), and GAMESS (Schmidt et al. 1993) have been developed to make it possible for non-specialists to create quantum-mechanical models of molecules and crystals.

Two common approximation methods are the Hartree-Fock (HF) method (Roothaan 1951) and density functional theory (DFT) (Hohenberg and Kohn 1964; Kohn and Sham 1965). In Hartree-Fock theory, the interactions between different electrons in a molecule are treated in a simplified way. Each electron "sees" other electrons as a time-averaged distribution, ignoring correlations between the positions of electrons in interacting orbitals. More accurate methods that take account of electron correlation have also been developed (Pople et al. 1976). HF calculations are much faster than correlation methods, however, and are often able to provide reasonably accurate descriptions of molecular structures and vibrational properties. HF methods work best for molecules with simple bond structures made up of low atomic-number atoms, and less well for molecules with exotic electronic structures (like O_3) and transition elements. DFT methods are distinguished from HF and correlated methods in their focus on determining electronic energies from the electron density. DFT methods are particularly well suited to crystals and solutions, and often provide a better description of transition-element bearing molecules than HF theory. Hybrid DFT-HF calculations such as the Becke three-parameter Lee-Yang-Parr (B3LYP) method (Becke 1993) are possible, and are commonly used to model molecular structures and vibrational frequencies. In fact B3LYP has become the method of choice for studying large molecules and atomic clusters in chemistry and earth science. Although the HF and DFT methods were developed separately and have different strengths and weaknesses, they have been implemented together in numerous commercial and open-source software packages, i.e. Jaguar (Schrödinger, Inc.), Gaussian (Gaussian, Inc.), and GAMESS (Schmidt et al. 1993), and it is natural to group them together in describing vibrational models. Note that some DFT methods are not strictly *ab initio* because a few empirical parameters (other than fundamental constants like the charge and mass of an electron and the speed of light) are used; these parameters are fixed within a given method however, and are so far removed from the user that the practical distinction is minor. In all of these methods, molecular electronic wavefunctions are typically built up using a set of orbital-like basis functions called a basis set. As with electronic correlation, the choice of basis set is governed by a compromise between accuracy and computational expense. Commonly used basis sets include 6-31G(d) (Francl et al. 1982) and 6-311G(d,p) (Krishnan et al. 1980).

Density functional theory has been extensively used to calculate vibrational properties of minerals and other crystalline phases in addition to molecules and molecule-like substances. This method has recently begun to be used to calculate isotope fractionation factors (Schauble et al. in press; Anbar et al. in press), and shows great potential for future research. Programs such as ABINIT (Gonze et al. 2002), pwSCF (Baroni et al. 2001)—both freely available—and the commercial package CASTEP (Accelrys, Inc.) can be used to calculate vibrational properties of crystals.

The Schrödinger equation can also be solved semi-empirically, with much less computational effort than *ab initio* methods. Prominent semi-empirical methods include MNDO, AM1, and PM3 (Dewar 1977; Dewar et al. 1985; Stewart 1989a; Stewart 1989b). The relative computational simplicity of these methods is accompanied, however, by a substantial loss of accuracy (Scott and Radom 1996), which has limited their use in geochemical simulations. Historically, semi-empirical calculations have also been limited by the elements that could be modeled, excluding many transition elements, for example. Semi-empirical calculations have been used to predict Si, S, and Cl isotopic fractionations in molecules (Hanschmann 1984), and these results are in qualitative agreement with other theoretical approaches and experimental results.

There are typically three steps in creating an *ab initio* quantum-mechanical model of a molecule. In the first step, the minimum-energy static structure of the molecule is determined via geometric relaxation. From an initial guess geometry, often the experimentally determined structure of the molecule, the forces on each atom are calculated, and a refined guess structure is determined. This procedure continues iteratively until the residual forces acting on each atom are sufficiently small—typically on the order of 10^{-10} Newtons or less. HF, DFT, and hybrid HF-DFT *ab initio* methods can be expected to reproduce experimental bond lengths and angles to within approximately 2 pm (0.02 Å) and 1°–2°, respectively, when the chosen method and basis set are appropriate. Once the minimum-energy configuration has been calculated, the second step is the determination of force constants for displacements of the atomic nuclei from their minimum energy positions. Finally, unknown frequencies are determined by a calculation with the appropriate isotopic masses. Since the force constants are not affected by isotopic substitution, this final step involves much less computational effort than the preceding two. Vibrational frequencies for numerous isotopic configurations of a molecule can be calculated quickly once the matrix of force constants has been determined, even for relatively complex substances.

It important to note that *ab initio* force fields of all types tend to make systematic errors in calculating vibrational frequencies (Pople et al. 1993; Scott and Radom 1996; Wong 1996). With HF calculations using the 6-31G(d) basis set, for instance, vibrational frequencies of gas-phase molecules are typically overestimated by about 12%, with both high- and low-frequency vibrations off by roughly the same scale factor (Scott and Radom 1996). Some gas-phase scale factors for other *ab initio* methods are given in Table 2. In cases where the *ab initio* molecular structure is close to the observed structure (angles within 1–2°, bond lengths within 2 pm) and calculated vibrational frequencies are related to observed frequencies by a uniform scaling factor, the ratios of frequencies of isotopic molecules should be accurately predicted (Schauble et al. 2003). For substances with relatively well-known vibrational spectra, model frequency ratios can be normalized to observed frequencies in the same way as MUBFF model frequencies.

Example calculation: $^{37}Cl \leftrightarrow {}^{35}Cl$ substitution in methyl chloride. Here we will create an *ab initio* vibrational model of CH_3Cl, a tetrahedral molecule with a less symmetric structure (point group C_{3v}) than chromate. Low-symmetry molecules can be tedious and difficult to model with empirical methods because of the large number of force-constant parameters that need to be constrained (although methyl chloride is small enough that such calculations

Table 2. Scale factors for *ab initio* model vibrational frequencies adapted from (Scott and Radom 1996). Please note that these scale factors are determined by comparing model and measured frequencies on a set gas-phase molecules dominated by molecules containing low atomic-number elements (H-Cl). These scale factors may not be appropriate for dissolved species and molecules containing heavier elements, and it is always a good idea to directly compare calculated and measured frequencies for each molecule studied. The root-mean-squared (rms) deviation of scaled model frequencies relative to measured frequencies is also shown, giving an indication of how reliable each scale factor is.

Method	Basis set	Scale factor	rms deviation (cm^{-1})
HF	3-21G	0.9085	87
HF	6-31G(d)	0.8953	50
HF	6-31G(d,p)	0.8992	53
HF	6-311G(d,p)	0.9051	54
B3LYP	6-31G(d)	0.9614	34

are feasible). Low-symmetry molecules are ideal candidates for *ab initio* modeling, which eliminates the time-consuming and error-prone process of fitting empirical parameters to measured frequencies. Calculations in this example are made using the open-source software package GAMESS (Schmidt et al. 1993), which can be downloaded without cost by academic and commercial users (*http://www.msg.ameslab.gov/GAMESS/GAMESS.html*). Hartree-Fock calculations appear to provide reasonably accurate models of the effects of isotopic substitution on vibrational frequencies of chlorocarbons and hydrochlorocarbons (Schauble et al. 2003), and are computationally fast. Here we will perform a Hartree-Fock calculation using the medium-accuracy 6-31G(d) basis set. As an initial guess for the structural relaxation step we can take the experimentally determined structure (Fig. 6), which has a C-Cl bond length of 177.6 pm, C-H bond lengths of 108.5 pm, and H-C-H bond angles of 110.4° (Jensen 1981). Geometry optimization takes <1 minute on a Macintosh desktop computer with a 400 MHz G4 processor, and yields a structure very close to what is observed: r(C-Cl) = 178.5 pm, r(C-H) = 107.8 pm, and ∠(H-C-H) = 110.5°. The total calculated electronic energy is –499.093153 Hartrees (–1,310,367.75 kJoules/mole).

Using the optimized structure, the vibrational frequencies are calculated. By default, GAMESS assumes that the dominant isotopic form of CH_3Cl ($^{12}C^{1}H_3{}^{35}Cl$) is present. GAMESS does not automatically classify vibrations by symmetry. However, this task can be accomplished by visual inspection using molecule animation software like MacMolPlt (Bode

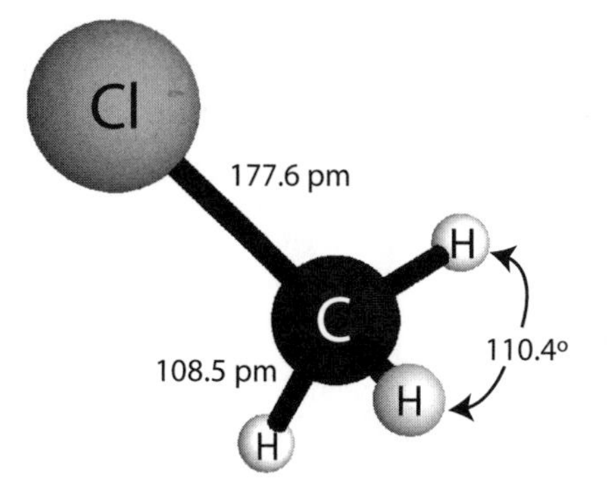

Figure 6. Measured molecular structure of methyl chloride (CH_3Cl), taken from Jensen (1981). CH_3Cl is a nearly tetrahedral molecule with C_{3v} symmetry. All C-H bond lengths, H-C-H angles and H-C-Cl angles are identical.

and Gordon 1998) and symmetry character tables found in many spectroscopy and inorganic chemistry textbooks (e.g., Nakamoto 1997; Shriver 1999). Methyl chloride is particularly simple because all doubly degenerate vibrations have E symmetry, and all remaining vibrations have A_1 symmetry. Calculated and measured frequencies are compared in Table 3. Measured frequencies are consistently ~10% lower than *ab initio* frequencies, in good agreement with the standard HF/6-31G(d) scale factor (Scott and Radom 1996).

^{37}Cl is sufficiently common that some vibrational frequencies of isotopically heavy methyl chloride have been measured spectroscopically, so it is possible to compare measured and model frequency shifts. It is found that ν_3 and ν_6 are the only frequencies that shift significantly when ^{37}Cl is substituted for ^{35}Cl; the model estimates that the ratio of ν_3 in $^{12}C^1H_3{}^{37}Cl$ divided by the frequency in $^{12}C^1H_3{}^{35}Cl$ is 0.9919, for ν_6 the ratio is 0.9996. These ratios are in excellent agreement with measured ratios of 0.9920 and 0.9996, respectively. Using normalized model frequencies (Eqn. 6), $\alpha_{\text{Methyl Chloride–Cl}} = 1.0088$ at 298 K.

Table 3. Measured (Black and Law 2001) and *ab initio* vibrational frequencies for methyl chloride, $^{12}C^1H_3{}^{35}Cl$. *Ab initio* frequencies are calculated with GAMESS, using the Hartree-Fock method and 6-31G(d) basis set. The ratio of each measured and model frequency is also shown.

Vibration	Measured frequency (cm^{-1})	*Ab initio* frequency (cm^{-1})	Meas./*Ab init.* Ratio	ν_{37}/ν_{35} (*ab initio*)
ν_1 (A_1)	2953.9	3267.33	0.9041	1.00000
ν_2 (A_1)	1354.88	1538.08	0.8809	0.99991
ν_3 (A_1)	732.84	782.60	0.9364	0.99194
ν_4 (E)	3039.29	3371.02	0.9016	1.00000
ν_5 (E)	1452.18	1628.96	0.8915	0.99999
ν_6 (E)	1018.07	1138.36	0.8943	0.99961

Theoretical applications of Mössbauer spectroscopy

The first theoretical calculations of stable isotope fractionation factors in the iron isotope system were based on Mössbauer spectroscopy rather than traditional vibrational spectroscopy (Polyakov 1997; Polyakov and Mineev 2000). These early studies successfully predicted that measurable equilibrium fractionations occur between different iron-bearing minerals, and that Fe^{3+}-bearing phases will tend to have higher $^{57}Fe/^{54}Fe$ than coexisting Fe^{2+}-bearing phases (Beard et al. 2003). However, predicted fractionations between different Fe^{2+}-bearing oxides appear to be larger than is observed in experimentally equilibrated phases and suites of natural samples. The source of disagreement between predicted Fe-isotope fractionations and experimental results has not yet been identified, but may lie in the extensive data modeling and processing required to predict fractionations from measured Mössbauer spectra.

Mössbauer spectroscopy involves the measurement of minute frequency shifts in the resonant gamma-ray absorption cross-section of a target nucleus (most commonly ^{57}Fe; occasionally ^{119}Sn, ^{197}Au, and a few others) embedded in a solid material. Because Mössbauer spectroscopy directly probes the chemical properties of the target nucleus, it is ideally suited to studies of complex materials and Fe-poor solid solutions. Mössbauer studies are commonly used to infer properties like oxidation states and coordination number at the site occupied by the target atom (Hawthorne 1988). Mössbauer-based fractionation models are based on an extension of Equations (4) and (5) (Bigeleisen and Mayer 1947), which relate α to either sums of squares of vibrational frequencies or a sum of force constants. In the Polyakov (1997) formulation, the fractionation factor is related directly to the average vibrational kinetic energy, $<KE>$, of atoms of the element of interest (the brackets indicate an averaged atomic thermodynamic quantity):

$$\ln\left(\alpha_{XR-X}\right)=\frac{\Delta m_X}{m_{heavy_X}}\left(\frac{\langle KE_X\rangle}{\mathrm{k}T}-\frac{3}{2}\right)$$

where the figure of 3/2 in parentheses represents classical thermal kinetic energy. Mössbauer spectroscopy can be used to estimate $<KE_X>$ via the second-order Doppler shift (SOD), an extremely small but measurable shift of the observed resonant frequency. Unlike conventional Doppler shifts, which are caused by line-of-sight motion, SOD results from the relativistic time dilation of the Mössbauer nucleus as it vibrates perpendicular to the line of sight between the gamma-ray source and the absorbing atom. In any iron-bearing crystal, ^{57}Fe atoms bound in the crystal lattice vibrate about their equilibrium positions, each with a finite velocity, v. Because it is moving, special relativity dictates that each atom has a slightly slower internal clock than an observer at rest. A photon that appears to a ^{57}Fe nucleus to have a frequency ν will appear to the observer as a photon with a frequency equal to $\nu+\delta\nu$, where

$$\frac{\delta\nu}{\nu}=\frac{1}{\sqrt{1+\frac{\mathrm{v}^2}{c^2}}}-1\approx-\frac{\mathrm{v}^2}{2c^2}$$

So by measuring the second-order Doppler shift of the Mössbauer nuclei in a material it is possible to determine their average velocity $<\mathrm{v}^2>$ and thus their average vibrational kinetic energy, $<KE_X> = m_{Mossbauer_X}<\mathrm{v}^2>/2$, where $m_{Mossbauer_X}$ is the mass of the Mössbauer nucleus. The magnitude S of a measured SOD is typically reported in terms of an equivalent conventional Doppler shift (relative velocity between gamma-ray source and absorber),

$$S=\frac{\delta\nu}{\nu}c$$

Using this convention, the Mössbauer model expression for equilibrium isotopic fractionation becomes:

$$\ln\left(\alpha_{XR-X}\right)=\frac{\Delta m_X}{m_{heavy_X}}\left(\frac{m_{Mossbauer_X}Sc}{\mathrm{k}T}-\frac{3}{2}\right)$$

(Polyakov 1997). Because the second-order Doppler shift is not the only factor controlling Mössbauer absorption frequencies, it is generally necessary to process data taken at a variety of temperatures, and to make a number of assumptions about the invariance of other factors with temperature and the form and properties of the vibrational density of states of the Mössbauer atom. Principles involved in analyzing temperature dependencies in Mössbauer spectra are extensively discussed in the primary literature (Hazony 1966; Housley and Hess 1966; Housley and Hess 1967) and reviews (e.g., Heberle 1971).

Integrating theoretically estimated fractionations with measurements

It should be clear from the preceding discussions that practical application of equilibrium stable isotope fractionation theory often requires a certain amount of simplification of complex and poorly studied systems. Given this reality, one should not be surprised to find that theoretically determined equilibrium fractionations rarely achieve accuracies approaching the nominal precisions of measurements made with modern analytical techniques. It should noted that future developments in vibrational modeling, particularly in *ab initio* force-field calculations and the evaluation of solvent effects, are likely to narrow the accuracy gap. The most important point, however, is that even simplified theoretical calculations have a number of powerful advantages. For instance, rough estimates can be particularly useful in developing a wide-ranging theoretical framework to help guide initial experimental investigations. In addition, theoretical calculations can greatly enhance the impact of limited experimental data, even when the former is much less accurate. They help constrain fractionation mechanisms in natural samples and experimental run products where it may not be clear that equilibrium has been attained (Fig. 7)—qualitative agreement between theoretical and observed fractionations, particularly when observed over a range of temperatures, constitutes a powerful argument in favor of the experimental attainment of equilibrium. Finally, theoretical calculations can be used to extend experimentally calibrated fractionations to low temperatures where isotopic exchange is too slow to reach equilibrium on a reasonable laboratory timescale (Clayton and Kieffer 1991).

BASIC KINETIC STABLE ISOTOPE FRACTIONATION THEORY

Kinetic fractionations can occur when there is incomplete isotopic exchange between the different phases present in a system. A thorough introduction to kinetic stable isotope fractionation theory is unfortunately beyond the scope of the present review. However, it is useful to include a brief discussion of some basic aspects, particularly in comparison to equilibrium fractionation theory. A simple example of kinetic fractionation is the evaporation of a liquid water droplet into a vacuum, in this example H_2O molecules entering the gas phase are physically removed from the vicinity of the droplet, so there is no chance for isotopic equilibration between vapor-phase molecules and the residual liquid. Isotopic fractionation in this case is determined by a one-way reaction path, and will not, in general, be the same as the fractionation in a system where vapor-phase molecules are able to equilibrate and exchange with the liquid. In other reactions, isotopic exchange is limited by an energy barrier—an

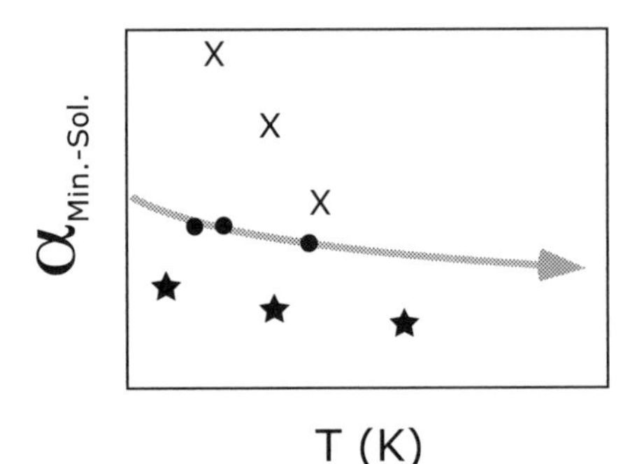

Figure 7. Using a theoretically determined equilibrium fractionation to interpret measured isotopic fractionations in a hypothetical mineral-solution system. Three sets of data are shown. The theoretical equilibrium fractionation for this system is indicated by the gray arrow. The first set of data, indicated by circles, closely follow the calculated fractionation, suggesting a batch equilibrium fractionation mechanism. The second set of data (stars) is displaced from the theoretical curve. This may either indicate a temperature-independent kinetic fractionation superimposed on an equilibrium-like fractionation, or that the theoretical calculation is somewhat inaccurate. The third set of data (crosses) shows much greater temperature sensitivity than the equilibrium calculation; this provides evidence for a dominantly non-equilibrium fractionation mechanism. For the first data set, the theoretical fractionation curve can be used to extrapolate beyond the measured temperature range. The second data set can also be extrapolated along a scaled theoretical curve (Clayton and Kieffer 1991).

exchange activation energy—that must be surmounted in order for two atoms with different masses to swap positions (Fig. 8).

The size of the activation energy for isotopic exchange reflects the need to break bonds and rearrange atoms in order to effect exchange. The rate at which atoms are exchanged decreases with increasing height of the energy barrier, increases with temperature, and also depends on geometric constraints. In natural systems, and particularly at low temperatures, kinetic stable isotope fractionations are common. With non-traditional stable isotope systems we are typically interested in heavier elements and condensed phases; for these the most common types of kinetic fractionations are likely to be those driven by the effects of isotopic mass on molecular and atomic velocities and diffusivity, and by isotopic effects on activation energies. Other mechanisms can be important in gas-phase and very low-temperature reactions, but are largely beyond the scope of the present work.

Effects of isotopic mass on molecular and atomic velocities and diffusivity

In a gas with a well defined temperature, the translational kinetic energies of all molecules are the same, on average,

$$\langle KE \rangle = \frac{3}{2}\mathrm{k}T = \frac{1}{2}m\langle \mathrm{v}^2 \rangle$$

where m here is the mass of the molecule and v is its velocity. It is apparent that molecules having different isotopic compositions will have different average velocities,

$$\frac{\langle v_{heavy}{}^2 \rangle}{\langle v_{light}{}^2 \rangle} = \frac{m_{light}}{m_{heavy}}$$

For fractionation processes that are chiefly dependent on molecular velocity, this relationship can be rearranged into a simple translational isotopic fractionation factor,

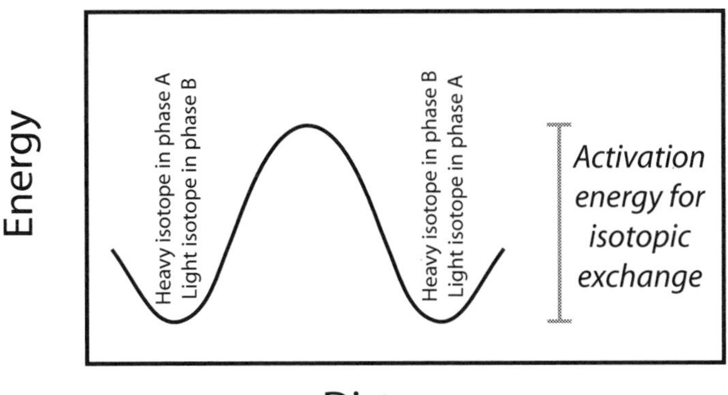

Figure 8. Schematic diagram of the potential energy trajectory of an isotope exchange reaction. In general, isotopic exchange requires the rearrangement or breaking of chemical bonds, which involves an increase in the potential energy of the molecule. This activation energy limits the rate at which exchange can take place, particularly at low temperatures. Equilibrium isotopic fractionation requires isotopic exchange, kinetic fractionations can occur if molecules are unable to surmount the exchange energy barrier. For simplicity isotopic effects on exchange energy are ignored.

$$\alpha_{XR(translation)} = \sqrt{\frac{m_{heavy\,XR}}{m_{light\,XR}}}$$

A good example of translational fractionation is one-way diffusion through an orifice that is smaller than the mean-free path of the gas. Related, but somewhat more complex velocity-dependent fractionations occur during diffusion through a host gas, liquid, or solid. In these fractionations the isotopic masses in the translational fractionation factor are often replaced by some kind of effective reduced mass. For instance, in diffusion of a trace gas *XR* through a medium, *Y*, consisting of molecules with mass m_Y,

$$\frac{D_{light\,XR}}{D_{heavy\,XR}} = \sqrt{\frac{\mu_{heavy\,XR-Y}}{\mu_{light\,XR-Y}}} = \sqrt{\frac{\dfrac{1}{m_{light\,XR}} + \dfrac{1}{m_Y}}{\dfrac{1}{m_{heavy\,XR}} + \dfrac{1}{m_Y}}}$$

where $D_{light_{XR}}$ and $D_{heavy_{XR}}$ are the diffusivities of the isotopically light and heavy forms of *XR*, respectively (Reid et al. 1977). Such diffusivity ratios will always be closer to unity than the simple translational fractionation factor.

Effects of isotopic mass on rates of activation

In many chemical processes the rate of reaction depends chiefly on one particular transformative step. Under conditions where reaction products are not able to back-exchange with the reactants, this rate-limiting step also can control the isotopic fractionation between reactants and products. For simple reactions where one molecule dissociates, a transition state theory of isotopic fractionation has been developed (Eyring 1935; Bigeleisen 1949). In this simplified method of calculating isotopic effects on reaction rates, only the energy barrier between the reactants and a single rate-limiting transition-state is considered. A reaction coordinate, defined as the lowest-energy path from the ground states of the reactants to the transition state, describes the mechanical and electronic distortions necessary before a reaction can take place (Fig. 9). In this theory, the rate at which reactant molecules achieve the transition-state (and thus react to form products) is determined by the statistical probability of a molecule possessing enough energy to reach the summit of the energy barrier,

$$p(E_A) = \exp\left(\frac{-E_A}{\mathrm{k}T}\right)$$

In theory, E_A is the activation energy (the height of the rate-limiting energy barrier in Joules/molecule). This probability is then multiplied by the rate at which reactant molecules sample the reaction coordinate. At the top of the energy barrier, the potential energy curve becomes flat, so that motion across the summit is rather similar to a free translation, rather than a vibrational mode. As a result, the activated molecule has three translational degrees of freedom, one reactive translational degree of freedom, and $3N$–7 vibrational degrees of freedom ($3N$–6 for a linear molecule). The energies of the reactants and the activated molecules can be determined by evaluating partition functions, using a translational partition function to evaluate the reaction coordinate in the activated molecule. Instead of using the molecular masses, however, an effective reduced mass of molecular motion along the coordinate is used,

$$Q_{Rxn\ Coordinate} = V\left(\frac{2\pi\mu_{Rxn}\mathrm{k}T}{\mathrm{h}^2}\right)^{3/2}$$

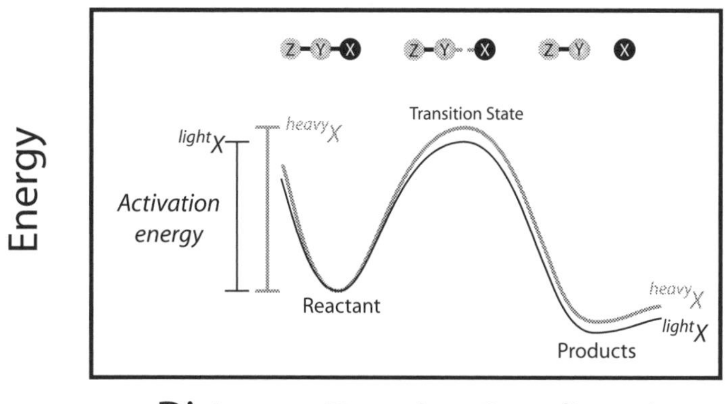

Figure 9. Schematic diagram of the potential energy trajectory of a molecular dissociation reaction. As with the isotopic exchange reaction, the molecule must acquire enough energy in the reaction coordinate to surmount an energy barrier. In this case the dissociating molecule, *ZYX*, splits into two fragments *ZY* and *X*, and the reaction coordinate is roughly equivalent to the length of the *X-Y* bond. Typically, the size of this activation energy barrier is slightly affected by isotopic substitution due to differences between the zero-point vibrational energies in the reactant molecule in the ground state and activated state. Here the potential energy curve for a molecule containing the light isotope of element X (^{light}X) is shown in black, and the corresponding curve for a molecule containing a heavier isotope of element X (^{heavy}X) is shown in grey. For clarity, both trajectories are set so that the potential energies of the reactant molecules are the same. In this simple example, the activation energy for the isotopically light molecule is smaller than for the isotopically heavy molecule. In addition, the isotopically light molecule will tend to sample the trajectory faster than the isotopically heavy molecule. For both reasons, the ^{light}X-bearing molecule will tend to react faster than the ^{heavy}X-bearing molecule.

For a simple decomposition reaction of a diatomic molecule, $AB \rightarrow A + B$, μ_{Rxn} is simply

$$\mu_{Rxn} = \frac{m_A m_B}{m_A + m_B}$$

In more complex reactions, it may not be as straightforward to determine the reduced mass along the reaction coordinate because the exact reaction mechanism is poorly known. This task can be facilitated by *ab initio* and empirical force-field software packages with built-in capacities to predict and evaluate reaction coordinates (Schmidt et al. 1993; Gale and Rohl 2003). Once the relevant reduced masses are known, a derivation similar to Urey's (1947) equation for equilibrium isotopic fractionation can be followed, obtaining:

$$\alpha_{transition-XR} = \frac{\mu_{^{heavy}XR(Rxn)}}{\mu_{^{light}XR(Rxn)}} \frac{\alpha_{XR(activated)-X}}{\alpha_{XR-X}}$$

where $\alpha_{transition-XR}$ is the ratio of reaction rates of $^{heavy}XR$ and $^{light}XR$, $\mu_{^{heavy}XR(Rxn)}$ and $\mu_{^{light}XR(Rxn)}$ are the effective reduced masses for motion along the reaction coordinate for the two isotopic forms of *XR*, $\alpha_{XR\text{-}X}$ for the reactant is calculated using Equation (3), and $\alpha_{XR(activated)}$ for the activated molecule is calculated using Equation (3) using the $3N$−7 vibrational modes of the activated complex. This method has been applied in theoretical studies of isotopic fractionations in non-traditional systems (e.g., Krouse and Thode 1962; Paneth 2003). The Paneth (2003) study predicts kinetic chlorine-isotope fractionations in reasonable agreement with experimental measurements.

MASS-DEPENDENCE OF EQUILIBRIUM AND KINETIC ISOTOPE FRACTIONATIONS

Recent discoveries of oxygen and sulfur fractionations (e.g., Farquhar et al. 2000; Thiemens et al. 2001) that appear to have unusual mass dependence has renewed interest in variations in the mass dependence of different fractionation mechanisms (Gao and Marcus 2002; Young et al. 2002). Usually, mass-dependent fractionations scale in proportion to differences in isotopic mass:

$$\frac{^{(^iX/^kX)}\alpha - 1}{^{(^jX/^kX)}\alpha - 1} \approx \frac{m_{^iX} - m_{^kX}}{m_{^jX} - m_{^kX}}$$

where $^{(^iX/^kX)}\alpha$ is the fractionation factor separating isotopes ^iX and ^kX in some reaction or process, $^{(^jX/^kX)}\alpha$ is the fractionation factor for isotopes ^jX and ^kX in the same reaction, and $m_{^iX}$, $m_{^jX}$, and $m_{^kX}$ are the masses of isotopes ^iX, ^jX, and ^kX respectively. Recent measurements in non-traditional stable isotope systems, including those described in this volume, appear to confirm that mass-dependent fractionation is the norm in geochemical processes and typical chemical reactions. Accurate mass-scaling laws can help to verify whether an observed fractionation has a mass-independent component. In addition, theoretical calculations suggest that different mass-dependent fractionation mechanisms will follow slightly different mass-scaling laws, so that it may be possible, for instance, to distinguish kinetic and equilibrium fractionations (Young et al. 2002). Young et al. (2002) derived and tabulated mass-scaling laws for several types of kinetic and equilibrium fractionations. For equilibrium fractionations,

$$\frac{\ln\left(^{(^iX/^kX)}\alpha\right)}{\ln\left(^{(^jX/^kX)}\alpha\right)} = \frac{\left(\frac{1}{m_k} - \frac{1}{m_i}\right)}{\left(\frac{1}{m_k} - \frac{1}{m_j}\right)}$$

This equation follows from the Bigeleisen and Mayer (1947) expression for equilibrium fractionation (Eqn. 4), and is appropriate at high temperatures ($h\nu/kT \leq 5$). This relation may not be accurate at low temperatures, especially near "crossovers" where fractionations switch direction (Matsuhisa et al. 1978; Deines 2003). As the equilibrium fractionation between two substances becomes small near a crossover, the mass-scaling law becomes strongly temperature dependent, and reduced partition function ratios calculated using Equation (3) should be used to estimate mass scaling. For elements other than H, C, N, and O, however, high atomic mass and relatively low vibrational frequencies are expected to make the high-temperature mass-scaling law accurate under most conditions. Anomalous equilibrium mass scaling has never been definitively observed, it may be most likely to occur when one equilibrated phase is a small, highly covalent molecule (e.g., SiC, SeO_3).

For translational kinetic fractionations dependent on molecular velocities, a slightly different relation holds (Young et al. 2002),

$$\frac{\ln\left(^{(^iX/^kX)}\alpha\right)}{\ln\left(^{(^jX/^kX)}\alpha\right)} = \frac{\ln\left(\frac{m_{^kXR}}{m_{^iXR}}\right)}{\ln\left(\frac{m_{^kXR}}{m_{^jXR}}\right)}$$

where $m_{^kXR}$, etc. are the masses of isotopically substituted molecules. Using the simplified transition state theory of Bigeleisen (1949), a somewhat more complex expression is obtained for fractionations controlled by activation rates,

$$\frac{\ln\left({}^{(i_X/k_X)}\alpha\right)}{\ln\left({}^{(j_X/k_X)}\alpha\right)} = \frac{\ln\left(\dfrac{\mu_{k_{XR(Rxn)}}}{\mu_{i_{XR(Rxn)}}}\right)}{\ln\left(\dfrac{\mu_{k_{XR(Rxn)}}}{\mu_{j_{XR(Rxn)}}}\right)}$$

where $\mu_{i_{XR(Rxn)}}$, $\mu_{j_{XR(Rxn)}}$ and $\mu_{k_{XR(Rxn)}}$ are the isotopic reduced masses along the relevant reaction coordinate. This formulation assumes that the activation energy is not significantly affected by isotopic substitution.

CONCLUSIONS

The purpose of this chapter is to provide a concise, comprehensible introduction to the theory of stable isotope fractionations. While many of the fundamental principles discussed here have been understood for decades, applications have been limited by incomplete knowledge of the vibrational properties and reaction mechanisms for all but the simplest gas-phase molecules. Recent developments in theoretical chemistry and spectroscopy are widening the scope of materials and processes that can be studied, however. In particular, *ab initio* methods and transferable empirical potentials are now routinely used to model the vibrational properties of molecules, crystals, and aqueous species with ever increasing precision and reliability; these methods hold great promise for estimating unknown vibrational frequencies of isotopically substituted materials, and for determining potential energy surfaces relevant to chemical reactions. At present these developments are just beginning to be used to full advantage, so that there is a major gap between our ability to measure stable isotope fractionations and our understanding of the processes that cause fractionations to occur. New studies are particularly important now, as novel measurement techniques rapidly expand the scope of stable isotope geochemistry.

ACKNOWLEDGMENTS

This chapter was improved by thoughtful suggestions from E.A. Johnson, T. Chacko, A. D. Anbar, J. R. O'Neil and C. M. Johnson. I would also like to thank H. P. Taylor, Jr. for introducing me to stable isotope fractionation theory, and G. R. Rossman for making vibrational spectroscopy fun.

REFERENCES

Anbar AD, Jarzecki A, Spiro T (in press) Theoretical investigation of iron isotope fractionation between $Fe(H_2O)_6^{3+}$ and $Fe(H_2O)_6^{2+}$. Geochim Cosmochim Acta

Angus WR, Bailey CR, Hale JB, Ingold CK, Leckie AH, Raisin CG, Thompson JW, Wilson CL (1936) Structure of benzene. Part VIII. Assignment of vibration frequencies of benzene and hexadeuterobenzene. J Chem Soc (London):971-987

Barling J, Arnold GL, Anbar AD (2001) Natural mass-dependent variation in the isotopic composition of molybdenum. Earth Planet Sci Lett 193:447-457

Baroni S, de Gironcoli S, Dal Corso A, Giannozzi P (2001) Phonons and related crystal properties from density-functional perturbation theory. Rev Modern Phys 73:515-562

Basile LJ, Ferraro JR, LaBonville P, Wall MD (1973) A study of force fields for tetrahedral molecules and ions. Coord Chem Rev 11:21-69

Beard BL, Johnson CM, Skulan JL, Nealson KH, Cox L, Sun H (2003) Application of Fe isotopes to tracing the geochemical and biological cycling of Fe. Chem Geol 195:87-117

Becke AD (1993) Density-functional thermochemistry. III. The role of exact exchange. J Chem Phys 98: 5648-5652

Bigeleisen J (1949) The relative velocities of isotopic molecules. J Chem Phys 17:675-678

Bigeleisen J (1955) Statistical mechanics of isotopic systems with small quantum corrections. I. General considerations and the rule of the geometric mean. J Chem Phys 23:2264-2267

Bigeleisen J (1998) Second-order correction to the Bigeleisen-Mayer equation due to the nuclear field shift. Proc National Acad Sci 95:4808-4809

Bigeleisen J, Mayer MG (1947) Calculation of equilibrium constants for isotopic exchange reactions. J Chem Phys 15:261-267

Black GM, Law MM (2001) The general harmonic force field of methyl chloride. J Mol Spectrosc 205:280-285

Bochkarev AV, Trefilova AN, Tsurkov NA, Klinskii GD (2003) Calculations of beta-factors by *ab initio* quantum-chemical methods. Russian J Phys Chem 77:622-626

Bode BM, Gordon MS (1998) MacMolPlt: a graphical user interface for GAMESS. J Mol Graphics and Modeling 16:133-138

Bottinga Y (1968) Calculations of fractionation factors for carbon and oxygen isotope exchange in the system calcite-carbon dioxide-water. J Phys Chem 72:800-808

Bullen TD, White AF, Childs CW, Vivit DV, Schulz MS (2001) Demonstration of significant abiotic iron isotope fractionation in nature. Geology 29:699-702

Burkholder JB, Hammer PD, Howard CJ, Maki AG, Thompson G, Chackerian CJ (1987) Infrared measurements of the ClO radical. J Mol Spectrosc 124:139-161

Catlow CRA, Freeman CM, Islam MS, Jackson RA, Leslie M, Tomlinson SM (1988) Interatomic potentials for oxides. Phil Mag A 58:123-141

Chacko T, Cole DR, Horita J (2001) Equilibrium oxygen, hydrogen and carbon isotope fractionation factors applicable to geologic systems. Rev Mineral Geochem 43:1-82

Chacko T, Mayeda TK, Clayton RN, Goldsmith JR (1991) Oxygen and carbon isotope fractionation between CO_2 and calcite. Geochim Cosmochim Acta 55:2867-2882

Clayton RN, Kieffer SW (1991) Oxygen isotopic thermometer calculations. *In*: Stable Isotope Geochemistry: A Tribute to Samuel Epstein, Special Publication No. 3. Taylor HPJ, O'Neil JR, Kaplan IR (eds) The Geochemical Society, San Antonio, Texas, p 1-10

Cotton FA (1971) Chemical Applications of Group Theory. Wiley-Interscience, New York

Criss RE (1991) Temperature dependence of isotopic fractionation factors. *In*: Stable Isotope Geochemistry: A Tribute to Samuel Epstein, Special Publication No. 3. Taylor HPJ, O'Neil JR, Kaplan IR (eds) The Geochemical Society, San Antonio, Texas, p 11-16

Criss RE (1999) Principles of stable isotope distribution. Oxford University Press, New York

Cygan RT, Kubicki JD (2001) Molecular Modeling Theory: Applications in the Geosciences. Reviews in Mineralogy and Geochemistry, Vol. 42, Mineralogical Society of America, Washington, D. C.

Cygan RT, Wright K, Fisler DK, Gale JD, Slater B (2002) Atomistic models of carbonate minerals: bulk and surface structures, defects, and diffusion. Mol Simul 28:475-495

Davis AM, Hashimoto A, Clayton RN, Mayeda TK (1990) Isotope mass fractionation during evaporation of Mg_2SiO_4. Nature 347:655-658

Decius JC (1948) A tabulation of general formulas for inverse kinetic energy matrix elements in acyclic molecules. J Chem Phys 16:1025-1034

Deines P (2003) A note on intra-elemental isotope effects and the interpretation of non-mass-dependent isotope variations. Chem Geol 199:179-182

Demiralp E, Cagin T, Goddard WA (1999) Morse stretch potential charge equilibrium force field for ceramics: Application to the quartz-stishovite phase transition and to silica glass. Phys Rev Lett 82:1708-1711

Dewar MJS (1977) Ground states of molecules. The MNDO method. Approximations and parameters. J Am Chem Soc 99:4899-4907

Dewar MJS, Zoebisch EG, Healy EF, Stewart JJP (1985) The development and use of quantum-mechanical molecular-models. 76. AM1 - a new general-purpose quantum-mechanical molecular-model. J Am Chem Soc 107:3902-3909

Dove MT, Winker B, Leslie M, Harris MJ, Salje EKH (1992) A new interatomic potential model for calcite: Applications to lattice-dynamics studies, phase-transformations, and isotope fractionation. Am Mineral 77:244-250

Driesner T, Ha TK, Seward TM (2000) Oxygen and hydrogen isotope fractionation by hydration complexes of Li^+, Na^+, K^+, F^-, Cl^-, and Br^-: A theoretical study. Geochim Cosmochim Acta 64:3007-3033

Driesner T, Seward TM (2000) Experimental and simulation study of salt effects and pressure/density effects on oxygen and hydrogen stable isotope liquid-vapor fractionation for 4-5 molal aqueous NaCl and KCl solutions to 400 degrees C. Geochim Cosmochim Acta 64:1773-1784

Eggenkamp HGM, Kreulen R, Van Groos AFK (1995) Chlorine stable isotope fractionation in evaporites. Geochim Cosmochim Acta 59:5169-5175

Elcombe MM, Pryor AW (1970) The lattice dynamics calcium fluoride. J Phys C 3:492-499

Eyring H (1935) The activated complex in chemical reactions. J Chem Phys 3:107-117

Farquhar J, Bao HM, Thiemens M (2000) Atmospheric influence of Earth's earliest sulfur cycle. Science 289: 756-758

Foresman JB, Frisch A (1996) Exploring chemistry with electronic structure methods. Gaussian, Inc., Pittsburgh, PA

Francl MM, Pietro WJ, Hehre WJ, Binkley JS, Gordon MS, DeFrees DJ, Pople JA (1982) Self-consistent molecular-orbital methods. 23. A polarization-type basis set for 2nd row elements. J Chem Phys 77:3654-3665

Fujii T, Suzuki D, Gunjii K, Watanabe K, Moriyama H, Nishizawa K (2002) Nuclear field shift effect in the isotope exchange reaction of chromium(III) using a crown ether. J Phys Chem A 106:6911-6914

Gale JD (2001) Simulating the crystal structures and properties of ionic materials from interatomic potentials. Rev Mineral Geochem 42:37-62

Gale JD, Rohl. AL (2003) The General Utility Lattice Program (GULP). Mol Simul 29:291-341

Galy A, Bar-Matthews M, Halicz L, O'Nions RK (2002) Mg isotopic composition of carbonate: insight from speleothem formation. Earth and Planet Sci Lett 201:1-11

Gao YQ, Marcus RA (2002) On the theory of the strange and unconventional isotopic effects in ozone formation. J Chem Phys 116:137-154

Gillet P, McMillan P, Schott J, Badro J, Grzechnik A (1996) Thermodynamic properties and isotopic fractionation of calcite from vibrational spectroscopy of ^{18}O-substituted calcite. Geochim Cosmochim Acta 60:3471-3485

Gonze X, Beuken J-M, Caracas R, Detraux F, Fuchs M, Rignanese G-M, Sindic L, Verstraete M, Zerah G, Jollet F, Torrent M, Roy A, Mikami M, Ghosez P, Raty J-Y, Allan DC (2002) First-principles computation of material properties: the ABINIT software project. Comput Mat Sci 25:478-492

Hanschmann G (1984) Reduzierte Zustandssummenverhältnisse isotoper Moleküle auf quantenchemischer Grundlage V. Mitt. MNDO-MO-Berechnungen zur ^{28}Si/^{30}Si-, ^{32}S/^{34}S- und ^{35}Cl/^{37}Cl-Substitution. Isotopenpraxis 12:437-439

Harrison NM, Leslie M (1992) The derivation of shell model potentials for $MgCl_2$ from *ab initio* theory. Mol Simul 9:171-174

Hawthorne FC (1988) Mössbauer spectroscopy. Rev Mineral 18:255-340

Hazony Y (1966) Effect of zero-point motion on the Mössbauer spectra of $K_4Fe(CN)_6$ and $K_4Fe(CN)_6{\bullet}3H_2O$. J Chem Phys 45:2664-2668

Heath DF, Linnett JW (1948) Molecular force fields. II. The force fields of the tetrahalides of the group IV elements. Trans Faraday Soc 44:561-568

Heberle J (1971) The Debye integrals, the thermal shift and the Mössbauer fraction. *In*: Mössbauer Effect Methodology, Vol 7. Gruverman IJ (ed), Plenum, p 299-308

Herzberg G (1945) Molecular Spectra and Molecular Structure. II. Infrared and Raman Spectra of Polyatomic Molecules. Von Nostrand Reinhold, New York

Hohenberg P, Kohn W (1964) Inhomogeneous electron gas. Phys Rev 136:B864-871

Horita J, Cole DR, Polyakov VB, Driesner T (2002) Experimental and theoretical study of pressure effects on hydrogen isotope fractionation in the system brucite-water at elevated temperatures. Geochim Cosmochim Acta 66:3769-3788

Housley RM, Hess F (1966) Analysis of Debye-Waller-factor and Mössbauer-thermal-shift measurements. 1. General theory. Phys Rev 146:517-526

Housley RM, Hess F (1967) Analysis of Debye-Waller-factor and Mössbauer-thermal-shift measurements. II. Thermal-shift data on Fe. Phys Rev 164:340-344

Johnson CM, Skulan JL, Beard BL (2002) Isotopic fractionation between Fe(III) and Fe(II) in aqueous solutions. Earth Plan Sci Lett 195:141-153

Kakihana H, Kotaka M, Satoh S, Nomura M, Okamoto M (1977) Fundamental studies on the ion-exchange separation of boron isotopes. Bull Chem Soc Japan 50:158-163

Kieffer SW (1982) Thermodynamics and lattice vibrations of minerals: 5. Applications to phase equilibria, isotopic fractionation, and high-pressure thermodynamic properties. Rev Geophys Space Phys 20:827-849

Kohn W, Sham LJ (1965) Self-consistent equations including exchange and correlation effects. Phys Rev 140: A1133-A1138

Kotaka M, Kakihana H (1977) Thermodynamic isotope effect of trigonal planar and tetrahedral species. Bull Research Lab Nuc Reactors 2:13-29

Kotaka M, Shono T, Ikuta E, Kakihana H (1978) Thermodynamic isotope effect of some octahedral hexahalo complexes. Bull Research Lab Nuc Reactors 3:31-37

Krishnan R, Binkley JS, Seeger R, Pople JA (1980) Self-consistent molecular-orbital methods. 20. Basis set for correlated wave-functions. J Chem Phys 72:650-654

Krouse HR, Thode HG (1962) Thermodynamic properties and geochemistry of isotopic compounds of selenium. Can J Chem 40:367-375

Krynauw GN (1990) The estimation of ν_6 of octahedral XY_6 species. Spectrochim Acta 46A:741-745

Le Roux H, Glasser L (1997) Transferable potentials for the Ti-O system. J Materials Chem 7:843-851

Lindemann FA (1919) Note of the vapor pressure and affinity of isotopes. Philosoph Mag 38:173-181

Lindemann FA, Aston FW (1919) The possibility of separating isotopes. Philosoph Mag 37:523-535

Mayer JE, Mayer MG (1940) Statistical Mechanics. John Wiley & Sons, Inc., New York

McMillan P, Hess AC (1988) Symmetry, group theory and quantum mechanics. Rev Mineral 18:11-62

Müller A, Königer F (1974) Schwingungsspektren von $^{50}CrO_4^{2-}$, $^{53}CrO_4^{2-}$, $Cr^{18}O_4^{2-}$, $^{92}MoO_4^{2-}$, $^{100}MoO_4^{2-}$ und $Ru^{18}O_4^{-}$. Zur Berechnung exakter Kraftkonstanten von Ionen. Spectrochim Acta 30A:641-649

Nakamoto K (1997) Infrared and Raman Spectra of Inorganic and Coordination Compounds. John Wiley & Sons, Inc., New York, NY

O'Neil JR (1986) Theoretical and experimental aspects of isotopic fractionation. Rev Mineral 16:1-40

Oi T (2000) Calculations of reduced partition function ratios of monomeric and dimeric boric acids and borates by the *ab initio* molecular orbital theory. J Nuclear Sci Tech 37:166-172

Oi T, Nomura M, Musashi M, Ossaka T, Okamoto M, Kakihana H (1989) Boron isotopic composition of some boron minerals. Geochim Cosmochim Acta 53:3189-3195

Oi T, Yanase S (2001) Calculations of reduced partition function ratios of hydrated monoborate anion by the *ab initio* molecular orbital theory. J Nuclear Sci Tech 38:429-432

Paneth P (2003) Chlorine kinetic isotope effects on enzymatic dehalogenations. Accounts Chem Res 36:120-126

Patel A, Price GD, Mendelssohn MJ (1991) A computer simulation approach to modeling the structure, thermodynamics and oxygen isotope equilibria of silicates. Phys Chem Min 17:690-699

Polyakov VB (1997) Equilibrium fractionation of the iron isotopes: estimation from Mössbauer spectroscopy data. Geochim Cosmochim Acta 61:4213-4217

Polyakov VB, Mineev SD (2000) The use of Mössbauer spectroscopy in stable isotope geochemistry. Geochim Cosmochim Acta 64:849-865

Pople JA, Binkley Js, Seeger R (1976) Theoretical models incorporating electron correlation. Int J Quantum Chem Y-10(Suppl.):1-19

Pople JA, Scott AP, Wong MW, Radom L (1993) Scaling factors for obtaining fundamental vibrational frequencies and zero-point energies from HF/6-31G* and MP2/6-31G* harmonic frequencies. Israel J Chem 33:345-350

Reid RC, Prausnitz JM, Sherwood TK (1977) The Properties of Gases and Liquids. McGraw-Hill, New York

Richet P, Bottinga Y, Javoy M (1977) A review of hydrogen, carbon, nitrogen, oxygen, sulphur, and chlorine stable isotope fractionation among gaseous molecules. Ann Rev Earth Planet Sci 5:65-110

Roe JE, Anbar AD, Barling J (2003) Nonbiological fractionation of Fe isotopes: evidence of an equilibrium isotope effect. Chem Geol 195:69-85

Roothaan CCJ (1951) New developments in molecular orbital theory. Rev Modern Phys 23:69-89

Rudolph W, Brooker MH, Pye CC (1995) Hydration of the lithium ion in aqueous-solution. J Phys Chem 99: 3793-3797

Russell WA, Papanastassiou DA, Tombrello TA (1978) Ca isotope fractionation on the Earth and other solar system materials. Geochim Cosmochim Acta 42:1075-1090

Schachtschneider JH (1964) Report 231/64. Shell Development Co., Houston, Texas

Schauble EA, Rossman GR, Taylor HP, Jr. (2001) Theoretical estimates of equilibrium Fe-isotope fractionations from vibrational spectroscopy. Geochim Cosmochim Acta 65:2487-2497

Schauble EA, Rossman GR, Taylor HPJ (2003) Theoretical estimates of equilibrium chlorine-isotope fractionations. Geochim Cosmochim Acta 67:3267-3281

Schauble EA, Rossman GR, Taylor HPJ (In review) Theoretical estimates of equilibrium chromium-isotope fractionations. Chem Geol

Schmidt MW, Baldridge KK, Boatz JA, Elbert ST, Gordon MS, Jensen JJ, Koseki S, Matsunaga N, Nguyen KA, Su S, Windus TL, Dupuis M, Montgomery JA (1993) General atomic and molecular electronic-structure system. J Computational Chem 14:1347-1363

Scott AP, Radom L (1996) Harmonic vibrational frequencies: An evaluation of Hartree-Fock, Møller-Plesset, quadratic configuration interaction, density functional theory, and semiempirical scale factors. J Phys Chem 100:16502-16513

Shriver DF (1999) Inorganic Chemistry. W. H. Freeman and Co., New York

Simanouti (Shimanouchi) T (1949) The normal vibrations of polyatomic vibrations as treated by Urey-Bradley field. J Chem Phys 17:245-248

Skulan JL, Beard BL, Johnson CM (2002) Kinetic and equilibrium Fe isotope fractionation between aqueous Fe(III) and hematite. Geochim Cosmochim Acta 66:2995-3015

Skulan JL, DePaolo DJ, Owens TL (1997) Biological control of calcium isotopic abundances in the global calcium cycle. Geochim Cosmochim Acta 61:2505-2510

Stewart JJP (1989a) Optimization of parameters for semiempirical methods. 1. Method. J Computational Chem 10:209-220

Stewart JJP (1989b) Optimization of parameters for semiempirical methods. 2. Applications. J Computational Chem 10:221-264

Thiemens M, Savarino J, Farquhar J, Bao HM (2001) Mass-independent isotopic compositions in terrestrial and extraterrestrial solid and their applications. Accounts Chem Res 34:645-652

Thijssen JM (1999) Computational Physics. Cambridge University Press, Cambridge, UK

Urey HC (1947) The thermodynamic properties of isotopic substances. J Chem Soc (London) 562-581

Urey HC, Greiff LJ (1935) Isotopic exchange equilibria. J Am Chem Soc 57:321-327

Wilson EBJ, Decius JC, Cross PC (1955) Molecular Vibrations: The Theory of Infrared and Raman Vibrational Spectra. Dover, New York

Wong MW (1996) Vibrational frequency prediction using density functional theory. Chem Phys Lett 256: 391-399

Yamaji K, Makita Y, Watanabe H, Sonoda A, Kanoh H, Hirotsu T, Ooi K (2001) Theoretical estimation of lithium isotopic reduced partition function ratio for lithium ions in aqueous solution. J Phys Chem A 105: 602-613

Young ED, Galy A, Nagahara H (2002) Kinetic and equilibrium mass-dependent isotope fractionation laws in nature and their geochemical and cosmochemical significance. Geochim Cosmochim Acta 66:1095-1104

APPENDIX: ANNOTATED BIBLIOGRAPHY OF THEORETICAL EQUILIBRIUM FRACTIONATIONS

A number of theoretical studies of equilibrium stable isotope fractionations in non-traditional systems have been published, including several in journals that may be less familiar to interested geochemists. Here I present an annotated bibliography, which is no doubt incomplete but should provide a good start. To maintain some consistency in the face of the many formats that have been used over the years in reporting theoretically calculated fractionations, most results have been converted to fractionation factors in α_{XR-X} form. One study does not contain tabulated results (Hanschmann 1984), but is worth mentioning because of the quantum-mechanical force-field method used to estimate vibrational frequencies. Most of the theoretical results reproduced here are tabulated at a few representative temperatures. Since fractionation factors are typically nearly linear vs. $1/T^2$, interpolations and extrapolations to higher temperatures made on this basis should be reasonably accurate. In some cases only a selection of the published results are given here, focusing on what appear to be the most relevant substances studied. Although these studies cited here use reasonable methods and appear to be carefully executed, critical evaluation of possible procedural, numerical, and typographical errors is encouraged.

GENERAL STUDIES

Urey and Greiff (1935)

This study is one of the earliest attempts to calculate equilibrium fractionation factors using measured vibrational spectra and simple reduced-mass calculations for diatomic molecules. For the sake of consistency I have converted reported single-molecule partition function ratios to α_{XR-X} units.

		α_{XR-X}	
	273.1 K	**298.1 K**	**600 K**
^{7}Li-^{6}Li			
LiH	1.0284	1.0251	1.0084
Li(g)	1.0000	1.0000	1.0000
^{37}Cl-^{35}Cl			
Cl_2	1.0088	1.0075	1.0020
HCl	1.0050	1.0046	1.0019

Bigeleisen and Mayer (1947)

In this study the authors develop simplified equations relating equilibrium fractionations to mass-scaling factors and molecular force constants. Equilibrium isotopic fractionations of heavy elements (Si and Sn) are predicted to be small, based on highly simplified, one-parameter empirical force-field models (bond-stretching only) of SiF_4, $[SiF_6]^{2-}$, $SnCl_4$, and $[SnCl_6]^{2-}$.

	$\alpha_{XR\text{-}X}$ **300 K**
^{30}Si-^{28}Si	
SiF_4	1.111
$[SiF_6]^{2-}$	1.109
^{120}Sn-^{118}Sn	
$SnCl_4$	1.00256
$[SnCl_6]^{2-}$	1.0028

Urey (1947)

Fractionation factors are calculated using measured vibrational spectra supplemented by simplified empirical force-field modeling (bond-stretching and bond-angle bending force constants only).

			$\alpha_{XR\text{-}X}$		
	273.1 K	**298.1 K**	**400 K**	**500 K**	**600 K**
^{7}Li-^{6}Li					
LiH	1.0281	1.0249	1.0161	1.0113	1.0083
Li(g)	1.0000	1.0000	1.0000	1.0000	1.0000
^{37}Cl-^{35}Cl					
$[ClO_4]^-$	1.0972	1.0847	1.0521	1.0353	1.0253
$[ClO_3]^-$	1.0551	1.0478	1.0291	1.0196	1.0140
ClO_2	1.0360	1.0313	1.0185	1.0130	1.0094
Cl_2	1.0086	1.0074	1.0043	1.0028	1.0019
HCl	1.0050	1.0046	1.0032	1.0024	1.0019
^{81}Br-^{79}Br					
$[BrO_3]^-$	1.0093	1.0080	1.0048	1.0032	1.0022
Br_2	1.0014	1.0012	1.0007	1.0004	1.0003
HBr	1.0009	1.0008	1.0006	1.0004	1.0003

Kotaka and Kakihana (1977)

Fractionation factors are calculated for a large variety of trigonal-planar (XY_3) and tetrahedral (XY_4) molecules and molecule-like complexes, with a particular focus on metal halides. Empirical force-field models (MUBFF) are used to estimate vibrational frequencies for the rarer isotopic forms of the substances studied, and aqueous and crystalline molecule-like species are modeled as gas-phase molecules. In the tabulation below the original equilibrium constants have been converted to fractionation factors ($\alpha_{XR\text{-}X}$).

	$\alpha_{XR\text{-}X}$					
	200 K	**300 K**	**400 K**	**500 K**	**700 K**	**1000 K**
^{30}Si-^{28}Si						
$[Si^{16}O_4]^{4-}$	1.1469	1.0760	1.0459	1.0305	1.0161	1.0081
SiF_4	1.1675	1.0886	1.0544	1.0365	1.0195	1.0098
$Si^{35}Cl_4$	1.0930	1.0455	1.0266	1.0174	1.0090	1.0045
$Si^{79}Br_4$	1.0755	1.0359	1.0207	1.0134	1.0069	1.0034
$Si^{127}I_4$	1.0577	1.0270	1.0155	1.0100	1.0051	1.0025
^{37}Cl-^{35}Cl						
$^{11}BCl_3$	1.0201	1.0100	1.0060	1.0039	1.0021	1.0010
$[^{11}BCl_4]^-$	1.0158	1.0074	1.0043	1.0028	1.0014	1.0007
$^{12}CCl_4$	1.0204	1.0097	1.0057	1.0037	1.0019	1.0009
$^{28}SiCl_4$	1.0178	1.0085	1.0050	1.0032	1.0017	1.0008
$^{76}GeCl_4$	1.0162	1.0076	1.0044	1.0028	1.0015	1.0007
^{76}Ge-^{70}Ge						
GeF_4	1.0744	1.0381	1.0229	1.0151	1.0080	1.0040
$Ge^{35}Cl_4$	1.0447	1.0211	1.0122	1.0079	1.0041	1.0020
$Ge^{79}Br_4$	1.0352	1.0162	1.0092	1.0059	1.0030	1.0015
$Ge^{127}I_4$	1.0271	1.0123	1.0070	1.0045	1.0023	1.0011
^{81}Br-^{79}Br						
$^{11}BBr_3$	1.0036	1.0017	1.0010	1.0007	1.0003	1.0002
$[^{11}BBr_4]^-$	1.0028	1.0013	1.0007	1.0005	1.0002	1.0001
$^{12}CBr_4$	1.0034	1.0016	1.0009	1.0006	1.0003	1.0001
$^{28}SiBr_4$	1.0030	1.0014	1.0008	1.0005	1.0003	1.0001
$^{76}GeBr_4$	1.0027	1.0012	1.0007	1.0004	1.0002	1.0001

Kotaka et al. (1978)

This study uses empirical force-field methods similar to those used by Kotaka and Kakihana (1977), with a focus on octahedral (XY_6) molecules and molecule-like complexes. The reader is cautioned that there appear to be typographical errors in the original tabulation.

	$\alpha_{XR\text{-}X}$					
	200 K	**300 K**	**400 K**	**500 K**	**700 K**	**1000K**
^{30}Si-^{28}Si						
$[SiF_6]^{2-}$	1.1432	1.0719	1.0427	1.0281	1.0147	1.0073
^{50}Ti-^{46}Ti						
$[Ti^{35}Cl_6]^{2-}$	1.0170	1.0078	1.0043	1.0028	1.0014	1.0007
$[Ti^{79}Br_6]^{2-}$	1.0194	1.0088	1.0050	1.0032	1.0016	1.0008
^{76}Ge-^{70}Ge						
$[GeF_6]^{2-}$	1.0895	1.0438	1.0256	1.0167	1.0087	1.0043
$[Ge^{35}Cl_6]^{2-}$	1.0358	1.0164	1.0093	1.0060	1.0031	1.0015
^{104}Ru-^{96}Ru						
RuF_6	1.0731	1.0369	1.0220	1.0145	1.0000	1.0067
^{130}Te-^{120}Te						
TeF_6	1.0598	1.0301	1.0179	1.0117	1.0062	1.0031
$[Te^{35}Cl_6]^{2-}$	1.0093	1.0042	1.0024	1.0015	1.0008	1.0004
$[Te^{79}Br_6]^{2-}$	1.0122	1.0055	1.0031	1.0020	1.0010	1.0005
^{180}Hf-^{174}Hf						
$[Hf^{35}Cl_6]^{2-}$	1.0031	1.0014	1.0008	1.0005	1.0003	1.0001
$[Hf^{79}Br_6]^{2-}$	1.0022	1.0001	1.0005	1.0003	1.0002	1.0001
^{192}Os-^{186}Os						
OsF_6	1.0158	1.0080	1.0047	1.0031	1.0016	1.0008
$[Os^{35}Cl_6]^{2-}$	1.0076	1.0035	1.0020	1.0013	1.0006	1.0003
^{193}Ir-^{191}Ir						
IrF_6	1.0062	1.0031	1.0019	1.0012	1.0006	1.0003
$[Ir^{35}Cl_6]^{2-}$	1.0017	1.0008	1.0004	1.0003	1.0002	1.0001
^{198}Pt-^{192}Pt						
PtF_6	1.0208	1.0106	1.0063	1.0042	1.0022	1.0011
$[PtF_6]^{2-}$	1.0092	1.0044	1.0026	1.0017	1.0009	1.0004
$[Pt^{35}Cl_6]^{2-}$	1.0049	1.0023	1.0013	1.0008	1.0004	1.0002
$[Pt^{79}Br_6]^{2-}$	1.0054	1.0024	1.0014	1.0009	1.0004	1.0002

Hanschmann (1984)

S, Cl and Si-isotope fractionations for gas-phase molecules and aqueous molecule-like complexes (using the gas-phase approximation) are calculated using semi-empirical quantum-mechanical force-field vibrational modeling. Model vibrational frequencies are not normalized to measured frequencies, so calculated fractionation factors are somewhat different from fractionations calculated using normalized or spectroscopically determined frequencies. There is no table of results in the original publication.

Bochkarev et al. (2003)

Li, Mg and Cl-isotope fractionations for gas-phase molecules and aqueous molecule-like complexes (using the gas-phase approximation) are calculated using *ab initio* vibrational modeling. The results below are calculated using Hartree-Fock quantum mechanical modeling. Model frequencies have not been normalized to spectroscopically measured frequencies, resulting in a probable overestimate of fractionation factors—compared for instance with Urey (1947). For consistency, results have been converted from the original format (lnβ) to $\alpha_{XR\text{-}X}$.

	$\alpha_{XR\text{-}X}$ **300 K**
^{7}Li-^{6}Li	
LiH	1.0252
LiF	1.0761
^{26}Mg-^{24}Mg	
MgO	1.0164
MgH	1.0041
MgF	1.0148
^{37}Cl-^{35}Cl	
$[ClO_4]^-$	1.0934
$[ClO_3]^-$	1.0637
ClO_2	1.0423
Cl_2	1.0087
HCl	1.0050
NaCl	1.0024
LiCl	1.0030

ELEMENT-SPECIFIC STUDIES

Lithium

Yamaji et al. (2001)

Fractionation factors for Li-H_2O clusters are calculated using *ab initio* vibrational models, in the gas-phase approximation. Vibrational frequencies in this system are largely unknown, and the few that have been measured are contentious. In the absence of reliable experimental constraints, Hartree-Fock model *ab initio* vibrational frequencies are normalized using a scaling factor of 0.8964. It is generally thought that aqueous lithium is coordinated to four water molecules (Rudolph et al. 1995). The authors speculate that 6-coordinate lithium in adsorbed or solid phases will have lower $^7Li/^6Li$ than coexisting aqueous Li^+.

	$\alpha_{XR\text{-}X}$ 298 K
^{7}Li-^{6}Li	
$[Li(H_2O)_4]^+$	1.064
$[Li(H_2O)_4]^+ \cdot 2H_2O$	1.072

Chromium

Schauble et al. (in press)

Fractionations (β-factors) calculated for aqueous molecule-like complexes and anhydrous crystalline Cr-metal and Cr_2O_3, using multiple empirical and *ab initio* force-field vibrational models as well as measured vibrational spectra of isotopically substituted species. Only best-estimate results are listed here. The authors predict high $^{53}Cr/^{52}Cr$ ratios in oxidized $[CrO_4]^{2-}$ relative to Cr^{3+}-bearing species like $[Cr(H_2O)_6]^{3+}$ and Cr_2O_3, and lower $^{53}Cr/^{52}Cr$ in species with Cr-N or Cr-Cl bonds than in structurally similar species with Cr-O bonds. The results also suggest a strong correlation between short bond lengths and high $^{53}Cr/^{52}Cr$ ratios at equilibrium.

	$\alpha_{XR\text{-}X}$			
	273.15 K	**298.15 K**	**373.15 K**	**573.15 K**
^{53}Cr-^{52}Cr				
$[CrCl_6]^{3-}$	1.0040	1.0034	1.0022	1.0009
$[Cr(NH_3)_6]^{3+}$	1.0061	1.0052	1.0034	1.0015
$[Cr(H_2O)_6]^{3+}$	1.0080	1.0069	1.0045	1.0020
$[CrO_4]^{2-}$	1.0156	1.0135	1.0092	1.0042
Cr-Metal	1.0041	1.0034	1.0022	1.0009
Cr_2O_3	1.0076	1.0065	1.0043	1.0019

Chlorine

Schauble et al. (2003)

Fractionations for gas-phase molecules and aqueous perchlorate (gas-phase approximation) calculated using *ab initio* force-field vibrational models normalized to measured frequencies. Fractionation factors are also calculated for crystalline chlorides using empirical force fields. Includes an indirect model of aqueous Cl^- ($\alpha_{Cl-(aq)-Cl} \approx$ 1.0021–1.0030 at 295K) based on measured NaCl–Cl^-(aq) and KCl–Cl^-(aq) fractionations (Eggenkamp et al. 1995) and the theoretically estimated α_{XR-X} for NaCl and KCl.

	α_{XR-X}									
	153 K	**173 K**	**193 K**	**233 K**	**273 K**	**298 K**	**373 K**	**473 K**	**673 K**	**1273 K**
^{37}Cl-^{35}Cl Molecules										
Cl_2	1.0227	1.0187	1.0157	1.0114	1.0087	1.0074	1.0049	1.0031	1.0016	1.0005
HCl	1.0099	1.0087	1.0077	1.0062	1.0052	1.0047	1.0036	1.0027	1.0017	1.0006
C_2Cl_4	1.0334	1.0273	1.0227	1.0164	1.0124	1.0106	1.0070	1.0045	1.0023	1.0007
C_2HCl_3	1.0310	1.0254	1.0213	1.0154	1.0117	1.0100	1.0067	1.0043	1.0022	1.0006
CCl_3F	1.0337	1.0274	1.0228	1.0163	1.0123	1.0104	1.0069	1.0044	1.0022	1.0006
CCl_4	1.0325	1.0264	1.0218	1.0156	1.0117	1.0099	1.0065	1.0041	1.0021	1.0006
$CHCl_3$	1.0294	1.0241	1.0201	1.0146	1.0110	1.0094	1.0063	1.0040	1.0020	1.0006
CH_2Cl_2	1.0267	1.0221	1.0187	1.0138	1.0106	1.0091	1.0061	1.0040	1.0020	1.0006
CH_3Cl	1.0243	1.0205	1.0175	1.0131	1.0102	1.0088	1.0060	1.0039	1.0020	1.0006

	273 K	**298 K**	**373 K**	**573 K**
Crystals				
NaCl	1.0039	1.0033	1.0021	1.0009
KCl	1.0030	1.0025	1.0016	1.0007
RbCl	1.0028	1.0023	1.0015	1.0006
$FeCl_2$	1.0067	1.0056	1.0036	1.0015
$MnCl_2$	1.0062	1.0052	1.0034	1.0014

Iron

Polyakov (1997), Polyakov and Mineev (2000)

$^{57}Fe/^{54}Fe$ fractionations (β-factors) are calculated for a range of iron-bearing minerals and crystal phases by modeling Mössbauer spectroscopic data. The ability of Mössbauer methods to characterize trace amounts of iron allows modeling of solid-solution effects and complex mineral species. The authors predict measurable fractionations between common mineral species, including a general enrichment of heavy iron isotopes in Fe^{3+} bearing minerals. Selected results from the more recent paper are reproduced here—these appear to recapitulate the earlier results. Results are tabulated in the form of a polynomial expansion vs. inverse temperature,

$$1000 \bullet \ln\alpha_{XR\text{-}R} = B_1x - B_2x^2 + B_3x^3,\ x = 10^6/T^2$$

^{57}Fe-^{54}Fe	B_1	B_2	B_3
Iron metal	0.69964	0.001476	0.000004845
Hematite Fe_2O_3	0.98684	0.002937	0.000013597
Magnetite Fe_3O_4 (A site)	1.74563	0.009190	0.000075260
(B site)	1.28250	0.004961	0.000029846
Goethite FeOOH	0.78168	0.001843	0.000006758
Akaganeite FeOOH (A site)	1.35089	0.005504	0.000034880
(B site)	0.94776	0.002709	0.000012050
Lepidocrocite FeOOH	0.72987	0.001607	0.000005501
Spinel $Mg_{0.9}Fe_{0.1}Al_2O_4$	0.45632	0.000628	0.000001344
Ilmenite $FeTiO_3$	0.37934	0.000434	0.000000772
Ferrochromite $FeCr_2O_4$	0.42987	0.000557	0.000001124
Periclase MgO:Fe	0.60395	0.001087	0.000003062
Diopside $Ca_{1.03}Mg_{0.64}Fe_{0.31}Si_{1.94}O_6$ (A site)	1.17247	0.004146	0.000022804
(B site)	0.48355	0.000705	0.000001560
Hedenbergite $CaMg_{0.15}Mn_{0.03}Fe_{0.76}Al_{0.03}Si_3O_6$	0.46984	0.000666	0.000001467
Aegirine $NaFeSi_2O_6$	1.15105	0.003996	0.000021577
Enstatite $Mg_{1.95}Fe_{0.05}Al_{0.05}Si_{1.96}O_6$ (A site)	0.49747	0.000746	0.000017420
(B site)	0.49747	0.000746	0.000017420
Enstatite $Mg_{1.65}Fe_{0.27}Al_{0.03}Si_{2.02}O_6$ (A site)	0.42987	0.000557	0.000001124
(B site)	0.66355	0.001328	0.000004134
Olivine $(Mg,Fe)_2SiO_4$	0.57000	0.000980	0.000002620
Monoclinic Chloritoid $Al_{1.98}Fe_{0.94}Mn_{0.10}SiO_5(OH)_2$	0.42987	0.000557	0.000001124
Triclinic Chloritoid $Al_{1.95}Fe_{0.75}Mn_{0.23}Mg_{0.13}SiO_5(OH)_2$	0.45632	0.000628	0.000001344
Pyrite FeS_2	1.46882	0.006506	0.000044835
Nickel Sulfide $Ni_{1-x}Fe_xS_2$	1.23789	0.004621	0.000026839
Ankerite $CaFe_{0.5}Mg_{0.5}(CO_3)_2$	0.35526	0.000381	0.000000634
Ankerite $Ca_{1.1}Fe_{0.3}Mg_{0.5}Mn_{0.1}(CO_3)_2$	0.48355	0.000705	0.000001560
Siderite $FeCO_3$	0.55510	0.000929	0.000002420

Schauble et al. (2001)

$^{56}Fe/^{54}Fe$ fractionations (β-factors) are calculated for aqueous molecule-like complexes, using empirical force-field (MUBFF) vibrational models. Materials studied include Fe-H_2O and Fe-Halide complexes with tetrahedral and octahedral geometries, cyano-complexes, and iron metal. These calculations predict measurable fractionations between aqueous species, governed by the oxidation state of iron (Fe^{3+} – high $^{56}Fe/^{54}Fe$, Fe^{2+} – low $^{56}Fe/^{54}Fe$), and covalent bonding strength of bond partners (CN^-, H_2O – high $^{56}Fe/^{54}Fe$, Cl^-, Br^- – low $^{56}Fe/^{54}Fe$). The calculations also suggested significant coordination-number effects (4-fold coordination – high $^{56}Fe/^{54}Fe$, 6-fold coordination – low $^{56}Fe/^{54}Fe$). Results for Fe-metal and ferrocyanide complexes (not shown here) generally agree with earlier Mössbauer-based models (Polyakov 1997; Polyakov and Mineev 2001).

			$\alpha_{XR\text{-}X}$		
	273.15 K	**298.15 K**	**373.15 K**	**473.15 K**	**573.15 K**
^{56}Fe-^{54}Fe					
$[Fe(H_2O)_6]^{3+}$	1.0137	1.0117	1.0076	1.0048	1.0033
$[FeCl_4]^-$	1.0085	1.0072	1.0047	1.0029	1.0020
$[FeBr_4]^-$	1.0076	1.0064	1.0041	1.0026	1.0018
$[Fe(H_2O)_6]^{2+}$	1.0073	1.0062	1.0040	1.0025	1.0017
Fe-metal	1.0063	1.0053	1.0034	1.0021	1.0015
$[FeCl_4]^{2-}$	1.0048	1.0040	1.0026	1.0016	1.0011
$[FeCl_6]^{3-}$	1.0046	1.0038	1.0025	1.0015	1.0011

Anbar et al. (in press)

$^{56}Fe/^{54}Fe$ fractionations (β-factors) are calculated for the aqueous Fe^{2+} and Fe^{3+} complexes $[Fe(H_2O)_6]^{2+}$ and $[Fe(H_2O)_6]^{3+}$, using hybrid density functional theory (DFT). *Ab initio* model frequencies are not scaled. The high electric charge of these complexes and the strong tendency of bonded water molecules to form hydrogen bonds can make gas-phase models of aquo-complexes like these inaccurate. Anbar et al. correct for these effects using an implicit model of the bulk solvent. In the implicit solvation models, the solvent (i.e., water not directly bonded to the central metal cations) is treated as a polarizable, dielectric continuum. This polarizable continuum model (PCM) improves the match between model and experimental molecular structures and vibrational frequencies, and probably yields more reliable fractionation estimates. However, there is still substantial disagreement (>100 cm^{-1}) with measured O-Fe-O bending frequencies in $[Fe(H_2O)_6]^{3+}$, possibly suggesting an error in the original spectroscopic interpretation. Calculated fractionations involving $[Fe(H_2O)_6]^{3+}$ appear to agree more closely with experiments than the MUBFF-based calculation of Schauble et al. (2001), mainly because of the different O-Fe-O vibrational frequencies used by these authors.

	$\alpha_{XR\text{-}X}$						
	273 K	**283 K**	**298 K**	**323 K**	**373 K**	**473 K**	**573 K**
^{56}Fe-^{54}Fe							
DFT (PCM solvent)							
$[Fe(H_2O)_6]^{3+}$	1.0108	1.0101	1.0092	1.0079	1.0060	1.0038	1.0026
$[Fe(H_2O)_6]^{2+}$	1.0079	1.0074	1.0067	1.0058	1.0044	1.0028	1.0019
DFT (gas phase)							
$[Fe(H_2O)_6]^{3+}$	1.0112	1.0104	1.0095	1.0082	1.0062	1.0040	1.0028
$[Fe(H_2O)_6]^{2+}$	1.0077	1.0072	1.0066	1.0057	1.0043	1.0028	1.0019

Selenium

Krouse and Thode (1962)

Fractionations are calculated for a number of gas-phase molecules and aqueous ions (treated as gases) using measured vibrational spectra and empirical force field calculations.

	$\alpha_{XR\text{-}X}$			
	273.15 K	**298.15 K**	**373.15 K**	**523.15 K**
^{82}Se-^{76}Se				
SeF_6	1.059	1.051	1.034	1.019
$[SeO_4]^{2-}$	1.044	1.038	1.023	1.014
Se_2	1.012	1.011	1.007	1.004
SeO(g)	1.009	1.008	1.005	1.003
H_2Se(g)	1.005	1.005	1.003	1.002
PbSe(g)	1.005	1.004	1.003	1.001
Se^-(g)	1.000	1.000	1.000	1.000

Reviews in Mineralogy & Geochemistry
Vol. 55, pp. 113-152, 2004
Copyright © Mineralogical Society of America

4

Analytical Methods for Non-Traditional Isotopes

Francis Albarède

Ecole Normale Supérieure de Lyon
46 Allée d'Italie, 69007 Lyon, France

Brian Beard

Department of Geology and Geophysics
University of Wisconsin
Madison, Wisconsin 53706, U.S.A.

INTRODUCTION

This chapter is devoted to the analytical methods employed for making high precision isotope ratio measurements that preserve naturally occurring mass-dependent isotopic variations. The biggest challenge in making these types of measurements is deconvolving mass-dependent isotopic fractionation produced in the laboratory and mass spectrometer, from naturally occurring mass-dependent isotopic fractionation, because the patterns of isotope variation produced by these processes are identical. Therefore, the main theme of this chapter is the description and mathematical treatment of mass-dependent isotopic variations and the possible pitfalls in deconvolving instrumental mass bias from naturally occurring mass-dependent isotopic variations. This chapter will not attempt to catalog methods for isotopic analysis of different elements. These details are better discussed in later chapters where 'element specific' analytical issues are covered. Rather, the effort in the chapter will be to focus on those specific items that make mass analysis of non-traditional isotopes challenging and unique, and the methods that can be employed to make precise and accurate isotope ratios.

INSTRUMENTATION USED FOR ISOTOPE RATIO ANALYSIS

Isotopic analysis of non-traditional isotopes is made using three main types of mass spectrometers. Elements that can easily be introduced as gases, such as Cl or Br, are typically analyzed using a gas source mass spectrometer. In contrast, metal elements are analyzed using either a thermal-ionization mass spectrometer (TIMS) or a multi-collector inductively coupled plasma mass spectrometer (MC-ICP-MS). The main difference between these three types of instruments is how the sample is introduced into the instrument, and how the sample is ionized. In contrast, the analyzer part of each instrument is similar. All three types of instruments use a stack of lenses with variable potential to focus the ion beam, a magnet to resolve the ion beam into different masses, and a series of collectors to measure ion currents of different isotopes simultaneously. The core of the mass analyzer is the magnet that is used to bend an ion beam, which has been accelerated through a potential, V, based on its mass to charge ratio (M/e). When the magnetic field, B, of the mass analyzer is varied around its initial position, the mass spectrometer scans the mass range according to the law:

1529-6466/04/0055-0004$05.00

$$\frac{M}{e} = \frac{\rho^2 B^2}{2V} \tag{1}$$

where ρ is the magnet's turning radius. Let us call N_k the number of atoms of isotope k, of an element of interest introduced into the mass spectrometer, and n_k the number of charges forming the electronic signal of the isotope. The transmission $T(M_k)$ of the isotope of mass M_k through the mass spectrometer is the ratio n_k/N_k and is observed to vary not only from element to element but also from one isotope of a given element to the next (Fig. 1).

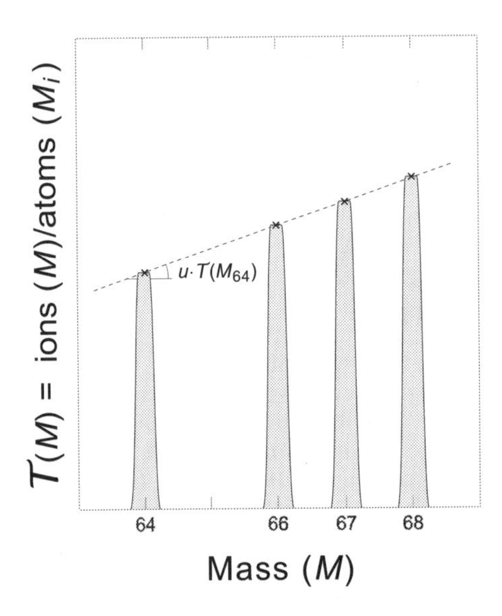

Figure 1. Transmission of Zn isotopes as a function of the collector position along the focal plane. This position is parameterized by mass M. Transmission is the ratio between the number of ions of mass M_i arriving at the collector at mass M divided by the number of atoms of mass M_i introduced in the mass spectrometer. The parameter u is the linear mass bias coefficient.

Mass bias, or the instrumental mass fractionation, is the variable transmission of the ion beam into the mass spectrometer. A variety of phenomena create conditions that lead to variable transmission of ion beams. For modern instruments, the transmission in the flight tube and the efficiency of ion conversion to electrons at the collector are almost quantitative. Most fractionation processes, therefore, take place within the source, namely in the area where the analyte is introduced into the mass spectrometer and ionized, or at the interface between the source and the mass analyzer.

Because there is so little mass bias in the mass analyzer, a discussion of ion transfer optics and collectors is not presented. The ion transfer optics of the magnetic sector mass analyzer, and the collectors used for isotope ratio measurements are critical design elements in all isotope ratio mass spectrometers and recent reviews of these items can be found in Habfast (1997) and Turner et al. (1998).

Gas source machines

As the name implies, the sample is introduced into the mass spectrometer as a gas (Nier 1940). There are two types of sources, the classic viscous flow source and the continuous flow source. The viscous flow source typically consists of two identical inlet systems that are coupled to the mass spectrometer by a change-over valve, which allows rapid switching for comparison of isotope ratios measured for sample and standard gases. In the continuous flow source, samples gas is introduced as a 'bubble' in a non-reactive carrier gas stream.

The continuous flow inlet system is typically able to work with smaller quantities of gas (e.g., Habfast 1997), as compared to the viscous flow inlet system. Ionization in gas source mass spectrometers is accomplished using an electron impact process, by which an electron stream produced by a heated filament (usually W) is shot into the gas, resulting in production of ions (e.g., Platzner 1997). Causes of mass fractionation are complex, normally entailing small differences in ion trajectories in the source, and from non-viscous flow regimes (Habfast 1997). Gonfiantini et al. (1997) discussed the effects of adsorption on mass fractionation in the mass spectrometer inlet. Mass fractionation corrections are made by comparing the isotope ratio of a sample gas to a standard gas, thus one of the most critical designs elements in gas source mass spectrometers is introducing the sample and standard gas into the instrument identically.

TIMS

Traditionally thermal ionization mass spectrometry was the instrument of choice for the isotopic analysis of metals because thermal ionization produced an ion beam with a very small kinetic energy spread (≈0.5 eV). Therefore only a magnetic mass analyzer is needed to resolve one isotope from another. Moreover, ionization of unwanted material, such as atmospheric contaminates, hydrocarbons from pump oil, or production of doubly ionized particles is almost non existent, thus background counts are minimized and signal-to-noise ratio is maximized.

Ionization in a TIMS involves using a single, double, or triple filament where the sample is loaded onto a filament, typically as a chloride or nitrate salt. Filaments are made of refractory non-reactive metal such as platinum, rhenium, tungsten or tantalum. The filament is put into the mass spectrometer under vacuum and a potential of 8 to 10 kV is applied to the filament. A current is passed through the filament causing the thin ribbon to heat, and as the temperature of the filament surpasses the vaporization temperature of the element/salt, neutral species and ions of the element are emitted from the hot surface of the filament (Langmuir and Kingdon 1925). For single filament loads, the filament acts as both the vaporization and ionization filament. In a double or triple filament source, one filament acts as an ionization filament and the other filaments are loaded with sample and act as vaporization filaments. The neutral metal atoms emitted by the ionization filament are absorbed on the surface of the vaporization filament where ionization of the analyte takes place. The separation of the ionization and vaporization filaments allows for a finer control of evaporation rate as well as allowing the ionization filament to be at an elevated temperature, which typically results in a higher ionization ratio for most elements. For example, if the element to analyze is volatile or if its ionization energy is too large, the analyte forms a cloud of neutral atoms above the filament ribbon.

In general the production of positive ions relative to neutral species can be predicted from the Saha-Langmuir equation:

$$\frac{n_+}{n_0} \propto \exp\left(e\frac{W - I_p}{\mathrm{k}T}\right) \tag{2}$$

where n_+/n_0 is the ratio of positive ions to neutral species, e the electronic charge, k is the Boltzmann constant, T is temperature in Kelvin, W is the work function of the filament in electron volts, and I_p is the first ionization potential in electron-volts. The ionization potential is a measure of the work required to remove the most loosely bound orbital electron, and the work function is the energy that must be applied to the filament to cause free electrons from the filament to escape their association with the bulk metal. For those elements with an ionization potential less than that of the work function of the filament, a low filament temperature will maximize the ratio of positive ions to neutral ions. In contrast, if the ionization potential is greater than the work function of the filament, high filament temperatures favor ionization. Typical work functions for the most common filament materials range from 4.3 to 5.1 eV.

In practice, the ionization efficiency of the element is of prime concern because it is the best measure of the quantity of material needed for an isotopic analysis. A working definition of ionization efficiency is the measurable ions that reach a collector relative to the number of atoms loaded onto the filament, and this ratio varies from a maximum of 80% for Cs to values <0.01%, depending on the element and the precision required for the isotopic analysis (e.g., DePaolo 1988). In general ionization efficiency is low for elements that have high ionization potentials and are highly volatile. However, for many elements, workers have developed specialized loading techniques to maximize ionization efficiency where, for example, sample requirements for a Pb isotope analysis were decreased to nanogram levels with the development of the silica gel and phosphoric acid technique (e.g., Cameron et al. 1969). Indeed, numerous materials and specially shaped filaments are used to enhance the ionization efficiency of the analyte (e.g., Smith et al. 1980; Dixon et al. 1993), and samples are laboriously purified using ion exchange chromatography to maximize ionization of the analyte. The materials used and design of the filament are typically done to enhance the work function of the filament, control the vaporization temperature of the analyte, or change how the analyte interacts with the filament substrate. The interaction of the analyte with the filament substrate by the formation of metal alloys or recrystallization is critical for stable ion emission. This curing time, during which the filament current is ramped slowly up to the temperature required for maximum ionization, can lengthen an individual isotopic analysis by several hours and often makes thermal ionization isotope analyses highly laborious, from the time a filament begins to be warmed up to the time the isotope analysis is completed.

During thermal ionization the light isotope is preferentially evaporated thus producing an instrumental mass-dependent fractionation where in the initial stages of the analysis the analyzed material is isotopically light relative to the true isotope composition, and at the end of the analysis, the analyzed isotope composition is isotopically heavy. An example of the change in isotope composition caused by the preferential vaporization of light relative to heavy isotopes during a thermal ionization mass spectrometry analysis of Fe is shown in Figure 2. The plotted curves show mathematical predictions of the changes in isotope composition using either a linear or exponential mass fractionation law (see section "A phenomenological description of the mass-dependent instrumental bias").

The majority of instrumental mass fractionation that takes place during TIMS occurs on or around the filaments used to produce ions. As a simple example, the ionization of Ca at the surface of a Ta filament may be represented by the formal reactions:

$$\mathrm{Ta}^+ + {}^{40}\mathrm{Ca} \rightarrow \mathrm{Ta} + {}^{40}\mathrm{Ca}^+ (k_1) \tag{3}$$

$$\mathrm{Ta}^+ + {}^{42}\mathrm{Ca} \rightarrow \mathrm{Ta} + {}^{42}\mathrm{Ca}^+ (k_2) \tag{4}$$

with the reaction rates k_1 and k_2 in parentheses. The actual reaction path is certainly more complicated but these equations serve their illustration purpose. If $k_1 \neq k_2$ then an instrumental mass fractionation is produced. Of course the actual reactions are more complicated because Ca is present as a nitrate or chloride salt and these reactions are unlikely to take place at equilibrium. Some of these complications are discussed by Habfast (1998), who developed an evaporative mass fractionation model for surface filament ionization, while Rameбäck et al. (2002) investigated the isotopic effect of diffusion in the sample film.

Typically, the relative instrumental mass-dependent fractionation is on the order of 1 to 5‰ per mass unit, but this fractionation is variable during the course of the analysis, as well as variable from filament to filament. The degree of instrumental mass bias can be minimized by use of a double or triple filament ionization source, as compared to a single filament source. In the double and triple filament source the temperature of the evaporation filament

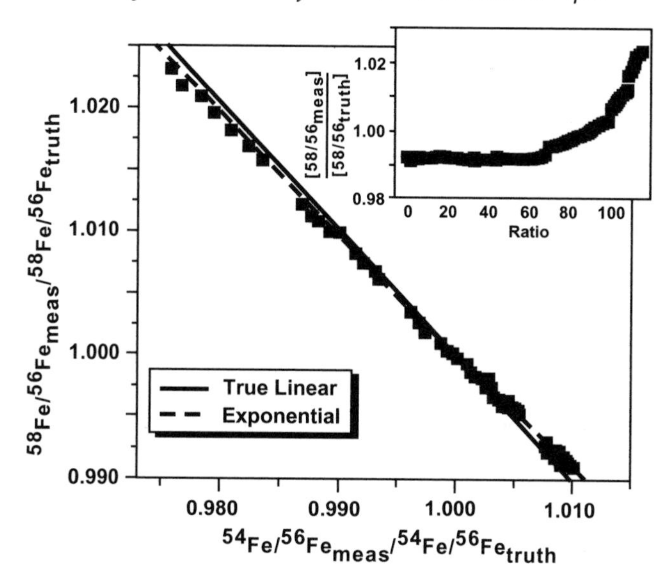

Figure 2. Plot of the measured $[^{54}Fe/^{56}Fe]_{measured}/[^{54}Fe/^{56}Fe]_{truth}$ versus the $[^{58}Fe/^{56}Fe]_{measured}/[^{58}Fe/^{56}Fe]_{truth}$ of a representative TIMS analysis of an equal atom Fe standard (e.g., $Ab^{54}Fe \approx Ab^{56}Fe \approx Ab^{57}Fe \approx Ab^{58}Fe$; Beard and Johnson 1999). Lines show the linear and exponential mass fractionation curves. The TIMS instrumental mass fractionation is best approximated using an exponential law for the first 2/3 to 3/4 of the data taken in a typical analysis. Inset shows the $[^{58}Fe/^{56}Fe]_{measured}/[^{58}Fe/^{56}Fe]_{truth}$ from the beginning to the end of the run. Modified from Beard and Johnson (1999).

can be minimized because the sample is isolated from the hot ionization filament. The low temperature of the evaporation filament minimizes the kinetic isotopic fractionation associated with vaporization. Moreover, for some elements, ionization efficiency can be increased and instrumental mass bias decreased, by using negative ions. Negative ions are typically some type of molecular species, as opposed to positive metal ions, and the increase in the overall mass of the analyte ion, can diminish mass fractionation, as shown by Walczyk (1997) who analyzed Fe as a FeF_4^- species and was able to control mass bias to 1‰ per mass as compared to the 2‰ per mass when Fe is run as a positive Fe^+ species (e.g., Dixon et al. 1993; Beard and Johnson 1999).

Corrections for instrumentally-produced mass fractionation that preserve natural mass dependent fractionation can be approached in one of two ways; a double-spike method, which allows for rigorous calculation of instrumental mass fractionation (e.g., Dodson 1963; Compston and Oversby 1969; Eugster et al. 1969; Gale 1970; Hamelin et al. 1985; Galer 1999; see section "Double-spike analysis"), or an empirical adjustment, based on comparison with isotopic analysis of standards (Dixon et al. 1993; Taylor et al. 1992; 1993). The empirical approach assumes that standards and samples fractionate to the same degree during isotopic analysis, requiring carefully controlled analysis conditions. Such approaches are commonly used for Pb isotope work. However, it is important to stress that the precision and accuracy of isotope ratios determined on unknown samples may be very difficult to evaluate because each filament load in a TIMS analysis is different.

Rigorous correction for instrumental mass bias is required if the precision of an isotope ratio measurement needs to be greater than 1‰ per mass unit. This concept is well illustrated by the definitive Ca isotope work of Russell et al. (1978), which used a double-spike approach. Prior to the Ca isotope investigation of Russell et al. (1978), natural mass-dependent Ca isotope variations were estimated to be large (12‰ per amu based on an empirical method for instrumental mass-fractionation correction; e.g., Miller et al. 1966). However, Russell et al. (1978) showed that naturally-occurring, mass-dependent variations in Ca isotopes are much smaller (total range of 0.6‰ per amu), and that the previously reported mass-dependent variations were entirely a result of imprecise corrections for instrumental mass fractionation.

MC-ICP-MS

The multi-collector inductively-coupled plasma mass spectrometer is becoming the instrument of choice for non-traditional stable isotope analysis because of the high ionization efficiency of the Ar plasma and rapid sample throughput. Moreover, because instrumental mass bias varies smoothly during the course of an analytical session, corrections for instrumental mass bias can be made by bracketing standards and by comparison to the mass bias inferred from an element of similar mass added to the analyte. However, MC-ICP-MS does have several drawbacks as compared to TIMS that include: 1) ions have a range of kinetic energies that requires using an electrostatic filter or collision cell to produce flat topped peaks for isotope ratio measurements, 2) the plasma ion source produces unwanted molecular ions from atmospheric and plasma gases as well as some doubly charged ions, which can create problems in the form of isobaric interferences, and 3) although instrumental bias is nearly constant during the course of an analysis, the overall magnitude of the mass bias is much greater than TIMS, and sample matrix effects can have a significant effect on instrumental mass bias, thus complicating assessment of the accuracy of isotope ratios.

The MC-ICP-MS consists of four main parts: 1) a sample introduction system that inlets the sample into the instrument as either a liquid (most common), gas, or solid (e.g., laser ablation), 2) an inductively coupled Ar plasma in which the sample is evaporated, vaporized, atomized, and ionized, 3) an ion transfer mechanism (the mass spectrometer interface) that separates the atmospheric pressure of the plasma from the vacuum of the analyzer, and 4) a mass analyzer that deals with the ion kinetic energy spread and produces a mass spectrum with flat topped peaks suitable for isotope ratio measurements.

Sample inlet system. The most common method for introduction of sample into a MC-ICP-MS is by aspiration of a liquid into the plasma as an aerosol. The size of the aerosol particles delivered to the plasma is critical because particles that are too large are not efficiently ionized in the plasma source. In order to assist in controlling the particle size, the aerosol is typically passed through a spray chamber which forces the largest particles to coalesce as droplets that are pumped away as waste. The aerosol delivery system is relatively inefficient, where for example, only 2% of an aspirated sample using a pneumatic nebulizer is of the correct aerosol size to be ionized in the plasma source (Platzner 1997). The various types of nebulizers and spray chambers that are used is reviewed by Montaser et al. (1998a), and the critical aspects of aerosol formation is reviewed in Montaser et al. (1998b). Sample delivery as a gas for most non traditional isotopes is not important because most metals do not easily form gasses. However, volatile hydride formation has been explored for As, Bi, Ge, Pb, Sb, Se, Sn, and Te.

Introduction of solid analytes is not common for most non-traditional isotopes but as laser ablation techniques are developed, the delivery of solid particulates to the Ar plasma will become increasingly important. The Mg isotope compositions of extra-terrestrial material were measured by Young et al. (2002a) with a 193 nm laser coupled with a MC-ICP-MS with a precision of ≈0.2‰. Using a similar laser, Jackson and Günther (2003) found that the grain size of sputtered particles is a crucial factor affecting the bias on Cu isotope analyses, while Hirata et al. (2003) emphasized the electronic settling of the amplification stage as perturbing the measurement of Cu and Fe isotopic ratios. Beyond the inherent limitations imposed by the limited counting statistics obtained on small beams, obtaining in situ laser-ablation data of the same quality as that of solution runs must await a better understanding of the fundamental instrumental mass biases.

Ar plasma. The analyte aerosol is typically introduced into the plasma using a Fassel type torch that is usually made of quartz glass and consists of three concentric tubes (e.g., Turner and Montaser 1998; Fig. 3). The outer tube typically has an Ar gas flow rate of 12–15 liter min^{-1} which is used as a cooling agent to keep the torch from melting. The intermediate tube typically has a gas flow rate of 0.5 to 1.0 $liter^{-1}$ that is used to shape, form, and stabilize the Ar plasma. The Ar gas into the outer and intermediate tube is introduced off the axis of the torch ensuring that the gas spirals along the axis of the torch (Fig. 3). The inner tube carries the aerosol using a gas flow rate of 0.4 to 1.0 $liter^{-1}$ and the gas + sample are introduced along the axis of the torch. The torch is concentrically placed in a load coil (Fig. 3) that typically is made with three turns of copper tubing (tubing is used so cooling water can be pumped through the load coil to keep it from melting). An RF power of 1 to 1.5 kW at a frequency of 27 or 40 MHz is applied to this load coil (note the choice of frequencies is driven by the fact that these are frequencies set aside for scientific instrumentation in order to avoid interferences from communication devices), which produces an intense electro magnetic field along the outside of the torch. Electrons are seeded into the torch using a tesla coil; these electrons are accelerated in the magnetic field so that some electrons have enough energy to ionize Ar gas during a collision. The electrons supplied from the ionized Ar gas lead to additional collisions, and ultimately a self-sustained plasma consisting of atomized Ar neutral species, positive Ar ions, and electrons is formed. The RF generator accelerates all the charged species of the plasma, but because of the much smaller mass of electrons, the electrons are accelerated to higher velocities. Hence, processes involving electrons dominate energy transfer into the plasma. The continuous transfer of energy from the RF generator through the load coil sustains the ICP discharge. The frequency of the RF generator may be controlled using a crystal oscillator, in which case the matching network compensates for the change in impedance of the plasma due to different matrices at constant frequency. Alternatively, if a self-oscillating generator is used, the frequency can be changed slightly to compensate for impedances changes, but much more rigorous shielding of the torch is required to reduce RF emissions which may be different from the permitted frequencies. The Ar plasma has a temperature of 8000 K at the site where ionization takes place. The ionization process includes dehydration of aqueous material producing solid particles. The solid particles

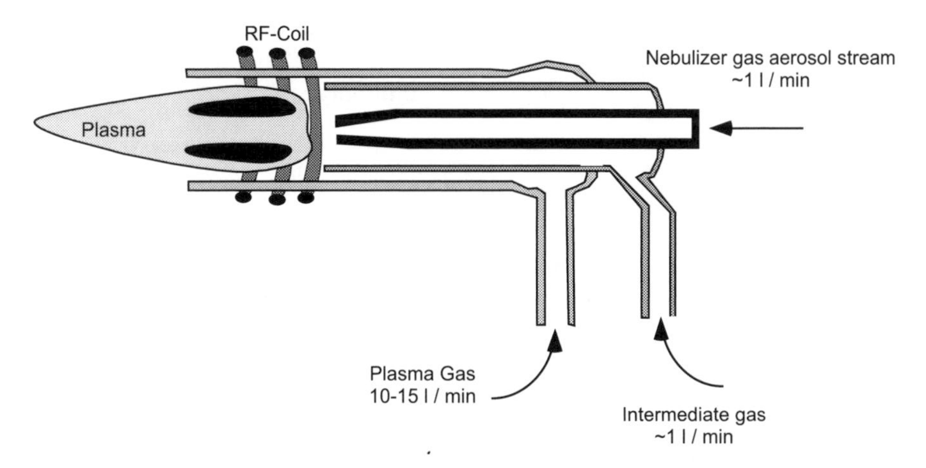

Figure 3. Fassel type torch that is typically used in MC-ICP-MS. Approximate Ar flow rates for the different plasma gasses are shown and the relative spatial relationships between the intermediate and sample lines relative to the RF-coil where the Ar plasma is generated are shown.

are vaporized to a gas, and the vaporized material is atomized. Collisions from the atomized particles with electrons from the Ar plasma produce positive ions. If dissociation of the analyte and ionization of the dissociated elements is not complete, mass dependent fractionation may be produced. In terms of ionization, the Ar plasma is a highly efficient ionizing source; elements with an ionization potential (I_n) less than 7 eV are completely ionized (Fig. 4), and high ion yields (≈90%) are produced for elements with I_n of up to 10.5 eV . Essentially, the only elements that are poorly ionized by the ICP Ar source are those that have an ionization potential greater than that of Ar (15.76 eV). Therefore, mass-dependent fractionation in the torch is unlikely to be a result of ionization but rather associated with vaporization and atomization processes in the Ar plasma.

The efficiency of Ar plasma ionization can introduce some unwanted isobars from the production of doubly ionized elements, as well as from the ionization of atmospheric gasses, the Ar plasma gasses, and formation of molecular ions such as metal oxides. Such unwanted molecular isobars can produce difficulties for some elements such as ^{40}Ca from $^{40}Ar^+$, ^{56}Fe from $^{40}Ar^{16}O^+$, or ^{80}Se from $(^{40}Ar^{40}Ar)^+$.

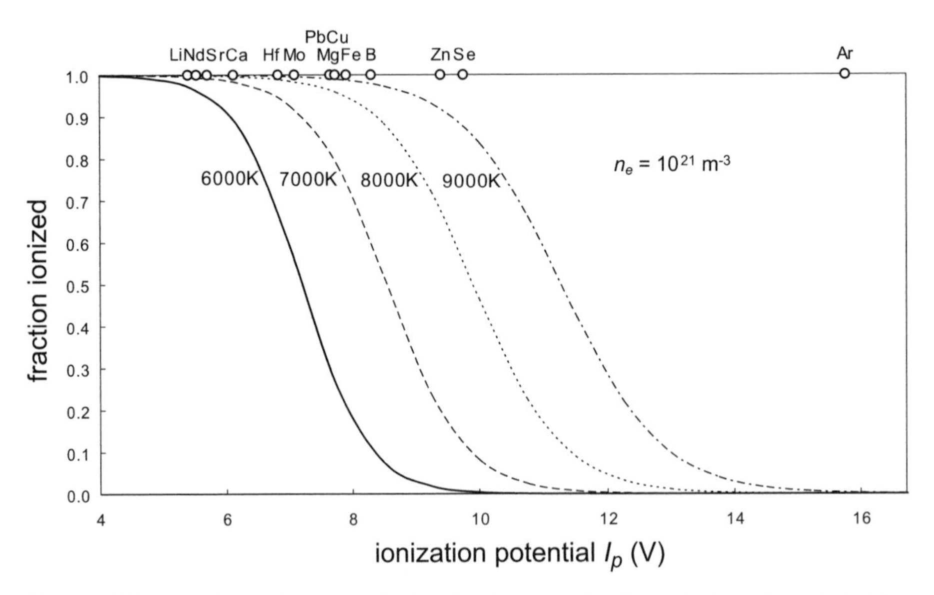

Figure 4. Efficiency of ionization as described by Saha's law as a function of the ionization potential for a representative electronic density of 10^{21} m^{-3}. The ionization potentials of some elements for which the isotopic composition is commonly analyzed are shown at the top.

Mass spectrometer interface. The plasma discharge is sampled through the MS interface which consists of two metal discs with small holes in the center through which the plasma and ions travel (Fig. 5). The pressure is dropped from atmosphere at the torch, to a pressure of 1–2 torr after the first disc (sampler cone), to a pressure of 10^{-4} torr behind the second disc (skimmer cone). The sampler cone is typically made of nickel with an orifice approximately 0.8 to 1.2 mm in diameter that the plasma travels through (Fig. 5). The sampler cone is typically kept cool by circulating cooling water through the metal to which the sampler cone is flanged. Behind the sampler cone is the skimmer cone (Fig. 5), which has an orifice with a diameter of 0.4 to 0.8 mm. Typically the skimmer cone has steeply sloped sides relative to the sampler cone and it also is usually made of Ni. The mass bias is drastically influenced by changes in the

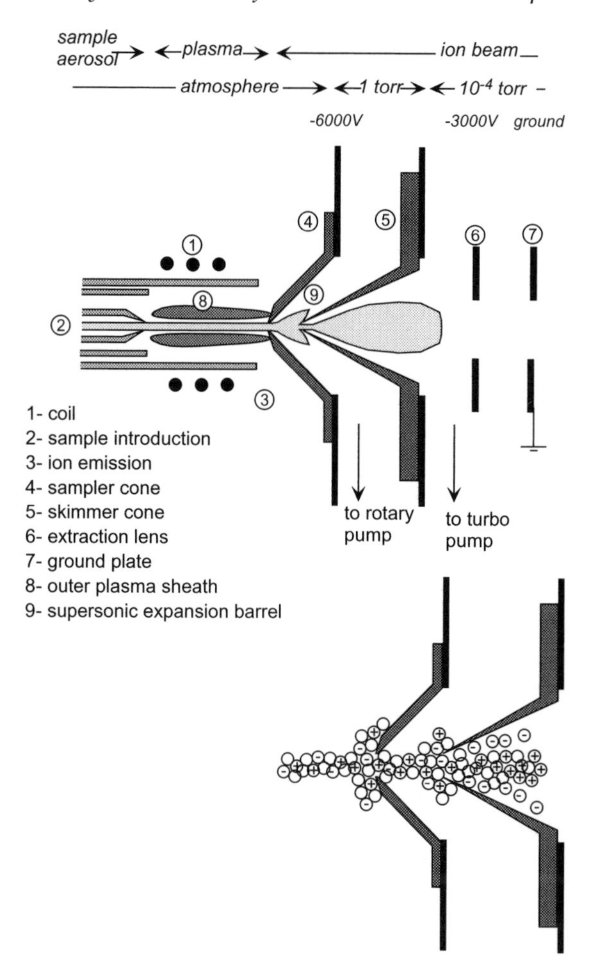

Figure 5. The inductively-coupled plasma source (inspired by Niu and Houk 1996). The original figure has been modified to show the electrical potentials, the vacuum cascade (top), and the distribution of ions and neutral (bottom) in an MC-ICP-MS similar to the VG Plasma 54. The zone with incipient voltage acceleration right behind the skimmer show maximum space-charge effect with the lighter ions being most efficiently driven off by the strong axial current of positive ions.

skimmer cone geometry (Campargue et al. 1984). The area between the sample and skimmer is referred to as the expansion chamber, and the pressure is maintained at a vacuum of 1–2 torr using a high pumping capacity rough pump such as a two-stage rotary vane pump.

The pressure difference across the sampler cone creates a barrel-shaped supersonic jet expansion of the plasma into the interface region (Fig. 5). This barrel shaped shock zone surrounds a zone of silence, which terminates in a Mach disc (e.g., Douglas and Tanner 1998). The skimmer cone is positioned such that the orifice is in the zone of silence thus achieving optimum ion extraction. It is the fixed positioning of the sampler and skimmer cone that dictates the vacuum required in the interface region (Douglass and Tanner 1998). The sensitivity of an ion signal can be increased by using a higher capacity rough pump on the expansion chamber, but this must be carefully balanced to eliminate transfer of additional atmospheric gasses into the MS interface. Based on this observation of an increase in sensitivity with higher pumping capacity it appears that some mass-dependent fractionation can be produced by preferential loss of light isotopes to the expansion chamber rough pump.

The sampler cone, in some instruments, has a positive high voltage potential applied to it such that ions are accelerated to achieve a resolving power of 400 or greater by the magnetic analyzer, which is held at a ground potential. In contrast, other instruments apply a negative high voltage potential to the mass analyzer and the sampler cone is held at ground, which serves an identical purpose; acceleration of ions through the mass analyzer.

Argon neutrals are the dominant species in the plasma and collisions with these species will cause all the ions to travel at the same velocity, approximately 2.5×10^5 cm s^{-1} through the MS interface. Therefore the kinetic energy of the ion beam will be controlled by the mass of each ion based on the relation:

$$E = \frac{1}{2}mv^2 \tag{5}$$

This corresponds to an energy spread from 0.5 to 10 eV for a mass range from 6 to 240 amu. An additional source of kinetic energy will be from the plasma potential relative to that of the sampler cone. The plasma potential is a result of capacitive coupling between the RF load coil and the plasma. During the initial development of ICP-MS high plasma potentials of 100 to 200 V resulted in arcing of the plasma to the sample cone. The arcing of the plasma to the sampler, which was referred to as a pinch discharge, resulted in significant changes in ion kinetic energy, and increases in the proportion of doubly charged ions. Plasma potential arcing problems were largely solved by grounding the load coil, thereby decreasing the plasma potential to 10 to 20 eV (Douglas and French 1986). Alternative methods of designing balanced load coils include putting opposite ends of the coil 180° out of phase from one another, or by interlacing load coils (Turner and Montaser 1998). Further reduction of the plasma potential, and hence kinetic energy spread, can be accomplished via shielding the load coil from the plasma via a screen. All of these methods of producing Ar plasmas with a minimum potential have largely decreased the plasma potential effect to approximately 5 eV (Turner and Montaser 1998). However, the shielding of the plasma from the load coil can make it difficult to ignite the plasma. A final source of variable kinetic energy in the ion beam is from thermal agitation which follows a relationship approximated by:

$$E = \frac{5}{2}\mathrm{k}T \tag{6}$$

where k is Boltzmann's constant and T is plasma gas kinetic temperature.

Maréchal et al. (1999) observed that instrumental mass bias was proportional to the square of the mass difference between isotopes of the analyte. These workers related this effect to the area of the ion-beam cross section and concluded from this analysis that the main locus for the instrumental mass bias is between the sampler cone and the grounding plate (e.g., Fig. 5). Part of this mass-bias is thought to be controlled by space charge affects which are believed to largely be a result of processes during the ion beam sampling through the skimmer cone and just behind it. The rapid gas driven jet expansion of the plasma into the MS interface preserves charge neutrality and the plasma is largely sampled, without change, through the sampler cone (e.g., Douglas and Tanner 1998). However, as the ion beam passes through the skimmer cone into the lower pressure mass analyzer, a net charge imbalance develops because electrons do not propagate very far down the negative voltage gradient (Niu and Houk 1994). The charge imbalance results in ions in the central portion of the ion beam repelling one another. Those ions with lowest kinetic energy (e.g., lowest mass to charge ratio) are more strongly repulsed than heavier ions. Importantly, in Ar plasmas, the ion current that is attributed to the ionization of Ar is high enough (≈1.5 mA; Gillson et al. 1988), so that such space charge effects will be developed regardless of analyte concentration. The net result is that heavy mass elements are more efficiently transferred relative to lighter mass elements. The preferential loss of light isotopes implies that measured isotopic compositions are heavy relative to true isotope ratio measurements. Overall, the sensitivity of a MC-ICP-MS can be described by a logarithmic function

$$S = \mathrm{a} \ln A + \mathrm{c} \tag{7}$$

where S is sensitivity (ion count rate normalized to analyte concentration), a and c are constants, and A is element mass number (e.g., Al-Ammar and Barnes 2001). Scatter about this curve is largely a reflection of ionization potentials. In general, the slope of this sensitivity response curve is relatively flat for high mass elements (e.g., U) to approximately a mass of Cu. At masses less than that of Cu, the sensitivity curve has significant curvature and a steep slope. The result of such mass dependent controls on sensitivity and mass bias, is that one cannot use the mass bias correction factor inferred for one element and directly apply this correction factor to another element (Maréchal et al. 1999; Al-Ammar and Barnes; 2001; Nonose and Kubota 2001).

Space charge effects are believed to largely develop as ions are transferred across the skimmer cone and into the low pressure of the analyzer. Experiments, using an electron gun located just behind the skimmer cone in which electrons are injected into the ion beam strongly diminish space charge effects, presumably because charge neutrality in the ion beam is maintained by addition of electrons (e.g., Parphairakist and Houk 2000a,b). Space charge effects have also been diminished by using a three chamber mass spectrometry interface in which the third chamber skims the ion current reducing overall total ion transmission and aids in preserving charge neutrality (Tanner et al. 1994). The reduction in transmission decreases space charge effects so overall sensitivity of the 3 chamber interface is similar to that of traditional two chamber interfaces. Space charge effects are very important for non-traditional isotope analysis, because they exert a strong control over instrumental mass bias. Moreover, because many non-traditional isotopes under investigation are relatively light in mass (e.g., lighter or similar in mass to Cu), space charge effects can be severe because of the strong curvature of the sensitivity response curve at low mass (e.g., Eqn. 7). In addition, space charge effects appear to dominate matrix effects (see section "Matrix effect").

Analyzer. Behind the skimmer cone there is series of ion lenses used to extract the ions from the skimmer, and to shape and steer the ion beam through the flight tube. The steering lens stack also provides a pressure differential where the pressure on the analyzer side of the final lens is lower (≈10^{-8} torr). Maintaining high-vacuum conditions in the flight tube is critical for achieving high abundance sensitivity (see section "Abundance sensitivity").

The mass analyzer of the MC-ICP-MS differs from that of gas source mass spectrometers and TIMS, because the Ar plasma ionization results in an isotope having a range of kinetic energies, typically on the order of 2–5 eV as measured at full width half maximum (Douglas and Tanner 1998). The range of ion energies is a result of thermal energy spread of ions (e.g., Eqn. 1) and from the plasma potential discharge. In order to achieve a high mass resolution with ion beams that have flat topped peaks, the MC-ICP-MS mass analyzer must control this kinetic ion energy range. Many MC-ICP-MS instruments use a double focusing instrument with a first-stage electrostatic field and a second-stage magnetic sector whose combination improves ion focusing at the collector, thus producing flat topped peaks suitable for isotopic ratio analysis (e.g., Walder 1997). Alternatively, a single focusing instrument that uses an RF-only collision cell with a collision gas, such as Ar or He, to produce ion beams with ≈1 eV energy spread has been utilized (e.g., Douglas and French 1992; Turner et al. 1998).

In addition to thermalizing the ion beam, reactive gasses can be added to the collision cell to eliminate or greatly reduce unwanted molecular species. For example, addition of H_2 to the collision cell can greatly diminish Argide molecular isobars that are isobaric at, for example ^{40}Ar on ^{40}Ca, $^{40}Ar^{16}O$ at ^{56}Fe, or $^{40}Ar_2^+$ at ^{80}Se (Turner et al. 1998).

A PHENOMENOLOGICAL DESCRIPTION OF THE MASS-DEPENDENT INSTRUMENTAL BIAS

This topic has been covered before in a number of papers (Hofmann 1971; Russell et al 1978; Hart and Zindler 1989; Habfast 1998; Maréchal et al. 1999; Beard and Johnson 1999; Kehm et al. 2003; Albarède et al. 2004). Regardless of the physical process responsible for a mass-dependent transmission in mass spectrometers, the analytical mass bias can be described using a general phenomenological theory which will be quickly reviewed here. Assuming that transmission is a function of mass, it will be expanded to the first order in the vicinity of a reference mass. Different mass fractionation laws are obtained by expanding different functions of the transmission with respect to different functions of the mass. An important property of a mass fractionation law is whether it is consistent or not: for instance, if the ratios $^{57}Fe/^{54}Fe$ and $^{56}Fe/^{54}Fe$ are fractionated according to a particular fractionation law, this law will be consistent if the ratio $^{57}Fe/^{56}Fe$ is fractionated according to the same law.

The *linear law*. This law (Fig. 1) is probably the most intuitive of all the mass-fractionation laws. Let us expand the transmission $\mathcal{T}(M)$ of isotopic beams at mass M as a function of the mass difference with a reference mass M_k:

$$\mathcal{T}(M) = \mathcal{T}(M_k) + \frac{\partial \mathcal{T}(M_k)}{\partial M}(M - M_k) + \mathcal{O}(M - M_k)^2 \tag{8}$$

where $\mathcal{O}$ stands for "of the order of." Let us evaluate this expression for $M = M_i = M_k + \Delta M$ and divide the results by by $\mathcal{T}(M_i)$:

$$\frac{\mathcal{T}(M_i)}{\mathcal{T}(M_k)} = 1 + \frac{\partial \ln \mathcal{T}(M_k)}{\partial M}(M_i - M_k) + \mathcal{O}(M_i - M_k)^2 \tag{9}$$

Let us now call u the derivative on the right-hand side evaluated at $M = M_k$ and consider the first order term only. The expression:

$$\frac{\mathcal{T}(M_i)}{\mathcal{T}(M_k)} \approx \frac{n_i/n_k}{N_i/N_k} \approx \frac{r_i}{R_i} \approx 1 + u(M_i - M_k) \tag{10}$$

gives the common form of the linear mass fractionation law. In this expression, r_i and R_i stand for the measured and true isotopic ratio, respectively, with the implicit convention that the ratios have the reference isotope k at the denominator. The parameter u is referred to as the linear mass bias. A typical mass bias for MC-ICP-MS may vary from a few percent per amu at low mass (>10% for Li and B) to one percent per amu for high masses, such as Hg. This bias is typically an order of magnitude larger than for the characteristic bias incurred in TIMS measurements. Dividing this equation by the same equation for a different isotopic ratio $R_j = N_j/N_k$ of the same element, we obtain:

$$\frac{r_i - R_i}{r_j - R_j} = \frac{R_i}{R_j} \frac{M_i - M_k}{M_j - M_k} \tag{11}$$

This expression shows that the replicates of a sample which fractionates according to the linear law define a linear alignment with slope given by Equation (11). It also explains why the isotopic dispersion created by natural magmatic, metamorphic, sedimentary, and biogeochemical processes does not create an alignment different from the alignment created by the analytical mass bias: although mass discrimination varies from one process to another, u does not enter the expression for the slope of the alignment when two isotopic ratios are plotted against each other. By introducing the conventional notation δ defined as

$$\delta_i = 1000\left(\frac{r_i}{R_i} - 1\right) \tag{12}$$

we obtain the useful equation:

$$\frac{\delta_i}{\delta_j} = \frac{M_i - M_k}{M_j - M_k} \tag{13}$$

which describes the linear relationship between the isotopic ratios in the delta notation predicted by the linear mass fractionation law. In Figure 6, we show the $^{66}Zn/^{64}Zn$ and $^{68}Zn/^{64}Zn$ ratios plotted against each other in the delta notation and observe that the line of slope two goes through all the analytical points. Assuming that all the Zn in the solar system has evolved from a single stock with homogeneous isotopic composition, all the terrestrial and planetary samples should plot on the same $\delta^{68}Zn$ vs. $\delta^{66}Zn$ fractionation line, regardless of whether these ratios have been adequately corrected for instrumental fractionation or not. This feature is well-known for oxygen isotopes, and the unique alignment formed by the $\delta^{17}O$ vs. $\delta^{18}O$ values of all terrestrial and lunar samples (Clayton 1976) is a strong case in favor of the impact origin of the Moon.

Unfortunately, the linear law is not consistent: if two ratios fractionate according to the linear law, the ratio of these ratios does not. For this reason, two additional mass fractionation laws are usually considered, the power law and the exponential law. It has been shown by Maréchal et al. (1999) that these two laws are two members of a broad family of laws, which they refer to as the generalized power law. The mass fractionation law predicted from quantum mechanics for isotopic equilibrium between vibrating molecules (Bigeleisen and Mayer 1947;

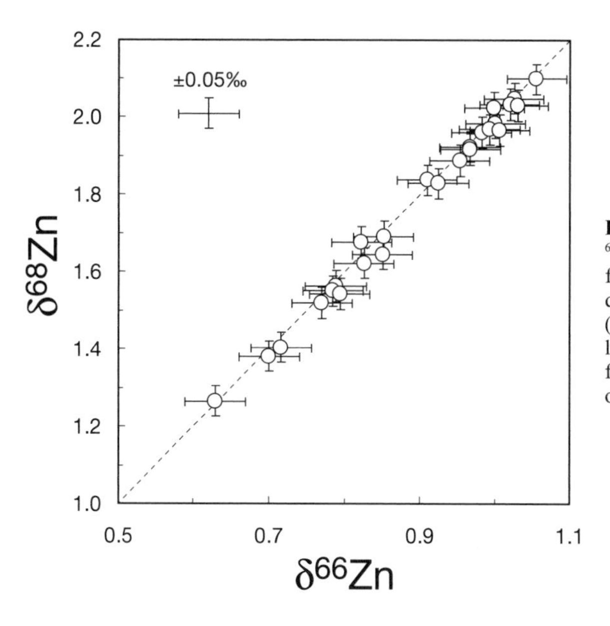

Figure 6. Plot of the $^{66}Zn/^{64}Zn$ and $^{68}Zn/^{64}Zn$ ratios in delta notation for ferromanganese nodules from different localities in the ocean (Maréchal et al. 2000). The dashed line represents the linear mass fractionation line going through the origin and has a slope of two.

Urey 1947; Criss 1999; Young et al. 2002b) is also a particular form of the generalized power law. It is demonstrated in Albarède et al. (2004) that the law used by Kehm et al. (2003) is an alternative form of the generalized power law.

The *power law.* Let us expand the logarithm of the transmission $\mathcal{T}(M)$ of isotopic beams at mass M as a function of the mass difference with a reference mass M_k:

$$\ln \mathcal{T}(M_k + \Delta M) = \ln \mathcal{T}(M_k) + \frac{\partial \ln \mathcal{T}(M_k)}{\partial M}\Delta M + \mathcal{O}\left(\Delta^2 M\right) \tag{14}$$

To the first order, the mass bias on the isotopic ratio N_i/N_k can be evaluated from:

$$\ln \mathcal{T}(M_i) - \ln \mathcal{T}(M_k) = \ln \frac{n_i/n_k}{N_i/N_k} \approx u(M_i - M_k) \tag{15}$$

We define the mass bias factor $g = e^u$ and finally obtain the expression for the so-called mass-fractionation power-law as:

$$r_i \approx R_i g^{M_i - M_k} \tag{16}$$

The *exponential law.* Let us now expand the logarithm of the transmission $\mathcal{T}(M)$ of isotopic beams at mass M as a function of $\ln M$:

$$\ln \mathcal{T}(M_i) = \ln \mathcal{T}(M_k) + \frac{\partial \ln \mathcal{T}(M_k)}{\partial \ln M}(\ln M_i - \ln M_k) + \mathcal{O}\left(\left[\ln M_i - \ln M_k\right]^2\right) \tag{17}$$

To the first order, the mass bias on the isotopic ratio N_i/N_k is evaluated from:

$$\ln \mathcal{T}(M_i) - \ln \mathcal{T}(M_k) = \ln \frac{n_i/n_k}{N_i/N_k} \approx \frac{\partial \ln \mathcal{T}(M_k)}{\partial \ln M} \ln \frac{M_i}{M_k} \tag{18}$$

The mass bias factor β is defined as:

$$\beta = \frac{\partial \ln \mathcal{T}(M_k)}{\partial \ln M} = M_k u \tag{19}$$

where u is again the linear mass bias factor, from which we get the mass-fractionation exponential as:

$$r_i \approx R_i \left(\frac{M_i}{M_k}\right)^{\beta} \tag{20}$$

For thermal ionization mass spectrometry the exponential law appears to be the best model that describes the mass-dependent fractionation taking place in the TIMS source for a wide variety of elements (e.g., Russel et al. 1978; Wasserburg et al. 1981; Hart and Zindler 1989; Beard and Johnson 1999).

Both the power law and the exponential law are consistent. They also share a second remarkable property: linear alignments are obtained in log-log plots in which one measured isotopic ratio is plotted against another. Taking the logarithms of Equation (16) and dividing this equation by the same equation for a different isotopic ratio R_j of the same element, we obtain for the power law:

$$\frac{\ln r_i - \ln R_i}{\ln r_j - \ln R_j} = \frac{M_i - M_k}{M_j - M_k} \tag{21}$$

while dividing Equation (20) by its equivalent for R_j gives

$$\frac{\ln r_i - \ln R_i}{\ln r_j - \ln R_j} = \frac{\ln M_i/M_k}{\ln M_j/M_k} \tag{22}$$

The right hand-side of these two equations is simply the slope $s_{j/k}^{i/k}$ of the isotopic $\ln(r_i)$ vs. $\ln(r_j)$ array for a same sample fractionated to different extents. For instance, we obtain the expression ofthe slope $s_{56/54}^{57/54}$ of the alignment of Fe isotopic ratios of a same sample in the plot of log $^{57}Fe/^{54}Fe$ vs. log $^{56}Fe/^{54}Fe$ as:

$$s_{56/54}^{57/54} = \frac{\ln\left(^{57}\mathrm{Fe}/^{54}\mathrm{Fe}\right)_2 - \ln\left(^{57}\mathrm{Fe}/^{54}\mathrm{Fe}\right)_1}{\ln\left(^{56}\mathrm{Fe}/^{54}\mathrm{Fe}\right)_2 - \ln\left(^{56}\mathrm{Fe}/^{54}\mathrm{Fe}\right)_1} = \frac{M_{^{57}\mathrm{Fe}} - M_{^{54}\mathrm{Fe}}}{M_{^{56}\mathrm{Fe}} - M_{^{54}\mathrm{Fe}}} = 1.5014 \tag{23}$$

for the power law and:

$$s_{56/54}^{57/54} = \frac{\ln\left(^{57}\mathrm{Fe}/^{54}\mathrm{Fe}\right)_2 - \ln\left(^{57}\mathrm{Fe}/^{54}\mathrm{Fe}\right)_1}{\ln\left(^{56}\mathrm{Fe}/^{54}\mathrm{Fe}\right)_2 - \ln\left(^{56}\mathrm{Fe}/^{54}\mathrm{Fe}\right)_1} = \frac{\ln\left(M_{^{57}\mathrm{Fe}}/M_{^{54}\mathrm{Fe}}\right)}{\ln\left(M_{^{56}\mathrm{Fe}}/M_{^{54}\mathrm{Fe}}\right)} = 1.4881 \tag{24}$$

for the exponential law. The slopes predicted by the two laws are very similar yet different enough to affect the corrected isotopic ratios by several tenths of a per mil.

For mixed solutions of two elements, such as Zn and Cu (Maréchal et al. 1999), Fe and Cu (Kehm et al. 2003; Roe et al. 2003) or Mo and Zr (Anbar et al. 2001; Barling et al. 2001), this linear property still holds provided their fractionation factors, either g or β, vary smoothly with respect to each other such as, for example, proportional $\ln(g^{Cu})$ and $\ln(g^{Zn})$ for the power law and proportional β_{Cu} and β_{Zn} for the exponential law (Fig. 7). For Cu and Zn, we can write the slope of the mass-fractionation line in a $\ln(^{65}Cu/^{63}Cu)$ vs. $\ln(^{66}Zn/^{64}Zn)$ diagram as:

$$s_{66/64}^{65/63} = \frac{\ln\left(^{65}\mathrm{Cu}/^{63}\mathrm{Cu}\right)_2 - \ln\left(^{65}\mathrm{Cu}/^{63}\mathrm{Cu}\right)_1}{\ln\left(^{66}\mathrm{Zn}/^{64}\mathrm{Zn}\right)_2 - \ln\left(^{66}\mathrm{Zn}/^{64}\mathrm{Zn}\right)_1} = \frac{\ln g^{\mathrm{Cu}}}{\ln g^{\mathrm{Zn}}} \frac{M_{^{65}\mathrm{Cu}} - M_{^{63}\mathrm{Cu}}}{M_{^{66}\mathrm{Zn}} - M_{^{64}\mathrm{Zn}}} \tag{25}$$

for the power law and:

$$s_{66/64}^{65/63} = \frac{\ln\left(^{65}\mathrm{Cu}/^{63}\mathrm{Cu}\right)_2 - \ln\left(^{65}\mathrm{Cu}/^{63}\mathrm{Cu}\right)_1}{\ln\left(^{66}\mathrm{Zn}/^{64}\mathrm{Zn}\right)_2 - \ln\left(^{66}\mathrm{Zn}/^{64}\mathrm{Zn}\right)_1} = \frac{\beta^{\mathrm{Cu}}}{\beta^{\mathrm{Zn}}} \frac{\ln\left(M_{^{65}\mathrm{Cu}}/M_{^{63}\mathrm{Cu}}\right)}{\ln\left(M_{^{66}\mathrm{Zn}}/M_{^{64}\mathrm{Zn}}\right)} \tag{26}$$

for the exponential law.

From a series of Zn isotope measurements, Maréchal et al. (1999) deduced in this way that the patterns of the Zn mass bias on the Lyon Plasma 54 are best accounted for by the exponential law.

Second-order correction can be implemented in a similar way. Let us illustrate a simple method for the exponential law. Retaining the second-order term of Equations (17) and (19), we obtain:

$$\ln \mathcal{T}(M_i) = \ln \mathcal{T}(M_k) + \beta' \ln \frac{M_i}{M_k} + \frac{\gamma}{2} \ln^2 \frac{M_i}{M_k} + \mathcal{O}\left(\ln^3 \frac{M_i}{M_k}\right) \tag{27}$$

where β' and γ are the first and second-order coefficients, respectively, of the $\ln \mathcal{T}$ expansion

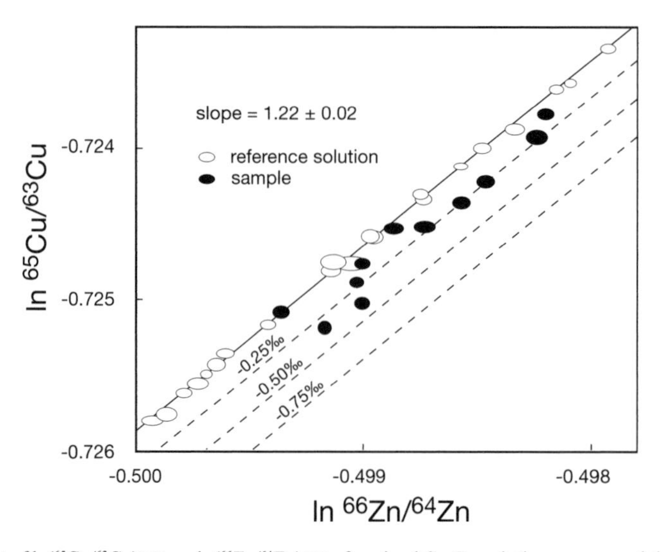

Figure 7. Plot of $\ln(^{65}Cu/^{63}Cu)^{meas}$ vs. $\ln(^{66}Zn/^{64}Zn)^{meas}$ of a mixed Cu-Zn solution run several times between samples during the same day (open 2σ error ellipses). The solid ellipses represent Cu_2S samples run in between the samples. Data acquired on the Micromass Isoprobe of University of Arizona (courtesy Steve Young). The linear relation holds even when the elements appearing on each axis are different. Although the slope $s^{65/63}_{66/64}$ of the alignment (1.2) is close to the value expected from the mass relationship (1.0) of the exponential law, assuming equal β_{Cu} and β_{Zn} and would result in errors of several tenths of a per mil. The figures on the dashed lines represent the difference in $\delta^{65}Cu$ values of the samples with respect to the standard solution.

with respect to M in the neighborhood of M_k. From the way it is evaluated, the first-order coefficient β' of the second-order expansion is different from the equivalent coefficient β in the first-order expansion. In terms of ratios, this equation can be simplified as:

$$\ln\frac{r_i}{R_i} \approx \beta'\mu_i + \gamma\frac{\mu_i^2}{2} \tag{28}$$

where u_i stands for $\ln(M_i/M_k)$. The correction formula is

$$R_i = r_i\left(\frac{M_i}{M_k}\right)^{-\beta'-\frac{1}{2}\gamma\mu_i} \tag{29}$$

Two normalization ratios R_i and R_j are required to solve this equation in β and γ and the solution is:

$$\begin{aligned} \beta' &= \frac{\mu_j}{\mu_i(\mu_j-\mu_i)}\ln\frac{r_i}{R_i} - \frac{\mu_i}{\mu_j(\mu_j-\mu_i)}\ln\frac{r_j}{R_j} = \frac{\mu_j\beta_i-\mu_i\beta_j}{\mu_j-\mu_i} \\ \gamma &= -\frac{2}{\mu_i(\mu_j-\mu_i)}\ln\frac{r_i}{R_i} + \frac{2}{\mu_j(\mu_j-\mu_i)}\ln\frac{r_j}{R_j} = 2\frac{\beta_j-\beta_i}{\mu_j-\mu_i} \end{aligned} \tag{30}$$

where β_i and β_j are the coefficients calculated with the standard first-order exponential law from the ratios r_i and r_j. The last expression for γ indicates that a second-order correction is appropriate whenever the first order β_i values increase smoothly with the mass difference.

Diagnostic of second-order effects is a residual correlation between isotopic ratios corrected for first-order fractionation (Albarède et al. 2004).

STANDARD BRACKETING METHODS

Standard bracketing consists in interpolating the mass bias of an unknown sample between the biases inferred from two standard runs, one preceding and one following the sample analysis (less stringent orders of interpolation are also used). Let us divide Equation (20) for the sample by the same equation for a standard (1) run just before the sample. The mass discrimination law used is the exponential law but the principle is easily adapted to the power law. We get:

$$\frac{(R_i)_{\text{sple}}}{(R_i)_{\text{std}}} = \frac{(r_i)_{\text{sple}}}{(r_i)^1_{\text{std}}}\left(\frac{M_i}{M_k}\right)^{\beta^1_{\text{std}}-\beta_{\text{sple}}} \tag{31}$$

Let us then write the same equation for a standard (2) run just after the sample and now assume that

$$\beta_{\text{sple}} = \frac{\beta^1_{\text{std}} + \beta^2_{\text{std}}}{2} \tag{32}$$

We now get

$$\frac{(R_i)_{\text{sple}}}{(R_i)_{\text{std}}} = \frac{(r_i)_{\text{sple}}}{(r_i)^1_{\text{std}}}\left(\frac{M_i}{M_k}\right)^{\beta^1_{\text{std}}-\beta^2_{\text{std}}} \tag{33}$$

$$\frac{(R_i)_{\text{sple}}}{(R_i)_{\text{std}}} = \frac{(r_i)_{\text{sple}}}{(r_i)^2_{\text{std}}}\left(\frac{M_i}{M_k}\right)^{\beta^2_{\text{std}}-\beta^1_{\text{std}}} \tag{34}$$

Multiplying the last two equations by each other, we obtain:

$$(R_i)_{\text{sple}} = (R_i)_{\text{std}}\frac{(r_i)_{\text{sple}}}{\sqrt{(r_i)^1_{\text{std}}\times(r_i)^2_{\text{std}}}} \tag{35}$$

This interpolation scheme is the most general and amounts to linearly interpolating the logarithms of the isotopic ratios. This procedure is easily adapted to procedures in which more than one sample is run between the bracketing standard solutions.

Equation (32) shows that mass fractionation by simple interpolation between standards, i.e., without internal or external isotopic normalization, is only correct when mass fractionation changes smoothly upon alternating between standards and samples, which requires extremely strict sample purification. A heavy sample matrix will likely result in β_{sple} falling outside the range $\beta^1_{\text{std}} - \beta^2_{\text{std}}$ and thus nullify the basic assumption of the method.

Standard bracketing is not applicable to TIMS measurements since each sample is on a different filament, which is successively positioned by a rotating turret: the mass bias depends on the temperature of the sample film loaded on the filament, as well as on its structure and composition. The run conditions for each sample are therefore variable from one sample to the next and the mass bias cannot be accurately controlled. Electron impact and ICP sources, in contrast, work at steady state: a well-purified sample and a reference can be considered as chemically equivalent while the conditions under which the mass spectrometer is operated may be kept constant. The method is by far the most common method to correct the experimental bias during the measurement of the isotopic compositions of H, N, O, C, S, etc. using electron impact ion sources but is also becoming a method of choice for the composition of transition elements measured by MC-ICP-MS (Zhu et al. 2002; Beard et al. 2003) (Fig. 8). The most visible difference between samples is the peak intensity but essentially identical signals can be achieved by changing the introduction volume for electron impact sources and dilution for MC-ICP-MS.

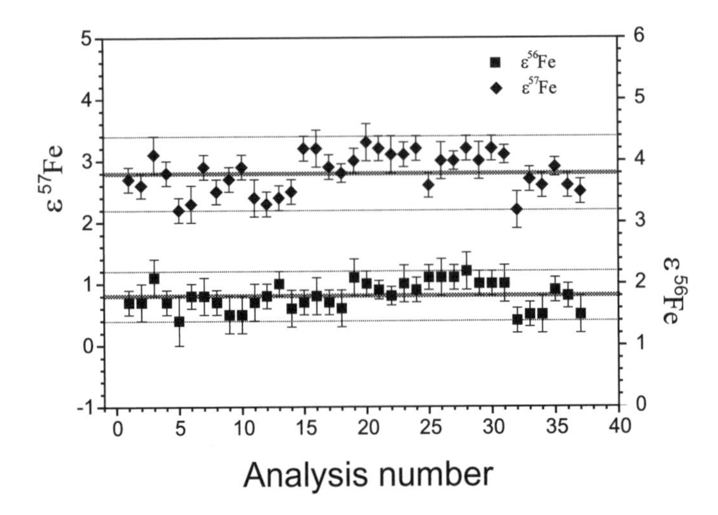

Figure 8. Deviations of the ^{56}Fe/^{54}Fe ratios (right axis) and ^{57}Fe/^{54}Fe (left axis) in parts per 10,000 of the IRMM-14 Fe standard with respect to an in-house reference solution for a series of 37 runs determined by the standard bracketing technique (Zhu et al. 2003). A precision of 0.04‰ and 0.06‰ is obtained on each ratio, respectively.

DOUBLE-SPIKE ANALYSIS

The double-spike technique was first proposed by Dodson (1963) as a way of evaluating the instrumental mass bias on isotope measurements. Many aspects of this technique have been discussed over the past three decades and the abundant literature on the subject has been adequately reviewed by Galer (1999) and Johnson and Beard (1999). Applications of double-spike analysis to Fe (Johnson and Beard 1999) and Mo (Siebert et al. 2001) isotopes have been presented. This method is ill-named, however, because the principle of double spike works for triple or multiple spikes as well (e.g., Galer 1999). As explained below, the technique can only be applied to elements with four isotopes and its geometry is simple (Fig. 9). It is based on the property that the mass bias line going through a given isotope composition has a direction that varies with the value of this composition. The straight-lines connecting the spike composition to all possible fractionated mixtures with the same sample/spike ratio form a surface, and the mixing line of the sample-spike mixture is a unique line for which there is no mass bias. The intersection of this surface with the fractionation line of the pure sample represents the unfractionated composition of the sample.

Given an element with four isotopes, we select one of the masses as the reference (ref). Let us call $R_i = N_i/N_{\text{ref}}$ the natural ratio of the isotope i to the reference isotope and r_i the ratio

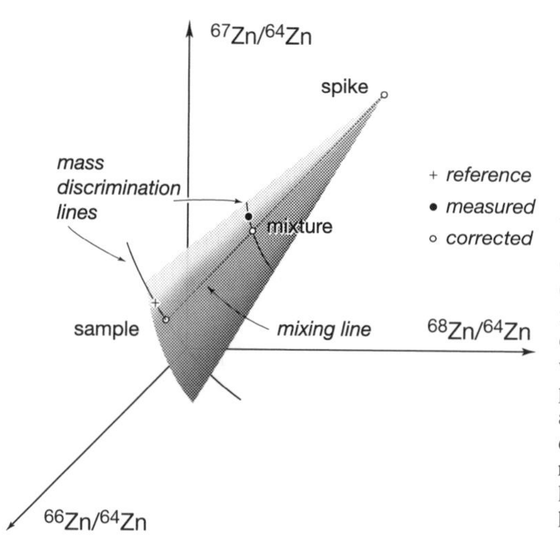

Figure 9. Sketch of the double spike ^{67}Zn-^{68}Zn method. The surface is constructed by drawing an infinite number of straight-lines through the point representing the spike composition (supposed to be known with no error) and each point of the mass fractionation line going through the point representing the measured mixture. One of these straight-lines, which is to be determined from the calculations, is the sample-spike mixing line (stippled line). Each determination of the Zn isotope composition of a sample involves only one run for the mixture of the sample with the spike. Since all natural samples plot on the same mass fractionation line, any reference composition will adequately determine isotope composition of the sample. note that, since the instrumental bias is not linear with mass, the mass discrimination lines are curved.

fractionated by the measurement. For illustration purposes, we assume that the mass bias follows an exponential law. Let us call $\varphi_{\text{ref}}^{\text{sp}}$ the proportion of the reference isotope contributed by the spike to the mixture, e.g., $^{64}\text{Zn}_{\text{sp}}/^{64}\text{Zn}_{\text{mix}}$ for the triplet of ^{66}Zn/^{64}Zn, ^{67}Zn/^{64}Zn, and ^{68}Zn/^{64}Zn ratios. Mass conservation requires

$$\varphi_{\text{ref}}^{\text{sp}} R_i^{\text{sp}} + \left(1-\varphi_{\text{ref}}^{\text{sp}}\right) R_i^{\text{sple}} = R_i^{\text{mix}} \tag{36}$$

Using the properties of the exponential law, this equation can be recast as:

$$\mathcal{F}_i\left(\varphi_{\text{ref}}^{\text{sp}}, \beta_{\text{sple}}, \beta_{\text{mix}}\right) = \varphi_{\text{ref}}^{\text{sp}} R_i^{\text{sp}} + \left(1-\varphi_{\text{ref}}^{\text{sp}}\right) r_i^{\text{sple}} \left(\frac{M_{\text{ref}}}{M_i}\right)^{\beta_{sple}} - r_i^{mix}\left(\frac{M_{\text{ref}}}{M_i}\right)^{\beta_{mix}} = 0 \tag{37}$$

This equation has three unknowns, $\varphi_{\text{ref}}^{\text{sp}}$, β_{sple}, and β_{mix}, and therefore three independent equations. Hence three isotope ratios of four isotopes are needed to solve for these unknowns. The system can be optimally solved by quadratically-converging Newton-Raphson steps, which involve the derivatives of $\mathcal{F}_i$ with respect to each unknown (Albarède 1995). For an element with one or more radiogenic isotopes such as Pb, the isotopic compositions of both the unspiked sample and the spiked mixture need to be measured. The example of Zn is developed in the Appendix. This scheme also provides excellent error propagation for double and multiple spike experiments. Alternatively, manual adjustment of the parameters on a spreadsheet also leads to the solution at the expense of rapidity, precision, and error propagation. A theory of error in double spiking has been formulated by a number of authors (Hamelin et al. 1985; Powell et al. 1998), but a detailed analysis involving the optimization of the spike composition for any element may be tailored from the model of the analysis of Pb isotopes by Galer (1999). In order to show how the choice of an optimum isotope composition of the spike conditions the precision on the isotope composition of the sample, we can simply evaluate the derivative of R_i^{sple} with respect to R_i^{sp}, which is

$$\frac{dR_i^{\text{sple}}}{dR_i^{\text{sp}}} = -\frac{\varphi_{\text{ref}}^{\text{sp}}}{1-\varphi_{\text{ref}}^{\text{sp}}} \tag{38}$$

In order to minimize the errors on the sample isotope composition, the variable $\varphi_{\text{ref}}^{\text{sp}}$ should be kept as small as possible. One consequence, previously arrived at by Galer (1999) in his numerical experiments, is that the reference isotope should not be chosen as a minor isotope when this isotope is abundant in the spike.

In addition to the rigorous mass dependent correction that can be made using the double-spike technique, it is also possible to use this technique to correct for mass-dependent fractionation associated with chemical purification of an analyte. Chemical purification, typically by ion exchange chromatography, is critical for both TIMS and MC-ICP-MS to eliminate spectral and non-spectral interferences (see section "Matrix effect"). However, if yields are not quantitative during purification a mass-dependent fractionation can be produced. If the double-spike is added prior to chemical purification, and the sample and spike achieve isotopic equilibrium, then the mass-dependent fractionation that can be produced during chemical purification can be rigorously corrected (e.g., Russell et al. 1978). Application of such techniques may be very important for the analysis of samples that have low analyte concentrations in high ion-strength solutions, for example analysis of transitions metal isotope compositions in seawater.

THE EFFECT OF PEAK SHAPE

When the peak tops are not perfectly flat, in particular as a consequence of energy spread and beam clipping, the isotopic ratios no longer depend on the mass only, but also on the position of the collectors. This is particularly critical for instruments with rotated focal planes which have imperfect z-focusing properties. Peak tops may be round, sloping, or both, although these defects remain hard to detect as a result of the plasma instabilities. Even small deviations from a flat peak top strongly affect the isotopic ratios. Let us call $M = M_i + \delta M_i$ the actual position of the cup with respect to its ideal position at mass M_i. The measured ratio must now be expanded as a function of the cup position rather than a function of mass. Assuming that the mass fractionation law is exponential and that the peak top is sloping, a first-order expansion is sufficient. We obtain:

$$r_i = R_i\left(\frac{M_i + \delta M_i}{M_k + \delta M_k}\right)^{\beta} = R_i\left(\frac{M_i}{M_k}\right)^{\beta}\left(\frac{1+\delta M_i/M_i}{1+\delta M_k/M_k}\right)^{\beta} \tag{39}$$

and the exponentially-corrected ratio as:

$$r_i^e = R_i\left(\frac{1+\delta M_i/M_i}{1+\delta M_k/M_k}\right)^{\beta} \tag{40}$$

A deviation from the ideal fractionation law will appear as a residual correlation between the ratios corrected for mass bias using the exponential law. It can be verified that even very small $\delta M/M$ in Equation (40) produces a potentially important isotopic effect on the order of $1 + \beta(\delta M_i - \delta M_k)/M_k$. The alignment of the correlation between the corrected ratios $x = \ln r_i^e$ and $y = \ln r_j^e$ produced by sloping peaks in a log-log plot has a slope of $\approx \delta M_j/\delta M_i$. For a round peak, the second-order term should be included.

COUNTING STATISTICS

The central assumption of ion counting is that ions arrive at the detector "at random," i.e., that the probability of arrival of an ion is the same for any time interval of a same length. The number n_i of ions i arriving at any collection device during the time interval δt is therefore subject to Poisson statistics: n_i is proportional to δt, the average count rate is $n_i/\delta t$, and its variance is also equal to $n_i/\delta t$. The standard deviation of an ion beam I_i measured in ions per second is proportional to the square-root of the beam intensity and the relative error is $\sqrt{I_i}/I_i = 1/\sqrt{I_i}$. Smaller beams therefore fluctuate more than larger beams. The noise, known as shot noise, due to counting statistics accounts for a 0.25‰ (2σ level) error on a typical 1 V signal collected in one second through a 10^{11} Ω resistor or in 10 seconds on a 100 mV signal. This noise, which is not to be confused with the thermal or Johnson noise of the resistor, imposes strong correlations on some isotopic plots that may easily be mistaken for mass-dependent fractionation effects. This is a familiar effect for Pb (e.g., Hamelin et al. 1985; Powell et al. 1998) because of the common usage of the minor ^{204}Pb isotope as the reference isotope, but it exists to a variable extent for other elements such as Zn (Fig. 10). Let us consider two measured isotopic ratios n_i/n_k and n_j/n_k. For the sake of clarity, we will neglect the thermal noise of the resistor. Taking the logarithms and differentiating, we obtain the approximate expression:

$$\frac{n_i/n_k - \overline{n_i/n_k}}{n_i/n_k} \approx \frac{n_i - \bar{n}_i}{n_i} - \frac{n_k - \bar{n}_k}{n_k} \tag{41}$$

in which the overbar indicates the mean value of the parameter. A similar expression can be written for n_j/n_k. By squaring this expression, taking its expectation and noting that statistical noise on different peaks is independent, we obtain:

$$\frac{\operatorname{var}(n_i/n_k)}{(n_i/n_k)^2} \approx \frac{\operatorname{var}(n_i)}{n_i^2} + \frac{\operatorname{var}(n_k)}{n_k^2} = \frac{1}{n_i} + \frac{1}{n_k} \tag{42}$$

with a similar expression for n_j/n_k. The covariance between the ratios is similarly:

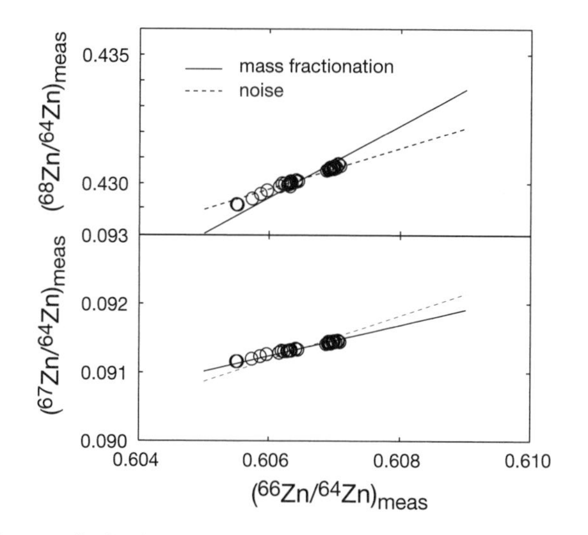

Figure 10. On this unusually fractionated block of 40 cycles, the effect of the noise-induced correlation observed on Zn isotopes can be separated from the mass fractionation effects. The solid line corresponds to mass-dependent fractionation, while the dashed line is defined by counting statistics. When the larger ^{68}Zn/^{64}Zn and ^{66}Zn/^{64}Zn ratios are plotted against each other (top), counting statistics tend to pull the results away from the mass fractionation line. When the smaller ^{67}Zn/^{64}Zn is considered (bottom), counting statistics has essentially no effect. Data acquired using the VG Plasma 54 of Lyon.

$$\frac{\operatorname{cov}\left(n_i/n_k, n_j/n_k\right)}{n_i n_j/n_k^2} \approx \frac{\operatorname{var}\left(n_k\right)}{n_k^2} = \frac{1}{n_k} \tag{43}$$

For ratios with the same denominator, e.g., ^{54}Fe for ^{56}Fe/^{54}Fe and ^{57}Fe/^{54}Fe, the slope of the noise correlation line in a ratio-ratio plot is the ratio of the standard deviations $\sigma_{j/k}$ and $\sigma_{i/k}$ along each axis:

$$\frac{\sigma_{j/k}}{\sigma_{i/k}} = \sqrt{\frac{n_j}{n_i}\frac{1+n_j/n_k}{1+n_i/n_k}} = \sqrt{\frac{n_j/n_k}{n_i/n_k}\frac{1+n_j/n_k}{1+n_i/n_k}} \tag{44}$$

while the correlation coefficient between the errors on the two ratios is:

$$\rho_{i/k}^{j/k} = \sqrt{\frac{n_i/n_k}{1+n_i/n_k} \times \frac{n_j/n_k}{1+n_j/n_k}} \tag{45}$$

Typical values of the correlation coefficients are 0.5 for two isotopic ratios equal to unity and quickly tend to unity for large ratios. Correlations between errors due to counting statistics are clearly minimized by using ratios with values <1.

The slope $s_{i/k}^{j/k}$ of the noise line in a log-log plot (or equivalently in an ε–ε plot) is similarly evaluated as

$$s_{i/k}^{j/k} = \sqrt{\frac{1+n_j/n_k}{n_j/n_k} / \frac{1+n_i/n_k}{n_i/n_k}} \tag{46}$$

It is important to separate the correlations between isotopic ratios introduced by mass discrimination from those introduced by counting fluctuations. Residual correlations between the mass-bias corrected isotopic ratios and the mass-bias index is expected. A correlation with a slope equal to the difference between those of the noise line and the fractionation line is expected between the mass bias-corrected ratios, which should not necessarily be taken as an indication that the instrumental mass-fractionation law should be amended. In a ln(^{56}Fe/^{54}Fe) vs. ln(^{57}Fe/^{54}Fe) plot, the slope of the noise correlation (1.87) is very similar to the slope of mass fractionation (1.49). The correlation coefficient of the counting statistics is 0.5, so that the risk of some noise being mistaken for natural fractionation is real. A test for unambiguously separating the effect of counting statistics from that of instrumental mass discrimination is to plot isotopic ratios with no common isotope, e.g., ^{67}Zn/^{68}Zn vs. ^{66}Zn/^{64}Zn (Fig. 11): the hypothesis of a null correlation coefficient is easily judged using standard statistical tests, so any significant correlation in these diagrams must be due to instrumental mass fractionation. Thus, the absence of correlations in plots of mass-fractionation corrected ratios with no common isotope shows that mass fractionation was corrected efficiently.

IMPLEMENTATION OF THE MASS BIAS CORRECTION FOR MC-ICP-MS

As shown by Maréchal et al. (1999), the linearity property of the isotopic array formed in log-log plots by all the measurements of a same solution can be used to obtain the isotopic ratio of a sample with respect to the same ratio in a standard solution of known isotopic properties of a different element. Combining either Equations (21) or (22) for both the sample and the standard, we obtain:

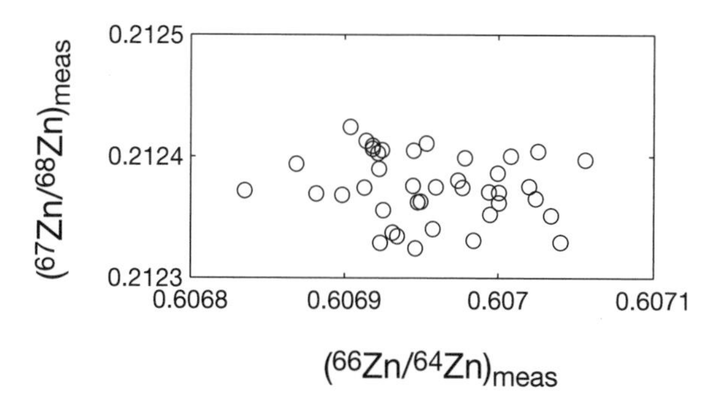

Figure 11. When the ratios on each axis have no common isotope and in the absence of mass-dependent fractionation, they are statistically independent. These data represent a run of 40 cycles in which $^{66}Zn/^{64}Zn$ and $^{67}Zn/^{68}Zn$ were measured. The horizontal array (uncorrelated variables) indicates that, in the present case, the variability of the isotopic ratios can be ascribed to counting statistics and not to mass-dependent fractionation. Data acquired using the VG Plasma 54 of Lyon.

$$\frac{\ln(R_i)_{\text{sple}} - \ln(r_i)_{\text{sple}}}{\ln(R_j)_{\text{sple}} - \ln(r_j)_{\text{sple}}} = \frac{\ln(R_i)_{\text{std}} - \ln(r_i)_{\text{std}}}{\ln(R_j)_{\text{std}} - \ln(r_j)_{\text{std}}} = s_{j/k}^{i/k} \tag{47}$$

Let us define the intercept I_{sple} of the mass fractionation line with the y-axis as:

$$I_{\text{sple}} = \ln r_i - s_{j/k}^{i/k} \ln r_j \tag{48}$$

We can re-arrange Equation (47) as:

$$\ln(R_i)_{\text{sple}} = s_{j/k}^{i/k} \ln(R_j)_{\text{sple}} + I_{\text{sple}} \tag{49}$$

and

$$\ln(R_i)_{\text{std}} = s_{j/k}^{i/k} \ln(R_j)_{\text{std}} + I_{\text{std}} \tag{50}$$

With the condition that $(R_j)_{\text{std}}$ and $(R_j)_{\text{sple}}$ are identical and by subtracting Equation (50) from Equation (49), we get the central expression:

$$\ln(R_i)_{\text{sple}} = \ln(R_i)_{\text{std}} + I_{\text{sple}} - I_{\text{std}} \tag{51}$$

The previous equations do not require that the isotopic ratio used for normalization (x-axis) and the ratio to be measured (y-axis) have to be for the same element. It is therefore possible to normalize the isotopic composition of Cu to that of a standard Zn solution without any assumption made on the particular mass-fractionation law. The original formulation of this property by Longerich et al. (1987) calls for identical isotopic fractionation factors for the two elements, but this is not at all a necessary constraint and Albarède et al. (2004) show that, in fact, this very assumption may lead to significant errors. For a Cu sample mixed with a Zn standard, in which the $^{66}Zn/^{64}Zn$ ratio of the standard solution is used for normalization, we obtain the expression:

$$\ln\left(^{65}\text{Cu}/^{63}\text{Cu}\right)_{\text{sple}}^{\text{corr}} = \ln\left(^{65}\text{Cu}/^{63}\text{Cu}\right)_{\text{std}}^{\text{true}} + I_{\text{sple}} - I_{\text{std}} \tag{52}$$

where

$$I = \ln\left({}^{65}\mathrm{Cu}/{}^{63}\mathrm{Cu}\right)^{\mathrm{meas}} - s_{66/64}^{65/63} \ln\left({}^{66}\mathrm{Zn}/{}^{64}\mathrm{Zn}\right)^{\mathrm{meas}} \tag{53}$$

In this equation, $s_{66/64}^{65/63}$ is the slope of the mixed Cu-Zn standard solution in a $y = \ln(^{65}\mathrm{Cu}/^{63}\mathrm{Cu})$ vs. $x = \ln(^{66}\mathrm{Zn}/^{64}\mathrm{Zn})$ array. Again, the ^{65}Cu/^{63}Cu of the sample is evaluated with respect to this ratio in the standard through the difference in elevation between the parallel lines drawn in the log-log plot between the sample and the standard (Fig. 7).

The slope $s_{j/k}^{i/k}$ is best estimated from the isotopic $\ln r_i$ vs. $\ln r_j$ array for a shelf solution and should be carefully determined for the interval during which the samples are measured. Standards bracketing the measured samples do not necessarily provide the best estimate of this slope: their isotope compositions are either very similar, in which case the error on the slope is large, or very different, which usually reflects a heavy matrix in the intercalated sample solution that abruptly changed the mass bias over an unknown period of time. Provided the alignment is good, we therefore rather suggest to use the slope determined on a large set of standards run over a period of several hours in alternation with samples. In the ideal case for which the precision on the slope $s_{j/k}^{i/k}$ is good, the knowledge of the particular fractionation law (power or exponential) applying to the process is immaterial to the correction.

The implicit assumption in this correction is that both mass fractionation factors β, e.g., for Cu and Zn, are different, but will be changed by the same constant upon switching between sample and standard solutions. Again, this assumption has better chances of being correct if purification of the sample is efficient. For isotopic ratios not too far from unity (e.g., ^{65}Cu/^{63}Cu, ^{66}Zn/^{64}Zn, ^{57}Fe/^{54}Fe), a precision of 0.05‰ can be achieved (Maréchal et al. 1999; Zhu et al. 2000; Beard et al. 2003). For larger ratios, the obtainable precision is 0.2–0.4‰.

So far, second-order corrections have only found their application for radiogenic isotopes (see a more extensive treatment in Albarède et al. 2004). The linear changes in the apparent mass bias of Nd with mass observed by Vance and Thirlwall (2002) is certainly an indication that high precision may benefit from such an elaborate scheme on at least some instruments.

MATRIX EFFECT

Matrix effects are typically divided into spectral (isobaric) and non-spectral types. The spectral or isobaric effects include 1) elemental isobaric interferences such as ^{54}Cr at ^{54}Fe, 2) molecular interferences such as ^{40}Ca^{16}O at ^{56}Fe and ^{40}Ar^{14}N at ^{54}Fe, 3) double charge interferences such as $^{48}\mathrm{Ca}^{2+}$ at ^{24}Mg. Non-spectral matrix effects are largely associated with changes in the sensitivity of an analyte due to the presence of other elements (Olivares and Houk 1986). Changes in sensitivity correspond to a change in instrumental mass bias, and therefore non-spectral matrix effects can have a significant impact on the accuracy of isotope measurements.

Spectral (isobaric) effects

For elemental interferences, mass difference between overlapping species is normally too small to be resolvable. An efficient separation chemistry is most commonly the solution, but the utmost care must be exercised to maintain quantitative yields because ion exchange chromatography can produce mass-dependent fractionations as shown for the elution of Ca (Russell and Papanastassiou 1978), Cu and Zn (Maréchal et al. 1999; Maréchal and Albarède 2002), and Fe (Anbar et al. 2000; Roe et al. 2003).

Oxides are a common issue since they can make up to several percent of the metal beam, while compounds of argon and atmospheric gasses, such as ArN, are ubiquitous. The abundance of molecular species is a function of their bond energy, of their ionization potential, but also of plasma composition and temperature. The mass difference is often large enough for the interfering species to be separated by positioning a faraday bucket to only intersect the analyte portion of a complicated multi-ion peak by using a full width faraday bucket and narrow ion beam defining slit (Weyer and Schwieters 2003). Molecular species are significantly reduced upon desolvation of the sample aerosol by letting a sweep gas (Ar) through a hydrophobic membrane (Montaser et al. 1991; Belshaw et al. 2000; Roe et al. 2003). Alternatively, ions can be introduced into a multiple (quadrupole, hexapole) collision cell filled with a light gas such as argon or helium: radio frequencies are used to focus the ions and induce collisions with the neutral gas (Douglas 1989; Beard et al. 2003). In the process, most molecular species are dissociated. Removal of interfering species using a collision cell may be preferable to the use of high resolution because there is no loss in ion beam transmission. Finally, the composition of the plasma may be altered and molecular species greatly reduced either by decreasing the power dispensed by the RF generator—the "cold plasma" of Jiang et al. (1988) and Kehm et al. (2003)—by shielding the plasma from the work coil (Sakata and Kawabata 1994) or by using mixtures of plasma gasses (Montaser and Zhang 1998; Lam and Horlick 1990).

Doubly-charged ions exist because the potential of second ionization of many metals is relatively low with respect to the plasma thermal energy. For instance, 11.9 eV are needed to remove two electrons, in contrast with 6.1 eV for one electron, from a calcium atom. At 8000K, a little less than 0.1% of Ca would be in the Ca^{2+} form and overlap with Mg^{+} isotopes. Likewise, Ba^{2+} is rather easily formed and overlaps with Zn^{+} isotopes. Isobaric interferences with doubly-charged ions are easily identified as odd-mass atoms will produce peaks at half masses, such as $^{43}Ca^{2+}$ at mass 21.5.

As long as isobaric interferences do not represent a large fraction of the signal, a correction step is always recommended. A simple rule-of-thumb is that, provided the isotopic composition of an interfering species accounting for 1% of a peak is known to a precision of 1%, it will contribute an uncertainty of ca. 10^{-4} on the corrected abundance of the main species. Peak "stripping" involves the resolution of a set of linear conservation equations, often by least-squares. It is a well-established straightforward technique (e.g., Albarède 1995), which can be handled on a spreadsheet and produce reliable results.

Non-spectral effects

In general, non-spectral matrix issues appear to be a result of space charge effects, where for example, the sensitivity of an analyte decreases as the mass to charge ratio of a matrix element increases (Gillson et al. 1988; Praphairaksit and Houk 2000a).

Alternatively, some non-spectral matrix issues may be a result of changes in vaporization position in the plasma (Carlson et al. 2001). There appears to be three main controls on non-spectral matrix effects, that include: 1) a dependence on the mass to charge ratio and the ionization potential of the analyte 2) a dependence on the mass to charge ratio and ionization potential of the matrix element and 3) a function of the high voltage acceleration potential (e.g., Olivares and Houk 1986; Nonose and Kubota 2001). Most of the studies that have conducted tests on non-spectral interferences have made measurements where the concentration ratio of matrix to analyte was 10 or greater (e.g., Olivares and Houk 1986; Date et al. 1987; Gillson et al. 1988; Nonose and Kubota 2001). At this high matrix to analyte ratio, the change in sensitivity of the analyte element can be on the order of 10 percent or more. In general the strongest matrix control on analyte sensitivity (and hence instrumental mass bias) is for matrix elements that have a mass greater than the analyte (e.g., Gillson et al. 1988). Changes in sensitivity for matrix elements that are lighter than the analyte are not as well established. Moreover, at lower matrix to analyte concentrations significant changes in analyte sensitivity were not observed.

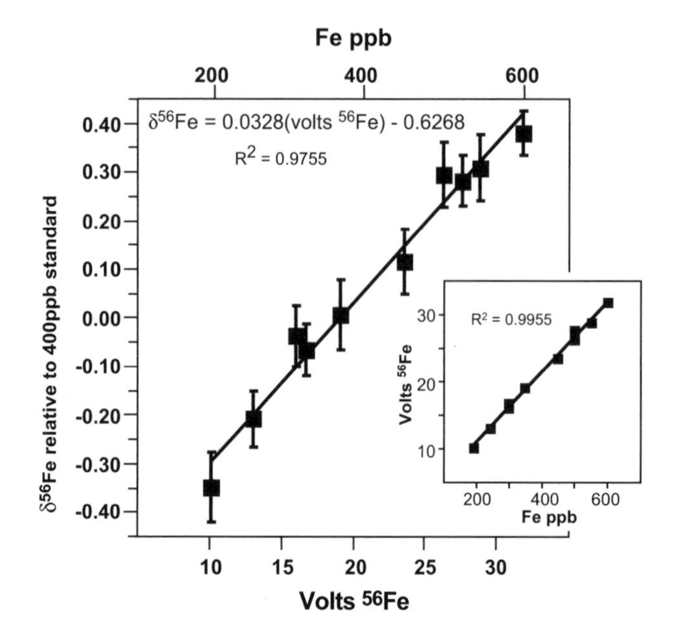

Figure 12. Plot of the $^{56}Fe/^{54}Fe$ ratio of an Fe standard, analyzed at 200 to 600 ppb concentrations, relative to the average $^{56}Fe/^{54}Fe$ of bracketing 400 ppb an Fe standard, versus the measured ^{56}Fe volts ($10^{10}\,\Omega$ resistor). The measured Fe isotope composition varies relative to Fe concentration, which reflects differences in instrumental mass bias as a function of concentration. Data were taken over a 24 hour period using the University of Wisconsin-Madison Micromass *IsoProbe*.

Small changes in mass bias can be produced with no appreciable change in sensitivity. There is a matrix effect associated solely with the concentration of the analyte (e.g., Carlson and Hauri 2001; Kehm et al. 2003). For example, in Fe isotope measurements an increase in mass bias is associated with high Fe concentrations. Figure 12 plots volts of ^{56}Fe vs. the $\delta^{56}Fe$ values of an Fe standard relative to the Fe isotope composition of the 400 ppb standard. In this example, the concentration effect may be a result of space charge effects because the increase in mass bias is positively correlated with an increase in ion current, which presumably results in stronger ion repulsion behind the skimmer cone. However, this slope of this line varies from day to day (although the slope is always positive) and hence there may also be matrix effects associated with plasma vaporization position (as noted by Kehm et al. 2003). Note, that there is no detectable change in sensitivity because a plot of total ion current versus Fe concentration defines a linear relationship over this concentration range. Corrections for changes in instrumental mass bias caused by concentration can easily be handled by carefully controlling the concentration of samples, or by constructing a working curve like that shown in Figure 12 and applying a correction to samples based on the measured ion intensity.

Although mass bias effects relative to total ion intensity may be corrected using a working curve (Fig. 12), anomalous mass bias in natural samples due to the presence of other elements cannot be corrected. Figure 13 shows the effects of anomalous mass bias in a 400 ppb Fe ultra pure standard that has been doped with varying concentrations (up to 75 ppb) of Mg, Al, or La. Carlson et al. (2001) noted a similar effect of Al on the isotope composition of Mg standard solutions. These matrix elements were chosen because they would not produce isobars on the Fe mass spectrum (Fig. 13D) and for their spread in atomic mass; Mg and Al both have masses less than Fe and La is greater. Additionally Mg and Al are major elements (7th and 3rd most

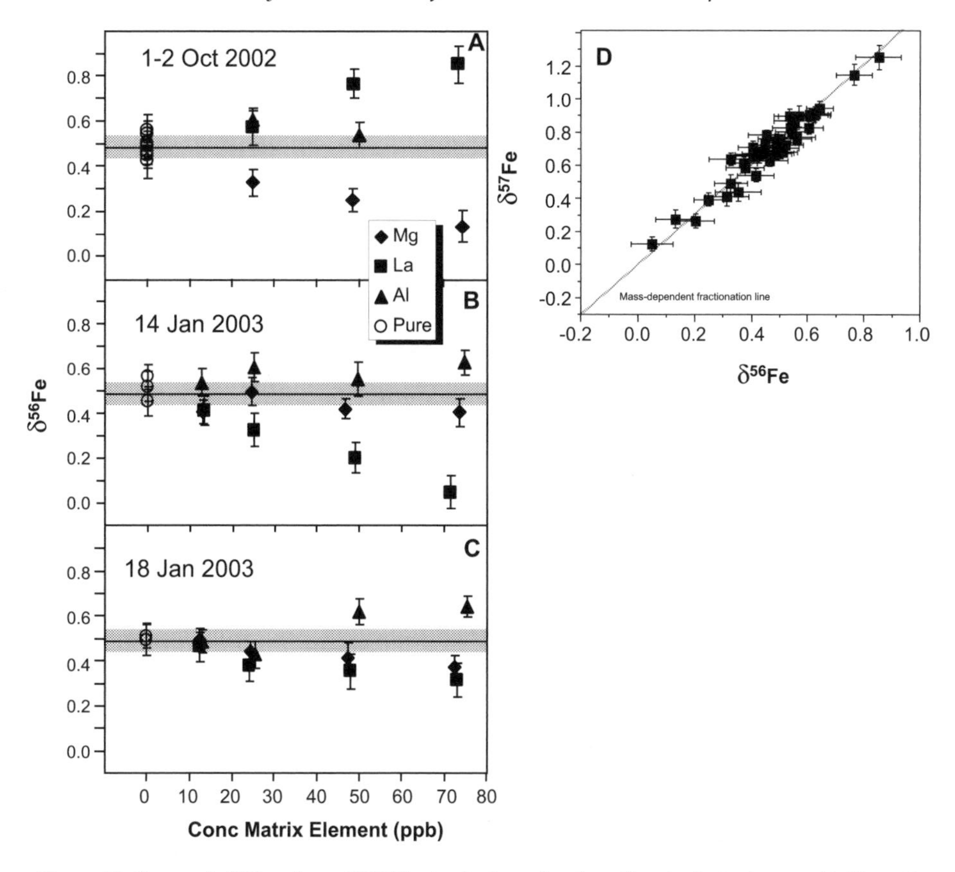

Figure 13. Changes in $\delta^{56}Fe$ values of HPS Fe standard as a function of contaminate elements Al, Mg, or La (12.5 to 75 ppb). All solutions were 400±2 ppb Fe. The Fe isotope compositions of the solutions are shifted from those in the pure Fe standard ($\delta^{56}Fe$ = +0.49±0.05‰) as a function of the impurity concentration. The magnitude of the this shift with impurity concentration is variable, as shown by data collected during three analytical sessions (parts A, B, and C). These impurity matrix elements do not produce molecular isobars, as evidenced by the fact that $\delta^{57}Fe$ and $\delta^{56}Fe$ values plot along a mass-dependent array (part D). Note that an important conclusion of these tests is that accuracy of Fe isotope measurements cannot be demonstrated by preservation of mass-dependent trends in $^{57}Fe/^{54}Fe$ and $^{56}Fe/^{54}Fe$. Data were collected using the Univ. of Wisconsin-Madison Micromass *IsoProbe*.

abundant elements in the crust, respectively) and hence likely to be present at trace levels in samples, even if the sample has been purified by ion exchange chromatography. In general, the Fe isotope composition shifts most strongly as the concentration of the contaminant is increased. The magnitude of the Fe isotope shift, however, is variable from day to day, which may be attributable to differences in nebulizer gas flow rates and sampler and skimmer cones that differed in their amount of wear during the three different days the tests were conducted. An example of the different behavior is the Fe isotope shift associated with La addition that shows an increase in $\delta^{56}Fe$ values with increasing La content for one analysis period, whereas for the other two analysis sessions, increasing La contents decrease the $\delta^{56}Fe$ value of the doped standard solution. All three-matrix elements had an impact on instrumental mass bias. Lanthanum produced the largest magnitude changes that may be a reflection of space charge effects. Magnesium and Al, which have similar masses, produced similar magnitude changes in instrumental mass bias for two of the three analysis sessions. Importantly, however, it appears that matrix elements with masses less than the analyte are able to impart changes in instrumental mass bias. Similar magnitude isotopic offsets have been measured for Ag isotopes, where for example, there is a 0.3‰ difference in $^{109}Ag/^{107}Ag$ ratios for pure Ag standard solutions purified by ion exchange as compared to the same Ag standard solutions that were not processed by ion exchange (Carlson and Hauri 2001). The differences in $^{109}Ag/^{107}Ag$ ratios may be a result of non-spectral matrix effects associated with the purity of the standard solutions (Carlson and Hauri 2001). However, it is difficult to uniquely determine if this is indeed a matrix effect because mass dependent fractionation on the ion exchange column could produce a similar shift in isotopic composition (Carlson and Hauri 2001).

Another type of non-spectral matrix effect, associated with the oxidation state of the analyte, was proposed by Zhu et al. (2002). Figure 14 plots the relative Fe(II) to total Fe ratio of ultra pure Fe standard solutions versus the difference between the $\delta^{56}Fe$ value of the mixed valence state standard and the $\delta^{56}Fe$ value of the Fe(III) only standard. The oxidation state of these standards was not quantified by Zhu et al. but based on colorimetric methods using 2,2'-bipyridine the relative Fe(II) to total Fe ratios of these standards are well known. This matrix effect appears to exert a significant control on isotope accuracy, where for example if a reduced ferrous solution was compared to an oxidized ferric standard, the accuracy of the $\delta^{56}Fe$ value could be affected by up to 1‰. This matrix effect associated with oxidation state is unlikely to be a result of space charge effects because the mass of an electron is unlikely to produce a large change in the mass of the ion beam. Perhaps this matrix effect may be associated with ionization properties in the plasma.

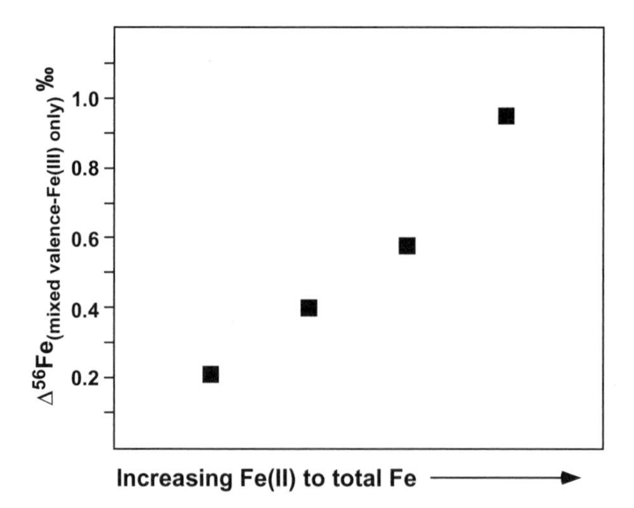

Figure 14. Plot of difference of the $\delta^{56}Fe$ value of an Fe standard that was partially reduced minus the $\delta^{56}Fe$ value of an Fe standard that was partially oxidized versus the relative oxidation state of the standard as determined colorimetrically using 2,2'-bipyridine. Data from Zhu et al. (2002) collected using an Nu-Plasma MC-ICP-MS.

INSTRUMENTAL BIAS: MEMORY EFFECTS

Background from previous runs vary from instrument to instrument, from laboratory to laboratory, and from day to day. Memory has an adverse effect on the accuracy and precision of isotopic compositions in the 0.1‰ range and the so-called on-peak zero (OPZ) methods, which amount to correct the signal for a background measured before the analysis, should be treated with utmost caution: it is always assumed that the background intensity and isotope composition are the same for a cleaning solution and for a sample solution. At least for Pb isotopes, this assumption is known to be incorrect as a result of multiple sources of memory effect (Woodhead 2002; Albarède et al. 2004). The intensity of the background peaks may vary as a function of the recent history, notably the isotopic composition of the samples run in the recent past, of a particular mass spectrometer. Additional care should be exercised when double-spike methods are used: these techniques tend to use the less abundant isotopes of the element, which has serious effects on the isotopic composition of the background signal (Albarède et al. 2004). Most modern MC-ICP-MS, however, should be able to run with a negligible background and the operator should be able to check that this background does not vary in intensity and composition when traces of other metals, e.g., Ca, or U, are introduced in the instrument, when the solutions contains anionic leftovers of the separation chemistry (such as perchlorate and chloride ions or organic acids), or even when the nature and strength of the acid used to dilute the sample is changed.

In order to quantify the effect of background on isotopic ratios, let us illustrate the problem with the case of Zn isotope measurements when samples doped with Zn double spike with a high ^{67}Zn abundance are routinely run on the same instrument. A 2×10^{-16} A beam equivalent to a 20 μV signal on a standard $10^{11}\Omega$ resistor would still be within the thermal (Johnson) noise of the resistor and therefore escape detection on a Faraday cup. In comparison, a 2.5 V total signal of normal Zn would come with about 100 mV at mass 67 and memory would contribute a bias of ca. 0.2‰ to ^{67}Zn. Since the isotope of mass 67 would be abundant in Zn double spike, the bias may rapidly become significant. For instance, expressing that the raw (total) ^{67}Zn and ^{64}Zn signals are the sum of a sample and a background signal, we get

$$\left(\frac{^{64}\mathrm{Zn}}{^{67}\mathrm{Zn}}\right)^{\mathrm{total}}=\left(\frac{^{64}\mathrm{Zn}}{^{67}\mathrm{Zn}}\right)^{\mathrm{sple}}+\left[\left(\frac{^{64}\mathrm{Zn}}{^{67}\mathrm{Zn}}\right)^{\mathrm{bkgd}}-\left(\frac{^{64}\mathrm{Zn}}{^{67}\mathrm{Zn}}\right)^{\mathrm{sple}}\right]\frac{^{67}\mathrm{Zn}^{\mathrm{bkgd}}}{^{67}\mathrm{Zn}^{\mathrm{total}}} \tag{54}$$

Not correcting the apparently negligible contribution of the background would create a bias of

$$\frac{\Delta^{64}\mathrm{Zn}/^{67}\mathrm{Zn}}{^{64}\mathrm{Zn}/^{67}\mathrm{Zn}}=\left[\frac{\left(^{64}\mathrm{Zn}/^{67}\mathrm{Zn}\right)^{\mathrm{bkgd}}}{\left(^{64}\mathrm{Zn}/^{67}\mathrm{Zn}\right)^{\mathrm{sple}}}-1\right]\frac{^{67}\mathrm{Zn}^{\mathrm{bkgd}}}{^{67}\mathrm{Zn}^{\mathrm{total}}} \tag{55}$$

In the particular case we just discussed, the requirement of a <0.05‰ bias necessitates that the $^{67}Zn/^{64}Zn$ ratio of the background be known *for this particular run* with a precision of 25%, which is a difficult goal to achieve on such a small signal. When isobaric interferences add up on the same masses, e.g., if minute traces of $^{51}V^{16}O$ are present, accuracy and precision on the $^{67}Zn/^{64}Zn$ ratio may become disastrous.

ABUNDANCE SENSITIVITY

For a nuclide of mass *M*, abundance sensitivity is the ratio between the signal at mass *M*+1 arising from the same species to the signal at mass *M*. Off-peak ions are present because of collisions behind the magnetic filter, of reflections on the tube wall, or of space-charge effects. As a result of the collisions, the energy of these ions is different from the energy of the main beam. They alter the apparent peak baseline in a continuous way. Abundance sensitivity decreases with the mean free path of ions, i.e., when pressure near the collector assembly increases. It should not be confused with mass resolution, which is a measure of how well separated two adjacent peaks are and which for a well-focused instrument reflects the widths of the entrance and exit slits. A high abundance sensitivity is necessary to measure very large ratios (>10,000) and has found its main applications in uranium disequilibrium series. Single-focusing TIMS and MC-ICP-MS instruments typically have abundance sensitivities of 2×10^{-6} which can be reduced by 2–3 orders of magnitude upon fitting a quadrupole filter or an electrostatic sector, which both act as energy filters, after the magnet. A wide-angle retarding potential analyzer (WARP) is particularly effective in plasma mass spectrometry when the ion energy has been thermalized using a collision cell interface. For all practical purposes, this effect can be ignored for stable isotope measurements.

CORRECTION OF THE NON-MASS DEPENDENT INSTRUMENTAL BIAS IN STATIC MODE

Because electronic components cannot be manufactured with perfect reproducibility and because the positioning of Faraday cups with respect to the instrument optical axis is usually imperfect, non-mass dependent biases arise as a consequence of a number of problems: (1) secondary electrons may not be suppressed with a 100 percent efficiency, (2) ions of various masses may bounce around the collection system and be reflected into the wrong cup, (3) the response of some components, such as resistors and capacitors, may be non-linear, (4) and last, but foremost, the ion beams may have cross sections so broad that that they are partially clipped along their trajectory in the flight tube, especially the top and the bottom of off-axis beams when *z*-focusing is imperfect. Although for modern mass spectrometers, these problems usually involve only a very small fraction of the ion beams, they may nevertheless alter the precision of the measurements in a way that is not a linear function of the mass difference between the beams collected in the Faraday cups. If the isotopic composition of a standard solution of the element under consideration is known, non-mass dependent isotopic fractionation can be conveniently dealt with by introducing correction factors (also know as efficiencies), which are determined by measuring the standard and comparing its isotopic abundances with the known values. In the following, we demonstrate, however, that although these non mass-dependent effects change the measured isotopic compositions, they do so in a way that preserves the slope of the linear alignments in log-log plots and therefore allows isotopic variations among standards to be determined with an excellent precision.

Let the transmission $\mathcal{T}_k^{\gamma}$ at mass k in collector γ be the product of a mass-dependent bias $\mathcal{Y}(M_k)$ and a cup-dependent factor (efficiency) $\mathcal{A}_{\gamma}$

$$n_k/N_k = \mathcal{A}_{\gamma}\mathcal{Y}(M_k) \tag{56}$$

$\mathcal{Y}(M_k)$ represents the mass bias when all the masses are peak-jumped in the same cup, such as with the standard procedure still employed on all single-collector mass spectrometers. If we assume that mass i is collected in cup α and the bias $\mathcal{Y}(M)$ follows the exponential law, we obtain the expression for the measured ratio of masses i to k as

$$r_i = \frac{\mathcal{A}_{\alpha}}{\mathcal{A}_{\gamma}} R_i \left(\frac{M_i}{M_k}\right)^{\beta} \tag{57}$$

Writing a similar equation for the ratio R_j with the assumption that mass j is collected in cup β, we can write

$$\frac{\ln R_i - \ln r_i - \ln A_\alpha / A_\gamma}{\ln R_j - \ln r_j - \ln A_\beta / A_\gamma} = \frac{\ln M_i / M_k}{\ln M_j / M_k} = s_{i/k}^{j/k} \tag{58}$$

In other words, variable cup efficiencies change the intercept, not the slope, of the mass-fractionation curves, and the procedure for correcting mass fractionation with respect to a standard is identical to that with no cup efficiency involved.

CORRECTION OF THE INSTRUMENTAL BIAS IN DYNAMIC MODE

With radiogenic isotopes, non-mass dependent isotopic fractionation can be evaluated by running the samples in dynamic mode, i.e., by switching the electromagnetic field and therefore the masses across the cups, which allows the efficiencies to cancel out between different beam configurations (Dodson 1963; Ludwig 1997; Wendt and Haase 1998). This technique cannot be applied to elements of light molar weight. By switching the magnetic field, perfect alignment cannot be achieved for all masses (Fig. 15). The absolute misfit for one mass difference at mass M is exactly $-1/M$, which approximately represents −0.02 for Fe, −0.01 for Mo, and −0.005 for Hg. In practice, because of the finite width of the peak flat top, dynamic measurements cannot be implemented below mass 80 and the potential for elements such as Mo, Cd, Hg remains to be explored. Fractionation conditions vary substantially during a same run on TIMS instruments. This technique cannot therefore be used to measure stable isotope abundances without resorting to a double-spike (see section "Double-spike analysis") or to heavy molecular beams such as $Cs_2BO_2^+$ for boron isotopes (Spivack and Edmond 1986). In contrast, the steady-state conditions of electron-impact and ICP sources make them suitable for the dynamic analysis of stable isotopes. Alternating an unknown sample with a reference (standard bracketing) makes it possible to obtain the relative deviation of the isotopic compositions between the two, but we are going to show that this technique can also be optimally combined with normalization with an external element.

Luais et al. (1997) and Blichert-Toft et al. (1997) published solutions for the dynamic

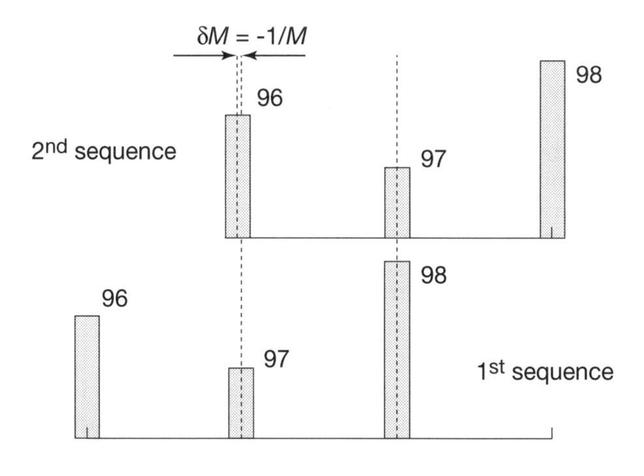

Figure 15. Mass misfit for two sequences of dynamic mode shifted by one mass. It is assumed that coincidence is obtained between ^{98}Mo for the first sequence and ^{97}Mo for the second sequence. The parameter δM (exaggerated) represents the misfit one mass down this peak.

analysis of $^{143}Nd/^{144}Nd$ and $^{176}Hf/^{177}Hf$, respectively, using an exponential law for mass bias, but these solutions are rather cumbersome. Similarly, Wendt and Haase (1998) presented a solution to the problem of dynamic analysis with cup efficiencies, but again the calculations are heavy to implement. Therefore, we here develop new systems of fully explicit equations that are versatile, lend themselves to the use of non-linear mass fractionation laws and to easy error assessment through matrix analysis. Let us start by rewriting Equation (56) in a log form:

$$\ln r_i = \ln \mathcal{A}_\alpha - \ln \mathcal{A}_\gamma + \beta \ln \frac{M_i}{M_k} + \ln R_i \quad (59)$$

When written in full, the number of unknowns is the sum $c + e + r$, where c is the number of active cups, e the number of elements for which the value h is needed, and r the number of ratios to measure. We first show how to use a standard solution (or a mixture of standards of different elements) of known isotopic compositions N_i/N_k to determine the cup efficiencies. The unknowns are the efficiencies A for each cup and the mass fractionation factors f (or h for other laws). The system of equations is particularly compact since Equation (59) now reduces to:

$$\ln \frac{r_i}{R_i} = \ln \mathcal{A}_\alpha - \ln \mathcal{A}_\gamma + \beta \ln \frac{M_i}{M_k} \quad (60)$$

This system can be conveniently written in a matrix form:

$$\mathbf{y} = \mathbf{Ax} \quad (61)$$

where the lower case symbols stand for vectors ($\mathbf{x}$ for the unknowns $\mathbf{x} = \ln\mathcal{A}_\alpha \ldots \ln\mathcal{A}_\gamma$, β, and $\mathbf{y}$ for the data) and the upper case symbol $\mathbf{A}$ for a rectangular matrix. In most cases, the number of measurements (dimension of $\mathbf{y}$) is larger than the number of unknowns (dimension of $\mathbf{x}$) and Equation (61) can be solved by standard least-square methods giving $\mathbf{x} = (\mathbf{A}^T\mathbf{A})^{-1}\mathbf{A}^T\mathbf{y}$. The errors on the unknown values $\mathbf{x}$ can simply be computed from the diagonal entry of $\mathbf{W_x} = (\mathbf{A}^t\mathbf{W_y}^{-1}\mathbf{A})^{-1}$, where $\mathbf{W_y}$ is the covariance matrix of the measurements $\mathbf{y}$. In most cases, an approximation of $\mathbf{W_y}$ by a diagonal matrix, in which the entries are the squared "errors," is sufficient.

Let us illustrate this method with the three-sequence dynamic run of a Hg-Tl mixture as described in Table 1 and assume for brevity that the exponential fractionation law holds. Efficiencies are only known as relative values and we will assume that the efficiency factor of the axial cup is unity. For the best possible precision, the ratios to be measured are $^{205}Tl/^{203}Tl$, and $^{199}Hg/^{202}Hg$, $^{200}Hg/^{202}Hg$, $^{201}Hg/^{202}Hg$, and $^{204}Hg/^{202}Hg$. The efficiencies will be evaluated for the cups L3, L2, L1, H1, and H2. The mass fractionation coefficients β^{Hg} and β^{Tl} are supposed to be different. The isobaric interference of Pb at mass 204 is supposed to be negligible or to have been perfectly corrected.

Table 1. Sample configuration of dynamic Hg isotope analysis in three sequences. Faraday cups are labeled L for low masses, H for high masses and Ax for the axial collector. The axial cup is the reference so its efficiency is assumed to be unity.

Sequence	*L3*	*L2*	*L1*	*(Ax)*	*H1*	*H2*
1	199	200	201	(202)	203	204
2	200	201	202	(203)	204	205
3	201	202	203	(204)	205	

$$\left|\begin{array}{c}
\ln\dfrac{\left({}^{199}\mathrm{Hg}/{}^{202}\mathrm{Hg}\right)_1^{\mathrm{meas}}}{\left({}^{199}\mathrm{Hg}/{}^{202}\mathrm{Hg}\right)^{\mathrm{ref}}} \\
\ln\dfrac{\left({}^{200}\mathrm{Hg}/{}^{202}\mathrm{Hg}\right)_1^{\mathrm{meas}}}{\left({}^{200}\mathrm{Hg}/{}^{202}\mathrm{Hg}\right)^{\mathrm{ref}}} \\
\ln\dfrac{\left({}^{201}\mathrm{Hg}/{}^{202}\mathrm{Hg}\right)_1^{\mathrm{meas}}}{\left({}^{201}\mathrm{Hg}/{}^{202}\mathrm{Hg}\right)^{\mathrm{ref}}} \\
\ln\dfrac{\left({}^{204}\mathrm{Hg}/{}^{202}\mathrm{Hg}\right)_1^{\mathrm{meas}}}{\left({}^{204}\mathrm{Hg}/{}^{202}\mathrm{Hg}\right)^{\mathrm{ref}}} \\
\ln\dfrac{\left({}^{200}\mathrm{Hg}/{}^{202}\mathrm{Hg}\right)_2^{\mathrm{meas}}}{\left({}^{200}\mathrm{Hg}/{}^{202}\mathrm{Hg}\right)^{\mathrm{ref}}} \\
\ln\dfrac{\left({}^{201}\mathrm{Hg}/{}^{202}\mathrm{Hg}\right)_2^{\mathrm{meas}}}{\left({}^{201}\mathrm{Hg}/{}^{202}\mathrm{Hg}\right)^{\mathrm{ref}}} \\
\ln\dfrac{\left({}^{204}\mathrm{Hg}/{}^{202}\mathrm{Hg}\right)_2^{\mathrm{meas}}}{\left({}^{204}\mathrm{Hg}/{}^{202}\mathrm{Hg}\right)^{\mathrm{ref}}} \\
\ln\dfrac{\left({}^{205}\mathrm{Tl}/{}^{203}\mathrm{Tl}\right)_2^{\mathrm{meas}}}{\left({}^{205}\mathrm{Tl}/{}^{203}\mathrm{Tl}\right)^{\mathrm{ref}}} \\
\ln\dfrac{\left({}^{201}\mathrm{Hg}/{}^{202}\mathrm{Hg}\right)_3^{\mathrm{meas}}}{\left({}^{201}\mathrm{Hg}/{}^{202}\mathrm{Hg}\right)^{\mathrm{ref}}} \\
\ln\dfrac{\left({}^{204}\mathrm{Hg}/{}^{202}\mathrm{Hg}\right)_3^{\mathrm{meas}}}{\left({}^{204}\mathrm{Hg}/{}^{202}\mathrm{Hg}\right)^{\mathrm{ref}}} \\
\ln\dfrac{\left({}^{205}\mathrm{Tl}/{}^{203}\mathrm{Tl}\right)_3^{\mathrm{meas}}}{\left({}^{205}\mathrm{Tl}/{}^{203}\mathrm{Tl}\right)^{\mathrm{ref}}}
\end{array}\right|
=
\left|\begin{array}{ccccccc}
1 & 0 & 0 & 0 & 0 & \ln\dfrac{M_{^{199}\mathrm{Hg}}}{M_{^{202}\mathrm{Hg}}} & 0 \\
0 & 1 & 0 & 0 & 0 & \ln\dfrac{M_{^{200}\mathrm{Hg}}}{M_{^{202}\mathrm{Hg}}} & 0 \\
0 & 0 & 1 & 0 & 0 & \ln\dfrac{M_{^{201}\mathrm{Hg}}}{M_{^{202}\mathrm{Hg}}} & 0 \\
0 & 0 & 0 & 0 & 1 & \ln\dfrac{M_{^{204}\mathrm{Hg}}}{M_{^{202}\mathrm{Hg}}} & 0 \\
1 & 0 & -1 & 0 & 0 & \ln\dfrac{M_{^{200}\mathrm{Hg}}}{M_{^{202}\mathrm{Hg}}} & 0 \\
0 & 1 & -1 & 0 & 0 & \ln\dfrac{M_{^{201}\mathrm{Hg}}}{M_{^{202}\mathrm{Hg}}} & 0 \\
0 & 0 & -1 & 0 & 1 & \ln\dfrac{M_{^{204}\mathrm{Hg}}}{M_{^{202}\mathrm{Hg}}} & 0 \\
0 & 0 & 0 & -1 & 0 & 0 & \ln\dfrac{M_{^{205}\mathrm{Tl}}}{M_{^{203}\mathrm{Tl}}} \\
1 & -1 & 0 & 0 & 0 & \ln\dfrac{M_{^{201}\mathrm{Hg}}}{M_{^{202}\mathrm{Hg}}} & 0 \\
0 & -1 & 0 & 1 & 0 & \ln\dfrac{M_{^{204}\mathrm{Hg}}}{M_{^{202}\mathrm{Hg}}} & 0 \\
0 & 0 & -1 & 0 & 1 & 0 & \ln\dfrac{M_{^{205}\mathrm{Tl}}}{M_{^{203}\mathrm{Tl}}}
\end{array}\right|
\left|\begin{array}{c}
\ln \mathcal{A}_{\mathrm{L3}} \\
\ln \mathcal{A}_{\mathrm{L2}} \\
\ln \mathcal{A}_{\mathrm{L1}} \\
\ln \mathcal{A}_{\mathrm{H1}} \\
\ln \mathcal{A}_{\mathrm{H2}} \\
\beta_{Hg} \\
\beta_{Tl}
\end{array}\right| \quad (62)$$

In the present case, the system has more equations (11) than unknowns (7) and may be conveniently solved for $\mathbf{x} = \mathcal{A}_{\mathrm{L3}}\ldots\mathcal{A}_{\mathrm{H2}}$, β_{Hg}, and β_{Tl} by the least-square solution alluded to above. The system is, in general, ill-conditioned and extended precision should be used for the inversion.

The next step is optional and consists in correcting the measured ratios for cup efficiencies. For instance, $(^{200}Hg/^{202}Hg)_2$ will be multiplied by $\mathcal{A}_{L1}/\mathcal{A}_{L3}$. Then, the slope $s^{200/202}_{205/203}$ in the $\ln(^{200}Hg/^{202}Hg)$ vs. $\ln(^{205}Tl/^{203}Tl)$ is calculated as:

$$s^{200/202}_{205/203} = \frac{\beta_{Hg}}{\beta_{Tl}} \frac{\ln\left(M_{^{200}Hg}/M_{^{202}Hg}\right)}{\ln\left(M_{^{205}Tl}/M_{^{203}Tl}\right)} \tag{63}$$

The rest of the correction procedure is identical to what is described above, notably the calculation of the intercepts I. The advantage of this procedure is that it can be applied without the a priori determination of the slope of the fractionation trends involving the Hg and Tl isotopic ratios.

A FEW SIMPLE RULES FOR PRECISE AND ACCURATE MC-ICP-MS MEASUREMENTS

The main advantages of MC-ICP-MS with respect to TIMS instruments are the efficient ionization of most elements and the operation of the instrument at steady-state, which allows a full control of mass fractionation. Its main disadvantages are a larger mass fractionation, common isobaric interferences, and a less-than-perfect peak shape due to the energy spread. Rather constant values of $\beta = 2.1 \pm 0.4$ are observed over the whole mass range of some MC-ICP-MS instruments, while others have a wider range. It is inevitable, however, that as improved transmission gets close to unity, mass fractionation diminishes ($\beta \rightarrow 0$). The bad news then is that in the process, mass fractionation will vary with transmission, at least more so than on the present-generation instruments. If, however, the signal intensities of samples and standards are kept within a narrow range, MC-ICP-MS should become the instrument of choice. Here, we assert that a few simple rules help achieving high precision and accuracy:

1. Because of its isotopic variability, background must be reduced at any cost. The matrix of samples and standards must be reduced by appropriate chemistry to trace amounts, typically to a total concentration far smaller than the element to be analyzed. This requirement is most critical when the mass bias is inferred not internally from the sample itself, but externally from bracketing standards or from a different element used for isotopic normalization. Even the most dilute heavy species may drastically affect mass discrimination.

2. A correlation between isotopic ratios corrected for mass fractionation may reveal (i) rounded or slopping peak tops (ii) second-order fractionation effects. The necessity of a second-order correction should be established by showing that the bias left after a first-order correction still depends smoothly on the mass.

3. Cup efficiencies, which largely reflect a trajectory-dependent transmission as a result of imperfect *z*-focusing, must be evaluated regardless of the instrument.

4. When mass discrimination is evaluated using a different element (e.g., Cu on Zn, Mo on Zr), the assumption that the two elements fractionate to the same extent is incorrect and results in systematic errors of several tenths of a per mil of even more. Forcing an arbitrary isotopic composition on the normalizing element in order to reduce the analytical bias on samples has no physical basis and may lead to long-term errors (Albarède et al. 2004). In contrast, if, as a result of efficient separation chemistry, the mass bias of standard solutions changes too little over long periods of time, the correlation between isotopic ratios is inadequate and standard bracketing is the method of choice.

5. In principle, double-spike techniques represent the most suitable approach to determine the isotope composition of elements with four isotopes or more (Fe, Zn). In most cases, these techniques involve the addition of an isotope which is usually minor in natural samples, such as ^{67}Zn or ^{58}Fe, implying that the risk introduced by memory effects on these spike isotopes must be carefully weighed against the added gain in precision from using the double spike. Such a risk is clearly more present with MC-ICP-MS than with TIMS.

ACKNOWLEDGMENTS

FA thanks Philippe Télouk for years of fruitful and friendly interaction, and all those who participated into the early development of the Lyon ICP-MS, notably, Janne Blichert-Toft, Maud Boyet, Béatrice Luais, Chloé Maréchal, Chantal Douchet, Jean-Marc Luck, and Dalila Ben-Othman. This manuscript was written during a leave of absence of FA at Caltech as a Moore Foundation scholar. Reviews by an anonymous reviewer, by Ariel Anbar, Rick Carlson, Ed Young, and by Clark Johnson are gratefully acknowledged. Janne Blichert-Toft obliged with careful English editing.

REFERENCES

Al-Ammar AS, Barnes RM (2001) Improving isotope ratio precision in inductively coupled plasma quadrupole mass spectrometry by common analyte internal standardization. J Anal At Spectrom 16:327-332

Albarède F (1995) Introduction to Geochemical Modeling. Cambridge University Press

Albarède F, Télouk P, Blichert-Toft J, Boyet M, Agranier A, Nelson B (2004) Precise and accurate isotopic measurements using multiple-collector ICPMS. Geochim Cosmochim Acta, in press

Anbar AD, Roe JE, Barling J, Nealson KH (2000) Nonbiological fractionation of iron isotopes. Science 288: 126-128

Anbar AD, Knab, KA, Barling J (2001) Precise determination of mass-dependent variations in the isotopic composition of molybdenum using MC-ICPMS. Anal Chem 73:1425-1431

Barling J, Arnold GL, Anbar AD (2001) Natural mass-dependent variations in the isotopic composition of molybdenum. Earth Planet Sci Lett 193:447-457

Beard BL, Johnson CM (1999) High precision iron isotope measurements of terrestrial and lunar materials. Geochim Cosmochim Acta 63:1653-1660

Beard BL, Johnson CM, Skulan JL, Nealson KH, Cox L, Sun H (2003) Application of Fe isotopes to tracing the geochemical and biological cycling of Fe. Chem Geol 195:87-117

Belshaw NS, Zhu XK, Guo Y, O'Nions RK (2000) High precision measurement of iron isotopes by plasma source mass spectrometry. Int J Mass Spectrom 197:191-195

Bigeleisen J, Mayer MG. (1947) Calculation of equilibrium constants for isotopic exchange reactions. J Chem. Phys 15:261-267

Blichert-Toft J, Chauvel C, Albarède F (1997) Separation of Hf and Lu for high-precision isotope analysis of rock samples by magnetic sector-multiple collector ICP-MS. Contrib Mineral Petrol 127:248-260

Cameron AE, Smith DH, Walker RL (1969) Mass spectrometry of nanogram-size samples of lead. Anal Chem 41:525-526

Campargue R (1984) Progress in overexpanded supersonic jets and skimmed molecular beams in free zones of silence. J Phys Chem A 88: 4466-4474

Carlson RW, Hauri EH, Alexander CMO'D (2001) Matrix induced isotopic mass fractionation in the ICP-MS. *In:* Plasma source mass spectrometry: The new Millennium. Holand G, Turner SD (eds), Cambridge: Roy Soc Chem, p 288-297

Carlson RW, Hauri EH (2001) Extending the ^{107}Pd-^{107}Ag chronometer to low Pd/Ag meteorites with multicollector plasma-ionization mass spectrometry. Geochim Cosmochim Acta 65:1839-1848

Clayton, RN, Onuma N, Mayeda TK (1976) Distribution of the presolar component in Allende and other carbonaceous chondrites. Earth Planet Sci Lett 30:10-18

Compston W, Oversby VM (1969) Lead isotopic analysis using a double spike. J Geophys Res 74:4338-4348

Criss RE (1999) Principles of Stable Isotope Distribution. University Press, Oxford

Date AR, Ying Y, Stuart ME (1987) The influence of polyatomic ion interferences in analysis by inductively coupled plasma source mass spectrometry (ICP-MS). Spectrochim Acta 42B:3-20

De Laeter JR (2001) Applications of Inorganic Mass Spectrometry. Wiley, New York 496 p

DePaolo DJ (1988) Neodymium isotope geochemistry an introduction. Spinger Verlag Berlin Heidelberg, Germany

Dixon PR, Perrin RE, Rokop DJ, Maeck R, Janecky DR, Banar JP (1993) Measurements of iron isotopes (^{54}Fe, ^{56}Fe, ^{57}Fe, and ^{58}Fe) in sub microgram quantities of iron. Anal Chem 65:2125-2130

Dodson MH (1963) A theoretical study of the use of internal standards for precise isotopic analysis by the surface ionization technique: Part I General first-order algebraic solutions. J Sci Instrum 40:289-295

Douglas DJ (1989) Some current perspectives on ICP-MS. Canad J Spectrosc 34:38-49

Douglas DJ, French JB (1986) An improved interface for inductively coupled plasma-mass spectrometry (ICP-MS). Spectrochim Acta 41B:197-204

Douglas DJ, French JB (1992) Collisional focusing effects in radio frequency quadropoles. J Am Soc Mass Spectrom 3:398-407

Douglas DJ, Tanner SC (1998) Fundamental considerations in ICP-MS. *In:* Inductively Coupled Plasma Mass Spectrometry. Montaser A (ed), Wiley-VCH, New York, p 615-679

Eugster O, Tera F, Wasserburg GJ (1969) Isotopic analyses of barium in meteorites and in terrestrial samples. J Geophys Res 74:3897-3908

Gale NH (1970) A solution in closed form for lead isotopic analysis using a double spike. Chem Geol 6:305-310

Galer S J (1999) Optimal double and triple spiking for high precision lead isotopic measurement. Chem Geol 157:255-274

Gillson GR, Douglas DJ, Fulford JE, Halligan KW, Tanner SD (1988) Nonspectroscopic interelement interferences in inductively coupled plasma mass spectrometry. Anal Chem 60:1472-1474

Gonfiantini R, Valkiers S, Taylor PDP, De Bievre P (1997) Adsorption in gas mass spectrometry. II Effects on the measurement of isotope amount ratios. Int J Mass Spectrom Ion Proc 171:231-242

Habfast K (1997) Advanced isotope ratio mass spectrometry. I: magnetic isotope ratio mass spectrometers. *In:* Modern Isotope Ratio Mass Spectrometry. Chemical Analysis Vol. 145. Platzner IT (ed). John Wiley and Sons, Chichester UK, p 11-82

Habfast K (1998) Fractionation correction and multiple collectors in thermal ionization isotope ratio mass spectrometry. Int J Mass Spec 176:133-148

Hamelin B, Manhès G, Albarède F, Allègre CJ (1985) Precise lead isotope measurements by the double spike technique: a reconsideration. Geochim Cosmochim Acta 49:173-182

Hart SR, Zindler A (1989) Isotope fractionation laws: A test using calcium. Int J Mass Spectr Ion Proc 89: 287-301

Hirata T, Hayano Y, Ohno T (2003) Improvements in precision of isotopic ratio measurements using laser ablation-multiple collector-ICP-mass spectrometry: reduction of changes in measured isotopic ratios. J Anal At Spectrom 18:1283-1288

Kehm K, Hauri EH, Alexander CMO'D, Carlson RW (2003) High precision iron isotope measurements of meteoritic material by cold plasma ICP-MS. Geochim Cosmochim Acta 67:2879-2891

Hofmann A (1971) Fractionation corrections for mixed-isotope spikes of Sr, K, and Pb. Earth Planet Sci Lett 10:397-402

Jackson SE, Günther D (2003) The nature and sources of laser induced isotopic fractionation in laser ablation-multicollector-inductively coupled plasma-mass spectrometry. J Anal At Spectrom 18:205-212

Jiang S-J, Houk RS, Stevens MA (1988) Alleviation of overlap interferences for determination of potassium isotope ratios by Inductively-Coupled Plasma Mass Spectrometry. Anal Chem 60:1217-1220

Lam JWH, Horlick G (1990) A comparison of argon and mixed gas plasmas for inductively coupled plasma-mass spectrometry. Spectrochim Acta Part B 45:1313-1325

Langmuir I, Kingdon KH(1925) Thermionic effects caused by vapours of alkali metals. Phil Trans R Soc A107:61-79

Johnson, CM, Beard BL (1999) Correction of instrumentally produced mass fractionation during isotopic analysis of Fe by thermal ionization mass spectrometry. Int J Mass Spectrom 193:87-99

Kehm K, Hauri EH, Alexander CMOD, Carlson RW. (2003) High precision iron isotope measurements of meteoritic material by cold plasma ICP-MS. Geochim Cosmochim Acta 67:2879-2891

Longerich HP, Fryer BJ, Strong DF (1987) Determination of lead isotope ratios by inductively coupled plasma-mass spectrometry (ICP-MS). Spectrochim Acta Part B 42:39-48

Ludwig KR (1997) Optimization of multicollector isotope-ratio measurement of strontium and neodymium. Chem Geol 135:325-334

Luais B, Télouk P, Albarède F (1997) Precise and accurate neodymium isotopic measurements by plasma-source mass spectrometry. Geochim Cosmochim Acta 61:4847-4854

Maréchal C, Télouk P, Albarède F (1999) Precise analysis of copper and zinc isotopic compositions by plasma-source mass spectrometry. Chem Geol 156:251-273

Maréchal CN, Douchet C, Nicolas E, Albarède F (2000) The abundance of zinc isotopes as a marine biogeochemical tracer. Geochem Geophys Geosyst 1:1999GC-000029

Maréchal C, Albarède F (2002) Ion-exchange fractionation of copper and zinc isotopes. Geochim Cosmochim Acta 66:1499-1509

Miller YM, Ustinov VI, Artemov YM (1966) Mass spectrometric determination of calcium isotope variations. Geochem Inter 3:929-933

Montaser A, Tan H, Iishi II, Nam SH, Cai M (1991) Argon inductively coupled plasma mass spectrometry with thermospray, ultrasonic, and pneumatic nebulization. Anal Chem 63:2660-2665

Montaser A, Minnich MG, Liu H, Gustavsson AGT, Browner RF (1998) Fundamental aspects of sample introduction in ICP spectrometry. *In*: Inductively Coupled Plasma Mass Spectrometry. Montaser A (ed), Wiley-VCH, New York, p 335-420

Montaser A, Minnich MG, McLean JA, Liu H, Caruso JA, McLeod CW (1998) Sample introduction in ICPMS. *In*: Inductively Coupled Plasma Mass Spectrometry. Montaser A (ed) Wiley-VCH, New York, p 83-264

Montaser A, Zhang H (1998) Mass spectrometry with mixed-gas and Helium ICPS. *In:* Inductively Coupled Plasma Mass Spectrometry. Montaser A (ed), Wiley-VCH, New York, p 809-890

Nier AO (1940) A mass spectrometer for routine isotope abundances measurement. Rev Sci Instrum 11:212-216

Niu H, Houk RS (1994) Langmuir probe measurements of the ion extraction process in inductively coupled plasma mass spectrometry-I. Spatially resolved determination of electron density and electron temperature. Spectrochim Acta 49B:1283-1303

Niu H, Houk RS (1996) Fundamental aspects of ion extraction in inductively coupled plasma mass spectrometry. Spectrochim Acta Part B 51:779-815

Nonose N, Kubota M (2001) Non-spectral interferences in inductively coupled plasma high-resolution mass spectrometry. Part 2. Comparison of interferences in quadrupole and high-resolution inductively coupled plasma mass spectrometers. J Anal At Spectrom 16:560-566

Olivares JA, Houk RS (1986) Suppression of analyte signal by various concomitant salts in inductively coupled plasma mass spectrometry. Anal Chem 58:20-25

Platzner IT (1997) Ion formation processes. *In:* Modern Isotope Ratio Mass Spectrometry. Chemical Analysis Vol. 145. Platzner IT (ed), John Wiley and Sons, Chichester UK, p 150-170

Praphairaksit N, Houk RS (2000a) Attenuation of matrix effects in inductively coupled plasma mass spectrometry with a supplemental electron source inside the skimmer. Anal Chem 72:2351-2355

Praphairaksit N, Houk RS (2000b) Reduction of space charge effects in inductively coupled plasma mass spectrometry using a supplemental electron source inside the skimmer: ion transmission and mass spectral characteristics. Anal Chem 72:2356-2361

Powell R, Woodhead J, Hergt J (1998) Uncertainties on lead isotope analyses: deconvolution in the double-spike method. Chem Geol 148:95-104

Rameбäck H, Berglund M, Kessel R, Wellum R (2002) Modeling isotope fractionation in thermal ionization mass spectrometry filaments having diffusion controlled emission. Int J Mass Spectrom 216:203-208

Roe JE, Anbar AD, Barling J (2003) Nonbiological fractionation of Fe isotopes: evidence of an equilibrium isotope effect. Chem Geol 195:69-85

Russel WA, Papanastassiou DA, Tombrello TA (1978) Ca isotope fractionation on the Earth and other solar system materials. Geochim Cosmochim Acta 42:1075-1090

Sakata KI, Kawabata K (1994) Reduction of fundamental polyatomic ions in inductively coupled plasma mass spectrometry. Spectrochim Acta Atom Spectrosc Part B 49:1027-1038

Siebert, C, Nägler, TF, Kramers JD (2001) Determination of molybdenum isotope fractionation by double-spike multicollector inductively coupled plasma mass spectrometry. Geochem Geophys Geosyst 2: 2000GC000124

Smith DH, Christie WH, Eby RE (1980) The resin bead as a thermal ion source: A SIMS study. Int Jour Mass Spectrom Ion Proc 36:301-316

Spivack AJ, Edmond JM (1986) Determination of boron isotope ratios by thermal ionization mass spectrometry of the dicesium metaborate cation. Anal Chem 58:31-35

Tanner SD, Cousins LM, Douglass DJ (1994) Reduction of space charge effects using a three-aperture gas dynamic vacuum interface for inductively coupled plasma mass spectrometry. Appl Spectrosc 48:1367-1372

Taylor PDP, Maeck R, De Bièvre P (1992) Determination of the absolute isotopic composition and Atomic Weight of a reference sample of natural iron. Int J Mass Spectrom Ion Processes 121:111-125

Taylor PDP, Maeck R, Hendrickx F, De Bièvre P (1993) The gravimetric preparation of synthetic mixtures of iron isotopes. Int J Mass Spectrom Ion Processes 128:91-97

Thirlwall MF (2002) Multicollector ICP-MS analysis of Pb isotopes using a ^{207}Pb-^{204}Pb double spike demonstrates up to 4000 ppm/amu systematic errors in Tl-normalization. Chem Geol 184:255-279

Turner IL, Montaser (1998) Plasma generation in ICPMS. *In*: Inductively Coupled Plasma Mass Spectrometry. Montaser A (ed),. Wiley-VCH, New York, p 265-334

Turner PJ, Mills DJ, Schröder E, Lapitajs G, Jung G, Iacone LA, Haydar DA, Montaser A (1998) Instrumentation for low- and high-resolution ICPMS. *In*: Inductively Coupled Plasma Mass Spectrometry. Montaser A (ed), Wiley-VCH, New York, p 421-501

Urey HC (1947) The thermodynamic properties of isotopic substances. J Chem Soc (London) 562-581

Vance D, Thirlwall MF (2002) An assessment of mass discrimination in MC-ICP-MS using Nd isotopes. Chem Geol 185:227-240

Walczyk T (1997) Iron isotope ratio measurements by negative thermal ionization mass spectrometry using FeF_4^- molecular ions. Int J Mass Spectrom Ion Proc 161:217-227

Wendt I, Haase G (1998) Dynamic double collector measurement with cup efficiency factor determination. Chem Geol 146:99-110

Weyer S, Schwieters JB (2003) High precision Fe isotope measurements with high mass resolution MC-ICP-MS. Int J Mass Spectrom 226:355-368

Woodhead J (2002) A simple method for obtaining highly accurate Pb isotope data by MC-ICP-MS. J Anal At Spectrom 17:1381-1385

Young ED, Ash RD, Galy A, Belshaw NS (2002a) Mg isotope heterogeneity in the Allende meteorite measured by UV laser ablation-MC-ICPMS and comparisons with O isotopes. Geochim Cosmochim Acta 66:683-698

Young ED, Galy A, Nagahara H (2002b) Kinetic and equilibrium mass-dependent isotope fractionation laws in nature and their geochemical and cosmochemical significance. Geochim Cosmochim Acta 66:1095-1104

Zhu XK, Guo Y, Williams RJP, O'Nions RK, Matthews A, Belshaw NS, Canters GW, de Waal EC, Weser U, Burgess BK, Salvato B (2002) Mass fractionation processes of transition metal isotopes. Earth Planet Sci Lett 200:47-62

APPENDIX A: TABLE OF SYMBOLS

$\mathbf{A}$: matrix of coefficients used for dynamic runs
$\mathcal{A}_k$: cup-dependent factor (efficiency) of Faraday cup k
E: ion energy
e: charge of the electron
$\mathcal{F}_i$: closure condition for the isotopic ratio N_i/N_{ref} in sample-spike mixtures (must be zero)
g: mass bias factor for the power law ($= e^u$)
h: mass bias factor for the generalized power law
I_p: ionization potential
k: Boltzmann's constant
M_i: Atomic mass of nuclide i
N_i: number of nuclides i introduced into the mass spectrometer
n_i: number of nuclides i detected by the mass spectrometer
R: gas constant
R^2: correlation coefficient of a regression line
R_i: true isotopic ratio N_i/N_{ref}
r_i: measured isotopic ratio n_i/n_{ref}
r_i^e : isotopic ratio n_i/n_{ref} corrected for an exponential bias f
$s_{i/k}^{j/k}$: slope of the mass fractionation curve in a log-log plot
S: sensitivity of the instrument
T: absolute temperature
$\mathcal{T}_i^\gamma$: transmission for isotope i in cup γ
u: mass bias factor for the linear law
v: particle velocity
W: metal work function
$\mathbf{W}$: covariance matrix
$\mathbf{x}$: vector of unknown parameters calculated from dynamic runs
$\mathbf{y}$: vector of data used in dynamic runs
$\mathcal{Y}_i$: mass bias for isotope i
β: first-order mass bias factor for the exponential law ($= M_{\mathrm{ref}}u$)
γ: second-order mass bias factor for the exponential law
μ_i: $\ln(M_i/M_{\mathrm{ref}})$
ε_i: deviation of the isotopic ratio from the reference value in part per 10,000$(r_i/R_i - 1)\times 10{,}000$
ρ: magnet turning radius
φ_i^{sp} : atomic proportion of the isotope i in a mixture contributed by the spike

APPENDIX B: EXAMPLE OF DOUBLE-SPIKE CALCULATION USING A NEWTON-RAPHSON ITERATION

For Zn, in which ^{64}Zn is abundant (49%), a double ^{67}Zn-^{68}Zn spike is a natural choice, which gives the following equation of $\mathcal{F}_{66}\left(\varphi_{64}^{sp},\beta_{sple},\beta_{mix}\right)$:

$$\mathcal{F}_{66}\left(\varphi_{64}^{sp},\beta_{sple},\beta_{mix}\right)=$$

$$\varphi_{64}^{sp}\left(\frac{^{66}Zn}{^{64}Zn}\right)^{sp}+\left(1-\varphi_{64}^{sp}\right)\left(\frac{^{66}Zn}{^{64}Zn}\right)_{ref}^{sple}\left(\frac{64}{66}\right)^{\beta_{sple}}-\left(\frac{^{66}Zn}{^{64}Zn}\right)_{meas}^{mix}\left(\frac{64}{66}\right)^{\beta_{mix}}=0 \quad (64)$$

where the "ref" subscript refers to a reference value of the un-spiked sample, normally a standard solution. The other two equations $\mathcal{F}_{64}=0$ and $\mathcal{F}_{68}=0$ make the system complete (least-square methods could be used for elements with more than four isotopes). The derivatives of $\mathcal{F}_{66}$ are written row-wise as:

$$\begin{aligned}
\frac{\partial \mathcal{F}_{66}}{\partial \varphi_{64}^{sp}} &= \left(\frac{^{66}Zn}{^{64}Zn}\right)^{sp}-\left(\frac{^{66}Zn}{^{64}Zn}\right)_{ref}^{sple}\left(\frac{64}{66}\right)^{\beta_{sple}} \\
\frac{\partial \mathcal{F}_{66}}{\partial \beta_{sple}} &= \beta_{sple}\left(1-\varphi_{64}^{sp}\right)\left(\frac{^{66}Zn}{^{64}Zn}\right)_{ref}^{sple}\left(\frac{64}{66}\right)^{\beta_{sple}-1} \\
\frac{\partial \mathcal{F}_{66}}{\partial \beta_{mix}} &= -\beta_{mix}\left(\frac{^{66}Zn}{^{64}Zn}\right)_{meas}^{mix}\left(\frac{64}{66}\right)^{\beta_{mix}-1}
\end{aligned} \quad (65)$$

Two similar sets of derivatives can be written by substituting 67 and 68 to 66. In order to solve the system for φ_{64}^{sp}, β_{sple}, and β_{mix}, we select a first guess of these variables, which we group into a 3×1 vector $\mathbf{x}_0$. Inserting this guess into each equation produces a 3×1 vector $\mathbf{y}_0$ out of the three $\mathcal{F}_i$ values and a 3×3 matrix $\mathbf{J}_0$ out of the nine derivatives. The updating Newton-Raphson step uses the first-order expansion of the $\mathcal{F}$ values about the current value $\mathbf{x}_0$ of $\mathbf{x}$ as:

$$\mathbf{x}_1=\mathbf{x}_0-\mathbf{J}_0^{-1}\mathbf{y}_0 \quad (66)$$

The variables are updated and the step is repeated until convergence is achieved. Errors can be propagated either linearly in a matrix form by computing the derivatives of the unknowns with respect to the observations or by Monte-Carlo (brute force) techniques (e.g., Albarède 1995).

Reviews in Mineralogy & Geochemistry
Vol. 55, pp. 153-195, 2004
Copyright © Mineralogical Society of America

Developments in the Understanding and Application of Lithium Isotopes in the Earth and Planetary Sciences

Paul B. Tomascak

Department of Geology
University of Maryland
College Park, Maryland, 20742, U.S.A.

INTRODUCTION

The significant relative mass difference (c. 16%) between the two stable isotopes of Li (approximately ^{6}Li 7.5%, ^{7}Li 92.5%), coupled with broad elemental dispersion in Earth and planetary materials, makes this a system of considerable interest in fingerprinting geochemical processes, determining mass balances, and in thermometry. Natural mass fractionation in this system is responsible for c. 6% variation among materials examined to date (Fig. 1). Although the "modern era" of Li isotope quantification has begun, there are still many questions about the Li isotopic compositions of fundamental materials and the nature of fractionation by important mechanisms that are unanswered (e.g., Hoefs 1997).

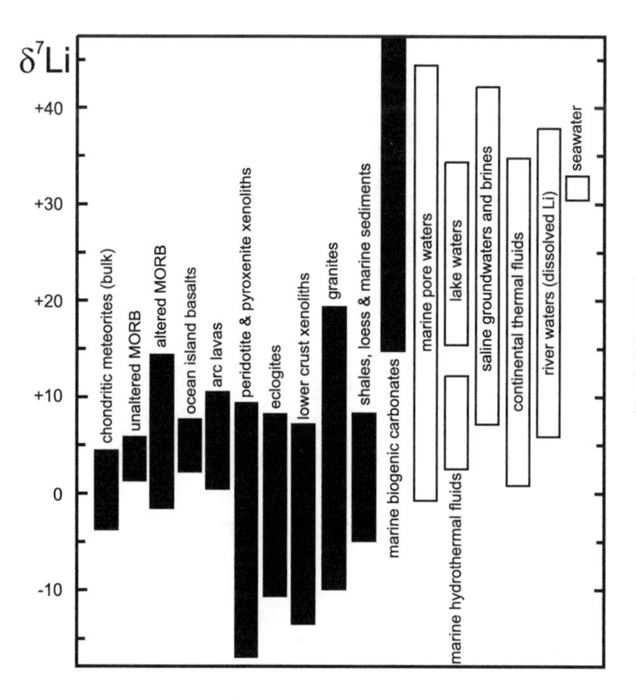

Figure 1. Summary of lithium isotopic compositions of Earth and planetary materials. Filled bars are solid samples, open bars are liquids. See text for references and details.

1529-6466/04/0055-0005$05.00

The purpose of this chapter is to summarize the current understanding of Li isotopes in geo- and cosmochemical systems and to indicate (1) where Li isotopes have a high probability of adding new understanding of these systems; (2) where some of the more significant deficits in knowledge exist. The small but burgeoning Li isotope community has not yet compiled the volume of peer-reviewed literature needed to adequately assess even that which has been studied to date. As a result, significant portions of this chapter are based on data reported in abstracts, and as such are more than normally subject to revisions over time. This chapter is anticipated to serve as a starting point for those interested in research incorporating Li isotope geochemistry, or in understanding the state of extant research.

BACKGROUND: LITHIUM ISOTOPES IN THE SCIENCES

Experiments in Li isotope fractionation

Knowledge of significant Li isotopic fractionation during basic chemical processes is long-standing. The early experiments by Taylor and Urey (1938), in which Li isotopes were fractionated by incomplete extraction of an aqueous solution from a zeolite exchange column, demonstrated clearly that the degree of induced isotopic fractionation was considerable, with ^{6}Li preferentially retained by the exchange medium (Fig. 2). This indicated that natural chemical exchange processes, such as water-rock or water-soil interaction, had a strong potential to generate isotopically distinct reservoirs in the Earth. Nevertheless, a long interregnum followed this early discovery, leading up to the development of routine accurate and precise methods for Li isotope analysis nearly half a century later.

Lithium isotopes in fields outside geochemistry

Lithium is an important element in many industries (Bach 1985). Lithium is used medically as a treatment for bipolar disorders (Schou 1988). Lithium toxicity, especially to the renal system, is problematical. Estimating systemic elemental mass balance, especially for patients receiving oral Li dosing, is important, and is one area in which Li isotope ratios are

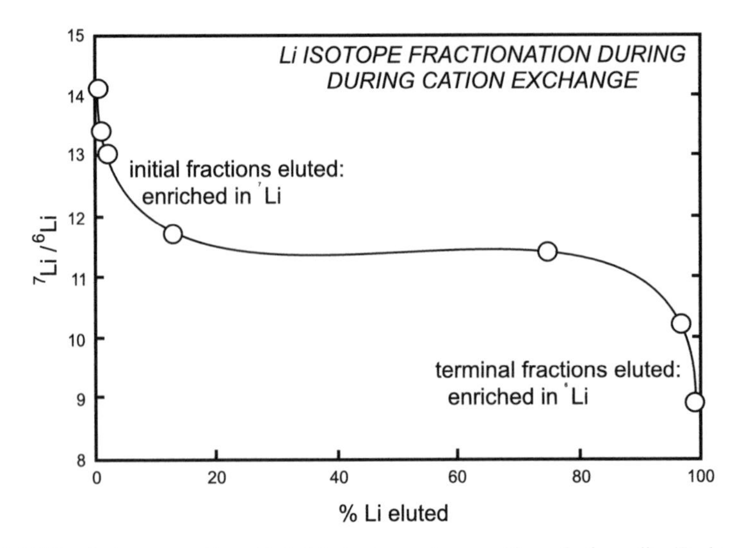

Figure 2. Lithium isotope separation effected during ion exchange with synthetic zeolite (Taylor and Urey 1938). As has been demonstrated repeatedly since that study, in both natural and synthetic experiments, ^{6}Li is fixed more effectively in the exchanger than ^{7}Li.

sought. Interest in the quantification of Li isotopes in biochemistry also stems from differential effects of ^{6}Li vs. ^{7}Li in kidney health (Stoll et al. 2001).

Lithium is an element of interest in astrophysics, owing to the marked variations in Li abundance and isotopic composition of materials emanating from different sources in the cosmos. The development of the "terrestrial" or "Solar system" Li isotopic compositions, which, to the limit of our current understanding, are quite different from values predicted for nucleosynthetic or interstellar Li, remains a puzzle for active astronomical research.

Large scale enrichment of lithium for thermonuclear uses took place at the Oak Ridge National Laboratory in the 1950's. The enrichment primarily employed ion exchange between aqueous/organic solutions and amalgam, commonly mercury-based (Palko et al. 1976). Electrochemical separation has also been employed for such operations (Umeda et al. 2001). These practices have not been taken up in academic laboratories in the intervening years, partly as they tend to be most effective only with relatively pure starting materials, partly because of the difference in scales involved. Enrichment factors of ^{6}Li of 1–7% are typical for these techniques (Symons 1985).

One interesting side-effect of the industrial isotopic enrichment of Li is the development of isotopically anomalous materials that make their way into other industries. For example, Qi et al. (1997a) found that commercial shelf standards for Li concentration had compositions that were over 300% enriched in ^{7}Li relative to known terrestrial materials (Fig. 3).

LITHIUM ISOTOPIC ANALYSIS

From early Li isotope measurement to the "modern era" of analysis

Shortly after the development of the early mass spectrometers, Li isotopes were identified by Francis Aston (1932). Although mass spectrometric techniques are those most commonly applied to the measurement of isotope ratios in geochemistry, attempts to quantify Li isotopes have been made using non-mass based emission methods (e.g., atomic absorption: Zaidel and Korennoi 1961; various nuclear methods: Kaplan and Wilzbach 1954; Brown et al. 1978;

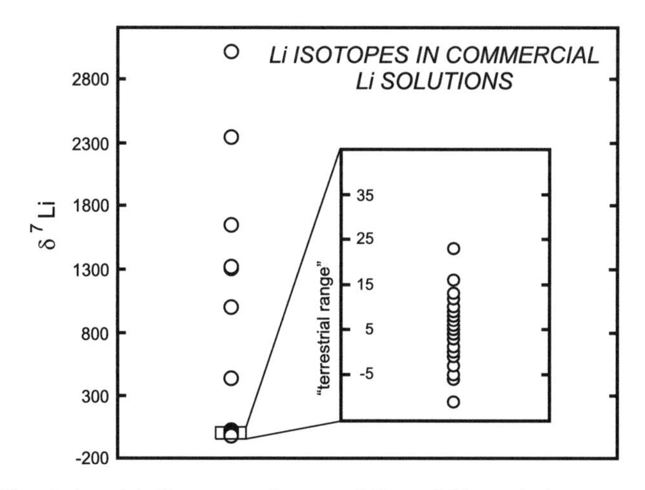

Figure 3. Lithium isotope data for a range of commercially-available synthetic concentration standards (Qi et al. 1997a). The inset expands the c. 60‰ range of reported natural samples. Although most anthropogenically-processed Li retains a broadly "terrestrial" value, nearly 20% of the samples examined show enormous isotopic enrichment in the heavy isotope.

Franklin et al. 1986). None of these techniques has enjoyed long term success. Measurement of Li isotopes by mass spectrometry faces the primary problem of controlling mass fractionation from the emitter. Ironically, the very property that makes Li geochemically interesting makes quantifying its isotopic composition with precision extraordinarily challenging. For this reason, mass spectrometric measurements of Li must be compared directly to a standard material. As long as all laboratories make use of the same standard material, its isotopic composition is academic, as the measured isotopic composition of the standard drops out of the arithmetic of normalization.

It is reasonable to assert that the "modern era" of Li isotope determination for geological materials began with the creation of an international standard for Li isotope measurement (NBS L-SVEC, now NIST SRM-8545; Flesch et al. 1973). The Li_2CO_3 standard was fabricated from "virgin" lithium ores (mainly spodumene) from near Kings Mountain, North Carolina. The homogeneity of the standard has been recently verified by analysis of aliquots used in four different laboratories (T Magna, written communication 2003). Since the manufacture of the L-SVEC standard, isotope ratios are presented in permil deviations (δ units) from the ratio measured for this material. The International Union for Pure and Applied Chemistry subsequently recommended the use of $^7Li/^6Li$ (Coplen et al. 1996), indicating that δ^7Li (= $[(^7Li/^6Li_{sample}/^7Li/^6Li_{L\text{-}SVEC}) - 1] \times 1000$) is the preferred normalized isotopic notation for lithium. Nevertheless, some authors continue to use the reverse notation (δ^6Li), following Chan (1987). At values within c. 7‰ of zero deviation from L-SVEC, a simple sign change is all that is required to switch between notations. At the extreme range of natural fractionation a value of $\delta^6Li = -38.5$ is equivalent to $\delta^7Li = +40.0$.

After the establishment of the L-SVEC standard, the development of accurate and precise quantification of Li isotope ratios was still more than a decade from being actualized. The initial publication of Chan (1987), using thermal ionization mass spectrometry (TIMS), was the first significant demonstration of routine precise determination of Li isotope ratios, although coeval work was ongoing in Belgium, at what would become the Institute for Reference Materials and Measurements (IRMM). The Li standard provided by the IRMM (IRMM-016; Michiels and DeBièvre 1983) has been assayed to have a Li isotopic composition indistinguishable from L-SVEC (Lamberty et al. 1987; Qi et al. 1997b).

In the late 1980's, methods for precise Li isotope determination (i.e., <2‰ long-term reproducibility) by a variety of TIMS methods became established. These techniques developed different strategies for dealing with the significant thermal mass fractionation of Li from the surface of a filament: from measurement of heavier ionic species (e.g., $Li_2BO_2^+$: Chan 1987, Sahoo and Masuda 1995; Li_2F^+: Green et al. 1988; $NaLiBO_2^+$: Chan et al. 1992) to evaporation of a Li molecule and subsequent ionization and measurement of Li as metal ions (Michiels and DeBièvre 1983; Xiao and Beary 1989; Moriguti and Nakamura 1993; Clausen 1995; You and Chan 1996; Sahoo and Masuda 1998).

Plasma-source mass spectrometry

Given considerations like availability and cost, isotope ratio measurement using single-collecting (quadrupole magnet) plasma source mass spectrometry (ICP-MS) has been an attractive pursuit, and Li isotopes have been measured using this equipment with improving success and precision over time (Sun et al. 1987; Koirtyohann 1994; Grégoire et al. 1996; Košler et al. 2001). Although the estimated precision is generally c. 2× poorer than mass analytical methods using sector-magnet deflection, quadrupole ICP-MS techniques are an attractive alternative, particularly in suites that show large isotopic fractionations and for surveying larger groups of samples.

The recent and meteoric development of plasma source mass spectrometers equipped with high stability sector-field magnets and multiple collectors (MC-ICP-MS) has lead to a colossal diversification in precise isotopic measurement, particularly with respect to the stable isotopes (Halliday et al. 1998). For Li isotope analysis, this instrumental advance has lead to the development of a set of techniques that have rapidly become widespread (Tomascak et al. 1999a; McDonough et al. 2001; Bouman et al. 2002; Magna et al. 2002; Nishio and Nakai 2002; Bryant et al. 2003a). These techniques have the common feature that Li isotope ratios are measured relative to bracketing standard analyses. Matrix effects, brought on by significant differences between the dissolved solid content of sample versus standard solution, perturb the essential smooth change in instrumental mass bias over time. The result of aspirating samples with unacceptable matrix loads into the plasma is erroneous sample/standard offsets. Unlike in TIMS analysis, however, the effect is obvious and suspect samples can be immediately culled.

Technical and theoretical aspects of MC-ICP-MS measurement were discussed in an earlier chapter of this volume (Albarède and Beard 2004). There are a few factors specific to Li isotope measurement by MC-ICP-MS that need to be assessed before analytical success can be achieved. Whereas measurement in the lightest part of the mass spectrum allows interference from molecular isobars to be avoided, both $^{12}C^{++}$ (on $^{6}Li^{+}$) and $^{14}N^{++}$ (on $^{7}Li^{+}$) interferences are possible, and need to be considered.

As the L-SVEC standard is ordinarily analyzed frequently among sample analyses by MC-ICP-MS, it is not useful to use this measurement uncertainty to estimate analytical precision (Tomascak et al. 1999a), but rather the reproducibility of independent measurements of samples in replicate or duplicate can be used. Parameters of accuracy and precision are discussed below.

Requirements in chemical separation

None of the bulk methods of Li isotope analysis permit high performance without isolation of Li from the sample matrix, although the degree to which other elements must be eliminated from an analyte solution differs among analytical methods. The techniques in common use for Li isolation involve chromatography with a variety of ion exchange media. The combined need for both good elemental purification and maximum Li yield has lead to the development of a plethora of techniques for Li separation chemistry in Earth materials (Chan 1987; Oi et al. 1997, 1999; Moriguti and Nakamura 1998a; Tomascak et al. 1999a).

Unlike exchange chemistry of, for example, long-lived radiogenic isotope systems (e.g., Sr, Nd, Pb), isolation chemistry for stable isotope quantification requires essentially 100% yield. Unless a multiple isotope spike is added, deviation from 100% yield introduces unquantifiable isotopic fractionation. Similarly, during digestion of solid samples, no residue can be tolerated, for the concern that Li in the residue will not be isotopically identical to that in the analyte solution. As shown by Oi et al. (1991), cationic solutions interacting with various types of commonly-used acidic exchange media show fractionation such that ^{6}Li has a higher apparent affinity for the stationery phase than ^{7}Li (analogous to the synthetic experiments of Taylor and Urey 1938; Fig. 2). Erroneously heavy compositions indicate a component of sample Li was left on the column, whereas erroneously light values can be manufactured by not collecting the elution from initial breakthrough of Li. A graphic demonstration of the potential for disaster in Li separation is given by Chan et al. (1999, 2002c), in which the Li elution from cation exchange resin was partly missed, resulting in measured δ^{7}Li up to 14‰ too low relative to later measurements with complete recovery during chemistry (Chan et al. 2002c). In many ion exchange procedures it is clear that the nature of the sample material added to a column has an effect on the specific volume needed to completely elute Li (e.g., Moriguti and Nakamura 1998a; Chan et al. 2002c; James and Palmer 2000a; Nishio et al. 2004), meaning that a single calibration of an exchange column may not be adequate for the entire range of materials a laboratory might process. This appears to be a particular problem for large samples with high Mg and Fe concentrations.

In situ analysis

Both for avoiding the rigors of chemical separation and for improving the spatial resolution of analysis, it has been desirable to perfect *in situ* isotope ratio measurement techniques. Although microsampling by laser ablation and introduction of sample material to a plasma source for ionization and mass analysis remains a "future direction" for the science, analysis by secondary ion mass spectrometry (SIMS) has been in practice for a number of years, with constant improvement (Chaussidon and Robert 1998; Hervig and Moore 2003; Layne 2003). Although this technique obviates the pitfalls of chemical removal of Li from samples, it is not without problems that historically limit its application in high precision isotope ratio measurement, in particular unpredictable shifts in mass fractionation caused by both instrumental and matrix effects.

A practical difficulty of SIMS Li isotope measurement is the paucity of appropriate mineral and rock standards with well characterized isotopic compositions. Although the estimated uncertainties are larger than for the more widespread TIMS and MC-ICP-MS techniques, the capacity to measure at very restricted spatial scales makes this a promising area of exploration, particularly for samples of restricted size and those with complex small-scale structure. The complimentary nature of *in situ* and bulk data in Li isotope studies shows promise for better understanding complex processes, e.g., fluid-rock interaction (Decitre et al. 2002).

Precision and accuracy

External precision is the ability to demonstrate analytical repeatability with multiple preparations and analyses of a material over a long period of time. The MC-ICP-MS techniques and the more widespread TIMS methods either demonstrate or claim external precisions in the range ±0.5 to 1.0‰ (2σ). The stated precision for most TIMS methods is estimated from the reproducibility of the L-SVEC standard. In many cases the analysis of individual samples prepared multiple times yields precisions poorer than this estimate. This is in part due to the heterogeneity of natural samples and in part due to effects introduced during preparation and analysis that are not experienced by the standard. Zhang et al. (1998) cite reproducibility of the L-SVEC standard of ± <1.0‰ (2σ), but their duplicate measurements of individual pore water samples vary from ±0.1‰ to ±6.1‰ (mean ±2.3‰; all 2σ). Later studies using refined TIMS procedures appear to achieve superior replicate precision (e.g., ±0.4‰ to ±1.1‰ for multiple replicates in Chan et al. 2002c).

Most MC-ICP-MS devices can demonstrate external precision in the range ±0.2 to 0.9‰ for synthetic standards (e.g., 10 month average of ±0.24 for 52 measurements of standard IRMM-016; Millot et al. 2004). However, natural samples, which require careful Li separation and purification from geological matrices, pose a significant and generally underestimated impediment to the achievement of long-term reproducibility much in excess of ±1.0‰. Nevertheless, given careful work, extremely high precision is now attainable (e.g., replicate reproducibility of 27 seawater aliquots of ±0.5‰; Millot et al. 2004).

As the achievable precision improves, more and more detailed geochemical problems can be realistically explored. That is, they may be explored if, at the new levels of precision, accurate measurement can be demonstrated. It is embarrassing to note the general absence of Li isotope analyses of international standard materials until fairly recently. It is only with the cross-reference of accepted standard compositions that comparisons of data among different laboratories can be accomplished. Table 1 lists the reported isotopic compositions of a variety of standard materials in the literature. Whereas the accumulated data for some materials agree quite well among laboratories (e.g., JR-2), others show poor agreement (e.g., JB-2, with a total range of 2.5‰). Lithium isotopic analysis of seawater is another means by which accuracy and precision have been estimated. This will be considered below under, "Lithium isotopic fractionation in the oceans."

Table 1. Reported Li isotopic compositions for international rock standards.

Name	*Producer**	*Material*	δ^7Li	*Source***
BHVO-1	USGS	basalt, Hawaii	+5.2 ± 0.5 (4)	1
BHVO-1			+5.1 ± 0.4 (2)	2
BHVO-1			+5.0 ± 1.9 (8)	3
BHVO-1			+5.8 ± 1.9 (3)	4
JB-2	GSJ	basalt, Japan	+6.8 ± 0.3 (3)	4
JB-2			+5.1 ± 1.1 (4)	5
JB-2			+4.3 ± 0.3 (5)	6
JB-2			+4.9 ± 0.7 (5)	7
JB-2			+5.1 ± 0.4 (3)	8
JB-3	GSJ	basalt, Japan	+3.9 ± 0.3 (3)	6
SRM-688	NIST	basalt, Nevada	+2.8 ± 1.1 (1)	5
JGb-1	GSJ	gabbro, Japan	+6.2 ± 2.5 (3)	4
DR-N	ANRT	diorite, France	+2.3 ± 1.1 (3)	4
JA-1	GSJ	andesite, Japan	+5.8 ± 0.7 (5)	6
G-2	USGS	granite, Rhode Island	−0.3 ± 0.4 (3)	2
G-2			−1.2 ± 0.6 (3)	4
JR-2	GSJ	rhyolite, Japan	+3.9 ± 0.4 (3)	1
JR-2			+3.9 ± 0.4 (1)	2
JR-2			+3.8 ± 1.0 (3)	4
JG-2	GSJ	granite, Japan	−0.7 ± 0.4 (2)	2
JG-2			+0.4 ± 1.0 (4)	3
JG-2			−0.4 ± 0.2 (3)	4
UB-N	ANRT	serpentine, France	−2.7 ± 1.1 (4)	4
UB-N			−2.6	9
SCo-1	USGS	shale, Wyoming	+5.3 ± 1.5 (3)	4

Note: uncertainties are those listed in the cited source or recalculated from data therein, 2σ population, where available; number in parentheses is number of analyses factored into this precision, where given

*Producers: USGS = United States Geological Survey; GSJ = Geological Survey of Japan; NIST = National Institute of Standards and Technology (USA); ANRT = Association Nationale de la Recherche Technique (France)

**Sources: 1 = Chan and Frey (2003); 2 = Pistiner and Henderson (2003); 3 = Bouman et al. (2002); 4 = James and Palmer (2000a); 5 = Tomascak et al. (1999a); 6 = Nishio and Nakai (2002); 7 = Moriguti and Nakamura (1998a); 8 = Chan et al. (2002c); 9 = Benton et al. (in review)

INTERNAL-PLANETARY SYSTEMS

Mass fractionation in igneous systems

At temperatures germane to melting and crystallizing mantle magmas, Li isotopes do not show permil-level mass fractionation (Fig. 4; Tomascak et al. 1999b). This has since been corroborated by examination of bulk rocks and olivine separates from basaltic lavas, which yield consonant isotopic values (Chan and Frey 2003). Also, whole rocks and omphacite mineral separates from alpine eclogite with metamorphic peak temperatures approximately 650°C (Zack et al. 2003) show no consistent Li isotopic difference.

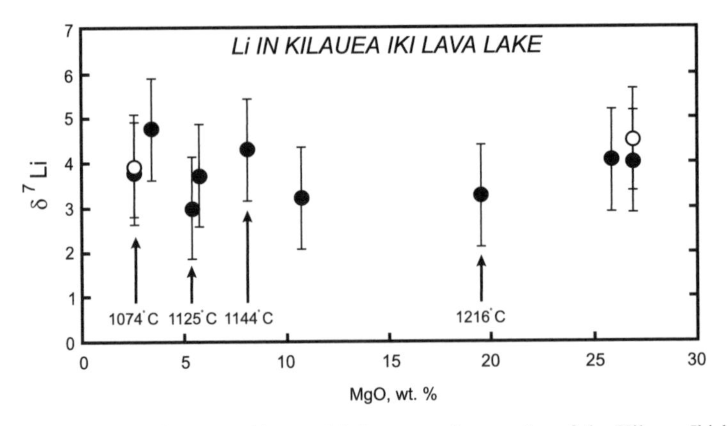

Figure 4. Plot of Li isotopic composition vs. MgO content for samples of the Kilauea Iki lava lake, Hawaii (Tomascak et al. 1999b). Cored basalt samples show a range of crystallization temperatures (estimated for four of the samples). The absence of permil-level variation in δ^7Li indicates that Li isotopes do not fractionate appreciably during crystallization in mantle systems. Open symbols (○) are replicate measurements.

However, evidence for mass fractionation at near-solidus temperatures in granitic systems exists. Early work on Li isotopes in crustal systems (Plyushin et al. 1979) suggested that significant isotopic effects could be seen in an evolving granitic system—effects that should be measurable using modern analytical methods. Similarly, permil-level variations in mineral isotopic compositions from different zones of the Tin Mountain pegmatite, South Dakota, have been interpreted to indicate the effects of crystallization of subsequent assemblages in the 500-600°C range (Tomascak et al. 1995c). Although these studies focus on mass fractionation introduced during crystallization processes, the potential for measurable isotopic fractionation during alkali diffusion has also been investigated (Lundstrom et al. 2001; Richter et al. 2003). These studies suggest that, because the greater diffusivity of ^{6}Li relative to ^{7}Li, contact zones where magmas flow through materials of contrasting concentration will show ^{6}Li enrichment. The magnitude of these effects in nature and their general importance in the interpretation of data from igneous systems awaits detailed study.

With the exception of these fractionation pathways, studies of igneous systems chiefly focus on the potential of Li isotopes as geochemical tracers: fingerprinting the cycling of Li derived from specific (low-temperature) sources through the solid Earth. The sections below deal with observations of Li isotopes in high-temperature systems, and the mechanisms for low-temperature fractionation processes are discussed after, under the heading, "Planetary surface systems."

Mantle processes

The mantle, although Li-poor relative to the continents (5–6 ppm in normal MORB and c. 1 ppm in depleted peridotites; Ryan and Langmuir 1987; Eggins et al. 1998), is a significant reservoir due to its large volume (discussed below under, "Significance of lithium isotopes in the bulk Earth").

Mantle reservoirs. The only quasi-systematic studies of igneous materials have centered on the mantle; in particular mid-ocean ridge basalts (MORB), ocean island basalts, and mantle peridotites. After reporting one MORB analysis in Chan and Edmond (1988), the first full study of MORB (Chan et al. 1992) reported three apparently unaltered Atlantic basalts and one from the East Pacific Rise, with a range in δ^7Li of +3.4 to +4.7 (Fig. 5). Subsequent studies have increased the global range of samples, the diversity of bulk compositions analyzed,

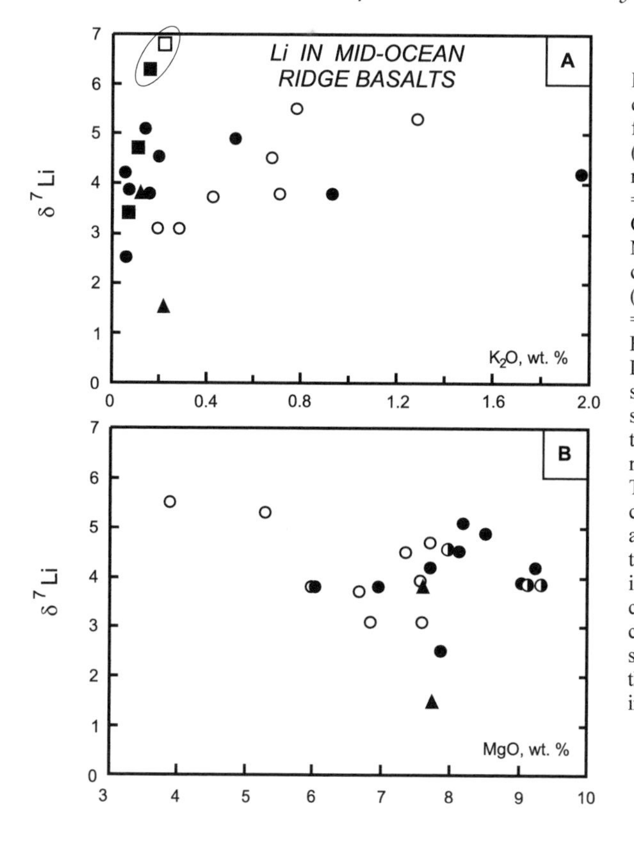

Figure 5. Plot of Li isotopic composition vs. (a) K_2O, (b) MgO for glassy mid-ocean ridge basalt (MORB) samples from various ridge segments. Symbols: square = Chan and Edmond (1988) and Chan et al. (1992); triangle = Moriguti and Nakamura (1998a); circle = Tomascak and Langmuir (1999). For all symbols, filled = Mid-Atlantic Ridge, open = East Pacific Rise, half-filled = Indian Ridge. The two encircled samples ($\delta^7Li > +6$) were considered by Chan et al. (1992) to be altered, although this was not petrographically obvious. The lack of an indication of correlation between Li isotopes and major elements (as well as trace elements and radiogenic isotopes) in a global sample set cannot be explained by a simple contamination process with near-sea floor material, suggesting the isotopic variation may be intrinsic to the upper mantle.

and the isotopic range (δ^7Li = +1.5 to +5.6; Moriguti and Nakamura 1998b; Tomascak and Langmuir 1999; Nishio et al. 2002). Although relative to typical analytical uncertainties of c. ±1‰ this range is relatively restricted, the 5‰ range manifest in the current global data set indicates that this reservoir is not homogeneous with respect to Li isotopes.

The lack of correlation between Li isotopes and other geochemical attributes (e.g., major elements; Fig. 5) indicates this an area where more detailed studies employing superior precision would be useful. The global data set may, however, unfairly average away geochemical features of Li in the MORB mantle. Samples from portions of the East Pacific Rise showed correlated variations in δ^7Li (+2.7 to +6.0) and Sr isotopic composition (Elliott et al. 2003), which was interpreted as reflecting heterogeneity in mantle sources of these lavas. Similar slight enrichment in δ^7Li was seen in an enriched MORB sample from the Juan de Fuca ridge (+6.8; Benton et al. 1999). The implications for a singular pristine upper mantle Li isotopic composition are unknown without further research, but could be significant.

Given the difference in Li concentration between MORB and sea floor altered MORB (which are commonly >12 ppm; Stoffyn-Egli and Mackenzie 1984; Chan et al. 1992), small contributions of altered oceanic crust could have a substantial impact on the measured Li isotopic composition of sea floor lavas. Indeed, two basalts that were petrographically "fresh" were excluded from consideration as pristine MORB by Chan et al. (1992) by virtue of slightly higher K_2O contents (>0.13 wt. %; Fig. 5). Within the global MORB data set no correlation between K_2O and δ^7Li is observed, suggesting that the variations in Li isotopes in MORB are not controlled primarily by near-surface contamination.

Mantle composition can also be assessed through examination of lavas from other oceanic settings. In a study of well-characterized Hawaiian lavas, Chan and Frey (2003) found a total range in δ^7Li that overlaps the range for MORB, +2.5 to +5.7. Other data from Kilauea volcano substantiate this range (Tomascak et al. 1999b). The Hawaii data, when compared with other isotopic and elemental parameters, suggested incorporation of recycled altered oceanic crust into the source of the Hawaiian plume (Chan and Frey 2003).

Nishio et al. (2003) analyzed basaltic rocks from six south Pacific islands to assess potential Li isotope contributions to magma sources in rocks with pronounced radiogenic isotope anomalies relative to MORB (e.g., high ^{206}Pb/^{204}Pb, "HIMU" end member; Zindler and Hart 1986). Whereas all of the lavas they studied had slightly enriched Li contents (up to 11.6 ppm), only the samples from Mangaia (δ^7Li = +7.4) and Ua Pou (+6.1) were particularly different from MORB in terms of Li, and there was not a straightforward correlation between δ^7Li and ^{206}Pb/^{204}Pb. The data suggest incorporation of an isotopically heavy, ultimately surficially-processed component (perhaps altered oceanic crust) in the sources of HIMU magmas. Nonetheless more complete data are required to elucidate the full extent of mantle Li isotopic heterogeneity. The only other report of Li isotope data from ocean island lavas (Ryan and Kyle 2000) showed slight enrichment in heavy Li relative to MORB in a minority of samples (δ^7Li up to +7.0). In addition to hot-spot lavas, samples from back-arc basins have been examined. Pacific back-arc basin basalts with MORB-like trace element contents examined by Chan et al. (2002c) and Nishio et al. (2002) have δ^7Li = +2.3 to +4.8, in agreement with the observed range for normal MORB.

Unmetasomatized peridotite xenoliths are additional potential sources of information on Li isotopes in the upper mantle. To date, the xenolith data set is very sparse, owing largely to the analytical challenges of these low Li-abundance, Mg,Fe-enriched materials. Benton et al. (1999) reported a rather light isotopic composition in a sample of continental peridotite from San Carlos, Arizona (δ^7Li = +1.1). However, Chan et al. (2002c) determined that an unaltered peridotite from Zabargad Island had δ^7Li = +5.0, identical to the value reported by Brooker et al. (2000) for peridotites from the same area. Oceanic peridotite xenoliths from La Palma showed a range of values (δ^7Li = +4 to +7; Bouman et al. 2000), overlapping and slightly higher than the normal MORB range. Peridotite xenoliths that show less pristine upper mantle elemental and radiogenic isotope characteristics are discussed below.

Xenoliths from Siberian continental lithosphere, with Archean model ages, had δ^7Li as low as +0.5 (Bouman et al. 2000). If these values accurately represent the Archean mantle, they suggest the potential for Li isotopic evolution in the Earth, from lighter compositions in the ancient mantle to what is seen in present-day MORB. In spite of the analytical challenges presented by ultramafic rocks, more data from these materials are crucial to an understanding of Li in the mantle, and in resolving questions about the appropriateness of the accepted MORB mantle range.

Subduction zones and related rocks. Lithium isotope characteristics of subduction zones have been considered in multiple thorough studies. If Li is transferred from a subducting slab into the overlying mantle along with other fluid-mobile elements, and if the slab-derived fluid has δ^7Li dissimilar to the mantle, then Li isotopes should provide a means of assessing, and perhaps quantifying this mass transfer. Given the differences in fluid/mineral partitioning of Li and B ($D_{\text{fluid/clinopyroxene}}$(B) ~ 10 × $D_{\text{fluid/clinopyroxene}}$(Li); Brenan et al. 1998), studies combining the elemental and isotopic compositions of Li and B should be particularly powerful in this quantification (e.g., Moriguti and Nakamura 1998b).

The isotopic composition of Li in fluids that leave a subducting slab are constrained to some extent by the current data base. As discussed below, low temperature sea floor alteration tends to generate minerals enriched in heavy Li relative to those present in pristine MORB.

Although the role of isotopically light sediments cannot be disregarded (also discussed below), altered oceanic crust is likely to be a primary source of Li to most subduction zone fluids (Tatsumi et al. 1986). Thus, these fluids should be isotopically heavier than MORB. The residue from this dehydration process may, therefore, be enriched in light Li relative to MORB.

The first studies of Li isotopes in subduction zones concentrated on young convergent margin lavas. Moriguti and Nakamura (1998b) reported correlated Li isotope and fluid-mobile element (notably boron) concentration variations in the Izu arc, southeastern Japan (δ^7Li = +1.1 to +7.6), consistent with significant incorporation of Li from altered oceanic crust into arc lava sources (Fig. 6). A similar trend has been reported in samples of basalts and basaltic andesites from Mt. Shasta, California (δ^7Li = +2.5 to +6.5; Magna et al. 2003).

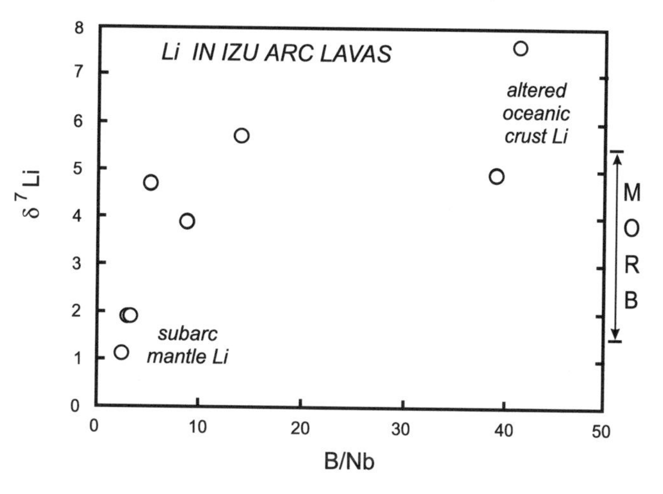

Figure 6. Plot of Li isotopic composition vs. B/Nb for a suite of samples from the Izu arc (Moriguti and Nakamura 1998b). The general correlation between high δ^7Li and elevated values of fluid mobile/immobile element ratios (as well as B and Pb isotopes) suggested incorporation of a Li component derived from the subducted slab, most likely from altered oceanic crust.

The majority of other studies on arcs, including true cross-arc evaluations, have not demonstrated strong correlations between Li isotopes and fluid mobile element contents of lavas or other geochemical parameters, although isotopic variations are apparent. Northeastern Japan arc lavas had compositions like MORB (δ^7Li = +1.5 to +5.5; Moriguti and Nakamura 2003). In the Marianas, the isotopically lightest sample (δ^7Li = +0.7) had elevated B/Nb and the isotopically heaviest sample (δ^7Li = +7.6) has somewhat lower B/Nb, but a systematic difference between Li isotopes and fluid mobile element enrichment is lacking (Benton and Tera 2000). Bouman and Elliott (1999) suggested that isotopic relations among a smaller set of Marianas samples could be satisfied by differences in inputs from isotopically light subducted sediments and isotopically heavier subducted altered oceanic crust. Chan et al. (2001; 2002b; 2002c) observed variability outside the MORB range in lavas from the South Sandwich Islands (δ^7Li = +3.4 to +7.4), part of the Aleutian chain (δ^7Li = +1.0 to +5.8) and regions within Central America (δ^7Li = +4.1 to +6.2), but the correlations relative to fluid mobile element enrichments were generally quite restricted (Fig. 7).

Tomascak et al. (2000, 2002) suggested that, for intra-oceanic arcs where such geochemical correlations were either restricted or non-existent, Li originating in the slab is likely removed from fluids/melts during mineral-fluid equilibration in the lower part of the subarc mantle. This reservoir of isotopically anomalous (heavy) slab-derived Li could be tapped in tectonically opportune situations, for instance in regions where young, warm crust is subducted (e.g., Panama: Tomascak et al. 2000; Cascades: Chan et al. 2001). Such processes

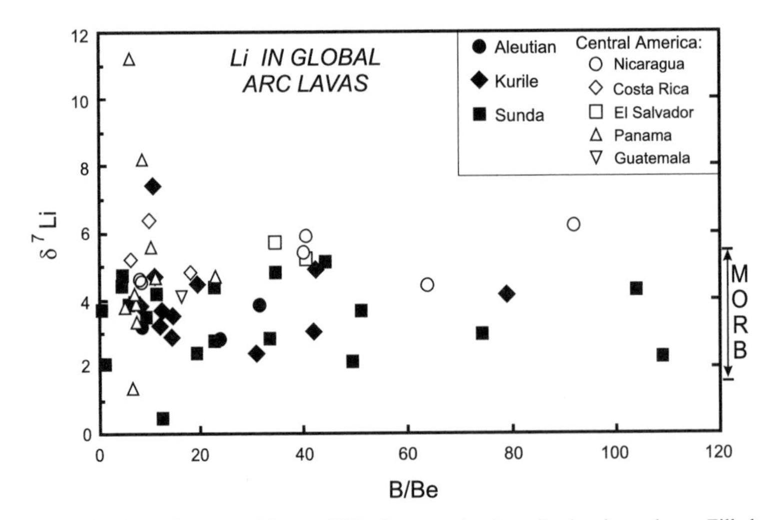

Figure 7. Plot of Li isotopic composition vs. B/Be for several suites of volcanic arc lavas. Filled symbols from Tomascak et al. (2002); open symbols circle from Tomascak et al. (2000) and Chan et al. (2002c). In these samples no simple strong correlation was seen between δ^7Li and indicators of slab-derived components. The lack of correlation and predominance of values around those seen in MORB suggested Li removal from slab derived fluids or melts in the subarc mantle prior to arc magma genesis (Tomascak et al. 2002).

would inevitably lead to the decoupling of Li isotopic systematics from fluid-mobile element contents (Tomascak et al. 2000). Slab Li incorporated into the subarc mantle might also persist there, perhaps later to be incorporated into the subcontinental lithosphere.

That dehydration of subducted oceanic crust leads to residues enriched in isotopically light Li is supported by a natural analog to the process. Alpine eclogites record, without significant later overprinting, the geochemistry of aborted subduction (Heinrich 1986). The eclogites of Trescolmen preserved δ^7Li as low as −11 (Zack et al. 2003) (Fig. 8). Calculation of the effect of distillation of altered oceanic crust during dehydration suggested that the more altered a basalt is prior to subduction, the lighter its isotopic composition after dehydration.

Eclogite xenoliths may yield complementary information to that gained from obducted eclogitic sections, while perhaps escaping some of the effects of pervasive retrograde re-equilibration. Eclogites with Archean Re-Os model ages from the Koidu kimberlite in Sierra Leone have δ^7Li that are both heavier and lighter than MORB (−2.3 to +8.0; Rudnick et al. 2003). This suggested that portions of subducted crust can retain geochemical records of both high and low temperature alteration histories on the ocean floor for geologically extensive durations.

Metasomatized mantle xenoliths, thought to derive ultimately from subduction-zone mantle that was affected by components (fluids and/or melts) discharged from the subducting lithosphere, are additional windows into mass transfer at convergent margins and into elemental cycling in the mantle. Pyroxenite veins from peridotites on Zabargad Island were dominantly isotopically heavier than their normal-mantle hosts (δ^7Li = +8.6 to +11.8), but also showed light values (−4.2) (Brooker et al. 2000). These signatures were interpreted to derive from metasomatism of subarc mantle by materials related to Pan-African subduction.

Metasomatized xenoliths studied by Nishio et al. (2004) show a diversity of Li isotopic compositions (δ^7Li = −17.1 to +6.8), some of these correlated with variations in radiogenic

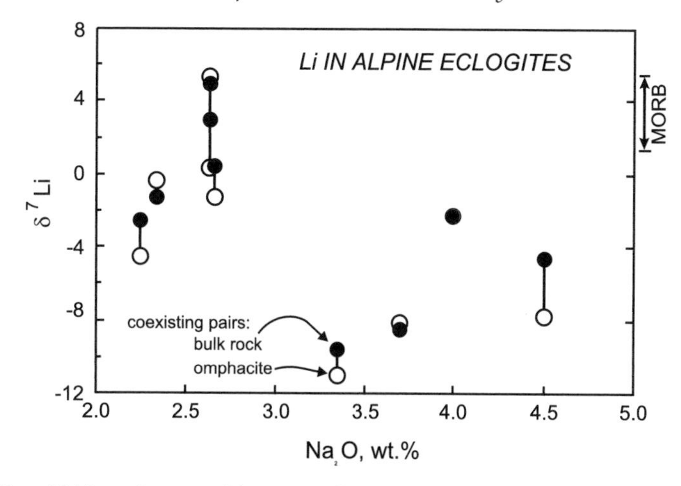

Figure 8. Plot of Li isotopic composition vs. Na_2O content of eclogites from Trescolmen, Swiss Alps (Zack et al. 2003). Data from pairs of bulk rock (●) and omphacite mineral separate (○) samples, indicating that bulk rocks generally preserve isotopic systematics of these rocks. Well-preserved, high pressure metamorphosed basaltic rocks were interpreted to retain much of their elemental character, and as such were good recorders of the residue remaining after subduction dehydration. The data indicated that originally isotopically heavy altered sea floor basalt could be transformed during subduction into some of the isotopically lightest materials in the Earth system.

isotopes. Lherzolite xenoliths from southeastern Australia retained essentially MORB-like Li isotopic compositions (δ^7Li = +5.0 to +6.0), but showed trends toward an end member composition with elevated δ^7Li and high $^{87}Sr/^{86}Sr$ (tentatively suggested to be EM2 of Zindler and Hart 1986; Fig. 9a,b). Xenoliths from three localities within the Sikhote-Alin district of eastern Russia possessed some of the lowest δ^7Li values measured. The data for these lherzolites defined arrays in Li isotope vs. $^{87}Sr/^{86}Sr$ and vs. $^{143}Nd/^{144}Nd$ space that suggested incorporation of a component with $\delta^7Li < -25$ (perhaps the EM1 end member of Zindler and Hart 1986; Fig. 9).

Regardless of the ultimate sources of these compositions, these results clearly show that strongly isotopically fractionated Li from crustal sources plays a role in the mantle. Processes active in subduction zones appear to be cardinal in the control of the Li isotopic composition of different parts of the mantle. The results to date imply that both isotopically enriched (δ^7Li > MORB) and depleted (δ^7Li < MORB) material are available for deep subduction, and that areas of the continental lithosphere may retain these records on long time scales.

Crustal processes

Crustal melting. No comprehensive study has yet to consider Li isotopes in igneous systems exclusive of purely mantle melting. Estimates of the Li isotopic composition of granitic rocks from various regions span a very wide range. Both Svec and Anderson (1965) and Isakov et al. (1969) showed large (percent-level) variability in Li isotopic compositions of minerals from granites and related deposits. As the analyses in these early works were not standardized, absolute values cannot be assessed and magnitudes may be questioned. Regardless of the vintage of these analyses, granitic systems (*sensu lato*) are likely to show substantial isotopic variability, the origins of which may prove to cement Li as an important tool in the study of petrogenesis or other processes.

Subsequent studies with improved analytical techniques include data from granitic rocks reveal values ranging from 15‰ lighter than MORB to 20‰ heavier. A number of granitic

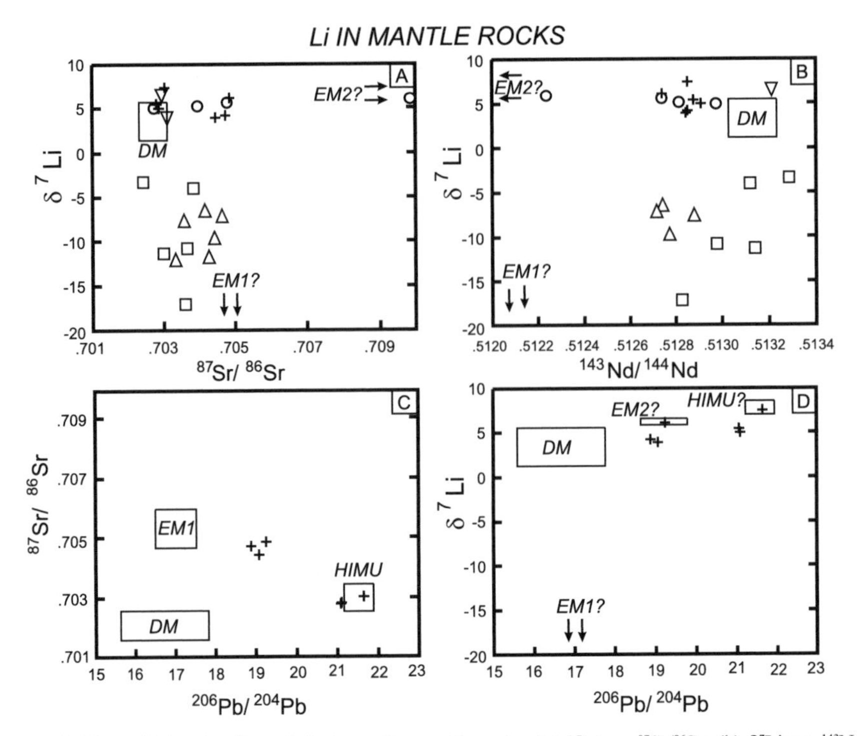

Figure 9. Plots of Li and radiogenic isotopes for mantle rocks. (a) δ^7Li vs. ^{87}Sr/^{86}Sr; (b) δ^7Li vs. ^{143}Nd/^{144}Nd; (c) ^{87}Sr/^{86}Sr vs. ^{206}Pb/^{204}Pb (d) δ^7Li vs. ^{206}Pb/^{204}Pb (Nishio et al. 2003, 2004). Symbols: + = south Pacific island basalts (six islands); ○ = lherzolite xenolith, Bullenmerri, Australia; □ = lherzolite xenolith, Sikhote-Alin, Russia (three localities); △ = dunite-peridotite-pyroxenite xenolith, Kyushu, Japan (two localities); ▽ = lherzolite xenolith, Ichinomegata, Japan. The ocean island data are from bulk rocks, the xenolith data are clinopyroxene separates. For explanations of the derivation of radiogenic isotope fields (DM, EM1, EM2, HIMU), see Zindler and Hart (1986). The estimate for Li isotopes in DM is based on MORB. The Li isotopic ranges for the other mantle reservoirs are based on Nishio et al. (2004) and Nishio et al. (2003), but these will require further examination (hence the use of question marks).

and rhyolitic rock standards have been analyzed, and they showed generally light but variable isotopic compositions: δ^7Li = −1.2 to +3.9 (Table 1; James and Palmer 2000a; Bouman et al. 2002; Chan and Frey 2003; Pistiner and Henderson 2003). The only large studies of granitic rocks to date examined rocks from eastern Australia, finding a range in δ^7Li of −1.4 to +8.0 (Bryant et al. 2003b; Teng et al. 2004). Of these, the granites derived from predominantly metasedimentary source rocks had lighter average compositions (S-type granites; δ^7Li = −1.4 to +2.1) than I-type granites (+1.9 to +8.0). Composite samples of granitic rocks from throughout China also had light isotopic compositions (δ^7Li = −3.4 to +3.0; Teng et al. 2004). Granites from the eastern Sierra Nevada, California, and Quaternary rhyolite from the Long Valley area had δ^7Li = 0.0 to +4.0 (Tomascak et al. 2003), overlapping values for least-hydrothermally altered rhyolite from the Yellowstone, Wyoming, area (+0.7; Sturchio and Chan 2003), and unaltered Liberty Hill granite, South Carolina (+2.3; Njo et al. 2003).

Individual samples of isotopically heavier granitic rocks have also been reported from the Canadian Shield (δ^7Li = +7.3 to +11.3; new and updated data from Bottomley et al. 2003) and from the Isle of Skye, Scotland (δ^7Li = +8.1; Pistiner and Henderson 2003). These are all rocks with long near-surface histories and modal chlorite or epidote, suggesting their isotopic compositions may not be pristine, igneous values (discussed below under, "Continental runoff and weathering processes").

Studies that have considered evolved granitic systems (associated with Li mineralization) have found both extremely isotopically light and highly variable isotopic compositions: Kings Mountain granite-pegmatite system, North Carolina: δ^7Li = −10 to +12 (Vocke et al. 1990); Harney Peak granite and related pegmatite, South Dakota: δ^7Li = 0 to +19 (Vocke et al. 1990; Tomascak et al. 1995c); Macusani high silica rhyolite, Peru: δ^7Li = −2.2 to −1.1 (Tomascak et al. 1999a). This dispersion of data may be an artifact of significant isotopic fractionation occurring at temperatures below the haplogranite granite solidus.

Metamorphism and ore formation. Variations in Li isotopes brought about regional metamorphism are essentially completely uncharacterized. Unattributed data in Huh et al. (1998) suggested values of δ^7Li > MORB for at least some regional metamorphic rocks, but a single analysis of a Precambrian schist yielded δ^7Li = +1.6 (Sturchio and Chan 2003). A variety of metamorphic rocks from the Canadian Shield had δ^7Li of +2.0 to +6.7, with a pegmatitic gneiss at +30.2 (Bottomley et al. 2003). Although Li isotope fractionation at granulite facies temperatures might be considered unimportant, highly fractionated compositions in lower continental crustal xenoliths, with δ^7Li as low as −13 (Teng et al. 2003), suggest that even at such conditions mass fractionation may take place, possibly governed by the presence (or composition) of a fluid phase, either as a part of lower crustal metamorphism or in the sources of these rocks.

Data on Li isotopic variations during local-scale metamorphism related to ore deposits are limited to analyses for which absolute values cannot be assessed. Morozova and Alferovskiy (1974) showed >1% variations in rocks/minerals of granite pluton contacts and amphibolite. Divis and Clark (1978) showed similar strong fractionation between granitic and volcanic rocks and their ore-related altered counterparts. The results of Divis and Clark (1978) suggest lighter isotopic compositions should be expected in higher-temperature ore zone rocks.

Differentiation of the highly geochemically evolved Tin Mountain pegmatite, South Dakota, led to fractionated isotope ratios in minerals during progressive crystallization, and suggested the possible viability of Li isotope thermometry (Tomascak et al. 1995c). In different internal zones of the lithium pegmatite, albite (δ^7Li = +7 to +11) was uniformly isotopically lighter than coexisting quartz and muscovite (δ^7Li = +16 to +19; Fig. 10). An increase of c. 2‰ in albite-quartz fractionation was observed from earlier- to later-crystallized assemblages which formed over an estimated temperature range from approximately 600°C to approximately 550°C, based on oxygen isotope data (Walker et al. 1986). This fractionation could be temperature-dependent equilibrium fractionation, where difference in magnitude of effect resulted from different available crystallographic sites in the crystallizing assemblage (considered in the next section). In light of the recent proposal of crystallization of this pegmatite at even lower temperatures (<400°C; Sirbescu and Nabelek 2003), the isotopic results should be considered cautiously.

Constraints on Li isotopic fractionation from equilibrium laboratory experiments in magmatic-hydrothermal systems have been examined in only one instance (Lynton 2003). This study examined Li in a quartz-muscovite-aqueous fluid system at conditions of formation of magmatic hydrothermal porphyry-type deposits (400–500°C, 100 MPa). In the presence of a fluid containing L-SVEC (δ^7Li = 0), quartz showed rapid shift from δ^7Li = +27 in the starting material to c. +10 at both run temperatures. Muscovite (initial δ^7Li = +9), shifted more sharply at 500°C (to c. +20) than at 400°C (to c. +13). Although the results are difficult to put directly into the context of a natural mineral deposit, they do indicate that over geologically relevant time scales, minerals in magmatic-hydrothermal systems should show appreciable Li isotopic fractionation, and that this may permit the composition and/or temperature of ore

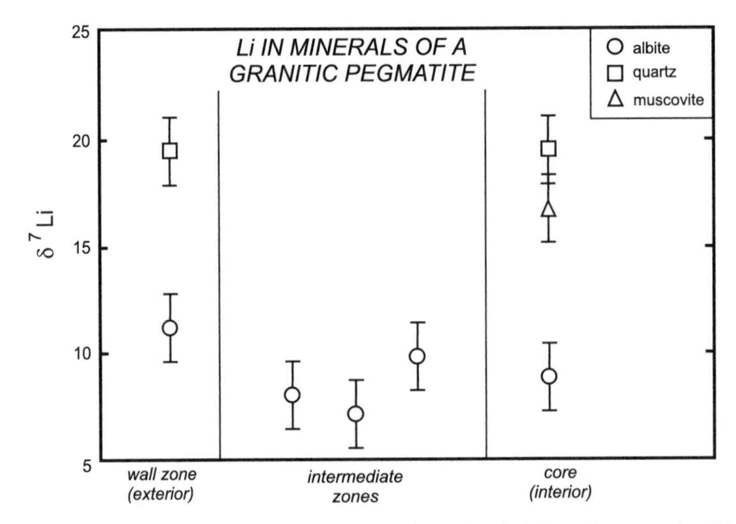

Figure 10. Lithium isotopic compositions of minerals from the Tin Mountain pegmatite, Black Hills, South Dakota (Tomascak et al. 1999c). Differences in the equilibrium quartz-feldspar oxygen isotopic compositions between the wall zone and intermediate zones suggested temperatures of 600 to 550°C (Walker et al. 1986). Despite poor precision of these data, they showed a consistent increase in the $\Delta^7Li_{quartz-albite}$ between the early-crystallizing wall zone and the later-crystallizing core.

fluids to be estimated. Despite the small number of data from natural and synthetic magmatic-hydrothermal systems, Li isotope research appears to offer the potential for new insights into the petrogenesis of magmatic and hydrothermal ore deposits.

PLANETARY SURFACE SYSTEMS

Mass fractionation in hydrologic cycle of Li

Most of the clearly identified isotopic fractionation in nature occurs close to the Earth's surface. The combination of low ionic charge, relatively small radius (approximately 0.068 nm; Shannon 1976), and high degree of covalency in its bonds affords Li potential for considerable adaptability in different geochemical environments. Lithium in aqueous solutions is highly hydrated relative to the other alkali metals, and ordinarily assumes tetrahedral hydrated coordination (Olsher et al. 1991). In most solids Li occupies either tetrahedrally- or octahedrally coordinated sites (Olsher et al. 1991; Wenger and Armbruster 1991). The occurrence of isotopic fractionation induced by recrystallization, in which an energetic advantage is gained by the incorporation of one isotope over another into a crystallographic site, is an important mechanism for near-surface isotopic fractionation (O'Neil 1986). Based on *ab initio* calculations, the isotopic effect of Li exchange into octahedral crystallographic sites from aqueous solution is preferential incorporation of ^{6}Li into the solid (Yamaji et al. 2001). This fractionation is borne out experimentally and in nature (examples summarized below), with a magnitude-dependent on reaction temperature and fluid/rock ratio.

The thermodynamic basis for considering Li isotopic fractionation during dissolution of primary minerals (equilibrium fractionation) takes into account small differences in bond energies of ions of different mass (O'Neil 1986). Breaking the higher energy bonds of ^{6}Li is energetically favorable, and so the equilibrium isotope effect would be to enrich the solution phase with the lighter isotope. However, Pistiner and Henderson (2003) suggested on the basis of laboratory dissolution experiments that incongruent mineral dissolution does not impart significant isotopic fractionation. It appears that this mechanism of isotopic fractionation is either not dominant in low temperature environments, or its effects are generally overprinted by later processes, as suggested by Huh et al. (2001) for continental weathering.

Fluid-particle interaction, in which aqueous ions are adsorbed by mineral surface sites, is another significant mechanism for the development of Li isotopic heterogeneity in the hydrosphere. Sorption of Li from aqueous solution by minerals at ambient temperatures has been demonstrated for a variety of materials (Taylor and Urey 1938; Anderson et al. 1989), despite the prediction that Li would tend to be the least sorbed of the alkalis, due to its high degree of hydration (Heier and Billings 1970). Sorption and retention of ions from solution depends largely on both mineral surface chemistry and the composition of the solution. Although it has been suggested experimentally that the magnitude of mass fractionation on desorption is equivalent to that on adsorption (Taylor and Urey 1938), natural and empirical data for both Li and Cs suggest that this fractionation is not reversible (Comans et al. 1991; James and Palmer 2000b). The irreversibility of this process is typically ascribed to incorporation of Li into crystallographic sites rather than surface ion exchange sites, suggesting that the magnitude of the effect will vary among minerals.

Zhang et al. (1998) examined the behavior of dissolved Li during sorption from seawater onto clay minerals and sediments. A similar degree of isotopic fractionation was observed for vermiculite, kaolinite and mixed Mississippi River suspended sediment ($\alpha \sim 0.978$; where $\alpha = {}^7Li/{}^6Li_{mineral}/{}^7Li/{}^6Li_{solution}$). Pistiner and Henderson (2003) conducted experiments on the isotopic effects of Li sorption on a variety of minerals. Their results suggested that sorption reactions in which Li was not structurally incorporated (physisorption) yielded no measurable isotopic fractionation. When the data indicated that Li was incorporated via stronger bonds (chemisorption), isotopic fractionation was observed, but was not homogeneous among minerals with different chemistry/structure, consistent with field studies (e.g., Anghel et al. 2002). The most significant effect was shown by gibbsite, with apparent $\alpha \sim 0.986$. It is clear that further experimental studies of isotopic fractionation effects of adsorption under geologically-relevant conditions are needed and will aid in understanding the hydrologic cycle of this element.

Transmission of solution through a semi-permeable membrane has been suggested as a means of inducing measurable stable isotope fractionation (Phillips and Bentley 1987); in principle, the concentration of ${}^7Li$ would be depleted on the high-pressure side of the membrane (Fritz 1992; Fritz and Whitworth 1994). This experimentally-derived separation has yet to be documented in nature (Remenda 1994), although deep sedimentary basins with clay-rich strata, elevated pore pressures and low flow rates are a setting where such an effect might be geochemically important.

The importance of equilibrium versus kinetic effects has yet to be addressed in any coherent way by the limited laboratory experiments conducted for Li isotope study. It is clear from the study of other stable isotope systems that kinetic effects may dominate the fractionation pathways under many circumstances (e.g., Johnson and Bullen 2004), especially in laboratory simulations of low-temperature natural phenomena (Beard and Johnson 2004). The clarification of how kinetic effects on Li isotopic compositions are manifested remains a major area for future study in natural, synthetic and theoretical systems.

Primarily clastic sediments and sedimentary rocks

Lithium isotope studies of sediments and sedimentary rocks have thus far concentrated on marine clastic and carbonate material. No systematic description of the effects of diagenetic processes on sediments has been made. Clay rich sediments are important to Li budgets in near-surface systems, as they concentrate Li relative to marine carbonates, which are among the most Li-impoverished materials in the crust (Ronov et al. 1970). Lacustrine carbonates have higher Li concentrations than their marine counterparts, and minerals generated in evaporitic environments are commonly enriched in Li (Tardy et al. 1972; Calvo et al. 1995). Marine carbonates are discussed below, and evaporites are mentioned under "Deep and shallow crustal fluids."

The first modern publication of Li isotope data from sediments came from Chan et al. (1994a), who reported the isotopic compositions of unaltered turbidites from DSDP Hole 477 and 477A, from the Gulf of California. These samples have been reanalyzed (LH Chan, written communication 2003), yielding lighter values. Similarly, Chan and Frey (2003) suggested a range for marine clastic sediments (δ^7Li = −1 to +5.6) based on new data and re-analysis of samples previously considered to be isotopically heavy.

Contact metamorphosed, hydrothermally altered and greenschist facies metamorphosed equivalents of these sediments showed a wide range of generally lighter compositions (δ^7Li as low as −1.3), and all altered samples had lower Li contents by >1.4×, consistent with removal of Li during >300°C alteration. At ODP Site 918, in the Irminger Basin, southeast of Greenland, Zhang et al. (1998) found that quartz-rich silts tend to have heavier compositions (δ^7Li = +8.5 to +11.8) than underlying chalk-dominated sediments (δ^7Li = +3.2 to +8.8).

Sediments ranging from clay-rich to carbonate-rich from ODP Sites 1039 and 1040, outboard of Costa Rica, had variable isotopic compositions, unrelated to their bulk chemistry (δ^7Li = +9.5 to +23.3; Chan and Kastner 2000). Turbidites and underlying ash-rich mudstones at ODP Site 808, southwest of Japan, defined a similarly large range in isotopic compositions (δ^7Li = −1 to +8; You et al. 1995). James et al. (1999) reported δ^7Li for variably altered sandy to silty turbidites to hemipelagic muds from ODP Site 1038, off the coast of northern California (−0.5 to +5.6).

Data for shales range from δ^7Li = +4.5 to +8.0 for formations in the western U.S. (James and Palmer 2000a; Sturchio and Chan, 2003), to light values for the Shimanto shale, Japan (δ^7Li = −2.7 to −1.5; Moriguti and Nakamura 1998b). These ranges overlap measurements from the one large-scale study of sedimentary materials to date (Teng et al. 2004). This study reported data for from five loess localities (δ^7Li = −3.1 to +4.8) and for nine shales used in the compilation of the post-Archean average Australian shale composition of Nance and Taylor (1976) (δ^7Li = −3.2 to +3.9). Composite samples of well-characterized sedimentary rocks from throughout China give a similar range of values (δ^7Li = −5.2 to +1.2; Teng et al. 2004). The isotopically light values from these sedimentary rocks contrast with a measurement of Vicksburg, Mississippi loess (δ^7Li = approximately +15; Huh et al. 1998)

Lithium isotopic fractionation in the oceans

The oceans have a Li concentration of c. 180 ppb, or 26 μM (Morozov 1968). Lithium in the oceans is well mixed vertically and between basins, and has not varied over the past >40 Ma by more than a factor of 1.4 (Delaney and Boyle 1986). Dissolved oceanic Li has a residence time estimated between 1.5 and 3 Ma (Stoffyn-Egli and Mackenzie, 1984; Huh et al. 1998). As a demonstration of this, Chan and Edmond (1988; same data reported in Chan 1987) found indistinguishable isotopic compositions of Atlantic and Pacific samples from both shallow (~100 m) and deep (>3700 m) levels (δ^7Li ~ +33). Analyses of seawater since this publication have maintained the isotopically heavy and consistent character of this reservoir, although a range beyond cited analytical uncertainties is evident: δ^7Li = +29.3 (Nishio and Nakai 2002); +29.6 (Sahoo and Masuda 1998; Pistiner and Henderson 2003); +30.0 (Moriguti and Nakamura 1998a); +31.0 (Millot et al. 2004); +30.4 to +32.0 (Bryant et al. 2003a); +31.8 (Tomascak et al. 1999a); +32.2 (James et al. 1999); +32.4 (You and Chan 1996); +32.5 (James and Palmer 2000a) (Fig. 11). It is unlikely that modern seawater truly shows isotopic variability on the order of 4‰. The range in measured values spans a variety of analytical methods and

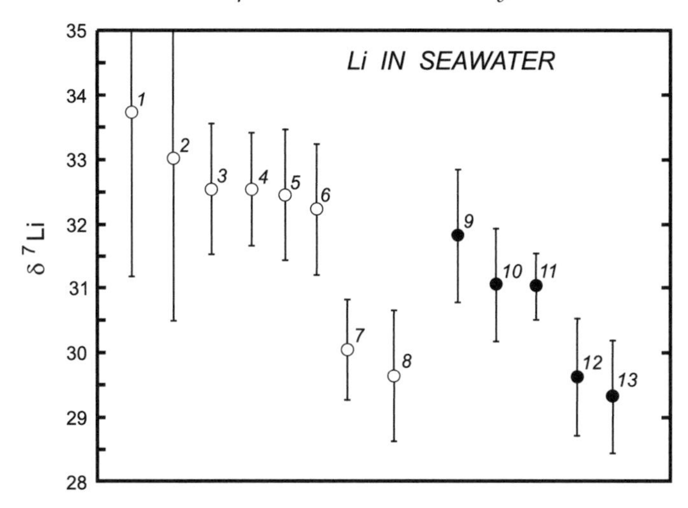

Figure 11. Reported Li isotopic compositions of seawater samples. ○ = analyses by TIMS, ● = analyses by MC-ICP-MS. Sources: 1 = Pacific average, Chan and Edmond (1988); 2 = Atlantic average, Chan and Edmond (1988); 3 = Gulf of Mexico, Zhang et al. (1998); 4 = location not given, James and Palmer (2000); 5 = Sargasso Sea, You and Chan (1996); 6 = Escanaba Trough, James et al. (1999); 7 = Mariana Trough, Moriguti and Nakamura (1998a); 8 = Japan Sea, Sahoo and Masuda (1998); 9 = central Pacific, Tomascak et al. (1999a); 10 = Great Barrier Reef, Bryant et al. (2003a); 11 = IRMM BCR-403, Millot et al. (in review); 12 = Pacific, off Hawaii, Pistiner and Henderson (2003); 13 = north Pacific, Nishio and Nakai (2002). Diagram uses conservative estimates of uncertainty (2σ) from the respective sources.

areas sampled in the oceans. Clearly analytical artifacts are present within this population and the true range for modern seawater remains to be conclusively determined.

The first substantive report of Li isotopes in any Earth materials (Chan and Edmond 1988) largely presaged what was to come in Li isotope research in the oceans. The interpretations therein, based on a handful of data from seawater, fresh and altered basalt, hydrothermal fluids and lake waters, laid out the foundation to what has come since, in terms of natural and laboratory-based studies of the marine geochemistry of Li isotopes.

As presently identified, only a few factors dominate the oceanic budget of Li and Li isotopes. Lithium is removed from the oceans via sea floor alteration reactions, high temperature sub-sea floor (hydrothermal) recrystallization, and adsorption on particle surfaces (discussed above). Lithium enters the oceans from rivers and from high and low temperature marine fluids. Crystallization of marine carbonates and biogenic silica are not important sinks (<5% total), given their low concentrations relative to their volumes (Chan et al. 1994b; Chan and Kastner 2000). Diffusion into and out of sediments are small components that roughly cancel one another (Stoffyn-Egli and Mackenzie 1984; Zhang et al. 1998; James and Palmer 2000b). The mass balance of oceanic Li has not yet been adequately resolved, due to the lack of a sink with sufficiently light Li isotopes. The components of the oceanic Li budget that have been examined in significant detail to date are summarized below.

Sea floor alteration. One of the most thoroughly studied settings in which Li isotope fractionation is observed is on the sea floor, during the alteration ("weathering" in many studies) of basalt (Chan 1987; Chan and Edmond 1988; Chan et al. 1992; Chan et al. 2002a) (Fig. 12). As oceanic basalt becomes altered, Li from seawater is taken up in the low temperature minerals that form (Stoffyn-Egli and Mackenzie 1984), most importantly clay minerals such as the smectites (Berger et al. 1988), such that the longer the exposure

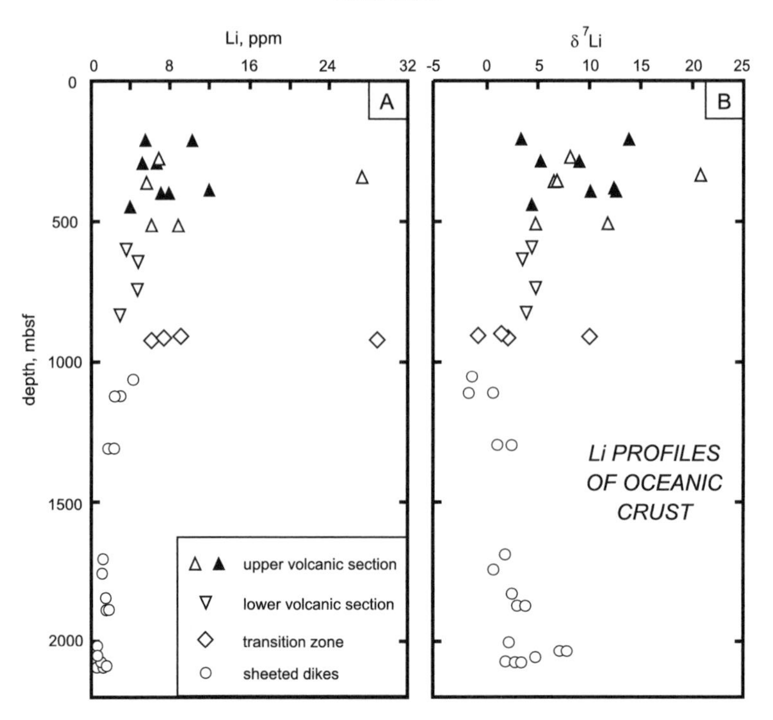

Figure 12. Depth profile of Li isotopic composition (a) and concentration (b) in drilled oceanic crust at ODP Sites 504B (open symbols) and 896A (filled symbols) off Costa Rica (Chan et al. 2002a). The transition zone exhibits mixing between hydrothermal fluids and seawater. Average oxygen isotopic ($\delta^{18}O$) composition of bulk samples decreases with depth: upper volcanic zone = +7.8, lower volcanic zone = +6.4, transition zone = +5.4, sheeted dikes = +4.3. However, despite many sheeted dike samples having δ^7Li less than unaltered MORB, there is no simple relationship between concentration and Li isotopes.

on the sea floor, the higher the Li concentration of the altered rock (to an "end member" altered basalt with δ^7Li ~ +14 and >75 ppm Li; Chan et al. 1992) (Fig. 13). Summarily, the generation of secondary minerals during sea floor alteration provides a major control on the isotopic composition of Li in the oceans (Chan et al. 1992). It is clear from the data for altered sea floor basalts that the process is not simply mixing between a pristine and a totally altered composition. Seawater Li enters newly-formed minerals in altered basalts with an apparent fractionation factor $\alpha \sim 0.981$ at 2°C for dredged basalts from the Atlantic and Pacific (Chan et al. 1992; recalculated to maintain consistency in this text: $\alpha = {}^7Li/{}^6Li_{mineral}/{}^7Li/{}^6Li_{solution}$). Uniformity in this fractionation for exclusively sea floor temperature altered crust sampled in a variety of locations suggests that although individual minerals may have distinct α values, the process as a whole produces a fixed average fractionation.

Marine hydrothermal fluids and alteration. At divergent plate boundaries, thinning of oceanic crust brings mantle magma close to the sea floor and engenders hydrothermal circulation of seawater through crust and overlying sediments. The upwelling fluids are a significant source of Li to the oceans, although the magnitude of the hydrothermal flux is not precisely quantified (Elderfield and Schultz 1996).

Lithium is enriched in high temperature (c. 350°C) vent fluids by a factor of 20–50 relative to seawater (Edmond et al. 1979; Von Damm 1995). The Li isotopic compositions of marine hydrothermal vent fluids ranged from MORB-like to heavier compositions (see

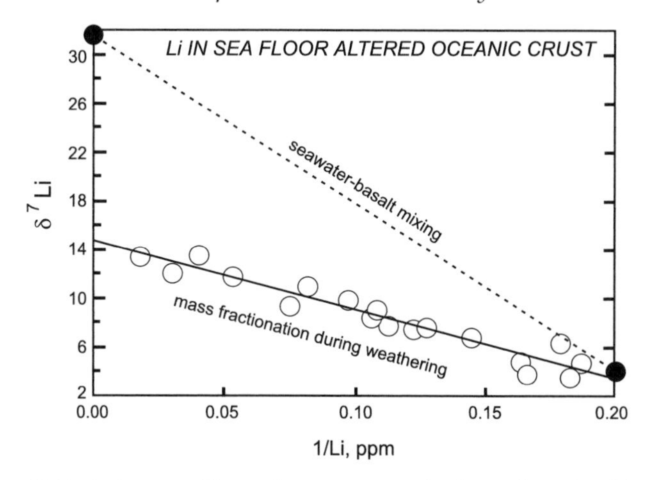

Figure 13. Plot of Li isotopic composition vs. inverse concentration for sea floor altered (weathered) mid-Atlantic ridge basalts (Chan et al. 1992). Solid line is the regression of the data (R^2 = 0.927), the dashed line shows the predicted relation of a pure mixture of seawater with unaltered basalt (Teng et al. 2004), underscoring that the altered basalts incorporated Li into secondary mineral with a fractionation factor α ~ 0.981.

Figs. 14, 15). Fluids from vents in various settings showed a degree of isotopic variability, with values extending from the MORB range to several permil heavier (δ^7Li = +2.6 to +11.6; Chan and Edmond 1988; Chan et al. 1993, 1994a; Bray 2001; Foustoukos et al. 2003) (Fig. 14). In view of the absence of correlation between δ^7Li and spreading rate, Chan et al. (1993) suggested that circulating mid-ocean ridge hydrothermal fluids derived most of their Li from young, unaltered basaltic crust, and not from older, sea floor altered basalts. The altered basalts have higher Li concentrations and markedly heavier isotopic signatures compared to unaltered basalts. High temperature fluids from systems capped with thick sediment cover showed similar concentration and isotopic ranges as those from ridge areas that lack sediment cover, requiring the development of the isotopic signature by deeper processes that are not vitally affected by shallow materials that are enriched in Li available for hydrothermal leaching (Chan et al. 1994a).

Rocks that have been altered during interaction with these kinds of hydrothermal fluids and retain the Li isotope record of the process may be somewhat uncommon because of the prominence of re-equilibration with low temperature fluids. Based on the empirical and experimental evidence, hydrothermal alteration preferentially removes heavy Li from the rock; in the absence of retrograde mineral formation, this generates altered rock with lower Li concentration and lower δ^7Li than in the unaltered state. This was first recognized in dredged ocean floor greenschist-facies metabasalt (4.6 ppm Li and δ^7Li = −2.1; Chan et al. 1992). Subsequent reports of hydrothermally or greenschist facies metamorphosed sediments and basalt in the oceanic crust demonstrate reduction in Li concentrations and a general trend toward lighter isotopic compositions (Chan et al. 1994a; You et al. 1995; James et al. 1999), although this process may be complex when complete sections are examined (Fig. 12; Chan et al. 2002a).

In oceanic fracture zones as well as in actively-upwelling forearc mud volcanoes, seawater or other marine fluids interact directly with mantle rock (Bonatti 1976; Fryer 1985). Serpentinization of mantle rocks at temperatures broadly <350°C involves influx of water, as well as general increase in Li concentration in the newly-formed hydrous assemblage. Considering the low temperatures involved, Li isotope exchange during seawater-mantle

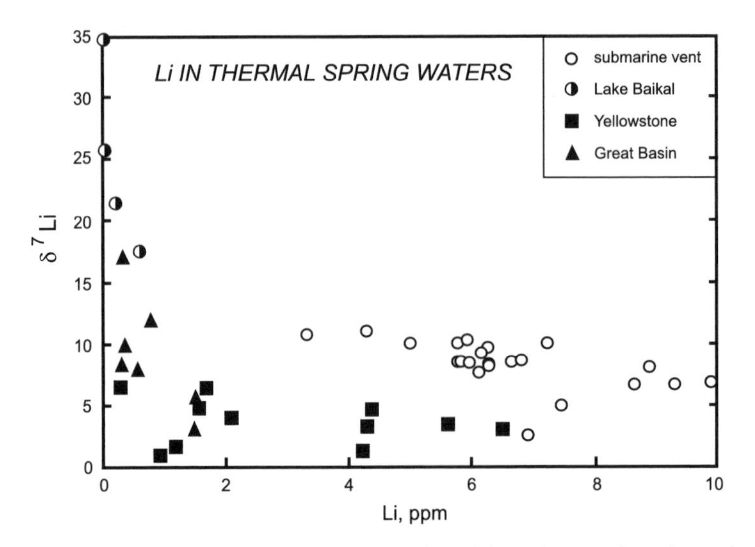

Figure 14. Plot of Li isotopic composition vs. concentration of thermal waters from the continents and the oceans (see text for references). Differences between the isotopic range of marine versus non-marine fluids emphasizes the variability in δ^7Li of continental rocks compared to oceanic basalt. The dilution of the continental fluids goes along with their lower temperatures; vent fluids are the only truly geothermal samples here, with temperatures in excess of 300°C.

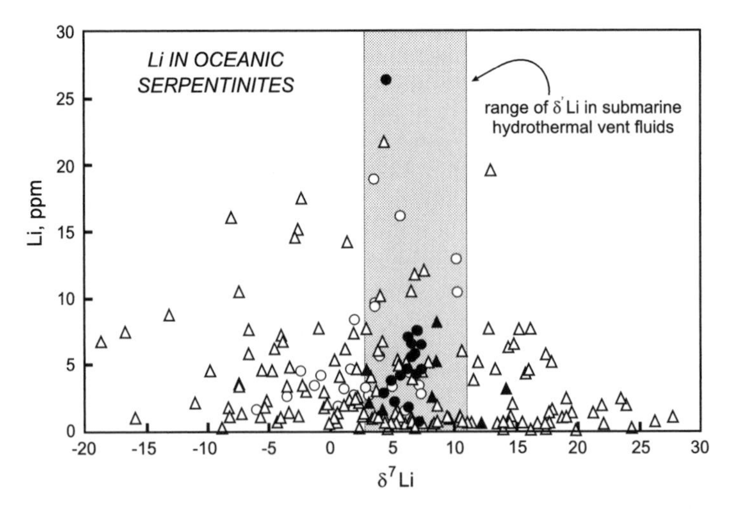

Figure 15. Plot of Li isotopic composition vs. concentration for serpentine and serpentinites samples (triangles = southwest Indian Ridge, Decitre et al. 2002; circles = Conical seamount, Marianas forearc, Benton et al. in review). For the Indian Ridge samples, open triangles represent *in situ* analyses (lower precision) of relict and alteration minerals, filled triangles are bulk samples. For the Conical seamount samples, open circles represent variably altered serpentinite clasts, filled circles are fine-grained homogenized serpentine muds that were the matrix to the clasts. The homogenized populations of these two geologically distinct settings display coherent systematics, with a mean value that overlaps the average for marine hydrothermal vent fluids (shaded region).

interaction can be expected. Results from dredged serpentinites (Decitre et al. 2002) and from forearc mud volcanoes (Savov et al. 2002; Benton et al. 1999) gave a consistent picture of fluid evolution during progressive interaction. Serpentinites in both studies were enriched in Li over unmetasomatized mantle rocks and were isotopically heterogeneous, ranging from much lighter than ordinary mantle rocks to slightly heavier (δ^7Li = −6.1 to +14.2 for bulk analyses, −19 to +28 for *in situ* analyses). Despite the differences in nature of formation of these two suites of serpentinites, the samples that showed the most coherent relations between elemental and isotopic data had a coincident mean δ^7Li = +7 ± 5‰, which was interestingly similar to the average composition of sea floor hydrothermal vent fluids (+8 ± 4‰) (Fig. 15).

Decitre et al. (2002) interpreted Li isotopic compositions of Indian Ocean serpentinites to represent internal recycling of Li from the basaltic crust into progressively-formed serpentinites, rather than Li extracted mainly from seawater. They envisioned the generation of the variable isotopic signatures in serpentinites (both lighter and heavier than the unaltered rocks) by interaction of mantle with a hydrothermal fluid that evolved, as it cooled and reacted with mantle rock along its flow path, toward heavier isotopic compositions. This model is consistent with the preponderance of compositions only slightly fractionated from MORB values, although better correlation between δ^7Li and Li concentration would be expected in the least complex scenario. The forearc serpentinites are considered to have been generated by infiltration of a fluid derived from the subducting Pacific plate (Benton et al. 1999), hence rather different in elemental contents than fluids expected to have interacted with fracture zone serpentinites.

Laboratory experiments have been carried out in order to gain insight on Li isotope behavior during interaction of high temperature fluids and various materials. Chan et al. (1994a) reacted Li-free aqueous fluid with a natural marine quartzofeldspathic mud (δ^7Li = +5.0) at an initial fluid/solid ~3. Fluids were sampled as temperatures were raised to 350°C over c. 600 h in the first experiment, and in a second run they were also sampled during cooling, after peak temperatures of approximately 350°C were reached (c. 540 h total). Both experiments demonstrated leaching of Li into solution, even at low temperatures, but more significantly as hydrothermal conditions are reached. The earliest fluids were isotopically heavy relative to the starting sediment, but under persistent hydrothermal conditions, fluid isotopic composition approached equilibrium with the starting solid. Fluids extracted in the second experiment during cooling also showed enrichment in heavy Li relative to the initial solid, and Li concentration in the fluid increased over the steady-state peak temperature value. The results are in general agreement with model predictions using empirical fractionation factors and assuming chlorite is the primary hydrothermal alteration mineral.

In the laboratory experiments of Seyfried et al. (1998), naturally altered sea floor basalt (δ^7Li = +7.4) was reacted with Li-free alkali-chloride aqueous fluid at 350°C for 890 hours (initial fluid/solid mass ratio ~2). Samples of the fluid were taken throughout the experiment, and showed initial rapid influx of isotopically heavy-enriched Li released by early-dissolving alteration minerals. However, with progressive reaction, isotopic composition of the fluid decreased and Li concentration reached apparent steady state. Although an equilibrium model applies best to the synthetic results, Rayleigh distillation was considered most likely to apply in hydrothermal reactions occurring in nature.

James et al. (2003) performed laboratory experiments in which aqueous fluids were reacted with natural samples of unaltered MORB (δ^7Li = +4.6) and a quartzofeldspathic mud/turbidite mixture (δ^7Li = +1.3). The starting fluids had elemental compositions generally intermediate between seawater and hydrothermal fluids, and were isotopically light (δ^7Li = −5.5). The initial fluid/solid was ~4. Fluid samples were extracted as the experiments were raised to 350°C over 400–500 h, and then during cooling (c. 740 h total). In both experiments Li was removed from solution at low temperatures, but the isotopic composition of the fluids rapidly shifted to heavier values. The Li isotopic composition of the fluid in the sediment experiment progressively changed toward that of the starting solid as the temperature reached 350°C, whereas that of the basalt changed more sluggishly, plausibly as result of coincident fractionation from secondary mineral formation at lower temperature (James et al. 2003).

All of the experimental evidence shows extraction of Li from basalt, altered basalt and sediments even at temperatures <100°C. Lithium was incorporated into clays at temperatures up to 150°C during basalt alteration (Seyfried et al. 1984; James et al. 2003), with affinity proportional to the fluid/rock ratio, such that Li was simultaneously added to and removed from fluids. Under high temperature, solid-dominated hydrothermal conditions (c. 350°C), Li was universally removed from materials and kept in solution, rather than in alteration minerals. The limited isotopic fractionation at these conditions has been estimated as $\alpha \sim 0.994$ (= $(^7Li/^6Li_{secondary\ minerals})/(^7Li/^6Li_{solution})$); Chan et al. 1993; Seyfried et al. 1998).

Marine pore fluid-mineral processes. During the diagenetic process (primarily compaction and mineral dehydration) fluids may be released from marine sediments, contributing waters of diverse chemical and isotopic compositions to the oceans. Pore fluids carry a record of fluid-sediment interaction. In the simplest case, fluids derived from seawater will possess lower concentrations of Li than seawater and $\delta^7Li > +31$, as a result of fractionation during fluid-sediment interaction or crystallization of low temperature minerals. However, depending on local geology, pore fluids can embody much more complex evolutionary histories. In order to derive meaningful conclusions, studies of these fluids have to contend with the potentially overprinting effects of variable mineral reactions and fractionation factors from strata of differing mineralogy and changing temperature in the sedimentary stack, *in situ* diffusion of mobile elements through the permeability network, and the vertical and lateral mixing of fluids of variable compositions (e.g., You et al. 2004). Whether or not fluids reach steady state compositions (at any stage in their geochemical evolution) is another complicating factor in these studies.

You et al. (1995) studied bulk samples of fluids incorporated in sediments from ODP Site 808, in the Nankai Trough, southwest of Japan. Pore fluids have somewhat variable isotopic compositions (δ^7Li = +10 to +21), with a spike of light compositions near the basal decollement. These authors interpreted the decollement zone geochemical anomaly to represent influx of waters with Li derived from leaching of sediments at high temperatures.

Zhang et al. (1998) considered Li isotopes in pore fluids of marine sediments at ODP Sites 918 (Fig. 16a) and 919, southeast of Greenland. They found fluids with isotopic compositions reflecting a complex range of diagenetic processes and δ^7Li = +24.5 to +43.7. They identified the importance of NH_4^+ concentrations in pore fluids as a counterion for Li exchange on sediments. The peak in NH_4^+ concentration at Site 918 coincided with Li isotopic compositions lighter than seawater, suggesting that desorbed Li was isotopically lighter than its seawater starting composition. Zhang et al. (1989) estimated a fractionation factor for this process of $\alpha \sim 0.975$ (= $(^7Li/^6Li_{mineral})/(^7Li/^6Li_{solution})$). This provides an interesting biogeochemical slant to the interpretation of Li isotope data from some pore waters, since NH_4^+ is a product of the biological degradation of organic matter. Lithium enrichment in deeper pore waters at Site 918 was interpreted to derive from influx of sediment-derived Li from lower horizons at higher temperatures (Zhang et al. 1998), a similar interpretation as has been applied since, where deep increases in pore water Li concentration have been observed.

James et al. (1999) reported findings from ODP Site 1038 (Fig. 16b), adjacent to the Gorda Ridge, an area of active hydrothermal venting through a thick sedimentary pile. Pore fluids there had a range in isotopic compositions (δ^7Li = −0.6 to +27.5) that were modeled in terms of initial hydrothermal removal of Li from sediments followed by down-temperature fluid fractionation during crystallization of secondary minerals, and ultimately dilution with

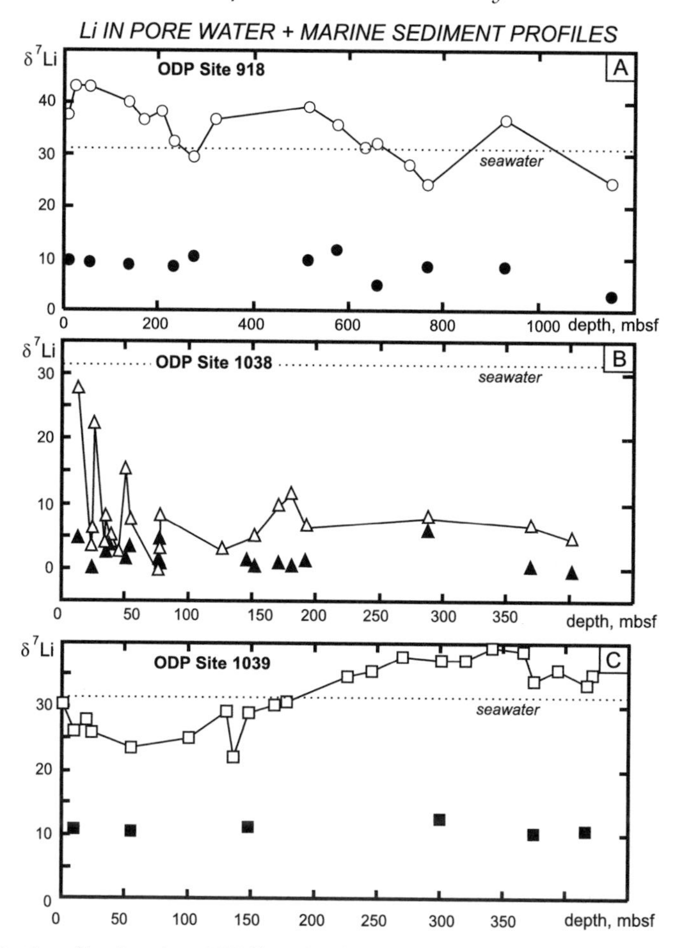

Figure 16. Depth profiles from three ODP Sites, showing Li isotopic composition variations in pore waters (open symbols) and associated sediments (filled symbols). (a) Site 918, Irminger Basin, north Atlantic (Zhang et al. 1998); (b) Site 1038, Escanaba Trough, northeastern Pacific (James et al. 1999); (c) site 1039, Middle American Trench off of Costa Rica (Chan and Kastner 2000). The average composition of seawater is noted on each profile with dashed line (note different scales). Whereas sediments have relatively monotonous compositions, pore waters have compositions reflecting different origins and processes in each site. Interpretations of the data are summarized in the text under, "Marine pore fluid-mineral processes."

seawater. This model fits the data more closely than simple mixing between hydrothermal fluid and sediment-derived Li, but fails to account for the isotopically lightest samples.

Chan and Kastner (2000) examined pore fluids from ODP Sites 1039 (Fig. 16c), 1040, and 1042, outboard of Costa Rica, with a total range in δ^7Li of +22.2 to +39.3. At Site 1039 they observed a pronounced correlation between Li concentration and Li isotopes, such that the pore waters with the highest concentrations (approximately equal to seawater) had the lowest δ^7Li. Variations in data from shallow depths at Site 1039 were interpreted to reflect superimposed effects of alteration of volcanic ash and ion exchange between Li^+ and NH_4^+ (Chan and Kastner 2000). The former consumes Li from fluids, the latter releases, but the isotopic fractionation upon desorption exceeds that of smectite formation (0.976 vs. 0.981), so the net process is a decrease in δ^7Li. Pore fluids deep at Site 1039 show increase in Li concentrations and decrease in δ^7Li, toward seawater values. This was interpreted as indicating mixing with a fluid derived from seawater infiltration, perhaps along the basement-cover boundary as a result of lithological control or non-uniform permeability structure (Chan and Kastner 2000). These are factors that should be common in deep sea sedimentary accumulations. The isotopically rather light pore fluids at Site 1040 have Li concentrations 10× greater than seawater and were interpreted to contain a component derived from mineral dehydration deeper in the sedimentary mass (at temperatures 100–150°C; Chan and Kastner 2000).

James and Palmer (2000b) analyzed pore fluids from a set of ODP locations in the Pacific northwest: Site 1037, adjacent to the Gorda Ridge, and Site 1034, in Saanich Inlet, BC, Canada. At the coastal Site 1034, pore fluids were more dilute than seawater with respect to Li and had δ^7Li less than seawater (+24.2 to +30.0). Moderate ammonium concentrations mimicked the Li concentration profile, suggesting that, particularly in the upper section of the Site, in spite of dilution by fresh water, some of the light isotopic signature came from isotopic fractionation during desorption. Pore fluids at Site 1037 showed a wide range of isotopic values (δ^7Li = +11.6 to +42.8). Controls on Li behavior were interpreted to be dominated by thermal alteration effects, perhaps up to c. 110°C. It was estimated that Li in pore fluids may have derived substantially (>65%) from Li initially adsorbed on particles in the water column (James and Palmer 2000b).

James et al. (2003) inferred, on the basis of comparison between experimental results and natural data, that upwelling rate is another parameter that is critical to interpretation of Li isotope signatures of pore fluids. At low temperatures (<100°C), Li may be lost to fluids by sediments (Chan et al. 1994a). However, enrichment in Li of pore fluids to concentrations greater than that of seawater near basement contacts may more prevalently reflect slow rate of upwelling, as fluid-sediment interaction is thereby favored. This interpretation is consistent with data from a variety of samples from ridge flanks (e.g., Elderfield et al. 1999; Wheat and Mottl 2000).

Marine biogenic and inorganic precipitates. Marine precipitates, ranging from biologically-secreted carbonate to the inorganic crystallization of ferromanganese minerals on the sea floor, have been of great interest in the quest for geochemical proxies of ocean chemistry and temperature. For Li isotopes, these studies have focused on either assessing temperature dependence in carbonate crystallization or on developing better tools for understanding weathering on the continents. Although but a few studies have thus far tantalized us with significant results, these will become more widespread as analytical barriers resultant from low concentrations (2 ppm average, with abundant biogenic carbonate shells possessing <1 ppm Li; Delaney et al. 1985; Hoefs and Sywall 1997; Huh et al. 1998) crumble. For example, studies dealing with small, single-species foraminiferal samples, for example, have recently become analytically feasible.

Pertinent laboratory experiments cast doubt on the viability of Li isotopes as paleotemperature tools (Marriott et al. 2004). This study showed no analytically resolvable isotopic difference among inorganic calcite precipitated from seawater composition over a range of 25°C (Fig. 17). The calcite was approximately 8.5‰ lighter than the solution from which it was precipitated, similar to the offset from natural *Acropora* coral. A sample of *Porites* coral was offset by approximately 11‰ from its seawater medium (Marriott et al. 2004), suggesting a difference in the metabolic effect in different species. Hence, use of Li isotopes as a seawater composition proxy in ancient corals, and perhaps other carbonate-secreting organisms, requires consideration of the specific creature being sampled, at least at the genus level. The only other examinations of dominantly inorganic carbonate precipitation were natural as opposed to synthetic studies, were less systematic, and arrived at different

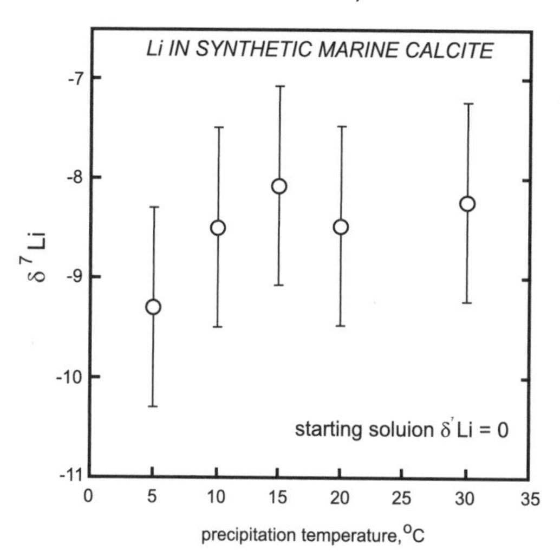

Figure 17. Plot of Li isotopic composition vs. temperature of growth for synthetic calcite crystallized from a solution containing Li from L-SVEC (Marriott et al. 2004). The results are most consistent with temperature not being a significant control on mass fractionation of Li during crystallization from aqueous solution, thus essentially eliminating Li isotopes as a paleotemperature proxy in marine carbonates.

conclusions. Chan (unpublished data, cited in Huh et al. 1998) indicated a 20‰ fractionation effect in the crystallization of Dead Sea aragonite, whereas Tomascak et al. (2003) observed no analytically-resolvable isotopic effect of crystallization of carbonate at Mono Lake, California. Clearly, additional laboratory studies, like that of Marriott et al. (2004), would be valuable in order to further map out the future of carbonate Li isotope proxies. The possible impact of crucial factors such as elemental composition of solution, carbonate ion content, and growth rate have yet to be revealed.

Foraminifera are also important in marine geochemistry studies, and Li isotopes have been measured in the shells of a variety of these organisms. The first report of this kind (You and Chan 1996) gave data for four Pleistocene samples of *P. obliquiloculata*: δ^7Li = +19.3 to +23.0 for two glacial period samples and +26.6 to +42.4 for two interglacial samples (data without blank correction). The species effect interpreted by Marriott et al. (2004) in corals has also been suggested for forams. In the study of Košler et al. (2001), core-top (i.e., Holocene) *P. obliquiloculata* samples yielded isotopic compositions close to modern seawater (δ^7Li = +27.8 to +31.1), whereas samples of *G. tumida* from the same samples had values of up to +50.5.

Samples of mixed foram species reported by Hoefs and Sywall (1997) showed a large isotopic range in samples as old as 57 Ma (δ^7Li = +3 to +41; recalculated assuming seawater with δ^7Li = +31.5). These data should be viewed skeptically, however, due to the large sample sizes and cleaning methods employed, which are not consistent with standard practices. Lithium isotope data for modern foraminifera show no such large variability. Forams of different genera (*Neogloboquadrina* and *Globorotalia*) from core tops at a single Indian Ocean site gave identical isotopic compositions of δ^7Li ~ +30 (Bruns and Elliott 1999).

Lithium isotope data from carbonate shells of other marine invertebrates have been reported. Hoefs and Sywall (1997) determined isotopic compositions of seven different species of modern bivalves from the North Sea coast. These samples had a relatively small range in δ^7Li (+15 to +21), which corresponds well to the seawater-carbonate offset from inorganic calcite and modern corals (Marriott et al. 2004).

Bulk carbonate samples have been analyzed in several studies. You and Chan (1996) determined the compositions of a pair of carbonate-rich sediment samples from ODP Site 851. The near-surface sample had δ^7Li = +6.2, whereas a sample from 196 m below the sea floor was +32.0. Carbonate-rich oozes from Hoefs and Sywall (1997) from the time period 81–9

Ma showed a large range in δ^7Li (−11 to +25). Although this study showed elemental data consistent with minimal incorporation of clays, the large sample size in this study makes the potential of contamination by small amounts of Li-rich clays a strong possibility, to say nothing of multi-species effects. The average δ^7Li of the oozes was considerably lower than of bulk foram samples over a similar time range (+8 for oozes vs. +23 for forams) and lower than oozes, marine chalk and oolites (+12.2 to +27.7) from Huh et al. (1998)—these are all consistent with the oozes incorporating variable proportions of isotopically light-enriched clays. Bulk Devonian to Cretaceous limestone samples from German localities had δ^7Li = −2 to +20 (Hoefs and Sywall 1997). Although low Li concentrations suggest a general lack of pervasive diagenetic alteration, this needs to be carefully assessed, particularly in bulk samples.

Non-carbonate marine precipitates are also used for the study of changes in ocean geochemistry over time. You and Chan (1996) reported data for three "hydrothermal" ferromanganese crusts with Li concentrations ranging from 9 to 30 ppm. The samples had δ^7Li = +25.7 to +36.6. Henderson and Burton (1999) reported a Li concentration of 4.62 ppm in a slow-accumulating Fe-Mn crust from the northwestern Atlantic and suggested that the low concentration, coupled with slow growth rate, indicates Li diffusion will be too rapid for a valid chemostratigraphy to be preserved.

Lithium isotopic fractionation on the continents

Although the processes that lead to Li isotope signatures of materials on the continents are somewhat similar to processes in the marine realm, Li isotopes in the latter are far more comprehensively studied at this point. The "starting materials" in the continents are more complex in terms of elemental and isotopic composition and history, so necessarily studies that consider materials on the continents have to compete with a larger initial level of complexity. Nonetheless, inroads have been made into understanding Li isotopes on the surfaces of the continents, particularly in quantifying the weathering process, and this remains among the most fertile areas for future study.

Continental runoff and weathering processes. Surface waters are generally dilute with respect to Li compared to groundwaters and other sub-surface fluids. Estimates of the global average Li concentration in river water range from 1.5 to 3 ppb for filtered samples (Morozov 1969; Edmond et al. 1979; Huh et al. 1998). Reported lithium contents of atmospheric precipitation range widely, and apparently depend somewhat on access to marine aerosols. Hence Li in Antarctic snow varies from approximately 0.05 ppb near the coast to $<$0.01 ppb inland (Ikegawa et al. 1999). Alpine snows have a lower range of reported concentrations (0.005–0.007 ppb; Bensimon et al. 2000). Limited data for Li contents of rainwater also show considerable variability. Filtered rainfall from Hawaii had 0.075 ppb Li compared to 1.01 ppb in unfiltered precipitation from the same area (Pistiner and Henderson 2003). By comparison, rainfall from eastern Maryland contained c. 0.1 ppb (Schwab et al. 1995). Such concentrations make precipitation an important consideration in the hydrologic budget of Li principally where rainfall is intensive and unweathered source rocks have low Li concentrations.

Two large studies on the geochemistry of rivers have been carried out in which Li isotope data are reported: an examination of rivers that make up c. 30% of total world discharge (Huh et al. 1998), and a detailed study of the Orinoco Basin (Huh et al. 2001). River data have been included in several other works (Chan et al. 1992; Falkner et al. 1997; Bottomley et al. 1999; Tomascak et al. 2003; Pistiner and Henderson 2003), and all are consistent with the larger, more comprehensive studies. Several important points about Li in continental runoff were manifest from these works. The global influx of Li to the oceans (δ^7Li = +6.0 to +37.5, with a flow-weighted average of +23.5) is isotopically lighter than seawater. Also, the suspended loads of rivers carry a substantial amount of the element, universally isotopically lighter than the associated dissolved loads.

How do these studies address the materials and processes that control the isotopic composition of Li in surface waters? From the works of Huh et al. (1998, 2001) a few details are clear. On a gross scale, the isotopic compositions of dissolved Li in world rivers showed no correlation with the lithology of the dominant exposed rocks of each basin. The Orinoco Basin data showed differences in the isotopic compositions of dissolved loads of tributaries draining the young Andes mountains (δ^7Li = +31.0 to +37.5) versus the ancient, low-relief craton (δ^7Li = +13.5 to +22.8) (Fig. 18). The elegant explanation from Huh et al. (2001) linked Li isotope signatures of dissolved loads not to lithology, but to fundamentally different weathering character: reaction-limited in the juvenile, high-relief Andes, transport limited in the stable craton. In the cratonic regions, weathering reactions were able to go closer to completion, and thus mean isotopic compositions were lighter. This hypothesis did not explain the isotopically heavy composition of water from the mouth of the Orinoco (δ^7Li = +33.3), however.

Similarly, rivers in Iceland (Vigier et al. 2002) showed a strong fractionation from the presumed-invariant isotopic composition of their solute sources (δ^7Li = +17.1 to +23.9, compared to approximately +4 for pristine basalts). Rivers draining terranes with the oldest exposed lavas yield low δ^7Li; a positive correlation exists between dissolved Li concentration and δ^7Li in Icelandic rivers. These observations are consistent with the data of Pistiner and Henderson (2003) for naturally weathered, historically-erupted Icelandic basalt. The outermost 4 mm of the basalt erupted in 1783 was shown to have δ^7Li 2‰ lighter than the interior of the sample, suggesting weathering over 200 yr, even in the arctic climate, lead to appreciable release of isotopically heavy Li to surface waters.

Soils are complex systems that require thorough examination in order to place data in a meaningful framework. For instance, Hawaiian soils from areas with essentially identical young basaltic bedrock and similar annual rainfall have been examined (Huh et al. 2002; Pistiner and Henderson 2003), but results are strikingly different. Huh et al. (2002) reported finding a 2× to 6× increase in Li concentration relative to unweathered basalt and variations in δ^7Li between +3 and +10 from lower to upper soil horizons. Pistiner and Henderson (2003) examined nine successive layers of volcanic soil that showed restricted variability in both

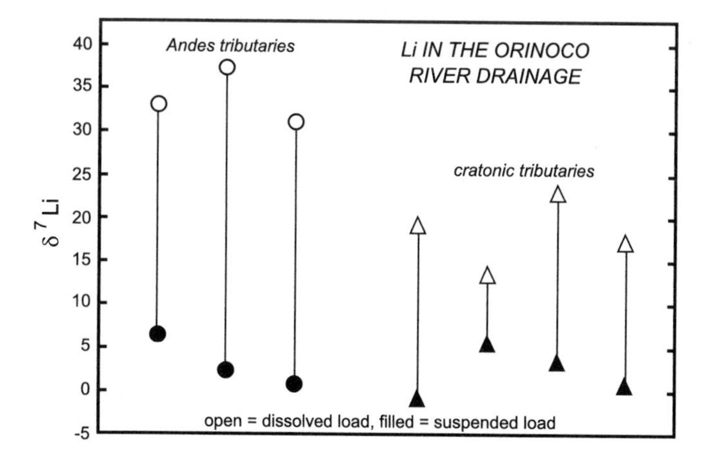

Figure 18. Orinoco River tributary Li isotope data (Huh et al. 2001). Tie lines connect samples pairs of dissolved (open symbol) and suspended load (filled symbol) for tributaries draining the Andes (circle) and those draining cratonic area (triangle). Isotopic compositions of suspended loads were similar in both cases, but the dissolved loads of Andes drainages were uniformly isotopically heavier. This difference was attributed not to lithology in the headwaters, but to transport- versus reaction-limited weathering conditions.

elemental content and isotopic composition relative to unweathered material. They suggested that heavy Li lost during weathering reactions was largely compensated for by the isotopic composition of rainfall (δ^7Li = +10.2 and +14.3, filtered and unfiltered, respectively). Both Hawaiian soil studies concurred that the influence of atmospheric Li (transporting both continental dust and seawater aerosols) was probably important to the Li isotope budget of weathered material in this environment. Also, the sequence examined by Huh et al. (2002) was considerably older than that in Pistiner and Henderson (2003) (400 ka versus 400 y).

Njo et al. (2003) presented results of a weathering study in a temperate, humid saprolite sequence in South Carolina. The unweathered metamorphosed basalt bedrock (δ^7Li = −4) showed correlated decrease in δ^7Li with higher degrees of cation leaching, to extremely light compositions (to δ^7Li = −20 in the most aluminous saprolite). Similarly, granite-bedrock saprolites in the same area exhibited strong isotopic fractionation in the same sense (from δ^7Li = +2.3 in granite to −6.8 in most weathered soil). These results suggest that soil sequences such as these may be an important missing link in understanding the continental weathering budget of Li, as they balance the heavy-Li-dominated dissolved loads in rivers, a role analogous to that played by isotopically light river suspended loads.

Pistiner and Henderson (2003) conducted laboratory experiments to examine Li isotope behavior during mineral dissolution. In the experiments, rocks were reacted with stronger acid than in natural weathering so as to minimize the potential for re-precipitation of secondary minerals. Solution recovered from partial dissolution of basalt in 0.8 M HNO_3 at 22°C and 45°C revealed no isotopic change from the initial rock. The same experimental process conducted with a Scottish granite (including chlorite, epidote, and calcite, suggesting it contained a non-igneous component from sub-magmatic alteration) reproducibly showed a removal of isotopically light Li to the solution (δ^7Li = +9.0 in starting granite, +1.6 in solution after reaction). Despite the acidity of the reacting solution, this may indicate that the precipitation of secondary minerals controlled the Li isotopic composition of the fluid.

The first report of the Li isotope geochemistry of lakes revealed the somewhat surprising result that some of the world's largest lakes had δ^7Li very similar to modern seawater (+32.1 to +34.4 in Lake Tanganyika, the Dead Sea and Caspian Sea; Chan and Edmond 1988) (Fig. 19). Later analyses of Lake Baikal found general agreement with these initial findings (δ^7Li = +28.7 to +33.4; Falkner et al. 1997), with the bulk of Baikal inputs having isotopic compositions similar to the lake itself. This is probably not generally the case: like the oceans, these lakes most commonly have inputs of Li that are isotopically lighter than their steady-state compositions. This suggests that the same processes that control the Li isotopic composition of the oceans may similarly govern the systematics of Li in these large lakes.

Hydrographically closed lakes are much more dynamic geochemical systems than the more persistent lakes, owing to the influences of climate and hydrology over short time scales. Compared to the lakes outlined above, closed lakes of the western U.S. Great Basin show highly variable modern Li isotopic compositions (δ^7Li = +16.7 to +23.7 for 5 lakes; Tomascak et al. 2003) (Fig. 19). In a detailed study of Mono Lake, California, Tomascak et al. (2003) constructed a general mass balance between fresh surface water inputs and saline spring inputs of both ambient temperature (tending toward heavy isotopic compositions) and elevated temperature (with light isotopic compositions). The data suggested the isotopic composition of the lake could be variable on short time scales as the balance between spring inputs changed. This hypothesis was tested in a set of samples from the Great Salt Lake, Utah (Tomascak et al. 2001). The lake waters, in which $^{87}Sr/^{86}Sr$ change correlated with decadal drought fluctuation, showed restricted Li isotope variability, weakly anticorrelated with radiogenic Sr. The study suffered from a lack of quantification of the isotopic compositions of solute source end members. Taken as a whole the lake data suggest a potential for Li isotopes to track hydrologic (and hence climatic) changes over lake histories.

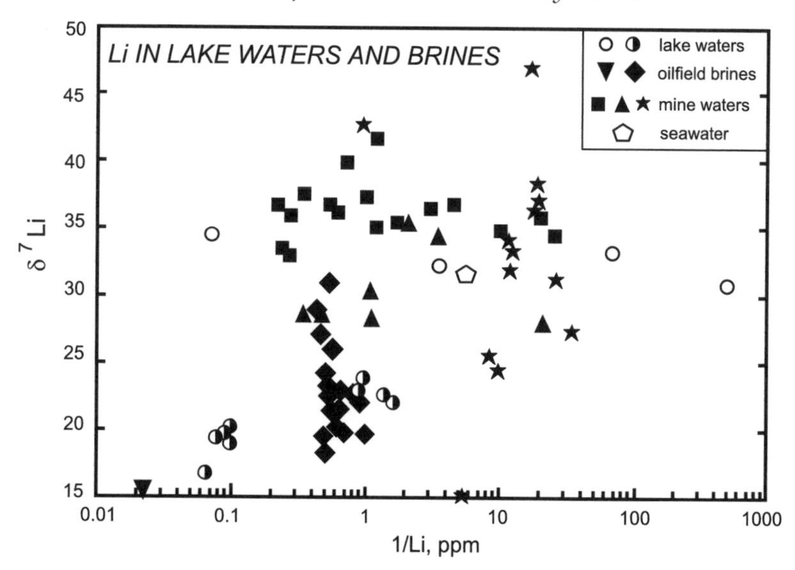

Figure 19. Plot of Li isotopic composition vs. inverse Li concentration for lakes and basinal/oilfield brines. Lakes: open circle = major global lakes (Chan and Edmond 1988; Falkner et al. 1997); semi-open circle = western U.S. closed basin lakes (Tomascak et al. 2003). Oilfield brines: inverted triangle = Williston basin, Saskatchewan (Bottomley et al. 2003); diamond = Israeli oilfields (Chan et al. 2002d). Mine waters (Canadian Shield basinal brines): square = Yellowknife, NWT (Bottomley et al. 1999); triangle = Sudbury, Ontario, area (Bottomley et al. 2003); star = Thompson, Manitoba, area (Bottomley et al. 2003). Average composition of seawater is included for reference.

Deep and shallow crustal fluids. Although comprehensive studies have yet to be undertaken, "ordinary" groundwater—ambient temperature water withdrawn from near-surface wells—exhibits a wide range in Li isotopic compositions, from extremely light (δ^7Li < -10; Gellenbeck and Bullen 1991) to similar to continental sedimentary rocks ($-10 < \delta^7$Li $< +20$; Tomascak et al. 1995a, 1999b; Tomascak and Banner 1998) to very heavy (δ^7Li $> +20$; Hogan and Blum 2003; Tomascak et al. 2003). This range no doubt reflects the variety of solute sources and fractionation processes that take place within aquifers. Saline groundwaters (0.01 to 16 ppm Li) are common in many arid to semi-arid regions. Where these have been examined, they exhibit relatively heavy Li isotopic compositions, evidently evolving from low temperature fluid-particle interactions (δ^7Li = +7.0 to +30.6; Vocke et al. 1990; Tomascak and Banner 1998; Tomascak et al. 2003).

Considering the sizeable contrast in Li isotopic characteristics between continental sediments and seawater, this system has been applied to geochemical problems related to the origins of solutes in deep sedimentary fluids, like basement and oilfield brines, or "formation waters." Bottomley et al. (1999) used Li isotopes to study Ca-chloride brines in the Miramar Con gold mine in Yellowknife, Northwest Territories, Canada. Shallow brines (c. 700 m below surface) had dissolved Li contents like saline groundwaters (39 to 99 ppb), whereas deep brines (>1000 m below surface) were considerably more saline and Li-rich (up to 4.3 ppm). The brines were isotopically fairly homogeneous, especially considering the range in Li concentration, with values equivalent to or heavier than modern seawater (δ^7Li = +33.2 to +41.8) (Fig. 19). Tritium concentrations indicate that all but the deepest mine fluids were in contact with the atmosphere within the last 40 years, but the light composition of local surface water (δ^7Li = +23.5) suggested it is not important to the process of brine formation. Combination of Li and Sr isotope and major ion concentrations suggested seawater, perhaps incorporated during the Devonian, was the ultimate source of solutes in these brines. The potential difference between Devonian and modern seawater δ^7Li was not considered.

In terms of Li content and isotopic signature, the Yellowknife mine waters are similar to waters from Sudbury (Ontario) area mines (0.05 to 2.9 ppm Li and $\delta^7Li = +27.9$ to $+42.6$; Bottomley et al. 2003). Compositions of the Sudbury brines are consistent with predominance of preserved marine Li. These waters contrast with samples from mines in central Manitoba, which are dilute (Li concentration ≤ seawater) and show a wide range of δ^7Li (+15.0 to +46.9; Bottomley et al. 2003). These waters are plausibly mixtures of isotopically heavy brine (developed through fluid-rock interaction in fracture zones) and waters that had interacted with isotopically light country rocks.

Oilfield brines have been suggested as a natural resource for Li owing to their very high concentrations (Collins 1976). Chan et al. (2002d) investigated Li isotopes in oilfield brines from Israel. The Ca-chloride brines from the Heletz-Kokhav oilfield were Li-rich (1.0 to 2.3 ppm) and showed a range in Li isotopic composition lighter than modern seawater, unlike the Yellowknife basement brines ($\delta^7Li = +18.2$ to $+30.8$) (Fig. 19). Evolution of these brines from seawater through a process of mineral reactions, evaporation and dilution was interpreted from the Li isotope and elemental data. The studies of Bottomley et al. (1999) and Chan et al. (2002d) succinctly illustrated the variable nature of Li in even broadly similar environments: in one case Li was taken up in secondary minerals, leaving a solution that was isotopically heavier than modern seawater; in the other case Li was interpreted to have been removed from sediments, leaving a solution isotopically lighter than seawater.

Continental hot spring waters, where essentially undiluted, have high Li concentrations (most in the range 0.1 to 10 ppm; Brondi et al. 1973; White et al. 1976; Shaw and Sturchio 1992). Although they showed a large range in Li isotopic composition ($\delta^7Li = +1.0$ to $+34.8$; Bullen and Kharaka 1992; Falkner et al. 1997; Sturchio and Chan 2003; Tomascak et al. 2003), the majority of these fluids were fairly similar to marine hydrothermal vent fluids in terms of Li concentration and isotopic composition (average $\delta^7Li = +5.1$ vs. +8.1 in vent fluids) (Fig. 14). As seen in a previous section, extensive hydrothermal alteration (that is not later overprinted by lower-temperature mineral reactions) will yield fluids with isotopic compositions that approach those of the materials from which Li is originally leached. Extant evidence from the eastern Sierra Nevada and Long Valley, California, and Yellowstone, Wyoming, are consistent with local geology having Li isotope characteristics congruent with the thermal fluids emanating from springs in these areas. The isotopically light sources of thermal waters at Yellowstone, lighter than any of the potential source materials of the area, would be satisfied by the high temperature dissolution of pre-existing alteration minerals in altered volcanic rocks, analogous to the laboratory results of granite dissolution by Pistiner and Henderson (2003).

Combinations of mineral reactions at lower temperatures and mixing with more dilute fluids are likely to result in the variations in concentration and isotopic composition in many of the continental thermal spring waters but not seen in their marine relatives. The extreme manifestation of this difference may have been generated in the dilute hot spring waters from around Lake Baikal, whose heavy isotopic compositions required extensive re-equilibration at temperatures 100–150°C (Falkner et al. 1997).

Minerals derived from evaporation of saline solutions have not been extensively examined. Evaporitic salts from the western U.S. Great Basin concentrated ^7Li ($\delta^7Li = +31.8$ to $+32.6$; Tomascak et al. 2003). Evaporation of thermal waters from Yellowstone yielded travertine with extremely low δ^7Li (−4.8; Sturchio and Chan 2003), offset from its coexisting fluid by −11.1‰. Such evaporites, where preserved in the sedimentary record, could afford an opportunity to examine variations in spring water compositions and hence information on hydrothermal processes over time.

Environmental tracers. Given the interest in the hydrological community in developing new geochemical tracers of near-surface water flow, it is perhaps unexpected that Li isotopes have seen such little exploitation in this field (Bullen and Kendall 1998), especially given the widespread use of Li as a conservative solute tracer (e.g., Bencala et al. 1990). The studies that have used Li isotopes in environmental capacities have demonstrated significant spread in composition between end-members, indicating that the system is well-suited for discerning solute sources in areas of geological and hydrological complexity. Also, as noted above, the strong anthropogenic signature of "processed" Li (Qi et al. 1997a) that is likely to be widespread among sources of industrial waste, suggests that Li isotopes could be well-suited in tracing point-source pollution (Bullen and Senior 1992; Hogan and Blum 2003).

Gellenbeck and Bullen (1991) showed that possible groundwater pollutants in an Arizona field site showed isotopic contrast of c. 40‰ (ground water with δ^7Li ~ −21; cow manure leachate δ^7Li ~ +19). In a study of groundwater flow, Bullen and McMahon (1992) used Li isotopes to gain insight into the nature of chemical reaction along a flow path. They noted variations in δ^7Li that were interpreted to result from dissolution of marine-derived minerals in one part of the aquifer (providing isotopically heavy Li relative to recharge), and from ion exchange in more distal parts of the aquifer (imparting isotopically light-enriched Li). Similarly, Tomascak et al. (1995a, b) examined variations in stream water Li isotopic composition from before, during and after a rain storm in eastern Maryland. The flushing of a pulse of isotopically heavy-enriched water on the increasing limb of the storm hydrograph indicated that easily-extracted Li, possibly anthropogenic in origin, was rapidly removed from adjacent agricultural fields at the onset of overland flow generation.

In a study of groundwater in and around the Fresh Kills landfill, New York, Hogan and Blum (2003) used Li isotopes to define mixtures of seawater and isotopically lighter fresh water. Waters extracted from wells in the landfill had variable δ^7Li (+22.0 to +53.9). Mass fractionation of diluted seawater-source groundwaters, presumably during fluid-particle interaction in unconsolidated sediments, was interpreted to have produced samples with isotope/concentration systematics that did not follow simple mixing of fresh water and seawater. Thus, in areas that have an initial isotopic difference between end-member compositions—and these many be commonplace—Li has the potential of demonstrating both solute sources and pathways of chemical evolution of groundwaters.

SIGNIFICANCE OF LITHIUM ISOTOPES IN THE BULK EARTH

Among the most glaring deficits in our present compilation of fundamental Li isotope data is in meteorites. To an extent this deficit results from the low Li concentrations of most stony meteorites (commonly <2 ppm; Heier and Billings 1972; Anders and Grevesse 1989), and hence the need to process large samples for bulk analysis. The lack of comprehensive studies of Li isotopes in meteorites is particularly striking in view of the progress made in understanding galactic Li through observational astronomy over the last two decades. Lithium isotopic fractionation during astration is considered to have produced the background heavy Li of the Sun and solar wind (^{7}Li/^{6}Li > 30; Boesgaard and Steigman 1985; Ritzenhoff et al. 1997; Chaussidon and Robert 1999). Certain observations of Li in interstellar space indicate a very light isotopic composition (^{7}Li/^{6}Li ~ 2; Knauth et al. 2000), consistent with the composition predicted from collisions between atoms in space and high energy galactic cosmic rays (Reeves et al. 1970). Meteorites and terrestrial materials appear to have intermediate Li isotopic compositions (^{7}Li/^{6}Li ~ 12; Krankowsky and Müller 1967; Chaussidon and Robert 1998). How this value developed is unclear, so more detailed Li isotopic studies of meteorites and their inclusions might be able to resolve key details about early Solar system evolution.

There have been many studies on Li isotopes in meteorites, but few since the development of high precision techniques. Of the earlier studies, most agree that meteorites do not show gross isotopic variations, but there are exceptions, especially in some studies of refractory inclusions (Klossa et al. 1981; Schirmeyer et al. 1997). The total extent in Li isotopic data from bulk meteorite samples falls within the range determined by Chaussidon and Robert (1998) for *in situ* analyses of three chondrules from the LL3 chondrite Semarkona (δ^7Li = +10 ± 20; recalculated assuming ^{7}Li/^{6}Li of L-SVEC = 12.175). James and Palmer (2000a) analyzed Li in the CI1 chondrite Orgeuil (δ^7Li = +3.9). McDonough et al. (2003) reported Li isotopic compositions for a range of petrologic types of chondrites. Petrologic type 1 – 2 carbonaceous chondrites (highly hydrated bodies) have δ^7Li in the terrestrial MORB range (+1 to +4), whereas type 3 – 4 carbonaceous and ordinary chondrites show a range of compositions (δ^7Li = −3.6 to +3) (Fig. 20). These data permit the interpretation that chondrites carry Li isotopic records of parent body processes involving fluid-solid interaction, generating isotopically heavy low temperature minerals, but they could equally be interpreted to reflect heterogeneity of source materials in the early Solar system. Using both bulk and spatially resolved analytical techniques it appears that Li isotope studies of meteorites may yield important information about early solar system and planetary processes.

As demonstrated by Teng et al. (2004), the mantle Li isotopic composition governs the bulk Earth composition, and hence an accurate quantification of the mantle is essential. Published estimates of Li isotopes in the bulk Earth have used values consistent with modern ocean floor basalts (δ^7Li ~ +6: Chaussidon and Robert 1998; δ^7Li ~ +4: Teng et al. 2004). However, the present day mantle isotopic composition is still rather murky, as described above. Basically, it is not clear if one small range exists to describe this reservoir. Although the MORB range may be as small as 5‰, it has been suggested that the actual uncontaminated MORB value has δ^7Li closer to +1.5, and that the range to higher values indicates subtle alteration (Moriguti and Nakamura 1998b). This would be consistent with the findings of

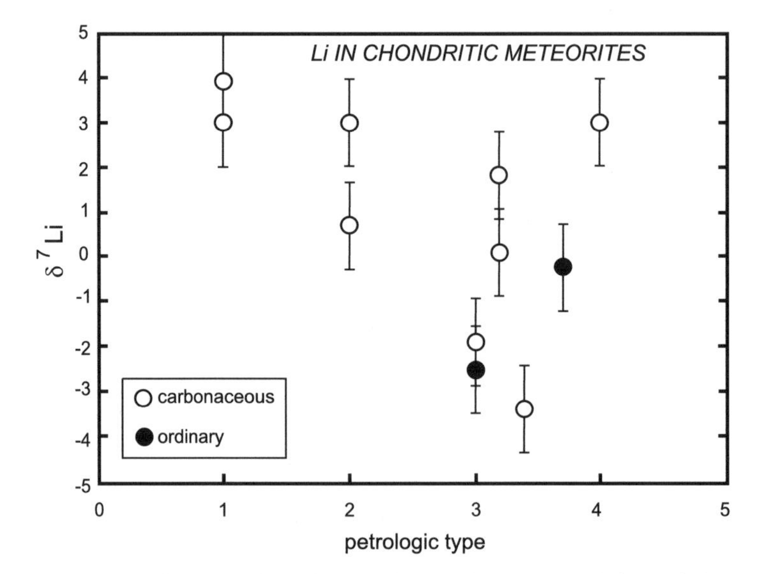

Figure 20. Lithium isotopic compositions of chondritic meteorites relative to their petrologic type (James and Palmer 2000a; McDonough et al. 2003). The extant data indicate lighter and more variable isotopic compositions in higher petrologic types of meteorites, those with the least record of hydrous parent body alteration.

Elliott et al. (2003). The data set for unmetasomatized peridotite xenoliths is too sparse to aid much in this controversy at present. As measurement reproducibility betters the ± 0.5‰ barrier, it is just a matter of analyzing the most appropriate samples before such questions can be answered and an accurate estimate can be made of the Li isotopic composition of the mantle, and hence of the bulk Earth.

CONCLUSIONS AND FUTURE DIRECTIONS

Although Li isotope study is still in its childhood in terms of mapping out the possibilities of the system, it is clear that many areas should benefit from further research. Based on the data generated to date, several general conclusions can be drawn and specific directions can be suggested:

1. There is no consensus on the specific mantle Li isotopic composition. Currently reported values for normal MORB suggest a range (δ^7Li = +1.5 to +5.6), but no single study has shed light on whether this range is an intrinsic feature of the upper mantle, to what extent cryptic contamination plays a role, or if bulk samples can only define an average for a reservoir with broad heterogeneity.
2. Given the fractionation of Li isotopes by processes near the Earth's surface, fractionated materials should be returned to the mantle throughout geologic time. There is apparent correlation between mantle heterogeneity in radiogenic isotopes and Li. Thus far the incorporation of material with δ^7Li < MORB, as predicted based on studies of subduction zone lavas and eclogites, is not evident in oceanic lavas. An initial attempt to track this recycling process over Earth history with peridotite xenoliths suggested an increase in δ^7Li by approximately 4‰ since the mid-Archean.
3. The extent to which isotopic fractionation occurs at sub-1000°C temperatures in magmatic and regional metamorphic systems is unstudied in detail, but limited natural and experimental data in granitic systems suggest that measurable fractionation may occur under these conditions. This fractionation might be applied in studies of magmatic and ore evolution, either for thermometry or mineral deposit exploration.
4. Clastic sediments are reservoirs of information about weathering processes, but are sufficiently complex that no study has yet to realize their potential. Despite a number of initial reports of relatively isotopically heavy samples, the majority of data for clastic sedimentary rocks have an average δ^7Li ~ 0, equivalent to the estimated average isotopic composition of the continental crust.
5. Although ocean water is well mixed with respect to Li, and many laboratories measure Li isotopes in seawater, the range in values reported in the literature (δ^7Li = +29.3 to +33) cast some uncertainty on this reservoir. As precision of analytical methods improves, a check of the viability of the "seawater standard" should be carried out.
6. Initial experiments on the viability of marine (as well as lacustrine) carbonates as geochemical proxies have produced encouraging results toward the goal of tracing temporal changes in the Li isotopic composition of the Earth's surface. Inorganic and biogene carbonate precipitation appears to favorably incorporate light Li, but this fractionation is not temperature sensitive. A better understanding of metabolic effects of invertebrate species will be necessary. More laboratory experiments and controlled natural studies should address the prospective of Li isotopes in assessing

compositional changes in waters over time. Modern analytical techniques, permitting precise analysis of small (≤1 ng Li) samples, will permit the examination of detailed populations of these low concentration materials.

7. Weathering releases isotopically heavy Li to the near-surface environment, regardless of the intensity of chemical weathering. Much of this Li goes to stream flow, contributing to the riverine average δ^7Li ~ +24. The detailed histories of weathering products on the continents--sediments and soils--can be complex, but studies of simple systems suggest retention of isotopically light material, and hence an evolution of the continents toward lower δ^7Li over Earth history.

8. Consensus should be reached on the interpretation of Li isotope data for chondritic meteorites. Can bulk planetary Li isotopic compositions be accurately estimated from meteorites, or do these objects preserve only parent body near-surface or metamorphic histories? Are either of these possibilities viable? Detailed, high precision studies should permit this assessment.

This is an exciting time for the study of Li isotopes, given the rapidly diminishing sample size boundary with a consequent increase in precision, allowing heretofore unachievable inroads into areas where isotopic variations are small. The capacity to make routine, high precision measurements is now possible by a number of techniques at a large number of laboratories. Although the community needs still be constantly vigilant that chemical preparation of samples imparts no laboratory fractionation effects, these problems are acknowledged and actively scrutinized by all engaged in the work. To continue to expand the analytical frontier, emphasis should be placed on the reproducibility of data for replicates (that is, complete reprocessing and reanalysis of a sample material) and on cross-checking and collaborating between laboratories and among different instrumental techniques.

Although certain areas still benefit from even modest-precision Li isotope analysis (i.e., where end-member compositions are sufficiently disparate), assessing small variations in Li isotopic composition will be important to a variety of future studies. Studies of xenoliths have the prospect of shedding light on: the nature and origin of chemical heterogeneity in the mantle over the course of Earth history, the process by which material is transferred from slab to mantle during subduction, and how the continental crust has grown. Analysis of marine carbonate samples (e.g., small single-species foram populations) has the potential of further defining the chemical evolution of the oceans and highlighting the style of continental contributions to the oceans, ultimately linked to climate change. Minerals generated during weathering on the continents (and by experimental simulations of these processes) will define in clearer detail the prospects for Li isotopes as tools in studying near-surface geochemical processes and fingerprinting these in the geological record.

ACKNOWLEDGMENTS

I became involved in Li isotope research thanks to a suggestion from Richard Walker in 1991. Work that followed has benefited from interactions with colleagues at the University of Maryland, the Department of Terrestrial Magnetism, and the Lamont-Doherty Earth Observatory. My Li isotope endeavors have been funded by several sources: the NSF (SGER EAR9404039 to Richard Walker in 1994; a Postdoctoral Research Fellowship in 1998; and most recently EAR0208012); The Carnegie Institution of Washington for postdoctoral research support; and the Lamont Investment Fund.

My thanks to Roberta Rudnick and Boswell Wing for comments on the manuscript and to Laurie Benton, Tom Bullen, Lui Chan, Sylvie Decitre, Tim Elliott, Gideon Henderson, Craig Lundstrom, Tomas Magna, Jeff Ryan, and Yoshiro Nishio for tips on works past or for sharing information on works-in-progress. Finally, the reviews of Lui Chan, Gideon Henderson, and James Hogan, as well as editing by Clark Johnson (and his invitation to submit this review to begin with), were essential and helpful, and I thank them for their considerable time and effort.

REFERENCES

Albarède F, Beard BL (2004) Analytical methods for non-traditional isotopes. Rev Mineral Geochem 55: 113-151

Anders E, Grevesse N (1989) Abundances of the elements: meteoritic and solar. Geochim Cosmochim Acta 53:197-214

Anderson MA, Bertsch PM, Miller WP (1989) Exchange and apparent fixation of lithium in selected soils and clay minerals. Soil Sci 148:46-52

Anghel I, Turin HJ, Reimus PW (2002) Lithium sorption to Yucca Mountain tuffs. Appl Geochem 17:819-824

Aston FW (1932) The isotopic constitution and atomic weights of cesium, strontium, lithium, rubidium, barium, scandium and thallium. Proc Roy Soc A 134:571

Bach RO (ed) (1985) Lithium—Current Applications in Science, Medicine, and Technology. Wiley-Interscience, New York

Beard BL, Johnson CM (2004) Fe isotope variations in the modern and ancient earth and other planetary bodies. Rev Mineral Geochem 55:319-357

Bencala KE, McKnight DM, Zellweger GW (1990) Characterization of transport in an acidic and metal-rich mountain stream based on a lithium tracer injection and simulations of transient storage. Water Resour Res 26:989-1000

Bensimon M, Bourquin J, Parriaux A (2000) Determination of ultra-trace elements in snow samples by inductively-coupled plasma source sector field mass spectrometry using ultrasonic nebulization. J Anal Atom Spectrom 15:731-734

Benton LD, Tera F (2000) Lithium isotope systematics of the Marianas revisited. J Conf Abst 5:210

Benton LD, Savov I, Ryan JG (1999) Recycling of subducted lithium in forearcs: Insights from a serpentine seamount. EOS Trans, Am Geophys Union 80:S349

Berger G, Schott J, Guy C (1988) Behavior of lithium, rubidium, and cesium during basalt glass and olivine dissolution and chlorite, smectite and zeolite precipitation from seawater: Experimental investigations and modeling between 50 and 300°C. Chem Geol 71:297-312

Boesgaard AM, Steigman G (1985) Big band nucleosynthesis: theories and observations. Ann Rev Astron Astrophys 23:319-378

Bonatti E (1976) Serpentinite protrusions in oceanic-crust. Earth Planet Sci Lett 32:107-113

Bottomley DJ, Katz A, Chan LH, Starinsky A, Douglas M, Clark ID, Raven KG (1999) The origin and evolution of Canadian Shield brines: Evaporation or freezing of seawater? New lithium isotope and geochemical evidence from the Slave craton. Chem Geol 155:295-320

Bottomley DJ, Chan LH, Katz A, Starinsky A, Clark ID (2003) Lithium isotope geochemistry and origin of Canadian Shield brines. Groundwater 41:847-856

Bouman C, Elliott TR (1999) Li isotope compositions of Mariana arc lavas: Implications for crust-mantle recycling. Ninth Goldschmidt Conf Abst, LPI Contribution 971, Lunar Planetary Institute, 35

Bouman C, Elliott TR, Vroon PZ, Pearson DG (2000) Li isotope evolution of the mantle from analyses of mantle xenoliths. J Conf Abst 5:239

Bouman C, Vroon PZ, Elliott TR, Schwieters JB, Hamester M (2002) Determination of lithium isotope compositions by MC-ICPMS (Thermo Finnigan MAT Neptune). Geochim Cosmochim Acta 66:A97

Bray AM (2001) The geochemistry of boron and lithium in mid-ocean ridge hydrothermal vent fluids. PhD thesis, University of New Hampshire, 125 p

Brenan JM, Ryerson FJ, Shaw HF (1998) The role of aqueous fluids in the slab-to-mantle transfer of boron, beryllium, and lithium during subduction: Experiments and models. Geochim Cosmochim Acta 62:3337-3347

Brondi M, Dall'Aglio M, Vitrani F (1973) Lithium as a pathfinder element in the large scale hydrogeochemical exploration for hydrothermal systems. Geothermics 2:142-153

Brooker R, Blundy J, James R (2000) Subduction-related mantle pyroxenites from Zabargad Island, Red Sea. J Conf Abst 5:249

Brown L, Rajan RS, Roberts RB, Tera F, Whitford DJ (1978) A new method for determining the isotopic composition of lithium. Nuc Inst Meth 156:541-546

Bruns P, Elliott T (1999) Investigating the potential of Li isotopes as a paleo-oceanologic proxy. EOS Trans, Am Geophys Union 80:547

Bryant CJ, McCulloch MT, Bennett VC (2003a) Impact of matrix effects on the accurate measurement of Li isotope ratios by inductively coupled plasma mass spectrometry (MC-ICP-MS) under "cold" plasma conditions. J Anal At Spectrom 18:734-737

Bryant CJ, Chappell BW, Bennett VC, McCulloch MT (2003b) Li isotopic variations in Eastern Australian granites. Geochim Cosmochim Acta 67:A47

Bullen TD, Kendall C (1998) Tracing of weathering reactions and water flowpaths: a multi-isotope approach. *In*: Isotope Tracers in Catchment Hydrology. Kendall, McDonnell JD (eds) Elsevier Science, Amsterdam, p 611-646

Bullen TD, Kharaka YK (1992) Isotopic composition of Sr, Nd, and Li in thermal waters from the Norris-Mammoth corridor, Yellowstone National Park and surrounding region. *In*: Water-Rock Interaction, Proceedings of the Seventh International Symposium on Water-Rock Interaction. Kharaka YK, Maest AS (eds) Balkema Publishers, Rotterdam p 897-901

Bullen TD, McMahon PE (1992) Evolution of $^{87}Sr/^{86}Sr$ and δ^7Li in groundwaters from the Black Creek Aquifer: confirmation of a model for the origin of Na^+-HCO_3^- waters in clastic aquifers. EOS Trans, Am Geophys Union 73:1719

Bullen TD, Senior LA (1992) Lithogenic Sr and anthropogenic Li in an urbanized stream basin: isotopic tracers of surface water-groundwater interaction. EOS Trans, Am Geophys Union 73:F130

Calvo JP, Jones BF, Bustillo M, Fort R, Alonso Zarza AM, Kendall C (1995) Sedimentology and geochemistry of carbonates from lacustrine sequences in the Madrid Basin, central Spain. Chem Geol 123:173-191

Chan LH (1987) Lithium isotope analysis by thermal ionization mass spectrometry of lithium tetraborate. Anal Chem 59:2662-2665

Chan LH, Edmond JM (1988) Variation of lithium isotope composition in the marine environment: A preliminary report. Geochim Cosmochim Acta 52:1711-1717

Chan LH, Kastner M (2000) Lithium isotopic compositions of pore fluids and sediments in the Costa Rica subduction zone: implications for fluid processes and sediment contribution to the arc volcanoes. Earth Planet Sci Lett 183:275-290

Chan LH, Edmond JM, Thompson G, Gillis K (1992) Lithium isotopic composition of submarine basalts: Implications for the lithium cycle in the oceans. Earth Planet Sci Lett 108:151-160

Chan LH, Edmond JM, Thompson G (1993) A lithium isotope study of hot springs and metabasalts from mid-ocean ridge hydrothermal systems. J Geophys Res 98:9653-9659

Chan LH, Gieskes JM, You CF, Edmond JM (1994a) Lithium isotope geochemistry of sediments and hydrothermal fluids of the Guaymas Basin, Gulf of California. Geochim Cosmochim Acta 58:4443-4454

Chan LH, Zhang L, Hein JR (1994b) Lithium isotope characteristics of marine sediments. EOS Trans, Am Geophys Union 75:314

Chan LH, Leeman WP, You CF (1999) Lithium isotopic composition of Central American volcanic arc lavas: implications for modification of subarc mantle by slab-derived fluids. Chem Geol 160:255-280

Chan LH, Leeman WP, Tonarini S (2001) Lithium isotope compositions of South Sandwich Arc and Southwest Washington Cascades: A comparative study of arc processes. EOS Trans, Am Geophys Union 82:1293

Chan LH, Alt JC, Teagle DAH (2002a) Lithium and lithium isotope profiles through the upper oceanic crust: as study of seawater-basalt exchange at ODP Sites 504B and 896A. Earth Planet Sci Lett 201:187-201

Chan LH, Leeman WP, Tonarini S, Singer B (2002b) Lithium and boron isotopes in the Aleutian Islands: Contributions of marine sediments to island arc magmas. EOS Trans, Am Geophys Union 83:1482

Chan LH, Leeman WP, You CF (2002c) Lithium isotopic composition of Central American volcanic arc lavas: implications for modification of subarc mantle by slab-derived fluids: Correction. Chem Geol 182:293-300

Chan LH, Starinsky A, Katz A (2002d) The behavior of lithium and its isotopes in oilfield brines: Evidence from the Heletz-Kokhav field, Israel. Geochim Cosmochim Acta 66:615-623

Chan LH, Frey FA (2003) Lithium isotope geochemistry of the Hawaiian plume: Results from the Hawaiian Scientific Drilling Project and Koolau Volcano. Geochem Geophys Geosyst 4:8707

Chaussidon M, Robert F (1998) $^7Li/^6Li$ and $^{11}B/^{10}B$ variations in chondrules from Semarkona unequilibrated chondrites. Earth Planet Sci Lett 164:577-589

Chaussidon M, Robert F (1999) Lithium nucleosynthesis in the Sun inferred from the solar-wind $^7Li/^6Li$ ratio. Nature 402:270-273

Clausen P (1995) Lithium Isotopenmessungen – konventionell vs. Toatalevaporation und Integration. Dissertation, Universität Göttingen, Göttingen, Germany

Collins AG (1976) Lithium abundances in oilfield waters. *In*: Lithium Resources and Requirements by the Year 2000. Vine JD (ed) U.S. Geol Surv Prof Pap 1005:116-123

Coplen TB (1996) Atomic weights of the elements 1995. Pure Appl Chem 68:2339-2359

Comans RNJ, Haller M, De Preter P (1991) Sorption of cesium on illite: Non-equilibrium behavior and reversibility. Geochim Cosmochim Acta 55:433-440

Decitre S, Deloule E, Reisberg L, James R, Agrinier P, Mével C (2002) Behavior of lithium and its isotopes during serpentinization of oceanic peridotites. Geochem Geophys Geosyst 3:10.1029/2001GC000178

Delaney ML, Boyle EA (1986) Lithium in foraminiferal shells: implications for high-temperature hydrothermal circulation fluxes and oceanic crustal generation rates. Earth Planet Sci Lett 80:91-105

Delaney ML, Bé AWH, Boyle EA (1985) Li, Sr, Mg and Na in foraminiferal calcite shells from laboratory culture, sediment traps and sediment cores. Geochim Cosmochim Acta 49:1327-1341

Divis AF, Clark JR (1978) Geochemical Exploration 1978: Proceedings of the Seventh International Geochemical Exploration Symposium, Golden, Colorado. Assoc Exploration Geochemists 233-241

Edmond JM, Measures C, McDuff RE, Chan LH, Collier R, Grant B, Gordon LI, Corliss JB (1979) Ridge crest hydrothermal activity and the balances of the major and minor elements in the ocean: The Galapagos data. Earth Planet Sci Lett 46:1-18

Eggins SM, Rudnick RL, McDonough WF (1998) The composition of peridotites and their minerals: a laser-ablation ICP-MS study. Earth Planet Sci Lett 154:53-71

Elderfield H, Schultz A (1996) Mid-ocean ridge hydrothermal fluxes and the chemical composition of the ocean. Ann Rev Earth Planet Sci 24:191-224

Elderfield H, Wheat CG, Mottl MJ, Monnin C, Spiro B (1999) Fluid and geochemical transport through ocean crust: a transect across the eastern flank of the Juan de Fuca Ridge. Earth Planet Sci Lett 172:151-165

Elliott T, Thomas A, Jeffcoate AB, Niu Y (2003) Li isotope composition of the upper mantle. EOS Trans, Am Geophys Union 84:1608

Falkner KK, Church M, Measures CI, LeBaron G, Thouron D, Jeandel C, Stordal MC, Gill GA, Mortlock R, Froelich P, Chan LH (1997) Minor and trace element chemistry of Lake Baikal, its tributaries, and surrounding hot springs. Limnol Oceanogr 42:329-345

Flesch GD, Anderson AR Jr., Svec HJ (1973) A secondary isotopic standard for $^{6}Li/^{7}Li$ determinations. Int J Mass Spect Ion Proc 12:265-272

Foustoukos DI, James RH, Seyfried Jr WE (2003) Lithium isotopic systematics of the Main Endeavor Field vent fluids, Northern Juan de Fuca Ridge. Geochim Cosmochim Acta 67:A101

Franklin KJ, Halliday JD, Plante LM, Symons EA (1986) Measurement of the Li-6 Li-7 isotope ratio for lithium-salts by FT NMR-spectroscopy. J Magnet Resonance 67:162-165

Fritz SJ (1992) Measuring the ratio of aqueous diffusion coefficients between $^{6}Li^{+}Cl^{-}$ and $^{7}Li^{+}Cl^{-}$ by osmometry. Geochim Cosmochim Acta 56:3781-3789

Fritz SJ, Whitworth TM (1994) Hyperfiltration-induced fractionation of lithium isotopes: Ramifications relating to the representativeness of aquifer sampling. Water Resour Res 30:225-235

Fryer P (1985) Origin and emplacement of Mariana fore-arc seamounts. Geology 13:774-775

Gellenbeck DJ, Bullen TD (1991) Isotopic characterization of strontium and lithium in potential sources of ground-water contamination in central Arizona. EOS Trans, Am Geophys Union 72:178

Green LW, Leppinen JJ, Elliot NL (1988) Isotopic analysis of lithium as thermal dilithium fluoride ions. Anal Chem 60:34-37

Grégoire DC, Acheson BM, Taylor RP (1996) Measurement of lithium isotope ratios by inductively coupled plasma mass spectrometry: application to geological materials. J Anal Atom Spectrom 11:765-772

Halliday AN, Lee DC, Christensen JN, Rehkämper M, Yi W, Luo X, Hall CM, Ballentine CJ, Pettke T, Stirling C (1998) Applications of multiple collector-ICPMS to cosmochemistry, geochemistry and paleoceanography. Geochim Cosmochim Acta 62:919-940

Heier KS, Billings GK (1970) Lithium. *In*: Handbook of Geochemistry. Wedepohl KH (ed) Vol II-1, Springer-Verlag, Berlin, p 3-A-1 – 3-O-1

Heinrich CA (1986) Eclogite facies regional metamorphism of hydrous mafic rocks in the Central alpine Adula nappe. J Petrol 27:123-154

Henderson GM, Burton KW (1999) Using ($^{234}U/^{238}U$) to assess diffusion rates of isotopic tracers in Mn crusts. Earth Planet Sci Lett 170:169-179

Hervig RL, Moore G (2003) Fractionation of boron (and lithium) between hydrous fluid and silicate melt: diffusion, contamination, and orphaned experiments. EOS Trans, Am Geophys Union 84:F163

Hoefs J (1997) Stable Isotope Geochemistry, 4 ed. Springer-Verlag, Berlin

Hoefs J, Sywall M (1997) Lithium isotope compositions of Quaternary and Tertiary biogene carbonates and a global lithium isotope balance. Geochim Cosmochim Acta 61:2679-2690

Hogan JF, Blum JD (2003) Boron and lithium isotopes as groundwater tracers: a study at the Fresh Kills Landfill, Staten Island, New York, USA. Applied Geochem 18:615-627

Huh Y, Chan LH, Edmond JM (2001) Lithium isotopes as a probe of weathering processes: Orinoco River. Earth Planet Sci Lett 194:189-199

Huh Y, Chan LH, Zhang L, Edmond JM (1998) Lithium and its isotopes in major world rivers: Implications for weathering and the oceanic budget. Geochim Cosmochim Acta 62:2039-2051

Huh Y, Chan LH, Chadwick O (2002) Lithium isotopes as a probe of weathering processes: Hawaiian soil climosequence. Geochim Cosmochim Acta 66:A346

Ikegawa M, Kimura M, Honda K, Akabane I, Makita K, Motoyama H, Fujii Y, Itokawa Y (1999) Geographical variations of major and trace elements in East Antarctica. Atmos Env 33:1457-1467

Isakov YA, Plyushin GS, Brandt SB (1969) Natural fractionation of lithium isotopes. Geochem Int 6:598-600

James RH, Palmer MR (2000a) The lithium isotope composition of international rock standards. Chem Geol 166:319-326

James RH, Palmer MR (2000b) Marine geochemical cycles of the alkali elements and boron: The role of sediments. Geochim Cosmochim Acta 64:3111-3122

James RH, Rudnicki MD, Palmer MR (1999) The alkali element and boron geochemistry of the Escanaba Trough sediment-hosted hydrothermal system: The role of sediments. Earth Planet Sci Lett 171:157-169

James RH, Allen DE, Seyfried Jr WE (2003) An experimental study of alteration of oceanic crust and terrigenous sediments at moderate temperatures (51 to 350°C): Insights as to chemical processes in near-shore ridge-flank hydrothermal systems. Geochim Cosmochim Acta 67:681-691

Johnson TM, Bullen TD (2004) Mass-dependent fractionation of selenium and chromim isotopes in low-temperature environments. Rev Mineral Geochem 55:289-317

Kaplan L, Wilzbach KE (1954) Lithium isotope determination by neutron activation. Anal Chem 26:1797-1798

Klossa B, Pierre A, Minster JF (1981) Mesures de la composition isotopique du lithium dans les inclusions refractaires d'Allende. Earth Planet Sci Lett 52:25-30

Knauth DC, Federman SR, Lambert DL, Crane P (2000) Newly synthesized lithium in the interstellar medium. Nature 405:656-658

Koirtyohann SR (1994) Precise determination of isotopic-ratios for some biologically significant elements by inductively-coupled plasma-mass spectrometry. Spectrochim Acta B-Atomic Spectrosc 49:1305-1311

Košler J, Kučera M, Sylvester P (2001) Precise measurement of Li isotopes in planktonic foraminiferal tests by quadrupole ICPMS. Chem Geol 181:169-179

Krankowsky D, Müller O (1967) Isotopic composition and abundance of lithium in meteoritic matter. Geochim Cosmochim Acta 31:1833-1842

Lamberty A, Michiels E, DeBièvre P (1987) On the atomic weight of lithium. Int J Mass Spect Ion Proc 79: 311-313

Layne GD (2003) Advantages of secondary ion mass spectrometry (SIMS) for stable isotope microanalysis of the trace light elements. EOS Trans, Am Geophys Union 84:F1635

Lundstrom CC, Chaussidon M, Kelemen P (2001) A Li isotope profile in a dunite to lherzolite transect within the Trinity Ophiolite: evidence for isotopic fractionation by diffusion. EOS Trans, Am Geophys Union 82:991

Lynton SJ (2003) An experimental study of the isotopic fractionation and partitioning of lithium among quartz, muscovite, and fluides. PhD Dissertation, University of Maryland, College Park, Maryland

Magna T, Wiechert U, Halliday AN (2002) Lithium isotopes and crust-mantle interaction. Geochim Cosmochim Acta 66:A474

Magna T, Wiechert UH, Grove TL, Halliday AN (2003) Lithium isotope composition of arc volcanics from the Mt. Shasta region, N California. Geochim Cosmochim Acta 67:A267

Marriott CS, Henderson GM, Belshaw NS, Tudhope AW (2004) Temperature dependence of δ^7Li, δ^{44}Ca and Li/Ca incorporation into calcium carbonate. Earth Planet Sci Lett (in press)

Michiels E, DeBièvre P (1983) Absolute isotopic composition and the atomic weight of a natural sample of lithium. Int J Mass Spect Ion Proc 49:265-274

McDonough WF, Rudnick RL, Dalpé C, Tomascak PB, Zack T (2001) Li isotopes as a tracer of Earth processes. EOS Trans, Am Geophys Union 82:F1410

McDonough WF, Teng FZ, Tomascak PB, Ash RD, Grossman JN, Rudnick RL (2003) Lithium isotopic composition of chondritic meteorites. Lunar Planet Sci XXXIV, Lunar Planet Inst, Houston 1931

Millot R, Guerrot C, Vigier N (2004) Accurate and high-precision measurement of lithium isotopes in two reference materials by MC-ICP-MS. Geostandards and Geoanalytical Res (in press)

Moriguti T, Nakamura E (1993) Precise lithium isotope analysis by thermal ionization mass spectrometry using lithium phosphate as an ion source. Proc Japan Acad Sci 69B:123-128

Moriguti T, Nakamura E (1998a) High-yield lithium separation and precise isotopic analysis for natural rock and aqueous samples. Chem Geol 145:91-104

Moriguti T, Nakamura E (1998b) Across-arc variation of Li isotopes in lavas and implications for crust/mantle recycling at subduction zones. Earth Planet Sci Lett 163:167-174

Moriguti T, Nakamura E (2003) Lithium and lead isotopes and trace element systematics of Quaternary basaltic volcanic rocks in Northeastern Japan. Geochim Cosmochim Acta 67:A305

Morozov NP (1968) Geochemistry of rare alkaline elements in the oceans and seas. Oceanology 8:169-178

Morozov NP (1969) Geochemistry of the alkali metals in rivers. Geochem Int 6:585-594

Morozova IM, Alferovskiy AA (1974) Fractionation of lithium and potassium isotopes in geological processes. Geochem Int 11:17-25

Nance WB, Taylor SR (1976) Rare earth element patterns and crustal evolution—I. Australian post-Archean sedimentary rocks. Geochim Cosmochim Acta 40:1539-1551

Nishio Y, Nakai S (2002) Accurate and precise lithium isotopic determinations of igneous rock samples using multi-collector inductively coupled plasma mass spectrometry. Anal Chim Acta 456:271-281

Nishio Y, Nakai S, Hirose K, Ishii T, Sano Y (2002) Li isotopic systematics of volcanic rocks in marginal basins. Geochim Cosmochim Acta 66:A556

Nishio Y, Nakai S, Kogiso T, Barsczus HG (2003) Lithium isotopic composition of HIMU oceanic island basalts: implications for the origin of HIMU component. *In:* XXXIII General Assembly of the International Union of Geodesy and Geophysics (IUGG 2003), 178 p

Nishio Y, Nakai S, Yamamoto J, Sumino H, Matsumoto T, Prikhod'ko VS, Arai S (2004) Li isotopic systematics of the mantle-derived ultramafic xenoliths: implications for EM1 origin. Earth Planet Sci Lett 217:245-261

Njo HB, Rudnick RL, Tomascak PB (2003) Lithium isotope fractionation during continental weathering. Geochim Cosmochim Acta 67:A340

O'Neil JR (1986) Theoretical and experimental aspects of isotopic fractionation. Rev Mineral 16:1-40

Oi T, Kawada K, Hosoe M, Kakihana H (1991) Fractionation of lithium isotopes in cation-exchange chromatography. Sep Sci Tech 26:1353-1375

Oi T, Odagiri T, Nomura M (1997) Extraction of lithium from GSJ rock reference samples and determination of their lithium isotopic compositions. Anal Chim Acta 340:221-225

Oi T, Shimizu K, Tayama S, Matsuno Y, Hosoe M (1999) Cubic antimonic acid as column-packing material for chromatographic lithium isotope separation. Sep Sci Tech 34:805-816

Olsher U, Izatt RM, Bradshaw JS, Dalley NK (1991) Coordination chemistry of lithium ion: a crystal and molecular structure review. Chem Rev 91:137-164

Palko AA, Drury JS, Begun GM (1976) Lithium isotope separation factors of some two-phase equilibrium systems. J Chem Phys 64:1828-1837

Phillips FM, Bentley HW (1987) Isotopic fractionation during ion filtration: I. Theory. Geochim Cosmochim Acta 51:683-695

Pistiner J, Henderson GM (2003) Lithium isotope fractionation during continental weathering processes. Earth Planet Sci Lett 214:327-339

Plyushin GS, Posokhov VF, Sandimirova GP (1979) Magmatic differentiation and the relationship of Li^7/Li^6 ratio to fluorine content. Doklady 248:223-225

Qi HP, Coplen TB, Wang QZ, Wang YH (1997a) Unnatural isotopic composition of lithium reagents. Anal Chem 69:4076-4078

Qi HP, Taylor PDP, Berglund M, DeBièvre P (1997b) Calibrated measurements of the isotopic composition and atomic weight of the natural Li isotopic reference material IRMM-016. Int J Mass Spect Ion Proc 171:263-268

Reeves H, Fowler WA, Hoyle F (1970) Galactic cosmic ray origin of Li, Be and B in stars. Nature 226:727-729

Remenda VH (1994) Distribution of 7Li/6Li with depth at an aquitard field site near Saskatoon. Geol Assoc Can – Min Assoc Can Abst Prog 19:A92

Richter FM, Davis AM, DePaolo DJ, Watson EB (2003) Isotope fractionation by chemical diffusion between molten basalt and rhyolite. Geochim Cosmochim Acta 67:3905-3923

Ritzenhoff S, Schröter EH, Schmidt W (1997) The lithium abundance in sunspots. Astron Astrophys 328: 695-701

Ronov AB, Migdisov AA, Voskresenskaya NT, Korzina GA (1970) Geochemistry of lithium in the sedimentary cycle. Geochem Int 7:75-102

Rudnick RL, McDonough WF, Tomascak PB, Zack T (2003) Lithium isotopic composition of eclogites: implications for subduction zone processes. Eighth Int Kimberlite Conf Abst

Ryan JG, Langmuir CH (1987) The systematics of lithium abundances in young volcanic rocks. Geochim Cosmochim Acta 51:1727-1741

Ryan JG, Kyle PR (2000) Lithium isotope systematics of McMurdo volcanic group lavas, and other intraplate sites. EOS Trans, Am Geophys Union 81:F1371

Sahoo SK, Masuda A (1995) Simultaneous measurement of lithium and boron isotopes as lithium tetraborate ion by thermal ionization mass-spectrometry. Analyst 120:335-339

Sahoo SK, Masuda A (1998) Precise determination of lithium isotopic composition by thermal ionization mass spectrometry in natural samples such as seawater. Anal Chim Acta 370:215-220

Savov IP, Ryan JG, Chan L, D'Antonio M, Mottl M, Fryer P, ODP Leg 195 Scientific Party (2002) Geochemistry of serpentinites from the S. Chamorro Seamount, ODP Leg 195, Site 1200, Mariana Forearc: Implications for recycling at subduction zones. Geochim Cosmochim Acta 66:A670

Schirmeyer S, Hoppe P, Stephan T, Bischoff A, Jessberger EK (1997) A lithium-bearing Ca,Al-rich inclusion from the CM-chondrite Cold Bokkeveld studied by TOF-SIMS and conventional SIMS. Lunar Planet Sci Conf 28:677-678

Schwab AB, O'Connell ME, Long SE (1995) The use of lithium concentration data and isotopic ratios as hydrologic tracers in a first-order catchment. Geol Soc Am Prog Abst 27:A97

Schou M (1988) Lithium treatment of manic-depressive illness—Past, present and perspectives. J Am Med Ass 259:1834-1836

Seyfried Jr WE, Janecky DR, Mottl M (1984) Alteration of the oceanic crust by seawater: implications for the geochemical cycles of boron and lithium. Geochim Cosmochim Acta 48:557-569

Seyfried Jr WE, Chen X, Chan LH (1998) Trace element mobility and lithium isotope exchange during hydrothermal alteration of seafloor weathered basalt: An experimental study at 350 degrees C, 500 bars. Geochim Cosmochim Acta 62:949-960

Shannon RD (1976) Revised effective ionic-radii and systematic studies of interatomic distances in halides and chalcogenides. Acta Crystallograph A 32:751-767

Shaw DM, Sturchio NC (1992) Boron-lithium relationships in rhyolites and associated thermal waters of young silicic calderas, with comments on incompatible element behavior. Geochim Cosmochim Acta 56:3723-3731

Sirbescu ML, and Nabelek PI (2003) Crustal melts below 400°C. Geology 31:685-688

Stoffyn-Egli P, Mackenzie FT (1984) Mass balance of dissolved lithium in the oceans. Geochim Cosmochim Acta 48:859-872

Stoll PM, Stokes PE, Okamoto M (2001) Lithium isotopes: differential effects on renal function and histology. Bipolar Disorders 3:174-180

Sturchio NC, Chan LH (2003) Lithium isotope geochemistry of the Yellowstone hydrothermal system. Soc Econ Geol Spec Pub 10:171-180

Sun XF, Ting BTG, Zeisel SH, Janghorbani M (1987) Accurate measurement of stable isotopes of lithium by inductively coupled plasma mass spectrometry. Analyst 112:1223-1228

Svec HJ, Anderson AR (1965) The absolute abundances of the lithium isotopes in natural sources. Geochim Cosmochim Acta 29:633-641

Symons EA (1985) Lithium isotope separation: a review of possible techniques. Sep Sci Tech 20:633-651

Tardy Y, Krempp G, Trauth N (1972) Le lithium dans les minéreaux argileux des sédiments et des sols. Geochim Cosmochim Acta 36:397-412

Tatsumi Y, Hamilton DL, Nesbitt RW (1986) Chemical characteristics of fluid phase released from a subducted lithosphere and origin of arc magmas: Evidence from high-pressure experiments and natural rocks. J Volc Geotherm Res 29:293-309

Taylor TI, Urey HC (1938) Fractionation of the lithium and potassium isotopes by chemical exchange with zeolites. J Chem Phys 6:429-438

Teng FZ, McDonough WF, Rudnick RL, Dalpé C, Tomascak PB, Gao S, Chappell BW (2004) Lithium isotopic composition and concentration of the upper continental crust. Geochim Cosmochim Acta (in press)

Teng FZ, McDonough WF, Rudnick RL, Tomascak PB, Saal AE (2003) Lithium isotopic composition of the lower continental crustal crust: A xenolith perspective. EOS Trans, Am Geophys Union 84:1608

Tomascak PB, Krogstad EJ, Prestegaard KL, Walker RJ (1995a) Li isotope variability in groundwaters from Maryland. EOS Trans, Am Geophys Union 76:S122

Tomascak PB, Krogstad EJ, Prestegaard KL, Walker RJ (1995b) Li isotope geochemistry of stormwater from Maryland. EOS Trans, Am Geophys Union 76:F207

Tomascak PB, Lynton SJ, Walker RJ, Krogstad EJ (1995c) Li isotope geochemistry of the Tin Mountain pegmatite, Black Hills, South Dakota. *In*: The Origin of Granites and Related Rocks. Brown M, Piccoli PM (eds) US Geol Surv Circ 1129:151-152

Tomascak PB, Banner JL (1998) Lithium isotope hydrogeochemistry: a multi-collector ICP-MS study. Geochemical Perspectives on Environmental Processes: New Theoretical and Analytical Approaches to Sources, Transport, and Bioavailability of Trace Elements, St. Louis, Conf Abst

Tomascak PB, Langmuir CH (1999) Lithium isotope variability in MORB. EOS Trans, Am Geophys Union 80:F1086-1087

Tomascak PB, Carlson RW, Shirey SB (1999a) Accurate and precise determination of Li isotopic compositions by multi-collector sector ICP-MS. Chem Geol 158:145-154

Tomascak PB, Tera F, Helz RT, Walker RJ (1999b) The absence of lithium isotope fractionation during basalt differentiation: New measurements by multi-collector sector ICP-MS. Geochim Cosmochim Acta 63: 907-910

Tomascak PB, Ryan JG, Defant MJ (2000) Lithium isotope evidence for light element decoupling in the Panama subarc mantle. Geology 28:507-510

Tomascak PB, Hemming NG, Pedone VA (2001) Lithium, boron, and strontium isotope constraints on solute sources for the Great Salt Lake, Utah. Eleventh VM Goldschmidt Conf Abst, LPI Contribution 1088, Lunar Planetary Institute, 3758

Tomascak PB, Widom E, Benton LD, Goldstein SL, Ryan JG (2002) The control of lithium budgets in island arcs. Earth Planet Sci Lett 196:227-238

Tomascak PB, Hemming NG, Hemming SR (2003) The lithium isotopic composition of waters from the Mono Basin, California. Geochim Cosmochim Acta 67:601-611

Umeda M, Tuchiya K, Kawamura H, Hasegawa Y, Nanjo Y (2001) Preliminary characterization on Li isotope separation with Li ionic conductors. Fusion Tech 39:654-658

Vigier N, Burton KW, Gislason SR, Roger NW, Schaefer BF, James RH (2002) Constraints on basalt erosion from Li isotopes and U-series nuclides measured in Iceland rivers. Geochim Cosmochim Acta 66:A806

Vocke RD, Beary ES, Walker RJ (1990) High precision lithium isotope ratio measurement of samples from a variety of natural sources. VM Goldschmidt Conf Prog Abst 89

Von Damm KL (1995) Controls on the chemistry and temporal variability of seafloor hydrothermal fluids. *In*: Seafloor Hydrothermal Systems: Physical, Chemical, Biologic, and Geologic Interactions. Humphris SE, Zierenberg RA, Mullineaux LS, Thomson RE (eds) Am Geophys Union, Geophys Monograph 91: 222-248

Walker RJ, Hanson GN, Papike JJ, O'Neil JR, Laul JC (1986) Internal evolution of the Tin Mountain pegmatite, Black Hills, South Dakota. Am Min 71:440-459

Wenger M and Armbruster T (1991) Crystal chemistry of lithium: oxygen coordination and bonding. Eur J Mineral 3:387-399

Wheat CG, Mottl MJ (2000) Composition of pore and spring waters from Baby Bare: Global implications of geochemical fluxes from a ridge flank hydrothermal system. Geochim Cosmochim Acta 64:629-642

White DE, Thompson JM, Fournier RO (1976) Lithium contents of thermal and mineral waters. *In*: Lithium Resources and Requirements by the Year 2000. Vine JD (ed) U.S. Geol Surv Prof Pap 1005:58-60

Xiao YK, Beary ES (1989) High-precision isotopic measurement of lithium by thermal ionization mass spectrometry. Int J Mass Spect Ion Proc 94:101-114

Yamaji K, Makita Y, Watanabe H, Sonoda A, Kanoh H, Hirotsu T, Ooi K (2001) Theoretical estimation of lithium isotopic reduced partition function ratio for lithium ions in aqueous solution. J Phys Chem A 105: 602-613

You CF, Chan LH, Gieskes JM, Klinkhammer GP (2004) Seawater intrusion through the oceanic crust and carbonate sediment in the Equatorial Pacific: Lithium abundance and isotopic evidence. Geophys Res Lett 30 (in press)

You CF, Chan LH (1996) Precise determination of lithium isotopic composition in low concentration natural samples. Geochim Cosmochim Acta 60:909-915

You CF, Chan LH, Spivack AJ, Gieskes JM (1995) Lithium, boron, and their isotopes in ODP Site 808, Nankai Trough sediments and pore waters: Implications for fluid expulsion in accretionary prisms. Geology 23: 37-40

Zack T, Tomascak PB, Rudnick RL, Dalpé C, McDonough WF (2003) Extremely light Li in orogenic eclogites: The role of isotope fractionation during dehydration in subducted oceanic crust. Earth Planet Sci Lett 208:279-290

Zaidel AN, Korennoi EP (1961) Spectral determination of isotopic composition and concentration of lithium in solutions. Optic Spectrosc 10:299-302

Zhang L, Chan LH, Gieskes JM (1998) Lithium isotope geochemistry of pore waters from Ocean Drilling Program Sites 918 and 919, Irminger Basin. Geochim Cosmochim Acta 62:2437-2450

Zindler A, Hart SR (1986) Chemical geodynamics. Ann Rev Earth Planet Sci 14:493-571

Reviews in Mineralogy & Geochemistry
Vol. 55, pp. 197-230, 2004
Copyright © Mineralogical Society of America

6

The Isotope Geochemistry and Cosmochemistry of Magnesium

Edward D. Young

Department of Earth and Space Sciences
and
Institute of Geophysics and Planetary Physics
University of California Los Angeles
Los Angeles, California 90095-1567, U.S.A.

Albert Galy

Department of Earth Sciences
University of Cambridge
Cambridge, CB2 3EQ, United Kingdom

INTRODUCTION

Magnesium is second only to oxygen in abundance among the rock-forming elements and is an important element in the oceans and in hydrological and biological systems. Differences in the relative abundances of its three stable isotopes, ^{24}Mg (78.99%), ^{25}Mg (10.00%), and ^{26}Mg (11.01%), are expected as a result of physicochemical processes because of the large relative mass differences of 4 and 8% between ^{25}Mg and ^{26}Mg, and ^{24}Mg, respectively. Although isotopes of Mg have been used for many years as tracers in artificially spiked systems (in which the abundance of one isotope is enriched) (Cary et al. 1990; Dombovari et al. 2000), reliable measurements of ^{25}Mg/^{24}Mg and ^{26}Mg/^{24}Mg in natural systems have been limited historically by the 1‰ (one part per thousand) reproducibility imparted by instrumental mass fractionation effects. In order to be useful for many geochemical and cosmochemical applications the isotope ratios of Mg must be resolved to ≤200 parts per million (ppm). As a result, with a few exceptions (e.g., Davis et al. 1990; Goswami et al. 1994; Russell et al. 1998), many past studies of Mg isotope ratios focused on detection of non-mass dependent, so-called "anomalous" Mg isotopic effects rather than on investigations of mass-dependent fractionation. The principle outcome of this focus was the discovery of radiogenic ^{26}Mg (^{26}Mg*) in primitive meteorites (Gray and Compston 1974; Lee and Papanastassiou 1974).

With the advent of multiple-collector inductively coupled plasma-source mass spectrometry (MC-ICPMS) it is now possible to measure ^{25}Mg/^{24}Mg and ^{26}Mg/^{24}Mg of Mg in solution with a reproducibility of 30 to 60 ppm or better (Galy et al. 2001). What is more, ultraviolet (UV) laser ablation combined with MC-ICPMS permits *in situ* analysis of Mg-bearing mineral samples with reproducibility of 100 to 200 ppm (Young et al. 2002a). These new analytical capabilities allow mass-dependent fractionations of the isotopes of Mg to be used as tracers in natural systems.

Perhaps the greatest advance afforded by the MC-ICPMS technology is the ability to measure ^{25}Mg/^{24}Mg and ^{26}Mg/^{24}Mg independently with precision many times smaller than the magnitude of the natural variations. Thermal ionization mass spectrometry (TIMS)

1529-6466/04/0055-0006$05.00

methods provide highly precise measurements of anomalies in $^{26}Mg/^{24}Mg$ if the data are normalized to a fixed "terrestrial" $^{25}Mg/^{24}Mg$ value (the latter compensates for instrumental mass fractionation). The MC-ICPMS technique does not require $^{26}Mg/^{24}Mg$ measurements to be corrected to a single $^{25}Mg/^{24}Mg$. Instead, both ratios are measured with a high level of precision. For samples of purified Mg in solution the precision of the MC-ICPMS measurements is sufficient to resolve the various mass-dependent fractionation relationships that exist between the two isotope ratios. We are therefore on the threshold of establishing an entirely new aspect of isotope geo-cosmochemistry based on the details of the mass-dependent fractionations of Mg isotopes.

At this writing the number of published papers reporting high-precision Mg isotope ratios of natural materials is small but growing rapidly. Since Mg isotope geo-cosmochemistry is a young field, our goal in writing this review is to identify for the reader those areas that we view as ripe for future exploitation, especially with respect to the significance of the observed patterns of $^{25}Mg/^{24}Mg$ and $^{26}Mg/^{24}Mg$ variation in terrestrial and extraterrestrial reservoirs.

REFERENCE MATERIALS, NOMENCLATURE, AND ANALYTICAL CONSIDERATIONS

Reference material

The National Institute of Standards and Technology (NIST, formerly the National Bureau of Standards, or NBS) developed an isotopic standard for Mg (Catanzaro et al. 1966). The NIST Standard Reference Material 980, or SRM 980, is a magnesium metal certified for $^{25}Mg/^{24}Mg = 0.12663 \pm 0.00013$ (±1.0‰) and $^{26}Mg/^{24}Mg = 0.13932 \pm 0.00026$ (±1.9‰) (Catanzaro et al. 1966). The former value is universally adopted while the latter value has been questioned (Galy et al. 2001; Schramm et al. 1970). A more accurate value for $^{26}Mg/^{24}Mg$ may be 0.139828 ± 0.000037 (±0.27‰, 2σ), a full 3.6‰ greater than the accepted value (Galy et al. 2001).

A more pressing problem is that the SRM 980 reference material appears to be heterogeneous in its isotopic composition. The heterogeneity is found at the scale of individual metal chips (~10 mg), is entirely mass dependent, and is comparable in magnitude to the reported precision of the certified isotope ratios determined by TIMS (Galy et al. 2003). As a result, and despite the attractiveness of reporting isotope ratios relative to a recognized and certified standard, the SRM 980 metal is inappropriate for use as an isotope standard in high-precision work where differences of 0.2‰ or less are significant.

Virtually all of the published Mg isotope data obtained by MC-ICPMS are referenced to the SRM 980 standard at the time of this writing. Fortunately, the vast majority were obtained using as the standard a single batch of dissolved SRM 980 housed in the Department of Earth Sciences, University of Oxford. This standard has been referred to as SRM 980_O (Galy et al. 2003). However, SRM 980_O was never intended to serve as a primary standard and there is the need to establish an alternative primary standard of uniform composition.

A new standard has been developed in the Department of Earth Sciences, University of Cambridge (Galy et al. 2003). It consists of approximately 10 g of pure magnesium metal (provided by Dead Sea Magnesium Ltd., Israel) dissolved in a liter of 0.3 N HNO_3. This material, referred to as DSM3, is a suitable reference because it is already in solution and therefore immune to heterogeneity. In addition, it has the added advantage that its isotopic composition is indistinguishable from MC-ICPMS measurements of carbonaceous chondrite materials, a rational reference point with cosmochemical and petrological significance.

All of the data reported in this review have been converted from the SRM 980_O scale to the DSM3 scale (with the exception of data from several previously published figures).

This was done as a first step towards eliminating the confusion that will ensue if laboratories continue to report their results relative to the heterogeneous SRM 980 metal, and because the Mg isotopic compositions relative to DSM3 are effectively identical to values reported relative to "chondrite."

The definitions of $\delta^{25}Mg_{SRM\,980\_O}$ and $\delta^{26}Mg_{SRM\,980\_O}$ are:

$$\delta^{x}\mathrm{Mg}^{\mathrm{sample}}_{\mathrm{SRM\,980\_O}} = \left(\frac{(^{x}\mathrm{Mg}/^{24}\mathrm{Mg})_{\mathrm{sample}}}{(^{x}\mathrm{Mg}/^{24}\mathrm{Mg})_{\mathrm{SRM\,980\_O}}} - 1\right)10^{3} \tag{1}$$

where x refers to either 25 or 26. From these values, conversion to the equivalent delta values on the DSM3 scale, $\delta^{25}Mg_{DSM3}$ and $\delta^{26}Mg_{DSM3}$, is accomplished with the expression

$$\delta^{x}\mathrm{Mg}^{\mathrm{sample}}_{\mathrm{DSM3}} = \delta^{x}\mathrm{Mg}^{\mathrm{sample}}_{\mathrm{SRM\,980\_O}} + \delta^{x}\mathrm{Mg}^{\mathrm{SRM\,980\_O}}_{\mathrm{DSM3}} + 0.001\,\delta^{x}\mathrm{Mg}^{\mathrm{sample}}_{\mathrm{SRM\,980\_O}}\,\mathrm{Mg}^{\mathrm{SRM\,980\_O}}_{\mathrm{DSM3}} \tag{2}$$

where

$$\delta^{25}\mathrm{Mg}^{\mathrm{SRM\,980\_O}}_{\mathrm{DSM3}} = -1.744 \tag{3}$$

and

$$\delta^{26}\mathrm{Mg}^{\mathrm{SRM\,980\_O}}_{\mathrm{DSM3}} = -3.405 \tag{4}$$

The values for the SRM 980_O relative to DSM3 are reported by Galy et al. (2003). The 0.001 factor arises when converting sample/standard ratios from one standard to another in δ notation.

Nomenclature

Early studies of Mg isotope ratios in geological materials used the notation $\Delta^{25}Mg$ to mean per mil deviations from a standard as expressed in Equation (1) above, a convention that persists today (e.g., Hsu et al. 2000). The values assigned to $\Delta^{25}Mg$ in those studies represent the level of mass-dependent isotopic fractionation relative to the standard. The same convention defined $\delta^{26}Mg$ as the per mil deviation from the standard after correction for the mass fractionation evidenced by $\Delta^{25}Mg$. In this system of nomenclature, Δ values refer to *mass dependent fractionations* while δ values refer to deviations from mass-dependent fractionation (i.e., the $\delta^{26}Mg$ defines excesses in ^{26}Mg relative to mass fractionation attributable to decay of the extinct nuclide ^{26}Al). In some cases $\Delta^{25}Mg$ has been replaced by the symbol F_{Mg} (Kennedy et al. 1997) where the F refers to "fractionation."

The nomenclature described above is common in the cosmochemistry literature for not only Mg but other systems as well (e.g., Ti; Fahey et al. 1987). It arose because naturally-occurring mass fractionation effects and instrumental effects were convolved and not easily distinguished. Instead it was the deviations from mass fractionation (i.e., nucleosynthetic effects) that were the primary focus. However, beginning with McKinney et al. (1950) the symbol δ has been used in the terrestrial literature to mean per mil deviations from a standard in the O, C, S, N, and D/H isotope systems. In this context the use of δ has no connotation with respect to the distinction between mass dependent and non-mass dependent fractionation. Following the discovery of non-mass dependent effects in oxygen in both extraterrestrial and terrestrial materials, the capital delta Δ became ensconced as the symbol that quantifies *deviation from mass-dependent fractionation* in that system. In the oxygen system $\Delta^{17}O$ refers to the intercept on a plot of $\delta^{17}O$ (ordinate) against $\delta^{18}O$ (abscissa) and $\delta^{17}O$ and $\delta^{18}O$ refer to per mil deviations in $^{17}O/^{16}O$ and $^{18}O/^{16}O$ from a standard (usually Standard Mean Ocean Water). Accordingly to this definition, $\Delta^{17}O$ should not change significantly with mass-dependent fractionation (within <0.3‰), although in detail small differences are expected (Young et al. 2002b). On the other hand, large differences in $\Delta^{17}O$ serve as indication that there are departures from mass-dependent fractionation. Similarly, sulfur is also now known to experience non-mass dependent fractionation and as a consequence $^{33}S/^{32}S$ and $^{34}S/^{32}S$ are reported as per mil deviations from a standard using $\delta^{33}S$ and $\delta^{34}S$ and $\Delta^{33}S$ is used, by analogy with $\Delta^{17}O$, as a measure of departure from mass-dependent fractionation curves in $\delta^{33}S$ vs. $\delta^{34}S$ space.

With regard to Mg isotopes, one is therefore forced to choose between the merit of precedence in the cosmochemistry community and that of conformity with the other light stable isotope systems when deciding upon a nomenclature for reporting $^{25}Mg/^{24}$ and $^{26}Mg/^{24}Mg$ data. We have chosen to come down on the side of conformity with the other light element isotope systems and will use $\delta^{25}Mg$ and $\delta^{26}Mg$ (defined in Eqn. 1) to mean simply the per mil deviations from a standard (usually DSM3). In keeping with the nomenclature of other stable isotope systems, $\Delta^{25}Mg$ is defined as the intercept on a plot of $\delta^{25}Mg$ (ordinate) vs. $\delta^{26}Mg$ (abscissa). A more precise definition of $\Delta^{25}Mg$ will be presented in the following section. The symbol $\delta^{26}Mg^*$ can then be reserved for deviations from mass fractionation attributable to $^{26}Mg^*$, consistent with the definition in many existing cosmochemistry papers.

We strongly recommend that these conventions be adopted at this early stage in the development of Mg isotope geochemistry and cosmochemistry to eliminate the current ambiguity in the meaning of the Δ symbol in the literature. Presumably, confusion will be kept to a minimum if the same meaning is assigned to the same symbols in different isotope systems where ever possible.

Analytical considerations

High-precision Mg isotope ratios can now be obtained by MC-ICPMS because instrumental fractionation (α_{inst}) can be precisely monitored. MC-ICPMS measurements of Mg isotope ratios are obtained by measuring ^{24}Mg, ^{25}Mg and ^{26}Mg signals simultaneously on three Faraday cup collectors. When using low-mass-resolution Ar-ICP-source mass spectrometers, the Mg mass spectrum can include spectral (i.e., isobaric) interferences. Potential interfering species include $^{48}Ti^{++}$, $^{50}Ti^{++}$, $^{48}Ca^{++}$, $^{50}Cr^{++}$, $^{52}Cr^{++}$, CN^+, CC^+ and various hydrides (e.g., C_2H^+). The H-bearing molecules, including Mg-hydride, become negligible in dry plasmas (Galy et al. 2001; Young et al. 2002a). The effects of N, although ubiquitous, can be rendered negligible by avoiding N_2 gas as an augmentation to Ar in the desolvating nebulizers. Effects of the other elements can be removed by measuring pure solutions of Mg (e.g., Chang et al. 2003). When purification is not possible the effects of the interfering species can be mitigated by matching analyte signal intensities. For laser ablation studies, interference effects must be determined in advance or monitored as part of the analytical scheme (e.g., Young et al. 2002a).

Higher mass resolution is becoming more common in MC-ICPMS technology. The result will be a reduction in the hindrances of isobaric interferences. With judicious use of narrow entrance slits and improvements in ion optics, even smaller radius instruments can resolve $^{50}Ti^{++}$ from $^{25}Mg^+$, for example. However, at this writing most studies have made use of low-mass-resolution instruments, and even with high mass resolution, care must still be taken to avoid changes in instrumental fractionation due the presence of elements other than the analyte in the plasma.

Solutions. Correction for α_{inst} is most often made using a standard-sample bracketing technique (e.g., Galy et al. 2001). In this protocol, standard and sample isotope ratios obtained by multiple measurement cycles are compared and the sample result expressed as a deviation from the standard. Cross contamination between the sample and the standard is avoided by washing the analytical instrumentation with dilute (usually about 0.1N) HNO_3 for several minutes between analyses. Introduction of Mg in dilute HNO_3 (e.g., 0.1N) into the MC-ICPMS via a desolvating nebulizer reduces O, N, C and H spectral interferences to an insignificant level. In addition, concentrations of Mg in the standard and the sample can be kept within 25% of each other to further minimize the potential influences of isobaric interferences.

Impure solutions of common minerals can induce isotopic biases of up to 0.5‰ per atomic mass unit (amu) (e.g., Galy et al. 2001). These so-called matrix effects are mass-dependent and are the result of changes in α_{inst}. Shifts in $\alpha_{inst.}$ are most sensitive to the presence of Ca and K followed by Zn, Al and Na. The matrix effect caused by the presence of Fe in solution is apparently less severe (Galy et al. 2001; Galy et al. in review). High precision and accuracy can still be obtained on multi-element solutions but only when the chemistry of the sample is identical to the chemistry of the standard. For accurate results, chemical purification is advisable. Chemical separation of Mg can be accomplished by liquid chromatographic methods but these introduce a mass-dependent isotopic fractionation of 1.25‰ per amu between the first and last 10% fractions (Chang et al. 2003). The preferred method makes use of Bio-Rad AG50W-X12 ion exchange resin through which >99% recovery of Mg is achieved (Galy et al. 2002; Chang et al. 2003). The size of the matrix effect related to alkali elements and Ca implies that [K+Ca+Na]/[Mg] must be lower than 0.05 (mol ratios) if the inaccuracy related to matrix effect is to be lower than the long term external reproducibility of the method. Therefore, K, Na and Ca blank levels in the eluent and acid used are the limiting factor of the sensitivity and accuracy of the method.

Laser ablation. Instrumental mass fractionation is generally accounted for in laser ablation analyses of Mg isotope ratios using sample-standard bracketing. The standard can be either a solid or a solution. The material liberated by laser ablation includes all of the elements comprising the target. As a result, investigations into the possible spectral interferences attributable to these elements are required. Young et al. (2002a) found that the interference effects of Ca and Ti were negligible for many minerals including Mg-rich olivine, clinopyroxene, Al-Ti-rich clinopyroxene, melilite, and spinel. For example, the concentration of Ca sufficient to bias measured magnesium isotope ratios was determined empirically by aspirating solutions of 1 to ~200 ppm Ca in dilute HNO_3 through the desolvating nebulizer and into the sample carrier gas during laser ablation of a standard consisting of a grain of forsterite. Because the mass resolution was too low to resolve $^{48}Ca^{++}$ and $^{24}Mg^{+}$, $^{44}Ca^{++}$ was used as a proxy for $^{48}Ca^{++}$ in order to calculate the amount of $^{48}Ca^{++}$ entering the source. Measurable interference on m/z = 24 signals due to the presence of $^{48}Ca^{++}$ occurred when the concentrations of Ca reaching the plasma source were sufficient to produce $^{44}Ca^{++}/^{24}Mg^{+}$ voltage ratios >1‰. This limit could then be used as a criterion for assessing the accuracy of analyses of unknowns that contain both Ca and Mg.

The "matrix effects" attributable to Fe, Ti, Al, Na, and K during laser ablation analyses of Mg isotope ratios have been investigated (Young et al. 2002a; Norman et al. 2004). Results indicate that these elements do not influence measured Mg isotope ratios at the 0.1‰ per amu level of detection for most common rock-forming minerals in which Mg is a significant component. However, effects greater than 1‰ have been observed when analyzing Mg as a minor element in feldspar and more work will be required to assess the accuracy obtainable by laser ablation of minerals containing small concentrations of Mg.

Matrix effects on measured Mg isotope ratios due to the presence of Fe in laser ablation targets require further study. Young et al. (2002a) found that adding Fe via solution to the plasma to yield $^{56}Fe^{+}/^{24}Mg^{+}$ up to 2.0 resulted in no measurable shifts in $^{25}Mg/^{24}Mg$ and $^{26}Mg/^{24}Mg$ of laser ablation products. Norman et al. (2004) showed that there is a shift in instrumental fractionation in measured $^{25}Mg/^{24}Mg$ on the order of +0.06‰ for every 1% decrease in Mg/(Mg+Fe) of the olivine target. These disparate results might be explained by ionization in different locations in the torch when samples are introduced by aspiration of solutions rather than by gas flow from a laser ablation chamber. Different matrix effects for solutions and laser ablation products will depend on the details of the instrument configuration, including the status of the guard electrode, sample load, and radio frequency power.

Inspection of ultraviolet laser ablation pits by backscattered electron imaging, micro-Raman spectroscopy, and characteristic X-ray analysis shows that extraction of the material from the laser pits is quantitative (Young et al. 2002a). Fractionation due to the laser sampling process must arise therefore from incomplete transfer of material after ablation rather than as a result of the ablation itself. Years of experience with UV laser ablation has shown that elemental and isotopic fractionation is minimized by ablating in an atmosphere of He (rather than Ar). The He gas permits the plasma plume that extends above the ablation pit to expand freely, enhancing entrainment into the carrier gas stream. An Ar makeup gas is added to the He flow prior to entering the plasma of the ICPMS in order to "punch" the sample into the torch. Studies to date indicate that fractionation attributable to transport of ablated materials is reproducible since accurate results (as compared with results from solutions of purified Mg) are obtained for a variety of mineral materials using sample-standard bracketing methods.

MAGNESIUM ISOTOPE RESERVOIRS

High-precision values of δ^{26}Mg for Mg obtained by acid digestion of bulk samples, purification of the analyte Mg (Galy et al. 2001), and analysis of dilute acid solutions of the extracted Mg are available for several meteorite samples, mineral grains from mantle nodules, a sample of loess from China representing continental crust, continental basalts, seawater, foraminifera, various carbonate rocks, speleothems and their associated waters, Mg metals, magnesia, and even spinach chlorophyl. These high-precision analyses are shown in Figure 1 and tabulated in Table 1.

Whole rock samples of CI meteorite, matrix of LL3 meteorite, and clinopyroxene from a single mantle nodule are indistinguishable from the DSM3 standard and suggest that δ^{26}Mg = 0.0‰ is a reasonable estimate of the chondritic reservoir for Mg (Galy and O'Nions 2000). Young et al. (2002a) report an analysis of olivine from the same sample of mantle nodule that is 0.8‰ lower in δ^{26}Mg than the clinopyroxene shown here. There is concern that this sample of olivine may have been altered. The olivine value has been verified using laser ablation several times on different instruments (Young et al. 2002a; unpub.) but the possibility for cryptic alteration remains and the olivine analysis may not be representative of mantle values.

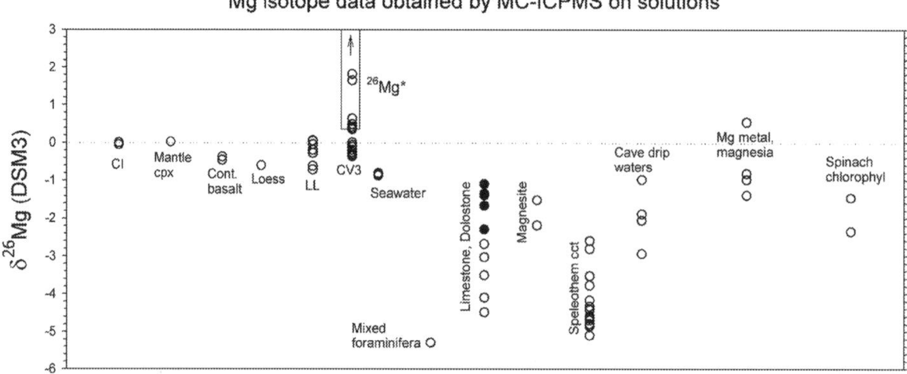

Figure 1. Compilation of δ^{26}Mg values (relative to DSM3) for various terrestrial and extraterrestrial reservoirs. Data and sources are tabulated in Table 1. For the carbonates, the black points represent dolomite while the open symbols represent calcite.

Table 1. Mg isotope ratios determined by MC-ICPMS analysis of purified Mg in weak acid solution. Three significant figures are shown to aid in the calculation of $\Delta^{25}Mg'$ values.

Sample	Source	$\delta^{26}Mg$	±2σ	$\delta^{25}Mg$	±2σ
Metal and oxide					
SRM 980 Mg metal	1	−3.405	0.07	−1.744	0.04
Romil Mg metal	1	−0.977	0.086	−0.508	0.049
Aldrich Mg metal	1	−0.816	0.172	−0.416	0.082
Dead Sea Mg metal	1	0.543	0.150	0.279	0.079
Petzl's magnesie	1	−1.383	0.042	−0.716	0.013
Organic matter					
Spinach chlorophyll (b)	1	−2.349	0.067	−1.204	0.026
Spinach chlorophyll (a)	1	−1.451	0.098	−0.741	0.062
Carbonate					
Magnesite, Piemont, Italy	1	−2.186	0.044	−1.146	0.035
Magnesite, Transvaal, South Africa	1	−1.508	0.143	−0.768	0.076
Dolomite	2	−1.771	0.040	−0.865	0.010
Dolostone, MSR	3	−1.089	0.036	−0.555	0.016
Dolostone, SR-D-1	3	−2.290	0.096	−1.184	0.033
Dolostone LH	3	−1.382	0.082	−0.710	0.108
Dolostone LH, AP867	3	−1.339	0.190	−0.676	0.094
Dolostone LH AP865	3	−1.660	0.119	−0.847	0.055
Marble TSS NL251	3	−3.496	0.036	−1.828	0.039
Marble TSS NL252	3	−2.676	0.018	−1.375	0.062
Marble TSS XP210	3	−3.024	0.076	−1.583	0.041
Limestone (Cen), PEK-HOST	3	−4.088	0.072	−2.123	0.003
Limestone (Tur), DDC	3	−4.472	0.015	−2.328	0.026
Speleothem, SR-2N-F	3	−4.628	0.115	−2.400	0.070
Speleothem, SR-2N-K	3	−4.623	0.083	−2.391	0.043
Speleothem SR-2-10-C1	3	−4.780	0.029	−2.465	0.019
Speleothem SR-2-10-C2	3	−4.820	0.001	−2.492	0.006
Speleothem SR-2-10-C2	3	−4.840	0.040	−2.498	0.045
Speleothem SR-2-10-D1	3	−4.826	0.022	−2.491	0.020
Speleothem SR-5-3-00-S	3	−4.636	0.116	−2.392	0.028
Speleothem SR-8-5-00-S	3	−4.617	0.019	−2.393	0.031
Speleothem SR-13-3-00-S	3	−2.594	0.025	−1.338	0.010
Spleothem, SR-2-8-B	3	−4.414	0.120	−2.282	0.065
Spleothem, SR-2-8-E3	3	−4.158	0.106	−2.154	0.074
Speleothem, SR-2-8-J	3	−4.579	0.126	−2.372	0.089
Speleothem, SR-2-8-G	3	−4.685	0.122	−2.414	0.027
Mixed-species foraminifera tests	2	−5.279	0.160	−2.732	0.100
Water					
Drip water, SR-5-3-00-W	3	−1.884	0.066	−0.965	0.034
Drip water, SR-8-5-00-W	3	−2.057	0.028	−1.048	0.060
Drip water, SR-13-3-00-W	3	−2.931	0.020	−1.500	0.047
Ground water, Har-Tuv4	3	−1.696	0.056	−0.870	0.014
Drip water, MSS	3	−0.975	0.057	−0.492	0.055
Drip water, TDGW	3	−2.433	0.119	−1.277	0.048
North Atlantic seawater	2	−0.824	0.040	−0.416	0.080
Mediterranean water	4	−0.864	0.122	−0.434	0.105
Atlantic seawater (SWOM)	4	−0.799	0.037	−0.407	0.007
Mediterranean surface water	4	−0.851	0.244	−0.422	0.121

Table continued on next page.

Table 1. *(continued from previous page)*

Sample	Source	$\delta^{26}Mg$	±2σ	$\delta^{25}Mg$	±2σ
Terrestrial silicate					
Continental basalt BR	4	−0.460	0.185	−0.250	0.047
Continental basalt BCR-1	4	−0.368	0.105	−0.186	0.070
Mantle nodule clinopyroxene Jl-1	5	0.022	0.093	0.017	0.073
Mantle nodule olivine Jl-1	5	−1.093	0.117	−0.558	0.096
Loess, China	4	−0.602	0.222	−0.308	0.148
Allende Meteorite (CV3)					
chondrule AG38	6	−0.333	0.092	−0.178	0.039
chondrule AG67II	6	−0.060	0.027	−0.027	0.015
chondrule A2	6	0.354	0.134	0.186	0.071
chondrule A3	6	−0.248	0.064	−0.118	0.056
chondrule A4	6	1.648	0.080	0.737	0.069
chondrule A5	6	0.487	0.053	0.223	0.085
chondrule A6	6	0.381	0.115	0.198	0.060
chondrule A7	6	−0.252	0.034	−0.118	0.018
chondrule A8	6	−0.362	0.113	−0.176	0.038
chondrule AH3	6	−0.192	0.043	−0.087	0.021
chondrule AH2	6	1.818	0.020	0.864	0.020
chondrule AH1	6	0.420	0.040	0.189	0.001
chondrule AH4	6	0.639	0.031	0.291	0.046
chondrule A9, rim	6	−0.116	0.043	−0.060	0.015
chondrule A9, core	6	0.009	0.042	0.004	0.028
CAI, AG178	6	11.917	0.024	5.117	0.005
Whole rock	6	−0.300	0.073	−0.160	0.031
Matrix	6	−0.287	0.044	−0.151	0.009
Bjurbole Meteorite (L/LL4)					
chondrule AG162	7	−0.194	0.159	−0.082	0.069
chondrule AG163	7	−0.275	0.043	−0.145	0.029
chondrule AG164	7	−0.714	0.068	−0.340	0.020
Orgueil Meteorite (CI)					
Whole rock, AG94	8	0.003	0.060	0.002	0.048
Whole rock, AG95	8	−0.050	0.079	−0.015	0.045
Chainpur Meteorite (LL3)					
Matrix, AG76	4	0.059	0.093	0.050	0.025
Matrix, AG77	4	0.046	0.110	0.023	0.066
Chondrule C11	4	−0.615	0.079	−0.307	0.043
Chondrule C12	4	−0.045	0.131	−0.027	0.062
Evaporation products					
chondrule AH3, re-melted	7	6.741	0.001	3.450	0.006
chondrule AG162, re-melted	7	0.861	0.060	0.459	0.014
chondrule AG163, re-melted	7	5.215	0.040	2.676	0.030
chondrule AG164, re-melted	7	0.899	0.227	0.472	0.107

Sources: (1) Galy et al. 2001; (2) Chang et al. 2003; (3) Galy et al. 2002; (4) This study; (5) Young et al. 2002; (6) Galy et al. 2000; (7) Young et al. 2002; (8) Galy et al. in press

Support for this conclusion comes from laser ablation analyses of mantle olivines recently reported by Norman et al. (2004). The loess and continental basalt samples suggest that evolved crustal materials may be on average approximately 0.4–0.6‰ lower in δ^{26}Mg than the primitive CI/mantle reservoir (Fig. 1).

Chondrules from Chainpur LL3 meteorite have somewhat lower δ^{26}Mg values than Chainpur LL3 matrix, CI, and terrestrial mantle clinopyroxene. Chondrules from the Bjurbole L/LL4 ordinary chondrite also have slightly lower δ^{26}Mg values than CI and LL3 matrix. Chondrules, matrix, and whole rock samples from the Allende CV3 meteorite span a larger range in δ^{26}Mg that overlaps the CI and ordinary chondrite data.

The CV3 δ^{26}Mg values greater than approximately 0.5‰ on the DSM3 scale reflect the presence of radiogenic ^{26}Mg (^{26}Mg *) formed by β+ decay of extinct ^{26}Al (half life = 0.72 m.y.). The origin of the ^{26}Mg* in the chondrules will be discussed further in the sections on extraterrestrial materials. Not shown in Figure 1 is a bulk CAI sample that has a δ^{26}Mg value of 11.85‰. This latter value reflects both mass fractionation and ^{26}Mg*. Excluding the ^{26}Mg*-bearing samples, the Allende data define a narrow spread in δ^{26}Mg values of about 1‰. The significance of the variability in δ^{26}Mg among the ^{26}Mg*-bearing samples and the uniformity in δ^{26}Mg among the ^{26}Mg*-free samples is discussed in sections describing the meteorite data that follow.

Carbonate rocks and foraminifera tests (a sample of mixed species) are consistently lower in δ^{26}Mg than Mg from seawater by several per mil. In addition, Mg in calcite is consistently lower in δ^{26}Mg than Mg in dolomite by approximately 2‰ (Fig. 1). These data together with the samples of coeval speleothem calcite and waters show that the heavy isotopes of Mg partition to water relative to carbonate minerals. In this respect the Mg isotopes behave much like the isotopes of Ca (Gussone et al. 2003; Schmitt et al. 2003). There is not yet sufficient data to assess with confidence the temperature dependence of the fractionation of Mg isotopes between carbonates and waters, although Galy et al. (2002) concluded that the evidence so far is that temperature effects are below detection in the range 4–18°C.

In summary, there is a clear mineralogical control on the partitioning of the Mg isotopes among carbonates and waters and, apparently, a weak dependence on temperature at low *T*. An important question is the extent to which these measured values for carbonates and waters reflect isotopic equilibrium. Hints to the answer come from comparing δ^{26}Mg to δ^{25}Mg, as shown in the section on terrestrial reservoirs.

Kinetic, or "vital," isotopic effects may be evidenced in the samples of mixed foraminifera and seawater reported by Chang et al. (2003). Again, the relationship between δ^{26}Mg and δ^{25}Mg help answer this question (see below), although the data are too few to reach firm conclusions at this stage.

On a standard plot of δ^{25}Mg vs. δ^{26}Mg all of the terrestrial data follow the expected mass-fractionation trends (Fig. 2). Deviations from mass fractionation are evident in several of the meteorite samples. The details of the relationships between δ^{25}Mg vs. δ^{26}Mg are the subject of several following sections.

RESOLVING MECHANISMS OF ISOTOPE FRACTIONATION WITH THREE ISOTOPE DIAGRAMS

The mass-dependent fractionation laws that describe the partitioning of three or more isotopes are different for kinetic and equilibrium reactions. The precision with which Mg isotope ratios can be measured by MC-ICPMS is sufficient to take advantage of these differences. Relationships between ^{25}Mg/^{24}Mg and ^{26}Mg/^{24}Mg are diagnostic of kinetic

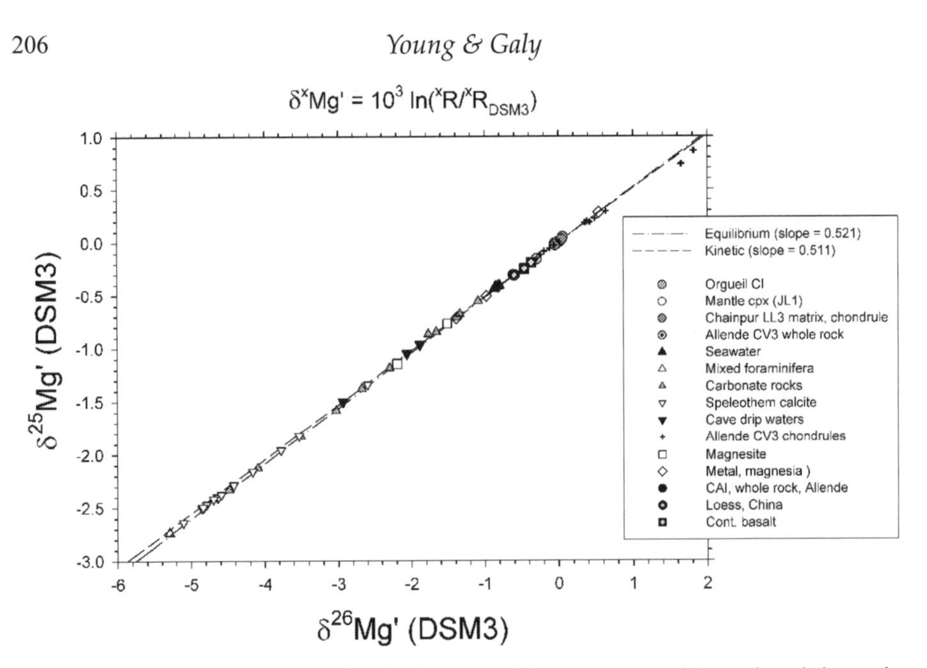

Figure 2. Magnesium three-isotope plot showing terrestrial and extraterrestrial samples relative to the predicted equilibrium and kinetic mass fractionation laws. The slopes in the caption refer to the β values that characterize the mass fractionation laws.

fractionation, equilibrium fractionation, and, in the case of extraterrestrial samples, even subtle departures from both. The theory and methods for applying this new technique to the Mg isotope system are described below.

Mass-dependent fractionation laws

The fractionation laws relating two or more isotope ratios are described by a relationship between the fractionation factors for the isotope ratios. Using Mg as an example, the three isotopes ^{24}Mg, ^{25}Mg, and ^{26}Mg define two fractionation factors

$$\alpha_{25/24} = \frac{(^{25}\mathrm{Mg}/^{24}\mathrm{Mg})_a}{(^{25}\mathrm{Mg}/^{24}\mathrm{Mg})_b} \tag{5}$$

and

$$\alpha_{26/24} = \frac{(^{26}\mathrm{Mg}/^{24}\mathrm{Mg})_a}{(^{26}\mathrm{Mg}/^{24}\mathrm{Mg})_b} \tag{6}$$

where a and b refer to either two different conditions or two different materials. At equilibrium the two fractionation factors are related by the expression

$$\frac{\ln \alpha_{25/24}}{\ln \alpha_{26/24}} = \frac{\left(\frac{1}{m_1} - \frac{1}{m_2}\right)}{\left(\frac{1}{m_1} - \frac{1}{m_3}\right)} \tag{7}$$

where m_1 is the atomic mass of ^{24}Mg (23.985042), m_2 is the atomic mass of ^{25}Mg (24.985837), and m_3 is the atomic mass of ^{26}Mg (25.982593) (Hulston and Thode 1965; Matsuhisa et al. 1978; Weston 1999; Young et al. 2002b). Equation (7) can be rearranged to give

$$\alpha_{25/24} = \left(\alpha_{26/24}\right)^{\beta} \tag{8}$$

where the exponent is

$$\beta = \frac{\left(\frac{1}{m_1} - \frac{1}{m_2}\right)}{\left(\frac{1}{m_1} - \frac{1}{m_3}\right)} \tag{9}$$

Equation (9) is a high-temperature limit for the exponent in the equilibrium fractionation law that is a good indication of the true equilibrium value (Young et al. 2002b). The value for the exponent obtained from the atomic masses of the Mg isotopes is 0.5210.

Kinetic isotope fractionation obeys a different fractionation law. Young et al. (2002b) showed recently that kinetic processes, again written here in terms of the two Mg isotope fractionation factors, obey the relation

$$\frac{\ln \alpha_{25/24}}{\ln \alpha_{26/24}} = \frac{\ln\left(\frac{M_1}{M_2}\right)}{\ln\left(\frac{M_1}{M_3}\right)} \tag{10}$$

where M_1, M_2, and M_3 are the masses in motion associated with the kinetic process. As before, Equation (10) can be rewritten as

$$\alpha_{25/24} = \alpha_{26/24}{}^{\beta} \tag{11}$$

where the exponent β is no longer given by Equation (9) but is instead

$$\beta = \frac{\ln\left(\frac{M_1}{M_2}\right)}{\ln\left(\frac{M_1}{M_3}\right)} \tag{12}$$

The masses in the kinetic β will be reduced masses if the rate limiting step involves vibrations (as during bond rupture), they will be molecular masses of isotopomers if the kinetic process involves transport of molecules, and they will be atomic masses if transport of atoms is involved. The value of the kinetic β goes down as the masses in (12) increase. As a result, an estimate of the maximum value for β during the kinetic fractionation of Mg isotopes is obtained by equating the M_i values with the atomic masses of Mg. The kinetic β obtained from the atomic masses of Mg is 0.5110.

In summary, the different mass-dependent fractionation laws that relate the two fractionation factors for the Mg isotopes are characterized by the value for β in the expression $\alpha_{25/24} = (\alpha_{26/24})^{\beta}$. For equilibrium processes β should be close to 0.521. For kinetic processes involving Mg atoms β should be close to 0.511.

Mass-fractionation in three-isotope space

The equilibrium and kinetic fractionation laws

$$\alpha_{25/24} = (\alpha_{26/24})^{\beta} \tag{13}$$

define curves on plots of δ^{25}Mg vs. δ^{26}Mg. These "mass fractionation curves" follow the general equation

$$\delta^{25}\text{Mg} = (10^3 + \delta^{25}\text{Mg}_{\text{ref}})\left(\frac{10^3 + \delta^{26}\text{Mg}}{10^3 + \delta^{26}\text{Mg}_{\text{ref}}}\right)^{\beta} - 10^3 \tag{14}$$

In this equation the reference composition (designated by the ref subscript) lies on the fractionation curve and might be, for example, the initial composition for an array of fractionated samples. All of the δ values in the equation refer to the same standard (e.g., DSM3) but this standard need not be on the fractionation curve itself (in other words, the fractionation curves described by Equation (14) do not have to pass through the origin on a plot of δ^{25}Mg vs. δ^{26}Mg).

One way to compare data to predicted fractionation laws is to plot the data on the three isotope plot in which δ^{25}Mg is the ordinate and δ^{26}Mg is the abscissa, and examine how closely the data fall to the different curves defined by the exponent β. However, the differences between the different β values are often evident only with careful attention to the statistics of the data. Ideally, the values of β should be obtained by a best fit to the data. This is most easily accomplished if the problem can be rewritten so that β is the slope in a linear regression.

Hulston and Thode (1965) showed that the relationship between δ values on a three isotope plot can be made linear if the definition of the δ's are modified so that the term $({}^xR/{}^xR_{std} - 1)$ is replaced by $\ln({}^xR/{}^xR_{std})$ where *x* refers to one of the minor isotopes, *R* is the ratio of the minor isotope to the major isotope and *std* refers to the standard ratio. In the case of Mg isotopes, the new definitions for δ^{25}Mg′ and δ^{26}Mg′ are:

$$\delta^{x}\text{Mg}' = \ln\left(\frac{({}^{x}\text{Mg}/{}^{24}\text{Mg})_{\text{sample}}}{({}^{x}\text{Mg}/{}^{24}\text{Mg})_{\text{DSM3}}}\right)10^3 \tag{15}$$

where *x* refers to either 25 or 26. The fractionation curves written in terms of these new, linearized δ' values are straight lines of the form

$$\delta^{25}\text{Mg}' = \beta\,\delta^{26}\text{Mg}' - \beta\,\delta^{26}\text{Mg}'_{\text{ref}} - \delta^{25}\text{Mg}'_{\text{ref}} \tag{16}$$

where the slope is the exponent β and the intercept is determined by the reference composition (if the reference composition is DSM3, then the intercept is the origin) .

When the Mg isotope data are cast in terms of the new δ' values, they can be subjected to well-established linear regression techniques to extract the best-fit exponent β. A handy equation that converts the original δ values to the new, linear δ' values is

$$\delta' = 10^3 \ln\left(\frac{(\delta + 10^3)}{10^3}\right) \tag{17}$$

The linear δ' values for the data from Table 1 are shown in Table 2.

By plotting the highly precise Mg isotope data collected by MC-ICPMS in terms of δ^{25}Mg′ vs. δ^{26}Mg′ (Table 2) it is possible to constrain the values for β from the best-fit slopes defined by the data and therefore the nature of the fractionation processes that lead to the distribution of data. This method is applied in sections below.

Table 2. Mg isotope ratios determined by MC-ICPMS analysis of purified Mg in weak acid solution expressed in terms of linear delta values.

Sample	Source	$\delta^{26}Mg'$	$\delta^{25}Mg'$	$\Delta^{25}Mg'$
Metal and oxide				
SRM 980 Mg metal	1	−3.415	−1.746	0.033
Romil Mg metal	1	−0.978	−0.509	0.001
Aldrich Mg metal	1	−0.816	−0.416	0.009
Dead Sea Mg metal	1	0.543	0.279	−0.004
Petzl's magnesie	1	−1.384	−0.716	0.005
Organic matter				
Spinach chlorophyll (b)	1	−2.352	−1.204	0.021
Spinach chlorophyll (a)	1	−1.452	−0.742	0.015
Carbonate				
Magnesite, Piemont, Italy	1	−2.189	−1.146	−0.006
Magnesite, Transvaal, South Africa	1	−1.509	−0.768	0.018
Dolomite	2	−1.772	−0.866	0.057
Dolostone, MSR	3	−1.090	−0.556	0.012
Dolostone, SR-D-1	3	−2.293	−1.185	0.010
Dolostone LH	3	−1.383	−0.711	0.010
Dolostone LH, AP867	3	−1.340	−0.676	0.022
Dolostone LH AP865	3	−1.661	−0.847	0.018
Marble TSS NL251	3	−3.502	−1.829	−0.005
Marble TSS NL252	3	−2.680	−1.376	0.020
Marble TSS XP210	3	−3.028	−1.585	−0.007
Limestone (Cen), PEK-HOST	3	−4.096	−2.126	0.009
Limestone (Tur), DDC	3	−4.482	−2.331	0.004
Speleothem, SR-2N-F	3	−4.638	−2.403	0.013
Speleothem, SR-2N-K	3	−4.634	−2.394	0.021
Speleothem SR-2-10-C1	3	−4.792	−2.468	0.029
Speleothem SR-2-10-C2	3	−4.832	−2.495	0.023
Speleothem SR-2-10-C2	3	−4.852	−2.501	0.027
Speleothem SR-2-10-D1	3	−4.837	−2.494	0.026
Speleothem SR-5-3-00-S	3	−4.647	−2.395	0.027
Speleothem SR-8-5-00-S	3	−4.627	−2.396	0.015
Speleothem SR-13-3-00-S	3	−2.598	−1.339	0.015
Spleothem, SR-2-8-B	3	−4.424	−2.285	0.020
Spleothem, SR-2-8-E3	3	−4.167	−2.156	0.015
Speleothem, SR-2-8-J	3	−4.590	−2.375	0.016
Speleothem, SR-2-8-G	3	−4.696	−2.416	0.030
Mixed-species foraminifera tests	2	−5.293	−2.736	0.021
Water				
Drip water, SR-5-3-00-W	3	−1.886	−0.965	0.017
Drip water, SR-8-5-00-W	3	−2.059	−1.049	0.024
Drip water, SR-13-3-00-W	3	−2.935	−1.501	0.028
Ground water, Har-Tuv4	3	−1.698	−0.870	0.014
Drip water, MSS	3	−0.975	−0.492	0.016
Drip water, TDGW	3	−2.436	−1.277	−0.008
North Atlantic seawater	2	−0.824	−0.416	0.013
Mediterranean water	4	−0.864	−0.434	0.016
Atlantic seawater (SMOW)	4	−0.800	−0.407	0.010
Mediterranean surface water	4	−0.852	−0.422	0.022

Table continued on next page.

Table 2. *(continued from previous page)*

Sample	Source	$\delta^{26}Mg'$	$\delta^{25}Mg'$	$\Delta^{25}Mg'$
Terrestrial silicate				
Continental basalt BR	4	−0.460	−0.250	−0.010
Continental basalt BCR-1	4	−0.368	−0.186	0.005
Mantle nodule clinopyroxene Jl-1	5	0.022	0.017	0.006
Mantle nodule olivine Jl-1	5	−1.094	−0.558	0.012
Loess, China	4	−0.602	−0.308	0.006
Allende Meteorite (CV3)				
chondrule AG38	6	−0.333	−0.178	−0.004
chondrule AG67II	6	−0.060	−0.027	0.004
chondrule A2	6	0.354	0.186	0.002
chondrule A3	6	−0.248	−0.118	0.011
chondrule A4	6	1.646	0.737	−0.121
chondrule A5	6	0.487	0.223	−0.031
chondrule A6	6	0.381	0.197	−0.001
chondrule A7	6	−0.252	−0.118	0.013
chondrule A8	6	−0.362	−0.176	0.012
chondrule AH3	6	−0.192	−0.087	0.013
chondrule AH2	6	1.817	0.863	−0.083
chondrule AH1	6	0.420	0.189	−0.030
chondrule AH4	6	0.639	0.291	−0.042
chondrule A9, rim	6	−0.116	−0.060	0.000
chondrule A9, core	6	0.009	0.004	−0.001
CAI, AG178	6	11.847	5.104	−1.068
Whole rock	6	−0.300	−0.160	−0.003
Matrix	6	−0.287	−0.151	−0.002
Bjurbole Meteorite (L/LL4)				
chondrule AG162	7	−0.194	−0.082	0.019
chondrule AG163	7	−0.275	−0.145	−0.002
chondrule AG164	7	−0.714	−0.340	0.032
Orgueil Meteorite (CI)				
Whole rock, AG94	8	0.003	0.002	0.000
Whole rock, AG95	8	−0.050	−0.015	0.011
Chainpur Meteorite (LL3)				
Matrix, AG76	4	0.059	0.050	0.019
Matrix, AG77	4	0.046	0.023	−0.001
Chondrule C11	4	−0.615	−0.307	0.013
Chondrule C12	4	−0.045	−0.027	−0.003
Evaporation products				
chondrule AH3, re-melted	7	6.718	3.444	−0.056
chondrule AG162, re-melted	7	0.861	0.459	0.010
chondrule AG163, re-melted	7	5.202	2.672	−0.038
chondrule AG164, re-melted	7	0.899	0.472	0.003

Sources: (1) Galy et al. 2001; (2) Chang et al. 2003; (3) Galy et al. 2002; (4) This study; (5) Young et al. 2002; (6) Galy et al. 2000; (7) Young et al. 2002; (8) Galy et al. in press

The definition of Δ^{25}Mg′

In order to facilitate the analysis of fractionation laws it is convenient to amplify small differences in β using the Δ^{25}Mg notation. We define Δ^{25}Mg in a fashion analogous to Δ^{17}O as

$$\Delta^{25}\text{Mg}' = \delta^{25}\text{Mg}' - 0.521\,\delta^{26}\text{Mg}' \tag{18}$$

where 0.521 refers to the equilibrium value for β. It is important to recognize that we have defined the "big delta 25" in terms of the linear forms of the δ values, δ' (data as represented in Table 2). We do this so that the β value used in the definition applies for all values of δ^{26}Mg′. Using the usual, non-linear definitions of δ values in Equation (18) would cause Δ^{25}Mg to vary with δ^{26}Mg due simply to the curvature of the fractionation relationships in δ^{25}Mg - δ^{26}Mg space. The reader is reminded that this Δ^{25}Mg is not at all related to the Δ^{25}Mg often reported in the cosmochemistry literature (see section on nomenclature), nor is it the Δ used in some of the terrestrial stable isotope literature where $\Delta \sim 10^3 \ln\alpha$, i.e. the difference in δ values between two materials at equilibrium.

The values of Δ^{25}Mg′ for a given datum are different on the SRM 980 and DSM3 scales. The SRM 980_O standard has a Δ^{25}Mg′ value of +0.0315‰ on the DSM3 scale (Galy et al. 2003). Two analyses of CI carbonaceous chondrite and the clinopyroxene from a terrestrial mantle nodule have Δ^{25}Mg′ values that are indistinguishable from 0.00 on the DSM3 scale (Table 2).

Uncertainties in Δ^{25}Mg′

Uncertainties in Δ^{25}Mg′ depend on the degree of covariance between δ^{25}Mg′ and δ^{26}Mg′ for each individual datum (i.e., the covariance for the internal precision defined by the variability among the measurement cycles on the mass spectrometer). Typical correlation coefficients between measured δ^{25}Mg′ and δ^{26}Mg′ on the MC-ICPMS instruments used by the authors are on the order of 0.8 (Young et al. 2002b), but variable. The corresponding covariance depends on the variances associated with the internal reproducibility of the measured δ^{25}Mg′ and δ^{26}Mg′ values. The reproducibility of individual measurements of δ^{25}Mg′ and δ^{26}Mg′ for solutions obtained in a single mass spectrometry session is on the order of ±0.015‰ and ±0.03‰ 1σ, yielding for the variances $\sigma^2_{\delta 25} = 0.00023$ and $\sigma^2_{\delta 26} = 0.0009$. These variances and the typical correlation coefficient in turn define a typical covariance $\sigma^2_{\delta 25,\delta 26}$ for δ^{25}Mg′ and δ^{26}Mg′ of 0.00036. Standard propagation of errors leads to the following expression for the standard deviation for an individual Δ^{25}Mg′ measurement:

$$\sigma^2_{\Delta^{25}\text{Mg}} = \sigma^2_{\delta^{25}\text{Mg}} + (-0.521)^2\,\sigma^2_{\delta^{26}\text{Mg}} + 2(-0.521)\,\sigma^2_{\delta^{25}\text{Mg},\delta^{26}\text{Mg}} \tag{19}$$

Substitution of the typical variances and covariance into Equation (19) suggests that the $1\sigma_{\Delta^{25}\text{Mg}}$ for the MC-ICPMS measurements of Mg in solutions is on the order of ±0.010‰. This is regarded as an internal precision for an individual solution measurement. We note, however, that the reported measurements represent averages of several replicate analyses of the same solution and so more realistic assessments of the *internal precision* for Δ^{25}Mg′ data presented here would be obtained from the uncertainties in the means (standard errors). For example, four analyses of the *same solution* yields a standard error for Δ^{25}Mg′ of ±0.005‰ (this is still regarded as an internal precision because the effects of column chemistry and sample dissolution are not included). No attempt has been made here to review all of the raw data sets to calculate standard errors for each datum in Table 1. However, the distribution of data indicates that ±0.010‰ 1σ is an overestimate of the internal precision of Δ^{25}Mg′ values and that a more realistic uncertainty is closer to a typical standard error, which in most cases will be $\leq$ ±0.005‰ (since the number of replicates is usually $\geq$4, e.g., Galy et al. 2001).

$\Delta' - \delta'$ Diagrams

The role that different fractionation laws, and departures from them, play in determining Mg isotopic compositions is evident when the data are shown on a plot of Δ^{25}Mg′ (ordinate) against δ^{26}Mg′ or δ^{25}Mg′ (abscissa). These Δ'-δ' plots magnify differences that are difficult or impossible to see on the more traditional three isotope plots in which δ^{25}Mg is plotted against δ^{26}Mg. Figure 3 shows how to interpret the Δ'-δ' diagrams in terms of mass-dependent fractionation. The dashed and dash-dot lines mark the kinetic and equilibrium mass fractionation paths on such a diagram. The shaded areas bounded by these lines correspond to the regions of the diagram accessible from the origin by a single stage of mass fractionation. This single stage can have any slope between the two extremes shown (in fact, in some circumstances involving large molecules kinetic slopes could be steeper than that shown in Fig. 3). The dark grey arrows show two examples of the effects of two episodes of mass fractionation. In these examples, a kinetic fractionation process is followed by an equilibrium process. The regions outside the shaded areas are accessible only by multiple fractionation steps involving different fractionation laws.

Figure 4 shows how the additions or subtractions of radiogenic ^{26}Mg* are manifested on the Δ'-δ' plot. It is relevant for interpreting the meteorite Mg isotope data. The grey arrows show addition or subtraction of ^{26}Mg* in the absence of mass fractionation relative to the origin (e.g., *in situ* growth of ^{26}Mg* by decay of ^{26}Al). Starting with an initial composition at the origin, combinations of mass fractionation and variations in ^{26}Mg* will define slopes

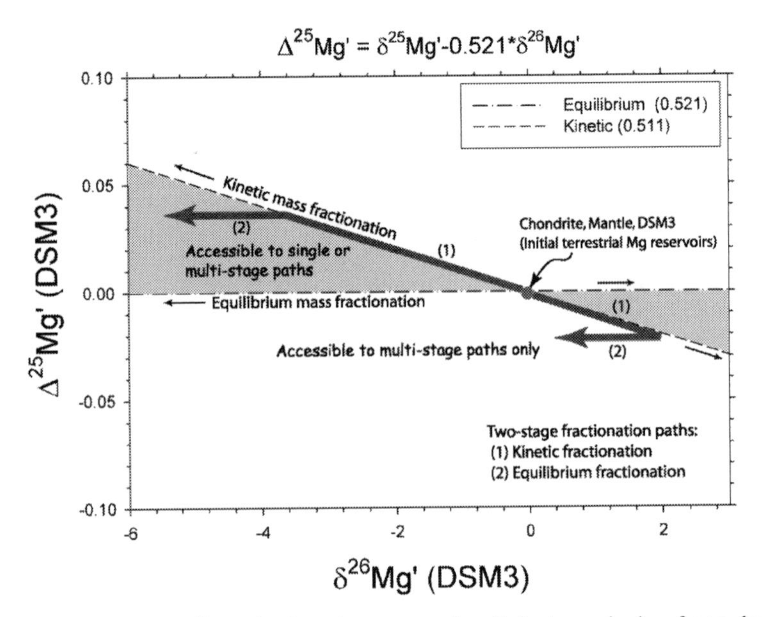

Figure 3. Schematic diagram illustrating how singe-stage and multiple-stage episodes of mass-dependent fractionation are expressed on a plot of Δ^{25}Mg′ vs. δ^{26}Mg′. The paths for pure equilibrium and pure kinetic fractionation relative to the origin are shown by the dashed-dot and dashed lines, respectively. The shaded region bounded by the kinetic and equilibrium fractionation lines is accessible from the origin by single stage fractionation. Other regions on the diagram are accessible by mass-dependent fractionation only if different fractionation laws are applied in succession (i.e., multiple steps with different β values). A variety of paths across the diagram are possible depending upon the nature and sign of the fractionation steps involved. Two multiple-stage paths are shown for illustration. In both examples a kinetic fractionation is followed by an equilibrium fractionation.

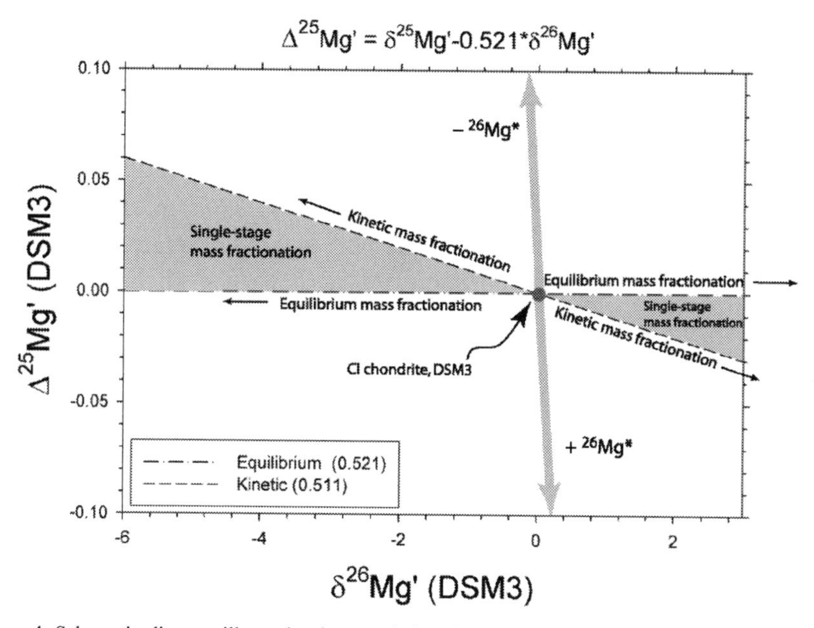

Figure 4. Schematic diagram illustrating how variations in the amount of radiogenic ^{26}Mg (^{26}Mg*) are expressed on a plot of Δ^{25}Mg′ vs. δ^{26}Mg′. Pure addition of ^{26}Mg* to a composition represented by the origin (e.g., CI chondrite) is manifest as a progression along the grey arrow pointing downward on the diagram. Single-step mass fractionation from the origin leads to compositions in the shaded regions bounded by the kinetic and equilibrium mass fractionation lines.

intermediate between the grey arrows and the shaded field bounded by the kinetic and equilibrium mass fractionation lines. On a plot of Δ^{25}Mg′ against δ^{25}Mg′ (rather than δ^{26}Mg′, not shown in Fig. 4) addition and subtraction of ^{26}Mg* results in a vertical path.

TERRESTRIAL MAGNESIUM ISOTOPE RESERVOIRS IN THREE-ISOTOPE SPACE: IMPLICATIONS FOR MECHANISMS OF FRACTIONATION

Measurements of terrestrial Mg isotope ratios on a plot of Δ^{25}Mg′ vs. δ^{26}Mg′ are all within the region bounded by the equilibrium and kinetic mass fractionation laws given expected uncertainties (Fig. 5). Apparently, all of the terrestrial reservoirs represented by the data thus far are related to the primitive chondrite/mantle reservoir by relatively simple fractionation histories. Adherence of the data to the regions accessible by simple mass fractionation processes in Figure 5 (the shaded regions in Fig. 3) is testimony to the veracity of the fractionation laws since there is no reason to suspect that Mg could be affected by any processes other than purely mass-dependent fractionation on Earth.

Casual inspection of Figure 5 suggests that Mg in waters is related to Mg from the primitive mantle (and chondrite) and evolved silicate crust (loess, continental basalt) by kinetic fractionation processes; most water data lie along the kinetic fractionation line that passes through mantle clinopyroxene with Δ^{25}Mg′ values greater than crust and mantle. The same is true for spinach chlorophyll. Carbonates tend to be related to waters by a significant equilibrium fractionation component since their greatest variance is along a trend parallel to the equilibrium fractionation line at constant Δ^{25}Mg′ (horizontal in Fig. 3). The fact that the carbonates tend to have positive Δ^{25}Mg′ values relative to mantle/chondrite and silicate

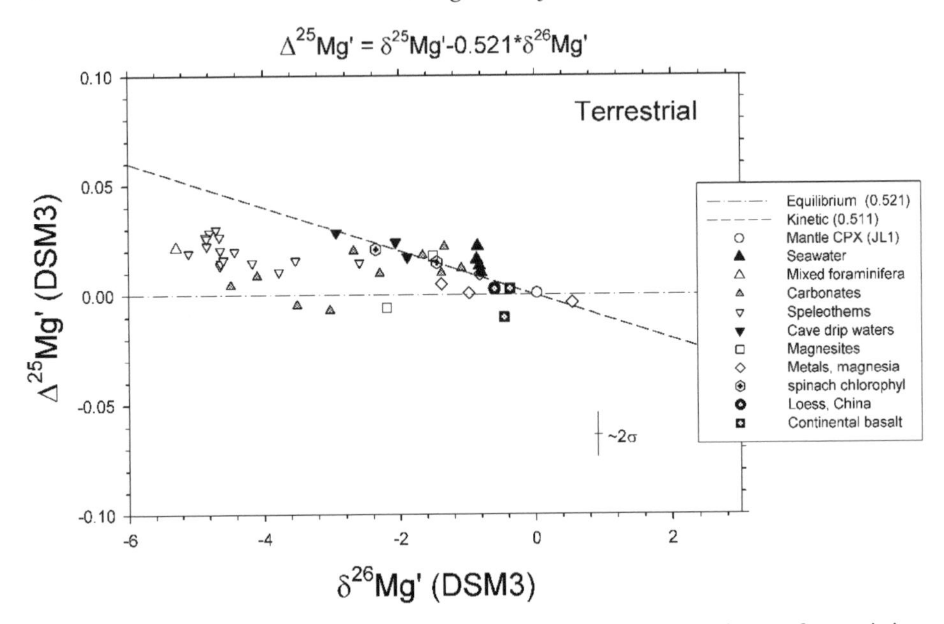

Figure 5. A plot of $\Delta^{25}Mg'$ vs. $\delta^{26}Mg'$ for terrestrial Mg materials. Within best estimates of uncertainties (cross) all of the data lie in the region bounded by equilibrium and kinetic mass fractionation laws. Waters, carbonates, and organic Mg (chlorophyll) have higher $\Delta^{25}Mg'$ values than mantle and crustal Mg reservoirs represented by mantle pyroxene, loess, and continental basalts. The difference in $\Delta^{25}Mg'$ values is attributable to episodes of kinetic mass fractionation.

crust can be explained by nearly equilibrium precipitation from waters that are themselves kinetically fractionated relative to mantle. These general observations can be quantified by extracting values for β from groups of genetically-related data.

For this purpose we applied the linear regression algorithms of Mahon (1996) to $\delta^{25}Mg'$ and $\delta^{26}Mg'$ values (Table 2) for subgroups of the data in Figure 5 to derive β values (β values are the slopes in $\delta^{25}Mg'$ -$\delta^{26}Mg'$ space). The results are listed in Table 3. The veracity of the technique was demonstrated recently by comparing the bulk Mg isotopic compositions of chondrules from the Allende and Bjurbole meteorites with residues left behind after the chondrules were partially evaporated in the laboratory (Young et al. 2002b). Evaporation of silicate liquids is a kinetic process that releases atomic Mg to the gas phase. The Mg isotopic compositions of the evaporation residues follow the predicted kinetic fractionation law for atomic Mg (Fig. 6), yielding a β value of 0.512 ± 0.002 (2σ) compared with the predicted

Table 3. Linear regression of $\delta^{25}Mg'$ and $\delta^{26}Mg'$ for groups of data from Table 1. Best-fit slopes are estimates of β for each data set.

Samples	Slope (β)	Low (95%)	High (95%)	Intercept ± 1σ	MSWD
Chlorophyll + Mg metal	0.512	0.510	0.514	0.031 ±0.002	1.49
Chondrule evaporation products	0.512	0.509	0.514	0.039 ±0.010	0.71
Mizpe Shlagim cave calcite + waters	0.522	0.517	0.527	0.013 ± 0.003	0.56
Soreq cave calcite + waters	0.519	0.517	0.520	0.020 ± 0.001	5.17
Dolostones + limestones + marble	0.524	0.518	0.531	0.004 ± 0.005	0.24

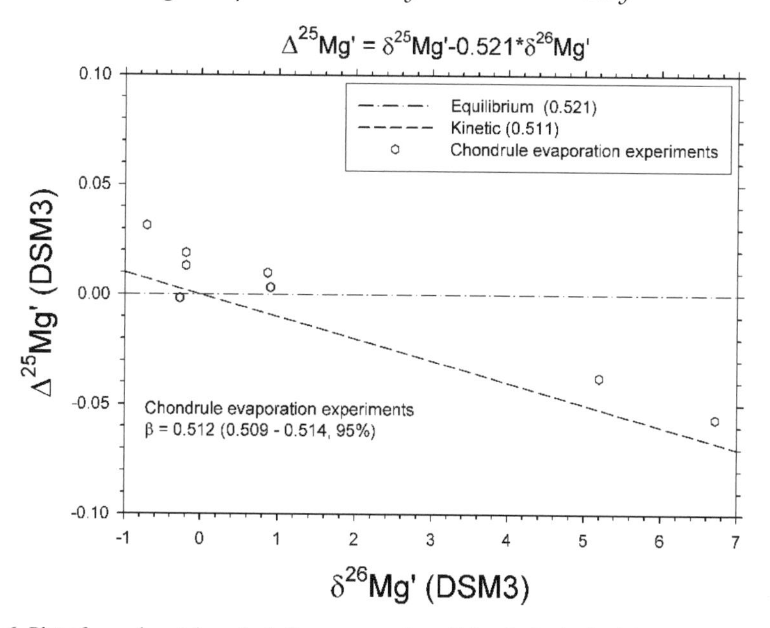

Figure 6. Plot of experimental products from evaporation of chondrules in the laboratory (Young et al. 2002b). The four data points below $\delta^{26}Mg' = 0.0$ are the starting materials. The evaporative residues exhibit heavy isotope enrichment along a kinetic mass fractionation path, verifying the prediction that evaporation should follow the kinetic mass fractionation law. The positive offset in $\Delta^{25}Mg'$ relative to DSM3 is a characteristic of the meteorite starting materials.

value of 0.511. The 95% confidence range in β for the evaporation experiments is 0.509–0.514 (Table 3) with an MSWD of 0.7, meaning that this kinetic β can be resolved with confidence from an equilibrium β value of 0.521. The intercept of the best-fit is 0.039‰, reflecting the positive $\Delta^{25}Mg'$ of the starting materials.

Extraction of β values from speleothems and waters apparently confirms the suspicion that Mg isotope partitioning during inorganic carbonate precipitation from waters is dominantly an equilibrium process (Fig. 7, Table 3). Linear regression of the $\delta^{25}Mg'$ and $\delta^{26}Mg'$ values for speleothem carbonates and associated drip waters and ground water from the Soreq cave site, Israel (Galy et al. 2002) yields a β of 0.519 ± 0.0015 (2σ) with the range being 0.517–0.520 at the 95% confidence level. Carbonate and drip water data for the Mitzpe Shlagim cave site, Israel, reported by Galy et al. (not shown in Fig. 7) yields a β of 0.522 ± 0.004 (2σ) and a 95% confidence range of 0.517–0.527. These values are near or indistinguishable from the equilibrium β of 0.521 and statistically far removed from the pure kinetic value of 0.511. Caution is warranted in the interpretation of these results because an equilibrium β does not exclude the possibility of a Rayleigh-like multi-step fractionation history in which the fractionation at each step is an equilibrium partitioning of the isotopes. The simplest interpretation, however, is that waters and their inorganic carbonate precipitates approach Mg isotopic equilibrium.

The β value defined by limestones, dolostones, marbles and seawater is consistent with equilibrium fractionation between aqueous Mg and carbonate Mg (Fig. 8, Table 3). The best-fit β is 0.524 ± 0.006 (2σ) with the 95% confidence range of 0.517–0.531. The data included in the regression represent an eclectic group of rocks, including three samples of marble (with the lowest $\Delta^{25}Mg'$ values near chondritic). However, there is some support for the equilibrium

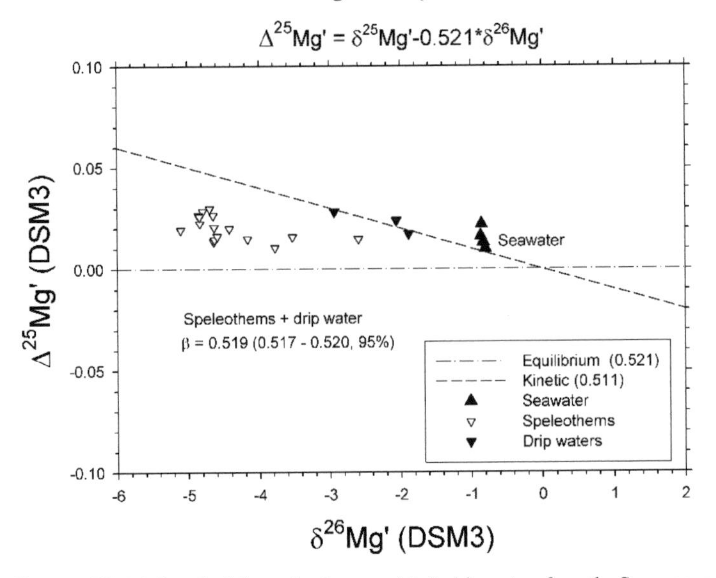

Figure 7. $\Delta^{25}Mg'$ vs. $\delta^{26}Mg'$ plot of calcite speleothems and their drip waters from the Soreq cave site, Israel (data from Galy et al. 2002) compared with seawater. The horizontal trend of the data suggests that Mg in carbonates is related to aqueous Mg by equilibrium fractionation processes. Results of a three-isotope regression, shown on the figure and in Table 3, confirm that the β value defined by the data is similar to the predicted equilibrium value of 0.521 and distinct from kinetic values. The positive $\Delta^{25}Mg'$ characteristic of the speleothem carbonates is apparently inherited from the waters. The positive $\Delta^{25}Mg'$ values of the waters appear to be produced by kinetic fractionation relative to primitive terrestrial Mg reservoirs (the origin).

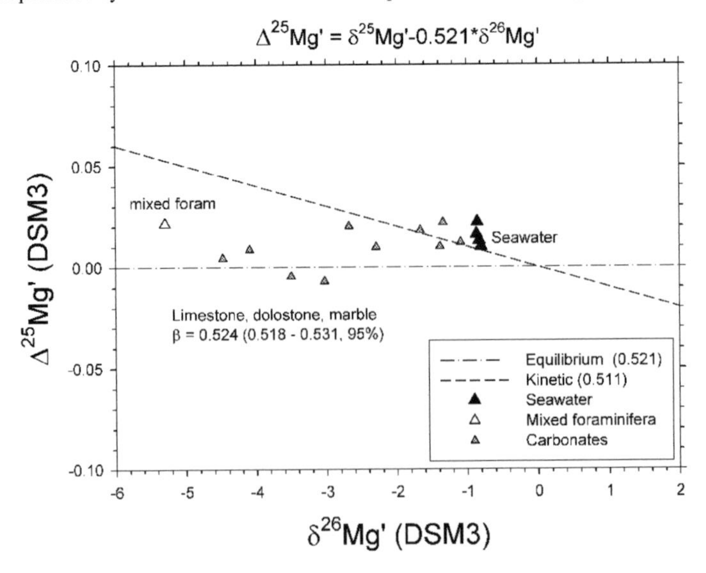

Figure 8. $\Delta^{25}Mg'$ vs. $\delta^{26}Mg'$ plot of limestone, dolostone, and marble samples (data from Galy et al. 2002) compared with a sample of foraminifera of various species (Chang et al. 2003) and seawater (Chang et al. 2003, this study). The broadly horizontal trend of the carbonates at elevated $\Delta^{25}Mg'$ suggests a component of equilibrium fractionation relative to seawater. The β value derived by regression of these $\delta^{25}Mg'$ and $\delta^{26}Mg'$ data is within the range for equilibrium fractionation and statistically distinguishable from purely kinetic fractionation.

relationship between marine carbonates and seawater from the mixed-species foraminifera analysis reported by Chang et al. (2003). Figure 8 shows that the mixed foraminifera and seawater have similar positive $\Delta^{25}Mg'$ values, meaning that the fractionation relationship between them is essentially equilibrium. However, it is also true that the foraminifera-seawater fractionation is the largest terrestrial difference observed to date and so vital effects might be expected. More work is required!

The analysis of fractionation law exponents quantifies the impression from the Δ'-δ' plots that aqueous Mg is related to primitive mantle and average crustal Mg by kinetic processes while carbonates precipitated from waters approach isotopic equilibrium with aqueous Mg. In any case, the positive $\Delta^{25}Mg'$ values of carbonates relative to the primitive chondrite/mantle reservoir and crust is a robust feature of the data and requires a component of kinetic Mg isotope fractionation prior to carbonate formation, as illustrated schematically in Figure 3.

Magnesium metal and chlorophyll Mg isotope ratios (lumped together because both are expected to exhibit kinetic effects) reported by Galy et al. (2001) define a best-fit β value of 0.512 ± 0.002 (2σ) and a 95% confidence range of 0.510 to 0.514 (Fig. 9, Table 3). This β is indistinguishable from the nominal kinetic value of 0.511 and statistically distinct from the equilibrium value of 0.521, and it is evidence for kinetic fractionation of the Mg isotopes among these materials, as one might expect from them. Again, more work is required, but these first results show that there is potential for distinguishing equilibrium from kinetic Mg exchange in natural systems using isotopes.

EXTRATERRESTRIAL MAGNESIUM ISOTOPE RESERVOIRS IN THREE-ISOTOPE SPACE

The meteorite data (Fig. 10) are not related by simple mass fractionation. They exhibit a clear negative trend in $\Delta^{25}Mg'$ -$\delta^{26}Mg'$ space that can be explained by a combination of the presence of excess $^{26}Mg^*$ due to decay of short-lived and extinct ^{26}Al and a mass-fractionation component (cf., Fig. 4). This is most easily seen on a plot of $\Delta^{25}Mg'$ vs. $\delta^{25}Mg'$ where decay of ^{26}Al in the absence of fractionation would result in a vertical trend (Fig. 11). The trend is dominated by the Allende chondrules with negative $\Delta^{25}Mg'$ in the data set but there are hints (e.g., the Bjurbole chondrules) that the trend may extend to positive $\Delta^{25}Mg'$ as well. A positive $\Delta^{25}Mg'$ may signify that bulk Earth has more radiogenic ^{26}Mg than these primitive samples. Alternatively, it may mean that there are small anomalies in the Mg isotopic system (e.g., excesses in ^{24}Mg) at the 0.03‰ level.

The simplest interpretation of the linear trend in Figure 10 (and 11) is that it is a mixing relationship. One endmember of the mixing trend is surely CAI material for which there is overwhelming evidence for *in situ* decay of ^{26}Al from an Al reservoir with an initial $^{26}Al/^{27}Al$ of about 5×10^{-5} (MacPherson et al. 1995), although the precise value for the initial $^{26}Al/^{27}Al$ recorded by CAIs is debated (see the section on laser ablation). The evidence for *in situ* decay in the other materials is not so clear. For all of the objects in the plot there must be just the right combination of *in situ* ^{26}Al decay and mass fractionation such that the combination of the two phenomena results in a strong correlation between $^{26}Mg^*$ and mass fractionation. Mixing appears to be the only plausible process capable of achieving this end unless one resorts to highly strained hypotheses. The single existing high-precision analysis of a bulk CAI (from Allende) published at the time of this writing has high $\delta^{26}Mg'$ and low $\Delta^{25}Mg'$ due to the presence of highly fractionated Mg and $^{26}Mg^*$ (Galy et al. 2000). The trend of the other chondrite data is in general from the origin towards the CAI datum (Fig. 12). Mixing between the highly fractionated CAI materials containing $^{26}Mg^*$ and more magnesian materials with no $^{26}Mg^*$ could produce the negative correlation between $\Delta^{25}Mg'$ and $\delta^{26}Mg'$. Mixing in this plot should produce a straight line. In detail the extension of the linear trend misses the one

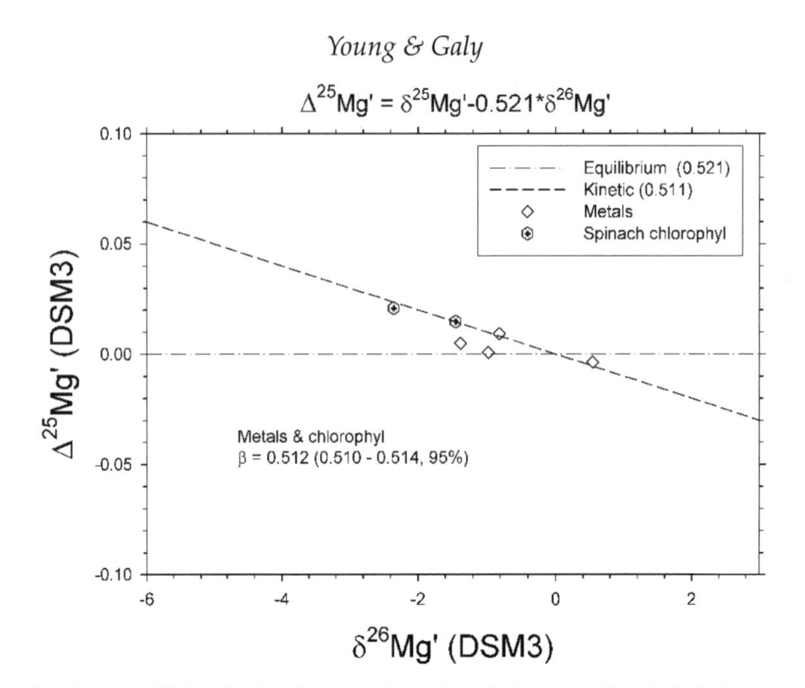

Figure 9. $\Delta^{25}Mg'$ vs. $\delta^{26}Mg'$ plot showing that Mg metals and Mg from chlorophyll, both representing Mg that is likely to be fractionated by kinetic processes relative to primitive terrestrial Mg reservoirs (the origin), follow a kinetic fractionation path relative to the origin. The β value derived by regression of these $\delta^{25}Mg'$ and $\delta^{26}Mg'$ data is statistically indistinguishable from the kinetic value of 0.511 and clearly resolved from the equilibrium value of 0.521.

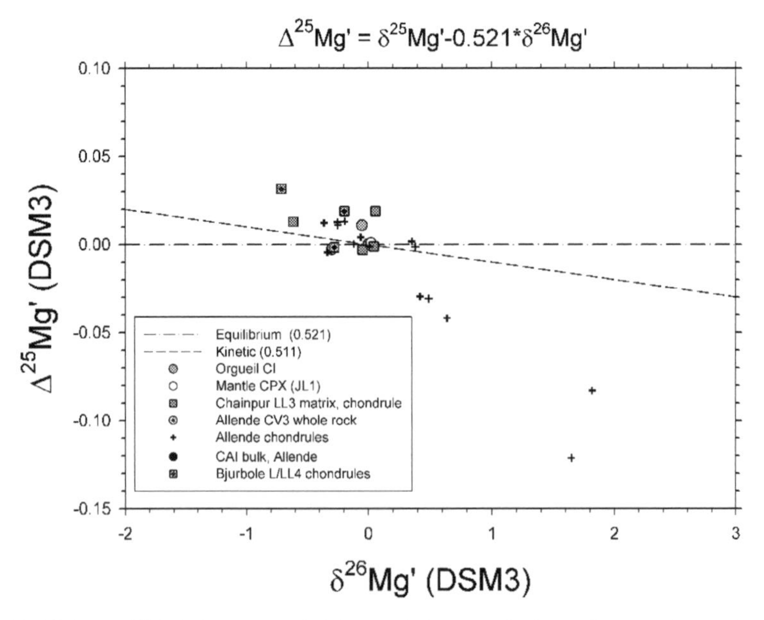

Figure 10. $\Delta^{25}Mg'$ vs. $\delta^{26}Mg'$ plot showing that the chondrite meteorite data define a trend inconsistent with mass-dependent fractionation. A single whole-rock CAI sample with the lowest $\Delta^{25}Mg'$ value plots off the diagram to the lower right.

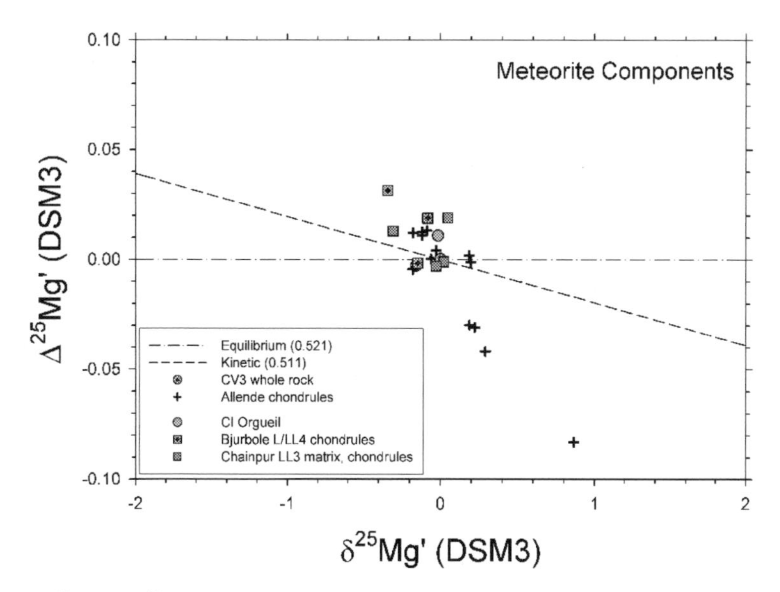

Figure 11. Δ^{25}Mg′ vs. δ^{25}Mg′ plot similar to Figure 10. In this diagram addition of δ^{26}Mg* by ^{26}Al β+ decay results in a vertical trend. The trend of the data shows that they can only be explained by a *combination* of mass-dependent fractionation and addition of ^{26}Mg*. The regularity of the trend among different chondritic components argues for a mixing line.

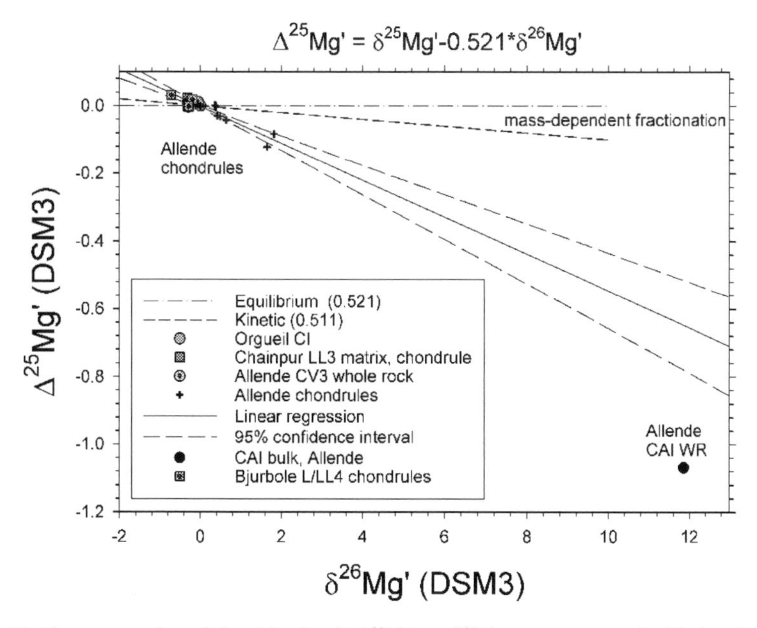

Figure 12. Linear regression of chondrite data in Δ^{25}Mg′ vs. δ^{26}Mg space compared with the whole-rock composition of an Allende CAI.

existing CAI datum. However, although we have only one whole-rock CAI sample analyzed by MC-ICPMS there are enough existing TIMS, secondary ion mass spectrometry (SIMS), and laser ablation MC-ICPMS data (published and unpublished) to safely say that CAIs do not have a unique bulk Mg isotopic composition (Clayton et al. 1988; Clayton and Mayeda 1977; Goswami et al. 1994; Kennedy et al. 1997; Russell et al. 1998; Wasserburg et al. 1977; Young et al. 2002a). In order to evaluate the likelihood that all of the variability in $\Delta^{25}Mg'$ seen in Figures 10–12 is due to mixing with CAIs, many more CAI whole rock analyses will be required.

If mixing between CAIs is the cause of the variable $\Delta^{25}Mg'$ among the Allende chondrules and other chondrite constituents shown here, then the departures of the chondrule and matrix data from mass fractionation is not indicative of *in situ* growth of extinct ^{26}Al in these objects but is instead due to inheritance of the product $^{26}Mg^*$ from ^{26}Al decay in the CAIs. The extent to which these high-precision Mg isotope data can be interpreted in terms of mixing rather than *in situ* decay of ^{26}Al was discussed by Galy et al. (2000). Indeed the origin of $^{26}Mg^*$ in Al-rich chondrules has been debated for years. The question is the extent to which the presence of $^{26}Mg^*$ in these non-CAI objects can be used to infer information about the timing of their formation. The problem is exacerbated by the fact that the mixing between Mg-rich chondritic materials and the Al-rich, Mg-poor and $^{26}Mg^*$-rich CAIs could result in correlations between $^{26}Mg^*$ and Al/Mg (Fig. 13). Such correlations are regarded as tell-tale signs of in situ decay of ^{26}Al; mixing lines resemble isochrons on ^{25}Mg-^{26}Al evolution diagrams and yet have no chronological meaning for the early Solar System.

A test of the hypothesis that the linear trends in Figures 10–12 are due to mixing comes from comparisons with oxygen isotopes. It has been conventional wisdom that $^{26}Mg^*$ does not correlate in detail with ^{16}O excesses in primitive chondrite components, including CAIs. However, combining the new highly precise MC-ICPMS data with isotope ratio data for the

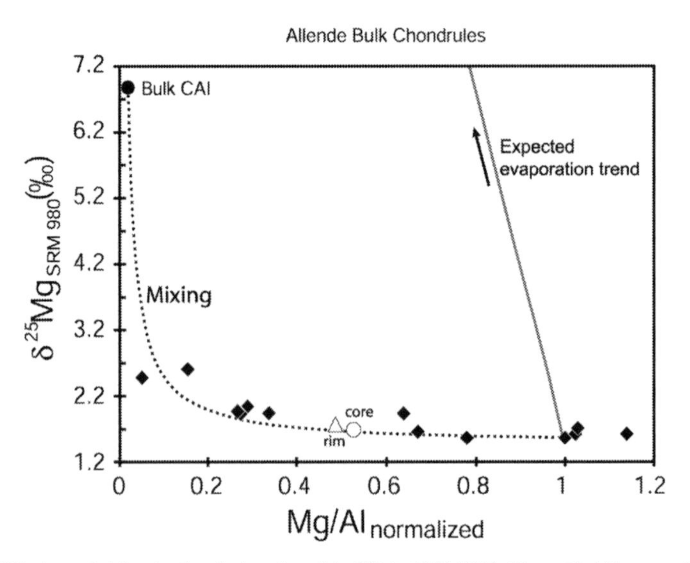

Figure 13. Whole-rock Allende chondrules plotted in $\delta^{25}Mg$ (SRM 980_O) vs. Mg/Al space showing that a mixing curve (dotted line) between the CAI whole-rock sample and the Mg-rich, Al-poor chondrules (i.e., "chondritic" chondrules) fits the trend defined by the bulk chondrule data (modified after Galy et al. 2000). The expected evaporation trend is shown for comparison and represents the consequences of free evaporation of chondrules in space. The evaporation curve is described in greater detail by Galy et al. (2000).

same materials suggests that such a correlation may exist. Figure 14 shows $\Delta^{25}Mg'$ data for components of various carbonaceous chondrite and ordinary chondrite meteorites, including Allende (CV3) matrix, Allende chondrules, Orgueil (CI) whole rocks, Chainpur (LL3) matrix/whole rock, and Bjurbole (L/LL4) chondrules collected by MC-ICPMS compared with $\Delta^{17}O$ (a measure of ^{16}O excess) obtained by laser ablation of the same materials in the author's laboratory or from the literature (Table 4). There is an overall correspondence between low $\Delta^{17}O$ (high concentrations of excess ^{16}O) and low $\Delta^{25}Mg'$ (high concentrations of excess ^{26}Mg) evident in these data. Addition of the CAI datum under the assumption that the CAI has a $\Delta^{17}O$ corresponding to the most enriched whole rocks (e.g., Clayton et al. 1977) (an assumption that may be justified if the $\Delta^{17}O$ of many CAIs was increased due to alteration either in the solar nebula or on the meteorite parent bodies) completes the trend (Fig. 15). The data resemble a hyperbolic curve in $\Delta^{25}Mg'$ - $\Delta^{17}O$ space. Hyperbolic mixing curves would result if the two endmember components had different Mg/O ratios.

The possibility for mixing between a CAI-like endmember characterized by a ^{16}O excess, a low Mg/O, and high $^{26}Mg^*$ (call it "CAI") with a "chondritic" endmember that is low in ^{16}O excess, higher in Mg/O, and has no $^{26}Mg^*$ is quantified by overlaying the data with hyperbolic mixing curves between these endmembers in Figure 15. The curvature of the mixing lines depends on $(Mg/O)_{(CAI)}/(Mg/O)_{(chondritic)}$. The calculation assumes a single composition for both endmembers, an oversimplification to be sure. Nonetheless, with this assumption the data are explained by $(Mg/O)_{(CAI)}/(Mg/O)_{(chondritic)}$ ranging from 0.2–0.02 (Fig. 15). These values are not unreasonable in view of the different chemical compositions of the Mg-rich "chondritic" objects like chondrules and matrix and the Mg-poor CAIs.

Many more high-precision whole-rock analyses of both Mg and O isotope ratios for chondrite components would help to establish whether or not the mixing trends in Figure 15 are valid, or even if the concept of mixing is useful. Bulk objects are desirable because their isotopic compositions are not affected by secondary inter-mineral exchange reactions that are know to be important for both the Mg and O isotopic systems in CAIs and chondrules.

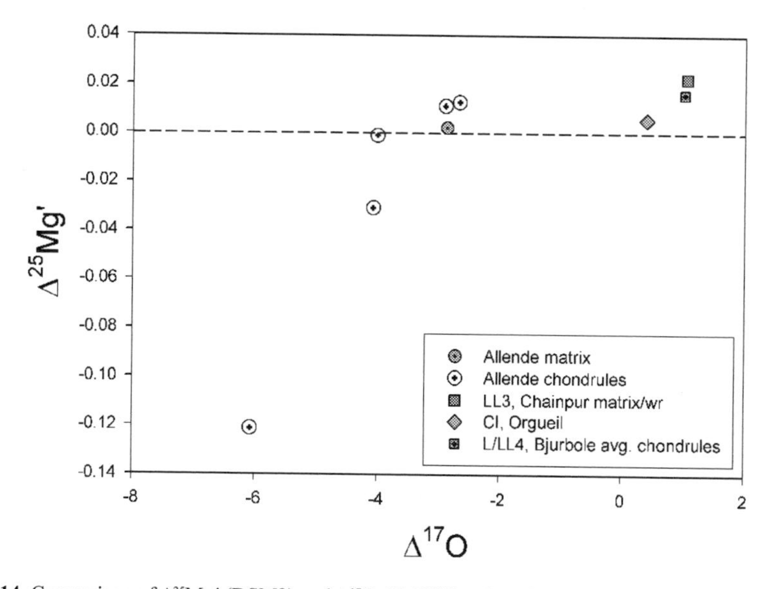

Figure 14. Comparison of $\Delta^{25}Mg'$ (DSM3) and $\Delta^{17}O$ (SMOW) values for various chondrules and whole-rock chondrite samples. The sources of the oxygen isotope data are given in Table 3. Aqueous alteration on this diagram will move points approximately horizontally on this diagram.

Table 4. Tabulation of $\Delta^{25}Mg'$ vs. $\Delta^{17}O$ for various chondrite components. $\Delta^{17}O$ laser ablation data are shown explicitly to demonstrate that averages are reflective of the bulk objects. Where data are few, the objects are small in comparison to laser beam diameter.

Sample	$\Delta^{25}Mg'$ [5]	$\Delta^{17}O$
Allende (CV3) matrix	0.002	−2.87 [1]
Allende chondrule A3	0.011	−2.48 −3.42 −2.58 −3.17 −3.02 −3.25 −3.23 −3.17 −2.31 −2.43 −2.92 −3.02 −2.65 −2.99 Avg = −2.90±0.35 1σ [2]
Allende chondrule A4	−0.121	−6.5 −5.7 −6 Avg = −6.07±0.40 1σ[2]
Allende chondrule A5	−0.031	−4.5 −4 −3.5 −4.3 Avg = −0.408±0.44 1σ [2]
Allende chondrule A6	−0.001	−3.8 −3.4 −4.5 −3.7 −4.3 −3.9 −3.9 −4.8 −3.8 Avg = −4.01±0.44 1σ [2]
Allende chondrule A7	0.013	−2.9 −2.7 −2.4 Avg = −2.67±0.25 1σ [2]
Allende CAI whole rock	−1.07	−19.2 [1]
Orgueil (CI)	0.006±0.001 1σ	0.39 [2]
Chainpur (LL3) matrix/wr	0.022	1.04 [4]
Bjurbole (L/LL4) avg. chondrules	0.016±0.017 1σ	1.00 [4]

[1] Data from Clayton and Mayeda 1999
[2] Data from Young et al. 2002a
[3] Value based on the assumption that the CAI has the maximum ^{16}O enrichment typical for Allende
[4] Data from Clayton et al. 1991
[5] Uncertainties on the order of 0.01 as discussed in text.

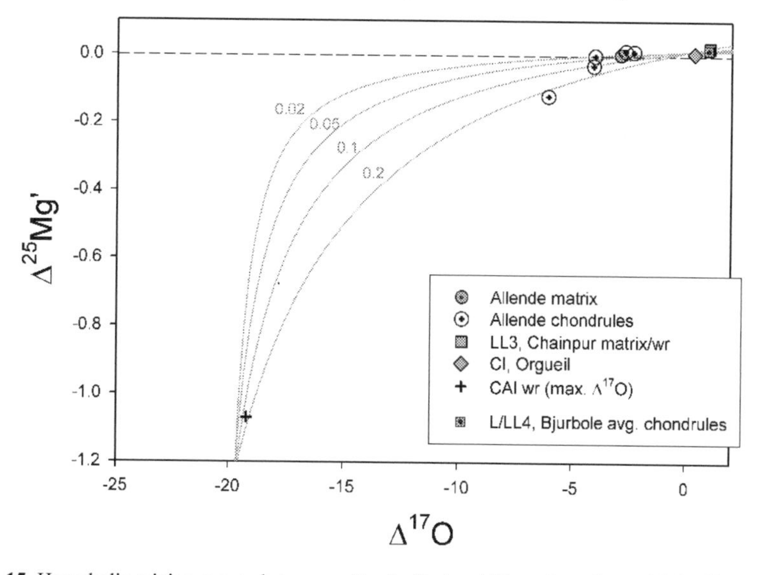

Figure 15. Hyperbolic mixing curves between a "typical" chondritic endmember at high Δ^{17}O (SMOW) and Δ^{25}Mg′ (DSM3) and a "typical" CAI composition at low Δ^{17}O and Δ^{25}Mg′ compared with existing high-precision data. The curvature of the mixing lines is a function of $(Mg/O)_{CAI}/(Mg/O)_{chondrite}$ where "CAI" and "chondrite" refers to the two endmember compositions. Each curve is labeled with the corresponding $(Mg/O)_{CAI}/(Mg/O)_{chondrite}$ values.

The whole-rock samples are not immune from other open-system exchange, however. For example, we do not yet know how aqueous alteration manifests itself on these diagrams because detailed studies of the Mg isotopic effects of water-rock reactions have not been carried out at the time of this writing.

ISOTOPIC COSMOBAROMETRY

The vast majority of the Mg isotope variability seen in the components of primitive meteorites and their constituents is apparently attributable to mixing with the highly mass fractionated and radiogenic Mg in CAIs (Fig. 10). Even if the mixing hypothesis proves to be incorrect, the amount of variability among chondrules is small in comparison to what one would expect. This is because the chondrules were once molten spheres in space and the expectation is that they should show evidence for evaporation. However, the high-precision Mg isotope data do not follow the trends expected if the molten spheres were present in a near vacuum. The lack of isotope fractionation in chondrules is not peculiar to Mg and is observed in K and Fe as well (Alexander et al. 2000; Galy et al. 2000; Alexander and Wang 2001; Zhu et al. 2001). It contrasts with the large fractionation effects seen in CAIs (Clayton et al. 1988). The general lack of light-element mass fractionation in chondrules has been a vexing problem because these objects show evidence for volatility-controlled variations in the concentrations of the very same elements lacking isotope fractionation, at odds with the expectations for free evaporation (Fig. 13) (Galy et al. 2000).

Two theories exist for the lack of isotope fractionation in chondrules. Both rely on the concept of quasi-statical (i.e., almost equilibrium) exchange of isotopes between liquid and gas followed by permanent loss of rock-forming elements to the escaping gas phase. One model (Alexander 2003) implies that the chondrules formed in such close proximity to one another that the partial pressures of the rock-forming elements in the surrounding ambient gas approached saturation (Fig. 16). The other (Richter et al. 2002) does not rely upon widespread saturation of the gas. Instead it allows that the liquid chondrules exchanged isotopes with the gas evaporating from the object. The local atmosphere of gas is bound to its chondrule by slow diffusion through the ambient gas (Fig. 16). The differences between these two physical situations can be seen with reference to an equation describing the net flux of a component i (e.g., Mg) volatilizing from a molten sphere (Richter et al. 2002):

$$J_{i,\text{net}} = \frac{J_{i,\text{evap}}\left(1 - \frac{P_{i,\infty}}{P_{i,\text{sat}}}\right)}{1 + \frac{\gamma_i r}{D_{i,\text{gas}}}\sqrt{\frac{RT}{2\pi m_i}}} \tag{20}$$

or

$$J_{i,\text{net}} = \frac{J_{i,\text{evap}}\left(1 - \frac{P_{i,\infty}}{P_{i,\text{sat}}}\right)}{1 + \Gamma} \tag{21}$$

where $J_{i,\text{net}}$ is the net difference between the evaporative and condensation fluxes for species i, $J_{i,\text{evap}}$ is the evaporative flux of i, $P_{i,\infty}$ is the partial pressure of species i far removed from the molten object, $P_{i,\text{sat}}$ is the saturation partial pressure of i, $D_{i,\text{gas}}$ is the diffusion coefficient of i through the gas phase, γ_i is the evaporation factor for i, r is the radius of the molten object, and m_i is the mass of the volatilizing species. In terms of this equation, isotope fractionation will occur when $J_{i,\text{net}}/J_{i,\text{evap}} \rightarrow 1$ while no fractionation occurs when $J_{i,\text{net}}/J_{i,\text{evap}} \rightarrow 0$. The latter occurs where the background pressure $P_{i,\infty}$ approaches the saturation pressure $P_{i,\text{sat}}$. This could be

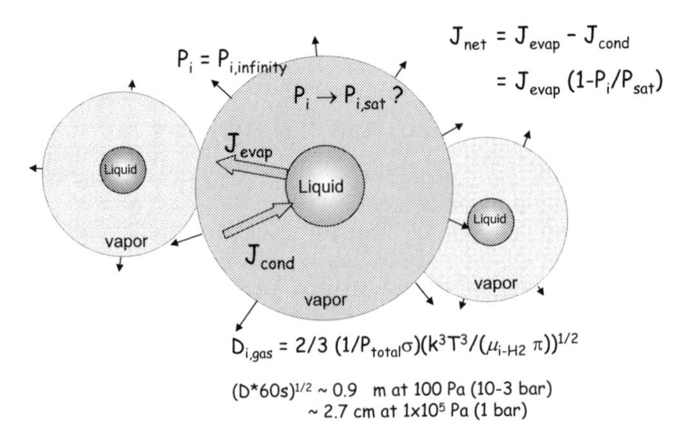

Figure 16. Schematic illustration of envelopes of gas species i, in this case Mg, surrounding a volatilizing molten chondrule in space. The size of the gas envelope is a function of ambient background pressure P_∞ by virtue of the effect that pressure has on the gas molecule diffusivity D_i. The diffusion coefficient can be calculated from the kinetic theory of gases, as shown. The level of isotopic fractionation associated with volatilization of the molten chondrule depends upon the balance between the evaporative flux J_{evap} and the condensation flux J_{cond}. When the fluxes are equal (i.e., when $J_{net} = 0$), there is no mass-dependent isotope fractionation associated with volatilization. This will be the case when the local partial pressure P_i approaches the saturation partial pressure $P_{i,sat}$

the case where a population of volatilizing chondrules contributes to an elevated background of partial pressure of rock-forming elements as envisioned by Alexander (2003). In the other extreme where $P_{i,\mathrm{sat}} >> P_{i,\infty}$, as would be the case where chondrules are not sufficiently close together to cause a pervasive rise in background partial pressures, the only way to prevent fractionation is for Γ to be large, as suggested by Galy et al. (2000). The kinetic theory of gases permits one to compute Γ as a function of pressure. Such calculations show that Γ is large enough to preclude fractionation when total pressures approach 1 bar (e.g., at T = 2000 K and $\gamma_i = 0.06$, $\Gamma = 0.062$ at 10^{-3} bar and 62 at 1 bar for a chondrule-sized object).

From the preceding discussion it should be clear that the level of isotopic fractionation for a major rock-forming element like Mg in an object that was once molten in the early Solar System is a barometer of either total pressure (large Γ) or partial pressures relative to saturation (small Γ). In order to use Mg as a "cosmobarometer," high-precision analyses of Mg isotope ratios in both chondrules and CAIs are required.

CAI's that were once molten (type B and compact type A) apparently crystallized under conditions where both partial pressures and total pressures were low because they exhibit marked fractionation of Mg isotopes relative to "chondritic" isotope ratios. But much remains to be learned from the distribution of this fractionation. Models and laboratory experiments indicate that Mg, O, and Si should fractionate to different degrees in a CAI (Davis et al. 1990; Richter et al. 2002) commensurate with the different equilibrium vapor pressures of Mg, SiO and other O-bearing species. Only now, with the advent of more precise mass spectrometry and sampling techniques, is it possible to search for these differences. Also, models predict that there should be variations in isotope ratios with growth direction and Mg/Al content in minerals like melilite. Identification of such trends would verify the validity of the theory. Conversely, if no correlations between position, mineral composition, and Mg, Si, and O isotopic composition are found in once molten CAIs, it implies that the objects acquired their isotopic signals prior to final crystallization. Evidence of this nature could be used to determine which objects were melted more than once.

From the high-precision Mg isotope data (and other K and Fe isotope data) available to date, it appears that chondrules crystallized under conditions of either high Γ, meaning high total pressure, or in close proximity to one another where $P_{i,\mathrm{sat}} \sim P_{i,\infty}$. There are some mass-dependent Mg isotopic heterogeneities in some chondrules when the objects are examined using laser ablation (see below) (Young et al. 2002a). We do not yet know how fractionation relates to mineralogy nor do we know if different populations of chondrules have different degrees of isotopic variability. More data are required before we can have confidence in the meaning of heterogeneity in Mg isotope fractionation in chondrules.

IN SITU LASER ABLATION

Young et al. (2002a) showed that ultraviolet laser ablation combined with MC-ICPMS (LA-MC-ICPMS) can offer advantages over other methods of spatially resolved Mg isotopic analysis of meteorite materials. They collected data for chondrules and a CAI from the Allende meteorite. Each datum in that study represents approximately 2.8 μg of material (based on a laser spot diameter of 100 μm and laser pit depth of 30 μm; depths are uncertain to ± 20%).

Allende chondrules exhibit internal mass-dependent variations in δ^{25}Mg values that deviate from the "canonical" chondritic value (we focus on ^{25}Mg/^{24}Mg here to avoid the ambiguities surrounding ^{26}Mg*) (Young et al. 2002a). One reason for this variation might be secondary alteration that appears to be associated with low ^{25}Mg/^{24}Mg relative to the matrix of the meteorite. The other is that portions of chondrules containing abundant relic grains of refractory olivine have high ^{25}Mg/^{24}Mg relative to matrix.

Alteration in Allende chondrule C6 is concentrated at the margins where δ^{25}Mg values are low (Fig. 17). Evaluation of the likelihood that low δ^{25}Mg values could be the result of aqueous alteration will require studies of Mg isotope fractionation during low-T alteration of terrestrial mafic rocks. An alternative explanation is that low δ^{25}Mg and alteration resulted from condensation (collision frequency of gaseous Mg is greater for the lighter isotopes).

Laser spots from unaltered parts of a CAI (USNM 3576-1) from Allende have the same ^{25}Mg/^{24}Mg ratio within analytical uncertainties with a mean δ^{25}Mg value of 5.3‰ ± 0.1 (2σ) on the SRM 980_O scale (Fig. 18), corresponding to +3.54‰ on the DSM3 scale. The laser ablation spot that sampled alteration at fassaite (Ti-Al-rich diopside)-anorthite grain boundaries yields a higher δ^{25}Mg of 5.6‰ (3.84‰ DSM3) (Fig. 18). These same laser spot samples vary in δ^{26}Mg by 1‰ (Fig. 18) as the result of in-growth of ^{26}Mg by decay of ^{26}Al.

Uniformity in δ^{25}Mg among unaltered phases in CAI USNM 3576-1 is noteworthy because these high-precision analyses were obtained without normalization to a single ^{25}Mg/^{24}Mg value. Previous measurements of δ^{26}Mg* (δ^{26}Mg* is the excess in δ^{26}Mg relative to mass fractionation shown by the dashed line in Fig. 18) in CAIs were all made either by TIMS or SIMS. A uniform δ^{25}Mg is imposed on the TIMS data because the high-precision δ^{26}Mg* values are obtained by normalizing to a single "terrestrial" ^{25}Mg/^{24}Mg. The detection limit for δ^{26}Mg* by SIMS is sufficiently high that δ^{26}Mg* can only be detected with certainty in the most aluminous phases, including gehlenitic melilite (Al/Mg > 10), hibonite (Al/Mg > 15), and feldspar (Al/Mg > 200) (Sahijpal et al. 1998). New innovations in SIMS techniques (e.g., multi-collection) show promise for lowering detection limits, but at present the laser ablation MC-ICPMS data are unique in that they show unequivocal δ^{26}Mg* in the low Al/Mg minerals, including fassaite (Ti-Al-rich diopside) and spinel. A spread in δ^{26}Mg values at constant δ^{25}Mg among the low-Al minerals sampled by the laser, together with correlation between δ^{26}Mg* and ^{27}Al/^{24}Mg (Fig. 19), is clear indication that ^{26}Mg was added independently of ^{25}Mg by decay of ^{26}Al in these minerals. Analyses with the greatest Al and highest δ^{26}Mg* values in this study come from the gehlenitic melilite typical of the margin of the CAI.

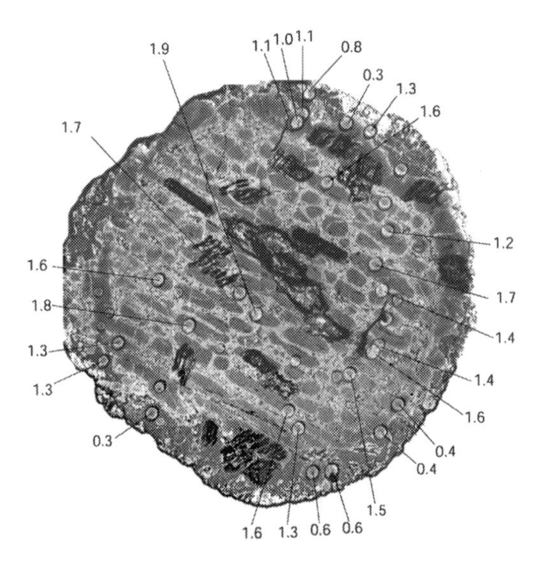

Figure 17. A backscattered electron image of a barred olivine chondrule from the Allende meteorite (chondrule C6l, after Young et al. 2002a) showing the locations of ultraviolet laser ablation pits (each pit is 100 μm in diameter) and the δ^{25}Mg values (SRM 980_O) obtained from the pits by LA-MC-ICPMS. Darkest areas around the margins are pyroxenes. Olivine bars appear as medium gray tones. The brighter mottled areas are altered mesostasis interstitial to the olivine bars. Large dark trenches are vestiges of oxygen isotope laser ablation pits.

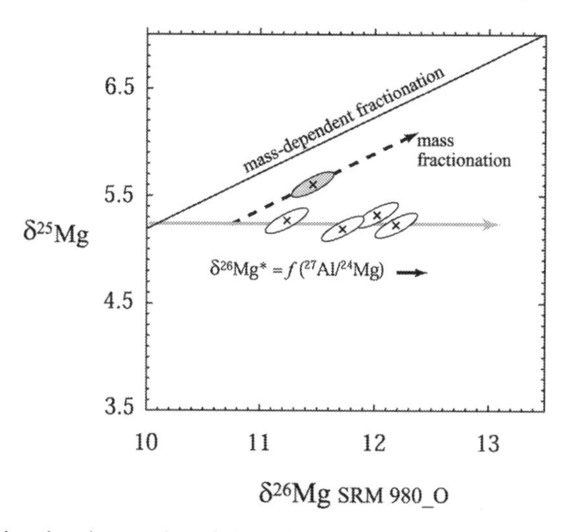

Figure 18. Magnesium three-isotope plot (relative to SRM 980_O) showing laser ablation data for Allende CAI 3576-1 (after Young et al. 2002a). Ellipses represent the 95% confidence for each datum. The shaded datum is the analysis that included alteration material in the CAI. This point is related to the others by mass fractionation (dashed line) at constant $\delta^{26}Mg^*$ where $\delta^{26}Mg^*$ is the horizontal deviation from a terrestrial mass-fractionation curve. The $\delta^{26}Mg^*$ value depends on the Al/Mg ratio, indicating *in situ* decay of ^{26}Al in the CAI. The high precision of the MC-ICPMS analyses makes it possible to resolve mass-dependent fractionation from excesses in ^{26}Mg at the sub-per mil level.

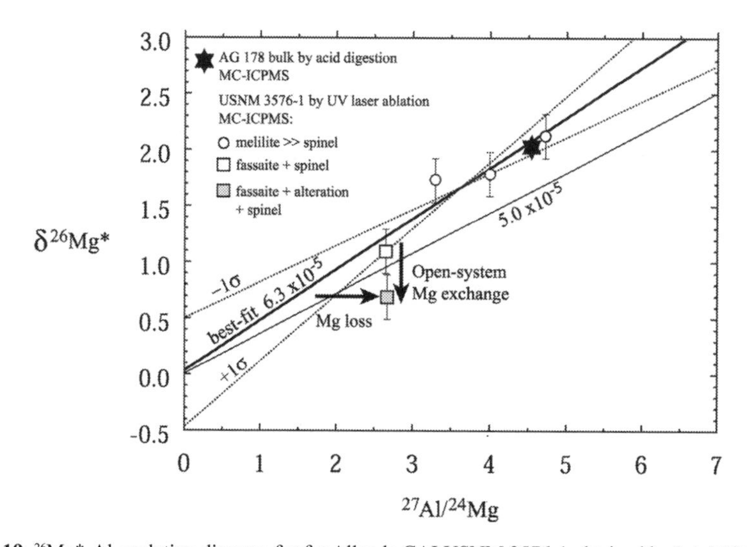

Figure 19. $^{26}Mg^*$-Al evolution diagram for for Allende CAI USNM 3576-1 obtained by LA-MC-ICPMS. Linear regression of the data in which alteration is absent (open symbols) yields an estimate of the initial $^{26}Al/^{27}Al$ for the sample (6.3×10^{-5}) based on the equation $\delta^{26}Mg^* = (^{26}Al/^{27}Al)_{initial}\ (10^3/(^{26}Mg/^{24}Mg\ )_{SRM\ 980})$ $(^{27}Al/^{24}Mg)$ and the $(^{26}Mg/^{24}Mg)_{SRM\ 980}$ value of Catanzaro et al. (1966). This estimate is indistinguishable from the value suggested by comparing the origin with a whole-rock sample of an Allende CAI measured by MC-ICPMS and reported by Galy et al. (2000). The uncertainties in the whole rock value are smaller than the star symbol. The position of the datum that includes alteration (grey symbol) requires either open-system exchange of Mg with an external reservoir or open-system loss of Mg from the alteration zone by leaching.

The distinct $\delta^{25}Mg$ value for the altered portion of the CAI suggests that alteration occurred in an open system. The data could be explained by closed-system Mg exchange if there was a reservoir within the CAI that had lower $^{26}Mg^*$ than the altered spot. However, there is no such reservoir; the altered material has the lowest Al/Mg present in the object. Open system deviations from a correlation line in Figure 19 are expected because Mg is generally more labile than Al, resulting in an increase in $^{27}Al/^{24}Mg$ at fixed $\delta^{26}Mg^*$ in Figure 19. Transition state theory predicts that leaching of Mg would likely be more efficacious for the light Mg isotope and a mass-dependent fractionation to higher $\delta^{25}Mg$ and $\delta^{26}Mg$ values at fixed $\delta^{26}Mg^*$ should occur as the isotopically lighter Mg is carried away.

On the ^{26}Al-$^{26}Mg^*$ evolution diagram analyses of unaltered material define a line with a slope corresponding to an initial $^{26}Al/^{27}Al$ of 6.3×10^{-5} and an intercept of $\delta^{26}Mg^* = 0.009$ (Fig. 19). The intercept is effectively zero despite the fact that the regression line was not forced to pass through the origin. The initial $^{26}Al/^{27}Al$, or $(^{26}Al/^{27}Al)^o$, of 6.3×10^{-5} defined by the unaltered minerals is identical to that suggested by the single acid-digestion MC-ICPMS measurement of a bulk CAI from Allende (Table 1) (Galy et al. 2000). Both the laser ablation data for unaltered minerals and the bulk CAI datum yield higher initial $(^{26}Al/^{27}Al)$ values than the canonical value of 5×10^{-5} (MacPherson et al. 1995). The uncertainty in the slope permits a lower $(^{26}Al/^{27}Al)^o$ consistent with the canonical value but only if the intercept approaches 0.5‰ rather than zero (Fig. 19). A non-zero intercept implies a significant departure from normal Mg isotope ratios in the Solar System (Fig. 2), and there is little evidence to support such a large intercept given the uniformity of Solar System materials (e.g., Fig. 2). For this reason the higher slope is the preferred interpretation of the data in Figure 19. Evaluation of the significance of the higher $(^{26}Al/^{27}Al)^o$ will require numerous measurements in a wide variety of CAIs (Young et al. 2002a).

These first experiments demonstrate the viability of using laser ablation in combination with MC-ICPMS as a tool for measuring Mg isotope ratios in-situ in extraterrestrial materials. The advantages are high analytical precision combined with in-situ sampling on the microgram scale, independent measurements of both $\delta^{25}Mg$ and $\delta^{26}Mg$ at the 0.2‰ (2σ) level of precision, and negligible matrix effects in Mg-rich minerals (Young et al. 2002a). In this particular application, it was shown that with this new tool it will be possible to determine the causes of dispersion on the ^{26}Al-$^{26}Mg^*$ evolution diagram with less ambiguity than with other methods, aiding the interpretation of the data in terms of relative chronology.

Future studies should make use of the combination of rapid analysis time and high precision afforded by laser ablation in order to compile larger data sets for the myriad of different objects found in meteorites.

SUMMARY

The high precision with which Mg isotope ratios can be measured using MC-ICPMS opens up new opportunities for using Mg as a tracer in both terrestrial and extraterrestrial materials. A key advance is the ability to resolve kinetic from equilibrium mass-dependent fractionation processes. From these new data it appears that Mg in waters is related to mantle and crustal reservoirs of Mg by kinetic fractionation while Mg in carbonates is related in turn to the waters by equilibrium processes. Resolution of different fractionation laws is only possible for measurements of Mg in solution at present; laser ablation combined with MC-ICPMS (LA-MC-ICPMS) is not yet sufficiently precise to measure different fractionation laws.

Variability in Mg isotope ratios among chondritic meteorites and their constituents is dominated by mixing between a radiogenic CAI-like reservoir and a reservoir resembling ordinary chondrites. The mixing is evident in $\delta^{25}Mg$ and $\delta^{26}Mg$, Al/Mg, and $\Delta^{17}O$ values but will require substantiation by collection of more MC-ICPMS Mg isotope data together with oxygen isotope and elemental compositions of bulk objects.

Laser ablation combined with LA-MC-ICPMS provides a new dimension to the analysis of Mg isotopes in calcium aluminum-rich inclusions from primitive meteorites. Dispersion in $^{26}Mg^*$-$^{27}Al/^{24}Mg$ evolution lines can be correlated with mass-dependent variations in $\delta^{25}Mg$ that distinguish open-system from closed-system processes. The ultimate product of such studies will be a better understanding of the chronological significance of variations in $^{26}Mg^*$ in these objects.

REFERENCES

Alexander CMOD (2003) Making CAIs and chondrules from CI dust in a canonical nebula. Lunar and Planetary Science Conference XXXIV:1391

Alexander CMOD, Grossman JN, Wang J, Zanda B, Bourot-Denise M, Hewins RH (2000) The lack of potassium-isotopic fractionation in Bishunpur chondrules. Meteor Planet Sci 35:859-868

Alexander CMOD, Wang J (2001) Iron isotopes in chondrules: implications for the role of evaporation during chondrule formation. Meteor Planet Sci 36:419-428

Cary EE, Wood RJ, Schwartz R (1990) Stable Mg isotopes as tracers using ICP-MS. J Micronutrient Anal 8(1):13-22

Catanzaro EJ, Murphy TJ, Garner EL, Shields WR (1966) Absolute isotopic abundance ratios and atomic weight of magnesium. J Res Natl Bur Stand 70A:453-458

Chang VT-C, Makishima A, Belshaw NS, O'Nions RK (2003) Purification of Mg from low-Mg biogenic carbonates from isotope ratio determination using multiple collector ICP-MS. J Anal At Spectrom 18: 296-301

Clayton RN, Hinton RW, Davis AM (1988) Isotopic variations in the rock-forming elements in meteorites. Philosoph Trans R Soc London, Ser A 325:483-501

Clayton RN, Mayeda TK (1977) Correlated oxygen and magnesium isotope anomalies in Allende inclusions, I: Oxygen. Geophys Res Lett 4(7):295-298

Clayton RN, Mayeda TK, Goswami JN, Olsen EJ (1991) Oxygen isotope studies of ordinary chondrites. Geochim Cosmochim Acta 55:2317-2337

Clayton RN, Mayeda TK (1999) Oxygen isotope studies of carbonaceous chondrites. Geochim Cosmochim Acta 63:2089-2103

Clayton RN, Onuma N, Grossman L, Mayeda TK (1977) Distribution of presolar component in Allende and other carbonaceous chondrites. Earth Planet Sci Lett 34:209-224

Davis AM, Hashimoto A, Clayton RN, Mayeda TK (1990) Isotope mass fractionation during evaporation of Mg_2SiO_4. Nature 347:655-658

Dombovari J, Becker JS, Dietze H-J (2000) Isotope ratio measurements of magnesium and determination of magnesium concentration by reverse isotope dilution technique on small amounts of ^{26}Mg-spiked nutrient solutions with inductively coupled plasma mass spectrometry. Int J Mass Spectrom 202:231-240

Fahey AJ, Goswami JN, McKeegan KD, Zinner E (1987) ^{26}Al, ^{244}Pu, ^{50}Ti, REE, and trace element abundances in hibonite grains from CM and CV meteorites. Geochim Cosmochim Acta 51(2):329-350

Galy A, Bar-Matthews M, Halicz L, O'Nions RK (2002) Mg isotopic composition of carbonate: insight from speleothem formation. Earth Planet Sci Lette 201:105-115

Galy A, Belshaw NS, Halicz L, O'Nions RK (2001) High-precision measurement of magnesium isotopes by multiple-collector inductively coupled plasma mass spectrometry. Int J Mass Spectrom 208:89-98

Galy A, O'Nions RK (2000) Is there a CHUR for Mg? Goldschmidt 2000, 424

Galy A, Yoffe O, Janney PE, Williams RW, Cloquet C, Alard O, Halicz L, Wadwha M, Hutcheon ID, Ramon E, Carignan J (2003) Magnesium isotopes heterogeneity of the isotopic standard SRM980 and new reference materials for magnesium-isotope-ratio measurements. J Anal At Spectrom 18:1352-1356

Galy A, Young ED, Ash RD, O'Nions RK (2000) The formation of chondrules at high gas pressures in the solar nebula. Science 290:1751-1753

Goswami JN, Srinivasan G, Ulyanov AA (1994) Ion microprobe studies of Efremovka CAIs: I. Magnesium isotope composition. Geochim Cosmochim Acta 58:431-447

Gray CM, Compston W (1974) Excess ^{26}Mg in the Allende meteorite. Nature 251:495-497

Gussone N, Eisenhauer A, Heuser A, Dietzel M, Bock B, Bohm F, Spero SJ, Lea DW, Bijma J, Nagler TF (2003) Model for kinetic effects on calcium isotope fractionation ($\delta^{44}Ca$) in inorganic aragonite and cultured planktonic foraminifera. Geochim Cosmochim Acta 67:1375-1382

Hsu W, Wasserburg GJ, Huss GR (2000) High time resolution by use of ^{26}Al chronometer in the multistage formation of a CAI. Earth Planet Sci Lett 182:15-29

Hulston JR, Thode HG (1965) Variations in the S^{33}, S^{34}, and S^{36} contents of meteorites and their relation to chemical and nuclear effects. J Geophys Res 70:3475-3484

Kennedy AF, Beckett JR, Edwards DA, Hutcheon ID (1997) Trace element disequilibria and magnesium isotope heterogeneity in 3655A: Evidence for a complex multi-stage evolution of a typical Allende type B1 CAI. Geochim Cosmochim Acta (7):1541-1561

Lee T, Papanastassiou DA (1974) Mg isotopic anomalies in the Allende meteorite and correlation with O and Sr effects. Geophys Res Lett 1:225-228

MacPherson GJ, Davis AM, Zinner EK (1995) The distribution of aluminum-26 in the early Solar System - a reappraisal. Meteoritics 30:365-386

Mahon KI (1996) The "new" York regression: application of an improved statistical method to geochemistry. Int Geol Rev 38:293-303

Matsuhisa J, Goldsmith JR, Clayton RN (1978) Mechanisms of hydrothermal crystallisation of quartz at 250 C and 15 kbar. Geochim Cosmochim Acta 42:173-182

McKinney CR, McCrea JM, Epstein S, Allen HA, Urey HC (1950) Improvements in mass spectrometers for the measurement of small differences in isotope abundance ratios. Rev Scientific Inst 21(8):724-730

Norman M, McCulloch M, O'Neill H, Brandon A (2004) Magnesium isotopes in the Earth, Moon, Mars, and Pallasite parent body: high precision analysis of olivine by laser ablation multi-collector ICPMS. Lunar and Planetary Science Conference XXXV:1447

Richter FM, Davis AM, Ebel DS, Hashimoto A (2002) Elemental and isotopic fractionation of type B calcium-aluminum-rich inclusions: Experiments, theoretical considerations, and constraints on their thermal evolution. Geochim Cosmochim Acta 66(3):521-540

Russell SS, Huss GR, Fahey AJ, Greenwood RC, Hutchison R, Wasserburg GJ (1998) An isotopic and petrologic study of calcium-aluminum-rich inclusions from CO3 meteorites. Geochim Cosmochim Acta 62(4):689-714

Sahijpal S, Goswami JN, Davis AM, Grossman L, Lewis RS (1998) A stellar origin for the short-lived nuclides in the early Solar System. Nature 391:559-560

Schmitt AD, Stille P, Vennemann T (2003) Variations of the $^{44}Ca/^{40}Ca$ ratio in seawater during the past 24 million years: evidence from $\delta^{44}Mg$ and $\delta^{18}O$ values of Miocene phosphates. Geochim Cosmochim Acta 67:2607-2614

Schramm DN, Tera F, Wasserburg GJ (1970) The isotopic abundance of ^{26}Mg and limits on ^{26}Al in the early solar system. Earth Planet Sci Lett 10(1):44-59

Wasserburg GJ, Lee T, Papanastassiou DA (1977) Correlated oxygen and magnesium isotopic anomalies in Allende inclusions: II. magnesium. J Geophys Res 4:299-302

Weston RE (1999) Anomalous or mass-independent isotope effects. Chemical Reviews 99:2115-2136

Young ED, Ash RD, Galy A, Belshaw NS (2002a) Mg isotope heterogeneity in the Allende meteorite measured by UV laser ablation-MC-ICPMS and comparisons with O isotopes. Geochim Cosmochim Acta 66(4): 683-698

Young ED, Galy A, Nagahara H (2002b) Kinetic and equilibrium mass-dependent isotope fractionation laws in nature and their geochemical and cosmochemical significance. Geochim Cosmochim Acta 66(6): 1095-1104

Zhu XK, Guo Y, O'Nions RK, Young ED, Ash RD (2001) Isotopic homogeneity of iron in the early solar nebula. Nature 412:311-313

Reviews in Mineralogy & Geochemistry
Vol. 55, pp. 231-254, 2004
Copyright © Mineralogical Society of America

7

The Stable-Chlorine Isotope Compositions of Natural and Anthropogenic Materials

Michael A. Stewart

Department of Geology
University of Illinois
Urbana, Illinois 61801, U.S.A.

Arthur J. Spivack

Graduate School of Oceanography
University of Rhode Island
Narragansett, Rhode Island 02882, U.S.A.

INTRODUCTION

The use of stable-isotopic ratios of a number of elements (e.g., H, Li, B, C, N, O, and S) is well established in the study of a broad spectrum of geological and environmental problems such as alteration of the oceanic crust, magmatic-crustal interactions, global chemical fluxes, the nature of the Precambrian crust and identifying the source and fate of pollutants (e.g., Taylor 1968; Muelenbachs and Clayton 1976; Muelenbachs 1980; Gregroy and Taylor 1981; Poreda et al. 1986; Spivack and Edmond 1986; Tanaka and Rye 1991; Mojzsis et al. 2001; Wilde et al. 2001; Numata et al. 2002). The fractionation of these light stable isotopes is a function of the relative mass differences between isotopes (e.g., Richet et al. 1977; Schauble, et al. 2003; Schauble 2004). The two stable isotopes of chlorine are ^{35}Cl and ^{37}Cl with a natural relative abundances of approximately 76% and 24%, respectively, and a relative mass difference of 5.7%, similar in magnitude to the relative mass differences between the isotopes of C and N. Hence, by analogy with these elements it is expected that stable isotopes of chlorine significantly fractionate and can be similarly exploited to understand and solve geological and environmental problems.

Early attempts at reproducibly determining ^{37}Cl/^{35}Cl ratios in natural samples were largely unsuccessful (e.g., Curie 1921; Owen and Schaeffer 1955; Hoering and Parker 1961), primarily, because of the limited effectiveness of the extraction and sample preparation techniques, the relatively poor precision of mass spectrometers at the time, and possibly the ^{37}Cl/^{35}Cl of the samples selected for analyses were inappropriate for the precision possible (e.g., Eggenkamp and Schuiling 1995). Taylor and Grimsrud (1969) developed a method for precisely measuring chlorine isotope ratios by mass spectrometry of methyl chloride (chloro-methane). Utilizing modifications of the this method, Kaufmann (1984) and Kaufmann et al. (1984) measured ^{37}Cl/^{35}Cl in seawater, groundwater, and chemical precipitates from evaporite deposits and hydrothermal systems with a precision of 0.24‰ based on the standard deviation of replicate standard analyses. These successes where followed by additional modifications that have now established this method and demonstrated that there are significant natural chlorine isotope variations (e.g., Eggenkamp 1994; Musashi et al. 1998). This method has been used to determine the δ^{37}Cl of groundwaters, and a variety of minerals and whole-rocks including evaporite deposits, fumarolic minerals, carbonatites, and granulites as well as organic compounds.

1529-6466/04/0055-0007$05.00

Despite these successes, the use of methyl chloride mass spectrometry is limited to relatively chlorine-rich samples (or samples with ample supply), thus, precluding many geologic and environmental samples with low chlorine abundance or limited supply. Xiao and Zhang (1992) and Magenheim et al. (1994) developed a high sensitivity method (micrograms of chlorine) suitable for stable-chlorine isotopic analysis of samples with low-chloride abundance by which, the $^{37}Cl/^{35}Cl$ ratios is measured by positive thermal ionization mass spectrometry of the relatively high-mass molecular ion, Cs_2Cl^+, with a reproducibility of 0.25‰. This method has been recently refined (e.g., Numata et al. 2001 2002). The pTIMS method has been used to investigate the Cl-isotopic composition of seafloor basalt glasses, silicate mineral phases from layered intrusions and altered oceanic crust, chondritic meteorites, sea floor sedimentary pore waters, atmospheric aerosols and organic compounds (e.g., Magenheim et al. 1994, 1995; Ransom, et al. 1995; Boudreau et al. 1997; Stewart et al. 1997, 1998; Volpe et al. 1998).

CHLORINE ISOTOPE NOMENCLATURE

Variations in chlorine isotopic composition are expressed as a per mil (‰) difference ($\delta^{37}Cl$) in a measured $^{37}Cl/^{35}Cl$ ratio (R) relative to the measured seawater chloride ratio (Standard Mean Ocean Chloride, SMOC),

$$\delta^{37}Cl = [R_{sample}/R_{seawater} - 1] \times 1000$$

Kauffmann (1984) reported analytical results for seawater samples taken from the Pacific and Atlantic oceans collected at depths ranging from 0 to 600 meters below sea level (mbsl) and showed that the stable-chlorine isotopic composition of modern seawater is analytically homogeneous. Recently, Godon et al. (2004) presented analytical results for 24 seawater samples from a variety of locations including the Atlantic, Pacific and Indian Oceans and a range of depths, exceeding 4000 mbsl. Their results confirm the constancy of the chlorine-isotopic composition of seawater in the modern ocean and that seawater is an appropriate and convenient reference material and standard. It is important to recognize that except for seawater, there are no natural or anthropogenic materials currently accepted as chlorine isotopic references or standards. Nor is there yet an accepted means for making interlaboratory comparisons and corrections although Long et al. (1993) propose that laboratories use National Institute of Standards and Technology's NaCl (NBS 975) and they report a $\delta^{37}Cl$ value of 0.52 ± 0.02‰ for NBS 975).

STABLE-CHLORINE ISOTOPE ANALYSES

Two methods are commonly used for stable-chlorine isotope analyses: Isotope ratio mass spectrometry (IRMS) of methyl chloride and positive thermal ionization mass spectrometer (pTIMS) of cesium chloride. Long et al. (1993) thoroughly detailed the IRMS method where AgCl is precipitated from aqueous solution, filtered and placed in vacuum with CH_3I to produce CH_3Cl by reaction. Solid samples are first converted to aqueous solution by dissolution, crushing and leaching, or hydrolysis. The CH_3Cl is purified by gas chromatography, and separated from CH_3I cryogenically prior to introduction to a gas-source mass spectrometer where mass numbers 50 and 52 ($CH_3{}^{35}Cl$ and $CH_3{}^{37}Cl$, respectively) are measured by peak-hopping and δ^{52} is calculated for SMOC and the sample relative to a laboratory reference gas. Differences in 52/50 are attributed entirely to differences in $^{37}Cl/^{35}Cl$ since neither carbon nor hydrogen have isotopes 2 mass numbers heavier than their most stable isotope (e.g., Long et al. 1993). The precision of this method was reported by Long et al. (1993) as ±0.09‰ based on 1σ standard deviation of a series of measurements, and for most accurate results, the procedure requires ~3 mg of chloride per measurement (e.g., Long et al. 1993, Godon et al. 2004). The relatively large abundance of chloride required for this method precludes the measurement of $\delta^{37}Cl$ in samples with low-chlorine abundance and short sample supply (e.g., seafloor basalts).

pTIMS of Cs_2Cl^+, requiring 3 to 6 μg chloride per analysis, is more suited to low-chlorine abundance samples with limited volume. In preparation of this method, chlorine is extracted form solid samples by pyrohydrolysis in an induction furnace and concentrated by evaporation (T < 100°C) under a heat lamp and filtered air (e.g., Magenheim et al. 1994; Stewart 2000). The concentrated aqueous solutions are purified and chloride balanced with cesium by cation exchange. The resulting $CsCl_{aq}$ is loaded onto a Ta filament for pTIMS (e.g., Xiao and Zhang 1992; Magenheim et al. 1994; Stewart 2000). Differences in mass numbers 170/168 ($Cs_2{}^{37}Cl^+/Cs_2{}^{35}Cl^+$) are attributed to differences in ^{35}Cl and ^{37}Cl as only one stable isotope of cesium exists (^{133}Cs). Based on 1σ standard deviation of a series of measurements, Magenheim et al. (1994) reported an external reproducibility of 0.25‰ and recently improved to 0.1‰ by Numata et al. (2002)—substantially greater than the IRMS method, however the pTIMS method is simpler and appropriate for Low-Cl abundance samples with limited supply.

CHLORINE ISOTOPE FRACTIONATION

The equilibrium isotopic fractionation of chlorine was first investigated theoretically in what are now classic papers by Urey and Greif (1935) and Urey (1947). Fractionation factors for a limited number of molecules have subsequently been calculated by other researchers (Kotaka and Kakihana 1977, Richet et al. 1977; Hanschmann 1984). Recently, however, Schauble et al. (2003) published theoretical estimates for a range of geochemically and environmentally important species that allow the prediction of general trends. These include, alkali salts, chlorinated aliphatic hydrocarbons, $FeCl_2$ and $MnCl_2$, as well as important atmospheric molecules and dissolved inorganic species. Silicate fractionation was estimated using $FeCl_2$ and $MnCl_2$ as model species. Fractionation factors were derived using classical statistical thermodynamics. Reduced partition function ratios were calculated using a combination of published vibrational spectra and force-field modeling. This approach is well described in Schauble (2004).

Schauble et al. (2003) summarize that the magnitude of fractionation is expected to systematically vary with the oxidation state of Cl, higher oxidation states being progressively enriched in the ^{37}Cl. Fractionation is also predicted to depend on the oxidation state of elements to which Cl is bound with greater fractionation for +2 cations than +1 cations. Silicates are predicted to be approximately 2‰ to 3‰ enriched compared to coexisting brines and organic molecules will be enriched by 5.8‰ to 8.5‰ at 295 K. Additionally, it is predicted that gas phase HCl will be enriched relative to dissolved Cl^- as was previously calculated by Richet et al. (1977). As described below, observed fractionations in silicates are generally consistent with these predictions as is the measured equilibrium fractionation between HCl and dissolved Cl^-. The observed fractionations associated with bio-mediated decomposition of chlorinated aliphatic hydrocarbons, however, are generally approximately one half of the calculated equilibrium values.

RESULTS AND IMPLICATIONS OF THE STABLE-CHLORINE ISOTOPIC COMPOSITION OF GEOLOGIC AND ANTHROPOGENIC MATERIALS

Meteorites

In order to develop a complete understanding of the distribution of chlorine-isotopes in Earth's exogenic (oceans, crust, and atmosphere) and mantle reservoirs it is necessary to determine the stable-chlorine isotope composition of the types of materials from which the earth may have formed. To date, however, the δ^{37}Cl for only four chondrites have been reported: Orgueil a C1-type, Murchison a C2-type, Allende a C3-type, and Zag an H4-Type chondrite (e.g., Magenheim et al. 1994, and Magenheim 1995; Bridges et al. 2001). The δ^{37}Cl values for carbonaceous chondrites range from 2.7‰ in the C1-type Orgueil meteorite to 4.0‰ in the C3-type Allende meteorite (Fig. 1). Interestingly, there is an apparent correlation between Cl-isotope composition and volatile concentration in carbonaceous chondrites: depletion of ^{37}Cl with decreasing volatile abundance from C1 through H4. However, given that C1- and C2-type meteorites acquired their volatile components through secondary aqueous alteration (e.g., McSween 1999), this correlation most likely does not reflect fractionation due to increasing "metamorphic" grade from C1 through C3 chondrites (e.g., increasing temperature, chemical equilibrium, decreasing volatile content; Van Schmus and Wood 1967; McSween 1999).

Bridges et al. (2001) analyzed Zag matrix and halite that crystallized from brine on the parent body (e.g., Bridges and Grady 2000; Whitby et al. 2000). These samples have relatively low δ^{37}Cl (−2.8‰ in the matrix, and ranging from −2‰ to −1‰ in the halite, Fig. 1) compared to the carbonaceous chondrites (2.7‰ to 4.0‰). This difference in δ^{37}Cl led Bridges et al. (2001) to suggest the existence of two distinct reservoirs, one associated with ice emplaced on the planetesimals surface (e.g., Whitby et al. 2000), and the other associated with silicates. The difference in Cl-isotopic composition of these two proposed reservoirs has profound implications for estimates of the initial Cl-isotopic composition of the Earth and its early exogenic and mantle reservoirs. However, the relative Cl-mass contribution of H-group chondrites to carbonaceous chondrites is not yet constrained. Nevertheless, this difference in Cl-isotopic composition in chondrites does suggest the possibility of using stable-chlorine isotopes to constrain the sources of terrestrial chlorine.

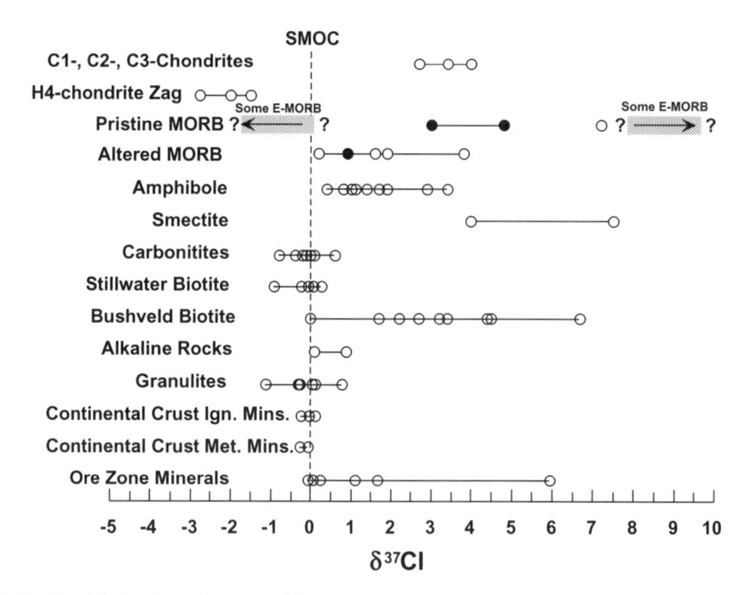

Figure 1. Stable-chlorine isotopic composition of meteorites, igneous and metamorphic rocks and minerals reported as δ^{37}Cl relative to ^{37}Cl/^{35}Cl in Standard Mean Ocean Chloride (SMOC) represented by the vertical dashed line at 0 ‰. Data from Magenheim et al. (1994), Magenheim et al. (1995), Bridges et al. (2001), Eggenkamp and Koster van Groos (1997), Boudreau et al. (1997), Willmore et al. (2002), Musashi et al. (1998), Markl et al. (1997), Eggenkamp and Schuiling (1995). Filled symbols are E-MORB and gray boxes represent preliminary data of E-MORB from Stewart (2000).

Igneous and metamorphic rocks and minerals

The igneous and metamorphic samples discussed below provide insights to the Cl-isotopic composition of both mantle and exogenic reservoirs. As noted above, currently there are relatively few Cl-isotope analyses of igneous and metamorphic samples. Therefore, the compositions of the Cl-reservoirs estimated from the few existing analyses are subject to modification by future analytical results. Estimates of the depleted mantle reservoir are relatively well constrained and range from 4.0‰ to 7.2‰ (Fig. 1; Magenheim et al. 1995; Stewart 2000). However, enriched mantle reservoirs such as un-degassed mantle, metasomatized mantle or mantle with a recycled crustal component require further investigation to constrain their Cl-isotopic compositions (e.g., Stewart 2000; see also E-MORB results below). Mantle enriched with a recycled crustal component (possibly through metasomatism by crustal-derived fluids) appears depleted in ^{37}Cl relative to SMOC (Fig. 1; Magenheim et al. 1995; Stewart et al. 1997). Notably, analyses have not yet revealed a reservoir enriched in ^{37}Cl derived from the residual subducted oceanic crust (see below). The Cl-isotopic composition of the un-degassed mantle reservoir is also relatively poorly constrained and requires further investigation, but preliminary results of Stewart et al. (1998) suggest it may be enriched in ^{37}Cl relative to SMOC.

Crustal reservoirs are also variable in Cl-isotope compositions (Figs. 1–6) due to fractionation of the Cl-isotope compositions inherited from their mantle source through fluid-mineral reactions, incorporation of Cl derived from the oceans and fractionation within fluid reservoirs by diffusion (see below). For example, the oceanic crust is enriched in ^{37}Cl (and pore fluids depleted in ^{37}Cl) through reaction of seawater with basaltic crust derived from the depleted mantle (Fig. 1; Magenheim et al. 1995). Undoubtedly, future investigations of Cl-isotopes in whole rocks and mineral separates will address the Cl-isotope compositions of these reservoirs and their evolution.

Mid-ocean ridge basalts. Measurements of stable-chlorine isotopes in mid-ocean ridge basalts (MORB) provide valuable insights into the origins of distinct mantle source compositions (e.g., Magenheim et al. 1995; Stewart et al. 1997; Stewart et al. 1998; Stewart 2000) and show the effects of assimilated seawater-derived chlorine components on the composition of basaltic magmas (e.g., Magenheim et al. 1995). Magenheim et al. (1995) first reported stable-chlorine isotope ratios in MORB. Their results show trace element depleted MORB (N-MORB) values range from 0.2‰ to 7.2‰ and trace element enriched MORB (E-MORB) values range from 0.9‰ to 4.8‰ (Fig. 1) with the lower values likely resulting from assimilation of seawater-derived chlorine

The effect of assimilation on the chlorine abundance of MORB magmas was thoroughly treated by Michael and Cornell (1992, and 1998). Their studies indicate that chlorine (and potassium) concentrations in MORB from slow-spreading ridges increase with indices of fractionation (e.g., MgO). These results are consistent with the highly incompatible behavior of Cl during mantle melting and basalt crystallization resulting from chlorine's large ionic radius and negative charge (e.g., Schilling et al. 1980). Michael and Cornell (1992 and 1998) also demonstrated that the Cl/K ratio increases with increasing indices of enrichment (e.g., K/Ti) in slow-spreading ridges, a result also consistent with chlorine's greater incompatibility relative to Large Ion Lithophile Elements with smaller ionic radii such as K and La (e.g., Schilling et al. 1980). This co-variation was interpreted as the result of mixing between depleted and enriched mantle components in the source of these MORB. The mixing relationship does not exceed Cl/K = 0.08 and Michael and Cornell (1998) concluded that this represents the maximum value for pristine, mantle-derived melts unaffected by assimilation of seawater-derived chlorine (i.e., pristine MORB). However, this may not be true of pristine mantle-derived melts affected by volatile-rich (i.e., Cl-rich) plume components such as those from the Reykjanes Ridge approaching Iceland (e.g., Schilling et al. 1980; Stewart 2000).

Michael and Cornell (1998) showed Cl/K increases beyond the maximum pristine mantle value (0.08) in MORB from fast-spreading ridges that fractionate at low-pressure (i.e., relatively shallow levels in the crust). They interpreted these results to indicate that these magmas assimilated altered crustal rocks containing excess chlorine gained from hydrothermal alteration with seawater-derived brines. The effect of assimilation of seawater-derived chlorine on the composition of MORB magma is to increase Cl/K with little change in K/Ti. This effect is greatest in N-MORB with generally lower initial Cl/K, and diminishes with the degree of trace element enrichment because both Cl and K concentrations are greater in E-MORB (e.g., Schilling et al. 1980; Jambon et al. 1995; Michael and Cornell 1998, Stewart 2000).

Magenheim et al. (1995) observed the relationships between δ^{37}Cl and Cl/K (and F/Cl) illustrating the effect of assimilation of seawater-derived chlorine on the δ^{37}Cl of MORB. They showed that MORB with Cl/K > 0.08 have F/Cl < 1 (i.e., are Cl-rich), and have δ^{37}Cl < 3.0‰. This relationship between trace element ratios and isotopic compositions led Magenheim et al. (1995) to conclude that contamination of the MORB magmas with seawater-derived chlorine results in lower δ^{37}Cl (<3.0‰). This was further demonstrated by Stewart (2000) where it was shown that N-MORB with Cl/K > 0.08 show a decreasing trend of δ^{37}Cl with increasing Cl/K, and that this trend is enveloped by calculated seawater-MORB mixing curves. Conversely, MORB with Cl/K < 0.08 (both N- and E-MORB), which are unaffected by assimilation of seawater-derived chlorine define δ^{37}Cl values for pristine MORB ranging from 4.0‰ to 7.2‰ (e.g., Magenheim et al. 1995; Stewart 2000).

Amphibole and smectite from the altered oceanic crust. The chlorine-isotopic composition of phases from the altered oceanic crust are key to understanding the effect of assimilation of seawater-derived chlorine on the composition of MORB magmas as discussed above (increasing Cl/K and decreasing F/Cl and δ^{37}Cl). Magenheim et al. (1995) analyzed the chlorine-isotopic composition of amphibole-bearing whole rocks and smectite veins from the oceanic crust recovered at Ocean Drilling Project holes 504B in the eastern Pacific Ocean and 735B in the southwestern Indian Ocean, and Mathematician Ridge. The smectite produced by low-temperature (<50°C) alteration in the upper crust have δ^{37}Cl of 7.5‰ and 4.0‰ (Fig. 1). The amphibole-bearing whole rocks produced at T > 350°C have δ^{37}Cl ranging from 0.4‰ to 3.4‰ (Fig. 1). The authors noted a possible decreasing trend in δ^{37}Cl with increasing depth of recovery of the alteration products: amphiboles from the base of the sheeted dike complex (hole 504B) have δ^{37}Cl ranging from 1.7‰ to 3.4‰. While those from the deeper plutonic sections (735B and Mathematician Ridge) range from 0.4‰ to 1.9‰. The correlation of trace element ratios (e.g., Cl/K and F/K) and stable-chlorine isotope ratios in MORB, with those of the residual amphibole and smectite from the altered oceanic crust led Magenheim et al. (1995) to conclude that assimilation of seawater-altered oceanic crust with excess chlorine incorporated in amphibole was likely responsible for the decrease in δ^{37}Cl with increasing Cl/K values observed in MORB with Cl/K > 0.08.

Implications from E-MORB results. Mantle domains with greater abundances of incompatible elements and distinct isotopic composition (i.e. enriched mantle) result from processes including recycling to the mantle though subduction and delamination of components derived from lithosphere and sediment (e.g., Hofmann and White 1982; Weaver et al. 1986; Zindler and Hart 1986; Ben Othman et al. 1989; Weaver 1991), or they may represent domains of relatively more un-degassed primitive mantle (e.g., Lupton 1983; White 1985; Poreda and Craig 1992). Incorporation of these enriched mantle components into the MORB mantle source results in melts (E-MORB) with relatively greater abundances of the more incompatible elements in addition to distinct isotopic signatures (possible including variations in δ^{37}Cl). Despite the relatively few (n = 3) E-MORB included in their study, Magenheim et al. (1995) noted a decrease in δ^{37}Cl with increasing source enrichment (e.g., K/Ti), suggesting that at least some types of enriched mantle may have δ^{37}Cl lower than the source of pristine N-MORB. Additional data from Stewart et al. (1997) suggest that some E-MORB (e.g., samples from the Chile Ridge and the Australian-Antarctic Discordance) may extend the range of observed $\delta^{37}Cl$ in MORB to negative values (Fig. 1). On the basis of trace element and radiogenic isotope systematics, enriched components in these MORB have been linked to a crustal component recycled to the mantle through subduction or possibly sub-continental delamination (e.g., Zindler and Hart 1986; Klein et al. 1988; Sturm et al. 1999). Thus, the inherent low (possibly negative) $\delta^{37}Cl$ composition of these samples may be linked to the recycling process. Such a link, if demonstrated, could serve as a powerful geochemical tool in aiding identification of recycled, enriched mantle components in mantle sources and in understanding the origin of these recycled signatures. Furthermore, preliminary results of Stewart et al. (1998), suggests it is also possible that E-MORB derived from relatively un-degassed, volatile-rich mantle (such as that underlying Iceland; Hart et al. 1973) may have $\delta^{37}Cl$ that is distinct (high positive) from E-MORB with recycled crustal signatures (Fig. 1).

Future research in stable-chlorine isotope geochemistry of MORB should focus on the nature of distinct mantle components (e.g., Stewart 2000). For example, analysis of glasses erupted from Loihi may provide insights to the high Cl (e.g., Kent and Stolper 1997), volatile-rich nature of the Hawaiian Plume with high $^{3}He/^{4}He$. Additional work on samples with recycled components in the source (e.g., the Azores and boninites) may reveal additional recycled signatures with distinct $\delta^{37}Cl$ inherited from the subducted oceanic lithosphere. As suggested by Stewart (2000), if such results are obtainable, then $\delta^{37}Cl$ could be used in conjunction with noble gas isotopic systems such as helium to distinguish un-degassed primordial mantle components from mantle enriched with recycled components from (1) intergranular or pore fluids derived from subducted slabs (e.g., see below the results of Ransom et al. 1995), or (2) melts and fluids derived from the residual lithosphere.

Carbonatites. Carbonatites offer further evidences and implications for the heterogeneity of stable-chlorine isotopes in the mantle. Carbonatites, containing at least 50% carbonate minerals and less than 10% silicates, represent magmas produced by melting of an anomalous mantle source enriched in carbonate fluids (e.g., Woolley and Kempe 1989). Eggenkamp and Koster van Groos (1997) reported the stable-Cl isotopic composition of carbonatites from Europe and Africa, and those from a primary mantle source (i.e., pristine and unaffected by post-emplacement alteration by crustal fluids) range in $\delta^{37}Cl$ from −0.4‰ to 0.1‰ (Fig. 1). The carbonatites altered by post-emplacement processes span the range of pristine carbonatites extending to both heavier (0.6‰) and lighter (−0.8‰) $\delta^{37}Cl$ values.

The isotopic compositions of carbonatites are relatively depleted in ^{37}Cl compared to pristine depleted mantle values (e.g., N-MORB, Fig. 1), with chlorine-compositions resembling E-MORB and Stillwater biotite (see below and Fig. 1). Eggenkamp and Koster van Groos (1997) suggested that the chlorine-isotopic composition of the pristine carbonatites is produced in the mantle source through fractionation occurring during accumulation of the carbonatite melts. They propose this occurs by one of two processes. Either preferential diffusion of ^{35}Cl from mantle originally of the assumed mantle composition (4.7‰, Magenheim et al. 1995), generating carbonatite melts that have a relatively light chlorine-composition. Alternatively, chlorine-rich carbonate-fluids exchange with chlorine-bearing mantle phases (e.g., amphiboles) producing isotopically light fluids that metasomatize the carbonatite mantle sources.

Layered intrusions. Chlorine isotope studies of the Stillwater Complex of Montana, USA (e.g., Stewart et al. 1996; Boudreau et al. 1997), and the Bushveld Complex in the Republic of South Africa (e.g., Willmore et al. 2002) show that both intrusions are anomalously rich in chlorine (e.g., Boudreau et al. 1986; Willmore et al. 2000), but that, their chlorine-isotopic compositions are distinct from one another (Fig. 1). The $\delta^{37}Cl$ values of Stillwater samples range from −0.93‰ to 0.27‰ with all but one value below 0.1‰ (Fig. 1). In contrast the

Bushveld samples have δ^{37}Cl ranging from 0.0‰ up to 6.7‰ with all but one sample greater than 1.0‰.

Likely Stillwater primary magmas are high-Mg basalt with boninite-like compositions, i.e., low TiO_2, but high K_2O and light rare earth element contents (e.g., Helz 1985). Boninites are likely generated by melting of previously depleted mantle that subsequently was metasomatized by fluid derived from subducted lithosphere (e.g., Hickey and Frey 1982; Wooden et al. 1991). This mechanism for boninite magma generation is the favored model for the generation of primary Stillwater magmas, and McCallum (1996) has demonstrated that it satisfies most of the isotopic constraints for the Stillwater parental magmas. Boudreau et al. (1997) and Stewart et al. (1996) suggest this mechanism may also explain the chlorine-isotopic compositions of Stillwater samples that are depleted in ^{37}Cl relative to pristine mantle (e.g., NMORB values, Fig. 1). They argued that the elevated chlorine concentration of the Stillwater magmas and their relatively light chlorine-isotopic character are imparted by the metasomatic fluid, which itself is derived from the subducted oceanic crust (i.e., chlorine-rich seawater-derived brines that have exchanged chlorine with amphibole during alteration of mafic crust resulting in low δ^{37}Cl).

In contrast to Stillwater Cl-isotope values, the relatively enriched δ^{37}Cl of the Bushveld Complex nearly covers the entire range of values in amphibole-containing whole rocks (0.4‰ to 3.4‰) and smectite veins (4.0‰ to 7.5‰) from the altered oceanic crust (e.g., Magenheim et al. 1995). As with the Stillwater Complex, the primary magmas of the Bushveld have been described as boninitic in character (e.g., Sharpe and Hulbert 1985; Hatton and Sharpe 1989) likely generated by metasomatism and melting of a previously depleted mantle in a subduction zone setting (e.g., Hickey and Frey 1982). However, the chlorine-rich fluid that metasomatized the mantle source of the Bushveld primary magmas could not have the same character (δ^{37}Cl $\leq$ 0‰) as that of the Stillwater samples. Willmore et al. (2002) propose a number of scenarios to account for the relatively ^{37}Cl-enriched isotopic composition in Bushveld samples. Common to each scenario is that the chlorine-rich metasomatic fluid is derived from the residual, subducted oceanic crust composed of alteration minerals enriched in ^{37}Cl (e.g., amphibole and smectite). The contrast between the Cl-isotopic composition of Stillwater and Bushveld samples, and yet both complexes possibly produced by metasomatism of the normal, depleted mantle, suggests that after further investigation, stable-chlorine isotopes may distinguish the agents of flux melting in subduction zones and possibly mantle enrichment processes in general (i.e., subducted pore fluids vs. fluids and melts derived from residual slab minerals).

Alkaline rocks. The only stable-chlorine isotopic data for alkaline igneous rocks are reported by Eggenkamp (1994) and Musashi et al. (1998). These analyses are from two samples of the Proterozoic Ilímaussaq alkaline intrusion of Greenland (~1.2 Ga; Blaxland et al. 1976) and described by Eggenkamp (1994). The average δ^{37}Cl values reported by Musashi et al. (1998) are 0.09‰ and 0.88‰, similar to values of other crustal samples (see below), Stillwater biotite separates, amphibole separates from the altered oceanic crust, and N-MORB with assimilated seawater-derived chlorine (Fig. 1). Comparing these few data to samples with similar chlorine isotopic compositions suggests that either the mantle source of these alkaline rocks has anomalously low δ^{37}Cl values relative to the pristine N-MORB mantle source and was possibly derived from recycled crustal pore-fluid derived chlorine, or the chlorine in these alkaline magmas was derived from a crustal source through assimilation as with the altered N-MORB values. This second scenario is more probable in light of the results of Stevenson et al. (1997) who showed that the Nd and Sr isotopic composition in Ilímaussaq samples support the incorporation of Proterozoic crust in parental magmas derived from a depleted mantle source.

Mid- to lower-crustal granulite facies rocks. Markl et al. (1997) reported δ^{37}Cl values of mid- to lower-crustal Precambrian aged granulite and retrograde amphibolite facies metamorphic rocks from the Lofoten Islands, Norway (Fig. 1). In part, their study shed light on the possible origins of CO_2-rich fluids involved in granulite facies metamorphism. As discussed by these authors, recent noble gas isotope studies of the Lofoten area (e.g., Dunai et al. 1995) and similar granulites (e.g., Dunai 1994; Dunai and Touret 1993) indicate the metamorphic fluids originate from degassing of the mantle as substantiated by some carbon isotope studies (e.g., Moecher et al. 1994; Santosh and Wada 1993). However, petrologic evidence exists that suggests crustal derived Na-K-Cl brines may be involved in the petrogenesis of some rocks commonly found in granulite facies terrains (e.g., Markl et al. 1997; Newton and Aranovich 1996).

In their study, Markl et al. (1997) analyzed whole rocks from two gabbroic anorthosites ($\delta^{37}Cl$ = −0.3‰ and 0.11‰) and one mangerite ($\delta^{37}Cl$ = 0.79‰) and amphibole-biotite separates from four whole rock samples with $\delta^{37}Cl$ ranging from −1.12‰ to 0.04‰ (Fig. 1). These values are all distinct from the assumed normal mantle values of 4.0‰ to 7.2‰ (e.g., Magenheim et al. 1995). They concluded that a mantle origin for the chlorine (and by inference the CO_2-rich fluids) is equivocally supported by these results, and instead, they favor the involvement of crustal-derived chlorine-rich brine in the petrogenesis of the Lofoten granulite facies rocks. However, Markl et al. (1997) note that the chlorine may be derived from mantle with anomalously low $\delta^{37}Cl$ values possibly derived from recycling of crustal chlorine to the mantle (e.g., Magenheim et al. 1995; Stewart 2000). Similar to studies of mantle derived rocks such as MORB and layered intrusions, analyses of $\delta^{37}Cl$ in granulite facies rocks from other terrains may enhance our understanding of the origins of chlorine in the lower crust in general (i.e. mantle- or crustal-derived).

Metamorphic, igneous and ore minerals from the continental crust. Eggenkamp and Schuiling (1995) reported $\delta^{37}Cl$ compositions near 0‰ in igneous apatite, sodalite and eudialite, and metamorphic amphiboles from the North American and northern European continental crust (Fig. 1). The relatively small variation in Cl-isotopic composition of the igneous minerals (Fig. 1) is thought to be the result from limited fractionation between fluids and minerals at high temperatures (e.g., Kaufmann 1989), suggesting that the average chlorine-isotopic composition of the continental crust is similar to seawater (e.g., Eggenkamp and Schuiling 1995; see also continental fluids below). The similar $\delta^{37}Cl$ values of the metamorphic minerals indicate that no significant fractionation occurs during metamorphism (e.g., Eggenkamp and Schuiling 1995).

The $\delta^{37}Cl$ values of oxidized ore minerals reported by Eggenkamp and Schuiling (1995) are nearly all enriched in ^{37}Cl ranging from –0.05‰ to 5.96‰ (Fig. 1). The distinctly positive values in some of these ore minerals led these authors to suggest that if chlorine-isotopic composition of the ore deposits was reflected in waters draining them, then the water compositions may be employed as a tool for economic exploration.

Fluids

Seawater and marine pore fluids. As discussed above, the chlorine isotopic composition of modern seawater does not vary measurably. This is not surprising in light of its long residence time (approximately 90 million years) and its conservative behavior in the water column. In contrast, marine pore fluids have been demonstrated to vary considerably. There is also the likelihood that hydrothermal fluids may be fractionated as a result of exchange with mineral phases, as phase separation under marine hydrothermal conditions does not appear to lead to measurable fractionation (e.g., Magenheim et al. 1995). However, to date no stable-chlorine isotopic compositions of marine hydrothermal fluids have been reported in the literature.

Ransom et al. (1995) and Godon et al. (2004) present $\delta^{37}Cl$ results for a variety of marine sedimentary pore fluids taken from the northeast African passive continental margin to pore fluids from the Nankai and Caribbean convergent margins. The measurement of these marine pore waters extend significantly the range of δ^{37}Cl values reported for natural waters (to −7.7‰).

Sediment pore waters from the Mazagan Plateau on the passive margin off northwest Africa were collected during Deep Sea Drilling Project Leg 79 at site 546 from clayey oozes overlying Jurassic evaporite deposits (e.g., Gieskes et al. 1984). Gieskes et al. (1984) demonstrated with major-element compositions that these pore waters were generated by mixing of seawater with brines derived from the dissolution of the subsurface evaporites. The δ^{37}Cl values of the Mazagan pore waters range from −0.41‰ to 0.06‰ (Fig. 2, Ransom et al. 1995). As demonstrated by Ransom et al. (1995), these values—similar to seawater—are consistent with an unfractionated mixture derived from evaporites and seawater. Marine evaporites (below) typically have δ^{37}Cl near 0‰ (e.g., Kaufmann et al. 1984) and only slight fractionation is expected during dissolution of the evaporite minerals (e.g., Long et al. 1993; Eggenkamp et al. 1995).

Pore waters from the convergent margins have stable-chlorine isotope composition distinct from seawater (Fig. 2) indicating significant fractionation of chlorine isotopes in their source regions or along their flow paths. Pore water samples from the Caribbean presented by Ransom et al. (1995) were collected from the uppermost meter of clay-rich volcaniclastic sediment in the Manon mud-volcano field of the Barbados accretionary prism during the Manon Cruise of the R/V L'Atlante (e.g., Martin 1993). Those from the Nankai margin include samples collected from below the subduction zone decollement in underthrust hemipelagic muds and altered volcanic ash (e.g., Kastner et al. 1993). The δ^{37}Cl of the pore waters from both convergent margins range to values significantly lower than seawater (Fig. 2) with the Barbados samples ranging from −5.29‰ to −0.89‰ and the depletion of ^{37}Cl in Nankai

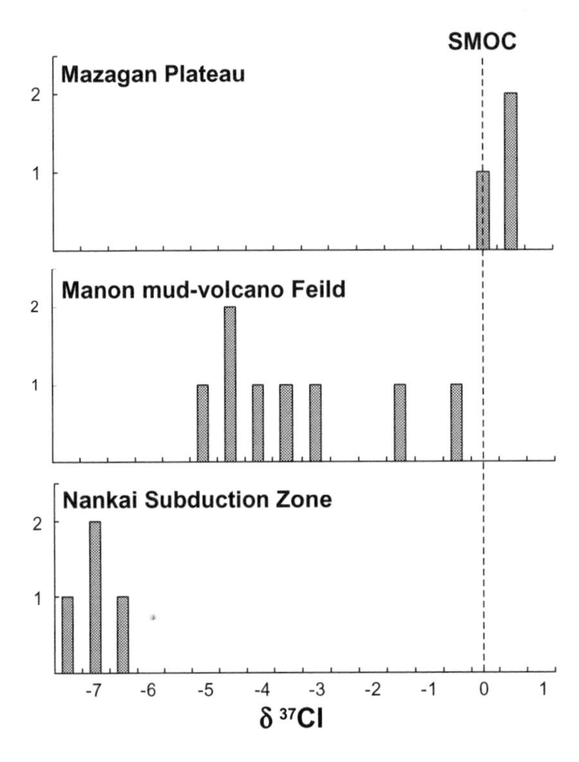

Figure 2. Histogram of stable-chlorine isotopic measurements in marine pore waters reported as δ^{37}Cl relative to ^{37}Cl/^{35}Cl in SMOC represented by the vertical dashed line at 0‰. Data from Ransom et al. (1995), and Godon et al. (2004).

samples even more extreme with δ^{37}Cl ranging from −7.71‰ to −6.78‰ (e.g., Ransom et al. 1995; Spivack et al. 2002; Godon et al. 2004). These observations demonstrate that chloride in these environments does not behave conservatively as has been assumed in a number of hydrologic models. It is necessary to understand the origin of these depletions in order to use chloride and the chlorine isotopes to constrain hydrologic models of subduction zones.

The chlorine-isotope fractionation of both the Barbados and Nankai pore waters most likely result from exchange of chlorine between formation waters and silicate minerals (rather than transport or methane hydrate transformations, See Ransom et al. 1995 and Spivack et al. 2002 and references therein). As discussed by Ransom et al. (1995), the generation of these isotopically light fluids is consistent with the alteration of the volcanic oceanic crust resulting in the positive δ^{37}Cl values of amphibole- and smectite-bearing samples discussed above and reported by Magenheim et al. (1995). However, it is not yet known whether reactions between the pore fluid and sediment may also fractionate chlorine isotopes in these environments. It has also been noted that the resultant isotopically light fluids may be recycled to the mantle during subduction and possibly involved in arc-magma genesis (see also discussion of volcanic gasses below). Of course, metasomatism of the mantle wedge by these fluids is only possible if they are not expelled from the slab prior to subduction into the mantle. Furthermore, in order for the ^{37}Cl-depleted signature of these fluids to be imparted to the volcanic arc source, the ^{37}Cl-depleted fluids and the ^{37}Cl-enriched component in the altered crust must migrate separately from the slab into the mantle wedge.

Continental fluids and evaporite minerals. Numerous studies have investigated the stable-chlorine isotopic composition of fluids and, to a lesser extent, evaporite minerals in continental settings (See for example references in Figs. 3 and 4 captions, also Martin 1993; Wirt 1988; Kaufmann et al. 1988; Frape and Bryant 1994; Lavastre et al. 2002). In a thorough investigation of the chlorine-isotopic composition of continental fluids, Eggenkamp (1994) presented results from a variety of settings, including groundwater, oil field formation water brines, volcanic springs and lakes, and volcanic gases. Long et al. (1993) briefly summarized many of these investigations, and here we can simply add recent results expanding the data set and confirming that stable-chlorine isotopes are useful in discriminating between continental fluids with distinct chloride sources.

The preponderance of the continental fluid data (e.g., groundwater, brines, etc.) and evaporite minerals (e.g., halite) have stable-chlorine isotopic compositions within 1‰ of seawater (Figs. 3 and 4). From these results it has been concluded that fractionation of chlorine isotopes occurs during the formation and dissolution of evaporite minerals (e.g., Eggenkamp and Coleman 1993; Eggenkamp et al. 1995). The greater degrees of fractionation observed in some continental fluids (e.g., North Sea, West-Nederland Basin, and Paris Basin formation waters; Fig. 3), and subduction zone pore waters (Fig. 2) likely result from diffusion (Eggenkamp 1994; Eggenkamp et al. 1994; Desauliniers et al. 1986), and exchange between fluids and chlorine-bearing phases (e.g., Ransom et al. 1995). Isotopic fractionation by diffusion is relatively uncommon in other stable isotope systems, however, diffusion commonly operates to fractionate stable-chlorine isotopes in continental fluids (e.g., Coleman et al. 2001; Figs. 3 and 4). Diffusive fractionation of isotopes results from the relatively greater mobility of the lighter isotope compared to the heavier isotope (e.g., ^{35}Cl vs. ^{37}Cl). For most elements, fractionation by diffusion is obscured by fractionations caused by chemical reactions and biological processes (e.g., Eggenkemp et al. 1994). However, in settings where chlorine is conservative and not involved in biological processes (e.g., crust), diffusion is a major process causing fractionation of chlorine isotope (e.g., Desaulniers et al. 1986; Coleman et al. 2001).

Sedimentary basins and oil fields. Despite their similarity with seawater chlorine-isotopic composition, enough variability between fluids exists locally (and the precision of the analytical method is fine enough) to distinguish the sources of chloride contributing to

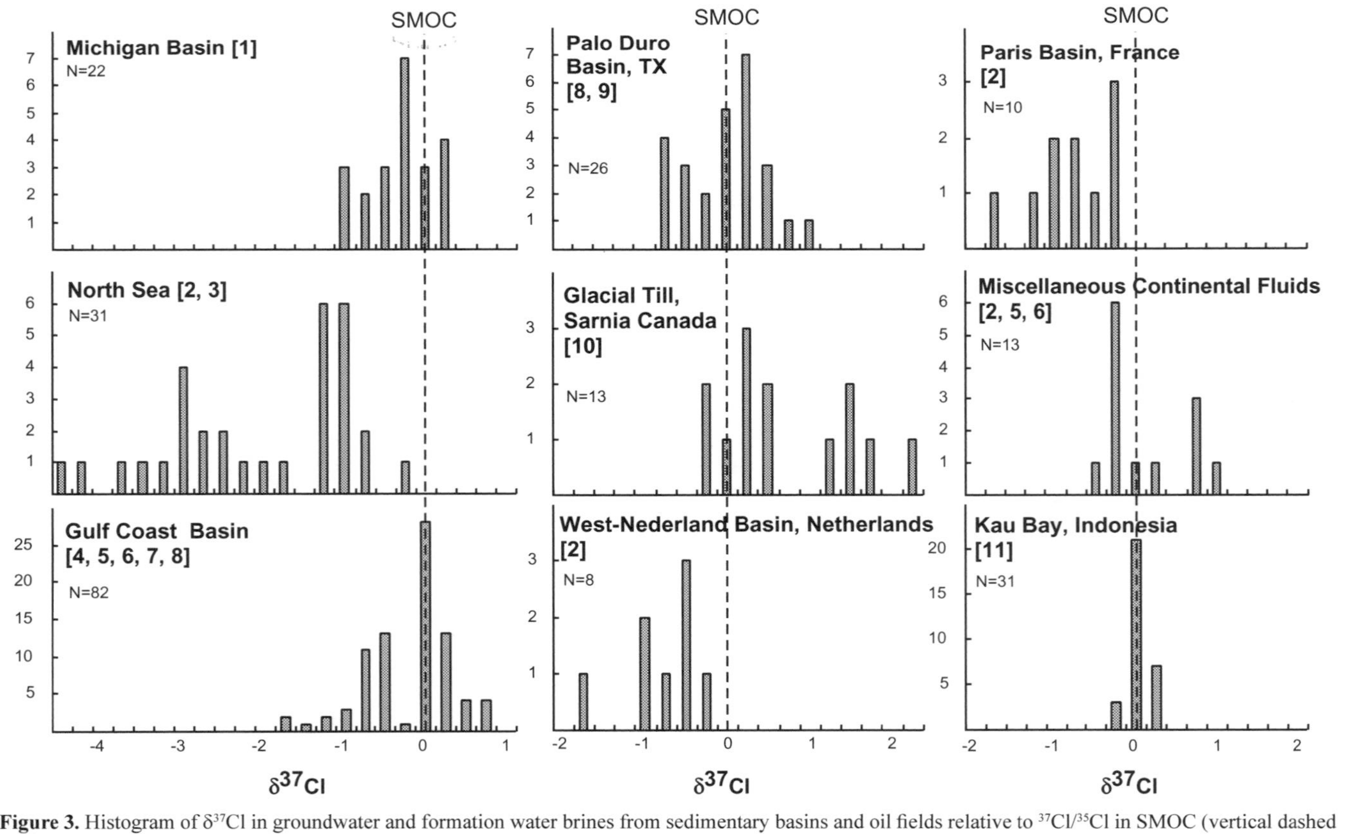

Figure 3. Histogram of $\delta^{37}Cl$ in groundwater and formation water brines from sedimentary basins and oil fields relative to $^{37}Cl/^{35}Cl$ in SMOC (vertical dashed line at 0‰). N is the number of analyses, and bracketed numbers identify references as follows: [1] Kaufmann et al. 1993; [2] Eggenkamp 1994; [3] Ziegleret al. 2001; [4] Eastoe et al. 2001; [5] Kaufmann 1984; [6] Kaufmann et al. 1984; [7] Kaufmann et al. 1988; [8] Eastoe and Guilbert 1992; [9] Eastoe et al. 1999; [10] Desauliniers et al. 1986; and [11] Eggenkamp et al. 1994.

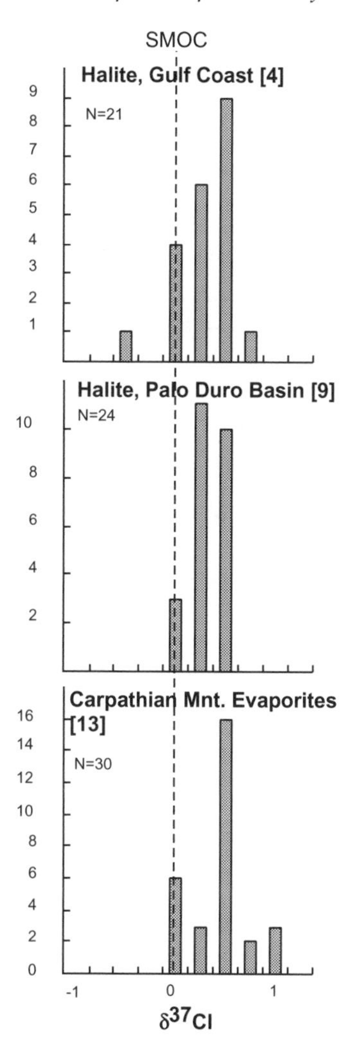

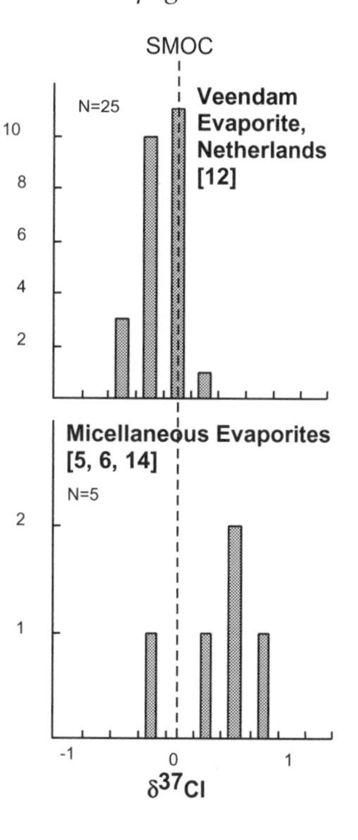

Figure 4. Histogram of δ^{37}Cl in evaporite deposits fields relative to ^{37}Cl/^{35}Cl in SMOC (vertical dashed line at 0‰). N is the number of analyses, and references (bracketed numbers) are as in Figure 3 and [12] Eggenkamp et al. 1995; [13] Eastoe and Peryt 1999; and [14] Eggenkamp and Schuiling 1995.

the fluid compositions. Many studies, including those referenced in Figure 3, use stable-Cl isotopic composition of fluids and their elemental concentrations compared with their associated evaporites from sedimentary basins to investigate the local sources of chlorine, the interaction and contributions of distinct fluids within or between basins, mixing of fluids, and fractionation of chlorine isotopes during fluid flow. For example, Kaufmann et al. (1993) determined that Michigan Basin formation water is a combination of formation water and local brines mixed with formation water infiltrating the Michigan Basin from the Canadian Shield. Similarly, Ziegler et al. (2001) determined the formation water in the Brent Group (North Sea). These have some of the lowest δ^{37}Cl values measured in basin fluids (−4.69‰ to −0.9‰) possibly from seawater mixed with brine from evaporated seawater and meteoric water. Based on the spatial distribution of waters with distinct δ^{37}Cl in the basin, they were also able to determine paleo-fluid flow paths. Not surprisingly, brines from the Gulf Coast Basin with δ^{37}Cl near seawater values (0‰) are best explained by mixing between seawater and fresh waters (e.g., Eastoe et al. 2001).

When present in a sedimentary basin, evaporite minerals can also supply chloride to fluids. For example, fluids from the Forties oil field (North Sea) range to extremely negative δ^{37}Cl values, −4.25‰ to −1.09‰ (Fig. 3, Coleman 1992; Eggenkamp 1994). The chloride in these formation waters is a mixture of two sources: Waters with δ^{37}Cl values near −1‰ (and high chlorine concentrations) likely derive their chlorine from evaporites, while those with relatively more negative δ^{37}Cl are likely derived from dehydration of clay mixed with connate water in the source rock (e.g., Coleman 1998). The negative δ^{37}Cl of these waters (−4.25‰) has been attributed to mass-dependent diffusional separation through the source rock with very low permeability (e.g., Eggenkamp and Coleman 1993; Eggenkamp 1994).

Furthermore, Eggenkamp et al. (1995) investigated the fractionation of chlorine isotopes between evaporite minerals (NaCl, KCl, $MgCl_2 \cdot 6H_2O$) and saturated solutions, and found small but significant variation in the fractionation factors ranging from –0.09 to 0.26. They concluded that δ^{37}Cl can be used to indicate evaporation cycles leading to precipitation of different chlorine-bearing salts and by inference the possible distinction of fluids with chlorine derived by the dissolution of evaporite deposits. For example, δ^{37}Cl of formation waters from the Palo Duro Basin of Texas (Fig. 3) discriminate between the contributions of evaporated seawater and mixed meteoric water and brine derived from dissolved evaporites (e.g., Eastoe et al. 1999). It follows that δ^{37}Cl can also discriminate sources of chloride incorporated into evaporite minerals, and in such a study Eastoe and Peryt (1999) determined that some chloride incorporated in Badenian evaporites (Carpathian mountains, Fig. 4) originates from two sources (brines); one enriched and the other depleted in ^{37}Cl and both distinctly not seawater.

Hydrothermal fluids and crystalline bedrock. A few investigations use stable-chlorine isotopes to characterize paragenetic stages in hydrothermally deposited ore zones. For example, Eastoe and Guilbert (1992) investigated the stable-chlorine isotopic composition of fluid inclusions in Mississippi Valley type deposits and porphyry-copper deposits. The Mississippi Valley type deposits of central Tennessee were precipitated in multiple paragenetic stages from saline brines at temperatures from 95°C to 135°C. Using stable-chlorine isotope compositions of fluid inclusions in the ore zone minerals (e.g., pyrite, quartz, biotite, sphalerite, calcite and barite, see Fig. 5), Eastoe and Guilbert (1992) were able to discriminate between the paragenetic stages by their chlorine-isotope composition and they conclude that alternating brines from at least two distinct sources were involved in the precipitation of the ore zone minerals.

These same authors applied stable-chlorine isotope techniques to the investigation of two porphyry-copper deposits in which the ore zone mineralization is related to salt-rich brines at high-temperature (near granitoid solidus). Although the chlorine-isotopic results did not distinguish the source of the brines (e.g., meteoric water – rock interaction vs. mixtures of magmatic and formation waters), they did, however, characterize the fluids as distinct from seawater (see Fig. 5). The ^{37}Cl-enriched biotites in the ore zones suggest fractionation of chlorine-isotopes during biotite precipitation.

A few studies have used stable-chlorine isotope techniques to characterize fluids in mining districts within the crystalline bedrock of Precambrian shields. Kaufmann et al. (1987) determined the δ^{37}Cl in Canadian Shield brines from the Sudbury, Thompson and Yellowknife mining districts where all but one brine sample were enriched in ^{37}Cl (Fig. 5). They hypothesized that the chloride is not of seawater origin, but rather is derived from multiple fluids of variable origin that have been modified by water-crystalline rock interactions. The Canadian Shield brines are distinct from brines in the Fennoscandian Shield (e.g., Frape and Bryant 1994, Fig. 5) that are mostly depleted in ^{37}Cl at shallow levels and become enriched in ^{37}Cl with depth. A few studies have investigated the chlorine-isotopic composition of Fennoscandian Shield brines (e.g., Frape et al. 1995; Frape et al. 1998; Sie and Frape 2002)

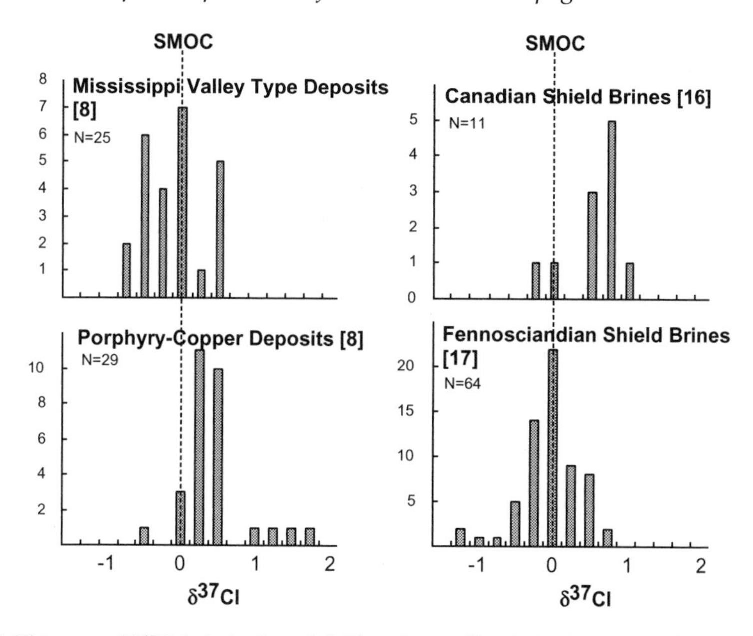

Figure 5. Histogram of $\delta^{37}Cl$ in hydrothermal fluids and crystalline bedrock relative to $^{37}Cl/^{35}Cl$ in SMOC (vertical dashed line at 0‰). N is the number of analyses, and references (bracketed numbers) are as in Figure 3 and [16] Kaufmann et al. 1987; [17] Sie and Frape 2002.

and concluded that brine chloride concentration and isotopic compositions are consistent with leaching from the crystalline basement.

Fumarolic minerals, volcanic gas condensates, and geothermal surface waters. There are relatively few stable-chlorine isotope measurements for volcanic gasses (N = 7) and fumarolic minerals (N = 3). The samples analyzed were collected from four volcanoes: Kudriavy in the Kuriles, Russian Federation, and Volcano, Etna and Vesuvius, Italy (e.g., Wahrenberger et al. 1997; Eggenkamp and Schuiling 1995; Godon et al. 2004). The $\delta^{37}Cl$ values of volcanic gas condensates range from −1.2‰ to 5.69‰ (Fig. 6), while the $\delta^{37}Cl$ of the two fumarolic minerals analyzed by Eggenkamp and Schuiling (1995) are −0.46‰ and −4.88‰. Interpretation of the chlorine-isotopic ratios in fumarolic minerals is complicated by multiple episodes of sublimation and condensation during precipitation of the minerals from volcanic gasses resulting in the substantial spread in measured $\delta^{37}Cl$ values (e.g., Eggenkamp and Schuiling 1995). The isotopic data of volcanic gases from Kudriavy have a linear correlation between chlorine content and isotopic composition. Wahrenberger et al. (1997) interpreted this to result from mixing of two components: a high $\delta^{37}Cl$, chlorine-rich magmatic (high-temperature) fluid and a low $\delta^{37}Cl$, Cl-poor meteoric-derived fluid. The data from Volcano are more complex and no simple interpretation was possible (e.g., Wahrenberger et al. 1997).

It is interesting to speculate on the implications of these and future analyses of volcanic gasses and geothermal surface waters from arc volcanoes (Fig. 6) in conjunction with the results from subduction zone pore-fluids (e.g., large negative $\delta^{37}Cl$ values, Fig. 2, Ransom et al. 1995): $\delta^{37}Cl$ values in volcanic gasses and geothermal fluids may reflect metasomatism of the mantle source by the subduction zone pore-fluids with negative $\delta^{37}Cl$, or fluid derived from minerals in the residual slab (e.g., amphibole with positive $\delta^{37}Cl$). However, the variability in $\delta^{37}Cl$ values from negative to positive values (Fig. 6) in geothermal waters from a single

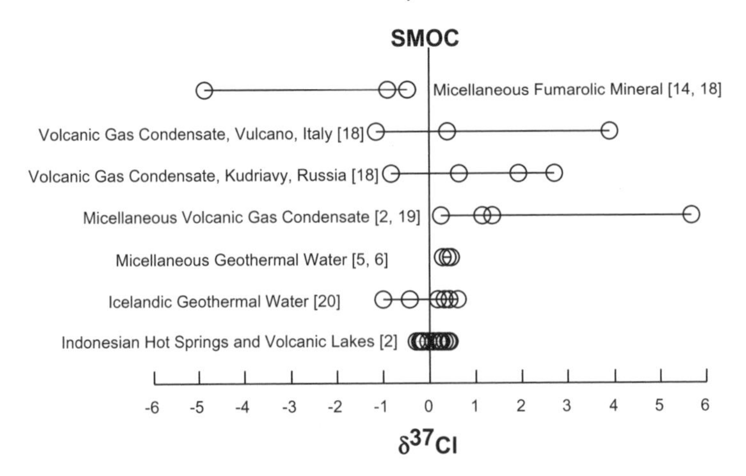

Figure 6. Stable-Cl isotopic composition of fumarolic minerals, volcanic gas condensates and geothermal waters. Bracketed numbers correspond to references listed in Figure 3, Figure 4, and, [18] Wahrenberger et al. 1997; [19] Godon et al. 2004; and [20] Kaufmann 1989.

setting (e.g., Iceland) and volcanic gasses from a single setting (e.g., Vulcano) complicate such interpretations and likely reflect exchange with hydrothermally altered rocks (e.g., Ransom et al. 1995) and fractionation of chlorine isotopes during vapor-solid phase separation (e.g., Kaufmann 1989). Note that these fractionations occur during phase separation between vapor-solid pairs and not vapor-brine pairs. Kaufmann (1989) used equilibrium-exchange models to calculate chlorine-fractionations from vapor-solid phase separations predicting measurable differences (~1‰) in $\delta^{37}Cl$ between vapor-solid pairs that are inversely correlated with temperature. These model calculations are not to be confused with the vapor-brine separation experiments referred to by Magenheim et al. (1995) and conducted at seafloor hydrothermal conditions (301–370 bar, 419–431°C) showing no measurable fractionation between coexisting vapor-brine pairs.

Summary of terrestrial chlorine isotope distributions

After consideration of the results from the limited number of analyses of chlorine-isotope composition in meteorites, rocks and fluids presented above, an understanding of the isotopic distribution of Earth's chlorine begins to emerge, but a number of fundamental questions need to be resolved before a complete picture is developed. It appears well established that the isotopic composition of the mantle is heterogeneous and that exogenous chlorine is distinct from the N-MORB mantle source. In principle, these differences may reflect an initial heterogeneity or may result from fractionation associated with the differentiation of the earth and the transfer of material between the mantle and exogenous reservoirs. Meteorites have a significant range in chlorine isotopic compositions suggesting the interesting possibility of an initially heterogeneous distribution of chlorine isotopes. However, the data are limited and the processes that produced this range of compositions are not well understood.

Magenheim et al. (1995) took the Orgueil value (2.7‰) to represent the undegassed mantle (bulk earth) in their calculation of the relative distribution of chlorine between the exogenic and degassed mantle reservoirs. Support for using the isotopic composition of Orgueil as an estimate of the bulk Earth composition comes from studies of halogen (Br, Cl, I) ratios, which show the bulk earth ratios are indistinguishable from those of CI-type meteorites (e.g., Dreibus et al. 1979; Schilling et al. 1980; Déruelle et al. 1992). Results of these calculations suggest that 40% of Earth's chlorine resides in exogenic reservoirs.

Magenheim et al. (1995) investigated two end-member possibilities to explain the mantle/seawater fractionation as a result of differentiation: (1) continuous mantle degassing with no recycling of exogenic chlorine and (2) complete degassing and recycling of exogenic chlorine by subduction. Their Rayleigh fractionation modeling results for (2) of 3.1‰ fractionation are indirectly supported by the observed amphibole-seawater fractionation. Spivack et al. (2002) further suggested that the light isotopic composition of seawater relative to the N-MORB source may result from the recycling to the mantle of ^{37}Cl-enriched complement of subduction zone pore fluids. Spivack et al. (2002) did point out however, that the apparent constancy of seawater isotopic composition (as evidenced by evaporates) indicates that a significant fraction of the ^{37}Cl-enriched chlorine isotope complement of the ^{37}Cl-depleted subduction zone pore-fluids must be recycled to seawater.

As well, intergranular pore fluids may be recycled to the mantle by subduction and released during early stages of slab devolatilization thereby metasomatizing the mantle with chlorine-rich, isotopically light fluids such as those proposed for genesis of Stillwater magmas and by inference boninites and arc magmas in general. Thus, recycling of chlorine to Earth's mantle by subduction (and possibly lower crustal delamination) may result in mantle domains either enriched or depleted in ^{37}Cl and later sampled by sub-ridge melting (E-MORB) or possibly by mantle plumes (e.g., Iceland).

Sea-salt aerosols

Sea-salt aerosols are a source of gaseous chlorine in the troposphere. Species include HCl, as well as photochemically reactive species. However, the chemical reactions involved are not well demonstrated and a number of reactions have been proposed. Volpe and Spivack (1994), Volpe (1998) and Volpe et al. (1998) examined δ^{37}Cl variations in marine aerosols and determined fractionation factors associated with HCl exchange. In their initial study of acidified bulk aerosols collected in the North Atlantic, it was observed that there was a positive correlation between δ^{37}Cl and the amount of chlorine lost from the aerosols. This relationship was reproduced in laboratory experiments designed to mimic irreversible volatilization of HCl due to acidification while the fractionation due to equilibrium HCl exchange has subsequently been demonstrated to be of opposite sign. In a follow up study of size segregated aerosols collected in the Pacific, distinct isotopic trends were observed for the small versus large aerosol particles. The smaller, acidified particles had positive δ^{37}Cl values as observed for the acidified Bermuda aerosols while the larger particles had negative δ^{37}Cl values that correlated with the amount of chlorine depletion. These data were used to infer that the various size fraction particles are not in isotopic equilibrium and that different mechanisms control the release of chlorine from acidified and unacidified aerosols. In particular it was argued that these data provided further evidence of the release of photochemically active chlorine. It was further suggested that experimental determination of the fractionation factors associated with the various suggested reactions that lead to reactive chlorine loss could be used with this type of field data to identify the release mechanism.

Environmental chemistry

Chlorine isotopic analyses have been proposed or utilized to understand the environmental chemistry of a number of natural and anthropogenic organic compounds. In general, these studies have the common goals of identifying and quantifying sources, transport and transformation of specific molecules or classes of compounds. To achieve these goals, these studies have addressed a variety of similar questions. These include, the isotopic composition of the various sources and the fractionations associated with physical processes (e.g., evaporation and adsorption) and chemical transformations (biologic and abiologic). Among the compounds that have been investigated are chlorinated aliphatic hydrocarbons (CAHs), polychlorinated biphenyls (PCBs), chloro-fluoro carbons (CFCs), Freons™ and other chlorinated organic compounds (COCs).

Tanaka and Rye (1991) reported the chlorine isotopic composition of a few compounds including Freon-12 and synthetic methyl chloride. Methyl chloride is the largest natural source of chlorine to the stratosphere. The $\delta^{37}Cl$ of Freon-12 and methyl chloride differed by nearly 8‰. Based on this, they suggested that if the isotopic composition of natural methyl chloride was also distinct from the Freons that the isotopic composition of stratospheric inorganic chlorine could be used to constrain the relative contributions of methyl chloride and the Freons™ as sources of stratospheric inorganic chlorine. Volpe (1998) further argued that the isotopic composition of methyl chloride could be used to help constrain its tropospheric budget. More recently, Harper et al. (2000) pointed out that the validity of such an approach is based on the conservative behavior of the bound chlorine. Based on the kinetics of inorganic exchange, the hydrolysis half-life of dissolved methyl chloride is 3–4 months. However, Harper et al. (2000) demonstrated in laboratory cultured soil bacterium that microbial transhalogenation occurs on the time scale of minutes, thus potentially compromising the use of the chlorine isotopes for constraining the atmospheric methyl chloride budget. Whether this is a significant effect in natural systems remains to be explored.

The CAHs (e.g., tetrachlorethene, trichloroethene, trichlorethane, and carbon tetrachloride) have been widely used since the 1940s as degreasers and due to their improper disposal now pollute many soils, and aquifers (e.g., Fischer et al. 1987). They are the most prevalent organic groundwater contaminant in the US that are known or suspected to be carcinogenic to humans. The PCBs are a class of toxic compounds that once had wide use in various applications notably in electrical capacitors and transformers. Although they are now banned from use, they are a pervasive environmental contaminant (e.g., Cairns et al. 1986). Considerable effort has gone into identifying the sources of both the PCBs and CAHs and their biogeochemical behaviors. In particular, there has been much effort aimed at identifying and quantifying their sources and in situ biodegradation based on concentrations. These efforts, however, are often confounded by adsorption, evaporation and dilution. Chlorine isotopic compositions of CAHs produced by a variety of manufacturers have been reported by a number of groups. Although there is no isotopic standard for any CAH and the groups have used different methodologies there is a general consistency in the data. The range of reported $\delta^{37}Cl$ is from approximately −5 to +6‰ with the $\delta^{37}Cl$ of each compound from a particular supplier having a distinct value. This has been argued to demonstrate that chlorine isotopes can help fingerprint CAH sources, in particular in conjunction with the determination of carbon and hydrogen isotope ratios (e.g., van Warmerdam et al. 1995; Beneteau et al. 1999; Jendrzjewski et al. 2001; Numata et al. 2002). Data from one study (e.g., Beneteau et al. 1999) indicated that the $\delta^{37}Cl$ of different batches of PCE and TCE produced by the same manufacturer were constant but that of TCA was variable.

Laboratory studies of the fractionation associated with the biodegradation of CAHs have been conducted by Heraty et al. (1999) and Numata et al. (2002), while Sturchio et al. (1998, 2003) inferred a fractionation factor based on field data. Heraty et al. (1999) determined carbon and chlorine fractionation factors for the aerobic decomposition of dichloromethane in batch cultures of a methylotrophic Bacteria. These values were 0.9576 and 0.9962 respectively, for carbon and chloride, demonstrating that bio-remediation should produce a characteristic $\delta^{37}Cl$–$\delta^{13}C$ trend. Numata et al. (2002) examined the reductive dechlorination of PCE and TCE to cis-1,2–dichloroethene (cDCE) using three different anaerobic cultures. A model was developed which considered that the chlorine in TCE are not chemically equivalent and thus may have different isotopic compositions. Fitting their data (isotopic composition of the three molecules combined) to the model they calculated that the fractionation factor for TCE dechlorination is 0.994 to 0.995 at 30°C and that for the first step of PCE dechlorination it is 0.987 to 0.991. Using a Rayleigh fractionation model and the observed correlation between the δ^{37}Cl and concentration of TCE of groundwaters from an aerobic aquifer, Sturchio et al. (1998) derive a fractionation factor of 0.999475 but note that this most likely represents an upper limit as the model does not account for mixing with TCE-free water.

Isotope fractionation associated with evaporation of two pure compounds has been investigated (e.g., Huang et al. 1999). Experiments were run to simulate irreversible and equilibrium evaporation. Derived irreversible fractionation factors were 0.9982 and 1.00031 for chlorine and carbon respectively for trichloroethene and 0.99952 and 1.00065 for dichloromethane. The authors report the equilibrium values as preliminary and give 0.99948 and 1.0009 for chlorine and carbon respectively for trichloroethene and 0.99987 and 1.0013 for dichloromethane. These results confirm the need to consider the fractionation associated with the evaporation of volatile organic carbon compounds such as the CAHs and that combined isotopic trends (δ^{37}Cl–δ^{13}C) may be diagnostic of particular processes such as evaporation. The isotopic effect of adsorption has been inferred from a study of a plume of dissolved PCE (e.g., Beneteau et al. 1999). In this plume, in which there is no evidence of biodegradation, δ^{37}Cl values are homogeneous and thus it is inferred that adsorption does not lead to significant fractionation.

PCB isotopic compositions have been reported for mixtures produced by Monsanto (USA), Bayer (Germany), and Caffaro (Italy) (e.g., Reddy et al. 2000; Drenzek et al. 2002). The Monsanto mixtures (Arochlors) were supplied from three separate environmental standard suppliers. The δ^{37}Cl values of all the samples fell within a limited range of –3.37 to –2.11‰. However, there were resolvable differences between the same Arochlor mixture supplied from different sources. A preliminary survey of contaminated sediments (e.g., Reddy et al. 2000) indicated that two samples which have PCBs that have been affected by alteration as evidenced by the lower amounts of less chlorinated congeners were isotopically distinct from the range of the source materials (assumed to be similar to the analyzed pure samples) while other samples were not distinguishable from the source material.

There are a number of COCs that have been detected in sediments, soils and animal tissue for which it is unknown whether they are natural products or anthropogenic. Reddy et al. (2002) suggested that if these compounds are natural they may be a useful analog for understanding how anthropogenic COCs behave. Alternatively if they are anthropogenic their source needs to be determined and their control should be considered. To test whether, natural compounds may be distinguished from anthropogenic compounds by their chlorine isotopic composition, Reddy et al. (2002) compared the isotopic fractionation associated with enzyme catalyzed chlorination to that of abiotic chlorination of trimethoxybenzene. The chlorinated products produced enzymatically were fractionated −10.2 to −11.2‰ relative to the starting chloride while the fractionation of the abiotic products was −2.7 to −3.1 from which Reddy et al. (2002) concluded that chlorine isotopes may be useful to distinguish natural from man made compounds.

Reported δ^{37}Cl work related to inorganic environmental problems is more limited. A method for the analysis of chlorates and perchlorates has been developed and used to demonstrate fractionation associated with the microbial degradation of perchlorate with a fractionation factor of 0.985 (Ader et al. 2001; Coleman et al. 2003). Additionally, δ^{37}Cl has been used to constrain the origin of a Cl-rich plume (Loomis et al. 1998).

ACKNOWLEDGMENTS

We appreciated the thoughtful and thorough reviews and comments by P. Michael, M. Coleman and the editor B. Beard. M. Stewart's Cl-isotope research from 1995 to 2002 was performed at Duke University, Durham NC, USA and generously supported by E. Klein and A. Spivack.

REFERENCES

Ader M, Coleman ML, Doyle SP, Stroud M, Wakelin D (2001) Methods for the stable isotopic analysis of chlorine in chlorate and perchlorate compounds. Anal Chem 73(20):4946-4950

Ben Othman D, White WM, Patchett J (1989) The geochemistry of marine sediments, island arc magma genesis, and crust-mantle recycling. Earth Planet Sci Lett 94:1-21

Beneteau KM, Aravena R, Frape SK (1999) Isotopic characterization of chlorinated solvents-laboratory and field results. Organic Geochemistry 30(8A):739-753

Blaxland AB, Van Breemen O, Steerfelt A (1976) Age and origin of agpaitic magmatism at Ilímaussaq, south Greenland: Rb-Sr study. Lithos 9:31-38

Boudreau AE, Mathez EA McCallum IS (1986) Halogen geochemistry of the Stillwater and Bushveld complexes: evidence for transport of the platinum-group elements by Cl-rich fluids. J Pet 27:967-986

Boudreau AE, Stewart MA, Spivack AJ (1997) Stable Cl isotopes and origin of high-Cl magmas of the Stillwater Complex, Montana. Geology 25:791-794

Bridges JC, Grady MM (2000) Petrography and fluid inclusion studies of Zag. Meteoritics Planet Sci 35: 33-34

Bridges JC, Banks DA, Grady MM (2001) Stable chlorine isotope reservoirs in chondrites. Meteoritics Planet Sci 36:A29-A30

Cairns T, Doose, GM, Froberg JE, Jacobson RA, Siegmund EG (1986) Analytical chemistry of PCBs. *In:* PCBs and the Environment. Waid JS (ed) CRC Press, Boca Raton, p 1-45

Coleman ML (1992) Water composition variation within one formation. *In:* Water-Rock Interaction Proceeding of 7th international symposium on water-rock interaction. Kharaka, Macst (eds) Park City, UT, p 1109-1112

Coleman ML (1998) Novel methods for measuring chemical composition of oil-zone waters: Implications for appraisal and production. PETEX 98 Conf Proc Publ Abs M3

Coleman ML, Ader M, Chaudhuri S, Coates JD (2003) Microbial isotopic fractionation of perchlorate chlorine. Appl Environ Microb 69(8):4997-5000

Coleman ML, Eggenkamp HGM, Aranyossy JF (2001) Chlorine stable isotope characterization of solute transport in mudrocks, ANDRA. Actes des Jounées Scientifiques: EDP Sciences, France, p 155-175

Curie I (1921) Sur le poids atomique du chloe dans quelques minéraux. CR Acad Sci (Paris) 172:1025-1028

Déruelle B, Dreibus G, Jambon A (1992) Iodine abundances in oceanic basalts: implications for Earth dynamics. Earth Planet Sci Lett 108:217-227

Desaulniers DE, Kaufmann RS, Cherry JA, Bently HW (1986) ^{37}Cl-^{35}Cl variations in a diffusion-controlled groundwater system. Geochim Cosmochim Acta 50:1757-1764

Dreibus G, Spettel B, Wanke H (1979) Halogens in meteorites and their primordial abundances. *In:* Origin and Distribution of the Elements, Proc and Symp. LH Ahrens (ed) Pergamon, Oxford, p 33-38

Drenzek NJ, Tarr CH, Eglinton CTI, Heraty LJ, Sturchio NC, Shiner VJ, Reddy CM (2002) Stable chlorine and carbon isotopic compositions of selected semi-volatile organochlorine compounds. Organic Geochemistry 33(4):437-444

Dunai TJ, Touret JLR (1993) A noble gas study of a granulite sample from the Nilgiri Hills, southern India: Implications for granulite formation. Earth Planet Sci Lett 119:271-281

Dunai TJ, Hilton DR, Touret JLR, Markl G (1995) Granulite formation and mantle CO_2 – evidence from the Lofoten and Vesterålen Islands, Norway. Eos Trans, Am Geophys Union 76:678

Dunai TJ (1994) Mantle derived CO_2 and granulite genesis: Evidence from noble gases. Mineral Mag 58: 245-246

Eastoe CJ, Long A, Knauth LP (1999) Stable chlorine isotopes in the Palo Duro Basin, Texas: Evidence for preservation of Permian evaporite basins. Geochim Cosmochim Acta 63:1375-1382

Eastoe CJ, Long A, Land LS, Kyle JR (2001) Stable chlorine isotopes in halite and brine from the Gulf Coast Basin: brine genesis and evolution. Chem Geol 176:343-360

Eastoe CJ, Guilbert JM (1992) Stable chlorine isotopes in hydrothermal systems. Geochim Cosmochim Acta 56:4247-4255

Eastoe CJ, Peryt T (1999) Multiple sources of chloride in Badenian evaporites, Carpathian Mountains: Stable chlorine isotope evidence. Terra Nova 11:118-123

Eggenkamp HGM, Coleman ML (1993) Extreme $\delta^{37}Cl$ variations in formation water and its possible relation to the migration from source to trap. AAPG Bull 77:1620

Eggenkamp HGM, Schuiling RD (1995) $\delta^{37}Cl$ variations in selected minerals: a possible tool for exploration. J Geoch Exploration 55:249-255

Eggenkamp HGM, Middelburg JJ, Kreulen R (1994) Preferential diffusion of ^{35}Cl relative to ^{37}Cl in sediments of Kau Bay, Halmahera, Indonesia. Chem Geol 116:317-325

Eggenkamp HGM, Kreulen R, Koster van Groos AF (1995) Chlorine stable isotope fractionation in evaporites. Geochim Cosmochim Acta 59:5169-5175

Eggenkamp HGM (1994) The geochemistry of chlorine isotopes. PhD Dissertation, Utrecht University, The Netherlands

Eggenkamp HGM, Koster van Groos AF (1997) Chlorine stable isotopes in carbonatites: evidence for isotopic heterogeneity in the mantle. Chem Geol 140:137-143

Fischer AJ, Rowan EA, Spalding RF (1987) VOCs in groundwater influenced by large-scale withdrawals. Groundwater 2:407-413

Frape SK, Bryant G, Durance P, Ropchan JC, Doupe J (1998) The source of stable chlorine isotopic signatures in groundwaters from crystalline shield rocks. *In:* Proceedings of the 9th International Symposium on Water-Rock Interaction, Arehart GB, Hulston JR (eds), Taupo, New Zealand p 223-226

Frape SK, Bryant G, Blomqvist R, Ruskeemiemi T (1995) Evidence from stable chlorine isotopes for multiple sources of chloride in groundwaters from crystalline shield environments. *In:* Proceedings of a Symposium on Isotopes in Water Resources Management Organized in Co-operation with The United nations Educational, Scientific and Cultural Organization and held in Vienna 20-24 March

Frape SK, Bryant G (1994) The stable chlorine isotopic signature of groundwaters in crystalline shield environments. Abst 8th International Conf Geoochron Cosmochron and Isotope Geol 8:105

Gieskes JM, Johnston K, Boehm M (1984) Interstitial water studies, Leg 79. *In:* Initial Reports of the Deep Sea Drilling Project. Vol 79. Hinz K, Winterer EL (eds) Washington, DC, U S Government Printing Office, p 825-834

Godon A, Jendrzejewski N, Eggenkamp HGM, Banks DA, Ader M, Coleman ML, Pineau F (2004) An international cross calibration over a large range of chlorine isotopic compositions. Chem Geol: in press

Gregory RT, Taylor Jr. HP (1981) An Oxygen isotope profile in a section of Cretaceous oceanic crust, Samail Ophiolite, Oman: evidence for ^{18}O-buffering of the oceans by deep (>5 km) seawater-hydrothermal circulation at mid-ocean ridges. J Geophys Res 86:2737-2755

Hanschmann G (1984) Reduzierte zustandssummenverhältnisse isotoper moleküle auf quantenchemischer grundlage v. mitt. MNDO-MO-berechnungen zur ^{28}Si/ ^{30}Si-, ^{32}S/ ^{34}S- und ^{35}Cl/ ^{37}Cl-substitution. Isotopenpraxis 12:437–439

Harper DB, Kalin RM, Larkin MJ, Hamilton JTG, Coulter C (2000) Microbial transhalogenation: A complicating factor in determination of atmospheric chloro- and bromomethane budgets. Environ Sci Technol 34(12):2525-2527

Hart SR, Schilling J-G, Powell JL (1973) Basalts from Iceland and along the Reykjanes Ridge: Sr isotope geochemistry. Nature 246:104-107

Hatton CJ, Sharpe MR (1989) Significance and origin of boninite-like rocks associated with the Bushveld Complex. *In:* Boninite*s* Crawford AJ (ed) Unwin Hyman, Boston, p 174-207

Helz RT (1985) Compositions of fine-grained mafic rocks from sills and dikes associated with the Stillwater complex. *In:* Stillwater complex, geology and guide. Special Pub 92. Czamanske GK, Zientek ML (eds) Montana Bureau of Mines and Geology, p 97-117

Heraty LJ, Fuller ME, Huang L, Abrajano T, Sturchio NC (1999) Isotopic fractionation of carbon and chlorine by microbial degradation of dichloromethane. Org Geochem 30(8A):793-799

Hickey R, Frey FA (1982) Geochemical characteristics of boninite series volcanics: implications for their sources. Geochim Cosmochim Acta 46:2099-2115

Hofmann AW, White WM (1982) Mantle plumes from oceanic crust. Earth Planet Sci Lett 57:421-436

Hoering TC, Parker PL (1961) The geochemistry of the stable isotopes of chlorine. Geochim Cosmochim Acta 23:186-199

Huang L, Sturchio NC, Abrajano T, Heraty LJ, Holt BD (1999) Carbon and chlorine isotope fractionation of chlorinated aliphatic hydrocarbons by evaporation. Org Geochem 30(8A):777-785

Jambon A, Deruelle B, Dreibus G, Pineau F (1995) Chlorine and bromine abundance in MORB: the contrasting behavior of the Mid-Atlantic Ridge and East Pacific Rise and implications for chlorine geodynamic cycle. Chem Geol 126:101-117

Jendrzejewski N, Eggenkamp HGM, Coleman ML (2001) Characterization of chlorinated hydrocarbons from chlorine and carbon isotopic compositions: Scope of application to environmental problems. Appl Geochem 16(9-10):1021-1031

Kastner M, Elderfield H, Jenkins WJ, Gieskes JM, Gano T (1993) Geochemical and isotopic evidence for fluid flow in the western Nankai subduction zone. Japan. *In:* Proceedings of the Ocean Drilling Program, Scientific Results, Leg 131. Hill IA, Taira A (eds) College Station, Texas, Ocean Drilling Program, p 397-416

Kaufmann RS (1984) Chlorine in ground water: Stable isotope distribution. PhD Dissertation, University of Arizona, Tucson, AZ

Kaufmann RS (1989) Equilibrium exchange models for chlorine stable isotope fractionation in high temperature environments. *In:* Proc 6th Int Symp Water-Rock Interaction. Miles DL (ed) p 365-368

Kaufmann RS, Frape SK, Fritz P, Bentley H (1987) Chlorine stable isotope composition of Canadian Shield brines. *In:* Saline Water and Gases in Crystalline Rocks, Fritz P, Frape SK (eds) Geological Association of Canada Special Paper 33:89-93

Kaufmann RS, Frape SK, McNutt R, Eastoe C (1993) Chlorine stable isotope distribution of Michigan Basin formation waters. Applied Geochem 8:403-407

Kaufmann RS, Long A, Bentley H, Davis S (1984) Natural chlorine isotope variations. Nature 309:338-340

Kaufmann RS, Long A, Campbell DJ (1988) Chlorine isotope distribution in formation waters, Texas and Louisiana. AAPG Bull 72:839-844

Kent AJR, Stolper EM (1997) Contamination of ocean-island basalt magmas by a seawater-derived component at Loihi Seamount, Hawaii. EOS Trans AGU 78:F806

Klein EM, Langmuir CH, Zindler A, Staudigel H, Hamelin B (1988) Isotope evidence of a mantle convection boundary at the Australian-Antarctic discordance. Nature 333:623-629

Lavastre V, Javoy M, Jendrzejewski N (2002) Stable chlorine isotopes as tracers of solute transport in clay-rock formation (Paris Basin, France). Goldschmidt Conf Abst: A435

Long AE, Eastoe CJ, Kaufmann RS, Martin JG, Wirt L, Finley JB (1993) High-precision measurement of chlorine stable isotope ratios. Geochim Cosmochim Acta 57:2907-2912

Loomis JL, Coleman M, Joseph J (1998) Use of stable chlorine isotopes to evaluate the origin of a Cl-rich plume in the Oxford Clay, England. Goldschmidt Conf. Toulouse. Mineral Mag 62A:901-902

Lupton JE (1983) Terrestrial inert gases: isotope tracer studies and clues to primordial components in the mantle. Annu Rev Earth Planet Sci 11:71-414

Magenheim AJ, Spivack AJ, Volpe C, Ransom B (1994) Precise determination of stable chlorine isotopic ratios in low-concentration natural samples. Geochim Cosmochim Acta 58:3117-3121

Magenheim AJ, Spivack AJ, Michael PJ, Gieskes JM (1995) Chlorine stable isotope composition of the oceanic crust: implications for Earth's distribution of chlorine. Earth Planet Sci Lett 131:427-432

Magenheim AJ (1995) Oceanic Borehole Fluid Chemistry and Analysis of Chlorine Stable Isotopes in Silicate Rocks. PhD dissertation, University of California, San Diego, San Diego, California

Markl G, Musashi M, Bucher K (1997) Chlorine stable isotope composition of granulites from Lofoten, Norway: implications for the Cl isotopic composition and for the source of Cl enrichment in the lower crust. Earth Planet Sci Lett 150:95-102

Martin JB (1993) Origins and Composition of Fluids at Convergent Margins, PhD Dissertation, Scripps Institution of Oceanography, La Jolla, California

McCallum IS (1996) The Stillwater complex, *In:* Layered Intrusions. Cawthorn RG (ed) Elsevier, Amsterdam, p 441-484

McSween HY (1999) Meteorites and their parent planets. Cambridge Univ Press, NY

Michael PJ, Cornell WC (1992) Spreading rate dependence on magma-hydrothermal systems. Goldschmidt Conf Program Abstr, A70-A71

Michael PJ, Cornell WC (1998) The influence of spreading rate and magma supply on crystallization and assimilation beneath mid-ocean ridges: evidence from chlorine and major element chemistry of MORB. J Geophys Res 103:18325-18356

Moecher DP, Valley JW, Essene EJ, (1994) Extraction and carbon isotope analysis of CO_2 from scapolite in deep crustal granulites and xenoliths. Geochim Cosmochim Acta 58:959-967

Mojzsis SJ, Harrison TM, Pidgeon RT (2001) Oxygen-isotope evidence from ancient zircons for liquid water at the Earth's surface 4,300 Myr ago. Nature 409:178-181

Muehlenbachs K, Clayton RN (1976) Oxygen isotope composition of the oceanic crust and its bearing on seawater. J Geophys Res 81:4365-4369

Muehlenbachs K (1980) The alteration and aging of the basaltic layer of the sea floor: oxygen isotope evidence from DSDP/IPOD Legs 51, 52, and 53. *In:* Initial Reports of the Deep Sea Drilling Project 53:1159-1167

Musashi M, Markl G, Kreulen R (1998) Stable chlorine-isotope analysis of rock samples: New aspects of chlorine extraction. Analytica Chim Acta 326:261-269

Newton R, Aranovich L (1996) Simple granulite melting in concentrated NaCl-KCl solutions at deep crustal conditions. Geol Soc Am Annu Meet Abstracts with Programs: 158

Numata M, Nakamura N, Koshikawa H, Terashima Y (2002) Chlorine isotope fractionation during reductive dechlorination of chlorinated ethenes by anaerobic bacteria. Env Sci Tech 36(20):4389-4394

Numata M, Nakamura N, Gamo T (2001) Precise measurement of chlorine stable isotopic ratios by thermal ionization mass spectrometry. Geochem J 35(2):89-100

Owen HR, Schaeffer OA (1995) The isotope abundances of chlorine from various sources. J Am Chem Soc 77:898-899

Poreda RJ, Craig H (1992) He and Sr isotopes in the Lau Basin mantle: Depleted and primitive mantle components. Earth Planet Sci Lett 113:487-493

Poreda R, Schilling J-G, Craig H (1986) Helium and hydrogen isotopes in ocean-ridge basalts north and south of Iceland. Earth Planet Sci Lett 78:1-17

Ransom B, Spivack AJ, Kastner M (1995) Stable Cl isotopes in subduction-zone pore waters: implications for fluid-rock reactions and the cycling of chlorine. Geology 23:715-718

Reddy CM, Heraty LJ, Holt BD, Sturchio NC, Eglinton TI, Drenzek NJ, Xu L, Lake JL, Maruya KA (2000) Stable chlorine isotopic compositions of aroclors and aroclor- contaminated sediments. Env Sci Tech 34(13):2866-2870

Reddy CM, Xu L, Drenzek NJ, Sturchio NC, Heraty LJ, Kimblin C, Butler A (2002) A chlorine isotope effect for enzyme-catalyzed chlorination. J Am Chem Soc 124(49):14526-14527

Richet P, Bottinga Y, Javoy M (1977) A Review of hydrogen, carbon, nitrogen, oxygen, sulphur, and chlorine stable isotope fractionation among gaseous molecules. Ann Rev Earth Planet Sci 5:65-110

Santosh M, Wada H (1993) A carbon isotope study of graphites from the Kerala Khondalite Belt, Southern India: Evidence for CO_2 infiltration in granulites. J Geol 101:643-651

Schilling J-G, Bergeron MB, Evans R (1980) Halogens in the mantle beneath the North Atlantic. Phil Trans R Soc Lond 297:147-178

Schauble EA (2004) Applying stable isotope fractionation theory to new systems. Rev Mineral Geochem 55: 65-111

Schauble EA, Rossman GR, Taylor Jr. HP (2003) Theoretical estimates of equilibrium chlorine-isotope fractionations. Geochim Cosmochim Acta 67:3267-3281

Sharpe MR, Hulbert LJ (1985) Ultramafic sills beneath the eastern Bushveld Complex: Mobilized suspensions of early Lower Zone cumulates in a parent magma with boninitic affinities. Econ Geol 80:849-871

Sie PMJ, Frape SK (2002) Evaluation of the groundwaters from the Stripa mine using stable chlorine isotopes. Chem Geol 182:565-582

Spivack AJ, Edmond JM (1986) Determination of boron isotope ratios by thermal ionization mass spectrometry of the dicesium metaborate cation. Anal Chem 58:31-35

Spivack AJ, Kastner M, Ransom B (2002) Elemental and isotopic chloride geochemistry and fluid flow in the Nankai Trough. Geophys Res Lett 29:1-4

Stevenson R, Upton BGJ, Steenfelt A (1997) Crust-Mantle interaction in the evolution of the Ilímaussaq Complex, South Greenland: Nd isotopic studies. Lithos 40:189-202

Stewart MA, Boudreau AE, Spivack AJ (1996) Cl isotopes of the Stillwater Complex, Montana, and the formation of high-Cl Bushveld and Stillwater parent liquids during mantle metasomatism by Cl-rich fluids. Eos Trans, AGU 77:F829

Stewart MA, Klein EM, Spivack AJ, Schilling J-G (1998) Stable Cl-isotope composition of MORB basalts from the North Atlantic. EOS Trans, AGU 79:F942

Stewart MA, Klein EM, Spivack AJ, Langmuir CH, Karsten JL, Bender JF, Magenheim AJ (1997) Stable Cl isotope compositions of N- and E-MORB glasses. EOS Trans, AGU 78:S323

Stewart MA (2000) Geochemistry of Dikes and Lavas from Hess Deep: Implications for Crustal Construction Processes Beneath Mid-Ocean Ridges and the Stable-Chlorine Isotope Geochemistry of Mid-Ocean Ridge Basalt Glasses. PhD Dissertation, Duke Univ, Durham, North Carolina

Sturchio NC, Clausen JL, Heraty LJ, Huang L, Holt BD, Abrajano TA (1998) Chlorine isotope investigation of natural attenuation of trichloroethene in an aerobic aquifer. Environ Sci Technol 32(20):3037-3042

Sturchio NC, Hatzinger PB, Arkins MD, Suh C, Heraty LJ (2003) Chlorine isotope fractionation during microbial reduction of perchlorate. Environ Sci Technol 37(17):3859-3863

Sturm ME, Klein EM, Graham DW, Karsten J (1999) Age constraints on crustal recycling to the mantle beneath the southern Chile Ridge: He-Pb-Sr-Nd isotope systematics. J Geophys Res 104:5097-5114

Tanaka N, Rye DM (1991) Chlorine in the stratosphere, Scientific Correspondence. Nature 353:707

Taylor Jr. HP (1968) The oxygen isotope geochemistry of igneous rocks. Contrib Mineral Pet 19:1-71

Taylor JW, Grimsrud EP (1969) Chlorine isotopic ratios by negative ion mass spectrometry, Anal Chem 41: 805-810

Urey HC (1947) The thermodynamic properties of isotopic substances. J Chem Soc 562–581

Urey HC, Greiff LJ (1935) Isotopic exchange equilibria. J Am Chem Soc 57:321–327

Van Schmus WR, Wook JA (1967) A chemical petrologic classification for the chondritic meteorites. Geochim Cosmochim Acta 31:747-765

van Warmerdam EM, Frape SK, Aravena R, Drimmie RJ, Flatt H, Cherry JA (1995) Stable chlorine and carbon isotope measurements of selected chlorinated organic solvents. Applied Geochem 10(5):547-552

Volpe C, Spivack AJ (1994) Stable chlorine isotopic composition of marine aerosol particles in the western Atlantic Ocean. Geophys Res Lett 21(12):1161-1164

Volpe C, Wahlen M, Spivack AJ (1998) Chlorine isotopic composition of marine aerosols: Implications for the release of reactive chlorine and HCl cycling rates. Geophys Res Lett 25(20):3831-3834

Volpe CM (1998) Stable Chlorine Isotope Variations in the Atmosphere. PhD Dissertation, University of California, San Diego, San Diego, California

Wahrenberger C, Eastoe CJ, Seward TM, Dietrich V (1997) Stable chlorine isotope composition of volcanic gas condensates. 7th Annual Goldschmidt Conf: 213

Weaver BL (1991) The origin of ocean island basalt end-member compositions: trace element and isotopic constraints. Earth Planet Sci Lett 104:381-397

Weaver Bl, Wood DA, Tarney J, Jornon JL (1986) Role of subducted sediments in the genesis of ocean-island basalts: geochemical evidence from south Atlantic Ocean islands. Geology 14:275-278

Whitby J, Burgess R, Turner G, Gilmour J, Bridges J (2000) Extinct ^{129}I in halite from primitive meteorite: evidence for evaporite formation in the early solar system. Science 288:1819-1821

White WM (1985) Sources of oceanic basalts: radiogenic isotopic evidence. Geology 13:115-118

Wilde SA, Valley JW, Peck WH, Graham CM (2001) Evidence from detrital zircons for the existence of continental crust and oceans on the Earth 44 Gyr ago. Nature 409:175-178

Willmore CC, Boudreau AE, Kruger FJ (2000) The halogen geochemistry of the Bushveld Complex, Republic of South Africa: Implications for chalcophile element distribution in the lower and critical zones. J Pet 41:1517-1539

Willmore CC, Boudreau AE, Spivack AJ, Kruger FJ (2002) Halogens of the Bushveld complex, South Africa: δ^{37}Cl and Cl/F evidence for hydration melting at the source region in a back-arc setting. Chem Geol 182: 503-511

Wirt L (1988) The origin of chloride in groundwater in the Stripa granite, Sweden. MS Thesis, Univ Arizona, Tucson, Arizona

Wooden JL, Czamanske GK, Zientek ML (1991) A lead isotopic study of the Stillwater Complex, Montana: Constraints on crustal contamination and source regions. Contrib Mineral Pet 107:80-93

Woolley AR, Kempe DRC (1989) Carbonatites: nomenclature, average chemical compositions, and element distribution. *In:* Carbonatites, Genesis and Evolution, Bell K (ed) Unwin Hyman, London, p 1-14

Xiao Y-K, Zhang C-G (1992) High precision isotopic measurement of chlorine by thermal ionization mass spectrometry of the Cs_2Cl^+ ion. Inter J Mass Spectrom Ion Proc 116:183-192

Ziegler KM, Coleman L, Howarth RJ (2001) Paleohydrodynamics of fluids in the Brent Group (Oseberg Field, Norwegian North Sea) from chemical and isotopic compositions of formation waters. Applied Geochem 16:609-632

Zindler A, Hart S (1986) Helium: Problematical primordial signals. Earth Planet Sci Lett 79:1-8

Reviews in Mineralogy & Geochemistry
Vol. 55, pp. 255-288, 2004
Copyright © Mineralogical Society of America

8

Calcium Isotopic Variations Produced by Biological, Kinetic, Radiogenic and Nucleosynthetic Processes

Donald J. DePaolo

Department of Earth and Planetary Science
University of California
Berkeley, California 94720, U.S.A.
and
Earth Sciences Division
E.O. Lawrence Berkeley National Laboratory
Berkeley, California 94720, U.S.A.

INTRODUCTION

Study of the isotopic variations of calcium is of interest because Ca is important in geochemical and biochemical processes, and is one of few major cations in rocks and minerals with demonstrated isotopic variability. Calcium is critical to life and a major component of the global geochemical cycles that control climate. Studies to date show that biological processing of calcium produces significant isotopic fractionation (4 to 5‰ variation of the $^{44}Ca/^{40}Ca$ ratio has been observed). Calcium isotopic fractionation due to inorganic processing at high (e.g., magmatic) temperatures is small. There are few studies of calcium isotope fractionation behavior for low-temperature inorganic processes. The Ca isotopic variations observed in nature, both biological and inorganic, are mostly attributed to kinetic effects, but this inference cannot be confirmed until equilibrium Ca isotope fractionation is more thoroughly investigated. Evaporation of silicate liquids into vacuum at high temperature is expected to produce kinetic Ca isotopic fractionation, as is diffusion of calcium in silicate liquids and aqueous solutions. The evaporation effects have been observed in meteorite samples. Diffusion effects have been observed in the laboratory but not yet in natural samples. The potential value of Ca isotopic studies has barely been tapped. Improvements in measurement precision would increase the attractiveness of Ca isotopes as a geochemical tool, but such improvements have been slow in coming.

Calcium is composed mainly of the isotope ^{40}Ca, which is a highly stable, doubly magic nuclide (both the number of neutrons and the number of protons represent closed nuclear shells). There are a total of six stable isotopes covering a mass range from 40 to 48. Ca has been studied for isotopic variations in three ways. The major isotope, ^{40}Ca is the primary radioactive decay product of radioactive ^{40}K. About 89.5% of the ^{40}K decays result in the production of ^{40}Ca; the other 10.5% produce ^{40}Ar and are the basis for the K-Ar and Ar-Ar dating techniques. Rocks and minerals that are both old and rich in potassium can have significant enrichments in ^{40}Ca relative to the other Ca isotopes. The 8-unit spread of isotopic masses also means that there can be significant mass-dependent isotopic fractionation in nature. Consequently there are both radiogenic- and thermodynamically generated Ca isotopic variations in terrestrial materials. In meteorites there is also evidence for isotopic effects reflecting incomplete mixing of materials from different nucleosynthetic sources, combined with processing in the early

1529-6466/04/0055-0008$05.00

solar nebula. Some of the mass-dependent variations found in meteorites are quite large (ca. 50‰ variation in the $^{44}Ca/^{40}Ca$ ratio).

CALCIUM ISOTOPE ESSENTIALS

The approximate abundances of the isotopes of calcium are shown in Table 1. Russell et al. (1978b) reported the first precise calcium isotope ratio determinations made with modern mass spectrometers. Their measured ratios are shown in Table 1 and constitute the basis for the isotopic abundances and most of the subsequent work described in this article. The Russell et al. (1978b) isotope ratios apply to terrestrial samples that have no significant additions of radiogenic ^{40}Ca beyond the small excesses (ca. 0.01%) expected for average crustal rocks and seawater as discussed below. Niederer and Papanastassiou (1984) and Lee et al. (1977, 1979) used the Russell et al. ratios as the terrestrial reference for assessing non-mass dependent isotopic effects in meteorites. Skulan et al. (1997) used the Russell et al. ratios as a basis for assessing natural mass dependent isotopic variations, and introduced the $\delta^{44}Ca$ designation (in preference to the $\delta^{40}Ca$ value used by Russell et al. 1978b). The use of the $^{44}Ca/^{40}Ca$ ratio and the $\delta^{44}Ca$ parameter is more consistent with the notation used for other stable isotope systems, where the heavier isotope is in the numerator and hence higher numbers for both the isotope ratio and the delta value represent relative enrichment of heavy isotopes.

Table 1. Natural abundances, measured isotope ratios, and enrichment factors used for Calcium isotopes.

Approximate isotopic abundances

Isotope	Abundance
^{40}Ca	96.98%
^{42}Ca	0.642%
^{43}Ca	0.133%
^{44}Ca	2.056%
^{46}Ca	0.003%
^{48}Ca	0.182%

Accepted "normal" isotopic ratios (Russell et al. 1978b; Niederer and Papanastassiou 1984)

$^{40}Ca/^{44}Ca = 47.153 \pm 3$ (can vary due to radiogenic ^{40}Ca)

$^{42}Ca/^{44}Ca = 0.31221 \pm 2$

$^{43}Ca/^{44}Ca = 0.06486 \pm 1$

$^{46}Ca/^{44}Ca = 0.001518 \pm 2$

$^{48}Ca/^{44}Ca = 0.088727 \pm 9$

$^{44}Ca/^{40}Ca = 0.0212076 \pm 13$

$^{40}Ca/^{42}Ca = 151.016^* \pm 0.008$ (Earth's mantle value)

*normalized to $^{42}Ca/^{44}Ca = 0.31221$

For stable isotope fractionation studies (Skulan et al. 1997)

$$\delta^{44}Ca = 1000\left(\frac{^{44}Ca/^{40}Ca_{sample}}{^{44}Ca/^{40}Ca_{std}} - 1\right) = 1000\left(\frac{^{44}Ca/^{40}Ca_{sample}}{0.0212076} - 1\right)$$

For radiogenic ^{40}Ca enrichment studies (Marshall and DePaolo 1982)

$$\varepsilon_{Ca} = 10000\left(\frac{^{40}Ca/^{42}Ca_{sample}}{^{40}Ca/^{42}Ca_{std}} - 1\right) = 10000\left(\frac{^{40}Ca/^{42}Ca_{sample}}{151.016} - 1\right)$$

Analysis methods and issues for radiogenic ^{40}Ca

Measurement of radiogenic ^{40}Ca enrichments is mainly done using thermal ionization mass spectrometry (TIMS). The procedures are analogous to those used for other radiogenic isotope measurements. Mass discrimination in the spectrometer must be corrected for using a natural isotope ratio that is not affected by radioactive decay. Because the amount of isotopic discrimination is roughly proportional to the mass difference between isotopes, it makes most sense to choose either the isotope ratio ^{40}Ca/^{42}Ca or ^{40}Ca/^{44}Ca to monitor radiogenic enrichments of ^{40}Ca. To measure the mass discrimination it can be advantageous to use an isotope ratio that encompasses a large mass difference between isotopes (such as ^{48}Ca/^{42}Ca or ^{48}Ca/^{44}Ca). However, there are other considerations that make these ratios unattractive for this purpose. The accuracy of the mass discrimination correction depends on the precision with which the isotope ratio can be measured, which is poorer for lower abundance isotopes. Hence ^{48}Ca is a poor choice, and the combination of ^{48}Ca and ^{42}Ca is particularly unattractive. Also, because the mass discrimination is mass dependent as discussed below, it is advantageous to use an isotope ratio that is similar in average mass to that of the target isotope ratio being measured.

The result of the various considerations is that it is generally best to use either the ^{40}Ca/^{42}Ca or ^{40}Ca/^{44}Ca ratios to monitor the radiogenic enrichments of ^{40}Ca, and to use ^{42}Ca/^{44}Ca to measure mass discrimination. Marshall and DePaolo (1982) chose to use ^{40}Ca/^{42}Ca as the target ratio and to use ^{42}Ca/^{44}Ca for the mass discrimination correction. They found that when using the value of ^{42}Ca/^{44}Ca = 0.31221 reported by Russell et al. (1978b) for the mass discrimination correction, the initial solar system value of ^{40}Ca/^{42}Ca is 151.016 ± 0.008.

Precise measurement of Ca isotope ratios is challenging because the range of isotopic abundances is large. The value of the ratio ^{40}Ca/^{42}Ca of ca. 150 means that achieving a typical strong beam intensity of 5×10^{-11} amp for the $^{40}Ca^{+}$ ion beam leaves the $^{42}Ca^{+}$ ion beam at an intensity of 3.3×10^{-13} amp. At these levels, counting statistics are a limitation on the measurement precision. Marshall and DePaolo (1982, 1989) used a single collector TIMS instrument to make their measurements, with one-second integration of the $^{40}Ca^{+}$ beam and four-second integration of the $^{42}Ca^{+}$ beam. At the ion beam intensities quoted above, this yields 3.1×10^{8} $^{40}Ca^{+}$ ions per measurement and 8.4×10^{6} $^{42}Ca^{+}$ ions per measurement. Ignoring any complications from background and electronic noise, this means that the accuracy of a single ratio measurement as determined by counting statistics is approximately the uncertainty associated with the measurement of ^{42}Ca, which is about $\pm(8.4\times10^{6})^{-1/2} \approx \pm0.3‰$. The theoretical accuracy of the ratio after correction for mass discrimination includes the uncertainty on the ^{42}Ca/^{44}Ca measurement, which is about ±0.5‰, so the overall limit imposed by counting statistics is about 0.6‰ at the 1σ level. By making ca. 200 measurements, this uncertainty can theoretically be reduced to about ±0.1‰ (2σ, or about 95% confidence limits) if there are no other sources of noise in the data. In practice, the precision that is achieved with the quoted beam intensities is about ±0.2 to 0.3‰. Although one might think it possible to improve the measurement by integrating the ion beams longer, this does not work for single collector measurements because it increases the amount of time between measurements of the reference isotope, and hence degrades the corrections for the changing beam intensity with time. Counting statistics can also theoretically be greatly improved by use of a multicollector mass spectrometer, where all of the isotopes can be collected at the same time. However, in practice it has been found that while multicollection does in fact improve the precision of individual measurements, there is unaccountable drift that worsens the reproducibility between mass spectrometer runs (e.g., Heuser et al. 2002). Russell et al. (1977,1978a,b) and Marshall and DePaolo (1982, 1989) achieved a ca. 50% improvement in measurement precision by increasing the beam intensities by 10× relative to those quoted above (i.e., $^{40}Ca^{+}$ beam intensity of 4×10^{-10} Amp. Hence the best that has so far been done is a reproducibility of about ±0.15‰, or ±1.5 units of ε_{Ca} at the 2σ level.

Analysis methods and issues for mass dependent Ca isotope fractionation

The limitations discussed above also apply approximately to measurements of mass dependent Ca isotope effects. The additional problem is to separate mass dependent fractionation in nature from mass dependent fractionation in the mass spectrometer. The maximum observed natural fractionation is about ±0.1% per mass unit, whereas instrumental fractionation is about ±0.5% per mass unit (for TIMS and much larger for ICPMS). The separation is accomplished with the use of a double spike (Russell et al. 1978b). The approach is illustrated here using the methods of Skulan et al. (1997), but other researchers have used slightly different algorithms and double spike isotopes (Zhu and MacDougall 1998; Heuser et al. 2002; Schmitt et al. 2003a).

Once geological samples are dissolved, a mixed ^{42}Ca–^{48}Ca tracer is added to the sample ($^{42}Ca/^{48}Ca \approx 1$). The isotopic abundances in the mixed sample-tracer solution are illustrated in Figure 1. When the mixed solution has roughly equal amounts of the isotopes ^{42}Ca, ^{44}Ca, and ^{48}Ca, then a near-optimal situation is achieved with regard to the corrections for the presence of the tracer (cf. Johnson and Beard 1999 for a detailed analysis of spike-sample ratios). The

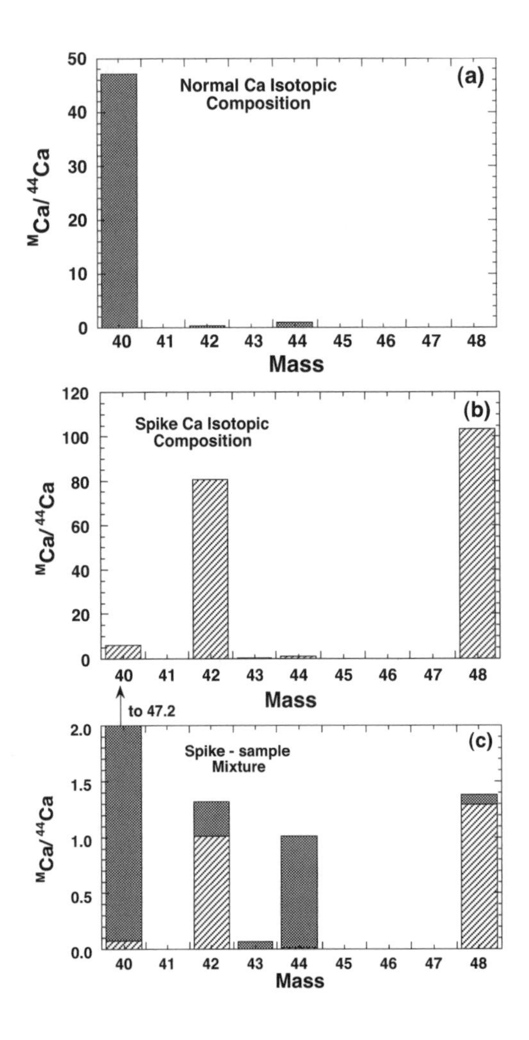

Figure 1. Schematic representation of the calcium mass spectrum in (a) natural materials, (b) a ^{42}Ca-^{48}Ca tracer solution used for separating natural mass dependent isotopic fractionation from mass discrimination caused by thermal ionization, and (c) a typical mixture of natural calcium and tracer calcium used for analysis. The tracer solution has roughly equal amounts of ^{42}Ca and ^{48}Ca. In (c) the relative isotopic abundances are shown with an expanded scale. Note that in the mixed sample, masses 42 and 48 are predominantly from the tracer solution, and masses 40 and 44 are almost entirely from natural calcium. This situation enables the instrumental fractionation to be gauged from the $^{42}Ca/^{48}Ca$ ratio, and the natural fractionation to be gauged from the sample $^{44}Ca/^{40}Ca$ ratio.

dissolved sample-tracer mixture is then loaded onto a cation exchange column and eluted with 2.5 N HCl. This step is needed to separate Ca from other major cations such as Mg, Ti, Fe, Na, K, and Al. The purified Ca is loaded onto Re filaments with tantalum oxide powder for mass spectrometric analysis. It is important to mix the sample and double-spike tracer solution prior to ion exchange separation of Ca because the ion exchange column can introduce substantial mass-dependent isotopic fractionation if the Ca yield of the column is not close to 100% (Russell et al. 1978a).

The mass spectrometric analysis uses Ca^+ ions and is carried out (at University of California, Berkeley) on a single-collector mass spectrometer: a modified VG354 design with one large Faraday bucket collector. Ca has been difficult to analyze reliably using a multi-collector mass spectrometer due to unspecified problems that cause poor reproducibility. Multi-collector runs typically result in excellent internal precision (≈0.05‰), but poor external precision (reproducibility of standards and repeat runs of samples can be as large as ±0.5‰), apparently because the measurements are affected by subtle differences in focusing from sample to sample (cf. Heuser et al. 2002). The single collector method has the disadvantage of requiring a very stable ion beam and long analysis times (several hours per sample), but the advantage of needing much less calibration, relative to the multi-collection method. There is an expectation that certain design improvements to the latest generation multi-collector instruments will make them more attractive for Ca isotope measurements. Fletcher et al. (1997b) and Heuser et al. (2002) report multi-collector TIMS measurements but do not report a large improvement in reproducibility. Halicz et al. (1999) report multi-collector ICPMS measurements with reproducibility similar to that achieved with other methods.

Normally it is possible to obtain stable $^{40}Ca^+$ ion beams with intensities up to about 10^{-10} amp. This intensity requires that about 3–5 μg of Ca be loaded onto the mass spectrometer filament. The amount of Ca needed for analysis is not generally a problem, because Ca is a major constituent of bone and shell material, and is present in natural waters in relatively high concentration as well. The intensity of the $^{40}Ca^+$ ion beam is adjusted so that it is between 6 and 9×10^{-11} amp. Beam intensity is normally kept within a narrow range for all measurements because large ion beams can produce a non-linear response of the amplifier feedback resistor. If the beam intensity is kept within a small range, the effect of any non-linearity on measured isotope ratios is minimized when samples and standards measured with the same instrument are compared. The large $^{40}Ca^+$ peak is integrated for one second whereas five second integrations are used on the smaller peaks $^{42}Ca^+$, $^{44}Ca^+$ and $^{48}Ca^+$, which are about 40 times smaller than $^{40}Ca^+$. For the smaller peaks this results in the counting of 5×10^7 ions per measurement, and for 200 measurements per sample per isotope, with the corrections involved, yields an analytical uncertainty of 0.1 to 0.2‰ at the 2σ level.

Using the three measured ratios, $^{42}Ca/^{40}Ca$, $^{44}Ca/^{40}Ca$ and $^{48}Ca/^{40}Ca$, three unknowns can be solved for: the tracer/sample ratio, the mass discrimination, and the sample $^{44}Ca/^{40}Ca$ ratio (see also Johnson and Beard 1999; Heuser et al. 2002). Solution of the equations is done iteratively. It is assumed that the isotopic composition of the ^{42}Ca–^{48}Ca tracer is known perfectly, based on a separate measurement of the pure spike solution. Initially it is also assumed that the sample calcium has a "normal" Ca isotopic composition (equivalent to the isotope ratios listed in Table 1). The $^{42}Ca/^{48}Ca$ ratio of the tracer is determined based on the results of the mass spectrometry on the tracer-sample mixture, by calculating the effect of removing the sample Ca. This yields a $^{42}Ca/^{48}Ca$ ratio for the tracer, which is in general different from that previously determined for the tracer. This difference is attributed to mass discrimination in the spectrometer ion source and is used to calculate a first approximation to the parameter "p" which describes the instrumental mass discrimination (see below). The first-approximation "p" is used to correct the *measured* isotope ratios for mass discrimination, and then a first-approximation tracer/sample ratio and a first-approximation sample $^{44}Ca/^{40}Ca$

ratio are calculated with standard spike-subtraction equations. This first-approximation sample $^{44}Ca/^{40}Ca$ ratio is in general different from the standard ratio, as a result of natural mass fractionation. Using the first approximation sample $^{44}Ca/^{40}Ca$ ratio, the $^{42}Ca/^{40}Ca$ and $^{48}Ca/^{40}Ca$ ratios are corrected for the natural mass fractionation, and these once-corrected sample Ca isotope ratios are then used to determine a second-approximation $^{42}Ca/^{48}Ca$ for the tracer, a second approximation "p," and second-approximation values of tracer/sample and sample $^{44}Ca/^{40}Ca$. This is continued until "p" no longer changes at the level of 1 ppm, which requires 3 to 5 iterations. The measured sample $^{44}Ca/^{40}Ca$ ratio is then converted to $\delta^{44}Ca$ using the relationship shown in Table 1.

The essential nature of the exponential parameterization of instrumental mass discrimination given by Russell et al. (1978b) is represented by the following two equations:

$$\frac{^{44}\mathrm{Ca}}{^{40}\mathrm{Ca}}_{\mathrm{frac}} = \frac{^{44}\mathrm{Ca}}{^{40}\mathrm{Ca}}_{\mathrm{unfrac}} \left(\frac{M_{44}}{M_{40}}\right)^{p} \tag{1a}$$

$$\frac{^{48}\mathrm{Ca}}{^{44}\mathrm{Ca}}_{\mathrm{frac}} = \frac{^{48}\mathrm{Ca}}{^{44}\mathrm{Ca}}_{\mathrm{unfrac}} \left(\frac{M_{48}}{M_{44}}\right)^{p} \tag{1b}$$

which gives the relationship between "fractionated" ratios and "unfractionated" ratios for two Ca isotope ratios. The parameters M_n are the exact atomic masses of the isotopes of mass number "n," and p is the discrimination parameter. These two equations show that for the same value of p, the amount of fractionation is greater for $^{44}Ca/^{40}Ca$ than it is for $^{48}Ca/^{44}Ca$; the discrimination is dependent not on the mass difference, but on the fractional mass difference (mass 44 is about 10% larger than mass 40, but mass 48 is only about 9.1% larger than mass 44). It is assumed that the fractionation of Ca isotopes in the mass spectrometer and the fractionation in nature are described by the same exponential law. If this were not the case it would make little difference to the results, because Ca isotopic fractionation in nature is sufficiently small.

Ca isotope standards and reference materials

The tracer-subtraction procedure adds negligible uncertainty to the measured $^{44}Ca/^{40}Ca$ ratios. However, it is in fact essentially impossible to entirely eliminate the effects of instrumental mass discrimination for the measurements of either the ^{42}Ca-^{48}Ca mixed tracer or for the "standard" Ca isotope ratios. Hence, it is necessary to have a standard material with an agreed-upon value of $\delta^{44}Ca$. At the time of writing of this article there is no such standard.

The isotope ratios listed in Table 1 are those of a CaF_2 purified salt used as an in-house standard by Russell et al. (1978b). Skulan et al. (1997) used a different purified Ca salt as a standard and showed that the $\delta^{44}Ca$ values measured for several terrestrial igneous rocks were close to zero on this scale. The Skulan et al. (1997) scale for $\delta^{44}Ca$ is used here. Zhu and MacDougall (1998) used a standard $^{44}Ca/^{40}Ca$ ratio (0.021747) that is 2.5% higher than the Russell et al. (1978b) value. However, they report all of their measured $\delta^{44}Ca$ values relative to seawater, which has a $\delta^{44}Ca$ value of +0.9 on the Skulan et al. (1997) scale. Hence the Zhu and MacDougall (1998) $\delta^{44}Ca$ values discussed here have 0.9 added to them to make them compatible with the Skulan et al. (1997) scale. Nägler et al. (2000) report $\delta^{44}Ca$ values relative to an in-house fluorite standard. Based on Figure 1 of Gussone et al. (2003), where seawater is plotted with $\delta^{44}Ca = +0.43$, it can be deduced that the $\delta^{44}Ca$ value of the fluorite standard used by Nägler et al. (2000) is close to +0.5 on the Skulan et al. (1997) scale. Hence, 0.5 has been added to the $\delta^{44}Ca$ values reported by Nägler et al. (2000) and Gussone et al. (2003). Schmitt et al. (2003a,b) report values for seawater and refer their measured values to seawater Ca as a standard (i.e., seawater $\delta^{44}Ca = 0$). Hence, 0.9 has been added to the $\delta^{44}Ca$ values reported by Schmitt et al.

As shown below, the $\delta^{44}Ca = 0$ as defined by Skulan et al. (1997) corresponds closely to the isotopic composition of calcium in igneous rocks (Table 2), and hence probably is close to the value for the bulk Earth. It is preferable to define $\delta^{44}Ca = 0$ as the bulk Earth value rather than that of seawater, because the seawater value is determined by geological processes that can change with time (discussed below), whereas the bulk Earth value represents the bulk of the Ca in the planet and does not change. Seawater also has a small radiogenic enrichment of ^{40}Ca relative to bulk Earth calcium (see discussion below). However, there is not yet a generally accepted rock standard, and there are small variations among volcanic rock samples (Table 2). An appropriate standard is needed, preferably a rock sample with high Ca and low K abundance, and with the mantle $^{40}Ca/^{42}Ca$ ratio (i.e., with negligible radiogenic ^{40}Ca). Some authors have measured the NIST calcium carbonate standard SRM915a (Schmitt et

Table 2. Calcium isotopic compositions of terrestrial volcanic rocks, with other related isotopic parameters.

Sample Number	CaO (wt.%)	$\delta^{18}O$	$\delta^{44}Ca$	±2s	ε_{Nd}	$^{87}Sr/^{86}Sr$	Sample Description (Ref.)
D54G	11.32	5.22	−0.33	0.07			Basalt, Marianas dredge (1)
KOO-10	8.13	5.71	−0.17	0.22			Tholeiitic Basalt, Koolau (1)
KOO-21	8.39	5.99	−0.10	0.19		0.70426	Tholeiitic Basalt, Koolau (1)
KOO-55	8.64	5.75	−0.09	0.14		0.70416	Tholeiitic Basalt, Koolau (1)
GUG-6	10.82	5.03	0.21	0.22	7.9		Basalt, Marianas (1)
ALV-1833	11.68	5.10	−0.17	0.23	8.3	0.70293	Basalt, Marianas (1)
HK-02	10.18		−0.16	0.26	7.7	0.70325	Alkali basalt, Haleakala (2)
HU-24	9.42		−0.32	0.23	5.7	0.70357	Alkali basalt, Hualalai (2)
HK-11	9.67		−0.37	0.27	8.2	0.70310	Alkali basalt, Haleakala (2)
HU-05	10.09		0.14	0.16	5.2	0.70360	Alkali basalt, Hualalai (2)
HSDP 452	10.46	4.80	−0.16	0.12	6.9	0.70358	Tholeiitic basalt Mauna Kea (1)
HSDP 160	10.45	4.87	0.34	0.25	7.3	0.70350	Alkali basalt, Mauna Kea (1)
92-12-29	4.89		−0.15	0.22	1.6	0.70446	Unzen Dacite (1992 eruption) (3)
94-02-05	4.91		0.02	0.24	1.4	0.70446	Unzen Dacite (1994 eruption) (3)
76DSH-8	5.35		0.12	0.21		0.70309	Shasta Dacite (Black Butte) (4)
76DSH-8	5.35		−0.11	0.21		0.70309	Shasta Dacite, Hornblende (4)
76DSH-8	5.35		−0.27	0.06		0.70309	Shasta Dacite, Plagioclase (4)
92DLV-113	0.53		−0.09	0.20	−0.3	0.70611	Inyo rhyolite
IO-14	11.82		−0.16	0.19			Basalt , Indian Ocean MORB
IO-38	11.30		−0.23	0.10			Basalt , Indian Ocean MORB
SUNY MORB	10.38		−0.22	0.03			MORB (5)
SUNY MORB	10.38		−0.27	0.12			MORB (5)
Lake Co. Obsidian	0.53		−0.23	0.05			Rhyolite (5)

References: (1) Eiler et al. (1996, 1997); (2) Sims et al. (1999); (3) Chen et al. (1999); (4) Getty and DePaolo (1995); (5) Richter et al. (2003)

al. 2003a,b; Gussone et al. 2003), and that standard could serve to facilitate inter-laboratory comparisons. The results from SRM915a suggest that the adjustments made here to the reported δ^{44}Ca values are appropriate and accurate within about ±0.1‰.

CALCIUM ISOTOPES IN METEORITES AND LUNAR SAMPLES

The Ca isotope ratios of meteoritic samples are of interest because they can give information on early solar system processes and because meteorites represent the materials from which the Earth accreted and hence relate to the expected values for the bulk Earth. Russell et al. (1978b) made the first measurements of stable Ca isotope variations in meteorites. They found variations of about ±1‰ for the ^{44}Ca/^{40}Ca ratio in samples from six different meteorites. Although some of these samples were spiked after having separated the Ca with an ion exchange column and hence may contain artifacts, it is clear from their data that bulk meteorites have some variability in δ^{44}Ca and that the average value is quite close to the terrestrial standard. No data on bulk meteorites have been reported since the Russell et al. (1978b) measurements, and since their one measurement of an ordinary chondrite had a poor Ca column yield, there exist no reliable measurements that can be used to verify the composition of typical chondritic meteorites.

Russell et al. (1977, 1978b) also measured two Apollo 17 and one Apollo 15 basalt samples and found small variations of δ^{44}Ca of ca. ±1‰, but again also spiked the samples after ion exchange separation so the results need to be verified. Russell et al. (1977) measured calcium obtained by lightly leaching an Apollo 15 soil sample and found it to have δ^{44}Ca of +3.3 ± 0.4. This heavy Ca is inferred to be associated with the grain surfaces and result from solar wind sputtering.

Lee et al. (1977, 1979), Niederer and Papanastassiou (1984), Jungck et al. (1984), Ireland et al. (1991) and Russell et al. (1998) reported Ca isotopic measurements on calcium-aluminum rich inclusions (CAI) and calcium-rich aggregates (CA) from the Allende meteorite. Both the CAI and CA show large mass-dependent fractionation effects (Fig. 2). The overall range of δ^{44}Ca is about 10 times larger than is found in terrestrial materials. The large mass fractionation effects are attributable to non-equilibrium evaporation from liquid or partially molten silicate droplets in the solar nebula. Light isotopes of Ca can be preferentially lost during evaporation into a near-vacuum, leaving residual Ca enriched in heavy isotopes and hence with high δ^{44}Ca. Condensation of light Ca from a vapor phase can generate solids with low δ^{44}Ca. Apparently both processes affected the calcium-aluminum rich materials. Similar fractionation effects are found also in Mg isotopes (Lee et al. 1977, 1979) and silicon isotopes (Clayton et al. 1988), although correlations between Ca isotope fractionation effects and Mg isotope fractionation effects are complex (Niederer and Papanastassiou 1984). Experimental studies have simulated the Si and Mg isotopic effects of high temperature evaporation into a low-pressure hydrogen medium similar to the solar nebula (Davis et al. 1990; Richter et al. 2002) and current models for the formation of CAI's involve cycling through the inner, high-temperature part of the nebula (Shu et al. 1996; Davis and MacPherson 1996). Consequently there are ample opportunities for heating and cooling to generate evaporation and recondensation.

The studies by Lee et al. (1977, 1979), Niederer and Papanastassiou (1984), as well those by Jungck et al. (1984), Ireland et al. (1991), Weber et al. (1995) and Russell et al. (1998), report small deviations of the abundances of ^{42}Ca, ^{46}Ca and ^{48}Ca relative to those of terrestrial materials. The most commonly observed deviation from normal terrestrial isotopic abundances is enrichment in ^{48}Ca, which can be quite large (Fig. 2b). These deviations are significant because they provide direct evidence for the existence of certain (in this case neutron-rich) nucleosynthetic environments in stars (cf. Cameron 1979 and references in the other papers listed above).

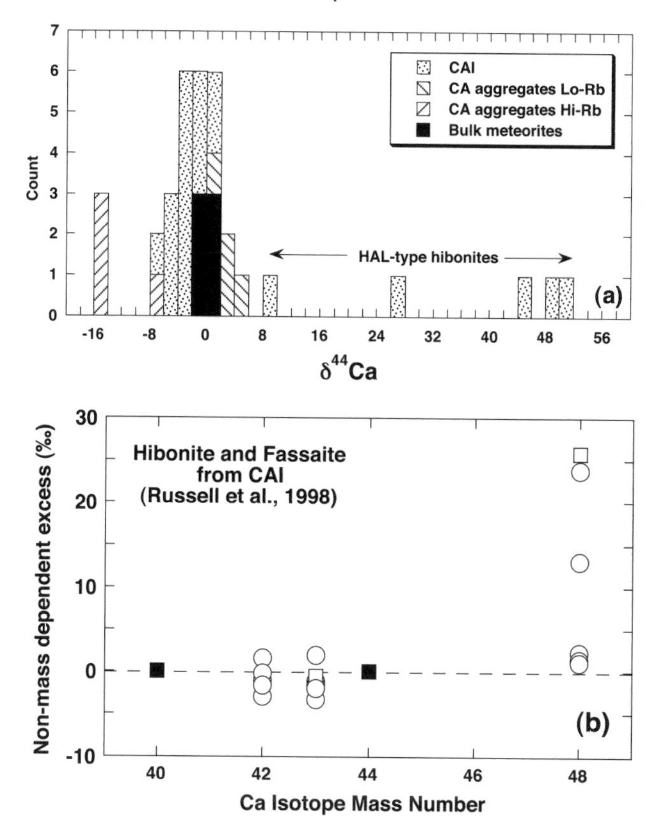

Figure 2. (a) Histogram of δ^{44}Ca values measured for samples from calcium-aluminum rich inclusions (CAI), calcium-aluminum rich aggregates (CA) from the Allende meteorite by Niederer and Papanastassiou (1984) and Lee et al. (1977, 1979), and bulk meteorites by these authors as well as Russell et al. (1978). Bulk meteorites cluster within about ±1‰ of δ^{44}Ca = 0, which is also the value that characterizes terrestrial igneous rocks (see Figs. 5 and 14). The large effects in Ca-Al rich samples have been interpreted as resulting from evaporation of silicate liquid in the early Solar Nebula at high temperature under non-equilibrium conditions, and recondensation of vapor enriched in light Ca isotopes. The large positive δ^{44}Ca values are from hibonite from CAI's and are most likely indicative of residual material that has been affected by partial evaporation. The Ca-Al aggregates have both high and low δ^{44}Ca, which tends to correlate with the amount of the volatile element Rb. (b) Examples of ^{48}Ca excess due to nucleosynthetic processes. These data were obtained by ion microprobe measurements of the minerals hibonite (circles) and fassaite (squares) from calcium-aluminum rich inclusions in type CO3 meteorites (Russell et al., 1998). The values shown represent isotopic ratios that have been corrected for mass dependent fractionation using the ^{44}Ca/^{40}Ca ratio (hence both ^{40}Ca and ^{44}Ca plot at zero); the other plotted values are residuals that cannot be explained by mass dependent fractionation. Typical uncertainty for these measurements is ±4‰.

TERRESTRIAL STUDIES OF RADIOGENIC CALCIUM

The radioactive, naturally occurring isotope of potassium—^{40}K—decays by β^- emission to ^{40}Ca. The accepted value for the total decay constant of ^{40}K is 5.543×10^{-10} yr^{-1}, which corresponds to a half-life of 1.25 billion years (Steiger and Jager 1977). As a consequence of radioactive decay, the abundance of ^{40}Ca increases with time in any substance that contains potassium, and this increase is independent of any mass-dependent isotopic fractionation of Ca isotopes. The equation describing the increase in ^{40}Ca with time can be written:

$$\frac{^{40}\text{Ca}}{^{n}\text{Ca}}(t) = \frac{^{40}\text{Ca}}{^{n}\text{Ca}}(t_o) + \frac{^{40}\text{K}}{^{n}\text{Ca}}(t) R_{\beta-}\left[e^{\lambda_{40}(t-t_o)} - 1\right] \quad (2)$$

where n is the mass number of one of the non-radiogenic Ca isotopes, λ_K is the total decay constant of ^{40}K, and $R_{\beta-}$ is the branching ratio (Steiger and Jaeger 1977):

$$R_{\beta-} = \frac{\lambda_{\beta-}}{\lambda_K} = 0.8952$$

To maximize analytical precision and reproducibility, Marshall and DePaolo (1982) chose $n = 42$, and hence report all data in terms of ^{40}Ca/^{42}Ca ratios normalized to ^{42}Ca/^{44}Ca = 0.31221. This choice allows one to use an isotope ratio spanning two mass units (^{42}Ca/^{44}Ca) to make a correction (for instrumental mass discrimination) to another isotope ratio spanning two mass units (^{40}Ca/^{42}Ca). The only other likely choice is to use ^{40}Ca/^{44}Ca (i.e., $n = 44$), which spans four mass units and hence would have twice as large a correction for instrumental mass discrimination.

Marshall and DePaolo (1982) demonstrated the use of the K-Ca method for geochronology by dating the Pikes Peak granite, which has also been dated by other methods (Fig. 3). Earlier work on K-Ca geochronology is reported by Inghram et al. (1950), Herzog (1956), Polevaya et al. (1958), and Coleman (1971). The earlier studies did not have the benefit of the more precise isotope ratio measurement techniques offered by automated mass spectrometers and were restricted to studies of minerals like sylvite and lepidolite which have extremely high K/Ca ratios. Baadsgaard (1987) also measured K-Ca ages of evaporite minerals and used them to clarify the prolonged history of diagenesis of the deposits. More recently Nägler and Villa (2000) measured both K-Ar and K-Ca ages on gem quality minerals in an attempt to measure the ^{40}K branching ratio on natural samples. They found that Ar ages are systematically younger

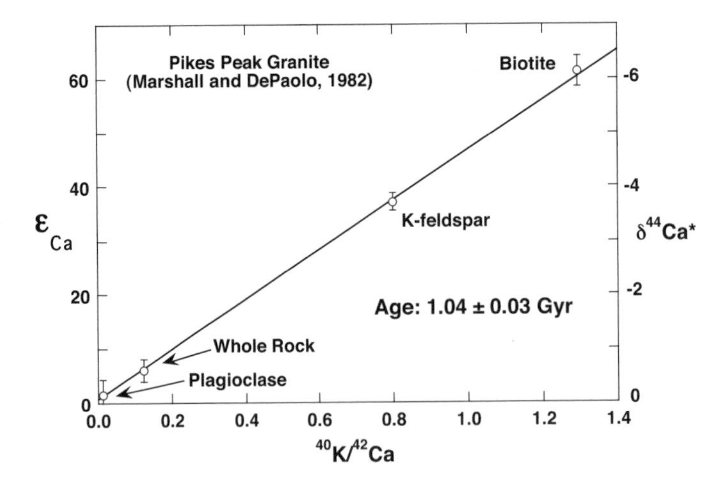

Figure 3. Variation of the ^{40}Ca/^{42}Ca value in minerals of the 1 billion year old Pikes Peak granite as measured by Marshall and DePaolo (1982). The currently accepted age is 1.08 Ga (Schärer and Allègre 1982; Smith et al. 1999). The left side scale shows the values as deviations from the mantle ^{40}Ca/^{42}Ca in units of ε_{Ca}, which is the parameter used to describe the radiogenic enrichments. The right side scale shows the equivalent value of δ^{44}Ca that would be inferred for a sample with radiogenic enrichment of ^{40}Ca, but analyzed according to the normal procedures for measuring mass dependent fractionation without correction for the radiogenic component. Whole rock ^{40}Ca enrichments typically are not greater than a few tenths of a unit of δ^{44}Ca, but mineral ^{40}Ca enrichments can be quite large.

by a few to several percent than K-Ca ages and suggest a small revision of the branching ratio based on one of their samples. However, the ultimate conclusion is that Ar loss from natural samples cannot be ruled out and hence any measurement of the branching ratio using geological samples is likely to have too much uncertainty to make one confident about revising the currently accepted value. Marshall et al. (1986) measured both K-Ca and K-Ar ages on Paleozoic authigenic sanidine and found that the K-Ar ages were systematically 8% younger. They interpreted the data to mean that Ar had diffused out of the sanidine, and that the diffusivity of Ar at ca. 20°C was several orders of magnitude higher than would be deduced from high temperature diffusion data. Fletcher et al. (1997a) measured both K-Ca and Rb-Sr ages in muscovites from Archean granitic rocks in Australia and found that the K-Ca dates are in general substantially younger than the Rb-Sr dates, and both are substantially younger than the accepted crystallization age of the rocks. Fletcher et al. (1997a) infer that the diffusivity of Ca in micas is about an order of magnitude greater than Sr in micas.

Because the ^{40}Ca isotopic enrichments in whole rocks are relatively small, Marshall and DePaolo (1982, 1989) used the ε_{Ca} value as defined in Table 1, which refers the $^{40}Ca/^{42}Ca$ value of rocks to the value expected in the Earth's mantle. Strictly speaking, the $^{40}Ca/^{42}Ca$ value of the mantle changes with time. However, taking the approximate value of K/Ca of the mantle to be 0.01, the $^{40}K/^{42}Ca$ ratio is about 0.0002, and the total change of $^{40}Ca/^{42}Ca$ over the age of the Earth is 0.002 or about 0.12 units of ε_{Ca}.

In general, magmas generated by melting of the mantle should have negligible enrichments of ^{40}Ca and hence have ε_{Ca} values that are zero within the analytical resolution of about ±1. However, granitic magmas that are melted from ancient K-rich crustal rocks can have significant enrichments of ^{40}Ca. The data reported by Marshall and DePaolo (1982, 1989) are shown in Figure 4. The highest ε_{Ca} values are found in granitic rocks with high K/Ca ratios (Fig. 4a) and with low ε_{Nd} values that indicate magma derived from crustal rocks (Fig. 4b). Based on the Sm-Nd model ages of the rocks, which are 1.8 to 3 Ga, the high ε_{Nd} values of the Tertiary and Cretaceous granites indicate that the magmas were melted from crustal rocks that had K/Ca ratios of about 0.5 to 2. Such high K/Ca values generally correspond to rocks with SiO_2 contents of about 65%. The Wyomingite lava measured from the Leucite Hills, which has a very high K/Ca ratio, shows very little ^{40}Ca enrichment as is expected for mantle rocks insofar as "enriched mantle" is unlikely to have a K/Ca ratio much higher than about 0.1. An unexpected aspect of the data, which has gone unconfirmed, is the slight enrichment of ^{40}Ca observed for island arc andesites and some low-K, high ε_{Nd} granitic rocks from the western Sierra Nevada. This effect would suggest that a large fraction of the calcium in island arc lavas comes from subducted carbonates, seawater-altered oceanic crust, and/or continental sediments. Nelson and McCulloch (1989) also reported a number of analyses of ε_{Ca} on mafic igneous rocks including carbonatites, kimberlites, ultrapotassic lavas, and island arc basalts. Some of their samples appear to have slightly elevated ε_{Ca}, but their interpretation was that none of the enrichments was clearly outside the analytical uncertainty of ±1.8 units.

Because of the 1.25 Gyr half-life of ^{40}K, a large amount of the Earth's ^{40}K has decayed over the past 4.55 Gyr. The $^{40}K/^{39}K$ ratio was higher by about 12.5 times at the time of formation of the earth relative to the present ratio, and was still enhanced by 4 to 9 times at 2.5 to 4.0 Ga. Consequently, ancient high-K granites can potentially have larger ε_{Ca} values if derived from yet older high-K crustal rocks. This aspect of the K-Ca system was employed by Shih et al. (1994) for dating of a ca. 3.6 Ga lunar "granite." However, the K-Ca age obtained on the granite was lower than the likely crystallization age.

The average age of continental crustal rocks is about 2 Ga. The typical upper crustal rock has K/Ca of 0.9 and the bulk continental crust has a ratio of about 0.35 (Rudnick and Fountain 1995, Taylor and Mclennan 1995, Shaw et al. 1986, Condie 1993, Wedepohl 1995). Using these values, the average continental crustal value for ε_{Ca} is 0.8 and that for upper continental

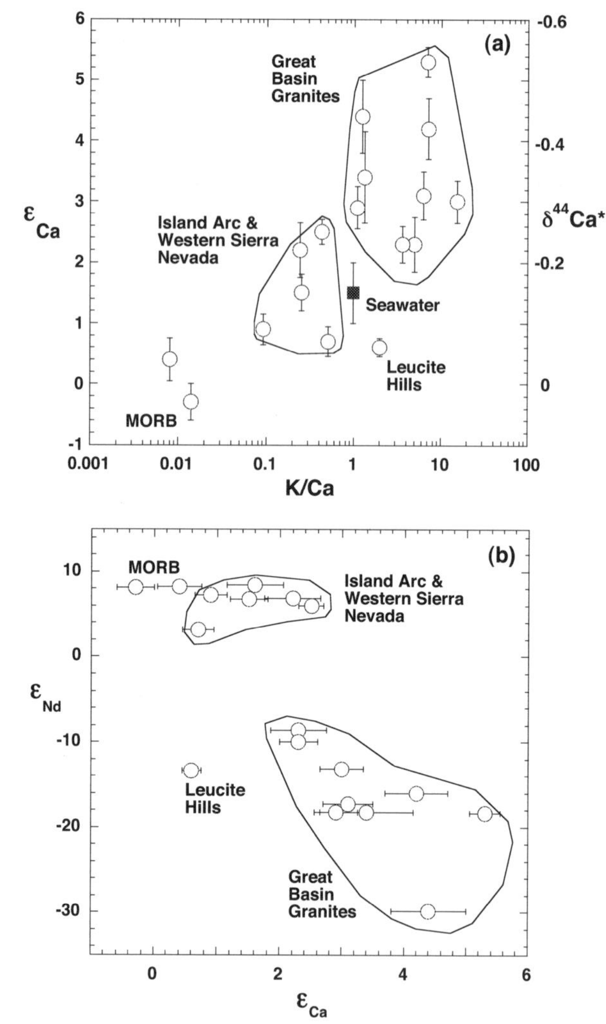

Figure 4. Radiogenic ^{40}Ca enrichments measured on whole rocks as reported by Marshall and DePaolo (1989), shown here as ε_{Ca} values with 1-sigma uncertainties and plotted versus (a) K/Ca and (b) ε_{Nd} value. The inferred value for seawater is also plotted in (a). Mid-ocean ridge basalts have no measurable enrichment of $^{40}Ca/^{42}Ca$ relative to the initial value for the Earth (151.016), as expected for magma derived from the Earth's mantle. Many granitic rocks, especially those with high K contents and low ε_{Nd} values, have significantly elevated ε_{Ca} values.

crust is 2.0. Marshall et al. (1986) measured modern shell material representative of seawater and found it to have an ε_{Ca} value of +1.5 ± 1.0 (Fig. 4a). The slight enrichment of ^{40}Ca in seawater is consistent with the estimated crustal ε_{Ca} values.

Effects of radiogenic calcium ^{40}Ca on measured $\delta^{44}Ca$ values

Radiogenic enrichments of ^{40}Ca can be accounted for in the double spike technique simply by changing the reference value of the $^{40}Ca/^{44}Ca$ ratio for "unfractionated" calcium (Table 1). Figures 3 and 4a show on the right hand scale the equivalent value of $\delta^{44}Ca$ that would be deduced from a double spike measurement that did not account for the radiogenic enrichment of ^{40}Ca. The $^{40}Ca/^{44}Ca$ values reported by Russell et al. (1978b), which were measured on purified CaF_2, probably represent continental crustal values and hence include an enrichment of ^{40}Ca of perhaps 0.1 to 0.2‰ relative to mantle samples and meteorites. This difference is only marginally significant because the typical analytical uncertainty is also about 0.1 to 0.2‰. However, the radiogenic component must be recognized (cf. Zhu and MacDougall

1998) and can be significant if not dominant in some cases. For example, in soils it is often found that biotite weathers faster than other silicate minerals (White et al. 1996; Blum and Erel 1997). Calcium derived from weathering biotite from the Pikes Peak granite (Fig. 3) would appear to have a δ^{44}Ca value of −6, and the Ca from the K-feldspar would have δ^{44}Ca of about −4. The radiogenic component of calcium from potassium minerals could be beneficial for tracing the fate of the calcium from these minerals during weathering and soil formation and during metamorphism (Marshall 1984). To be rigorous, the measurement of stable isotope fractionation effects should include determinations of the radiogenic ^{40}Ca enrichment as well (Zhu and MacDougall 1998; Schmitt et al. 2003b). In practice this is not necessary for many samples, and in addition the small ^{40}Ca enrichments are typically difficult to resolve. For example, Schmitt et al. (2003b) use seawater as a reference and claim to be able to resolve radiogenic ^{40}Ca effects at the level of about ±0.2‰. The seawater value of δ^{44}Ca, 0.15 ± 0.2, covers most of the observed range in whole rock samples due to radiogenic effects (Fig. 4a).

STUDIES OF TERRESTRIAL CALCIUM STABLE ISOTOPE FRACTIONATION

Igneous and metamorphic rocks and petrogenetic processes

The measurements that are available to assess high temperature fractionation of Ca isotopes in nature do not constitute a representative sampling of crystalline rocks; most of the samples measured are volcanic and are relatively common rock types (Table 2; Fig. 5). There are few or no measurements available for clastic sedimentary rocks, soils, high-K granites, and many other rock types that might be of interest. All of the measured δ^{44}Ca values of typical volcanic rocks are between −0.3 and +0.3. The mean and standard deviation of the values from the table is −0.11 ± 0.18. Since virtually all of the samples listed in Table 2 are mantle-derived rocks with no significant involvement of old continental rocks, the δ^{44}Ca values do not reflect any radiogenic ^{40}Ca enrichment. Hence the slightly negative average value may in fact be confirmation that the mantle ^{40}Ca/^{44}Ca ratio is slightly lower than the value for the Skulan et al. (1997) standard. The variation of the δ^{44}Ca values is larger than analytical uncertainty, which suggests that there are small variations in igneous rocks that are barely resolvable with current techniques. There are some hints that the δ^{44}Ca values may correlate with other isotopic parameters (Fig. 5), but there are too few data to be definitive.

Richter et al. (2003) have shown experimentally that there are physical processes that can fractionate Ca isotopes significantly (Fig. 6). A diffusion couple consisting of basaltic liquid (10.38% CaO) and rhyolitic liquid (0.5% CaO) was allowed to evolve for 12 hours at a temperature of 1450°C. Isotopic effects are generated during multi-component diffusion where Ca (as well as Mg, Fe, etc.) diffuses from the basalt liquid into the rhyolite liquid. The basalt and rhyolite start out with identical δ^{44}Ca values of −0.2. The lighter isotopes of Ca diffuse slightly faster than the heavier isotopes and consequently the basalt becomes enriched in heavy Ca while the rhyolite develops a negative δ^{44}Ca value. As shown in the figure, the experiment generates an isotopic contrast of about 6.5‰ in δ^{44}Ca where none existed at the outset. This type of effect has not yet been observed in natural samples, but the experiments confirm that diffusion within silicate melts can fractionate Ca isotopes. Richter et al. (2003) modeled the results and found that the ratios of the diffusivities of the two isotopes is described approximately by:

$$\frac{D_{44}}{D_{40}} = \left(\frac{m_{40}}{m_{44}}\right)^{0.075} \tag{3}$$

where D_{44}/D_{40} is the ratio of the diffusivities of the diffusing species containing ^{44}Ca and

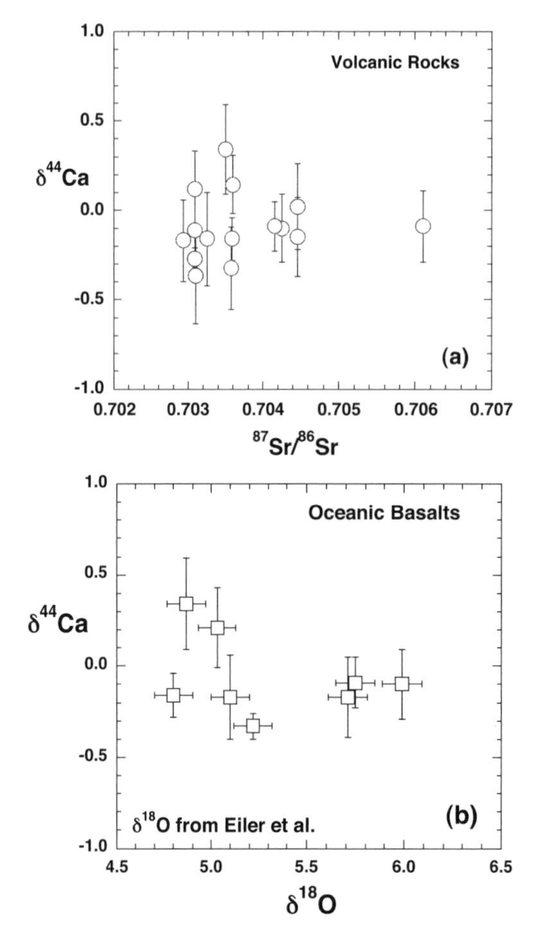

Figure 5. Available data on δ^{44}Ca values in oceanic basalts (Table 2) plotted against (a) ^{87}Sr/^{86}Sr and (b) δ^{18}O. The average δ^{44}Ca value for all igneous rocks measured so far is -0.05 ± 0.2. Small variations in mantle-derived igneous rocks could be expected as a result of recycling (subduction) of seawater-altered rocks, which would have relatively high δ^{44}Ca, recycling of weathering products such as clay-rich sediment, which are also expected to have high δ^{44}Ca (not confirmed by measurements), recycling of oceanic carbonates, which have relatively low δ^{44}Ca, and recycling of materials enriched in radiogenic ^{40}Ca.

^{40}Ca respectively, and m_{40} and m_{44} are the nuclidic masses of the two calcium isotopes. This result indicates that the masses of the diffusing species are considerably larger than the masses of the individual Ca atoms. Richter et al. (2003) did not comment on the nature of the diffusing species in the silicate liquid. However, it may be noteworthy that, for the ratio of the diffusivities to equal the square root of the mass of the diffusing species, the diffusing species containing ^{40}Ca needs to have a mass of about 278, exactly the mass of the anorthite formula unit ($^{40}CaAl_2Si_2O_8$). Hence, a model using the anorthite formula unit as the diffusing species would be consistent with the observations. This approach could be useful for estimating the polymer sizes associated with cations in silicate melts.

Biological fractionation through food chains

The observation suggesting that biological processes fractionate Ca isotopes is the systematic lowering of δ^{44}Ca values through food chains (Skulan et al. 1997). In both terrestrial and marine environments, carnivores at the end of the food chain have significantly lighter skeletal δ^{44}Ca values (Fig. 7). The starting point for the marine systems is seawater, which is about 1‰ higher in δ^{44}Ca than average igneous rocks (the reason for this is discussed further below). Organisms that obtain their Ca directly from seawater, such as foraminifera, mollusks, and fishes, have δ^{44}Ca values of about +0.5 to −1.0. Seals, which get their Ca from

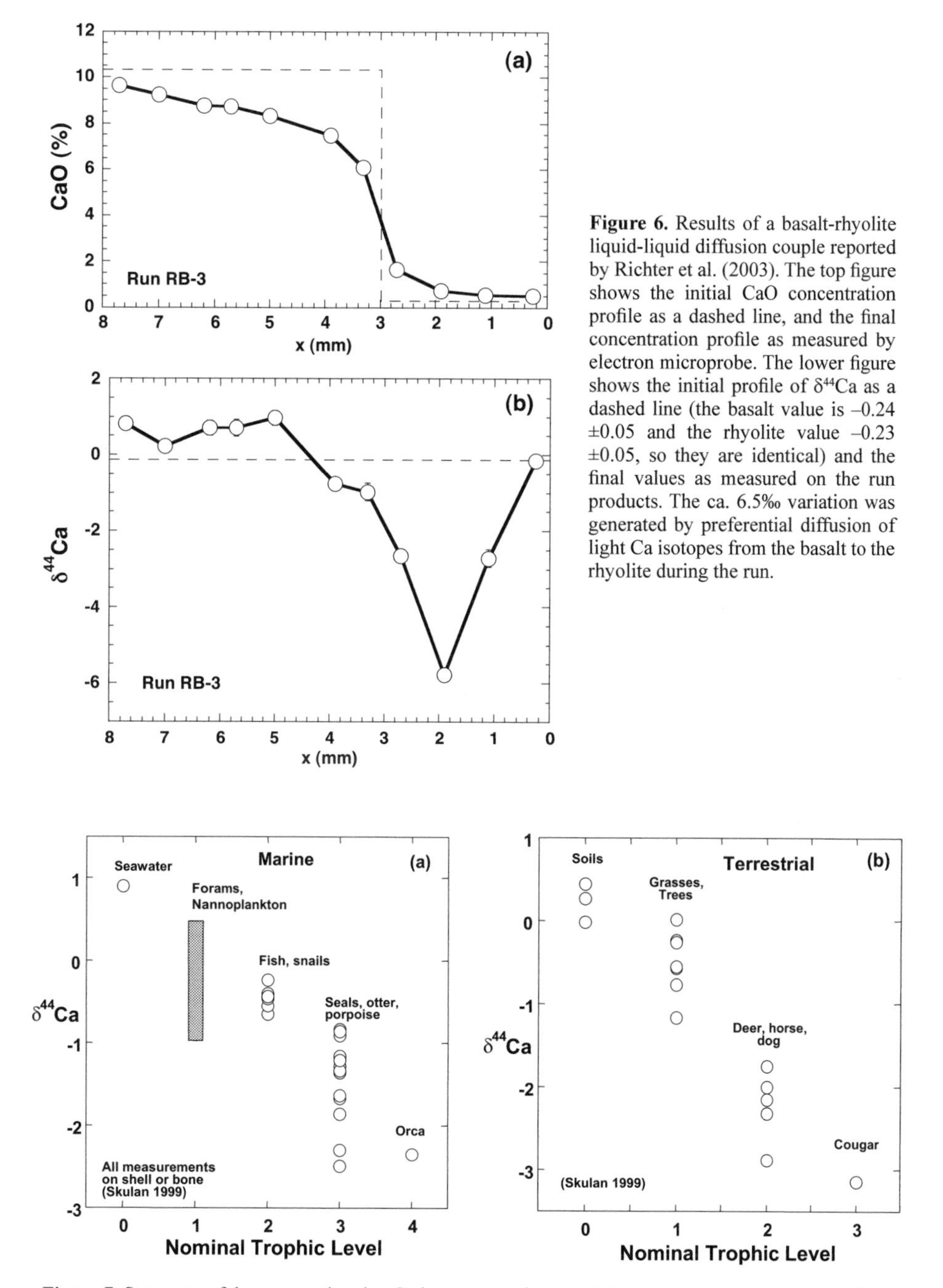

Figure 6. Results of a basalt-rhyolite liquid-liquid diffusion couple reported by Richter et al. (2003). The top figure shows the initial CaO concentration profile as a dashed line, and the final concentration profile as measured by electron microprobe. The lower figure shows the initial profile of δ^{44}Ca as a dashed line (the basalt value is –0.24 ±0.05 and the rhyolite value –0.23 ±0.05, so they are identical) and the final values as measured on the run products. The ca. 6.5‰ variation was generated by preferential diffusion of light Ca isotopes from the basalt to the rhyolite during the run.

Figure 7. Summary of data suggesting that Ca isotopes are fractionated as calcium proceeds through food chains. Data are from Skulan et al. (1997), Skulan and DePaolo (1999) and Skulan (1999). Additional data are reported by Clementz et al. (2003).

smaller organisms rather than seawater, have an average δ^{44}Ca value of about −1.3. Bone from an orca is the lightest material so far measured in a higher marine organism at −2.3. The total difference between orca bone and seawater is 3.2‰. For terrestrial organisms the range of δ^{44}Ca is similar to that found in the marine environment, but the δ^{44}Ca values are lower. In a small sampling of materials in Northern California and in New York (Skulan 1999; Skulan and DePaolo 1999) it is found that whole soils are close to or slightly higher than rock values for δ^{44}Ca, plants are about 1‰ lower, horse and deer have bone δ^{44}Ca values of about −2 and a cougar has a bone δ^{44}Ca value of −3.2. The total range of δ^{44}Ca values in the terrestrial environment is about 3.5‰.

The Ca isotope data from food chains indicate that the Ca fixed in mineral matter in organisms is typically lighter than the Ca available to them from their surroundings or through their diets. In succeeding levels of a food chain the δ^{44}Ca values decrease by about 1 unit per level. Terrestrial animals at analogous levels of the food chain tend to have δ^{44}Ca values that are about 1 unit lower than their marine counterparts. A recent report by Clementz et al. (2003) confirms the systematic relationship between δ^{44}Ca and trophic level in marine organisms, and suggests that the relationship is preserved in 15 million year old bone specimens as well. Skulan (1999) also analyzed a large number of fossil bone and shell samples and suggests that Ca isotopes can be used for paleodiet reconstruction.

Model for Ca isotope fractionation in vertebrates

In an effort to further elucidate the nature of Ca isotope fractionation in animals, Skulan and DePaolo (1999) studied tissues from living and recently deceased (naturally) organisms. For the four animals studied, it was found that bone calcium typically has δ^{44}Ca that is 1.3‰ lower than the value in the dietary calcium (Fig. 8). The soft tissue calcium, however, is quite variable, and has values closer to that of the dietary calcium. Skulan and DePaolo (1999) concluded that the primary source of Ca isotope fractionation was in the formation of bone, and that the δ^{44}Ca values of soft tissue were variable in time and dependent on the immediate status of the Ca balance in the organisms.

The model proposed for understanding the Ca balance, and both the soft tissue and bone δ^{44}Ca values (Fig. 9) suggests that the isotopic values are dependent on the ratio of the available Ca coming from dietary intake, and the rate of bone growth and/or bone loss. The assumption is that exchange of Ca between different soft tissues involves negligible isotopic fractionation. Fractionation occurs only when Ca is fixed into bone, and the fractionation factor is about −1.5‰ (Fig. 9a). If the dietary supply of Ca (V_d) is much larger than the rate of Ca use by bone growth (V_b), then bone should have δ^{44}Ca about 1.5 units lower than the dietary Ca, and the average δ^{44}Ca value of soft tissue is close to that of the dietary source (Fig. 9b). However, if a large fraction of the dietary supply is being used to produce new bone (i.e., $V_b/V_d \approx 1$), the bone δ^{44}Ca values will be closer to that of the dietary supply, and the soft tissue values will be higher than those of the diet. If there is a significant amount of bone loss by comparison with the dietary Ca supply ($V_l/V_d \geq 1$), then the low-δ^{44}Ca calcium coming from dissolving bone will make the soft tissue δ^{44}Ca values low (Fig. 9c).

The net result, according to the model, is that under circumstances that might be considered typical of healthy juvenile organisms—substantial bone growth but with ample dietary Ca supply—typical bone material is slightly less than 1.5‰ lower in δ^{44}Ca than dietary calcium. In healthy adults, where new bone growth is balanced by bone loss (remodeling), there would be little tendency for the average bone δ^{44}Ca values to change. If there is a shortage of dietary Ca while new bone is being made in juveniles, then the bone δ^{44}Ca values will be closer to the dietary values. If there is major bone loss in adults, then soft tissue (and remodeled bone) δ^{44}Ca values will be exceptionally low (Fig. 9c). The bone δ^{44}Ca values reflect conditions integrated over the residence time of bone Ca, which is many years. Soft tissue δ^{44}Ca values, however,

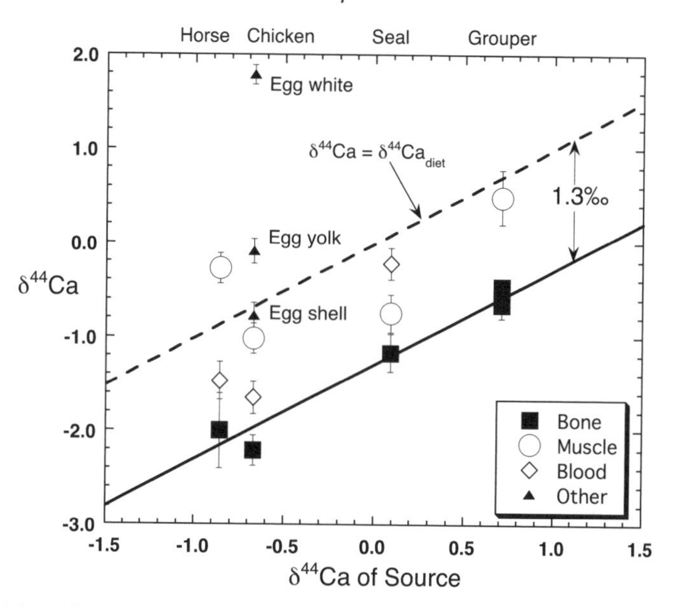

Figure 8. Calcium $\delta^{44}Ca$ values in vertebrate bone and soft tissue samples versus $\delta^{44}Ca$ in dietary source (Skulan and DePaolo 1999). Bone values are systematically about 1.3‰ lower than source values. Soft tissue values are more variable. All of the values are hypothesized to reflect the balance between Ca dietary intake and exchange with bone calcium (Fig. 9). The soft tissue values are variable largely because the residence time of Ca in the tissues is short. The high value of the egg white reflects Rayleigh-type distillation; the egg white loses light Ca to the shell as the shell forms. The small amount of Ca left in the egg white is highly fractionated. The low $\delta^{44}Ca$ value of the seal muscle is interpreted as a sign of distress; the seal may have had a dietary Ca deficiency for several days or longer before it died, and hence was deriving most of its Ca from bone dissolution.

may reflect the conditions of the Ca balance within hours or days of the sampling time. For example, the low $\delta^{44}Ca$ value of muscle Ca in the fur seal, which was found dead, suggests that it had a dietary Ca deficiency for at least several days before it died.

Ca isotope fractionation in marine shelled organisms

The calcium carbonate shells of marine microfauna are a large repository of terrestrial calcium and constitute a potential record of changes in the cycling of calcium at and near the earth's surface (Zhu and MacDougall 1998; De La Rocha and DePaolo 2000; Schmitt et al. 2003a,b). To understand the record held in deep sea carbonate sediments, it is necessary to document any Ca isotopic fractionation that occurs between dissolved seawater Ca and carbonate shell material.

Skulan et al. (1997) reported analyses of marine carbonate sediments of various ages and analyses of foraminifera tests and other marine organisms. They concluded that typical marine shell material had $\delta^{44}Ca$ values that are about 1‰ lower than seawater Ca, and that there might be a small temperature dependence of about 0.02 per degree. Zhu and MacDougall (1998) reported measurements of foraminifera that indicate that the carbonate $\delta^{44}Ca$ values are 0.56 to 1.45‰ lower than those of seawater, and suggested that there was a temperature dependence to the fractionation within individual species, at the level of about 0.05 to 0.1 per degree. Zhu and MacDougall (1998) also reported data on coccolith oozes that were 1.88 to 2.64‰ lighter than seawater. De La Rocha and DePaolo (2000) reported data on one species of foraminifera that showed an average fractionation relative to seawater of −1.3‰ and a hint

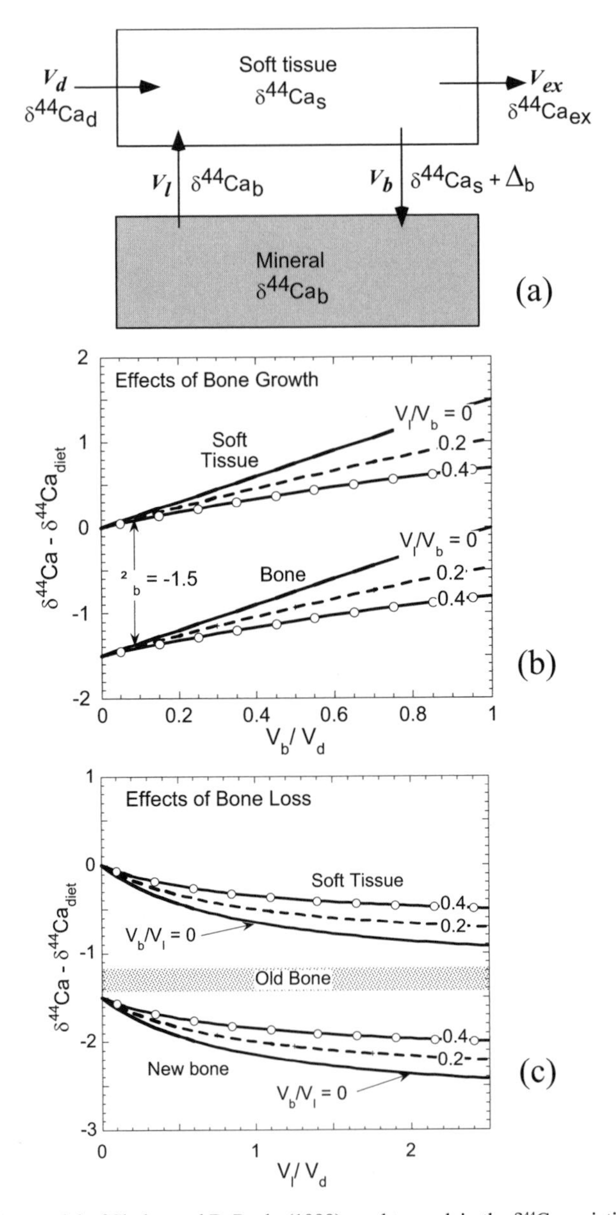

Figure 9. (a) Box model of Skulan and DePaolo (1999) used to explain the δ^{44}Ca variations found in the bone and soft tissue of vertebrates. The δ^{44}Ca values of bone and soft tissue are generally determined by the ratios of the dietary Ca flux (V_d), to the Ca fluxes due to bone formation (V_b), and bone dissolution or loss (V_l). It is hypothesized that only the step associated with bone growth involves isotopic fractionation ($1000\ln\alpha = \Delta_b \approx -1.5$). (b) Calculated effects on calcium isotopic fractionation due to variations of the ratio V_b/V_d, and (c) variations of the ratio V_l/V_d. In both cases, three curves are shown to describe the simultaneous variation of the third flux. Under healthy growth conditions, $V_b/V_d << 1$, and bone develops with δ^{44}Ca that is about 1.3‰ lower than the average value of the dietary calcium. When soft tissue Ca is heavily influenced by Ca derived from bone dissolution, then the soft tissue δ^{44}Ca values approach those of the average bone, and remodeled bone tends to be exceptionally low in δ^{44}Ca.

of a temperature dependence (Fig. 10). De La Rocha and DePaolo (2000) also reported data on cultured specimens of the coccolith *Emiliania huxleyi* and found the δ^{44}Ca values to be 1.3‰ lower than that of the seawater growth medium.

The most striking data yet published are those of Nagler et al. (2000), who provide evidence that cultured specimens of the foraminifer species *Globigerinoides sacculifer* exhibit a strongly temperature-dependent Ca isotope fractionation (Fig. 10). They proposed that this species could be used for ocean paleo-temperature determinations and demonstrated the application with analyses of separated *G. sacculifer* from a Late Pleistocene marine sediment core (Fig. 11a). Subsequent work by Gussone et al. (2003) on a different foraminifer species, *Orbulina universa* indicate that the temperature dependence of the Ca isotope fractionation in other foraminifera is much smaller than in *G. sacculifer* (Fig. 10). Data obtained by the author from bulk carbonate ooze from DSDP Site 590B (cf. Grant and Dickens 2002) show only small variations in δ^{44}Ca over the past 3 million years (Fig. 11b) and contrast with the *G. sacculifer* data. The Site 590B data may be recording small changes in the δ^{44}Ca of seawater or small shifts in local water temperature over this time period (see discussion below).

Gussone et al. (2003) were also able to demonstrate fractionation during inorganic precipitation of aragonite from a Ca-Mg-Cl solution and that the fractionation factor is temperature dependent (Fig. 10). The precipitation experiment involved diffusing CO_2 into a Ca-Mg-Cl solution, which generates $CaCO_3$ precipitation as aragonite. Gussone et al. (2003) claim that the isotopic fractionation observed is a kinetic effect and imply that it is a result of differences in the diffusivity of isotopic species in solution. Although not explained in the paper, they infer that precipitation of aragonite is limited by the rate that dissolved Ca can diffuse to the surfaces of growing aragonite crystals. It is surprising that this process should have temperature dependence, because isotopic fractionation caused by aqueous diffusion

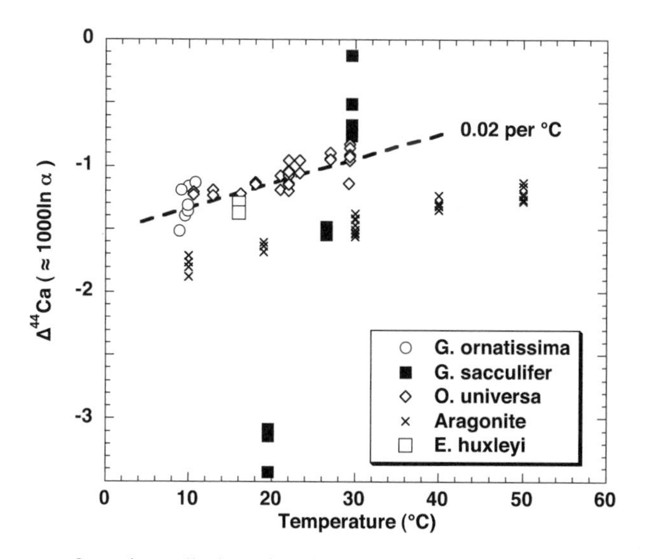

Figure 10. Summary of experimentally determined fractionation factors for Ca isotopes in the formation of foraminifera and coccolith shell carbonate, and for rapid inorganic precipitation of aragonite from an Mg-Ca-Cl solution. Data for the foraminifer *G. ornatissima* and the coccolith *E. huxleyi* are from De La Rocha and DePaolo (2000). Data on *G. sacculifer* are from Nagler et al. (2000). Data for *O. universa* and aragonite are from Gussone et al. (2003). Two of the forams and the coccolith *E. huxleyi* have similar fractionation behavior, with an overall fractionation factor of −1 to −1.5‰, and a small temperature dependence of about 0.02 per °C. The foram *G. sacculifer* appears to have a strongly temperature dependent fractionation factor.

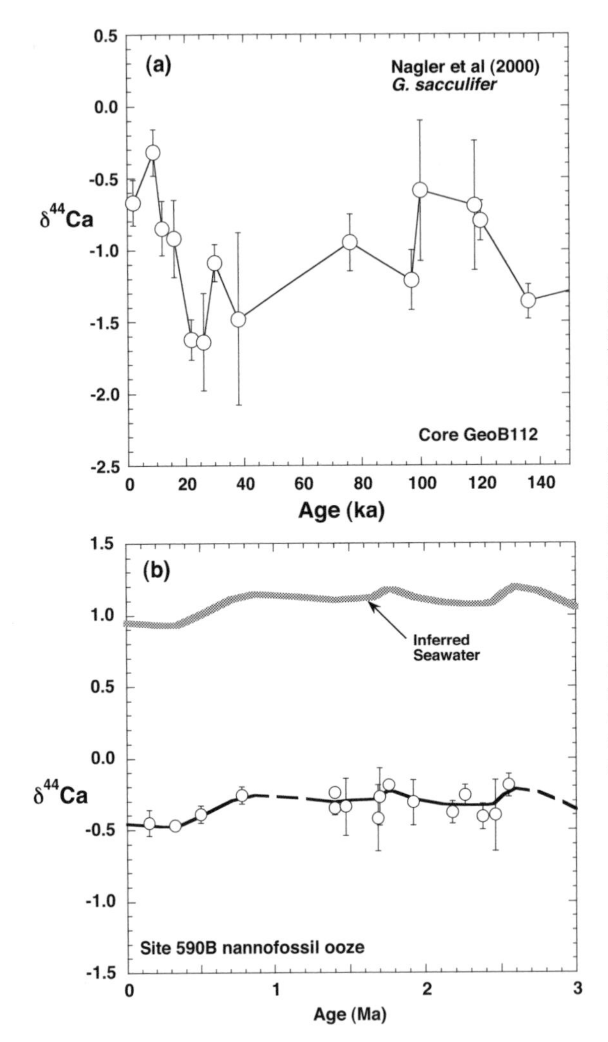

Figure 11. (a) Late Pleistocene $\delta^{44}Ca$ record based on measurements of *G. sacculifer* (from Nagler et al. 2000). The inferred variations of temperature are similar to those inferred from variations of Mg/Ca in the same sediment core. (b) Plio-Pleistocene record of seawater $\delta^{44}Ca$ based on bulk coccolith ooze from DSDP Site 590B in the southwestern Pacific (Tasman Sea). The 2.5 m.y. record shows only small variations of $\delta^{44}Ca$. The seawater curve is constructed assuming that the fractionation between seawater and bulk sediment remained constant. The decrease of $\delta^{44}Ca$ at ca. 0.7 Ma could reflect cooling rather than a change in the seawater $\delta^{44}Ca$.

should be a function only of the mass-dependent part of the diffusivity ratio as in Equation (3). Gussone et al. (2003) assume that they have measured a kinetic fractionation effect, but in fact they give no evidence that it is not an equilibrium fractionation. In any case, the results are important in that they demonstrate a behavior that mimics the fractionation in foraminifera with regard to temperature dependence, and a fractionation factor that is similar in size to that observed in both foraminifera and in vertebrate bone formation. For comparison, Skulan (1999) precipitated $CaCO_3$ from seawater at room temperature and estimated the fractionation factor to be smaller than that found by Gussone et al. (2003), only −0.2 to −0.4‰.

Gussone et al. (2003) argue that the difference in fractionation behavior between foraminifera species may be explained by the way in which Ca ions are transported across cell membranes or through aqueous media by diffusion. The relatively small fractionation observed in *O. universa* and in inorganic aragonite precipitation suggests that the diffusing Ca species is an aquocomplex with a mass of about 600. This would require that the Ca^{2+} ions be associated with about 30 water molecules, or that they are in complexes involving both water molecules and anions. Gussone et al. (2003) suggest the reason that the temperature dependence of the fractionation in *G. sacculifer* is so large is that this organism dehydrates the Ca^{2+} ions before calcification so that the diffusing species is the Ca^{2+} ion. These conclusions are interesting but difficult to understand insofar as the absolute value of the kinetic fractionation factor for Ca isotopes should theoretically be dependent on the mass of the diffusing species, but the fractionation factor should theoretically have no temperature dependence. Hence the difference in the temperature dependence of the fractionation factor for different foraminifera species is not likely to be due to differences in the mass of the diffusing complex.

Calcium in the weathering cycle

The global calcium cycle is intimately linked to the carbon and nitrogen cycles. The weathering of Ca from silicates and its eventual sequestration in marine carbonates plays a major role in the control of atmospheric CO_2 concentrations (Berner et al. 1983; Raymo and Ruddiman 1992; Sundquist 1993). Calcium concentrations in the ocean have varied over geologic time. Kempe and Degens (1985) theorize that Ca^{2+} concentrations in the ocean were near zero from 4–1 Ga, whereas the modern day value is 10mM (Holland 1978; Broecker and Peng 1982). Models suggest secular variations in the marine Ca^{2+} concentration of seawater ranging from 10 to more than 40 mM during the Phanerozoic (Hardie 1996; Stanley and Hardie 1998), and these expectations are confirmed to some degree by studies of fluid inclusions in evaporite minerals (Horita et al. 2002; Fig. 12). Deposition rates of carbonate also vary widely. Massive carbonate deposition occurred during the late Cretaceous, for instance, following an extended period of deposition of carbonate-poor sediments (Kennett 1982).

To relate the isotopic composition of marine calcium to variations in the calcium cycle requires characterization of the $\delta^{44}Ca$ values of the sources and sinks of Ca to the oceans, and estimates of the Ca fluxes. Neither is well documented presently. Estimates of the calcium fluxes from Milliman (1993) are shown in Table 3. There are six analyses of $\delta^{44}Ca$ from mid-ocean ridge hydrothermal vents, and these average +0.2 ± 0.2 (Zhu and MacDougall

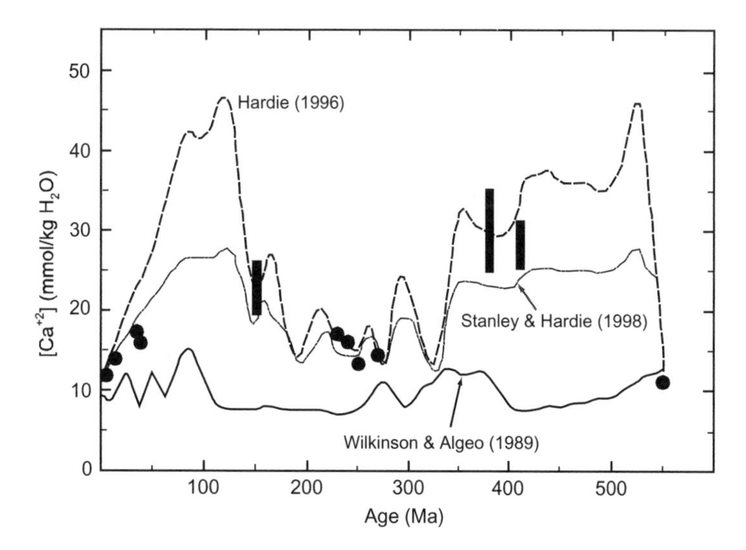

Figure 12. Models and data showing variations in the concentration of Ca^{2+} in the oceans over Phanerozoic time. Figure adapted from Horita et al. (2002), who used the mineralogy of evaporite deposits to infer the values shown as filled circles and bars.

Table 3. Modern marine Ca^{2+} budget*

Inputs	Flux (Tmol y^{-1})	δ^{44}Ca (‰)
Riverine	12-15	−0.2 ± 0.2
Diagenetic	3-5	−0.5 ±0.2
Groundwater	5-16	−0.4 ± 0.2
Hydrothermal	2-3	+0.2 ± 0.2
Aeolian	<0.05	−0.1 ?
Total	**23 to 40**	**−0.3 ± 0.2**
Outputs		
Biogenic Carbonate	**32**	**−0.4 ± 0.2****

*Fluxes are based on Milliman (1993). Estimates of δ^{44}Ca were compiled from Skulan et al. (1997), Zhu and MacDougall (1998), De La Rocha and DePaolo (2000), Schmitt et al. (2003b), and DePaolo (unpublished data).

**Based on Pleistocene and younger deep sea ooze data of De La Rocha and DePaolo (2000) and DePaolo (unpublished; see Fig. 11b). Zhu and MacDougall (1998) estimate this value as about −1.2 ± 0.3.

1998; Schmitt et al. 2003b). There are five analyses of groundwater reported by Schmitt et al. (2003b), and those average δ^{44}Ca = −0.4 ± 0.2. There are no data for pore fluids from deep-sea sediments, but the δ^{44}Ca values of the diagenetic flux from sediments is expected to resemble the average value of carbonate ooze, which as described below is probably about −0.5 ± 0.2.

The δ^{44}Ca values of riverine dissolved calcium show significant variability and correlate with $^{87}Sr/^{86}Sr$ (Fig. 13). The Eel River in Northern California has a high δ^{44}Ca value. The Ganges tributaries, which drain the Himalaya, have low δ^{44}Ca values. Insofar as the

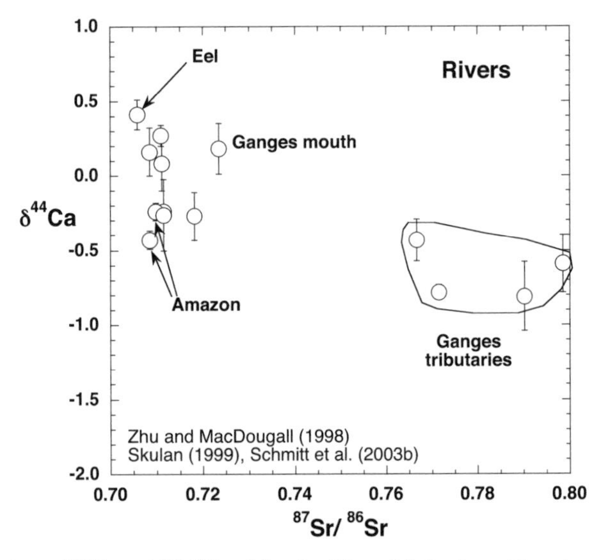

Figure 13. Variations of δ^{44}Ca and $^{87}Sr/^{86}Sr$ of dissolved Ca and Sr in rivers. There is variation of >1.0‰ in δ^{44}Ca although recently reported data (Schmitt et al. 2003) suggest that the low Ganges tributary values are not seen in samples from near the mouth of the Ganges.

Himalayan Rivers also have extremely high $^{87}Sr/^{86}Sr$ ratios, the low $\delta^{44}Ca$ values may be partly attributable to the presence of excess radiogenic ^{40}Ca. Judging from the Sr isotopic data and the low published values of ε_{Nd} for the Himalayan granites (Deniel et al. 1987), the radiogenic ^{40}Ca excess could account for a shift of −0.2 to −0.4 in the $\delta^{44}Ca$ value (see Fig. 4). Zhu and MacDougall (1998) state that their $\delta^{44}Ca$ measurements account for variations in ^{40}Ca but they do not provide the data to demonstrate this. The Ganges tributaries represent an instance where separate analysis of radiogenic ^{40}Ca enrichments would be useful. Overall, the average $\delta^{44}Ca$ value of the riverine flux is not accurately determined. The value given in Table 3 is the approximate average value measured for large rivers.

At steady state the Ca isotopic composition of the inputs of calcium to the oceans should be identical to the outputs. Calcium is removed from the oceans by carbonate sedimentation. Measurements of bulk carbonate sediments are summarized in histogram format in Figure 14. Virtually all of the samples measured at the University of California, Berkeley (Skulan et al. 1997; De La Rocha and DePaolo 2000; and unpublished data from DSDP Site 590B), fall in the range $\delta^{44}Ca$ = +0.2 to −0.9. The mean and standard deviation are −0.5 ± 0.2. Comparison to the river data (Fig. 12) shows that the deep sea carbonate mean value is slightly lower than river calcium, and slightly lower than the average value estimated for the total Ca input to the oceans (Table 3). The bulk carbonate ooze $\delta^{44}Ca$ data from Zhu and MacDougall (1998), however, are distinct from the Berkeley data, and the values are considerably lower than appears likely for the average input calcium.

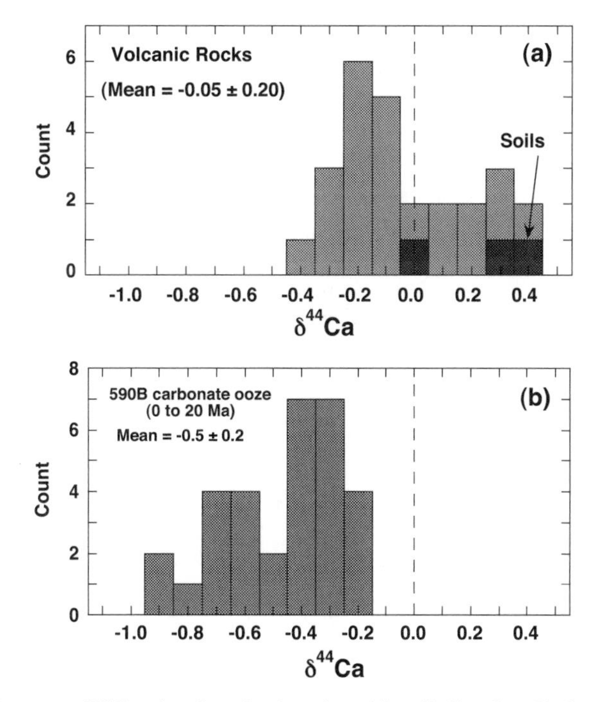

Figure 14. (a) Histogram of $\delta^{44}Ca$ values for volcanic rocks and for soils (data from Skulan et al. 1997; Skulan 1999; Zhu and MacDougall 1998; and Table 2). The average value of $\delta^{44}Ca$ is close to zero. Soils tend to have relatively high $\delta^{44}Ca$ suggesting that weathering preferentially removes light calcium isotopes. The average value of $\delta^{44}Ca$ in carbonate ooze from DSDP Site 590B is significantly lower than the average volcanic rock value, and similar to the average carbonate value reported by De La Rocha and DePaolo (2000). The data suggest that $\delta^{44}Ca$ of the overall weathering flux is lower than that of typical volcanic rocks.

The deduced average input $\delta^{44}Ca$ is not identical to the average volcanic rock value, but is lower by about 0.4‰ (Fig. 14). This suggests that weathering may discriminate between calcium isotopes. This inference is supported by the few available data on soils (Fig. 14; Skulan and DePaolo 1999; Skulan 1999; Schmitt et al. 2003b). The soils generally have $\delta^{44}Ca$ values that are higher than the volcanic rocks. Data to allow direct comparison of soils to parent rock material are not available.

For a simple model of Ca cycling, where the Ca sources for the ocean are weathering of continental rocks, pore fluids in the marine environment, and ocean floor basalt (Gieskes and Lawrence 1981; Berner et al. 1983; Elderfield et al. 1999), and the primary sink is the biological fixation of Ca into sediments, the rate of change of $\delta^{44}Ca$ (= δ_{SW}) of the oceans is given by:

$$N_{Ca}\frac{d\delta_{SW}}{dt} = F_C(\delta_C - \delta_{SW}) + F_H(\delta_H - \delta_{SW}) - F_{Sed}\Delta_{Sed} \tag{5}$$

where N_{Ca} is the number of moles of Ca in the oceans, F_C is the continental (riverine and groundwater) weathering flux modified by the contributions of the marine diagenetic flux, F_H is the hydrothermal flux from mid-ocean ridges and off-ridge oceanic basalt weathering, F_{Sed} is the rate of biological removal of Ca into sediments, and Δ_{Sed} is the average fractionation factor for $\delta^{44}Ca$ associated with biological removal of Ca from the oceans. The steady state $\delta^{44}Ca$ of the ocean is given by:

$$\delta_{SW}(s.s.) = \frac{F_C\delta_C + F_H\delta_H - F_{Sed}\Delta_{Sed}}{F_C + F_H} \tag{6}$$

If the Ca concentration of the ocean is unchanging, then $F_{Sed} \approx F_C + F_H$, and the solution is:

$$\delta_{SW} = \delta_W - \Delta_{Sed} \tag{7}$$

where δ_W is the average value for all rock weathering fluxes including the hydrothermal flux.

The overall Ca isotope cycle is diagrammed in Figure 15. The average igneous rock has $\delta^{44}Ca \approx 0$ and the average carbonate rock has $\delta^{44}Ca \approx -0.5$. The calcium released from weathering of carbonate and silicate rocks probably has $\delta^{44}Ca \approx -0.4$ to -0.5 and the hydrothermal flux modifies this only slightly. The fractionation associated with calcium removal from the oceans is estimated to be $\Delta_{Sed} = -1.4 \pm 0.1$. Hence the expected steady state

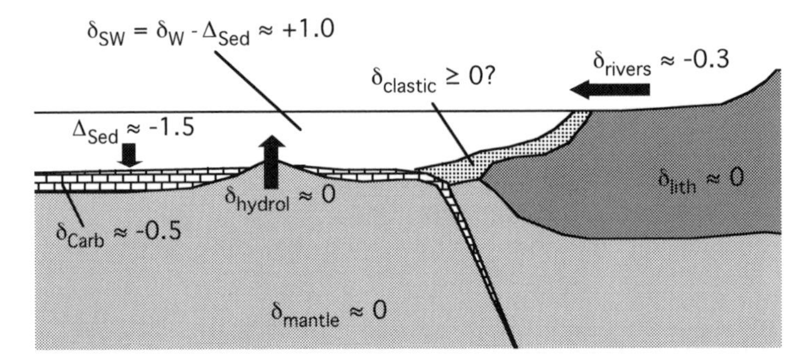

Figure 15. Diagram showing the major components of the global calcium cycle with $\delta^{44}Ca$ values (denoted as δ). The modern residence time of Ca in the oceans is about 1 million years (Holland 1978; 1984). Abbreviations used are: SW = seawater, Sed = sedimentation, clastic = clastic sediments, carb = marine carbonate sediments, hydrol = mid-ocean ridge hydrothermal systems, lith = continental lithosphere.

value for the δ^{44}Ca of the oceans is +0.9 to +1.0. Using these numbers, therefore, it appears that the current system is close to steady state, since the seawater value is in fact very close to +1.0 (De La Rocha and DePaolo 2000; Schmitt et al. 2003b). Zhu and MacDougall (1998), however, suggest that the value for Δ_{Sed} should be −2.2 ± 0.3. If this number were used, then it would have to be concluded that the system is far from steady state. The Zhu and MacDougall (1998) carbonate ooze data are the only data available that suggest that Δ_{Sed} should be much different from −1.4 ± 0.1. The fractionation studies (Fig. 10), as well as the studies of deep sea ooze and chalk by Skulan and DePaolo (1997), De La Rocha and DePaolo (2000), and the unpublished data shown in Figure 11b, all suggest that Δ_{Sed} is in the range −1.3 to −1.5.

Ca isotopic variations in the paleo-oceans

As noted above, it is likely that the calcium input fluxes to the oceans and the outputs fluxes are not always equal. According to Equation (5), this means that the seawater δ^{44}Ca ratio can vary with time. The rapidity with which the seawater δ^{44}Ca can change is dictated by the residence time of calcium in seawater. At present, the residence time is estimated to be about 1 million years (e.g., Holland 1978). In the past, the residence time could have been larger or smaller, perhaps by as much as a factor of 5 or even 10, depending on the Ca concentration in seawater (Fig. 12) and the riverine, diagenetic and hydrothermal fluxes of calcium to the oceans.

The best method for reconstructing the seawater δ^{44}Ca in the geologic past has not been demonstrated. The work done so far has been mainly on bulk carbonate ooze and chalk, mainly of Cenozoic age, and the general features of the data have been corroborated by studies of marine phosphate (Schmitt et al. 2003a). The assumption is that the fractionation factor applicable to the formation of the carbonate sediments from seawater does not vary with time. De La Rocha and DePaolo (2000) reported the data shown in Figure 16a for Cenozoic carbonates. The data indicate that carbonate δ^{44}Ca values are not constant with time, and if one accepts that these sediment samples are globally representative, then this would suggest that the seawater values have also changed with time. The average spacing of the De La Rocha and DePaolo (2000) data is about 3 m.y. between 0 and 45 Ma. This spacing is longer than the nominal residence time and hence the data may be aliased. Much more closely spaced data are shown for the past 3 million years in Figure 11b, all from a single carbonate ooze section cored at DSDP Site 590B in the Tasman Sea (DePaolo, unpublished data). The Site 590B data show that on timescales less than 1 million years, there are only very small changes in the δ^{44}Ca of the bulk ooze, as expected if they are reflecting seawater variations.

The interpretation of the data from bulk oozes and chalks, even if it is accepted that they are providing a global signal, is different depending on whether the timescales of interest are greater than or smaller than the residence time of Ca. Equation (5) can be written also in the following form:

$$\frac{N_{Ca}}{F_W}\frac{d\delta_{SW}}{dt} = (\delta_W - \delta_{SW}) - \frac{F_{Sed}}{F_W}\Delta_{Sed} \tag{8}$$

where all of the inputs to the ocean have been lumped into a single weathering term. The amount of dissolved calcium in the oceans is described by:

$$\frac{dN_{Ca}}{dt} = F_W - F_{Sed} \tag{9}$$

In the case that $F_W = F_{Sed}$, and δ_W and Δ_{Sed} are constant, then Equation (7) will be satisfied, and δ_{SW} will be unchanging with time. At steady state it will also be the case that:

$$\delta_{Sed} = \delta_{SW} + \Delta_{Sed} = \delta_W \tag{10}$$

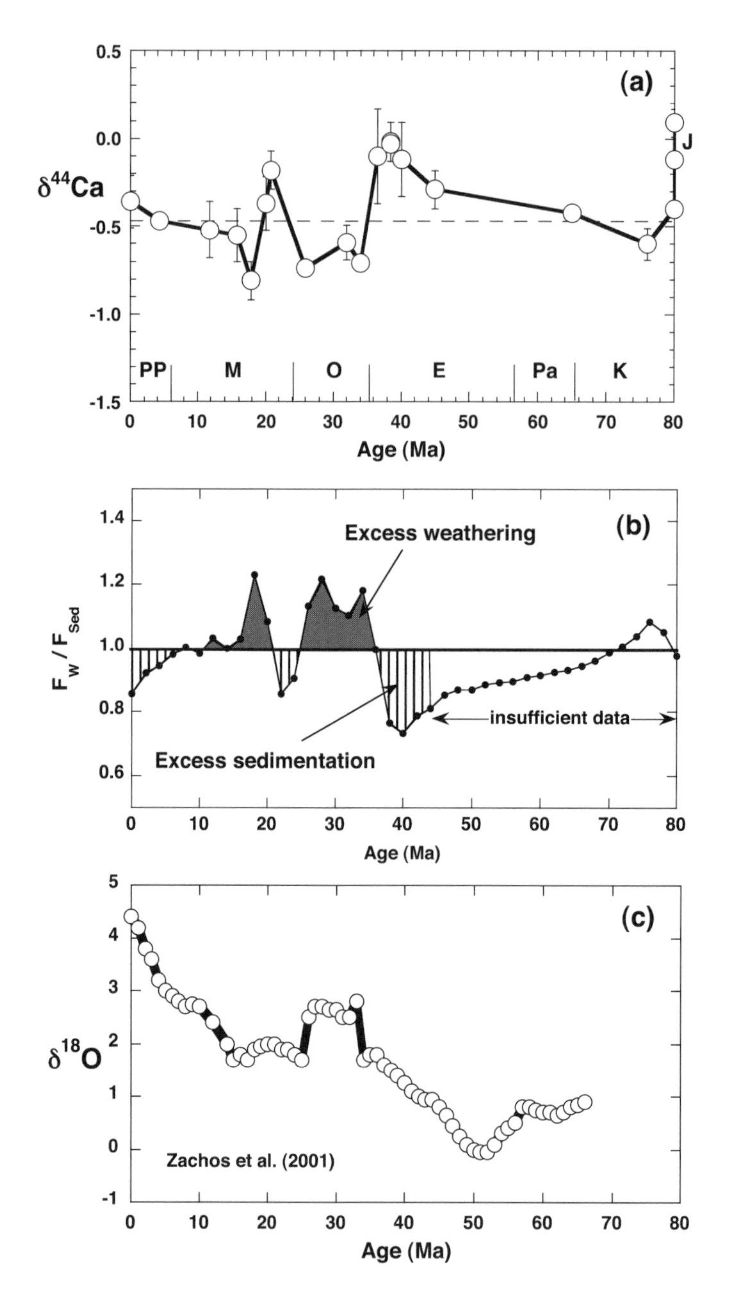

Figure 16. (a) Ca isotope record from marine carbonates (De La Rocha and DePaolo 2000). The variations are inferred to reflect variations in the isotopic composition of seawater (which is heavier by about 1.4‰). The small excursions of δ^{44}Ca reflect changes in the global weathering cycle; they are recast in (b) in terms of the ratio of the flux of calcium being delivered to the ocean by weathering (F_W) to the flux of Ca being removed from the ocean by carbonate sedimentation (F_{Sed}). (c) Smoothed record of benthic foraminiferal δ^{18}O for the Cenozoic time period from Zachos et al. (2001).

Hence, over the long term, the $\delta^{44}Ca$ of the sedimentation flux is not dependent on the fractionation factor, only on the $\delta^{44}Ca$ of the weathering flux. And overall, in the long term the $\delta^{44}Ca$ of the oceans and of sedimentary Ca is not expected to change unless the $\delta^{44}Ca$ of the weathering flux changes. The equations allow mainly for excursions from the weathering flux value due to shifts in Δ_{Sed} or F_W/F_{Sed}; excursions that last for periods of time up to a few times the residence time (generally a few million years).

One way to temporarily change δ_{SW} is to change Δ_{Sed}, which is possible if there is sufficient change in ocean temperature (Fig. 10). Based on the lower slope curves from Figure 10, a temperature change of 5°C might generate a change of $\delta^{44}Ca$ of about 0.1‰. Local or even global changes in ocean temperature can happen quickly (hundreds to thousands of years), which is much faster than the ocean can change $\delta^{44}Ca$. Hence temperature fluctuations may produce small $\delta^{44}Ca$ fluctuations in the sediment on timescales short in comparison to the residence time of calcium. However, on long timescales, the oceans will change $\delta^{44}Ca$ to adjust to any change in Δ_{Sed}, whether it is due to temperature or some other effect such as a change in the mode of calcium removal from deep sea to shelf carbonates.

Changes in δ_W can generate changes in δ_{SW} that persist for time periods longer than the residence time. The restricted range in $\delta^{44}Ca$ in rocks suggests that δ_W changes are likely to be small (Schmitt et al. 2003a; De La Rocha and DePaolo 2000), but the relatively large range in the river $\delta^{44}Ca$ values (Fig. 13) suggests that it is indeed possible and perhaps likely that relatively long-term but small changes in δ_W occur.

Considering that there is geologic evidence that the Ca concentration of the oceans change with time, it is necessary that F_W/F_{Sed} changes with time. As noted by De La Rocha and DePaolo (2000), if the steady state equation for the sedimentary values is written as:

$$\delta_{Sed}(t) = \delta_W(t) - \Delta_{Sed}\left[\frac{F_{Sed}}{F_W}(t) - 1\right] \tag{11}$$

then it can be estimated that variations of δ_{Sed} of ±0.4‰, as suggested by the data, can be generated by variations of F_W/F_{Sed} from about 0.7 to 1.2 (Fig. 16b). If F_W/F_{Sed} is 1.2 and the residence time of Ca in the oceans is 1 m.y., then the ocean Ca concentration will double in a period of about 4 residence times. The timescale for doubling could be lengthened or shortened depending on whether the absolute values of the fluxes change at the same time that the ratio changes. The long timescales of the variations shown by De La Rocha and DePaolo (2000) data could only be explained by changes in either or both of F_W/F_{Sed} and δ_W.

Paleoclimatology

The small changes in $\delta^{44}Ca$ may turn out to be useful for characterizing past climates and climate changes. This can be illustrated by "straw man" analyses of the data in Figures 11b and 16. The only potentially significant signal in the Pleistocene 590B record is the slight drop in $\delta^{44}Ca$ between 0.8 and 0.4 Ma. Because of the short timescale, it is possible that this represents cooling beginning at about 0.7 or 0.8 Ma. The longer-term variations shown in Figure 16a cannot be due to temperature changes. De La Rocha and DePaolo (2000) favored an interpretation in terms of changing F_W/F_{Sed}. An important implication of such changes is the change of seawater Ca concentration.

The CO_2 concentration in the earth's atmosphere is ultimately governed by the calcium carbonate equilibrium in the ocean (e.g., Berner et al. 1983). If the oceans are in equilibrium with calcite, which is usually the case, then to a reasonable approximation, the pCO_2 of the atmosphere is defined by the equilibrium:

$$CO_2(v) + Ca^{2+}(aq) + H_2O = CaCO_3(s) + 2H^+(aq) \tag{12}$$

And therefore, pCO_2 is related to ocean pH and Ca^{2+} concentration by:

$$\mathrm{pCO_2} = \frac{[H^+]^2}{K_{eq}(T)[Ca^{2+}]} \tag{13}$$

where K_{eq} is the equilibrium constant for equation (12). It has recently been proposed that boron isotopes allow reconstruction of past ocean pH values (Pearson and Palmer 2000). If Ca isotopes can help estimate changes in the ocean Ca concentration, then one can generate a fully constrained paleo-pCO_2 curve based on Equation (13).

The ratio of the fluxes of Ca into and out of the ocean can be reconstructed from the seawater δ^{44}Ca record using the relation (from Eqn. 8):

$$\frac{F_{\mathrm{Sed}}}{F_{\mathrm{W}}} = \frac{1}{\Delta_{\mathrm{Sed}}}\left[\frac{N_{\mathrm{Ca}}}{F_{\mathrm{W}}}\frac{d\delta_{\mathrm{SW}}}{dt} + (\delta_{\mathrm{W}} - \delta_{\mathrm{SW}})\right] \tag{14}$$

Starting with present conditions, where $N_{Ca}/F_W \approx 1$ million years, and assuming that both Δ_{Sed} and δ_W are constant, the past variations of F_{Sed}/F_W are deduced as shown in Figure 16b. This curve does not uniquely determine the magnitude of past changes in [Ca^{2+}], which can be seen from:

$$\frac{dN_{\mathrm{Ca}}}{dt} = F_{\mathrm{W}}\left(1 - \frac{F_{\mathrm{Sed}}}{F_{\mathrm{W}}}\right) \tag{15}$$

With sufficiently closely spaced data such as those from Site 590B, it may be possible to estimate past variations in [Ca^{2+}] by integrating Equation (15), and to combine that information with estimates of paleo-pH from B isotopes to calculate the expected changes in pCO_2 in the past. De La Rocha and DePaolo (2000) gave somewhat more general conclusions based on their data. For example, the high δ^{44}Ca in the Late Eocene corresponds to a period of relatively high global temperature as suggested by the low foraminiferal δ^{18}O values (Fig. 16c). The high δ^{44}Ca corresponds to excess sedimentation or a decreased weathering flux, which suggests decreasing [Ca^{2+}] and hence increasing pCO_2. The Oligocene is a period of lower global temperatures, and the relatively low δ^{44}Ca indicates increasing [Ca^{2+}] and hence decreasing pCO_2.

A bolder approach is to integrate the curve in Figure 16b to estimate past ocean [Ca^{2+}] variations and combine these with estimates of paleo-pH from boron isotope data. This exercise does not yield tight constraints on past atmospheric CO_2 levels because the [Ca^{2+}] estimates are too sensitive to the assumptions (and there are too few data), but helps to illustrate how the parameters of the Ca cycle are related to other observations. Figure 17a shows the variations of seawater [Ca^{2+}] over the past 60 million years based on the data of Figure 16a and for three different assumed values of the δ^{44}Ca value of the weathering flux. Varying the assumed values of Δ_{Sed} and the Ca residence time (see Eqn. 14) has much less effect on the results. The [Ca^{2+}] estimates become highly uncertain with increasing age. However, it is interesting to note that the estimates of paleo-seawater [Ca^{2+}] given by Horita et al. (2002; see Fig. 12) are most compatible with a relatively low value of δ_W. This observation can be taken as support for the idea that the weathering flux has δ^{44}Ca that is significantly different from average igneous rocks, and hence that light Ca is preferentially released from rocks by weathering processes. When the [Ca^{2+}] values are combined with the paleo-pH values of Pearson and Palmer (2000) (Fig. 17b), the result is the paleo-pCO_2 curves shown in Figure 17c. The curve for $\delta_W = -0.5$ is similar to the curve produced by Pearson and Palmer (2000), who assumed that [Ca^{2+}] is proportional to seawater alkalinity. This curve also shows similarities to what might be expected from δ^{18}O and other evidence (e.g., Fig. 16c). The early

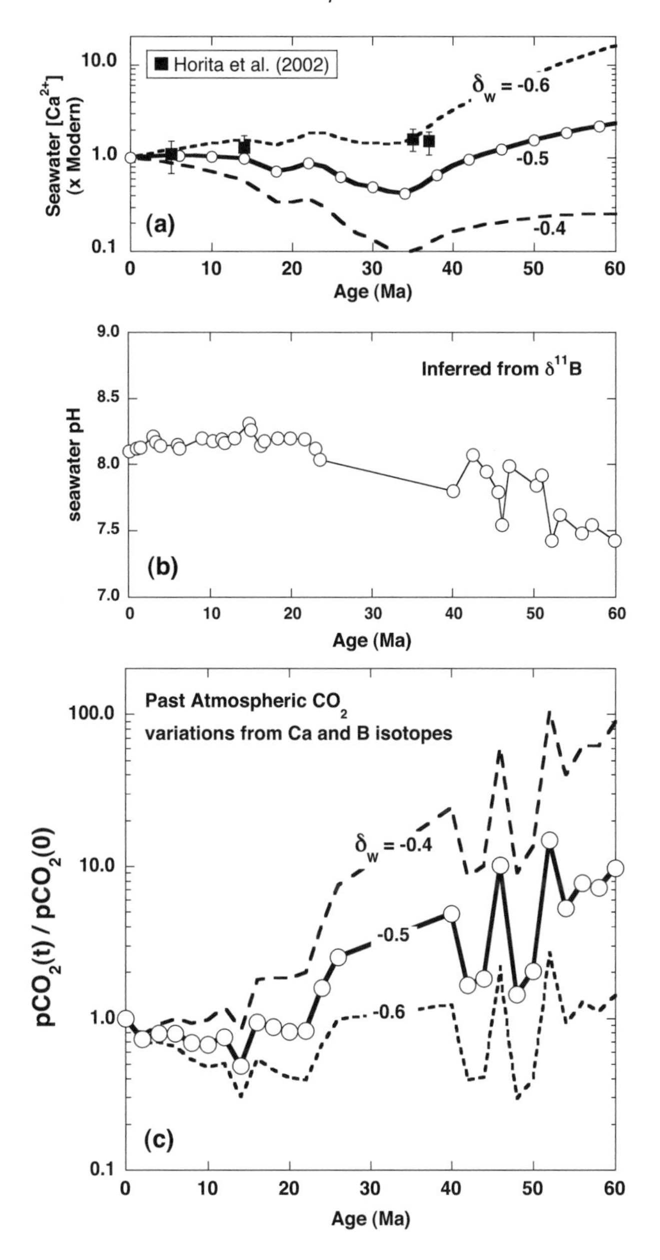

Figure 17. (a) Seawater calcium concentration versus geologic age calculated by integration of the F_W/F_{Sed} curve from Figure 16b under the assumptions that the residence time of Ca in seawater is constant and equal to 2 million years, and the δ^{44}Ca of the weathering flux is –0.5 ± 0.1. The results are highly sensitive to the choice of the value of δ^{44}Ca for the riverine input and become more uncertain at greater age. Also shown are estimates of paleo-seawater [Ca^{2+}] from Horita et al (2002) based on studies of fluid inclusions in evaporite minerals. (b) Inferred variation of ocean pH with geologic age based on the results of boron isotopic studies (Pearson and Palmer 2000). (c) Calculated atmospheric CO_2 concentration as function of geologic age using the curves from (a) and (b) and Equation (13). The value of *log K* is assumed to be constant.

Cenozoic is a time of high but variable pCO_2. By early Miocene time (ca. 22 Ma) CO_2 levels are comparable to the present (pre-industrial) values. The calculated pCO_2 for the period from the middle Miocene through the Pliocene comes out lower than the modern value, which does not correspond well with the $\delta^{18}O$ evidence for continued cooling between 15 Ma and the present (Fig. 16c). It may be noteworthy that higher assumed values for δ_W correspond to very high levels of atmospheric CO_2 in the early Cenozoic, whereas δ_W values in the range −0.5 to −0.6 correspond to the 1× to 10× higher CO_2 levels that are typically estimated with other approaches (e.g., Ekart et al. 1999).

CONCLUSIONS AND IMPLICATIONS

The research summarized here demonstrates that there are significant and systematic variations in the Ca isotopic composition of geological and biological materials. Most if not all of the natural isotopic fractionation has been interpreted in terms of kinetically controlled processes, but there has not yet been any measurement of the equilibrium isotopic partitioning. Laboratory studies demonstrate that some inorganic processes—diffusive transport in silicate liquids and precipitation of aragonite from aqueous solution—can fractionate Ca isotopes. In nature the effects of inorganic fractionation of Ca isotopes have not been demonstrated. Organic processes, however, clearly do fractionate Ca isotopes and hence the Ca isotope system may have its greatest geochemical value as a tracer of organic processes in the geologic record. There also appears to be a role for Ca isotopes in studies of the metabolism of higher organisms and for intracellular ion transport. The research to date suggests that Ca isotopes are fractionated within organisms mainly in the process of formation of mineral matter in bone or shell material.

Calcium also has isotopic variations stemming from the radioactive decay of ^{40}K to ^{40}Ca. These variations can be used for geochronology and may also be useful for studies of rock weathering, soil formation, magma genesis, diagenesis, and metamorphism.

The observed range of natural variations of $\delta^{44}Ca$ is about 4 to 5‰ in terrestrial materials and up to 50‰ in high temperature condensate minerals in carbonaceous chondrites. The typical reproducibility of measurements is about ±0.15‰. Broader application of Ca isotope measurements in geochemistry may be possible, particularly if the reproducibility can be improved to ±0.05‰ to 0.03‰. There is hope that this can be achieved either with inductively coupled plasma source mass spectrometry (Halicz et al. 1999) or with a new generation of multi-collector thermal ionization mass spectrometers (Heuser et al. 2002).

Studies of the global calcium cycle with isotopes have promise but require further characterization of the components of the cycle—rocks, soils, rivers, sediments, organic material—to be fully realized. Study of the calcium cycle is complementary to studies of the carbon cycle, and may contribute to understanding past Earth climates and oceans. The fractionation factors for Ca isotopes, their causes, magnitude, temperature dependence and ubiquity in metabolic processes, are still poorly understood. Further studies may yield information on metabolism, Ca complexing in aqueous solutions, and polymerization in silicate melts. Studies of vertebrates suggest that significant Ca isotopic variations may be generated in soft tissue during bone mass loss, and whenever the dietary Ca supply is small in comparison to Ca demand.

ACKNOWLEDGMENTS

The manuscript was significantly improved as a result of reviews by Brian Marshall and Tom Bullen. The author's recent work on Ca isotopes has been supported largely by the

Director, Office of Energy Research, Basic Energy Sciences, Chemical Sciences, Geosciences and Biosciences Division of the U.S. Department of Energy under Contract No. De-AC03-76SF00098 to the E.O. Lawrence Berkeley National Laboratory, and by the National Science Foundation, most recently by grant number EAR-9909639.

REFERENCES

Ahrens LH (1951) The feasibility of a calcium method for the determination of geological age. Geochim Cosmochim Acta 1:312-316

Baadsgaard H (1987) Rb–Sr and K–Ca isotope systematics in minerals from Potassium horizons in the Prairie Evaporite Formation, Saskatchewan, Canada. Chem Geol Isotope Geosci Sec 66:1-15

Beckinsale RD, Gale NM (1969) A reappraisal of the decay constants and branching ratio of ^{40}K. Earth Planet Sci Lett 6:289-294

Barker F, Hedge C, Millard H, O'Neil J (1976) Pikes Peak Batholith: Geochemistry of some minor elements and isotopes, and implications for magma genesis. *In:* Studies in Colorado Field Geology. Vol 8. Weimer RJ (ed) Colorado School of Mines Professional Contributions, Colorado, p 44-56

Berner RA, Lasaga AC, Garrels RM (1983) The carbonate-silicate geochemical cycle and its effect on atmospheric carbon dioxide over the past 100 million years. Amer J Sci 283:641-683

Blum JD, Erel Y (1997) Rb-Sr isotope systematics of a granitic soil chronosequence: The importance of biotite weathering. Geochim Cosmochim Acta 61:3193-3204

Broecker W, Peng TH (1982) Tracers in the Sea. Eldigio Press, Palisades NY

Cameron AGW (1979) From nucleosynthesis to the birth of the Earth; origin of the planets and of the Earth. *In:* The rediscovery of the Earth. Motz L (ed) Van Nostrand Reinhold New York, p 51-58

Chen CH, Nakada S, Shieh YN, DePaolo DJ (1999) The Sr, Nd and O isotopic studies of the 1991-1995 eruption at Unzen. Japan J Volc Geothermal Res 89:243-253 (1999)

Clayton RN, Hinton RW, Davis AM (1988) Isotopic variations in the rock-forming elements in meteorites. Phil Trans R Soc Lond A325:483-501

Clementz MT, Holden P, Koch PL (2003) Are calcium isotopes a reliable monitor of trophic level in marine settings? Int J Osteoarchaeology 13(1-2):29-36

Coleman ML (1971) Potassium–calcium dates from pegmatitic micas. Earth Planet Sci Lett 12:399-405

Condie KC (1993) Chemical composition and evolution of the upper continental crust: contrasting results from surface samples and shales. Chem Geol 104:1-37

Davis AM, Hashimoto A, Clayton RN, Mayeda TK (1990) Isotope mass fractionation during evaporation of Mg_2SiO_4. Nature 347:655-658

Davis AM, MacPherson GJ (1996) Thermal processing in the solar nebula: Constraints from refractory inclusions. *In:* Chondrules and the Protoplanetary Disk. Hewins RH, Jones RH, Scott ERD (eds) Cambridge University Press, New York, p 71–76

De La Rocha CL, DePaolo DJ (2000) Isotopic evidence for variations in the marine calcium cycle over the Cenozoic. Science 289(5482):1176-1178

Deniel D, Vidal P, Fernandez A, LeFort P, Peucat JJ (1987) Isotopic study of the Manaslu granite (Himalaya, Nepal): inference on the age and source of Himalayan leucogranites. Contrib Mineral Petrol 96:78-92

DePaolo DJ (1986) Detailed record of the Neogene Sr isotopic evolution of seawater from DSDP Site 590B. Geology 14:103-106

Edmond JM (1992) Himalayan tectonics, weathering processes, and the strontium isotope record in marine limestones. Science 258:1594-1597

Elderfield H, Wheat CG, Mottl MJ, Monnin C, Spiro B (1999) Fluid and geochemical transport through oceanic crust; a transect across the eastern flank of the Juan de Fuca Ridge. Earth Planet Sci Lett 172: 151-165

Eiler JM, Valley JW, Stolper EM (1996) Oxygen isotope ratios in olivine from the Hawaii Scientific Drilling Project. J Geophys Res-Solid Earth. 101(B5):11807-11813

Eiler JM, Farley KA, Valley JW, Hauri E, Craig H, Hart SR, Stolper EM (1997) Oxygen isotope variations in ocean island basalt phenocrysts. Geochim Cosmochim Acta 61:2281-2293

Ekart DD, Cerling TE, Montanez IP, Tabor NJ (1999) A 400 million year carbon isotope record of pedogenic carbonate: Implications for paleoatmospheric carbon dioxide. Amer J Sci 299:805-827

Elderfield H, Schultz, A (1996) Mid-ocean ridge hydrothermal fluxes and the chemical composition of the ocean. Ann Rev Earth Planet Sci 24:191-224

Fletcher IR, McNaughton NJ, Pidgeon RT, Rosman KJR (1997a) Sequential closure of K-Ca and Rb-Sr isotopic systems in Archaean micas. Chem Geol 138:289–301

Fletcher IR, Maggi AL, Rosman KJR, McNaughton NJ (1997b) Isotopic abundance measurements of K and Ca using a wide-dispersion multi-collector mass spectrometer and low-fractionation ionisation techniques. Int J Mass Spectrom Ion Proc 163(1-2):1-17

Getty SJ, DePaolo DJ (1995) Quaternary geochronology by the U-Th-Pb method. Geochim Cosmochim Acta 59:3267-3272

Gieskes JM, Lawrence JR (1981) Alteration of volcanic matter in deep sea sediments: Evidence from the chemical composition of interstitial waters from deep sea drilling cores. Geochim Cosmochim Acta 45: 1687-1703

Grant KM, Dickens GR (2002) Coupled productivity and carbon isotope records in the southwest Pacific Ocean during the late Miocene-early Pliocene biogenic bloom. Paleogeography, Paleoclimatology, Paleoceanography 187:61-82

Gussone N, Eisenhauer A, Heuser A, Dietzel M, Bock B, Bohm F, Spero H, Lea D, Buma J, Nagler, TF (2003) Model for kinetic effects on calcium isotope fractionation (δ^{44}Ca) in inorganic aragonite and cultured planktonic foraminifera. Geochim Cosmochim Acta 67:1375-1382

Halicz L, Galy A, Belshaw NS, O'Nions RK (1999) High precision measurement of calcium isotopes in carbonates and related materials by multiple collector inductively coupled plasma mass spectrometry (MC-ICP-MS). J Anal Atom Spectr 14:1835-1838

Hardie LA (1996) Secular variation in seawater chemistry: An explanation for the coupled secular variation in the mineralogies of marine limestones and potash evaporites over the past 600 m.y. Geology 24:279-283

Herzog LF (1956) Rb-Sr and K-Ca analyses and ages. *In:* Nuclear processes in geologic settings. Natl Research Council, Comm Nuclear Sci, Nuclear Sci Ser Rept 19:114-130

Heuser A, Eisenhauer A, Gussone N, Bock B, Hansen BT, Nägler TF (2002) Measurement of calcium isotopes (δ^{44}Ca) using a multicollector TIMS technique. Int J Mass Spec 220:387–399

Hippler D, Schmitt A-D, Gussone N, Heuser A, Stille P, Eisenhauer A, Nagler T (2003) Calcium isotopic composition of various reference materials and seawater. Geostandards Newsletter 27:13-19

Holland HD (1978) The Chemistry of the Atmosphere and Oceans. John Wiley and Sons, New York

Holland HD (1984) The Chemical Evolution of the Atmosphere and Oceans. Princeton University Press, Princeton NJ

Horita J, Zimmermann H, Holland HD (2002) Chemical evolution of seawater during the Phanerozoic: implications from the record of marine evaporates. Geochim Cosmochim Acta 66:3733–3756

Inghram MG, Brown H, Patterson C, Hess DC (1950) The branching ratio of K-40 radioactive decay. Phys Rev 80:916-917

Ireland TR, Fahey AJ, Zinner EK (1991) Hibonite-bearing microspherules; a new type of refractory inclusions with large isotopic anomalies. Geochim Cosmochim Acta 55:367-379

Johnson CM, Beard BL (1999) Correction of instrumentally produced mass fractionation during isotopic analysis of Fe by thermal ionization mass spectrometry. Int J Mass Spect 193:87-99

Jungck MHA, Shimamura T, Lugmair GW (1984) Calcium isotope variations in Allende. Geochim Cosmochim Acta 48:2651–2658

Kempe S, Degens ET (1985) An early soda ocean? Chemical Geology 53:95-108

Kennett J (1982) Marine Geology. Prentice-Hall, Englewood Cliffs NJ

Lee T, Papanastassiou DA, Wasserburg GJ (1977) Mg and Ca isotopic study of individual microscopic crystals from the Allende meteorite by the direct loading technique. Geochim Cosmochim Acta 41:1473-1485

Lee T, Russell WA, Wasserburg GJ (1979) Calcium isotopic anomalies and the lack of aluminum-26 in an unusual Allende inclusion. Appl J Lett 228(L93–L98):661–662

Marshall BD, DePaolo DJ (1982) Precise age determinations and petrogenetic studies using the K-Ca method. Geochim Cosmochim Acta 46:2537-2545

Marshall BD (1984) The potassium-calcium geochronometer. PhD Dissertation, University of California, Los Angeles

Marshall BD, Woodard HH, Krueger HW, DePaolo DJ (1986) K-Ca-Ar systematics of authigenic sanidine from Waukau, Wisconsin, and the diffusivity of argon. Geology 14:936-938

Marshall BD, DePaolo DJ (1989) Calcium isotopes in igneous rocks and the origin of granite. Geochim Cosmochim Acta 53:917-922

Milliman JD (1993) Production and accumulation of calcium carbonate in the ocean: budget of a non-steady state. Global Geochem Cycles 7:927-957

McCauley S, DePaolo DJ (1997) The marine ^{87}Sr/^{86}Sr and ∂^{18}O records, Himalayan alkalinity fluxes and Cenozoic climate models. *In:* Tectonic Uplift and Climate Change. Ruddiman WF (ed) Plenum, New York, p 427-467

Nägler TF, Villa IM (2000) In pursuit of the ^{40}K branching ratios: K-Ca and ^{39}Ar-^{40}Ar dating of gem silicates. Chem Geol 169:5–16

Nägler TF, Einsenhauer A, Müller A, Hemleben C, Kramers J (2000) The δ^{44}Ca-temperature calibration on fossil and cultured Globigerinoides sacculifer: New tool for reconstruction of past sea surface temperatures. Geochem Geophys Geosys 1(2000GC000091)

Nelson DR, McCulloch MT (1989) Petrogenic applications of the ^{40}K-^{40}Ca radiogenic decay scheme—a reconnaissance study. Chem Geol (Isot Geosci Sect) 79:275–293

Niederer FR, Papanastassiou DA (1984) Ca isotopes in refractory inclusions. Geochim Cosmochim Acta 48: 1279-1293

Platzner I, Degani N (1990) Fractionation of stable calcium isotopes in tissues of date palm trees. Biomed Environ Mass Spect 19:822-824

Pearson PN, Palmer MR (2000) Atmospheric carbon dioxide concentrations over the past 60 million years. Nature 406:695-699

Polevaya NI, Titov NE, Belyaer VS, Sprintsson VD (1958) Application of the calcium method in the absolute age determination of sylvites. Geochemistry 8:897-906

Raymo ME, Ruddiman WF (1992) Tectonic forcing of late Cenozoic climate. Nature 359:117-122.

Renne PR, Swisher DD, Deino AL, Karner DB, Owens TL, DePaolo DJ (1998) Intercalibration of standards, absolute ages and uncertainties in ^{40}Ar/^{39}Ar dating. Chem Geol 145:117-152

Richter FM, Rowley DB, DePaolo DJ (1992) Sr isotope evolution of sea water: the role of tectonics. Earth Planet Sci Lett 109:11-23

Richter FM, Davis AM, Ebel DS, Hashimoto A (2002) Elemental and isotopic fractionation of Type B calcium-, aluminum-rich inclusions: Experiments, theoretical considerations, and constraints on their thermal evolution. Geochim Cosmochim Acta 66:521-540

Richter FM, Davis AM, DePaolo DJ, Watson EB (2003) Isotope fractionation by chemical diffusion between molten basalt and rhyolite. Geochim Cosmochim Acta 67:3905-3923

Rudnick RL, Fountain DM (1995) Nature and composition of the continental crust—a lower crustal perspective. Rev Geophys 33:267-309

Russell SS, Huss GR, Fahey AJ, Greenwood RC, Hutchison R, Wasserburg GJ (1998) An isotopic and petrologic study of calcium-aluminum-rich inclusions from CO_3 meteorites. Geochim Cosmochim Acta 62:689-714

Russell WA, Papanastassiou DA, Tombrello TA, Epstein S (1977) Ca isotope fractionation on the moon. Proc. Lunar Sci. Conf. 8th, 3791-3805.

Russell WA, Papanastassiou DA (1978a) Calcium isotope fractionation in ion-exchange chromatography. Anal Chem 50:1151-1153

Russell WA, Papanastassiou DA, Tombrello TA (1978b) Ca isotope fractionation on the Earth and other solar system materials. Geochim Cosmochim Acta 42:1075-1090

Schärer U, Allègre CJ (1982) Uranium–lead system in fragments of a single zircon grain. Nature 295:585-587

Schmitt AD, Bracke G, Stille P, Kiefel B (2001) The calcium isotope composition of modern seawater determined by thermal ionisation mass spectrometry. Geostandard Newsletter 25:267–275

Schmitt A-D, Stille P, Venneman T (2003a) Variations of the ^{44}Ca/^{40}Ca ratio in seawater during the past 24 million years: evidence from δ^{44}Ca and δ^{18}O values of Miocene phosphates. Geochim Cosmochim Acta 67:2607-2614

Schmitt AD, Chabaux F, Stille P (2003b) The calcium riverine and hydrothermal isotopic fluxes and the oceanic calcium mass balance. Earth Planet Sci Lett 6731:1-16

Shaw DM, Cramer JJ, Higgins MD, Truscott MG (1986) Composition of the Canadian Precambrian Shield and the continental crust of the Earth. *In:* The Nature of the Lower Continental Crust. Dawson JB, Carswell DA, Hall J, Wedepohl KH (eds) Geological Society Special Publications 24:275-282

Shih C-Y, Nyquist LE, Wiesmann H, (1994) K–Ca and Rb–Sr dating of two lunar granites: relative chronometer resetting. Geochim Cosmochim Acta 58:3101-3116

Shu FH, Shang H, Lee T (1996) Toward an astrophysical theory of chondrites. Science 271:1545-1552

Sims KWW, DePaolo DJ, Murrell MT, Baldridge WS, Goldstein S, Clague D, Jull M (1999) Porosity of the melting zone and variations in solid mantle upwelling rate beneath Hawaii: inferences from ^{238}U-^{230}Th-^{226}Ra and ^{235}U-^{231}Pa. Geochim Cosmochim Acta 63:4119-4138

Skulan JL (1999) Calcium isotopes and the evolution of terrestrial reproduction in vertebrates. PhD Dissertation, University of California, Berkeley

Skulan J, DePaolo DJ, Owens TL (1997) Biological control of calcium isotopic abundances in the global calcium cycle. Geochim Cosmochim Acta 61:2505-2510

Skulan J, DePaolo DJ (1999) Calcium isotope fractionation between soft and mineralized tissues as a monitor of calcium use in vertebrates. Proc Nat Acad Sci 96:13,709-13,713

Smith DR, Noblett J, Wobus RA, Unruh D, Douglass J, Beane R, Davis C, Goldman S, Kay G, Gustavson B, Saltoun B, Stewart J (1999) Petrology and geochemistry of late-stage intrusions of the A-type, mid-Proterozoic Pikes Peak batholith (Central Colorado, USA): implications for petrogenetic models. Precambrian Res 98:271-305

Stanley SM, Hardie LA (1998) Secular oscillations in the carbonate mineralogy of reef-building and sediment-producing organisms driven by tectonically forced shifts in seawater chemistry. Palaeogeography Palaeoclimatology Palaeoecology 144:3-19

Steiger RH, Jager E (1977) Subcommission on geochronology: convention on the use of decay constants in geo- and cosmochronology. Earth Planet Sci Lett 36:359-362

Sundquist ET (1993) The global carbon dioxide budget. Science 259:934-941

Taylor SR, Mclennan SM (1995) The geochemical evolution of the continental crust. Rev Geophys 33:241-265

Weber D, Zinner E, Bischoff A (1995) Trace element abundances and magnesium, calcium, and titanium isotopic compositions of grossite-containing inclusions from the carbonaceous chondrite Acfer 182. Geochim Cosmochim Acta 59:803-823

Wedepohl KH (1995) The composition of the continental crust. Geochim Cosmochim Acta 59:1217-1239

White AF, Blum AE, Schulz MS, Bullen TD, Harden JW, Peterson ML (1996) Chemical weathering rates of a soil chronosequence on granitic alluvium: I. Quantification of mineralogical and surface area changes and calculation of primary silicate reaction rates. Geochim Cosmochim Acta 60:2533-2550

Wilkinson BH, Algeo TJ (1989) Sedimentary carbonate record of calcium-magnesium cycling. Am J Sci 289: 1158-1194

Zachos J, Pagani M, Sloan L, Thomas E, Billups K (2001) Trends, rhythms, and aberrations in global climate 65 Ma to present. Science 292:686-693

Zhu P, Macdougall JD (1998) Calcium isotopes in the marine environment and the oceanic calcium cycle. Geochim Cosmochim Acta 62:1691-1698

Reviews in Mineralogy & Geochemistry
Vol. 55, pp. 289-317, 2004
Copyright © Mineralogical Society of America

9

Mass-Dependent Fractionation of Selenium and Chromium Isotopes in Low-Temperature Environments

Thomas M. Johnson

Department of Geology
University of Illinois at Urbana-Champaign
Urbana, Illinois, 61801, U.S.A.

Thomas D. Bullen

Water Resources Division
United States Geological Survey
Menlo Park, California, 94025, U.S.A.

INTRODUCTION

Selenium (masses 74, 76, 77, 78, 80, and 82; Table 1) and chromium (masses 50, 52, 53 54; Table 1) are treated together in this chapter because of their geochemical similarities and similar isotope systematics. Both of these elements are important contaminants in surface and ground water. They are redox-active and their mobility and environmental impact depend strongly on valence state and redox transformations. Isotope ratio shifts occur primarily during oxyanion reduction reactions, and the isotope ratios should serve as indicators of those reactions. In addition to environmental applications, we expect that there will be geological applications for Se and Cr isotope measurements. The redox properties of Se and Cr make them promising candidates as recorders of marine chemistry and paleoredox conditions.

There are only about a dozen published studies on Se isotopes and only two on Cr isotopes. This chapter summarizes what has been learned thus far, and almost all of this work concentrates on aqueous reactions at earth surface temperatures. It also attempts to provide some geochemical background and reviews some relevant points from the sulfur isotope literature, which provides insight into the isotopic systematics of Se and Cr.

Table 1. Compositions of natural Se and Cr and currently used double spikes (atom %).

	^{74}Se	^{76}Se	^{77}Se	^{78}Se	^{80}Se	^{82}Se
Natural Se	0.889	9.366	7.635	23.772	49.607	8.731
82/74 Spike	31.757	1.9260	0.9159	2.2908	4.0870	59.023
	^{50}Cr	^{52}Cr	^{53}Cr	^{54}Cr		
Natural Cr	4.3450	83.785	9.5059	2.3637		
54/50 Spike	48.517	2.232	0.527	48.724		

1529-6466/04/0055-0009$05.00

Se geochemistry

Se is chemically similar to sulfur, which is immediately above it in the periodic table. Its concentration in the earth's crust is small, with most rocks containing less than 0.1 ppm Se except for shales, which span a wide range of concentration and average roughly 1ppm (Faure 1991). Coal is also relatively rich in Se, averaging 3 ppm (Cooper et al. 1974). Se can substitute extensively for S in pyrite (Coleman and Delevaux 1957). However, Se concentrations of many sulfide minerals are quite small, suggesting strong decoupling of Se from S in some systems. Se^0, SeO_2 and SeO_3 vaporize at 680°C, 350°C and 118°C, respectively, and thus Se is volatilized during volcanic activity and burning of coal in power plants. Volatile alkylselenides also comprise a significant atmospheric flux in the global Se cycle (Amouroux et al. 2001).

The chemical behavior of Se depends strongly on redox reactions (Elrashidi et al. 1987; McNeal and Balistrieri 1989). An Eh-pH stability diagram is given in Figure 1. In this diagram, oxic waters plot just below the uppermost diagonal line and conditions are increasingly reducing with decreasing Eh. Under oxidizing conditions, Se(VI) is thermodynamically favored and forms the selenate (SeO_4^{2-}) anion. Selenate is highly soluble and not strongly adsorbing (Neal and Sposito 1989). Under mildly reducing conditions (e.g., Eh = 300 to 500 mV at pH 7) Se(VI) may be reduced to Se(IV), which forms the highly soluble anions selenite (SeO_3^{2-}) and biselenite ($HSeO_3^-$), but which adsorbs strongly onto Fe and Al oxides (Bar-Yosef and Meek 1987). Under moderately reducing conditions, Se(IV) reduces to Se(0), an extremely insoluble non-metallic solid that is resistant to re-oxidation. Under strongly reducing conditions, Se(−II), as HSe^-, is the thermodynamically stable form, and it behaves much like sulfide. Thermodynamic calculations suggest that if metals such as Cu, Pb, Cd, Hg, and Ag are present, the metal selenides are extremely stable and may form instead of Se(0) at moderate to strongly reducing conditions (Elrashidi et al. 1987).

Although Se and S are similar chemically, their redox speciation is different enough so that decoupling of Se from S can occur. This is illustrated in the Eh-pH stability diagrams for Se and S given in Figure 1. Under moderately reducing conditions, Se is stable as Se(IV) or Se(0), whereas S(IV) is not stable at all, and S(0) is stable only under a restricted set of conditions. Thus, Se may be separated from S if it is precipitated as Se(0), for example.

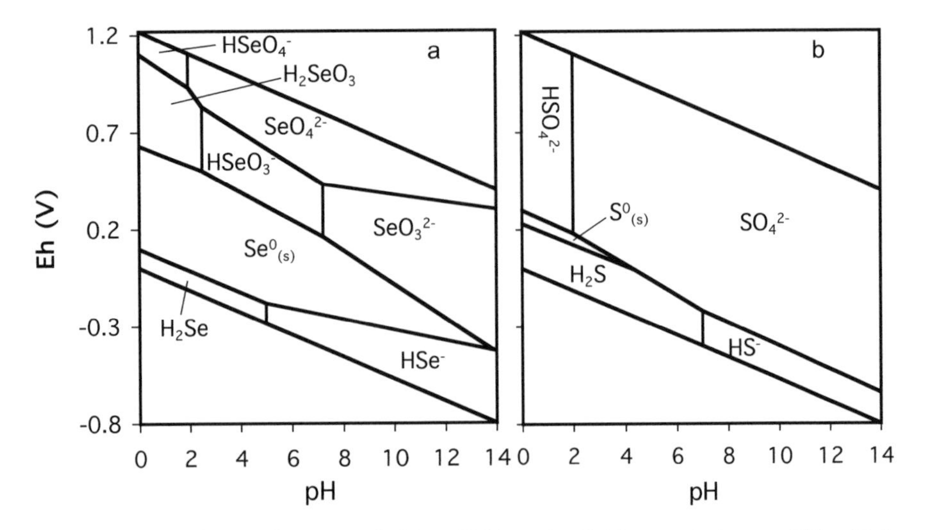

Figure 1. Stability diagrams for aqueous Se (a) and S (b) species. Total concentrations are 10^{-6} mol/L for Se and 10^{-3}mol/L for S.

Additionally, Se is more strongly concentrated by living organisms than S is (see below) and this also leads to decoupling of Se and S.

Biological role. Se is an essential nutrient at low concentrations but it becomes toxic to fish and waterfowl at concentrations as low as 120 nmol/L (Cooper and Glover 1974; Ganther 1974; Lemly 1998; Skorupa 1998). Se contamination of water has resulted mainly from irrigation of Se-rich soils developed on certain shales (Presser 1994; Seiler 1998). In western North America, Cretaceous to Eocene shales extending from North Dakota to central California contain elevated Se concentrations, and the arid environment of this region allows Se(VI) generated by weathering to accumulate in soils. Disposal of fly ash from coal-fired power plants is another problematic Se source (Lemly 1985), and some mining and smelter operations also release Se into the environment.

Biological action is very important in Se redox transformations. Rates of abiotic selenium redox reactions tend to be slow, and in soils and sediments, Se(VI), Se(IV), Se(0) and organically bound Se often coexist (Tokunaga et al. 1991; Zhang and Moore 1996; Zawislanski and McGrath 1998). Bacteria use Se(VI) and Se(IV) as electron acceptors (Blum et al. 1998; Dungan and Frankenberger 1998; Oremland et al. 1989), or oxidize elemental Se (Dowdle and Oremland 1998), and it is likely that most of the important redox transformations are microbially mediated.

Volatile organic compounds containing Se are commonly observed and are an important flux of Se in the environment. Algae and other organisms can emit significant masses of Se as volatile alkylselenides (Frankenberger and Karlson 1994; Fan and Higashi 1998). The flux of volatile Se from the ocean surface is estimate to be 35×10^9 g/yr; this is large enough to have a significant impact on the global Se cycle and terrestrial Se budgets in some areas (Amouroux et al. 2001).

Marine Se. Se concentrations in the oceans are very low, ranging from roughly 2 nmol/L in deep waters to much less in near-surface waters (Measures et al. 1980, 1983; Cutter and Bruland 1984; Cutter and Cutter 1995). Se(VI), Se(IV), and organically associated Se are all important species, and dissolved Se(VI) and Se(IV) are both greatly diminished in near-surface waters because of biological uptake. In reducing waters, organically associated Se dominates (Cutter 1982; Takayanagi and Wong 1985). The residence time of Se in the oceans is thought to be of order 10^4 years (Whitfield and Turner 1987).

Cr geochemistry

Chromium is a compatible element in the Earth's mantle, and tends be present in much greater concentrations in mafic igneous rocks than in felsic ones (Faure 1991). Ultramafic rocks often contain over 1,000 ppm Cr and can generate environmental problems when they weather (Robertson 1975; Robles and Armienta 2000). Granites may contain less than 20 ppm Cr, whereas shales contain roughly 90 ppm.

Chromium is a common anthropogenic contaminant in surface water and ground water, as it is used in electroplating, leather tanning, wood preservation, and cooling water preservation (Nriagu and Niebor 1988). Cr is redox-active; an Eh-pH stability diagram is given in Figure 2. The two common valences in natural waters are Cr(VI) and Cr(III). Cr(VI), the most common form of contaminant Cr, is highly soluble and mobile, and is a carcinogen (Davis and Olsen 1995; Fendorf 1995). Cr(III) is chemically similar to Fe(III), and in the neutral pH range is insoluble, immobile, and less toxic than Cr(VI). At pH 7.0, the transition from thermodynamic Cr(VI) dominance to Cr(III) dominance occurs at an Eh value of roughly 540 mV (Brookins 1988). Thus, Cr(VI) is not as oxidizing as nitrate or MnO_2, but is more oxidizing than Fe(III). In many aquifers, microbes and/or naturally occurring reductants such as Fe(II) and certain organic compounds remove Cr(VI) from solution and make active remediation of the contamination unnecessary (Blowes 2002). In others, Cr(VI) can migrate for kilometers without much attenuation (Perlmutter et al. 1963). In some cases, reductants are injected into aquifers to remove Cr(VI) from solution (Blowes 2002). Reduction of Cr(VI) to Cr(III) is thus a central issue in approaches to Cr(VI) contaminant remediation.

Marine Cr. In the oceans, Cr occurs as both Cr(VI) and Cr(III) (Jeandel and Minster 1987; Mayer 1988; Mugo 1997; Pettine 2000; Sirinawin et al. 2000). Total Cr concentrations generally range from 1 to 10 nmol/l. The residence time of Cr is thought to be approximately 10^4 years, and it is classified either as a "recycled" element that is taken up as a nutrient (Whitfield and Turner 1987) or as transitional between recycled and conservative (Sirinawin et al. 2000).

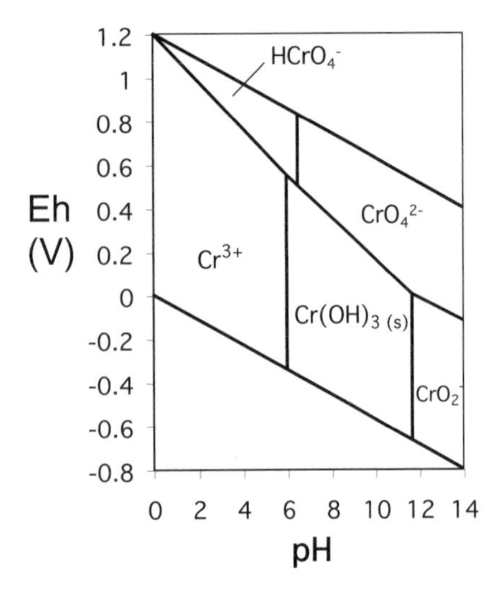

Figure 2. Stability diagram for aqueous Cr species. Total Cr concentration is 10^{-6} mol/L.

Reduction of sulfate, nitrate, selenate and chromate: Analogous isotopic behavior

Because selenate is chemically similar to sulfate, much of the work on isotopic fractionation during sulfate reduction is useful in approaching isotopic fractionation during selenate reduction. More broadly, sulfate, selenate, nitrate and chromate reduction may be expected to have some common isotope systematics, as reduction of these oxyanions involves transfer of multiple electrons and rearrangement of the oxygens bonded to the S, N, Se, or Cr.

Sulfate reduction converts S(VI), as SO_4^{2-}, to S(−II) as H_2S or HS^-. At earth surface temperatures (e.g., <40°C), abiotic sulfate reduction is kinetically sluggish in solution in the absence of enzymes or other catalysts. Presumably this reflects the strength of the tetrahedral cage of oxygens surrounding the S in sulfate. Calculations of equilibrium isotope fractionation between SO_4^{2-} and H_2S have yielded a value of $\alpha_{SO4\text{-}H2S} = 1.074$ for $^{34}S/^{32}S$ at 25°C (Tudge and Thode 1950). Also, experiments with abiotic sulfate reduction revealed a kinetic isotope effect whereby the product is enriched in the lighter isotope. An $\alpha_{SO4\text{-}H2S}$ value of 1.021 was determined, i.e., the product's $^{34}S/^{32}S$ ratio was approximately 21‰ less than the reactant (Harrison and Thode 1957). Although these abiotic reactions do not occur at less than 40°C and are of limited relevance in low-temperature aqueous environments, they provide a foundation for understanding S isotope systematics.

Sulfate reduction is dominated by bacterial processes in nature, and several studies have revealed strong variability in the isotopic fractionation (Thode et al. 1951; Harrison and Thode 1958; Kaplan and Rittenberg 1964; Kemp and Thode 1968; Rees 1973; Dickman and Thode 1990; Habicht and Canfield 1997; Brüchert et al. 2001). The fractionation appears to depend both on the microbial species and on the metabolic state of an individual species (Kaplan and Rittenberg 1964). This is discussed in greater detail below.

Microbial reduction of nitrate to N_2, known as denitrification, is similar. It is kinetically inhibited in the absence of bacteria and is known to induce a kinetic isotope effect (Blackmer and Bremner 1977; Kohl and Shearer 1978; Mariotti et al. 1981; Bryan et al. 1983; Hübner 1986; Mariotti et al. 1988). $^{15}N/^{14}N$ shifts ranging from 6.5‰ to 20‰ have been observed experimentally. As with sulfate, microbial fractionations appear to depend on the metabolic states of the microbes.

The body of research on isotopic fractionation induced by sulfate and nitrate reduction provides insight into selenate, selenite and chromate reduction. For sulfate and nitrate oxyanions, reduction is generally microbially mediated, is irreversible, and involves a fairly large but variable isotopic fractionation. As described below, Se and Cr oxyanion reduction follows suit, though abiotic reactions may have a greater role in some transformations.

ANALYTICAL METHODS

General information about analytical techniques can be found in another chapter of this volume (Albarède and Beard 2004). Details specific to Se and/or Cr are covered here. Most of the existing Se and Cr isotope measurements were made by TIMS, but MC-ICP-MS methods are supplanting TIMS methods for Se (Rouxel et al. 2002) and will probably do so for Cr within a few years. With both TIMS and MC-ICP-MS analyses, there is a measurement bias, or discrimination, that must be corrected for (Albarède and Beard 2004). For example, Se is particularly evaporation prone because of its volatility; lighter isotopes of Se evaporate from the filament faster than heavier isotopes, resulting in large changes in measured ratios over time (Johnson et al. 1999).

A "double spike" technique is essential for TIMS analyses of Se and Cr, and may also be useful in MC-ICP-MS analysis. Briefly, two spike isotopes with a known ratio are added to each sample, and the measured ratio of the spike isotopes is used to determine and correct for instrumental bias. Examples of Se and Cr double spikes currently in use are given in Table 1. The fact that small amounts of the spike isotopes are present in the samples and small amounts of nominally unspiked isotopes are found in the spikes is not a problem, as the measurements allow highly precise mathematical separation of spike from samples. Algorithms for such calculations are described by Albarède and Beard (2004) and, specifically for Se, by Johnson et al. (1999).

For MC-ICP-MS analysis of Se and Cr, the double spike approach is not essential, but has four potential advantages over sample-standard bracketing: 1) It tolerates and corrects for mass bias differences between samples and standards caused by matrix differences. 2) It can correct for any isotopic fractionation that occurs during sample preparation. 3) It is affected little by short-term drift in the mass bias. 4) Analytical throughput may be greater because fewer standards are analyzed. On the other hand, there are disadvantages. The initial effort involved in obtaining, characterizing, and calibrating the spikes is great. After that is done, the extra effort required per analysis is minimal. Spiking of samples take minutes and data reduction can be done as fast as the raw data can be imported in a spreadsheet. A more serious disadvantage is the potential for interferences on the spiked isotopes. Polyatomic interferences such as $ArCl^+$ and ArC^+ are one of the major challenges facing MC-ICP-MS analysts, and addition of two more measured masses can increase the difficulty and/or uncertainty of the measurements.

Selenium mass spectrometry

Gas-source mass spectrometry. Work on Se stable isotopes has a long history, dating back to the Ph.D. work of H. Roy Krouse around 1960. From 1960 to 1990, analyses were done by gas-source mass spectrometry using SeF_6 (Krouse and Thode 1962). The sample Se was converted to Se(0), then reacted with F_2 gas to produce SeF_6. This method required large quantities (e.g., tens of micrograms) of Se for measurements and thus was not widely applied. Recent continuous flow mass spectrometry methods could enable gas-source measurements on much smaller quantities, but will still use too much sample to compete with TIMS and MC-ICP-MS methods.

TIMS methods. In the 1990's, procedures for TIMS measurements were developed for Se (Wachsmann and Heumann 1992). A negative ion approach was used, as thermal ionization produces abundant Se^- ions but very few Se^+ ions. The volatility of Se (SeO_2 and Se^0 vaporize at 350°C and 684°C, respectively) presents a challenge for TIMS analysis. In this case, the Se was loaded with silica gel (to decrease volatilization) on one filament of a double Re filament assembly, and kept relatively cool. The second filament was heated to higher temperatures (920°C to 960°C) and was coated with a layer of ionization-enhancing BaO to ionize Se vapor exuding from the sample filament. This procedure used 500ng Se to obtain ion currents of 10^{-11} A, but had limited precision because of the lack of correction for instrumental bias.

This problem was corrected by applying the double spike method (Johnson et al. 1999). A $^{74}Se + ^{82}Se$ double spike was developed, yielding $^{80}Se/^{76}Se$ measurements with a precision of ±0.2‰. Filament loading was altered from the technique of Wachsmann and Heumann (1992) to improve ionization and stability and to simplify loading. One microliter of a saturated barium hydroxide solution (24 μg Ba) was placed on the filament and dried. The sample and 0.2 μg colloidal graphite (to control Se volatility) were placed on top of the dried $Ba(OH)_2$, and gently evaporated to dryness. Maximum signal intensity occurred between 950°C and 1000°C, though data were usually collected at lower temperature to minimize drift in instrumental discrimination. This TIMS method provided precise measurements using 100-500 ng Se, enabling measurement of Se isotopes in environmental studies and laboratory investigations of Se isotope systematics.

MC-ICP-MS method. Rouxel et al. (2002) developed an MC-ICP-MS method that allows precise analysis with as little as 20 ng Se. This development should allow analysis of seawater and uncontaminated fresh water. The great sensitivity of this method arises from "hydride generation"—injection of Se into the plasma as H_2Se. Very little of the sample Se is lost in the sample introduction system. The H_2Se is generated by mixing the Se(IV)-bearing sample in 1 to 3N HCl medium with a 1% $NaBH_4$ solution (both delivered by a peristaltic pump), then scrubbing the H_2Se from the solution in a stripping vessel and passing it to the plasma torch. This method also has the advantage of minimizing the amount of sample matrix entering the plasma; only H_2 (generated by the $BH_4^- + H^+$ reaction) and small amounts of volatile substances such as HCl vapor are passed from the hydride generator.

Polyatomic interferences are an important issue with Se isotope measurements by ICP-MS. $ArAr^+$ ions are generated in the plasma, creating a large isobaric interference at mass 80 and smaller, as well as significant interferences at masses 76 and 78. One instrument currently on the market, the IsoProbe, has a collision cell in which polyatomic ions are broken down if H_2 is introduced, resulting in negligible $ArAr^+$ effects. Rouxel et al. (2002) utilized H_2 in the collision cell for some measurements, but avoided this approach for most analyses, in order to avoid forming problematic AsH^+ ions at mass 76 and other hydrides. Sample-standard bracketing was used, with 1:1 alternation of sample and standard. This method yielded a precision of ±0.25‰ on $^{82}Se/^{76}Se$.

Selenium sample preparation

The complex chemistry of Se provides many options for purification, but also causes some difficulties in sample preparation. Notably, evaporation of an HCl matrix to dryness causes severe loss of Se (Chau and Riley 1965). Storage of Se(IV) samples for long periods can lead to losses due to reduction to Se(0). Standard ion exchange methods are not effective with acidic sample matrices, as Se(IV) is present as uncharged H_2SeO_3 at $pH < 2$ (Fig. 1).

Hydride generation provides a rapid method for separation of Se from most of the sample matrix (Ellis et al. 2003). Se(IV) is reacted with a $NaBH_4$ solution to produce H_2Se, then scrubbed from the sample solution by N_2 carrier gas bubbling through a gas-liquid separator and transferred to an oxidizing trap solution. The reaction is Se(IV)-specific, and can be used to separate Se species. This single step purifies the Se sufficiently for TIMS analysis. Several other elements that form hydrides (As, Ge, Sn, Pb, Te, Sb and a few others) are not separated from Se. In MC-ICP-MS analysis, Ge^+ and AsH^+ are isobaric interferences and thus Ge and As must be removed. More importantly, the hydride generation process does not produce 100% yields and has a strong potential to fractionate Se isotopes. One standard that was processed through hydride generation prior to double spike addition was shifted by +4.2‰ in $^{80}Se/^{76}Se$ (unpublished data). Accordingly, this purification technique has only been demonstrated effective when used in conjunction with the double spike.

Rouxel et al. (2002) purified Se using thiol-impregnated cotton fiber (TCF). Cotton is reacted with thioglycollic acid to create activated sites that attract certain species of Se, As, Te, Sb from solutions. The sample, in 1 N HCl matrix, is passed through a column containing 0.2 g of the TCF. The column is washed with 6 N HCl to remove the matrix elements, and Se(IV) is then recovered by oxidizing the cotton with 7 N HNO_3. This method recovers Se, along with a few percent of the sample's As, but gives complete separation from Ge. Quantitative recovery of Se ensures that samples are not isotopically shifted, and this enables analysis by MC-ICP-MS without a double spike.

Tanzer and Heumann (1991) used other methods for purification of Se prior to TIMS measurements of $^{82}Se/^{80}Se$ for isotope dilution determination of Se species in fresh water samples. These included anion exchange separation of Se(IV) from Se(VI), reduction of Se(IV) with ascorbic acid to form a Se(0) precipitate, and recovery of certain organic Se species via volatilization and trapping. Anion exchange is effective for some sample matrices, but separation of Se(IV) from other components is impossible in strongly acidic digests (it is uncharged at pH <2; Fig. 1) and difficult in general because a variety of weak acid anions such as phosphate and arsenate are eluted with Se(IV) by weak acid eluents. Precipitation of Se(0) via reduction of Se(IV) appears to be effective as a purification step, but has not been proven with the current high-precision methods. Purging and trapping volatile organic Se should be effective in recovering certain Se species, but this approach has also not been proven.

Chromium mass spectrometry and sample preparation

A TIMS method has existed for many years for measurement of radiogenic ^{53}Cr, the stable daughter of the extinct nuclide ^{53}Mn, in meteorites (Birck and Allegre 1988; Lee and Tera 1986; Shukolyukov and Lugmair 1998). However, this method measures only variations in radiogenic ^{53}Cr relative to the other isotopes. It removes the effects of mass-dependent isotope fractionation by normalizing measured $^{53}Cr/^{52}Cr$ ratios to an assumed $^{50}Cr/^{52}Cr$ ratio. A new method that measures mass-dependent fractionation via the double spike technique has been developed recently (Ellis et al. 2002). The natural choices for spike isotopes are ^{50}Cr and ^{54}Cr, leaving $^{53}Cr/^{52}Cr$ as the ratio to be determined for the sample. After purification, the sample, typically 250-500 ng Cr, is mixed with 20 μg colloidal silica and 0.6 μL of a saturated boric acid solution and loaded onto a rhenium filament. $^{50}Cr/^{52}Cr$, $^{53}Cr/^{52}Cr$, and $^{54}Cr/^{52}Cr$ ratios are measured, and delta values calculated using a double spike data reduction routine as described above. External reproducibility is ±0.2‰ on $\delta^{53}Cr$. MC-ICP-MS methods will certainly be developed, and will probably be similar to those used for Fe. The abundances of ^{50}Cr and ^{54}Cr are small (Table 1) and thus $^{53}Cr/^{52}Cr$ will probably be used as the measured ratio.

Purification of Cr can be achieved through ion exchange. Ellis et al. (2002) used a two step procedure in which the sample Cr, as Cr(VI), was first adsorbed onto an AG1-X8 anion exchange column. After rinsing with 0.1 N HCl, only strong acid anions such as sulfate remained. The Cr(VI) was then reduced to Cr(III) by sulfurous acid and eluted in 0.1 N HCl. After conversion of the sulfurous acid to sulfate, the sulfate was removed by passing the sample, in a weak HCl matrix, through a second AG1-X8 column. This method achieves complete removal of Fe and V, which can produce isobaric interferences.

Standards currently in use

Presently, no certified reference material for Se isotopic composition exists. A few analyses of Se recovered from the meteorite-derived S isotope standard Canyon Diablo Troilite (CDT) have been done, but this is not an acceptable standard. Large amounts of this meteorite -derived material must be destroyed to obtain the Se, and heterogeneity in sulfur isotopes in CDT suggest heterogeneity problems with Se (Beaudoin et al. 1994). NIST SRM3149 is an intermediate purity Se solution (nitric acid matrix) sold as a 10,000 ppm concentration standard. Analyses of three vials of this solution indicate that it is isotopically identical, at the ±0.2‰ level ($^{80}Se/^{76}Se$), to the 4 CDT analyses done to date by two independent groups (Rouxel et al. 2002; our unpublished data). Accordingly, we expect that SRM3149, or a supply of SRM3149 vials given a new number by NIST, may become an accepted standard. Rouxel et al. (2002) used a reagent Se called "Merck" as a standard in their work, whereas several other publications used another provisional standard, MH-495 (Hagiwara 2000; Johnson et al. 2000; Herbel et al. 2002). The differences between these standards and CDT are given in Rouxel et al (2002).

NIST SRM-979 is a high purity $Cr(NO_3)_3{\bullet}9H_2O$ crystalline solid sold as an isotopic reference material. It was prepared from a commercial material and its isotopic composition was analyzed by TIMS, along with gravimetric mixtures of purified isotopes to examine instrumental bias (Sheilds et al. 1966). The absolute $^{53}Cr/^{52}Cr$ ratio is reported as 0.11339, with a 95% confidence interval of ±1.3‰. The homogeneity of this material at the high precision attainable with the more recent methods has not been established, but it is unlikely that significant isotopic variability could have occurred during the precipitation of $Cr(NO_3)_3{\bullet}9H_2O$. As described below, existing data suggest that the $^{53}Cr/^{52}Cr$ ratio of SRM-979 is very close to that of mantle-derived basalts and chromite ores.

NOTATION

The notation we use here is identical to that used Canfield (2001) and by most of the sulfur isotope community. The fractionation factor, $\alpha_{A\text{-}B}$, is defined:

$$\alpha_{A\text{-}B} = R_A/R_B \tag{1}$$

where R_A and R_B are the isotope ratios of the reactant, A, and the product, B. For a kinetically controlled reaction, R_A and R_B are the ratios of the reactant pool and the product generated at one instant in time. We define epsilon, the per mil fractionation:

$$\varepsilon_{A\text{-}B} = 1000(\alpha_{A\text{-}B} - 1) \tag{2}$$

This notation is convenient because

$$\varepsilon_{A\text{-}B} \approx \delta_A - \delta_B \tag{3}$$

where δ_A and δ_B are the isotopic values of the reactant reservoir and of the product generated at one instant in time. This is a close approximation when $\delta_{reactant}$ and $\delta_{product}$ are close to 0‰. The definition given in (2) departs from that used in some earlier Se isotope publications, in which $\varepsilon_{A\text{-}B}$ was defined as $\delta_{product} - \delta_{reactant}$. With that definition, and the usual kinetic fractionation, with the product enriched in lighter isotopes, ε is negative, whereas it is positive as defined in this chapter. There is some inconsistency in the definition of $\varepsilon_{A\text{-}B}$ in the literature, but we recommend the form in Equation (2) above as it is the most widely used.

KINETIC ISOTOPE EFFECTS: THEORETICAL CONSIDERATIONS AND INSIGHTS FROM THE SULFUR ISOTOPE LITERATURE

Dominance of kinetic effects at earth-surface temperatures

Most of the reactions that involve significant fractionation of Se and Cr isotopes appear to be far from chemical or isotopic equilibrium at earth-surface temperatures. Redox disequilibrium is common among dissolved Se species. Dissolved Se(IV) and solid Se(0) are often observed in oxic waters despite their chemical instability (Tokunaga et al. 1991; Zhang and Moore 1996; Zawislanski and McGrath 1998). In one study of shallow groundwater, Se species were found to be out of equilibrium with other redox couples such as Fe(III)/Fe(II) (White and Dubrovsky 1994). The kinetics of abiotic Se(VI) reduction, like those of sulfate, are quite slow. In the laboratory, conversion of Se(VI) to Se(IV) requires one hour of heating to ca. 100°C in a 4 M HCl medium.

Even when forward reactions proceed rapidly at laboratory conditions, as is observed with Se(IV) and Cr(VI) reduction, evidence exists that chemical and isotopic equilibrium are not approached rapidly. Altman and King (1961) studied the kinetics of equilibration between Cr(III) and Cr(VI) at pH = 2.0 to 2.5 and 94.8°C. Radioactive ^{51}Cr was used to determine exchange rates, and Cr concentrations were greater than 1 mmol/L. Time scales for equilibration were found to be days to weeks. The mechanism of the reaction was inferred to involve unstable, ephemeral Cr(V) and Cr(IV) intermediates. Altman and King (1961) stated that the slowness of the equilibration was expected because the overall Cr(VI)-Cr(III) transformation involves transfer of three electrons and a change in cooordination (tetrahedral to octahedral). Se redox reactions also involve multiple electron transfers and changes in coordination.

Accordingly, isotopic equilibration for Cr and Se species is expected to be much slower than for the aqueous Fe(III)-Fe(II) couple, which reaches equilibrium within minutes in laboratory experiments (Beard and Johnson 2004). Additionally, Cr(III) and Se(0) are highly insoluble and their residence times in solution are small, which further decreases the likelihood of isotopic equilibration. In the synthesis below, isotopic fractionations are assumed to be kinetically controlled unless otherwise stated. However, definitive assessments of this assumption have not been done, and future studies may find that equilibrium fractionation is attained for some reactions or under certain conditions.

The basic principles of kinetic isotope fractionation can be understood in the context of kinetic theory. A review of this theory can be found in another chapter in this volume (Schauble 2004). For elementary (i.e., single step) reactions, lighter isotopes usually have greater rate constants because the activation energies are lesser than those of heavier isotopes. For example, breakage of one of the four S-O bonds in SO_4^{2-} is an important isotope-fractionating reaction step during SO_4^{2-} reduction, and this breakage requires less energy for ^{32}S than for ^{34}S (Harrison and Thode 1957).

Dependence of kinetic isotope effects on reaction mechanisms and rates

When isotopes are fractionated kinetically during chemical reactions, the isotope ratio shift of the reaction products relative to the reactants often depends on reaction mechanisms and rates. This contrasts with isotopic fractionations between phases in isotopic equilibrium, where the isotopic differences are thermodynamic quantities and thus do not depend on reaction mechanisms or rates. In this section, we briefly review the well-developed theory for kinetic isotope effects that appears in the S isotope literature. This background serves as a guide for interpreting and predicting Se and Cr isotope systematics.

The chemical reactions of interest in this chapter mostly involve multiple steps. For example, one can write a reaction for $HSeO_3^-$ being reduced directly to Se^0, but such a reaction is extremely unlikely in nature, as this would require transfer of 4 electrons simultaneously.

More likely, there are ephemeral intermediate species with short residence times, and the reaction proceeds in several steps with several intermediates. In such a reaction pathway, changes in the relative rates of the reaction steps can result in changes in the fractionation. Furthermore, there may be multiple pathways by which a chemical transformation can occur. For example, transformation of Se(IV) to Se(0) could proceed via simple abiotic reaction, or via uptake of $HSeO_3^-$ by a plant, reduction to Se(−II) within the plant, incorporation into amino acids, death and decay of the plant, release of the Se(−II), and oxidation to Se^0. The overall transformation, from Se(IV) to Se(0), is the same, but because the two reaction pathways differ greatly, the overall isotopic fractionation may be greatly different.

Model for multi-step reactions. A simple model has been put forward to explain variations in S isotope fractionation during microbial sulfate reduction (Canfield 2001; Rees 1973). This model provides a simple illustration of how kinetic isotope effects vary with changes in the reaction. Our intent here is to present the simplest possible version of the model to illuminate the basic cause of variability in a simplified version of the model. Step 1 of the reaction is diffusion or active transport of sulfate into the cell, which involves no change in the bonding environment around S and thus induces little isotopic fractionation. Step 2 is the enzyme-mediated breakage of one of the S-O bonds in the sulfate, and this step strongly favors lighter isotopes. After this step, several additional steps take place, involving, for example, intermediates such as S(IV). In this chain of steps, the rate of the initial S-O bond breakage is highly variable, as it depends on the supply of an activated enzyme in the cell.

If the cell is well supplied with nutrients, then the production of activated enzyme is great and this step is relatively fast. If the transport of sulfate into the cell cannot keep up with the reduction of sulfate, the concentration of sulfate within the cell becomes small, and very little of the isotopically fractionated sulfate inside the cell can leak back out of the cell. Thus, the effect of the internal isotopic fractionation on the outside world is minimal and the overall fractionation of the process is small. In a hypothetical extreme case, every sulfate anion entering the cell would be consumed by reduction. This would require a complete lack of isotopic fractionation, because when all S atoms entering are consumed, there can be no selection of light vs. heavy isotopes. The isotopic fractionation of the overall reduction reaction would be equal to that which occurs during the diffusion step only.

This may seem paradoxical, as the kinetic isotope effect induced by S-O bond breakage still exists. How can the overall reaction have little isotopic fractionation when one step within it has a large kinetic isotope effect? The key to understanding this is in the isotopic composition of the intermediate species in the reaction chain. An intermediate may become enriched in heavier isotopes if the next step in the reaction chain preferentially consumes lighter isotopes. In the hypothetical case described above, at steady state the sulfate within the cell is enriched in the heavy isotope by an amount equal to the kinetic isotope effect occurring at step 2. Thus, the isotopic composition of the flux of S through step 2 is the same as that of the flux of S into the cell and the kinetic isotope effect occurring at step 2 has no effect on the overall isotopic fractionation.

Implications of this model. A complete description of this phenomenon is beyond the scope of this chapter, but the implications can be summarized in a few rules:

If a single step of the reaction is much slower than all the others (i.e., it is the rate-limiting step), then the isotopic shift induced by the overall reaction is equal to the kinetic isotope effect occurring at the rate limiting step plus the sum of any equilibrium isotope effects occurring between the preceding species in the reaction chain.

Kinetic isotope effects occurring after the rate-limiting step do not contribute to the overall isotopic fractionation of the reaction. This is why the "later steps" after the initial S-O bond breakage are ignored in the simplified analysis above.

If two reaction steps both control the overall rate (e.g., they are equally slow), the isotopic fractionation of the overall reaction equals the sum of the equilibrium fractionations occurring before the rate limiting steps, plus the kinetic isotope effect occurring at the first rate limiting step (including back-reaction), plus part of the kinetic isotope effect occurring at the second rate limiting step.

Using these rules, observed variation in S isotope fractionation during sulfate reduction can be understood. If the cell is starved of nutrients, it cannot produce much of the activated enzyme that drives step 2. This step is then very slow, it is the rate-limiting step for the overall reaction, and the overall reaction is slow. The overall isotopic fractionation for sulfate reduction is equal to the large kinetic fractionation occurring at step 2 (S-O bond breakage) plus the small or zero equilibrium fractionation for step 1 (reversible diffusion). As the abundance of nutrients increases, the rate of step 2 increases, the diffusion step becomes partially rate-limiting, and the isotopic fractionation decreases. This trend has been observed experimentally. In several experiments, greater rates of reduction per bacterial cell result in less fractionation (Habicht and Canfield 1997; Kaplan and Rittenberg 1964). Similar trends have been observed with N isotope fractionation during nitrate reduction (Kohl and Shearer 1978). This theoretical framework provides some understanding of how fractionations vary, but the actual processes are more complex and at present there is no practical algorithm for calculations of actual isotope ratio shifts.

Other controls on isotopic fractionation. As one might expect, factors aside from reaction rate influence the isotopic shift. For example, when hydrogen was supplied by Kaplan and Rittenberg (1964) as the electron donor for sulfate reduction, the isotope fractionation was less than with other donors. Also, there appear to be differences in isotopic fractionation between different sulfate-reducing bacterial species (Brüchert et al. 2001; Detmers et al. 2001). Clearly, the simple model described above captures only one aspect of the variability and cannot be used to extrapolate beyond the conditions of the experiments, but it provides insight into the variations that can be expected in kinetic fractionation factors. The carbon isotope literature contains more involved models for the more complex biochemical pathways involved in photosynthesis. This body of work was reviewed by Hayes (2001) in Volume 43 of the Reviews in Mineralogy Series. The reactions considered include branching pathways and greater numbers of steps than the model described above, and may provide additional insight into complex kinetic reactions. Also, Farquhar (2003) similarly considers complex reaction pathways specifically for sulfate reduction. However, such complexity is not needed to make sense of the current data on Cr and Se isotope fractionation.

Startup effects. Startup effects must also be considered in the interpretation of laboratory experiments. For example, during sulfate reduction, the first small amount of sulfur to pass through the chain of reaction steps would be subject to the kinetic isotope effects of all of the reaction steps. This is because it takes some time for the isotopic compositions of the pools of intermediates to become enriched in heavier isotopes as described above for the steady-state case. Accordingly, the first H_2S produced would be more strongly enriched in the lighter isotopes than that produced after a steady state has been approached. This principle was modeled by Rashid and Krouse (1985) to interpret kinetic isotope effects occurring during abiotic reduction of Se(IV) to Se(0) (see below). Startup effects may be particularly relevant in laboratory experiments where Se or Cr concentrations are very small, as is the case in some of the studies reviewed below.

SELENIUM ISOTOPE SYSTEMATICS

The isotope geochemistry of Se is complex because it involves several inorganic oxidation states and organic forms, kinetic isotope effects, and microbial action (Fig. 3). Fortunately,

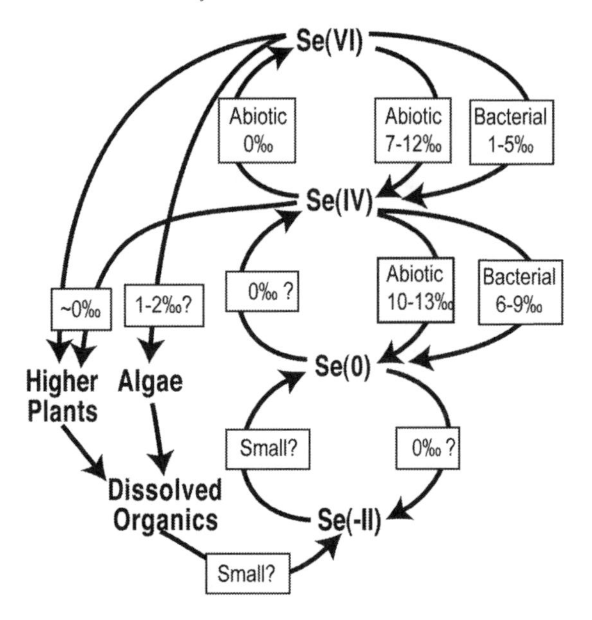

Figure 3. Isotopic effects for redox transformations of Se. Modified from Johnson (in press).

many of these transformations involve negligible Se isotope effects, and reduction of Se(VI) and Se(IV) oxyanions is the dominant cause of isotopic fractionation, according to work done to date (see below). However, this work provides only a reconnaissance understanding of Se isotope geochemistry; additional work will be needed to refine our understanding and support interpretation of field data.

In this section, all data are discussed in terms of $^{80}Se/^{76}Se$ ratios to simplify presentation and facilitate comparisons between various isotopic fractionations. Several of the studies reviewed here measured $^{82}Se/^{76}Se$ ratios, and in these cases, results have been converted to $^{80}Se/^{76}Se$ fractionations using $\varepsilon_{80/76} = 2/3\varepsilon_{82/76}$. The results are summarized in Tables 2 and 3, which express fractionations in term of both $^{80}Se/^{76}Se$ and $^{82}Se/^{76}Se$.

Theoretical estimates of equilibrium isotope fractionation

Although most of the studies reviewed here involve kinetic isotope effects, estimates of equilibrium fractionations are useful in obtaining an understanding of potential isotopic fractionations between species and insight into bonding issues without the added complications inherent in kinetic fractionations. Krouse and Thode (1962) used measured vibrational spectra

Table 2. Theoretical estimates of equilibrium Se isotope fractionations.

	$1000\ln\alpha_{A\text{-}B}$ ($^{82}Se/^{76}Se$)			$1000\ln\alpha_{A\text{-}B}$ ($^{80}Se/^{76}Se$)		
*Exchanging species**	*0°C*	*25°C*	*100°C*	*0°C*	*25°C*	*100°C*
SeO_4^{2-} - H_2Se	39	33	20	26	22	13
SeO_4^{2-} -Se_2	32	27	16	22	18	11
H_2Se - PbSe	0	1	0	0	1	0

*In each pair, the first species is A; the second is B. $\alpha_{A-B} = R_A/R_B$

Table 3. Kinetic isotopic fractionation caused by Se transformations.

Reference	*Transformation*	*Reacting Agent*	ε (‰) $^{80}Se/^{76}Se$	ε (‰) $^{82}Se/^{76}Se$
Rees and Thode (1966)	Se(VI) to Se(IV)	HCl, 25°C	12*	18
Johnson and Bullen (2003)	"	"Green rust"	7.4	11.1*
Ellis et al. (2003)	"	Sediment slurry (microbial)	2.8	4.2*
Herbel et al. (1998)	"	Bacterial cultures	1.1 to 4.8	1.7 to 7.2*
Krouse and Thode (1962), Rees and Thode (1966), Webster (1972), Rashid and Krouse (1985)	Se(IV) to Se(0)	NH_2OH or ascorbic acid	10 to 13*	15 to 19
Ellis et al. (2003)	"	Sediment slurry (microbial)	5.6	8.4*
Herbel et al. (1998)	"	Bacterial cultures	6.0 to 9.1	9.0 to 13.7*
Johnson (in press)	Se(IV) to Se(VI)	NaOH + H_2O_2	<0.2	<0.5*
Johnson et al. (1999)	Se(0) oxidation	Incubated soil	<0.3	<0.5*
Herbel et al. (2002)	Plant uptake	Wetland plants	<1.0	<1.5*
Hagiwara (2000)	Algal uptake	C. reinhardtii	1.0 to 2.6	1.5 to 3.9*
Johnson et al. (1999)	Se volatilization	Cyanobacteria	<1.1	<1.7*
"	"	Soil (Microbes)	<0.6	<0.9*
"	Se(IV) adsorption	$Fe(OH)_3{\bullet}nH_2O$	0.5	0.8*

* Converted to or from $^{80}Se/^{76}Se$ using the relation: $\varepsilon_{80/76} = 2/3\ \varepsilon_{82/76}$.

from Se-bearing compounds to estimate fractionation factors for several gas-phase and aqueous species. Se(IV) was not included, presumably because of a lack of spectroscopic data at that time. The methods and results are tabulated in another chapter in this volume (Schauble 2004). Relevant values of 1000lnα, converted from $^{82}Se/^{76}Se$ ratios to $^{80}Se/^{76}Se$ ratios for consistency in this chapter, are given in Table 2. Notably, the estimate for SeO_4^{2-}-H_2Se equilibrium isotope fractionation at 25°C is 22‰, whereas a similar estimate for SO_4^{2-}-H_2S calculated by Tudge and Thode (1950) is 75‰. This suggests that Se isotope fractionations should be much smaller than S isotope fractionations (e.g., 1000lnα more than a factor of three smaller) in cases where Se reactions are similar to those of sulfur. On the other hand, Se isotope fractionations still have the potential to be much larger than typical analytical uncertainties.

Se(VI) reduction: abiotic

Rees and Thode (1966) reduced Se(VI) to Se(IV) with room temperature ~8M HCl, and obtained a fractionation factor ($\varepsilon_{Se(VI)\text{-}Se(IV)}$) of 12‰ (±1). Interpretation of the results was complicated somewhat by the fact that the Se(IV) product was recovered by reduction to Se(0). The authors reported that some back-reaction of this precipitate may have occurred. They also presented some theoretical estimates of α, based on the theory of Bigeleisen (1949) and some assumptions about the nature of the reaction, that were consistent with the experimental results. In more recent experiments (Johnson et al. 1999), Se(VI) reduction by 4N HCl at 70°C yielded an $\varepsilon_{Se(VI)\text{-}Se(IV)}$ value 5.5‰ (±0.3). The difference between this result and that of Rees and Thode (1966) might be explained by the difference in temperature, but a difference in reaction mechanism cannot be ruled out.

These results from strong HCl matrices are not directly transferable to natural settings, but they provide some initial insights into the isotopic effects that are possible as Se(VI) is reduced. If we assume rupture of one Se-O bond is the dominant isotope-fractionating step in the strong HCl experiments, then we can tentatively project that other, naturally relevant processes may induce similar isotope effects.

Only one naturally relevant abiotic Se(VI) reduction process has been documented to date. Se(VI) can be reduced to Se(IV) and ultimately to Se(0) by "green rust", an Fe(II)- and Fe(III)-bearing phase with sulfate occupying interlayer spaces (Myneni et al. 1997). Johnson and Bullen (2003) obtained an $\varepsilon_{Se(VI)\text{-}Se(IV)}$ value of 7.4‰ (±0.2) for the Se(VI) reduction reaction. The result was not sensitive to changes in pH or solution composition within the ranges over which green rust is stable.

Because this reaction must involve two steps, diffusion of selenate into the interlayer spaces of the green rust followed by electron transfer from Fe(II) green rust, Johnson and Bullen (2003) interpreted this result using a two-step model similar to that discussed above. The diffusion step presumably has very little isotopic fractionation associated with it. Step 2 might be expected to involve a kinetic isotope effect similar to that observed in the HCl reduction experiments. As is discussed above, if the diffusion step is partially rate-limiting, the isotopic fractionation for the overall process should be less than the kinetic isotope effect occurring at the reduction step. This appears to be the case, as the $\varepsilon_{Se(VI)\text{-}Se(IV)}$ value of 7.4‰ is somewhat smaller than that observed for reduction by strong HCl (12‰).

Se(IV) reduction: abiotic

Se(IV) reduction is not kinetically inhibited to the extent that selenate and sulfate reduction are. For example, the strong reducing agent BH_4^- readily reduces Se(IV) to H_2Se at room temperature but leaves Se(VI) essentially untouched. Se(IV) can be reduced by mild reducing agents such as ascorbic acid and SO_2 without a catalyst, and presumably a variety of naturally occurring organic reducing agents can reduce Se(IV) to Se(0) in nature.

In the earliest work, Krouse and Thode (1962) found the Se isotope fractionation factor $\varepsilon_{Se(IV)\text{-}Se(0)}$ to be 10‰ (±1‰) with hydroxylamine (NH_2OH) as the reductant. Rees and Thode (1966) obtained a larger value, 12.8‰, for reduction by ascorbic acid. Webster (1972) later obtained 10‰ for NH_2OH reduction. Rashid and Krouse (1985) completed a more detailed study, and found that the fractionation factor varied with time over the course of the experiments. They explained the variations observed among the experiments in all four studies using a model in which reduction consists of two steps. With the rate constant of the second step two orders of magnitude smaller than the first, and kinetic isotope effects of 4.8‰ and 13.2‰ for the first and second steps, respectively, all the data (Table 3) were fit. Thus, kinetic isotope effects of apparently simple abiotic reactions can depend on reaction conditions.

Se(VI) reduction: microbial

Herbel et al. (2000) reduced Se(VI) to Se(IV) with growing, anaerobic batch cultures of a haloalkaliphile, *Bacillus arsenicoselenatis* strain E1H, and a batch cultures of a haloalkaliphile, *Bacillus arsenicoselenatis* strain E1H, and a freshwater bacterium, *Sulfurospirillum barnesii* strain SES-3. For *B. arsenicoselenatis*, $\varepsilon_{Se(VI)\text{-}Se(IV)}$ = 5.0‰ (±0.5), whereas that for *S. barnesii* increased from 0.2‰ (±0.2) early in the experiment to 4.0 (±1.0) as the Se(VI) was exhausted from the solution. Resting cell suspensions of *S. barnesii* produced $\varepsilon_{Se(VI)\text{-}Se(IV)}$ values of 1.1‰ (±0.3) at both 15°C and 30°C. All of the Herbel et al. (2000) experiments were conducted with media containing 20 mmol/L lactate and 10 mmol/L Se(VI). Because the Se(VI) and electron donor concentrations were much higher than would be found in almost all environments, the results do not translate directly to natural environments, but indicate that microbial reduction of Se(VI) induces isotopic fractionation and that the amount of fractionation depends on the conditions under which the bacteria respire Se(VI). Given the high concentrations of electron donors used, Herbel et al. (2000) speculated that isotopic fractionation for this reaction in natural settings would be close to the maximum values (5‰) observed in the experiments.

Ellis et al. (2003) reduced Se(VI) with anaerobic sediment slurries in order to approximate conditions in natural wetlands. Sediments and waters from the northern reach of the San Francisco estuary, the San Luis Drain, and a man-made wetland, all in California, were used. Reduction was apparently carried out by microbes, as autoclaved control experiments exhibited little reduction. Despite differences between the sediments and concentrations of Se(VI) used in the various experiments, $\varepsilon_{Se(VI)\text{-}Se(IV)}$ varied little, from 2.6‰ to 3.1‰. The starting Se(VI) concentrations of three experiments ranged from 230 nmol/L to 430 nmol/L; that of a fourth experiment was much greater, at 100 μmol/L. Thus, it appears based on these few data that significant Se isotope fractionations persist to very low concentrations, though extrapolation to seawater concentrations (e.g., 1 nmol/L) would be risky.

This lack of a concentration dependence contrasts with the sulfur isotope literature, which suggests that sulfur isotope fractionation induced by sulfate reduction decreases as the sulfate concentration decreases below 0.2 mmol/L (Canfield 2001; Habicht et al. 2002). This difference may reflect differences between S(VI) and Se(VI) reduction pathways or possible adaptations of bacteria to low Se concentrations, but at present no clear explanation has emerged.

The small size of this isotopic fractionation is broadly consistent with the theory, developed in the S isotope literature and outlined above, for isotope fractionation during dissimilatory reduction. If the Se-O bond breakage step has a kinetic isotope effect equal to that observed in the HCl reduction experiments (12‰), then the overall isotopic fractionation observed by Ellis et al. (2003) is approximately one fourth of the bond-breakage effect. This suggests that the transport of selenate into the cells was partially rate-limiting. This is not surprising, given the abundance of decaying organic matter in the sediments, which would provide electron donors to the cells and promote production of respiratory enzymes. Accordingly, the results probably do not apply to the electron donor-poor conditions of deep ocean sediments and oligotrophic aquifers.

Abiotic reduction of Se(VI) appears to induce greater isotopic fractionation than does microbial reduction. The two room-temperature abiotic experiments produced shifts greater than 7.4‰, whereas microbial reduction appears to induce less fractionation. If further work strengthens the evidence for this contrast, it may be possible to distinguish between microbial and abiotic Se(VI) reduction in natural settings using the size of the Se isotope fractionation. However, until more detailed knowledge is gained (e.g., how bacterial fractionations depend on Se concentration and other factors) such an approach would be highly speculative.

Se(IV) reduction: microbial

Se isotope fractionation occurring during microbial reduction of Se(IV) to Se(0) is greater than that occurring during Se(VI) reduction to Se(IV). Herbel et al. (2000) determined the size of the Se isotope fractionation during Se(IV) reduction by the two species used in their Se(VI) reduction experiments, *B. arsenicoselenatis* and *S. Barnesii*. A third species, *B. selenitireducens*, which respires Se(IV) and can grown with Se(IV) as an electron acceptor, was also studied. Starting Se(IV) and lactate concentrations were 10 mmol/L. As with the Se(VI) reduction experiments, in some cases, the isotopic fractionation appeared to be small (e.g., 1.0‰) in the early stages of some experiments. However, in the middle to later stages of most experiments, $\varepsilon_{Se(IV)\text{-}Se(0)}$ ranged from 7.9‰ to 9.1‰. The experiment with growing *B.*

arsenicoselenatis was the only exception, with a fractionation of 6.0‰ determined for a single sample taken after a relatively small amount of Se(IV) reduction.

Ellis et al. (2003) obtained an $\varepsilon_{Se(IV)-Se(0)}$ range for Se(IV) reduction of 5.5‰ to 5.7‰ in three experiments with two of the three sediment slurries used for the Se(VI) reduction experiments reviewed above. Se(IV) concentrations were 100 μmol/L and 240 nmol/L for the estuarine sediment experiments and 460 nmol/L for the wetland sediment. As with the Se(VI) reduction experiments from this study, there was no apparent dependence on Se concentration or sediment type, though the number of experiments was small.

This range of isotopic fractionation (5.5‰ to 9.1‰, excluding the early time points from Herbel et al. 2000) overlaps strongly with the range observed for abiotic reduction (10‰ to 13‰; Table 3). This suggests a fundamental difference between Se(IV) reduction and Se(VI) reduction, in which microbial Se isotope fractionation is much smaller than that caused by abiotic reduction.

Se oxidation

Results of experiments by Johnson (in press) suggest that oxidation of Se(IV) to Se(VI) does not fractionate Se isotopes measurably. The oxidation was carried out with a starting concentration of 0.5 mmol/L Se(IV) in a 0.8 M NaOH + 3% H_2O_2 solution, and the remaining Se(IV) was analyzed after approximately 60% reduction of the initial Se(IV). No $^{80}Se/^{76}Se$ shift was detected within the uncertainty of the measurements (±0.2‰). Whereas the breaking of Se-O bonds necessary to reduce Se(VI) or Se(IV) involves a significant kinetic isotope effect, the process of oxidizing Se(IV) and forming new Se-O bonds apparently does not. The simple experiment described above does not approximate natural conditions well, and microbes could mediate Se(IV) oxidation in nature, but the absence of abiotic fractionation suggests fractionation will be small regardless of the exact pathway.

Oxidation of Se(0) is unlikely to involve Se isotope fractionation. Se(0) is extremely insoluble and elemental Se precipitates are often found in moderately reducing environments. As solid Se(0) is consumed by an oxidation reaction, any kinetic isotope effect is ultimately negated by mass balance effects. For example, if a strong kinetic isotope effect preferentially removes lighter isotopes from the surface of the solid, the surface becomes enriched in heavier isotopes. Ultimately, the removal of successive layers from the solid requires 100% oxidation of the Se(0) and thus there is no opportunity for any kinetic isotope effect to be expressed.

It is possible that a branching reaction could cause isotopic fractionation of Se removed from a Se(0) precipitate, but no evidence exists for this at present. Elemental sulfur can be converted to sulfate and sulfide through a branching reaction mediated by disproportionating bacteria, and the produced sulfate's $^{34}S/^{32}S$ ratio is shifted +17‰ to +31‰ relative to the reacting elemental sulfur. However, this reaction is driven by the thermodynamic instability of S^0; bacteria can gain energy by mediating the reaction. In contrast, Se^0 is thermodynamically stable (Fig. 1) and thus we do not expect that Se disproportionation occurs.

No experiments have been completed to determine isotopic shifts induced by Se(-II) oxidation, but results from the S isotope literature suggest they should be small. Canfield (2001) reviewed previous work on oxidation of dissolved sulfide and concluded that little isotopic fractionation occurs in this process.

Se(VI) and Se(IV) uptake by algae and plants

Assimilation of Se(VI) and Se(IV) by algae appears to induce a small Se isotope fractionation. Hagiwara (2000) grew a fresh water alga, *Chlamydomonas reinhardtii,* in Se(VI)- and Se(IV)-bearing media. The Se in the algae was found to be enriched in the lighter isotopes by 1.1 to 2.6‰. The Se concentration in most of these experiments was in excess of 1.5 μmol/L; the smallest enrichment came from the experiment with the lowest Se concentration (190 nmol/L). There was no consistent difference in fractionation between the results of the Se(VI) and Se(IV) assimilation experiments. Also, algae recovered from the Panoche Treatment Plant, a pilot-scale Se remediation plant in the San Joaquin Valley of California, were found to be shifted by −1.0‰ relative to the saline water, containing 5.2 μmol/L Se(VI), in which it had grown. Thus, based on the preliminary evidence available, it appears that Se isotopic fractionation by algae is significant but small. The algae in these experiments were oven-dried at 70°C to begin the sample preparation, and there is some chance that the fractionation resulted from volatilization of organically bound Se. Future studies should employ freeze drying. Minor isotopic fractionation (0.9 to 2.8‰) is induced by algal assimilation of sulfate (Canfield 2001; Kaplan et al. 1963; Kaplan and Rittenberg 1964) and a small effect is thus expected for Se, at least in Se-rich waters. In Se-poor waters, including seawater, the effect may be different because Se is scavenged as a nutrient and occurs at extremely low concentrations in surface waters.

Se isotope effects associated with assimilation by higher plants have not yet been measured in well-controlled laboratory experiments, but field data suggest that the effects are small. Herbel et al. (2002) determined the isotopic composition of Se in five macrophytes from an artificial wetland receiving water with approximately 320 nmol/L Se(VI). The plants' isotope ratios varied little (0.52‰ total range) and the mean value was slightly (0.74‰) less than that of the surface waters in the wetland. This result is consistent with those in the S isotope literature. Trust and Fry (1992) reviewed data on assimilation of S by higher plants and found the plants to be an average of 1.5‰ enriched in the lighter isotope, with significant scatter. Accordingly, all evidence at present suggests that Se isotopes fractionate only slightly during assimilation by higher plants, and that the plant tissues' compositions should reflect the soil solutions from which the plants grew.

Volatilization

Johnson et al. (1999) measured $\delta^{80/76}$Se in volatilized Se, presumably in the form of alkylselenides, that was generated by cyanobacterial mats and incubated soils. For the cyanobacteria, an early sample of a vigorously growing culture yielded no measurable difference between the volatilized Se and the growth medium, whereas a later sample was enriched in the lighter isotope by 1.1‰. In the soil volatilization experiments, four samples from two different incubated soils were analyzed. All of the samples' $\delta^{80/76}$Se values were within 0.6‰ of the total Se in the soil. Both the cyanobacteria and soil experiments were not highly controlled, but they do suggest that isotopic fractionation related to volatilization is small.

Lack of isotopic fractionation between adsorbed and dissolved species

Isotopic fractionation between adsorbed and dissolved Se(IV) appears to be limited. Johnson et al. (1999) found Se(IV) adsorbed onto hydrous ferric oxides to be slightly (0.53‰) enriched in the lighter isotope relative to the coexisting solution. Given the analytical uncertainties, this effect is barely distinguishable at the 95% confidence level, and only one experiment was conducted. In the S isotope literature, one study of sulfate adsorption found a similar, but smaller effect (Van Stempvoort et al. 1990). Results from Cr(VI) adsorption experiments (Ellis et al. submitted) indicate that Cr isotope fractionation between adsorbed and dissolved Cr(VI) is less than 0.04‰ (see below). Taken together, these results suggest that sorption of oxyanions in general induces little isotopic fractionation. This result is not surprising, as the central atom in the oxyanion complex is bonded to the same number of oxygen atoms in the adsorbed as in the dissolved species. Distortion of this coordination environment does occur, however, and is likely responsible for the effect observed with Se(IV).

SELENIUM ISOTOPE APPLICATIONS AND DATA FROM NATURAL MATERIALS

Meteorites and igneous rocks

The $^{82}Se/^{76}Se$ ratios of CDT and 3 other iron meteorite samples were determined by Rouxel et al. (2002). CDT had the greatest ratio, and the other meteorites ranged from −0.2‰ to −0.6‰ relative to CDT. Four basaltic reference materials, two glassy MORB's, and one peridotite also analyzed by Rouxel et al. (2002) were within 0.2‰ of CDT. These results suggest that the earth's mantle is close in Se isotope composition to CDT, and that CDT is, tentatively, a reasonable proxy for the bulk composition of the earth.

Wetlands

The most notorious environmental threat from Se toxicity has occurred in wetland settings, most notably in the San Joaquin Valley of California (Presser 1994). Herbel et al. (2002) measured $^{80}Se/^{76}Se$ ratio variations in waters, plants, and sediment extracts from man-made wetlands and ponds designed to manage Se-rich agricultural drainage water from this region. Little isotopic contrast (mean difference 0.7‰) existed between the oxidized waters containing 320 nmol/L Se(VI) entering the systems and the reduced Se forms present in the sediments. Similar results were obtained by Johnson et al. (2000) for wetland sediments from the northern reach of the San Francisco Estuary, California, where dissolved total Se concentrations were ca. 2 nmol/L. Total sediment digests, and extracts designed to recover Se(0) from sediments, were only slightly enriched in the lighter isotope relative to the overlying water column (1.2‰ difference in mean $^{80}Se/^{76}Se$).

Although the sediments in these systems accumulate Se over time, the small isotopic contrast suggests that dissimilatory reduction is not the dominant accumulation mechanism. If dissimilatory reduction of Se(VI) and/or Se(IV) to Se(0) by bacteria were the dominant mechanism, one would expect the accumulated Se(0) to be enriched in the lighter isotope. In the San Francisco Estuary case, this assumes that the isotopic fractionations measured by Ellis et al. (2003) can be extrapolated to much lower concentrations. Incorporation of Se into algae and macrophyte tissues, followed by decay of some material and conversion of its Se to Se(0), is more consistent with the observed Se isotope data. Notably, the mean Se isotope composition of the Se(0) in the sediments of the Herbel et al. (2002) study was identical to that of the macrophytes.

Hydrothermal sulfides

Rouxel et al. (2002) measured $^{82}Se/^{76}Se$ ratios in 16 sulfide mineral samples and 2 hydrothermally altered basalts from four hydrothermal vent fields on the mid-Atlantic ridge. The alteration of the basalts apparently added Se that was enriched in the lighter isotope by about 1.7‰. Two sphalerite samples lining the central conduit of a vent chimney were isotopically heavy (+0.8‰) relative to CDT and also contained relatively little Se. All other sulfide minerals were enriched in the lighter isotopes, with $^{82}Se/^{76}Se$ ratios ranging from that of CDT to 5.1‰ less than CDT, with a mean value 1.3‰ less than CDT. These values suggest partial reduction of oxidized Se along the flow path of fluids through the crust (Rouxel et al. 2002).

Marine sediments

Hagiwara (2000) completed a reconnaissance survey of Se isotope variation in marine sediments and sedimentary rocks (Table 4). The most important observation was a lack of strong enrichment in lighter isotopes in most shale samples and three Black Sea sediments. It appears that near-surface alteration has altered Se isotope ratios in some cases. All of the Phosphoria formation samples were probably altered by deep groundwater or hydrothermal

Table 4. Se concentration and $\delta^{80/76}$Se of marine sediments and rocks (Hagiwara 2000).

Sample Name	*Location*	*Material Age and Type*	*[Se], mg/kg*	*(‰, SRM3149)*
11805 136-138	Mid-Atlantic	Holocene sediment- core	0.88	−0.95
STA 9 12-14	Black Sea- Anoxic	"	1.88	−0.43
STA 9 20-22	"	"	2.69	0.00
STA 14 20-22	"	"	1.97	−0.37
MV 27.6	Mancos Fm. Colorado, USA	Eocene Shale- outcrop	0.57	−0.42
MV 38.8	"	"	0.86	−1.05
GH1	Niobrara Fm. Kansas, USA	Cretaceous shale- outcrop	102	3.04
BHP-1	"	Cretaceous Chalk- core	4.17	0.97
Redbird	Pierre Fm. Wyoming, USA	Cretaceous Shale- outcrop	21.9	0.34
Mike	Pottsville Fm. Pennsylvania, USA	Pennsylvanian Shale- outcrop	4.03	1.94
Hermosa	Hermosa Fm. Colorado, USA	Pennsylvanian Shale- core	13.1	−1.11
13 IL 5074.4′	New Albany Fm. Illinois, USA	Devonian shale- core	4.85	−0.08
13 IL 5149.5′	"	"	2.29	0.35
WPSE 82	Phosphoria Fm. Idaho, USA	Permian Shale, altered	249	−1.24
WPSZ 75	"	"	5380	−4.8
WPSC 157	"	"	1070	−2.45
83FP215	"	"	242	0.44
FMC220-2	"	"	91	−0.96

fluids, and this probably led to the high Se concentrations observed (Hagiwara 2000). Some other samples were from outcrops in which redistribution of Se could have occurred during weathering. The phosphoria and outcrop samples give the greatest and least values in the data set. When they are removed, all of the $\delta^{80/76}$Se values are within ±1.1‰ of SRM-3149 (or CDT), and the mean is −0.2‰.

This suggests a major difference between the geochemical cycling of Se and that of S. The δ^{34}S values in phanerozoic shales are in most cases strongly enriched in lighter isotopes relative to CDT because of the kinetic isotope effect associated with sulfate reduction (Canfield 2001), and the δ^{34}S values of the Black sea sediments analyzed here range from −37 to −38‰ vs. CDT (Lyons 1997). The lack of a light-isotope enrichment for Se indicates that the biogeochemical cycling of Se differs from that of S. Se is strongly scavenged by phytoplankton in marine surface waters; dissolved Se(VI) and Se(IV) concentrations decrease toward the surface like those of Si and P (Measures et al. 1980, 1983; Cutter and Bruland 1984; Cutter and Cutter 1995). The marine Se cycle is greatly influenced by fluxes of organically-bound Se, whereas the marine S cycle is dominated by sulfate.

Se concentrations in seawater tend to be very small, and Se isotope measurements of the dissolved Se have not yet been published. Analysis of one Mn nodule by Rouxel et al. (2002) yielded a Se isotopic composition very close to CDT. This suggests that seawater is not highly fractionated relative to the bulk earth and basaltic rocks, though it is possible that the nodule composition reflects only a subset of the total Se species.

CHROMIUM ISOTOPE SYSTEMATICS

Equilibrium isotope fractionation

Schauble et al. (in press) calculated theoretical estimates of equilibrium Cr isotope fractionation. These are reviewed and tabulated in another chapter in this volume (Schauble 2004). Values of $1000\ln\alpha$ for the most relevant pairs are given in Table 5. The largest fractionations, up to 7.6‰, are between Cr(VI) and Cr(III) species, and those between Cr(III) and Cr metal are somewhat smaller than that. Notably, there is little fractionation between aqueous Cr(III) and solid Cr_2O_3. This reflects the similarity in octahedral bonding of Cr(III) to oxygen in solution to that in minerals. A moderate-size isotopic fractionation between $Cr(H_2O)_6^{3+}$ and $Cr(Cl)_6^{3-}$ suggests that complexing of aqueous Cr(III) that results in bonding between Cr and elements other than oxygen will influence isotopic fractionations. Overall, these results suggest that redox processes should be the dominant drivers of Cr isotope fractionation, but that there may be second-order effects related to aqueous speciation or precipitation of solids. At earth-surface temperatures, the redox reactions probably do not approach equilibrium (see discussion above) and thus experiments must be done to determine kinetic isotope fractionations for the various reactions mechanisms that operate in the systems of interest.

Table 5. Theoretical estimates of equilibrium Cr isotope fractionations (Schauble 2004).

*Exchanging species**	$1000\ln\alpha_{A\text{-}B}$ (0°C)	$1000\ln\alpha_{A\text{-}B}$ (25°C)	$1000\ln\alpha_{A\text{-}B}$ (100°C)
CrO_4^{2-} - $Cr(H_2O)_6^{3+}$	7.6	6.6	4.7
$Cr(H_2O)_6^{3+}$ - Cr Metal	3.9	3.5	2.3
$Cr(H_2O)_6^{3+}$ - Cr_2O_3	0.4	0.4	0.2
$Cr(H_2O)_6^{3+}$ - $Cr(Cl)_6^{3-}$	4.0	3.5	2.3

*In each pair, the first species is A; the second is B. $\alpha_{A-B} = R_A/R_B$

Fractionation during Cr(VI) reduction

Results from the two experimental studies of Cr isotope fractionation are summarized in Table 6. An initial set of experiments has proven that Cr(VI) reduction induces a kinetic isotope effect. Ellis et al. (2002) reduced Cr(VI) to Cr(III) in anaerobic serum bottles using

Table 6. Experimental determinations of Cr isotope fractionations.

Reference	*Process*	*Reacting agents*	ε (‰)
Ellis et al. (2002)	Cr(VI) reduction	Magnetite, Sediment Slurries	3.4
Ellis et al. (submitted)	Cr(VI) adsorption	γ- Al_2O_3, Geothite	<0.04

three different reductants: magnetite, a mildly reducing pond sediment, and a mildly reducing estuarine sediment. In all three experiments, it appeared that reduction occurred abiotically, as reduction rates in autoclaved duplicates were no different from unautoclaved bottles. As reduction proceeded, δ^{53}Cr values of the remaining unreduced Cr(VI) increased, and were consistent with simple Rayleigh systematics (Fig. 4). Fitting the data to a Rayleigh model yielded a fractionation factor of $\varepsilon_{Cr(VI)\text{-}Cr(III)} = 3.4‰$ for all three experiments.

Additional work is needed to determine how greatly isotopic fractionation varies between different mechanisms of Cr(VI) reduction and between different field settings. It is perhaps not surprising that abiotic reduction would produce the same degree of fractionation in three contrasting experiments, if one envisions that isotopic selection occurred during Cr-O bond cleavage, that the three experiments had little difference in the bond energy of the broken Cr-O bonds, and that if there were multiple steps in the reduction reactions, the Cr-O bond cleavage was the rate-limiting step. Presumably, Cr(VI) reduction in the magnetite experiment involved transfer of electrons from Fe(II) on the magnetite surfaces to adsorbed Cr(VI) ions. In the sediment slurries, the reduction mechanism was unknown, though abundant dissolved organic molecules were present, and certain organics are known to reduce Cr(VI), especially when sorption surfaces for Cr(VI) are present (Deng and Stone 1996). Future experiments will explore this further and determine if $\varepsilon_{Cr(VI)\text{-}Cr(III)} = 3.4‰$ for all abiotic Cr(VI) reduction reactions.

It is not known whether Cr(VI) reduction is dominantly microbial or abiotic in nature. Whereas SO_4^{2-} and SeO_4^{2-} are strongly resistant to abiotic reduction, the CrO_4^{2-} anion is known to undergo abiotic reduction by various reducing agents at room temperature. Notably, dissolved Fe(II) at micromolar concentrations at nearly neutral pH reduces Cr(VI) in a matter of minutes or hours, depending on solution chemistry specifics. On the other hand, several bacteria are known to reduce Cr(VI), and it is certainly possible that, in the absence of dissolved Fe(II), bacterial Cr(VI) reduction could be more rapid than the various abiotic mechanisms. Accordingly, experiments must be done to examine Cr isotope fractionation during Cr(VI) reduction by microbes.

Lack of isotopic fractionation between adsorbed and dissolved Cr(VI)

In groundwater settings, a small isotopic fractionation between adsorbed Cr(VI) and dissolved Cr(VI) could be magnified over time. This effect is analogous to that observed on chromatographic columns in the laboratory (Russell and Papanastassiou 1978; Anbar et al.

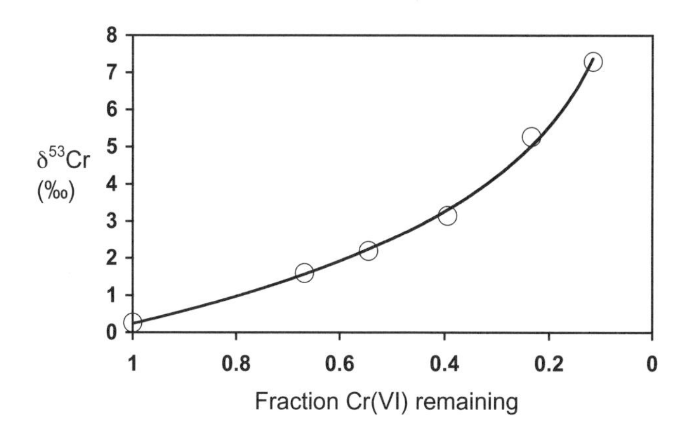

Figure 4. δ^{53}Cr as a function of the amount of Cr(VI) remaining in a batch slurry experiments with estuarine sediment. Line gives a Rayleigh fractionation model, with $\varepsilon = 3.4‰$. Data from Ellis et al. (2002).

2000). Simple models of Cr(VI) transport through aquifers suggest that if adsorbed Cr(VI) were enriched in the heavier isotope by 0.1‰, the leading edge of a Cr(VI) plume could have δ^{53}Cr values roughly 1.0‰ greater than the plume core. In such a case, the isotopic contrast could be attributed to Cr(VI) reduction when in fact it is caused by adsorption.

Recent experiments have determined that the isotopic fractionation between adsorbed and dissolved Cr(VI) is very small (Ellis et al. submitted). In batch experiments, dissolved Cr(VI) was equilibrated with finely ground alumina or goethite. The results of these experiments should be applicable to natural settings, as oxides chemically similar to alumina and goethite provide the main adsorption sites for Cr(VI) anions. Because the analytical precision of the measurements was ±0.2‰, it was necessary to amplify any isotopic shifts so that effects smaller than 0.1‰ could be detected. This was accomplished by repeating the sorption step ten or more times, so that a fractionation between dissolved and adsorbed Cr(VI) of 0.04‰ was detectable at the 95% confidence level.

No δ^{53}Cr differences were detected between the starting solutions and the final solutions. Thus, the equilibrium δ^{53}Cr offset between dissolved and adsorbed Cr(VI) is less than 0.04‰. If this small offset were amplified ten times by a chromatographic effect in a groundwater system, the resulting δ^{53}Cr shift would have a very small impact on interpretation of δ^{53}Cr variations in terms of Cr(VI) reduction. This impact would be limited to the leading edges of plumes or other areas where concentrations change strongly.

Other processes

Cr isotope shifts that might occur during Cr(III) oxidation have not yet been studied. However, in the neutral pH range, Cr(III) is insoluble like Fe(III), and thus, its oxidation would involve layer-by-layer removal of solid surface material. As described above for Se(0), isotopic fractionation is unlikely after the removal of a few layers of Cr(III) atoms, unless there is a branching of the reaction pathway after the Cr has been removed from the solid. However, if Cr(III) oxides are present as very thin coatings, the transient phase of isotopic fractionation during the early stages of oxidation could involve a significant fraction of the total Cr(III) pool. Also, Cr(III) might be somewhat soluble under low pH conditions, or in cases where it complexes with organic ligands, so this oxidation reaction should be examined.

Isotope effects during assimilation of Cr by plants and algae have not been studied. With higher plants we expect that, as with Se and S, the effects will be minimal because of the lack of communication between the plant interior, where reduction occurs, and the outside environment. With single-celled phytoplankton, there could be some fractionation, and this should be investigated as studies of Cr isotopes in the marine environment move forward.

Cr isotope data from igneous rocks, reagent Cr, and Cr plating baths

Measurements and theoretical considerations suggest that igneous rocks, Cr ores, and Cr reagents all have isotopic compositions close to NIST SRM-979, and that SRM-979 is close to the earth's mantle isotopically. Table 7 gives data reported by Ellis et al. (2002). Three basalt samples, from contrasting environments, were indistinguishable from SRM-979. Two reagent grade Cr compounds were slightly enriched in the heavier isotope.

Table 7. Isotopic composition of igneous rock-derived Cr (Ellis et al. 2002).

Material	*δ^{53}Cr (‰, SRM-979)*
BIR Basalt, Iceland	−0.04
BHVO Basalt, Kilauea	0.05
JB Basalt, Japan	−0.04
$K_2Cr_2O_7$ reagent	0.35
$Cr(NO_3)_3$ reagent	0.32
Industrial CrO_3 supply	−0.07
Plating bath 1	0.37
Plating bath 2	0.36
Plating bath 3	0.29

At the elevated temperatures of the earth's mantle, and with all Cr in the +3 valence, little equilibrium isotopic fractionation between minerals phases, or between extracted liquids and residual phases is possible. Thus, magmas ascending from the mantle should be isotopically similar to it. Recycling of sediments with potentially contrasting isotopic compositions into the earth's mantle should have relatively small effects because of high Cr concentrations in the mantle. Cr ores are derived from mafic igneous rocks, which generally have little admixture of crustal material. Even if crustal material is assimilated by these magmas, it has relatively small Cr concentrations. Finally, crystallization of chromite from the magmas should induce very little isotopic fractionation, as temperatures are still high during this process, there is no redox reaction involved, and the bonding environment of Cr is similar in silicate liquids and crystals.

Chromite ($FeCr_2O_4$) ores are the source for industrial Cr supplies. Cr is recovered from the ore by roasting with an alkali fluxing agent, which converts the Fe and Cr to Fe(III) and Cr(VI), respectively, which are then easily separated, for example by dissolving the soluble Cr(VI) into a basic solution. The roasting process involves an oxidation reaction, but it occurs at higher temperatures so little or no fractionation is expected. The overall Cr refining process involves very little loss of Cr and therefore the final product is unlikely to be isotopically shifted relative to the ore. Accordingly, we expect that most or all moderate-purity industrial Cr supplies, such as the chromic acid flakes that are used to make up Cr plating baths, are isotopically close to the earth's mantle.

Cr plating baths appear to be isotopically similar to the CrO_3 used to make them up, even after several years of use, during which significant amounts of Cr(VI) are consumed by the plating process and replaced by addition of more CrO_3. Two of the three plating baths in Table 7 were from a single facility; the $\delta^{53}Cr$ value for the CrO_3 supply at that facility is also given. The third plating bath was from a different facility with somewhat different plating conditions. Because Cr(VI) is reduced to Cr metal, one might expect that kinetic isotope effects would selectively remove lighter isotopes from solution, but the plating baths' $\delta^{53}Cr$ values are only slightly greater than that of the CrO_3 supply. For some reason, the expected kinetic effect is not expressed. This phenomenon may be related to the fact that the first step in the Cr(VI) reduction process should be migration of Cr(VI) across an electric field gradient. Assuming this step has little back reaction (i.e., little migration against the gradient), then the $\delta^{53}Cr$ of the Cr removed from solution is quite close to that of the bulk solution, despite the kinetic isotope effect occurring during a Cr(VI) reduction step (see above discussion of kinetic isotope effects of multi-step processes).

Purification of industrial grade Cr to produce reagent-grade Cr compounds could involve processes in which 1) Cr isotope fractionation occurs and 2) fractionated Cr is lost so that the final product is shifted in $\delta^{53}Cr$ from the initial Cr. Such an effect is suggested by the slightly positive $\delta^{53}Cr$ values of the two reagent-grade Cr compounds listed in Table 7. However, the differences are very small.

A larger data set is necessary to explore the Cr isotope variability of Cr sources. If the emerging pattern holds, this will have great impact on the use of $\delta^{53}Cr$ as an environmental tool. First, it is unlikely that different contaminant sources will have distinct $\delta^{53}Cr$ "signatures." But on the other hand, if all sources turn out to be isotopically similar, then any deviations from $\delta^{53}Cr = 0$ can be attributed to Cr(VI) reduction. This would be particularly useful in the many cases where the source of an existing groundwater plume has been removed and is no longer available for sampling.

Cr isotope data from groundwater systems

A data set from three groundwater systems supports the hypothesis that Cr(VI) reduction in groundwater systems causes enrichment of the remaining Cr(VI) in the heavier isotope, and provides assurance that the purification methodology is effective. Ellis et al. (2002) measured

δ^{53}Cr at two field sites where other studies had inferred Cr(VI) reduction from concentration data (Table 8). We are adding here unpublished data from one of those sites and three analyses from a third site where conditions are strongly oxidizing and Cr(VI) reduction should not occur. All of the groundwater samples are enriched in the heavy isotope relative to SRM-979, indicating that Cr(VI) reduction has taken place.

The Connecticut and Northern California sites appear to have substantial Cr(VI) reduction occurring, based on other studies using temporal and spatial patterns in concentration data. The general pattern in the isotope data is that the lowest concentration samples, which are usually distal samples near the plume fringes, have the greatest δ^{53}Cr values. This is consistent with the expectation that the reducing capacity of plume cores will tend to be depleted by extended contact with Cr(VI)-rich waters, whereas the plume fringes should have greater reduction rates. Using a Rayleigh model and $\varepsilon_{Cr(VI)-Cr(III)}$ value of 3.4‰ from the laboratory experiments, and assuming the initial plating bath Cr was +0.3‰, we estimate that between 20% and 80% of the originally present Cr(VI) has been removed by reduction.

Table 8. Cr isotope ratios in groundwaters.

Sample	*[Cr(VI)] (mg/L)*	*δ^{53}Cr (‰)*
Connecticut Site		
Plating baths, mean (n=2)	---	0.3
MW-8 groundwater	4.5	2.2
MW-9 groundwater	8.61	1.3
MW-11s groundwater	16.1	1.9
MW-11 groundwater	0.63	3.6
MW-12 groundwater	1.63	4.0
Northern California Site		
MW-3 groundwater	0.98	1.1
MW-7 groundwater	1.35	1.4*
MW-8 groundwater	1.97	1.2*
MW-10 groundwater	0.13	5.8
MW-11 groundwater	0.751	3.4*
MW-12 groundwater	3.10	3.4
MW-15 groundwater	1.04	3.2*
MW-16 groundwater	3.12	3.6*
Southern California Site		
MW-1 groundwater	0.238	1.6*
MW-3 groundwater	0.889	1.4*
MW-5 groundwater	186.1	1.7*

*Unpublished data; others from Ellis et al. (2002)

At the Southern California site, there is no isotopic contrast between the proximal, high concentration sample and the distal, low concentration samples. This suggests that there is no reduction occurring within the groundwater plume. However, δ^{53}Cr values of all three samples are elevated and suggest either the source itself contains Cr enriched in the heavy isotope or that roughly 20% of the original Cr(VI) was reduced to Cr(III) in the source zone before migration in the oxidizing aquifer. Locally reducing conditions at the source area could exist in soils or as a result of degradation of organic co-contaminants.

If remobilization of Cr(III) in an aquifer were to occur, this process may be detectable using Cr isotope ratios. When Cr(VI) reduction is occurring, Cr(III) in the aquifer matrix should be enriched in the lighter isotopes relative to the Cr(IV) in the water. If part of a Cr(VI) plume were to change over time from enrichment in heavier isotopes to enrichment in lighter isotopes, this would suggest remobilization of previously reduced Cr has begun.

Need for field calibration

The greatest uncertainty in interpreting Cr isotope data from groundwater systems is in the fractionation factor used to calculate the extent of reduction. It is not known at present if natural Cr(VI) reduction in aquifers is predominantly abiotic or microbially mediated. If microbial reactions dominate, then the isotopic fractionation is likely not the same as for abiotic reduction and may depend on the microbial species present and their metabolic state. It is also possible that variation exists between the various abiotic reduction mechanisms. It will thus be necessary to determine Cr isotope fractionation factors in groundwater systems without greatly altering the flow or water chemistry.

FUTURE DIRECTIONS

The basic isotope systematics of Se and Cr are now defined well enough to enable applications in several areas. As noted above, however, there are many uncertainties remaining and many more well-controlled laboratory experiments to be done. The interpretation of field data is only as good as our understanding of the isotope systematics, and the isotope systematics are quite complex with these elements.

As both elements are important environmental contaminants whose mobility and environmental impact depend on redox reactions, we expect that Cr and Se isotope analyses will be used widely as indicators of oxyanion reduction. It may also be possible to trace the origin of the Se or Cr via isotopic "signatures," but only if the possible confounding effects of reduction can be constrained.

Applications in studies of the modern and ancient oceans also look promising. Se and Cr are biologically active elements in the oceans, and changes in their cycling have almost certainly occurred over time in response to changes in oxygenation of the oceans. Se concentrations in seawater are extremely small, but with the advent of a highly sensitive MC-ICP-MS method (Rouxel et al. 2002), isotopic analysis of the various Se species is now thinkable. Cr concentrations are greater and the isotopic analyses may be somewhat less challenging, but just as informative. Ultimately, an array of isotopic analyses (e.g., S, Mo, Cr and Se), each of which has a unique redox threshold, may be used to better define the oxygenation history of the oceans and the atmosphere.

Se is concentrated in certain ore deposits, such as roll-front U complexes and Carlin-type gold deposits. Roll-front U deposits occur where U(VI)-bearing groundwater migrates into reducing zones which precipitate both U(IV) and Se(0) or Se(−II). It may be possible to constrain Se sources and/or ore-forming processes using Se isotope variations.

ACKNOWLEDGMENTS

The writing of this chapter and much of the work reviewed herein was supported by the National Science Foundation, Division of Earth Sciences, Hydrologic Sciences Program under grants EAR 97-25799, EAR 00-03381, EAR 00-01153, and EAR 02-29079. Thoughtful reviews by O. Rouxel, E. Schauble, and D. Canfield were very helpful in improving the manuscript.

REFERENCES

Albarède F, Beard BL (2004) Analytical methods for non-traditional isotopes. Rev Mineral Geochem 55: 113-151

Altman C, King EL (1961) The mechanism of the exchange of chromium(III) and chromium(VI) in acidic solution. J Amer Chem Soc 83:2825-2830

Amouroux D, Liss PS, Tessier E, Hemren-Larsson M, Donard OFX (2001) Role of oceans as biogenic sources of selenium. Earth Planet Sci Lett 189: 277-283

Anbar AD, Roe JE, Holman ES, Barling J, Nealson KH (2000) Non-biological fractionation of iron isotopes. Science 288:126-128

Bar-Yosef B, Meek D (1987) Selenium sorption by kaolinite and montmorillonite. Soil Sci (144):11-19

Beard BL, Johnson CM (2004) Fe isotope variations in the modern and ancient earth and other planetary bodies. Rev Mineral Geochem 55:319-357

Beaudoin G, Taylor BE, Rumble III D, Thiemens M (1994) Variations in the sulfur isotope composition of troilite from the Cañon Diablo iron meteorite. Geochim Cosmochim Acta 58:4253-4255

Birck JL, Allegre CJ (1988) Manganese-chromium isotope systematics and the development of the early solar system. Nature 331(6157):579-584

Blackmer AM, Bremner JM (1977) Nitrogen isotope discrimination in denitrification of nitrate in soils. Soil Biol Biochem 9:73-77

Blowes DW (2002) Tracking hexavalent Cr in groundwater. Science 295:2024-2025

Blum JS, Bindi AB, Buzzelli J, Stolz J, Oremland RS (1998) *Bacillus arsenicoselenatis,* sp. nov., and *Bacillus selenitireducens,* sp. nov.: two haloalkaliphiles from Mono Lake, California that respire oxyanions of selenium and arsenic. Arch Microbiol 171:19-30

Brookins DG (1988) Eh-pH Diagrams for Geochemistry. Springer-Verlag, New York

Brüchert V, Knoblauch C, Jørgensen BB (2001) Controls on stable sulfur isotope fractionation during bacterial sulfate reduction in Arctic sediments. Geochim Cosmochim Acta 65:763-776

Bryan BA, Shearer G, Skeeters JL, Kohl DH (1983) Variable expression of the nitrogen isotope effect associated with denitrification of nitrate. J Biol Chem 258:8613-8617

Canfield DE (2001) Biogeochemistry of sulfur isotopes. Rev Mineral Geochem 43:607-636

Chau YK, Riley JP (1965) The determination of selenium in sea water, silicates, and marine organisms. Anal Chim Acta 33:36-49

Coleman RG, Delevaux MH (1957) Occurrence of selenium in sulfides from some sedimentary rocks of the western United States. Econ Geol 52(5):499-527

Cooper WC, Bennett KG, Croxton FC (1974) The history, occurrence, and properties of selenium. *In:* Selenium. Zingaro RA, Cooper WC (eds), Van Nostrand Reinhold, New York, p 1-30

Cooper WC, Glover JR (1974) The toxicology of selenium and its compounds. *In:* Selenium. Zingaro RA, Cooper WC (eds), Van Nostrand Reinhold, New York, p 654-674

Cutter GA (1982) Selenium in reducing waters. Science 217(4562):829-831

Cutter GA, Bruland KW (1984) The marine biogeochemistry of selenium; a re-evaluation. Limnol Ocean 29(6):1179-1192

Cutter GA, Cutter LS (1995) Behavior of dissolved antimony, arsenic, and selenium in the Atlantic Ocean. *In:* IOC Contaminants Baseline Study. Marine Chemistry. Yeats PA, Shiller AM (eds), Elsevier, Amsterdam, p 295-306

Davis A, Olsen RL (1995) The geochemistry of chromium migration and remediation in the subsurface. Ground Water 33:759-768

Deng B, Stone A (1996) Surface-catalyzed chromium(VI) reduction: reactivity comparisons of different organic reductants and different oxide surfaces. Environ Sci Technol 30(8):2484-2494

Detmers J, Brüchert V, Habicht KS, Kuever J (2001) Diversity of sulfur isotope fractionations by sulfate-reducing prokaryotes. Appl Eviron Microbiol 67:888-894

Dickman M, Thode H (1990) Sulphur bacteria and sulphur isotope fractionation in a meromictic lake near Toronto, Canada. *In:* Facets of Modern Biogeochemistry. Ittekkot V, Kempe S, Michaelis W, Spitzy A (eds), Springer-Verlag, New York, p 225-241

Dowdle PR, Oremland RS (1998) Microbial oxidation of elemental selenium in soil slurries and bacterial cultures. Environ Sci Technol 32:3749-3755

Dungan RS, Frankenberger Jr. WT (1998) Reduction of selenite to elemental selenium by Enterobacter cloacae SLD1a-1. J Env Qual 27:1301-1306

Ellis AS, Johnson TM, Bullen TD (2002) Cr isotopes and the fate of hexavalent chromium in the environment. Science 295:2060-2062

Ellis AS, Johnson,TM, Bullen TD (submitted) Using chromium stable isotope ratios to quantify Cr(VI) reduction: lack of sorption effects. Environ Sci Technol submitted 9/2003

Ellis AS, Johnson TM, Bullen TD, Herbel MJ (2003) Stable isotope fractionation of selenium by natural microbial consortia. Chem Geol 195:119-129

Elrashidi MA, Adriano DC, Workman SM, Lindsay WL (1987) Chemical equilibria of selenium in soils; a theoretical development. Soil Sci 144(2):141-152

Fan TWM, Higashi RM (1998) Biochemical fate of selenium in microphytes; natural bioremediation by volatilization and sedimentation in aquatic environments. *In:* Environmental Chemistry of Selenium. Frankenberger Jr. WT, Engberg RA (eds), Marcel Dekker, New York, p 545-564

Farquhar J, Johnston DT, Wing BA, Habicht KS, Canfield DE, Airieau S, Thiemens MH (2003) Multiple sulphur isotopic interpretations of biosynthetic pathways: implications for biological signatures in the sulphur isotope record. Geobiol 1:27-36

Faure G (1991) Principles and Applications of Geochemistry. Wiley, New York

Fendorf SE (1995) Surface reactions of chromium in soils and waters. Geoderma 67:55-71

Frankenberger Jr. WT, Karlson U (1994) Microbial volatilization of selenium from soils and sediments. *In:* Selenium in the Environment. Frankenberger WT, Benson S (eds), Marcel Dekker, New York, NY, United States, p 369-387

Ganther HE (1974) The biochemistry of selenium. *In:* Selenium. Zingaro RA, Cooper WC (eds), Van Nostrand Reinhold Co., p 546-614

Habicht KS, Canfield DE (1997) Sulfur isotope fractionation during bacterial sulfate reduction in organic-rich sediments. Geochim Cosmochim Acta 61(24):5351-5361

Habicht KS, Gade M, Thamdrup B, Berg P, Canfield DE (2002) Calibration of sulfate levels in the Archean ocean. Science 298:2372-2374

Hagiwara Y (2000) Selenium isotope ratios in marine sediments and algae. A reconnaissance study. M.S. Thesis, Univ. of Illinois at Urbana-Champaign, Urbana, IL

Harrison AG, Thode HG (1957) The kinetic isotope effect in the chemical reduction of sulphate. Trans Faraday Soc 53:1648-1660

Harrison AG, Thode HG (1958) Mechanism of the bacterial reduction of sulphate from isotope fractionation studies. Trans Faraday Soc. 54:84-96

Hayes JM (2001) Fractionation of carbon and hydrogen isotopes in biosynthetic processes. Rev Mineral Geochem 43: 225-277

Herbel MJ, Johnson TM, Oremland RS, Bullen TD (1998) Selenium stable isotope fractionation during bacterial dissimilatory reduction of selenium oxyanions (Abstract). EOS, Transact. A.G.U. 79, Suppl.: F356

Herbel MJ, Johnson TM, Oremland RS, Bullen TD (2000) Selenium stable isotope fractionation during bacterial dissimilatory reduction of selenium oxyanions. Geochim Cosmochim Acta 64:3701-3709

Herbel MJ, Johnson TM, Tanji KK, Gao S, Bullen TD (2002) Selenium stable isotope ratios in agricultural drainage water systems of the western San Joaquin Valley, CA. J Environ Qual 31:1146-1156

Hübner H (1986) Isotope effects of nitrogen in the soil and biosphere. *In:* Handbook of Environmental Isotope Geochemistry: Vol. 2 The Terrestrial Environment, B. Fritz P, Fontes JC (eds), Elsevier, New York, p 361-425

Jeandel C, Minster JF (1987) Chromium behavior in the ocean; global versus regional processes. Global Biogeochem Cycles 1(2):131-154

Johnson TM (in press) A review of mass-dependent fractionation of selenium isotopes and implications for other "heavy stable isotopes." Chem Geol

Johnson TM, Bullen TD (2003) Selenium isotope fractionation during reduction by Fe(II)-Fe(III) hydroxide-sulfate (green rust). Geochim Cosmochim Acta 67:413-419

Johnson TM, Bullen TD, Zawislanski PT (2000) Selenium stable isotope ratios as indicators of sources and cycling of selenium: Results from the northern reach of San Francisco Bay. Environ Sci Tech 34: 2075-2079

Johnson TM, Herbel MJ, Bullen TD, Zawislanski PT (1999) Selenium isotope ratios as indicators of selenium sources and oxyanion reduction. Geochim Cosmochim Acta 63(18):2775-2783

Kaplan IR, Emery KO, Rittenberg SC (1963) The distribution and isotopic abundance of sulphur in recent marine sediments off southern California. Geochim Cosmochim Acta 27(4):297-331

Kaplan IR, Rittenberg SC (1964) Microbial fractionation of sulfur isotopes. Gen Microbiol 43:195-212

Kemp ALW, Thode HG (1968) The mechanism of the bacterial reduction of sulphate and of sulphite from isotope fractionation studies. Geochim Cosmochim Acta 32(1):71-91

Kohl DH, Shearer G (1978) Isotope effects in metabolic studies. *In:* Recent developments in mass spectrometry in biochemistry and medicine. Frigerio A (ed), Plenum, New York, p 623-641

Krouse HR, Thode HC (1962) Thermodynamic properties and geochemistry of isotopic compounds of selenium. Can J Chem 40(2):367-375

Lee T, Tera F (1986) The meteoritic chromium isotopic composition and limits for radioactive ^{53}Mn in the early solar system. Geochim Cosmochim Acta 50(2):199-206

Lemly AD (1985) Toxicology of selenium in a freshwater reservoir: Implications for environmental hazard evaluation and safety. Ecotoxicol Environ Saf 10:314-338

Lemly AD (1998) Pathology of Selenium Poisoning in Fish. *In:* Environmental Chemistry of Selenium. Frankenberger Jr. WT, Engberg RA (eds), Marcel Dekker, New York, p 281-296

Lyons TW (1997) Sulfur isotopic trends and pathways of iron sulfide formation in upper Holocene sediments of the anoxic Black Sea. Geochim Cosmochim Acta 61(16):3367-3382

Mariotti A, Germon JC, Hubert P, Kaiser P, Letolle R, Tardieux A, Tardieux P (1981) Experimental determination of nitrogen kinetic isotope fractionation: some principles; illustration for the denitrifcation and nitrification processes. Plant Soil 62:413-430

Mariotti A, Landreau A, Simon B (1988) ^{15}N isotope biogeochemistry and natural denitrification process in groundwater; application to the chalk aquifer of northern France. Geochim Cosmochim Acta 52(7): 1869-1878

Mayer LM (1988) Geochemistry of chromium in the oceans. *In:* Chromium in the Natural and Human Environments. Nriagu JO, Nieboer E (eds), Wiley & Sons, New York, p 173-187

McNeal JM, Balistrieri LS (1989) Geochemistry and occurrence of selenium; an overview. *In:* Selenium in Agriculture and the Environment. SSSA Special Publication. Jacobs LW (ed), Soil Science Society of America, Madison, WI, p 1-13

Measures CI, Grant BC, Mangum BJ, Edmond JM (1983) The relationship of the distribution of dissolved selenium IV and VI in three oceans to physical and biological processes. *In:* Trace Metals in Sea Water. Wong CS, Boyle EA, Bruland KW, Burton JD, Goldberg ED (eds), Plenum, New York, p 73-83

Measures CI, McDuff RE, Edmond JM (1980) Selenium redox chemistry at GEOSECS I re-occupation. Earth Planet Sci Lett 49(1):102-108

Mugo, RK (1997) The geochemistry of chromium in various marine environments. PhD Dissertation, University of British Columbia

Myneni SCB, Tokunaga TK, Brown Jr. GE (1997) Abiotic selenium redox transformations in the presence of Fe(II,III) oxides. Science 278(5340):1106-1109

Neal RH, Sposito G (1989) Selenate adsorption on Alluvial soils. Soil Sci Soc Amer J 53(1):70-74

Nriagu JO, Niebor E (eds) (1988) Chromium in the natural and human environments. John Wiley and Sons, New York, 462 pp

Oremland RS, Hollibaugh JT, Maest AS, Presser TS, Miller LG, Culbertson CW (1989) Selenate reduction to elemental selenium by anaerobic bacteria in sediments and culture: Biogeochemical significance of a novel, sulfate-independent respiration. Appl Env Microbiol 55(9):2333-2343

Perlmutter NM, Lieber M, Frauenthal HL (1963) Movement of waterborne cadmium and hexavalent chromium wastes in South Farmingdale, Nassau County, Long Island, U.S. Geological Survey Professional Paper 475C, p C170-C184

Pettine M (2000) Redox processes of chromium in sea water. *In:* Chemical Processes in Marine Environments. Gianguzza A, Pelizzetti E, Sammartano S (eds) Springer, Berlin, p 281-296

Presser TS (1994) The Kesterson effect. Environ Manage 18:437-454

Rashid K, Krouse HR (1985) Selenium isotopic fractionation during SeO_3 reduction to Se^0 and H_2Se. Can J Chem 63:3195-3199

Rees CB, Thode HG (1966) Selenium isotope effects in the reduction of sodium selenite and of sodium selenate. Can J Chem 44:419-427

Rees CE (1973) A steady-state model for sulphur isotope fractionation in bacterial reduction processes. Geochim Cosmichim Acta 37:1141-1162

Robertson FN (1975) Hexavalent chromium in the ground water in Paradise Valley, Arizona. Ground Water 13(6):516-527

Robles CJ, Armienta MA (2000) Natural chromium contamination of groundwater at Leon Valley, Mexico. J Geochem Explor 68(3):167-181

Rouxel O, Ludden J, Carginan J, Marin L, Fouquet Y (2002) Natural variations of Se isotopic composition determined by hydride generation multiple collector inductively coupled plasma mass spectrometry. Geochim Cosmochim Acta 66(18):3191-3199

Russell WA, Papanastassiou DP (1978) Calcium isotope fractionation in ion-exchange chromatography. Anal Chem 50(8):1151-1154

Schauble EA (2004) Applying stable isotope fractionation theory to new systems. Rev Mineral Geochem 55: 65-111

Schauble E, Rossman GR, Taylor HP (in press) Theoretical estimates of equilibrium chromium-isotope fractionations. Chem Geol

Seiler RL (1998) Prediction of lands susceptible to irrigation-induced selenium contamination of water. *In:* Environmental Chemistry of Selenium. Frankenberger Jr. WT, Engberg RA (eds), Marcel Dekker, New York, p 397-418

Sheilds WR, Murphy TJ, Catanzaro EJ, Garner J (1966) Absolute isotopic abundance ratios and the atomic weight of a reference sample of chromium. J Res Natl Bur Standards 70A(2):193-197

Shukolyukov A, Lugmair GW (1998) Isotopic evidence for the Cretaceous-Tertiary impactor and its type. Science 282(5390):927-929

Sirinawin W, Turner DR, Westerlund S (2000) Chromium(VI) distributions in the Arctic and the Atlantic oceans and a reassessment of the oceanic Cr cycle. Marine Chem 71(3-4):265-282

Skorupa JP (1998) Selenium poisoning of fish and wildlife in nature: Lessons from twelve real-world-examples. *In:* Environmental Chemistry of Selenium. Frankenberger Jr. WT, Engberg RA (eds), Marcel Dekker, New York, p 315-354

Takayanagi K, Wong GTF (1985) Dissolved inorganic and organic selenium in the Orca Basin. Geochim Cosmochim Acta 49(2):539-546

Tanzer D, Heumann KG (1991) Determination of dissolved selenium species in environmental water samples using isotope dilution mass spectrometry. Anal Chem 63:1984-1989

Thode HG, Kleerekoper H, McElcheran DE (1951) Isotope fractionation in the bacterial reduction of sulfate. Research Lond 4:581

Tokunaga TK, Lipton DS, Benson SM, Yee AW, Oldfather JM, Duckar EC, Johannis PW, Halvorsen KE (1991) Soil selenium fractionation, depth profiles and time trends in a vegetated site at Kesterson Reservoir. Water Air Soil Pollut 57:31-41

Trust BA, Fry B (1992) Stable sulphur isotopes in plants: a review. Plant Cell Environ 15:1105-1110

Tudge AP, Thode HG (1950) Thermodynamic properties of isotopic compounds of sulfur. Can J Res B28: 567-578

Van Stempvoort DR, Reardon EJ, Fritz P (1990) Fractionation of sulfur and oxygen isotopes in sulfate by soil sorption. Geochim Cosmochim Acta 54(10):2817-2826

Wachsmann M, Heumann KG (1992) Negative thermal ionization mass spectrometry of main group elements, part 2. 6th group: Sulfur, selenium, and tellurium. Int J Mass Spectrom Ion Proc 114:209-220

Webster CL (1972) Selenium isotope analysis and geochemical applications. Ph.D Dissertation, Colorado State University, Fort Collins, CO

White AF, Dubrovsky NM (1994) Chemical oxidation-reduction controls on selenium mobility in groundwater systems. *In:* Environmental Chemistry of Selenium. Frankenberger Jr. WT, Engberg RA (eds), Marcel Dekker, New York, p 185-221

Whitfield M, Turner DR (1987) The role of particles in regulating the composition of seawater. *In:* Aquatic Surface Chemistry. Stumm W (ed) Wiley, New York, p 457-493

Zawislanski PT, McGrath AE (1998) Selenium cycling in estuarine wetlands; overview and new results from the San Francisco Bay. *In:* Environmental Chemistry of Selenium. Frankenberger Jr. WT, Engberg RA (eds), Marcel Dekker, New York, p 223-242

Zhang Y, Moore JN (1996) Selenium fractionation and speciation in a wetland system. Env Sci Technol 30(8): 2613-2619

Reviews in Mineralogy & Geochemistry
Vol. 55, pp. 319-357, 2004
Copyright © Mineralogical Society of America

10A

Fe Isotope Variations in the Modern and Ancient Earth and Other Planetary Bodies

Brian L. Beard and Clark M. Johnson

Department of Geology and Geophysics
University of Wisconsin-Madison
1215 West Dayton Street
Madison, Wisconsin 53706, U.S.A.

INTRODUCTION

Iron, the fourth most abundant element in the Earth's crust, has four naturally occurring stable isotopes: ^{54}Fe (5.84%), ^{56}Fe (91.76%), ^{57}Fe (2.12%), and ^{58}Fe (0.28%), and the natural, mass-dependent isotope variations of Fe in the rock record span a range of ~4 per mil (‰) in ^{56}Fe/^{54}Fe ratios (Fig. 1). The field of Fe isotope geochemistry is relatively new but has received considerable attention because it may allow us to gain a better understanding of how Fe is cycled in different environments. Iron typically occurs as either reduced ferrous Fe in oxygen-poor environments, or as oxidized ferric iron in oxygen-rich environments. Notably, only the reduced species is soluble in oxygenated aqueous solutions, unless the pH is low. In the Archean and Early Proterozoic, the earth may have been relatively oxygen-poor (e.g., Kasting et al. 1979; Grandstaff 1980; Holland 1994), suggesting that there may have been significant quantities of Fe (0.9 millimolar) dissolved in the oceans as $Fe(II)_{aq}$ (e.g., Ewers 1983; Sumner 1997). The extensive iron formations of Archean to Early Proterozoic age may have been deposited from such Fe(II)-rich oceans (e.g., Beukes and Klein 1992). In the modern oxic oceans, however, Fe contents are exceedingly low, <1 nanomolar in the open oceans (e.g., Martin and Gordon 1988; Bruland et al. 1991; Martin 1992; Johnson et al. 1997), and it is now recognized that marine productivity is Fe-limited in parts of the open oceans (e.g., Martin and Fitzwater 1988; Martin et al. 1989, 1994). The differences in the behavior of Fe with redox state, and the significant isotope fractionations (1‰ or more in ^{56}Fe/^{54}Fe) that are associated with redox conditions, suggests that Fe isotope studies will be extremely useful for tracing the Fe geochemical cycle.

The largest variations in Fe isotope compositions are associated with chemically precipitated sediments such as Pliocene to recent Fe-Mn crusts (Zhu et al. 2000) and Archean and Proterozoic age Banded Iron Formations (BIFs; Johnson et al. 2003a). Moreover, there is a similarly large range in the Fe isotope compositions of black shales, where diagenesis occurred under anoxic conditions (Yamaguchi et al. 2003; Matthews et al. 2004; Fig. 1). In contrast, clastic sedimentary materials such as loess, turbidites, the suspended load of rivers, and grey shales which were deposited in oxic environments have a very narrow range in Fe isotope compositions (Fig. 1; Beard et al. 2003a). The variability in Fe isotope compositions that is seen in nature, however, is not limited to low temperature sedimentary environments, where for example, chondritic meteorites, in particular individual chondrules, define a spread of ^{56}Fe/^{54}Fe ratios of 2.5‰ (e.g., Zhu et al. 2001; Kehm et al. 2003; Mullane et al. 2003a; Fig. 1). In contrast, bulk-rock analyses of achondrite meteorites (i.e., meteorites from differentiated planetary bodies) and terrestrial igneous rocks define a relative narrow range of Fe isotope

1529-6466/04/0055-010A$05.00

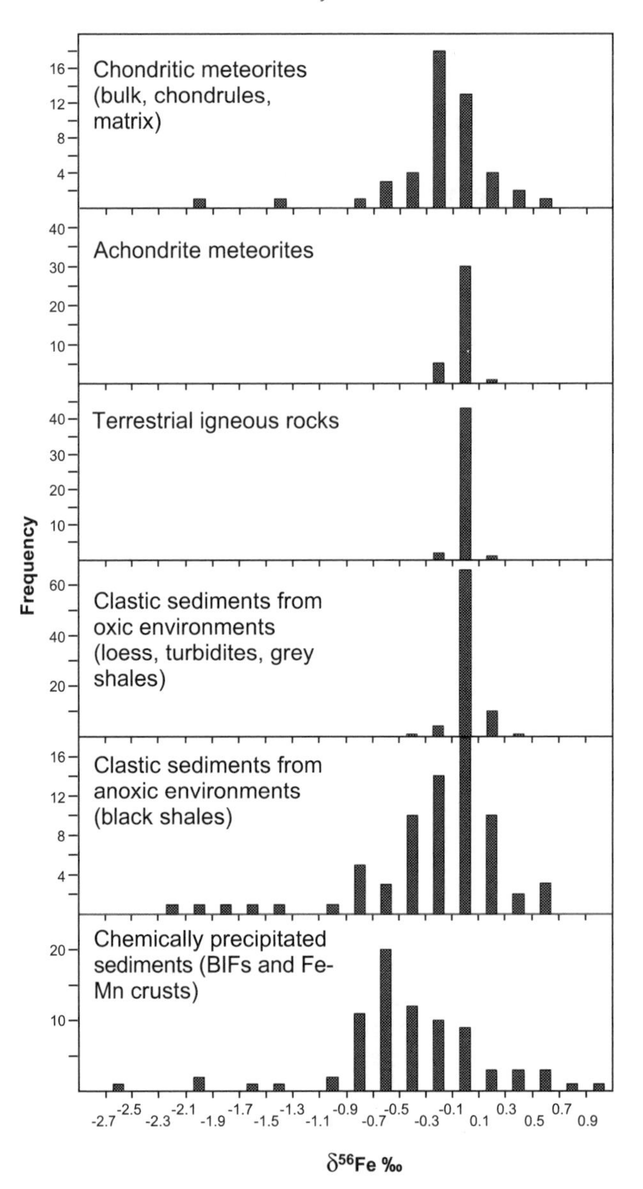

Figure 1. Frequency diagram of Fe isotope compositions measured in rocks from different environments. Interlaboratory bias in Fe isotope measurements have been corrected by normalizing to a δ^{56}Fe value of −0.09‰ for the IRMM-014 standard, which places Fe isotope measurements relative to the average of terrestrial igneous rocks. Data sources are: chondrite meteorite data from Zhu et al. (2001), Mullane et al. (2003a,b), and Kehm et al. (2003); achondrites including samples from the SNC, HED, Pallasite, and iron groups from Zhu et al. (2001), Mullane et al. (2003c), and Kehm et al. (2003); terrestrial igneous rocks from Beard et al. (2003b); clastic sediments from oxic environments including loess, turbidites, low-C_{org} shales, modern aerosols, and suspended river load sediments from Beard et al. (2003a); high-C_{org} black shales from Yamaguchi et al. (2003) and Matthews et al. (2004); chemically precipitated sediments including layers from Archean and Proterozoic Banded Iron Formations from Johnson et al. (2003a), and an Atlantic Ocean Fe-Mn crust from Zhu et al. (2000).

composition (Fig. 1). Indeed, terrestrial igneous rocks are very homogenous in Fe isotope compositions; the average of 43 igneous rocks that range in composition from peridotite to high-silica rhyolite are identical in $^{56}Fe/^{54}Fe$ ratios to within ±0.05‰ (1σ).

This chapter is divided into four main parts. In the first section, the nomenclature and analytical methods used in Fe isotope geochemistry are discussed. In the second section, we summarize the emerging data base of Fe isotope fractionation factors in abiological systems. The final two sections discuss Fe isotope variations in natural environments. In the third section, Fe isotope variability in high-temperature environments is reviewed and the variability of Fe isotope compositions in the solar system which may arise from planetary accretion processes and evaporation and condensation of Fe is highlighted. The final section discusses the variability of Fe isotope ratios measured on rocks and fluids from low-temperature environments. The origins of these Fe isotope variations may be evaluated based on Fe isotope fractionation factors, as well as mass-balance constraints that can be inferred from the growing database of Fe isotope compositions of natural samples. In the following chapter (Johnson et al. 2004), Fe isotope fractionations that are associated with biological processes are discussed and this provides a compliment to the review in the present chapter.

NOMENCLATURE AND ANALYTICAL METHODS

Nomenclature

Iron isotope data in the literature have been reported in both standard δ notation (parts per 10^3), and ε notation (parts per 10^4), and workers have reported data in terms of $^{57}Fe/^{54}Fe$, $^{57}Fe/^{56}Fe$, or $^{56}Fe/^{54}Fe$ ratios. The low abundance of ^{58}Fe limits its usefulness in describing mass-dependent variations, although ^{58}Fe may be important for evaluating non-mass dependent effects (e.g., Völkening and Papanastassiou 1989). In standard per mil notation, the $\delta^{56}Fe$ and $\delta^{57}Fe$ values are most commonly defined as:

$$\delta^{56}\text{Fe}=\left(\frac{^{56}\text{Fe}/^{54}\text{Fe}_{\text{sample}}}{\left(^{56}\text{Fe}/^{54}\text{Fe}_{\text{standard}}\right)-1}\right)\times 10^3 \tag{1}$$

and

$$\delta^{57}\text{Fe}=\left(\frac{^{57}\text{Fe}/^{54}\text{Fe}_{\text{sample}}}{\left(^{57}\text{Fe}/^{54}\text{Fe}_{\text{standard}}\right)-1}\right)\times 10^3 \tag{2}$$

Epsilon notation is defined similarly, although the deviations are in parts per 10,000. Comparison between these numbers is straightforward as shown in Figure 2, which plots the co-variations in $^{56}Fe/^{54}Fe$ and $^{57}Fe/^{54}Fe$ of layers from BIFs in both δ and ε notation. The mass-dependent fractionation line, based on a simple harmonic oscillator approximation (Criss 1999), lies close to a line of 1.5:1 for $\delta^{57}Fe$–$\delta^{56}Fe$ variations. For example, point A in Figure 2 has an $\delta^{57}Fe$ value of +15.0, which would be approximately equal to a $\delta^{56}Fe$ value of +1.00, as defined here, assuming normalization to an identical reference reservoir.

The choice of the reference reservoir for calculating δ or ε values is quite variable among different laboratories. Table 1 lists all the laboratories that have published Fe isotope data in an abstract or peer-reviewed journal as of September 2003, and include the form in which the data are reported (e.g., $\delta^{56}Fe$ or $\delta^{57}Fe$), as well as the reservoir or "standard" used to define δ or ε values. A common procedure of all labs currently conducting Fe isotope studies is to measure at least three Fe isotopes, which is done to provide a check on data quality, primarily evaluation of potential isobaric interferences. The choice of reference reservoir used in this chapter follows the approach of other stable isotope systems such as oxygen, which defines

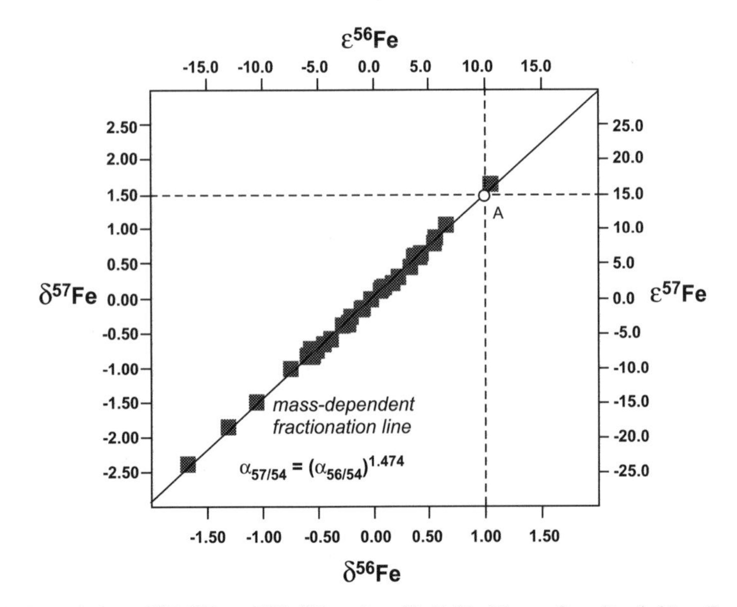

Figure 2. Co-variation of $^{56}Fe/^{54}Fe$ and $^{57}Fe/^{54}Fe$ ratios of individual layers from Banded Iron Formations, modified from Johnson et al. (2003a). The horizontal axis shows isotopic compositions in δ^{56}Fe (bottom) and ε^{56}Fe (top), and the vertical axis shows values in δ^{57}Fe (left) and ε^{57}Fe (right). Point A has a δ^{56}Fe of +1.00‰, which corresponds to an ε^{57}Fe value of +15.0. Figure illustrates the multiple ways in which the same Fe isotope composition may be reported when isotopic variations are due to mass-dependent processes. See text for discussion.

Table 1. Laboratories pursuing Fe isotope studies and methods of reporting data.

Laboratory	Data Reporting / reservoir		Example Reference
CNRS Nancy	δ^{57}Fe	IRMM-014	Rouxel et al. (2003)
DTM	δ^{56}Fe	igneous rock	Kehm et al. (2003)
ETH	δ^{57}Fe	IRMM-014	Poitrasson et al. (2002)
Frankfurt Min. Inst.	δ^{57}Fe	IRMM-014	Weyer and Schwieters (2003)
Geol. Surv. Israel	δ^{57}Fe	IRMM-014	Matthews et al. (2004)
London Royal Holloway	δ^{56}Fe	IRMM-014	Archer and Vance (2002)
Luleå Univ. Tech.	δ^{56}Fe	IRMM-014	Malinovsky et al. (2003)
MIT	δ^{56}Fe	igneous rock	Bergquist and Boyle (2002)
Nat. Hist. Mus., London	δ^{56}Fe	IRMM-014	Mullane et al. (2003a)
Oxford Univ.	ε^{57}Fe	IRMM-014	Zhu et al. (2000)
Rochester	δ^{56}Fe	Canyon Diablo Troilite	Sharma et al. (2001)
UC Berkley	δ^{56}Fe	igneous rock	Fantle and DePaolo (2002)
Univ. Bern	δ^{56}Fe	IRMM-014	Walczyk and von Blackenburg (2002)
Univ. Southampton	δ^{56}Fe	IRMM-014	Severmann et al. (2002)
Univ. WI-Madison	δ^{56}Fe	avg. igneous rock	Beard and Johnson (1999)
USGS Menlo Park	δ^{56}Fe	igneous rock BIR-1	Mandernack et al. (1999)

Note that many labs analyze at least three isotopes, and many data tables typically report both $^{56}Fe/^{54}Fe$ and $^{57}Fe/^{54}Fe$ isotope ratios. Conversion between igneous rock baselines and IRMM-014 baselines can be done using the IRMM-014 standard.

δ values relative to a significant planetary reservoir such as Standard Mean Ocean Water (SMOW). It is also an approach used in radiogenic isotopes where, for example, ε_{Nd}, ε_{Hf}, and γ_{Os} values are defined relative to a bulk earth or planetary reference value. Based on Table 1, the two main methods of reporting Fe isotope data are relative to terrestrial igneous rocks or relative to the IRMM-014 standard (available from the Institute of Reference Materials and Measurements, Belgium; Taylor et al. 1992, 1993). Regardless of the choice of reference reservoir, inter-laboratory comparison of Fe isotope ratios can be made by normalizing through a common standard such as IRMM-014, and this standard has rapidly become the accepted interlaboratory standard for the field. On an igneous rock scale, the δ^{56}Fe value of IRMM-014 is −0.09‰ and the δ^{57}Fe value is −0.11‰ (Beard et al. 2003b). All of the data discussed in this chapter are reported as δ^{56}Fe values relative to the average of igneous rocks (Beard et al. 2003b). Data from labs that only report ^{57}Fe/^{54}Fe values were converted by multiplying the δ^{57}Fe value by 0.667. For data reported relative to IRMM-014, −0.09‰ was subtracted from the δ^{56}Fe value to place it on the igneous rock scale.

Mass-dependent iron isotope fractionation between two phases, A and B, is noted as

$$\alpha_{A\text{-}B} = \frac{\left({}^{56}\mathrm{Fe}/{}^{54}\mathrm{Fe}\right)_A}{\left({}^{56}\mathrm{Fe}/{}^{54}\mathrm{Fe}\right)_B} \tag{3}$$

where, for example, an $\alpha_{A\text{-}B}$ of 1.001 means that phase A would have a δ^{56}Fe value 1‰ greater than phase B, based on the approximation that $10^3\ln\delta_{A\text{-}B}$ is ~ $\delta^{56}Fe_A - \delta^{56}Fe_B$.

Analytical methods

Iron isotope measurements are made by both the double-spike method using thermal ionization mass spectrometry (TIMS) (e.g., Johnson and Beard 1999; Beard and Johnson 1999; Beard et al. 1999; Mandernack et al. 1999; Bullen et al. 2001), as well as multi-collector inductively-coupled plasma mass spectrometry (MC-ICP-MS) (e.g., Belshaw et al. 2000; Sharma et al. 2001; Beard et al. 2003b; Kehm et al. 2003). The biggest benefit of TIMS-based double-spike analyses is that there are relatively few isobaric interferences associated with TIMS analyses (Albarède and Beard 2004), except for elemental isobars from ^{54}Cr on ^{54}Fe and ^{58}Ni on ^{58}Fe. The biggest difficulties using TIMS analysis are the lengthy amount of time required for each analysis (4-8 hours per analysis) and the low ionization efficiency of Fe, which limits TIMS Fe isotope analyses to working with microgram quantities of Fe. Most workers are pursuing Fe isotope studies by MC-ICP-MS because of the high ionization efficiency and rapid sample throughput. Instrumental mass bias corrections are made using bracketing standards (e.g., Belshaw et al. 2000; Beard et al. 2003b; Albarède and Beard 2004), or by comparison to mass bias inferred from an element spike, such as Cu or Ni, that has been added to the solution (e.g., Sharma et al. 2001; Kehm et al. 2003; Malinovsky et al. 2003; Albarède and Beard 2004).

Iron isotope analysis made using MC-ICP-MS presents special challenges because of isobars that are produced by the Ar plasma, including ^{40}Ar^{14}N on ^{54}Fe, ^{40}Ar^{16}O on ^{56}Fe, and ^{40}Ar^{16}OH on ^{57}Fe. Many workers have dealt with these isobars by using large quantities (~20 μg) of Fe during an isotopic analysis to minimize the relative intensities of Ar isobars, have employed cool-plasma techniques to eliminate Argides at the expense of decreased sensitivity, or used a narrow defining slit to allow Faraday collectors to be positioned to coincide with the Fe shoulder of a multi-species mass spectrum (e.g., Belshaw et al. 2000; Kehm et al. 2003; Weyer and Schwieters 2003). Other workers have utilized a collision cell to eliminate or greatly reduce Ar isobars, and this approach provides very high sensitivity, where high-precision Fe isotope analyses can be made on samples as small as 100 ng (Beard et al. 2003b). MC-ICP-MS analyses generally produce an external precision of 0.10 and 0.15‰ (2σ) for ^{56}Fe/^{54}Fe and ^{57}Fe/^{54}Fe ratios, respectively (e.g., Beard et al. 2003a;b).

The accuracy of Fe isotope compositions can be strongly affected by the purity of the sample. For example, there can be significant matrix effects that change instrumental mass bias for samples as compared to standards unless they are purified to similar degrees (Kehm et al. 2003; Albarède and Beard 2004). Indeed, in order to avoid these matrix affects the Fe concentrations of samples and standards must be carefully controlled (Kehm et al. 2003; Albarède and Beard 2004). Apparently the redox state of Fe in solution may be an issue as well (Zhu et al. 2002). Purification of the sample is important for minimizing matrix effects, as well as eliminating isobars from Cr and Ni, in addition to molecular isobars such as $^{40}Ca^{16}O$ on ^{56}Fe and $^{40}Ca^{16}OH$ on ^{57}Fe. It is important to stress, however, that during ion-exchange chromatography, yields must be quantitative in order to avoid laboratory-induced Fe isotope fractionations (e.g., Anbar et al. 2000; Roe et al. 2003). One of the best methods for demonstrating that the procedures used for purifying samples do not induce laboratory artifacts is to prepare artificial samples that match the chemical composition of the natural sample using Fe of known isotopic composition. Such a procedure was used by Beard et al. (2003a) to demonstrate the accuracy of Fe isotope compositions measured in Mid-Ocean Ridge (MOR) hydrothermal vent fluids.

ORIGIN OF Fe ISOTOPE VARIATIONS

Calculated fractionation factors

Calculation of stable isotope fractionation factors under equilibrium conditions provides important insight into expected isotopic fractionations, and the principles behind such calculations are discussed in detail in Schauble (2004). In the temperature range of 0–250°C, Fe isotope fractionations up to ~10‰ are predicted (Fig. 3), based on comparisons of reduced partition function ratios ($\delta_{56/54}$ factors; Schauble 2004). In general, calculated fractionation factors indicate that ferric Fe-bearing phases (aqueous or solid) will have higher $\delta^{56}Fe$ values than ferrous Fe-bearing phases (Fig. 3). The ferric Fe-bearing oxide minerals magnetite, hematite, goethite, and lepidocrocite, for example, are predicted to have higher $\delta^{56}Fe$ values than minerals that only contain ferrous Fe such as siderite, ankerite, and silicates such as olivine (Fig. 3A). In the case of aqueous Fe species, most ferric Fe-bearing species, such as $[Fe^{III}(H_2O)_6]^{3+}$, $[Fe^{III}(H_2O)_4Cl_2]^+$, and $[Fe^{III}Cl_4]^-$ are predicted to have higher $\delta^{56}Fe$ values than the hexaquo ferrous Fe complex $[Fe^{II}(H_2O)_6]^{2+}$ (Fig. 3B). Exceptions to the generalization that ferric Fe phases are expected to have the highest $\delta^{56}Fe$ values include pyrite, which is predicted to have the highest $\delta^{56}Fe$ values of the minerals illustrated in Figure 3A. In addition, in the case of the octahedral ferro- and ferri-cyanide complexes, the ferrous Fe-bearing species is predicted to have the higher $\delta^{56}Fe$ values (Fig. 3B). Finally, the octahedral ferric chloride complex is predicted to have some of the lowest $\delta^{56}Fe$ values (Fig. 3B). Of these exceptions, the only phase that is common in natural low-temperature environments is pyrite.

The calculated Fe isotope fractionation factors suggest that Fe isotope shifts will occur during several important geologic processes. At high temperatures, such as those commonly found in magmas (800–1000°C), only a few tenths per mil fractionation in $^{56}Fe/^{54}Fe$ ratios are predicted among the common silicate minerals, as well as between silicates and oxides such as ilmenite and spinel; slightly larger fractionations between silicates and magnetite are predicted at high temperatures (Polyakov and Mineev 2000). If we take, therefore, the $\beta_{56/54}$ factor for olivine as representative of silicate minerals in igneous and metamorphic rocks, the relatively high $\beta_{56/54}$ factors for weathering products such as ferric oxides or hydrous ferric Fe-bearing silicates such as celadonite suggest that chemical weathering should produce phases that have $\delta^{56}Fe$ values that are several ‰ greater than those of igneous or metamorphic protoliths (Fig. 3A). In contrast, Fe carbonates, which may form in relatively anoxic marine environments, are predicted to have $\delta^{56}Fe$ values that are 1–2‰ lower than those of igneous and metamorphic rocks (Fig. 3A).

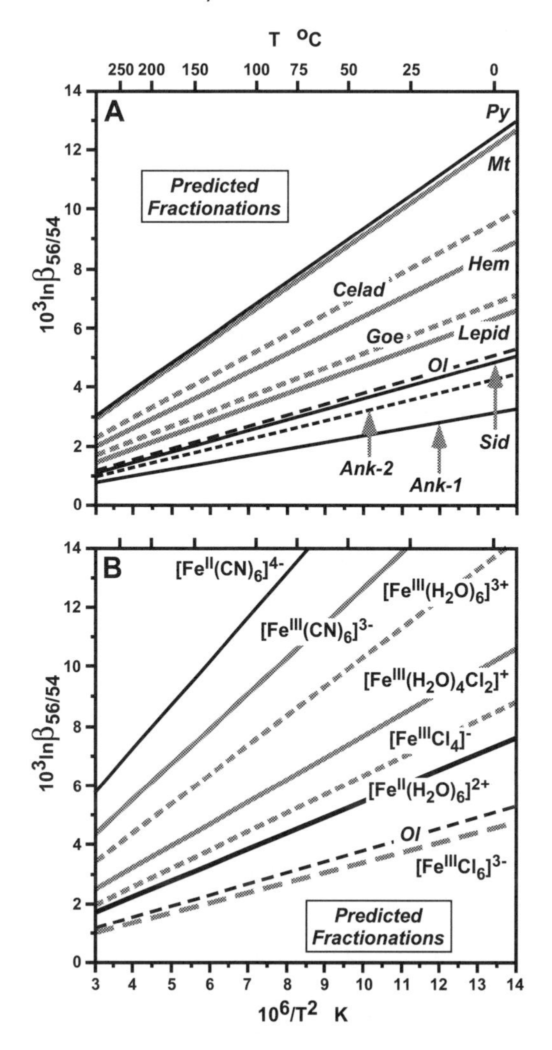

Figure 3. Plot of $10^3\ln\beta_{56/54}$ values versus $10^6/T^2$ (K) for (A) Fe minerals calculated from Polyakov and Mineev (2000), and for (B) aqueous Fe species calculated from Schauble et al. (2001). Temperature scale in °C shown at top. Mineral abbreviations are Py: pyrite, Mt: magnetite, Celad: celadonite, Hem: hematite, Goe: goethite, Lepid: lepidocrocite, Ol: olivine, Sid: siderite, Ank-2: ankerite ($Ca_{1.1}Mg_{0.5}Fe_{0.3}Mn_{0.1}(CO_3)_2$), Ank-1: ankerite ($Ca_1Mg_{0.5}Fe_{0.5}(CO_3)_2$).

In the case of fluid-mineral fractionation involving Fe-bearing aqueous solutions, the $\beta_{56/54}$ factors for hexaquo ferric Fe are significantly higher than those of ferric Fe oxides (Fig. 3). In addition, the $\beta_{56/54}$ factors for hexaquo ferrous Fe are significantly higher than those of Fe carbonates (Fig. 3). These observations suggest that fluid-mineral Fe isotope fractionation factors for common natural fluids and minerals will tend to be positive, assuming no change in redox state for Fe. The predicted $\beta_{56/54}$ factors, however, suggest that significant changes in Fe isotope fractionations may occur depending on the structure of the mineral phase involved, even within a single group such as ferric oxides/hydroxides and Fe carbonates. For example, at low temperatures, the δ^{56}Fe values for hematite, goethite, and lepidocrocite are predicted to be different by several per mil, as are the Fe carbonates, including siderite and ankerite (Fig. 3A). In the case of aqueous Fe species, significant Fe isotope effects are predicted for Cl^- substitution into the hexaquo ferric Fe complex (Fig. 3B). Determination of the exact solid phase, as well as the aqueous Fe species involved, may therefore be important in interpreting Fe isotope fractionations that are expected in natural systems.

Experimental determination of fractionation factors

The large Fe isotope fractionations predicted to occur between aqueous ferric and ferrous Fe species (Fig. 3) has been investigated experimentally at two temperatures for hexaquo Fe(III) and Fe(II), as well as the effect of Cl^- substitution (Johnson et al. 2002; Welch et al. 2003). The kinetics of isotopic exchange in these experiments was determined using an enriched ^{57}Fe tracer for the ferric Fe phase. The ^{57}Fe tracer experiments reported in Johnson et al. (2002) and Welch et al. (2003) indicate that 95% isotopic equilibrium between $Fe(III)_{aq}$ and $Fe(II)_{aq}$ occurs within ~60 seconds at room temperature (22°C), or within ~5 minutes at 0°C. The relatively slower isotopic exchange rates at lower temperatures are expected.

Although isotopic equilibrium will be quickly attained between the hexaquo ferric Fe, hexaquo ferrous Fe, and monochloro ferric Fe complexes, the relatively rapid exchange rates presents a challenge in terms of separation of the two species for isotopic analysis. Johnson et al. (2002) and Welch et al. (2003) rapidly (~1 s) precipitated the ferric Fe fraction through carbonate addition, during which time approximately 10–20% isotopic exchange occurred between the aqueous ferric and ferrous Fe species, based on the isotope exchange kinetics determined from the enriched ^{57}Fe tracer experiments. The errors introduced by partial re-equilibration during species separation are largest for experiments using low $Fe(II)/Fe_T$ ratios, and this is the likely explanation for the deviations in $\delta^{56}Fe$ values for $Fe(II)_{aq}$ at low $Fe(II)/Fe_T$ ratios relative to that expected from isotopic mass balance (Fig. 4).

Welch et al. (2003) determined that $Fe(III)_{aq}$-$Fe(II)_{aq}$ fractionation at 22°C (corrected for partial re-equilibration) was +3.02 ± 0.14‰, +3.05 ± 0.17‰, and +2.92 ± 0.08‰ for solutions that contained 0, 11, and 111 mM Cl^-, respectively, which are all the same within error. At 0°C, the $Fe(III)_{aq}$–$Fe(II)_{aq}$ fractionation (corrected for partial re-equilibration) for the three Cl^- contents was +3.50 ± 0.31‰, +3.57 ± 0.17‰, and +3.65 ± 0.15‰. The consistent $Fe(III)_{aq}$-$Fe(II)_{aq}$ fractionations at a given temperature determined for a variety of solution compositions, and at different Fe(III) to total Fe ratios suggests that variations in the hydroxide and chloride species studied by Johnson et al. (2002) and Welch et al. (2003) has no significant isotopic effect. For example, the proportions of $[Fe^{III}(H_2O)_6]^{3+}$, $[Fe^{III}(H_2O)_5OH]^{2+}$, and $[Fe^{III}(H_2O)_5Cl]^{2+}$ in these solutions varied from 0.1 to 64%, 0.0 to 47%, and 0.0 to 44%, respectively, over the range of pH values and Cl^- contents studied by Johnson et al. (2002) and Welch et al. (2003), and yet the $Fe(III)_{aq}$-$Fe(II)_{aq}$ fractionations are the same within error when corrected for small extents of partial isotopic re-equilibration during species separation. This suggests that the equilibrium Fe isotope fractionation between $[Fe^{III}(H_2O)_6]^{3+}$ and at least its mononuclear hydroxy complex is negligible.

The insignificant effect of Cl^- substitution on the $Fe(III)_{aq}$–$Fe(II)_{aq}$ fractionation is somewhat surprising. Chloride substitution lowers the vibrational frequency of the Fe–Cl bond pair as compared to the Fe-O bond (248 versus 505 cm^{-1} for v_3, and 184 versus 304 cm^{-1} for v_4; see Schauble et al. 2001 and references within). If we assume that the difference in $\beta_{56/54}$ factors for $[Fe^{III}(H_2O)_6]^{3+}$ and $[Fe^{III}(H_2O)_5Cl]^{2+}$ is half that calculated for $[Fe^{III}(H_2O)_6]^{3+}$ and $[Fe^{III}(H_2O)_5Cl_2]^+$ by Schauble et al. (2001), we would expect a 1.5‰ difference in $\delta^{56}Fe$ values between $[Fe^{III}(H_2O)_6]^{3+}$ and $[Fe^{III}(H_2O)_5Cl]^{2+}$ at 22°C. Because the $[Fe^{III}(H_2O)_5Cl]^{2+}$ species ranged from 0 to 44% in the experiments of Welch et al. (2003), such a fractionation effect should have produced $Fe(III)_{aq}$-$Fe(II)_{aq}$ fractionations that varied by up to 0.7‰ and correlated with Cl^- contents, but this was not observed. Although the reason for the relative insensitivity of $Fe(III)_{aq}$-$Fe(II)_{aq}$ fractionations to Cl^- contents is not clear, this result greatly simplifies application of the measured Fe isotope fractionations to natural fluids, which may be quite variable in their chloride and hydroxide species. For example, seawater contains ~0.5 M Cl^-, 0.03 M SO_4^{2-}, and 0.002 M total CO_2, where dissolved ferrous speciation is dominated by $[Fe^{II}(H_2O)_6]^{2+}$ and ferric speciation is dominated by $[Fe^{III}(H_2O)_4(OH)_2]^+$ (>99%), $[Fe^{III}(H_2O)_5(OH)]^{2+}$, and $[Fe^{III}(OH)_3]^0$ (e.g., Millero et al. 1995); the results suggest

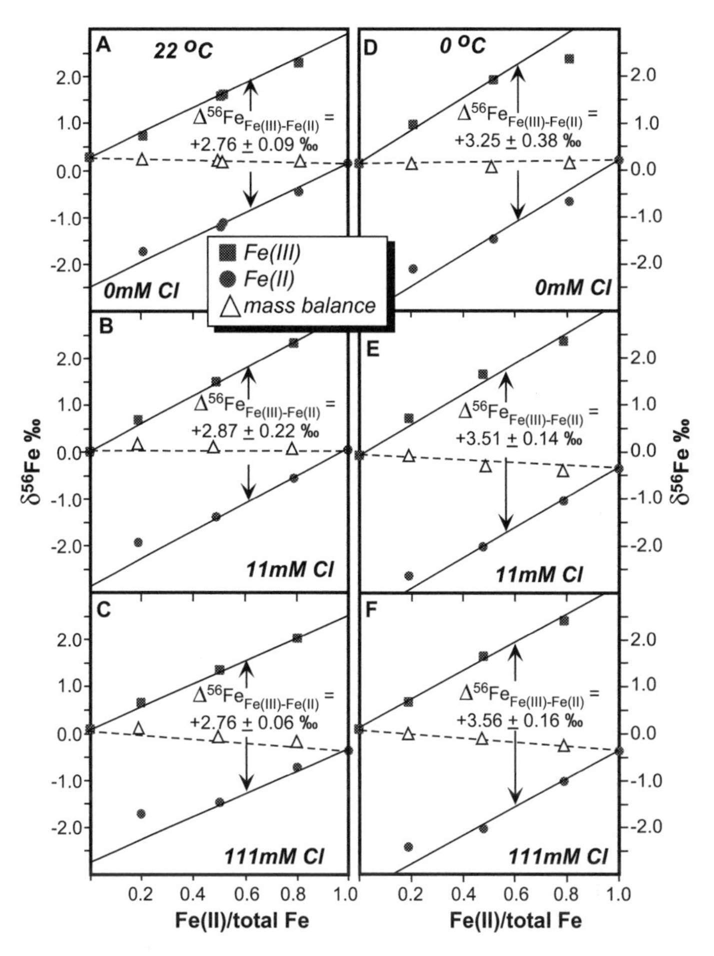

Figure 4. Isotopic mass-balance for measured δ^{56}Fe values of ferric (■) and ferrous (●) Fe in solution as a function of Fe(II)/Fe_T ratios. Although the initial Fe isotope compositions of the fluids have δ^{56}Fe values of ~0‰ (shown by dotted mixing line), the lighter Fe isotopes are portioned into the ferrous species following attainment of isotopic equilibrium. The triangles (△) represent the calculated isotopic mass balance of the different solutions. Fractionations noted are measured values based on data for equimolar and higher Fe(II)/total Fe ratio experiments. (A) and (D) are for experiments with zero Cl^- at 22 and 0°C, respectively. (B) and (E) are experiments done with at 11 mM Cl^- at 22 and 0°C, respectively. (C) and (F) are experiments done with 111 mM Cl^- at 22 and 0°C, respectively. Modified from Welch et al. (2003).

that these variations in speciation will have negligible isotopic effects relative to hexaquo Fe(III).

The experimentally measured temperature dependence of the $Fe(III)_{aq}$–$Fe(II)_{aq}$ fractionation at low temperatures is:

$$10^3 \ln\alpha_{Fe(III)_{aq}\text{-}Fe(II)_{aq}} = A \times \frac{10^6}{T^2} + B \tag{4}$$

where $A = 0.334 \pm 0.032$ and $B = -0.88 \pm 0.38$, based on the determinations at 0 and 22°C by

Welch et al. (2003). This produces a temperature dependence that is similar to that calculated by Schauble et al. (2001), but is displaced to $Fe(III)_{aq}$–$Fe(II)_{aq}$ fractionations that are approximately half that predicted by Schauble et al. (2001). This discrepancy lies outside the estimated uncertainties for the predicted and measured $Fe(III)_{aq}$–$Fe(II)_{aq}$ fractionations (Fig. 5). Based on comparison of measured $Fe(III)_{aq}$-hematite fractionations with those predicted from spectroscopic data, Skulan et al. (2002) noted that the predicted $Fe(III)_{aq}$–hematite and $Fe(III)_{aq}$–$Fe(II)_{aq}$ fractionations may be brought into closer agreement with the measured $Fe(III)_{aq}$–hematite and $Fe(III)_{aq}$–$Fe(II)_{aq}$ fractionations if the $\beta_{56/54}$ factor for $[Fe^{III}(H_2O)_6]^{3+}$ from Schauble et al. (2001) is reduced by 1–2‰ at low temperatures, and this suggestion has been supported by recent ab inito calculations of the $\beta_{56/54}$ factor for $[Fe^{III}(H_2O)_6]^{3+}$ by Anbar et al. (2004; Fig. 5).

There does appear to be small (~0.1‰) Fe isotope fractionations between some Fe(III) chloro-complexes at very high Cl concentrations (e.g., 7–2 M HCl; Anbar et al. 2000; Roe et al. 2003). Based on a series of anion-exchange resin experiments using aqueous Fe(III) solutions in 7 and 2 M HCl, Anbar et al. (2000) and Roe et al. (2003) inferred that there is a small (~0.1‰) fractionation between Fe(III) that is tetrahedrally complexed with Cl^- relative to Fe(III) that is octahedrally coordinated with water and chloride ions. Matthews et al. (2001) showed that aqueous Fe species that are complexed by organic ligands may record significant Fe isotope fractionation where $Fe(II)_{aq}$ and $Fe(III)_{aq}$ that was bound with 2, 2′- bipyridine and chloride ions, respectively, recorded fractionations up to 12‰. The isotopic fractionations were interpreted to reflect kinetic effects due to the relatively slow separation process, as well as breakdown of $[Fe^{II}(bipy)_3]^{2+}$ in the 6M HCl solution. Although, as noted by Matthews et al. (2001), the measured isotopic fractionations do not reflect equilibrium isotope partitioning, they do highlight the importance of highly covalent bonds, where, in this case, ferrous Fe is enriched in $^{56}Fe/^{54}Fe$ ratios relative to ferric Fe chloro complexes.

Skulan et al. (2002) investigated Fe isotope fractionation between aqueous Fe(III) and hematite at 98°C. During the initial stages of this reaction (12 hours), at a rapid hematite

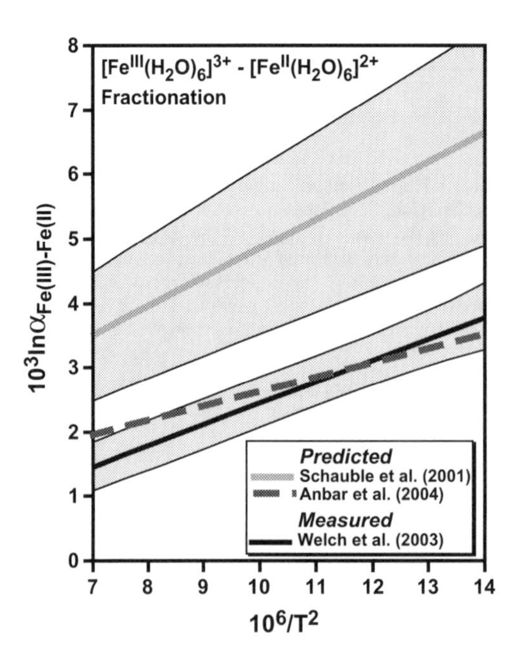

Figure 5. Plot of $10^3\ln\alpha_{Fe(III)aq\text{-}Fe(II)aq}$ versus $10^6/T^2$ (K) for the calculated fractionation factor between $[Fe^{III}(H_2O)_6]^{3+}$–$[Fe^{II}(H_2O)_6]^{2+}$ by Schauble et al. (2001) and Anbar et al. (2004), as compared to the experimentally-determined curve of Welch et al. (2003). The grey shaded fields are 2σ errors (field omitted for study of Anbar et al. (2004) for simplicity).

precipitation rate, the Fe isotope fractionation between hematite and Fe(III) can be modeled as a Rayleigh distillation process using $\Delta_{Fe(III)\text{-hematite}} = +1.3‰$ (Skulan et al. 2002). This rapidly synthesized hematite is isotopically zoned as determined by step-wise leaching of hematite after 98% precipitation. Skulan et al. (2002) inferred that initial rapid hematite synthesis resulted in kinetic Fe isotope fractionations because in longer-term experiments the measured Fe isotope fractionation factor between Fe(III) and hematite is a strong function of time (Fig. 6). Based on a series of long-term experiments (up to 203 days) that involved a variety of hematite synthesis rates and initial conditions, Skulan et al. (2002) extrapolated the Fe(III)-hematite fractionation to a zero precipitation rate (Fig. 7), and inferred this to be the best estimate of the equilibrium fractionation factor. Additionally, Skulan et al. (2002) applied a small correction (~0.2‰) to these data to subtract the effects of the large initial kinetic isotope fractionation (Fig. 6) that were not completely erased during the long-term experiments. This correction, which Skulan et al. (2002) define as the "terminal offset" is a small correction that is based on modeling of the extent of dissolution and re-precipitation based on parallel experiments that used enriched ^{57}Fe tracers. Extrapolation of the data corrected for the "terminal offset" to a zero hematite precipitation rate yields a $\Delta_{Fe(III)\text{-hematite}}$ fractionation of $-0.10 \pm 0.20‰$ at 98°C (Fig. 7).

Composite effects of equilibrium and kinetic fractionations in an experiment

The preceding discussion suggests that, in general, isotopic equilibrium is attained between aqueous Fe species within minutes, whereas kinetic isotope fractionation is likely to be significant during rapid precipitation of minerals from fluids, particularly where precipitation occurs over timescales of hours or even days. Oxidation of aqueous Fe(II), followed by precipitation of ferric oxide/hydroxides, is a common process in near-surface environments, where $Fe(II)_{aq}$ may be supplied by hydrothermal sources or anoxic groundwaters (Bullen et al. 2001), and consideration of combined equilibrium and kinetic effects during such processes is important to understanding the Fe isotope variations that may be produced. If we assume that precipitation of ferric Fe minerals from $Fe(II)_{aq}$ is preceded by formation of aqueous Fe(III), the overall process may be envisioned to occur in two steps:

$$Fe(II)_{aq} \rightarrow Fe(III)_{aq} \rightarrow Fe(OH)_3(s) \tag{5}$$

where the ferric Fe hydroxide precipitate, most likely ferrihydrite, is represented as $Fe(OH)_3$

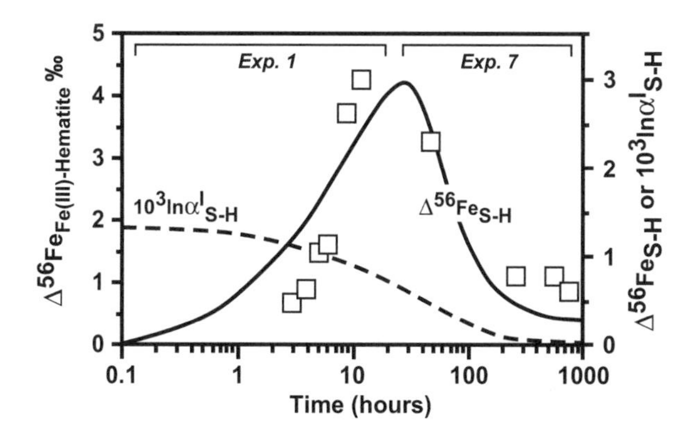

Figure 6. Superposition of a model for the $\Delta^{56}Fe_{Fe(III)\text{-hematite}}$ (solid line) and the assumed instantaneous $10^3 ln\alpha^I_{Fe(III)\text{-hematite}}$ fractionation (dashed line) with time relative to measured experimental data (□) from the study of $Fe(III)_{aq}$-hematite fractionation by Skulan et al. (2002). Over short timescales (≤12 h), kinetic isotope fractionation dominates, whereas over longer timescales, the isotopic fractionations move back toward a near-zero value. Scale for measured data on left and bottom, whereas scale for model curves is on right. Adapted from Skulan et al. (2002).

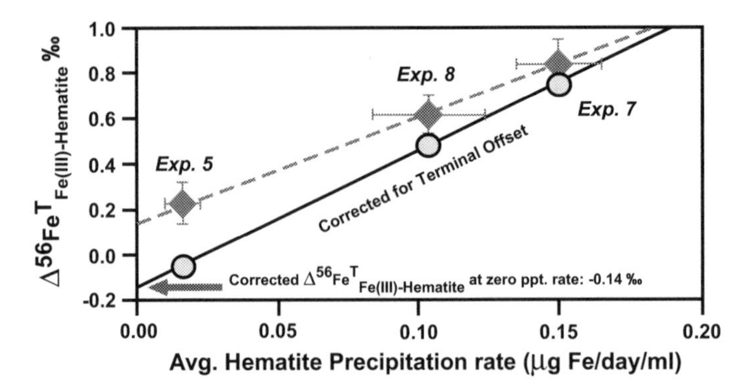

Figure 7. Measured and corrected $\Delta^{56}Fe^{T}_{Fe(III)\text{-}Hematite}$ values (◆ and ●, respectively) relative to average hematite precipitation rate for Experiments 5, 7, and 8 of Skulan et al. (2002). The $\Delta^{56}Fe^{T}_{Fe(III)\text{-}Hematite}$ values are defined as those measured at the termination of the experiments; the corrected $\Delta^{56}Fe^{T}_{Fe(III)\text{-}Hematite}$ values reflect the estimated correction required to remove any residual kinetic isotope fractionation that was produced early in experiments that was not completely removed by dissolution and re-precipitation over the long term. Extrapolation of the corrected $\Delta^{56}Fe^{T}_{Fe(III)\text{-}Hematite}$ values to zero precipitation rates yields an estimate for the equilibrium $Fe(III)_{aq}$–hematite fractionation, $\Delta^{56}Fe_{Fe(III)\text{-}hematite}$, of −0.14‰. Modified from Skulan et al. (2002).

for simplicity. Bullen et al. (2001) measured the overall Fe isotope fractionation produced by oxidation of $Fe(II)_{aq}$, followed by precipitation to ferrihydrite, and the overall reaction in their batch experiment is well fit by a first-order rate law (Fig. 8). It is important to note, however, that Bullen et al. (2001) interpret their data in a different conceptual framework than that of Equation (5), instead focusing on Fe isotope exchange between aqueous ferrous Fe hydroxide species. The motivation for re-evaluating this framework lies in the subsequent studies of Fe isotope exchange between $Fe(II)_{aq}$ and $Fe(III)_{aq}$ by Johnson et al. (2002) and Welch et al. (2003), both in terms of rates of exchange and magnitude of fractionation. Assuming that Equation (5) is applicable to oxidation of $Fe(II)_{aq}$ to $Fe(III)_{aq}$, followed by precipitation of ferric Fe hydroxide, we may define a series of first-order rate equations to describe the overall process:

$$\frac{d\left[\mathrm{Fe(II)_{aq}}\right]}{dt} = -k_1\left[\mathrm{Fe(II)_{aq}}\right] \tag{6}$$

$$\frac{d\left[\mathrm{Fe(III)_{aq}}\right]}{dt} = k_1\left[\mathrm{Fe(II)_{aq}}\right] - k_2\left[\mathrm{Fe(III)_{aq}}\right] \tag{7}$$

$$\frac{d\left[\mathrm{Fe(OH)_3}\right]}{dt} = k_2\left[\mathrm{Fe(III)_{aq}}\right] \tag{8}$$

where k_1, and k_2 are first-order rate constants. The size of the $Fe(III)_{aq}$ component will therefore be determined by the k_2/k_1 ratio, where increasing k_2/k_1 ratios will produce a decreasing proportion of $Fe(III)_{aq}$.

Equations (6)–(8) predict that the proportions of $Fe(II)_{aq}$, $Fe(III)_{aq}$, and $Fe(OH)_3(s)$ will change over time. However, if the rate of Fe isotope exchange is rapid between, for example, $Fe(II)_{aq}$ and $Fe(III)_{aq}$, Fe isotope equilibrium may still be maintained between aqueous Fe species, and this may be evaluated through comparison of the residence time of $Fe(III)_{aq}$ relative to the time required to attain isotopic equilibrium between $Fe(II)_{aq}$ and $Fe(III)_{aq}$. The residence time (τ) of $Fe(III)_{aq}$ may be defined as:

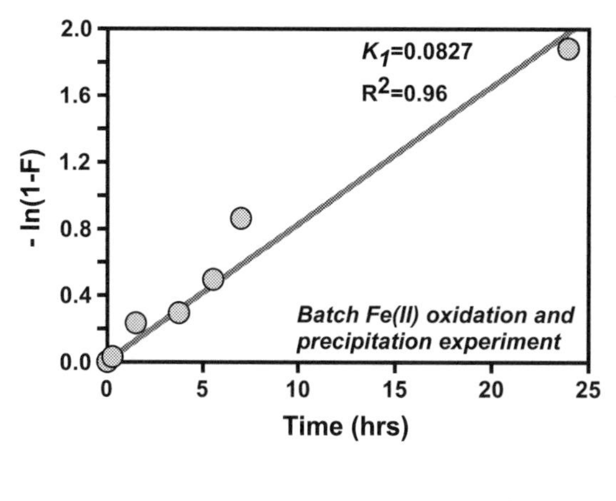

Figure 8. Regression of batch oxidation and precipitation experiment of Bullen et al. (2001), where $Fe(II)_{aq}$ was oxidized, followed by precipitation of ferrihydrite, over a 24 h period. The reaction progress (F) is well-fit by a first-order rate law, where the rate constant is 0.0827 F/h, with an R^2 of 0.96. In the model illustrated by Equations (5)–(8) in the text, this rate constant would be set to k_1.

$$\tau_{Fe(III)_{aq}} = \frac{M_{Fe(III)_{aq}}}{J_{Fe(III)_{aq}}} \tag{9}$$

where $M_{Fe(III)aq}$ is the total moles of Fe in the $Fe(III)_{aq}$ reservoir, and $J_{Fe(III)aq}$ is the flux of Fe through the $Fe(III)_{aq}$ reservoir. Substituting the rate equations into Equation (9), we obtain:

$$\tau_{Fe(III)_{aq}} = \frac{\left[Fe(III)_{aq}\right]}{k_1\left[Fe(OH)_3\right] - k_2\left[Fe(III)_{aq}\right]} \tag{10}$$

If the time to reach isotopic equilibrium through isotopic exchange is t_{ex}, isotopic equilibrium between $[Fe(II)_{aq}]$ and $[Fe(III)_{aq}]$ should be maintained when:

$$\tau_{Fe(III)_{aq}} \gg t_{ex} \tag{11}$$

Based on the ^{57}Fe tracer work of Johnson et al. (2002) and Welch et al. (2003), the time required to attain isotopic equilibrium between $Fe(III)_{aq}$ and $Fe(II)_{aq}$ is on the order of a few minutes or less in dilute aqueous solutions.

In Figure 9, the proportions of $Fe(II)_{aq}$, $Fe(III)_{aq}$, and $Fe(OH)_3(s)$ over time are illustrated, as well as the residence time of $Fe(III)_{aq}$, as calculated using Equations (6)–(11), and the rate for k_1 from Figure 8. The time required to reach steady-state conditions in terms of the $Fe(III)_{aq}/Fe(II)_{aq}$ ratio is, of course, dependent upon the k_2/k_1 ratio, and varies from ~2 hours for $k_2/k_1 = 50$, to ~20 hours for $k_2/k_1 = 5$ (Fig. 9). It is important to note, however, that attainment of isotopic equilibrium between $Fe(II)_{aq}$ and $Fe(III)_{aq}$ is not related to the time required to reach steady-state in terms of the concentration or molar ratios of Fe in various species, because t_{ex} is entirely independent of the rate equations and rate constants of Equations 6-8. For the range of k_2/k_1 ratios illustrated in Figure 9, the residence time for $Fe(III)_{aq}$ greatly exceeds that of the time required for isotopic exchange after one hour or less into the experiment (Fig. 9), indicating that isotopic equilibrium between $Fe(II)_{aq}$ and $Fe(III)_{aq}$ should have been maintained for essentially the entire duration of the batch precipitation experiment of Bullen et al. (2001).

The calculations illustrated in Figure 9 demonstrate that attainment of steady-state conditions in terms of elemental abundances may not be related to attainment of isotopic equilibrium. The converse may also be true, a system that is at equilibrium in terms of concentrations may be far from isotopic equilibrium. Moreover, a system at steady-state conditions in terms of concentrations may also be far from isotopic equilibrium. This case is well illustrated by the experiments on $Fe(III)_{aq}$-hematite fractionations discussed above (Skulan et al. 2002). In Skulan et al.'s (2002) experiments, approximately constant $Fe(III)_{aq}$ contents

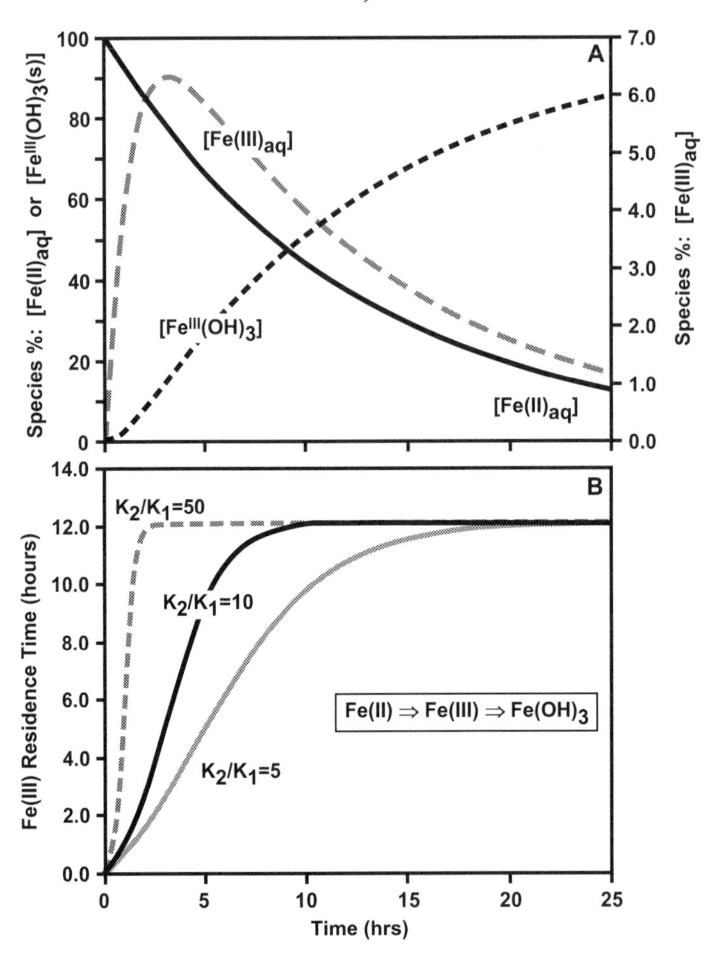

Figure 9. Proportions of species over time for the process of oxidation of $Fe(II)_{aq}$ to $Fe(III)_{aq}$, followed by precipitation of ferrihydrite (Eqn. 5), as calculated using the first-order rate laws of Equations (6)–(8), and the rate constant k_1 from Figure 8. (A) Proportion of species calculated from Equations (6)–(8), assuming a k_2/k_1 ratio of 10. Ferrihydrite represented as $Fe(OH)_3$ for simplicity. (B) The residence time of intermediate species $Fe(III)_{aq}$, as calculated using Equation (10), for various k_2/k_1 ratios.

were maintained through dissolution and reprecipitation processes—which were in approximate balance—but where large changes in $Fe(III)_{aq}$-hematite fractionations were measured over time as early Fe isotope fractionations produced by rapid hematite precipitation were progressively erased by slow dissolution and reprecipitation (Fig. 6). We note that our analysis of the process of oxidation and precipitation is different than that of Anbar (2004), where we explicitly consider the residence time of Fe relative to the time required for isotopic exchange.

We now turn to the Fe isotope fractionations that are predicted by a model where oxidation of $Fe(II)_{aq}$ to $Fe(III)_{aq}$ occurs, followed by precipitation of $Fe(III)_{aq}$ to ferrihydrite (FH) (Eqn. 5). In a closed system, the $\delta^{56}Fe$ values of the three components are constrained by simple mass balance as:

$$\delta_{SYS} = X_{Fe(II)aq}\delta_{Fe(II)aq} + X_{Fe(III)aq}\delta_{Fe(III)aq} + X_{FH}\delta_{FH} \quad (12)$$

where X represents the mole fractions of the three components (Eqn. 5).

In the case of the aqueous Fe components, the isotopic mass-balance relation is:

$$\delta_{(\mathrm{Fe_{aq}})_T} = \frac{X_{\mathrm{Fe(II)_{aq}}}}{X_{\mathrm{Fe(II)_{aq}}} + X_{\mathrm{Fe(III)_{aq}}}}\delta_{\mathrm{Fe(II)_{aq}}} + \frac{X_{\mathrm{Fe(III)_{aq}}}}{X_{\mathrm{Fe(II)_{aq}}} + X_{\mathrm{Fe(III)_{aq}}}}\delta_{\mathrm{Fe(III)_{aq}}} \tag{13}$$

If the fractionation factor between $\mathrm{Fe(III)_{aq}}$ and $\mathrm{Fe(II)_{aq}}$ is:

$$\Delta_{\mathrm{Fe(III)_{aq}-Fe(II)_{aq}}} = \delta_{\mathrm{Fe(III)_{aq}}} - \delta_{\mathrm{Fe(II)_{aq}}} \tag{14}$$

then Equation (13) can be simplified as:

$$\delta_{(\mathrm{Fe_{aq}})_T} = \delta_{\mathrm{Fe(III)_{aq}}} - \left(\Delta_{\mathrm{Fe(III)_{aq}-Fe(II)_{aq}}}\right)\left(\frac{X_{\mathrm{Fe(II)_{aq}}}}{X_{\mathrm{Fe(II)_{aq}}} + X_{\mathrm{Fe(III)_{aq}}}}\right) \tag{15}$$

where $\delta_{(\mathrm{Fe_{aq}})_T}$ is the δ^{56}Fe value of the total aqueous Fe pool ($\mathrm{Fe(II)_{aq}}$ and $\mathrm{Fe(III)_{aq}}$), and $X_{\mathrm{Fe(II)}}$ and $X_{\mathrm{Fe(III)}}$ are the mole fractions of $\mathrm{Fe(II)_{aq}}$ and $\mathrm{Fe(III)_{aq}}$ in the total system, respectively. As precipitation of $\mathrm{Fe(III)_{aq}}$ to ferrihydrite proceeds, the ratio of $\{X_{\mathrm{Fe(II)aq}}/(X_{\mathrm{Fe(II)aq}} + X_{\mathrm{Fe(III)aq}})\}$ increases, and becomes unity when precipitation of $\mathrm{Fe(III)_{aq}}$ is complete. Accordingly, if the value of $\Delta_{\mathrm{Fe(III)\text{-}Fe(II)}}$ is constant (i.e., isotopic equilibrium is maintained), large changes will occur in the δ^{56}Fe value of the remaining aqueous Fe components. If the proportion of $\mathrm{Fe(III)_{aq}}$ is small, as would be the case for high k_2/k_1 ratios (Fig. 9), the $\{X_{\mathrm{Fe(II)aq}}/(X_{\mathrm{Fe(II)aq}} + X_{\mathrm{Fe(III)aq}})\}$ ratio is close to unity, and $\delta_{(\mathrm{Fe_{aq}})_T}$ is equal to $\delta_{\mathrm{Fe(II)aq}}$. In this case Equation (15) may be re-arranged to be equal to

$$\delta_{\mathrm{Fe(II)aq}} - \delta_{\mathrm{FH}} \approx \Delta_{\mathrm{Fe(III)\text{-}FH}} - \Delta_{\mathrm{Fe(III)\text{-}Fe(II)}} \tag{16}$$

where we define the fractionation factor between $\mathrm{Fe(III)_{aq}}$ and ferrihydrite (FH) as:

$$\Delta_{\mathrm{Fe(III)aq\ -\ FH}} = \delta_{\mathrm{Fe(III)aq}} - \delta_{\mathrm{FH}} \tag{17}$$

Equation (16) notes that the difference in measured δ^{56}Fe values for $\mathrm{Fe(II)_{aq}}$ and ferrihydrite precipitate is equal to difference in the $\mathrm{Fe(III)_{aq}}$–ferrihydrite and $\mathrm{Fe(III)_{aq}}$–$\mathrm{Fe(II)_{aq}}$ fractionation factors, assuming that the proportion of $\mathrm{Fe(III)_{aq}}$ is very small ($\leq$5%). In cases where the proportion of $\mathrm{Fe(III)_{aq}}$ ratio is significant (>5%), the isotopic effects of combined oxidation and precipitation may still be calculated using an incremental approach and Equation (12), along with the pertinent fractionations between components (Eqns. 14 and 17).

The usefulness of Equation (16) lies in its ability to show that the measured fractionation between $\mathrm{Fe(II)_{aq}}$ and ferrihydrite precipitate of the experiments of Bullen et al. (2001), under conditions of high $\mathrm{Fe(II)_{aq}}/\mathrm{Fe(III)_{aq}}$ ratio and maintenance of isotopic equilibrium between $\mathrm{Fe(III)_{aq}}$ and $\mathrm{Fe(II)_{aq}}$, likely reflects the combined effects of an equilibrium $\mathrm{Fe(III)_{aq}}$-$\mathrm{Fe(II)_{aq}}$ fractionation and a fractionation between $\mathrm{Fe(III)_{aq}}$ and ferrihydrite. For example, the −0.9‰ fractionation between aqueous Fe (which existed as essentially all ferrous Fe; T. Bullen, pers. commun. 2002) and ferrihydrite precipitate early in the experiment probably reflects the combined effects of two fractionation steps: a +2.9‰ equilibrium fractionation between $\mathrm{Fe(III)_{aq}}$ and $\mathrm{Fe(II)_{aq}}$, and a +2.0‰ fractionation between $\mathrm{Fe(III)_{aq}}$ and ferrihydrite upon precipitation (e.g., Eqn. 16).

The combined effects of a +2.9‰ equilibrium fractionation between $\mathrm{Fe(III)_{aq}}$ and $\mathrm{Fe(II)_{aq}}$, and a +2.0‰ fractionation between $\mathrm{Fe(III)_{aq}}$ and ferrihydrite upon precipitation is illustrated in Figure 10. Although a k_2/k_1 ratio of 5 appears to fit the fractionations measured by Bullen et al. (2001) (Fig. 10), we note that at these relatively low k_2/k_1 ratios, Equation (16) cannot be used, but instead the calculations are made using an incremental approach and simple isotope mass-balance equations (e.g., Eqn. 12). As the k_2/k_1 ratio increases to ~20, Equation (16) may

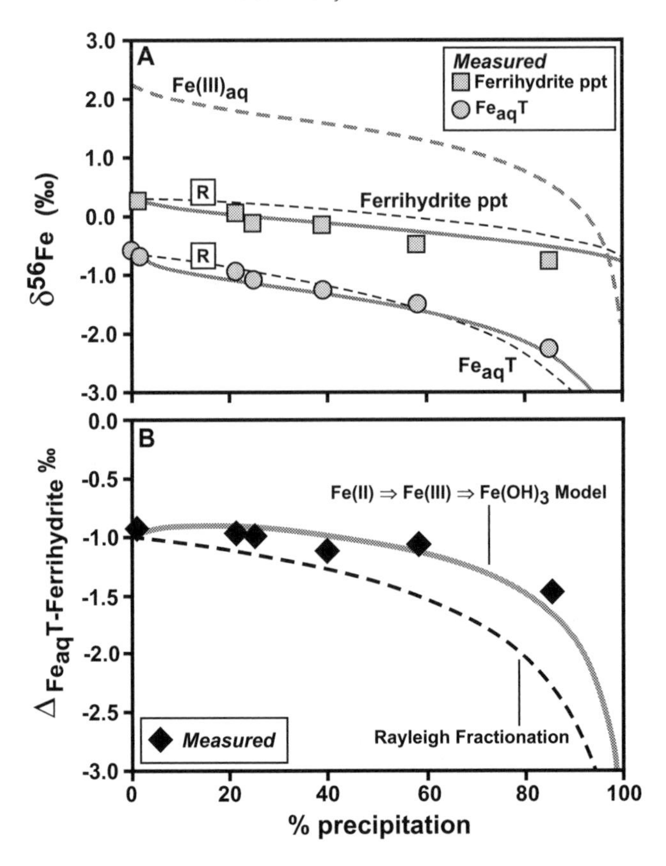

Figure 10. Fe isotope compositions for total aqueous Fe (Fe_{aq},T) and ferrihydrite (FH) precipitate and aqueous Fe-ferrihydrite fractionations from the batch oxidation and precipitation experiment of Bullen et al. (2001). (A) Measured $\delta^{56}Fe$ values from Bullen et al. (2001), compared to simple Rayleigh fractionation (short-dashed lines, noted with "R") using $10^3 \ln\alpha_{Feaq\text{-}FH} = -0.9‰$, as well as the two-step re-equilibration model discussed in the text (i.e., Eqn. 12), shown in solid gray lines for the aqueous Fe and ferrihydrite components; the predicted $\delta^{56}Fe$ value for $Fe(III)_{aq}$ is shown in the heavy dashed line, which reflects continual isotopic equilibrium between $Fe(II)_{aq}$ and $Fe(III)_{aq}$. Note that in the experiment of Bullen et al. (2001), aqueous Fe existed almost entirely as $Fe(II)_{aq}$. (B) Measured fractionation between total aqueous Fe and ferrihydrite precipitate, as measured, and as predicted from simple Rayleigh fractionation (black dashed line) and the two-step model where isotopic equilibrium is maintained between aqueous $Fe(II)_{aq}$ and $Fe(III)_{aq}$ (solid gray line).

be used, and the two-step fractionation model becomes equivalent to common models such as Rayleigh fractionation, where the overall fractionation is equal to the difference of the two fractionation factors (Fig. 10). Bullen et al. (2001) noted that their data did not strictly lie along a Rayleigh fractionation curve, and hypothesized this may reflect isotopic "back exchange" between ferrihydrite and aqueous Fe. This possibility now seems unlikely, based on $Fe(III)_{aq}$-ferrihydrite exchange experiments (Poulson et al. 2003), where isotopic exchange is limited to surface sites. The two-step fractionation model discussed here is an alternative explanation for the deviation from a Rayleigh model (Fig. 10), but we cannot discount changes in the $\Delta_{Fe(III)aq\text{-}FH}$ fractionation factor with precipitation as a possibility as well.

A $Fe(III)_{aq}$-ferrihydrite fractionation of +2.0‰, as suggested by the above discussion, seems likely to be a kinetic isotope fractionation when the rates of precipitation of the experiment of Bullen et al. (2001) are considered relative to the observations of Skulan et al. (2002), who measured a kinetic $Fe(III)_{aq}$-hematite fractionation of +1.3‰ at similar precipitation rates. That the precipitation experiment of Skulan et al. (2002) was run at 98°C, rather than at room temperature, where the Bullen et al. (2001) experiment was performed, may be one explanation for the larger inferred $Fe(III)_{aq}$-ferrihydrite fractionation, which would be expected to be larger at the lower temperature. However, it is difficult to compare these two studies directly, and it is possible that there are significant isotopic differences between hematite and ferrihydrite; Polyakov and Mineev (2000), for example, predict significant differences in $\beta_{56/54}$ factors for hematite, goethite, and lepidocrocite (Fig. 3), which remains to be tested experimentally. Fully understanding the kinetic and equilibrium fractionations involved during oxidation of $Fe(II)_{aq}$ and eventual precipitation to ferrihydrite will require experimental determination of kinetic fractionations between $Fe(III)_{aq}$ and ferrihydrite over a range of precipitation rates, as well as determination of the equilibrium $Fe(III)_{aq}$-ferrihydrite fractionation factor.

The two-step model discussed here, which is primarily motivated by the recognition that rapid isotopic exchange occurs between aqueous ferric and ferrous Fe species, contrasts with that proposed by Bullen et al (2001), who interpreted their data to reflect isotopic fractionation between ferrous Fe species such as $[Fe^{II}(OH)(H_2O)_5]^+$ and $[Fe^{II}(H_2O)_6]^{2+}$, given the rapid pathway through the ferrous Fe hydroxide species during oxidation of $Fe(II)_{aq}$. This theme was later echoed by Bullen et al. (2003), who suggested that the moderate pH experiments of Johnson et al. (2002) cannot be used to estimate the $Fe(III)_{aq}$–$Fe(II)_{aq}$ fractionation because $Fe(II)_{aq}$ is unstable at moderate pH. In reply, Johnson et al. (2003b) note that kinetic issues cannot influence the isotopic fractionation measured under equilibrium conditions, and point out that if isotopic exchange is rapid relative to the timescale of changing $Fe(II)_{aq}/Fe(III)_{aq}$ ratios in solution, isotopic equilibrium will still be maintained even if the solutions are out of chemical equilibrium. Although one might assume that as an intermediate species, $Fe(III)_{aq}$ has no effect on the overall measured Fe isotope fractionation between $Fe(II)_{aq}$ and ferrihydrite precipitate because essentially all $Fe(III)_{aq}$ is converted to ferrihydrite, this would only be true if $Fe(III)_{aq}$ was "inert" and not open to isotopic exchange, which is clearly not the case given the rapid isotopic exchange that occurs between $Fe(II)_{aq}$ and $Fe(III)_{aq}$. Consideration of the residence times of $Fe(III)_{aq}$ during oxidation of $Fe(II)_{aq}$, relative to the time required to attain isotopic equilibrium among aqueous species, indicates that, in fact, $Fe(III)_{aq}$ exerts a major control on the overall measured isotopic fractionation.

Summary observations on Fe isotope fractionation factors

A key component to developing the Fe isotope system is accurate determination of kinetic and equilibrium Fe isotope fractionation factors among a wide variety of aqueous Fe species and minerals. In the experimental systems studied to date, where equilibrium isotope fractionation factors can be confidently inferred for $Fe(III)_{aq}$–$Fe(II)_{aq}$ and $Fe(III)_{aq}$–hematite, comparison of the measured fractionations with those predicted from spectroscopic data (Polyakov and Mineev 2000; Schauble et al. 2001; Anbar et al. 2004) yields agreement in some instances but is inconsistent in others. For example, there is excellent agreement between the experimentally measured $Fe(III)_{aq}$–$Fe(II)_{aq}$ fractionation with that calculated by Anbar et al. (2004), but disagreement with Schauble et al. (2001). The greatest disparity between the two predicted fractionations lies in the reduced partition function ratio for $[Fe^{III}(H_2O)_6]^{3+}$. Use of the reduced partition function ratio for $[Fe^{III}(H_2O)_6]^{3+}$ by Anbar et al. (2004), in conjunction with that for hematite determined by Polyakov and Mineev (2000), yields a calculated $\Delta_{Fe(III)aq-hematite}$ of approximately +1.3‰ at 98°C, which is different than the measured equilibrium fractionation factor $\Delta_{Fe(III)aq-hematite}$ of −0.1 ± 0.2‰ at 98°C by Skulan et al. (2002). Considering the uncertainties in the predicted $\beta_{56/54}$ factors, however, which are

1‰ or greater at low temperature (Polyakov and Mineev 2000), this discrepancy may not be significant.

Table 2 summarizes experimentally determined Fe isotope fractionation factors for abiotic systems. These results confirm that the largest magnitude Fe isotope fractionations involve redox changes in Fe-bearing minerals and/or fluids. Many of the major Fe-bearing mineral groups (e.g., oxide, carbonate, and sulfide) have been studied to some degree, although the effect that mineralogy within a group plays has yet to be established. For example are the predicted fractionations between hematite and goethite real. The compositional effects predicted for Fe carbonates (Polyakov and Mineev 2000) have yet to be studied experimentally. Are there differences between Fe monosulfide and pyrite? It remains unknown if the differences in $Fe(II)_{aq}$–FeS and $Fe(II)_{aq}$–FeS_2 fractionations that have been experimentally determined or calculated reflect real differences, or problems with experimental and/or predicted fractionations. Finally we note that no experimental work has been conducted on Fe silicates, especially those silicates that form at low temperatures such as celadonite, whose predicted $Fe(II)_{aq}$–mineral fractionations are predicted to be significant.

Table 2. Summary of experimentally measured abiotic Fe isotope fractionation between Fe species.

Fe species	$\Delta^{56}Fe_{A-B}$	Reference
$Fe(II)_{aq}$–Fe(III) oxide	−0.9 ± 0.2	Bullen et al. (2001)
$Fe(II)_{aq}$–Hematite	−3.0 + 0.3	Skulan et al. (2002); Johnson et al. (2002); Welch et al. (2003)
$Fe(III)_{aq}$–Hematite	−0.1 ± 0.2	Skulan et al.(2002)
$Fe(II)_{aq}$–FeS	+0.3 ± 0.05	Butler et al. (2003)
$Fe(II)_{aq}$–Siderite	+0.5 ± 0.2	Wiesli et al. (2003b)
$Fe(III)_{aq}$–$Fe(II)_{aq}$	+2.9 ± 0.2	Johnson et al. (2002); Welch et al. (2003)

All results are reported for experiments conducted at room temperature (~20°C) except for $Fe(II)_{aq}$–hematite that was done at 98°C and the $Fe(II)_{aq}$–FeS which was done from 2 to 40°C.

IRON ISOTOPE VARIATIONS IN HIGH TEMPERATURE ENVIRONMENTS

Figure 11 is a histogram plot of the Fe isotope composition of high temperature rocks from the solar system. The average bulk-rock $\delta^{56}Fe$ value of 43 terrestrial igneous rocks is 0.00 ± 0.05‰ (on an IRMM-014 scale, the $\delta^{56}Fe$ of igneous rocks averages +0.09 ± 0.05‰), and there is no correlation between Fe isotope compositions and bulk-rock chemical compositions. For example, silicic continental volcanic and plutonic rocks have the same isotopic composition as oceanic and continental basalts. Lunar rocks define a range of Fe isotope compositions that is similar to those of terrestrial igneous rocks (Fig. 11). In contrast, the lunar regolith has higher $\delta^{56}Fe$ values as compared to those of crystalline lunar rocks (Fig. 11), which probably reflects space weathering processes and Fe loss during vaporization (Wiesli et al. 2003a; see below). The HED meteorite group (Howardites, Eucrites, Diogenites) define a range of Fe isotope composition that is similar to that of terrestrial samples (Fig. 11; Mullane et al. 2003c), although Poitrasson et al. (2002) suggests that Eucrites are skewed to lower $\delta^{56}Fe$ values by ~0.05‰. Metal from Iron meteorites has the same isotopic composition as terrestrial igneous rocks, where the average $\delta^{56}Fe$ value of 10 analyzed Iron meteorites is −0.03 ± 0.04‰ (Zhu et al. 2001; Kehm et al. 2003). In contrast, Pallasites (stony iron meteorites) have a wide range in isotopic compositions (Fig. 11; $\delta^{56}Fe$ from approximately −1.0 to +0.1‰) as measured in Fe metal and silicates (Zhu et al. 2001, 2002; Mullane et al. 2002; Kehm et al. 2003). It is

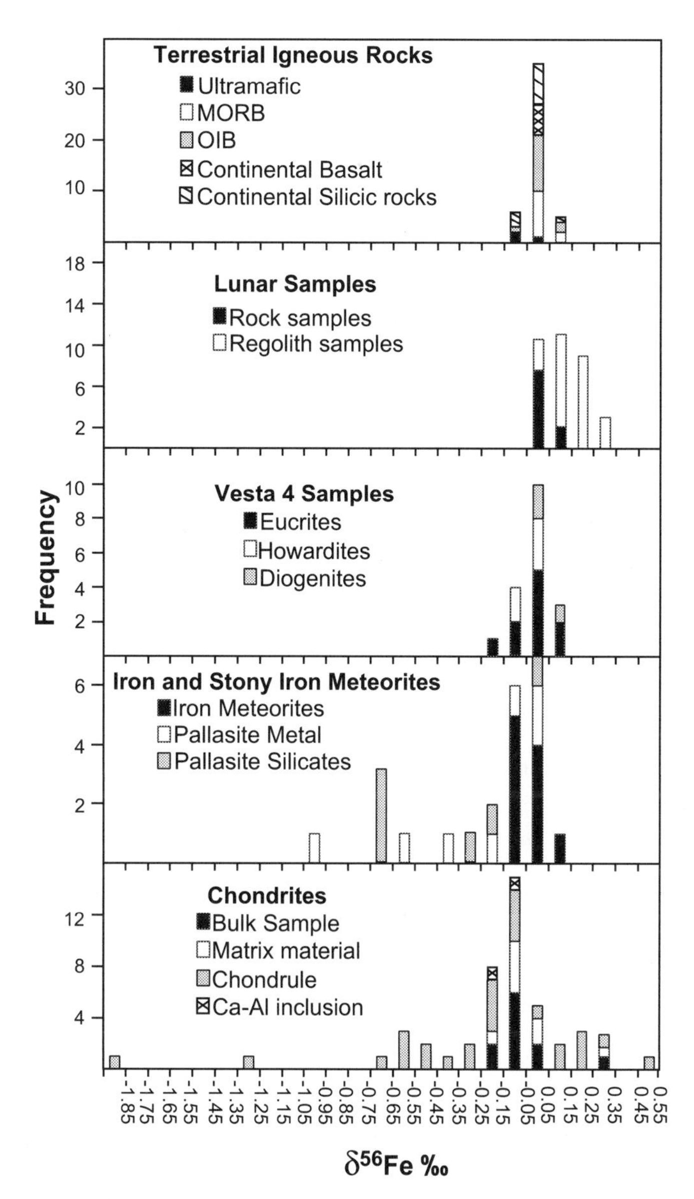

Figure 11. Frequency diagram of Fe isotope compositions measured in igneous rocks from terrestrial and extraterrestrial samples. The bin range is 0.1‰ in δ^{56}Fe values, which is the typical external precision (2σ) of an Fe isotope measurement. The data summarized in the diagram includes Fe isotope analyses from four laboratories and inter-laboratory bias has been corrected by normalization to a constant δ^{56}Fe value of −0.09‰ for IRMM-014, which places Fe isotope variations relative to the average of terrestrial igneous rocks (e.g., Beard et al. 2003b). Data sources are: terrestrial igneous rocks from Beard et al. (2003b); lunar rock and regolith samples from Wiesli et al. (2003a) HED meteorite data from Zhu et al. (2001), and Mullane et al. (2003c); Iron and Pallasite meteorite data from Zhu et al. (2001), Mullane et al. (2002), and Kehm et al. (2003); chondrite meteorite data from Zhu et al. (2001), Mullane et al. (2003a; b), and Kehm et al. (2003).

unknown if the wide range in isotopic compositions of Pallasites is due to high-temperature fractionation between Fe metal and silicate minerals (see below) or oxidation by vapors that are produced during crystallization of silicate liquids (e.g., Wasson and Choi 2003).

The largest variations in Fe isotope compositions occur in chondrites (Fig. 11). For example, individual chondrules define a range in δ^{56}Fe values from −1.9 to +0.5‰ (Zhu et al. 2001; Kehm et al. 2003; Mullane et al. 2003a,b). Bulk chondritic meteorites define a significant range in Fe isotope compositions, where 8 bulk chondritic meteorites vary from δ^{56}Fe values of −0.24 to +0.29‰ (Zhu et al. 2001; Kehm et al. 2003; Fig. 11). The matrix material in chondritic meteorites is also variable where δ^{56}Fe varies from −0.15 to +0.34‰ (Zhu et al. 2001; Mullane et al. 2003b; Fig. 11). Two bulk calcium aluminum inclusions (CAI) have been measured in chondrites and these have Fe isotope compositions that are similar to the matrix material analyzed in the same meteorite (Mullane et al. 2003a,b).

The Fe isotope variations measured in these terrestrial and extraterrestrial samples were all produce by mass-dependent processes, and it appears that the solar nebula was well homogenized with respect to variations in ^{54}Fe, ^{56}Fe, ^{57}Fe, and ^{58}Fe (Zhu et al. 2001; Kehm et al. 2003). This stands in marked contrast to the O isotope compositions of extraterrestrial materials, where different meteorite groups tend to have unique ^{17}O/^{16}O–^{18}O/^{16}O variations, including incomplete mixing with extrasolar ^{16}O-rich components. The only non-mass-dependent Fe isotope variations that have been reported for extraterrestrial samples are parts of some CAIs from chondritic meteorites, where there appears to be an excess of ^{58}Fe that likely reflects nucleosynthesis processes (Völkening and Papanastassiou 1989). In the one high precision Fe isotope study that made ^{54}Fe, ^{56}Fe, and ^{58}Fe isotope analyses in a manner that preserved natural, mass-dependent variations (Kehm et al. 2003), these workers found that all variations in bulk chondrites and individual chondrules are associated with mass-dependent processes. It appears that evidence for non-mass dependent Fe isotope variations can only be found in the most refractory portions of the oldest extraterrestrial materials (e.g., Völkening and Papanastassiou 1989; Birck 2004), or perhaps in presolar grains (Clayton 1999).

The range in δ^{56}Fe values in bulk chondrites and chondrite matrices is though to reflect alteration processes that occurred on the chondritic parent bodies, probably through interaction with aqueous solutions (Kehm et al. 2003). Of particular significance is the composition of the bulk analysis of the chondritic meteorite Orgueil. Both Zhu et al. (2001) and Kehm et al. (2003) have analyzed aliquots from this sample and the analysis by Zhu et al. (2001) is the only bulk chondrite that has a positive δ^{56}Fe value (+0.29‰); however, the aliquot analyzed by Kehm et al. (2003) had a δ^{56}Fe of −0.04‰, a value that is similar to the 7 other bulk chondrites that have been analyzed. This contrast does not appear to be an analytical artifact, since both studies analyzed bulk samples of Murchison and Allende, and the reproducibility between the two labs is good. The different isotopic compositions analyzed by both labs for the Orgueil meteorite is inferred to be a result of fine-scale heterogeneities in the meteorite. Mullane et al. (2003a,b) consider the similar Fe isotope compositions of aliquots of matrix material and CAI from the Allende meteorite to be evidence for redistribution of Fe because CAIs have exceedingly low Fe contents.

The 2.5‰ range in ^{56}Fe/^{54}Fe ratios of chondrules is interpreted to reflect evaporation and condensation during chondrule formation, alteration processes, and initial isotopic heterogeneities in the source materials. Iron isotope analyses conducted by ion microprobe on olivine from Chainpur chondrules place some constraints on possible Fe evaporation models during chondrule formation (Alexander and Wang 2001). Rapid calculated cooling rates are required to produce minimal Fe isotope fractionation, and these agree well with cooling rates inferred from experiments that produce the petrographic textures which are found in chondrules. The precision of individual Fe isotope analyses, however, were ~2‰ in ^{57}Fe/^{56}Fe, and were not sufficient to satisfactorily resolve distinct differences in Fe isotope compositions.

Based on high-precision $^{56}Fe/^{54}Fe$ ratios determined for 8 chondrules from the ordinary chondrite Tieschitz, Kehm et al. (2003) concluded that there was no evidence for positive $\delta^{56}Fe$ values relative to terrestrial igneous rocks. The lack of evidence for heavy Fe isotope compositions led Kehm et al. (2003) to conclude that evaporative loss of Fe was minimal, or if evaporation occurred it did not proceed via Rayleigh process. Instead, Kehm et al. (2003) suggest that the trend of Fe isotope compositions in chondrules that have negative $\delta^{56}Fe$ values is better explained by loss of Fe during condensation processes, assuming that gas and condensate do not maintain equilibrium. In another high-precision Fe isotope study of individual chondrules from the Allende and Chainpur chondritic meteorites, Mullane et al. (2003a,b) concluded that the limited Fe isotope variations in chondrules rules out Rayleigh evaporation processes. Mullane et al. (2003a,b) also highlight the lack of correlation between Fe isotope composition and chondrule size and texture, suggesting that mass-dependent isotopic heterogeneities in the chondrule starting materials is the most likely explanation for the range in Fe isotope compositions. Secondary processes are likely to homogenize Fe isotope compositions between matrix, chondrules, and CAI, and it is possible that initial isotopic variability in the solar nebula was much greater.

There may be some subtle differences in the Fe isotope composition of different planetary bodies. For example, the average $\delta^{56}Fe$ value of bulk chondritic meteorites is -0.11 ± 0.07‰, excluding the Orgueil analysis by Zhu et al. (2001; see above). Poitrasson et al. (2002) has noted that the SNC and HED meteorite groups have $\delta^{56}Fe$ values that are slightly lower than terrestrial igneous rocks, approximately -0.07‰, where as they note that lunar rocks have slightly higher $\delta^{56}Fe$ values that are 0.06‰ greater than terrestrial rocks. The similarity between SNC and HED and the average of chondritic meteorites, and the slightly higher $\delta^{56}Fe$ of the Earth-Moon system is interpreted as evidence for vaporization and loss of Fe from the Earth-Moon system during formation of the moon by a giant impactor (Poitrasson et al. 2002). The isotopic differences between the Earth-Moon system and the SNC and HED parent bodies, however, are near the limit of analytical precision, and these interpretations may change as the database is expanded. For example, using a different suite of lunar samples, Wiesli et al. (2003a) determined that the average $\delta^{56}Fe$ value of 9 lunar rock samples was 0.03 ± 0.05‰, and individual samples varied from -0.04 to $+0.13$‰. Similarly, in a more extensive Fe isotope survey of HED meteorites (5 Howardites, 9 Eucrites and 3 Diogenites), Mullane et al. (2003c) observed that there are no Fe isotope differences between Howardites, Eucrites, and Diogenites, and that the average $\delta^{56}Fe$ value of the HED meteorites analyzed by Mullane et al. (2003c) is -0.01 ± 0.06‰. Iron meteorites, which presumably represent material from differentiated planets, have an average $\delta^{56}Fe$ of -0.03 ± 0.04‰ for the 10 analyzed samples. In summary, based on the current data set, it is difficult to determine if different planets have resolvable differences in their Fe isotope composition, or if these planets have compositions that are unique relative to bulk chondrites. An important avenue for further study would be measurement of the isotopic compositions of elements of different volatility, including Fe, which may provide more robust constraints on accretion history of planetary bodies.

Fe isotope fractionation produced by vaporization

Loss of Fe during vaporization processes can lead to significant Fe isotope fractionation in a variety of environments. For example, deep sea iron spherules (micro iron meteorites) have $\delta^{56}Fe$ values that are up to +40‰ greater than the average of terrestrial igneous rocks and iron meteorites (Herzog et al. 1999). The isotopic compositions of Cr and Ni in deep sea spherules are similarly enriched in the heavy isotope relative to an average terrestrial value. These extreme isotopic compositions suggest that up to 95% of the Fe in the spherule was lost by evaporation and heating during passage through the Earth's atmosphere. Similar fractionations are produced in thermal ionization mass spectrometry, when Fe isotope analysis during heating results in the preferential loss of the light isotopes (Beard and Johnson 1999).

Vaporization processes that produce the high δ^{56}Fe values of deep sea spherules are, of course, more complicated, because a large portion of the heating and vaporization occurs in the atmosphere, and is accompanied by conversion of Fe metal to Fe oxides.

Vaporization and loss of Fe is the preferred explanation for the high δ^{56}Fe values measured in the lunar regolith, as compared to lunar rock samples (Wiesli et al. 2003a). The lunar regolith is formed by space weathering processes that produce changes in the primary mineralogy of crystalline rocks, largely through production of agglutinitic glass fragments and formation of regolith that is primarily controlled by micrometeorite impacts and solar wind sputtering processes (e.g., Keller and McKay 1993; Bernatowicz et al. 1994; Hapke et al. 1994; Hapke 2001). During formation of agglutinitic glass, nano-phase Fe metal is produced which has a strong influence on the magnetic properties of regolith as measured by I_S values (ferromagnetic resonance intensity; Morris 1976). High I_S values are correlated to a high proportion of nano-phase Fe metal. The relative exposure age of regolith to space weathering processes is measured using the I_S/FeO ratio (I_S values are normalized to the soil's total Fe in order to normalize for differences in bulk composition), where values of 30 and 60 delineate immature from submature, and submature from mature, respectively (Morris 1976). Within an individual soil sample, the finest fractions have the highest I_S/FeO values and the highest agglutinitic glass proportion (Taylor et al. 2001), and the finest size fractions have a greater proportion of nano-phase Fe metal relative to coarser size fractions.

Immature soil samples have δ^{56}Fe values that are indistinguishable from lunar rocks, whereas submature and mature soils have δ^{56}Fe values that are greater than those of lunar rocks, and δ^{56}Fe values are positively correlated with I_S/FeO values (Fig. 12). Lunar regolith samples in general tend to have heavy isotopic compositions as compared to lunar rock samples, as demonstrated by isotopic analyses of O, Si, S, Mg, K, Ca, and Cd (Epstein and Taylor 1971; Clayton et al. 1974; Russell et al. 1977; Esat and Taylor 1992; Humayun and Clayton 1995; Sands et al. 2001; Thode 1976). The origin of isotopic compositions that are enriched in the heavy isotopes has been presumed to reflect sputtering by solar wind and vaporization, where preferential loss of the lighter isotope to space occurs. In contrast to previous isotopic studies, the Fe isotope compositions measured in the Lunar Soil Characterization Consortium samples can be related to a specific phase based on the positive correlation in I_S/FeO and δ^{56}Fe values (Fig. 12).

High-temperature inter-mineral Fe isotope fractionation

The calculated Fe isotope fractionation factors of Polyakov and Mineev (2000) indicate that at high temperatures (>500°C) there may be analytically resolvable (e.g., 0.1‰ in ^{56}Fe/^{54}Fe) inter-mineral fractionations between magnetite and olivine and between clinopyroxene and olivine (Fig. 13). Berger and von Blanckenburg (2001) reported magnetite-silicate fractionations for metamorphic rocks that were consistent with the calculated values of Polyakov and Mineev (2000), although these workers suggested that Fe isotope fractionation was more strongly controlled by changing redox conditions as opposed to temperature dependency. Zhu et al. (2002) reported significant high-temperature, inter-mineral isotope fractionation based on analyses of 3 spinel peridotite mantle xenoliths, where olivine has δ^{56}Fe values that are ~0.2‰ lower than coexisting clino- and orthopyroxene, and concluded that these inter-mineral fractionations were a result of high-temperature equilibrium isotopic exchange (Fig. 13). Similar Fe isotope compositions measured on minerals from spinel peridotite were reported by Williams et al. (2002). In addition to the inter-mineral Fe isotope differences measured in spinel peridotites, Zhu et al. (2002) reported that olivine from two Pallasite meteorites has δ^{56}Fe values that are 0.14 and 0.19‰ lower than coexisting Fe metal (Fig. 13).

In contrast to the above findings, which suggest that there are equilibrium high-temperature inter-mineral fractionations, Mullane et al. (2002) reported that silicate minerals

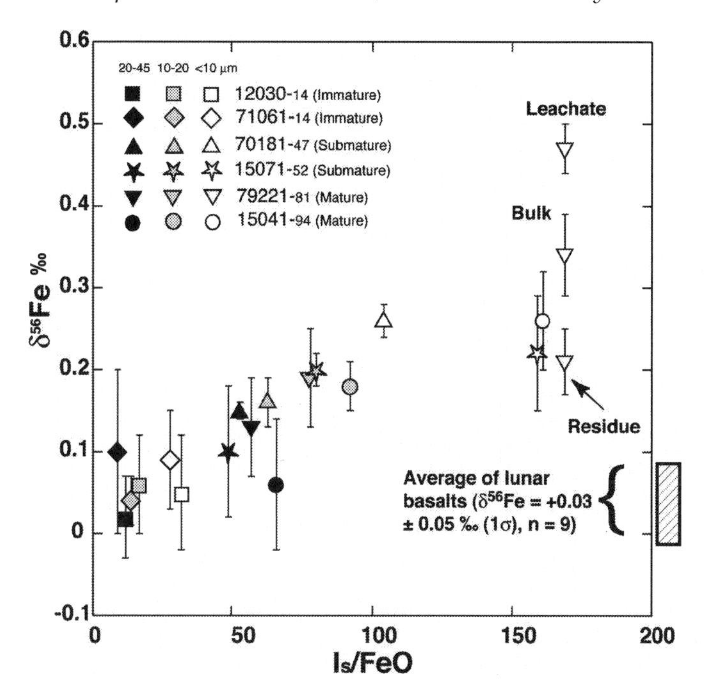

Figure 12. Plot of I_s/FeO versus δ^{56}Fe values of lunar regolith samples from the Lunar Soil Characterization Consortium. The sub-scripted numbers after the sample numbers are the I_s/FeO values measured for the <250 μm sized fraction. All analyses are for bulk samples of the different sized fractions; error bars are 2σ as calculated from 2 or more complete Fe isotope analyses. Modified from Wiesli et al. (2003a).

in Pallasites are homogenous in Fe isotope composition, and that coexisting Fe metal in some samples has a δ^{56}Fe value that is greater than the silicate mineral and in other samples the Fe-metal has a δ^{56}Fe that is less than the coexisting silicate mineral. Beard and Johnson (2004) analyzed coexisting olivine, clinopyroxene, and orthopyroxene from 8 spinel peridotite mantle xenoliths and found that there was not a significant Fe isotope fractionation between orthopyroxene and olivine (Fig. 13). Clinopyroxene-olivine fractionations for all but one sample were insignificant as well (Fig 13). Beard and Johnson (2004) noted that the measured Fe isotope differences between clinopyroxene and olivine were a function of Fe isotope composition as shown on a δ-δ plot (Fig. 14). The negative correlation between $\Delta^{56}Fe_{cpx\text{-}ol}$ and the δ^{56}Fe value of olivine (or clinopyroxene) is interpreted to be a result of open-system processes, drawing on analogous relations that have been observed in oxygen isotope studies (e.g., Gregory and Criss 1986). Specifically, Beard and Johnson (2004) hypothesized that metasomatic alteration by a fluid that had a low δ^{56}Fe value coupled with differences in the relative Fe diffusion rates of olivine, orthopyroxene, and clinopyroxene, may produce the trends in δ^{56}Fe values measured in mantle minerals.

Beard and Johnson (2004) also investigated Fe isotope fractionation between magnetite and silicate minerals in 4 volcanic rocks from Mt. Lassen, California. In these samples there was no resolvable difference in Fe isotope composition between magnetite, olivine, amphibole, and biotite, and all the minerals had δ^{56}Fe values that overlapped with the Fe isotope composition of igneous rocks. These results stand in contrast to the findings of Berger and von Blanckenburg (2001), who analyzed magnetite and silicate minerals from slowly cooled metamorphic rocks and found significant differences between their Fe isotope

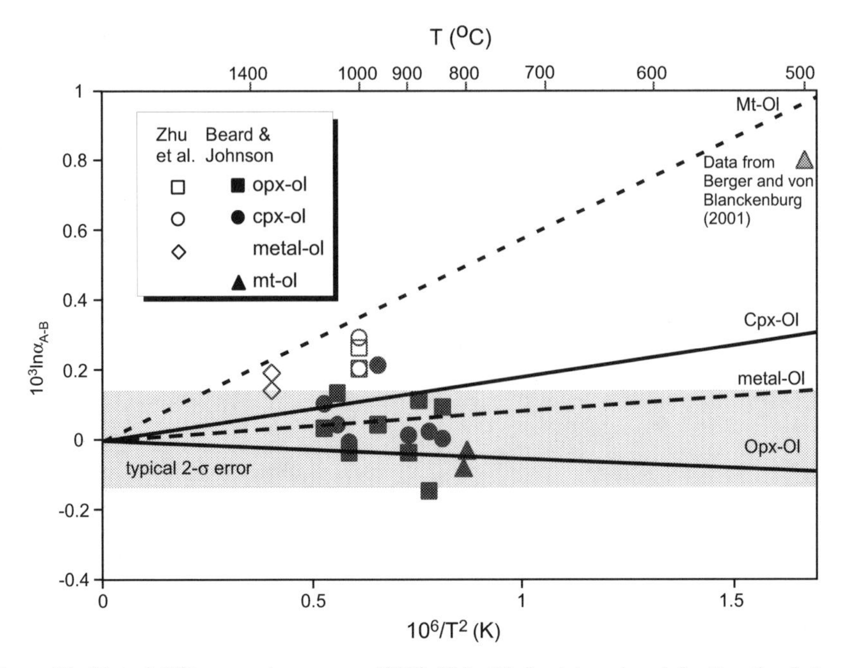

Figure 13. Plot of $10^3\ln\alpha_{A\text{-}B}$ values versus $10^6/T^2$ (T in K) for inter-mineral fractionation between magnetite-olivine, orthopyroxene-olivine, and clinopyroxene-olivine, and Fe metal-olivine as calculated from spectroscopic data by Polyakov and Mineev (2000), and as measured from natural samples by Zhu et al. (2002), Beard and Johnson (2004). Also shown is the Fe isotope fractionation factor between magnetite and Fe-silicates measured by Berger and von Blanckenburg (2001).

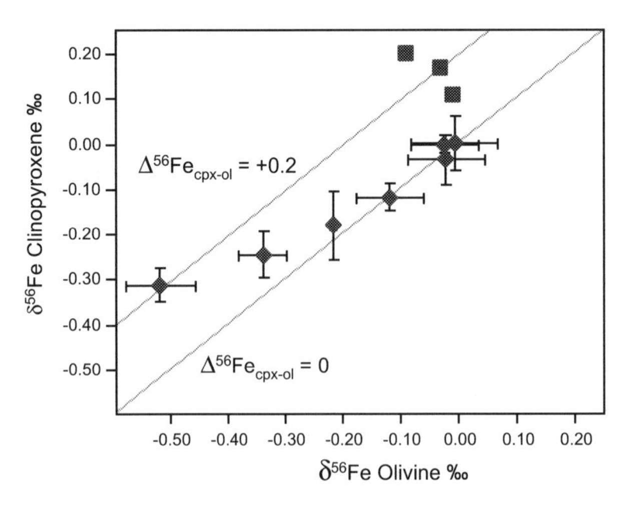

Figure 14. Inter-mineral Fe isotope fractionations among olivine and clinopyroxene from spinel peridotite mantle xenoliths. Data are from Zhu et al. (2002) (■) and Beard and Johnson (2004) (◆). In the study by Beard and Johnson (2004), the difference in the Fe isotope composition between clinopyroxene and olivine is larger as a function of their δ^{56}Fe values, suggesting disequilibrium fractionation.

compositions. The differences in Fe isotope fractionation between magnetite and silicate minerals analyzed in these two studies may reflect differences between the cooling rates of the samples. For example, open-system isotopic exchange between magnetite and silicates may have taken place during the slow cooling of the metamorphic rocks analyzed by Berger and von Blanckenburg (2001), whereas the rapidly cooled volcanic rocks would have been resistant to such open system behavior.

Iron isotope fractionation between garnet and clinopyroxene was measured for coexisting mineral pairs from 4 eclogites, 2 garnet peridotites and 1 garnet pyroxenite (Beard and Johnson 2004). Garnets have $\delta^{56}Fe$ values that range from −0.39 to +0.01‰, and clinopyroxenes have $\delta^{56}Fe$ values that range from −0.03 to +0.30‰ (Fig. 15). The Fe isotope compositions of coexisting garnet and clinopyroxene are positively correlated, and six of the seven samples have $\Delta^{56}Fe_{cpx\text{-}gt}$ of approximately +0.3‰, whereas the garnet pyroxenite sample has a significantly lower $\Delta^{56}Fe_{cpx\text{-}gt}$ fractionation of 0.07‰. Mass-balance calculations using the measured Fe contents and isotope compositions, as well as visually estimated modal proportions of garnet and clinopyroxene, indicate that the bulk Fe in these samples range from −0.10 to +0.07‰, in excellent agreement with analyzed bulk-rock Fe isotope analyses. Based on the Fe isotope variations between garnet and pyroxene on a δ-δ plot, Beard and Johnson (2004) inferred that there is resolvable high-temperature inter-mineral fractionation between garnet and pyroxene.

Perhaps the most surprising result of these mantle mineral studies is that the mantle has heterogeneities of 0.5‰ in $\delta^{56}Fe$ values (Fig. 14). These isotopic heterogeneities stand in marked contrast to the isotopic homogeneity of igneous rocks in oceanic and continental environments. Therefore it is believed that isotopically anomalous mantle is minor in abundance, and any contribution to basaltic melts must be small. Additionally, because there are no significant Fe isotope fractionations among the major Fe-bearing minerals (silicates and oxides) that are common liquidus phases in crustal magmas, the isotopic homogeneity recorded in basaltic magmas will be retained during differentiation. We are therefore left with the conclusion that although significant Fe isotope variability may be produced in parts of the mantle, such variability may be "left in the mantle" during magma genesis, and significant Fe

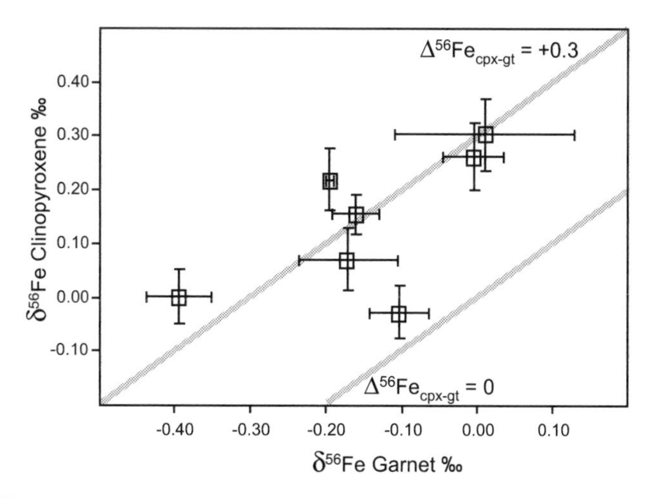

Figure 15. $\delta^{56}Fe$ of garnet versus that of coexisting clinopyroxene for eclogites, garnet peridotites, and garnet pyroxenite. Six of the seven samples scatter about a $\Delta^{56}Fe_{cpx\text{-}gt}$ of +0.3‰ over a wide range in $\delta^{56}Fe$ values measured in the minerals. Error bars are 1-SD of the mean as determined by two or more replicate analyses of the mineral. Modified from Beard and Johnson (2004).

isotope fractionations only re-appear in natural systems in surface environments where low-temperature fractionations between fluids minerals occurs, as well as biological cycling of Fe.

IRON ISOTOPE VARIATIONS IN LOW TEMPERATURE ENVIRONMENTS

The homogenous Fe isotope composition of igneous rocks (δ^{56}Fe = 0.00 ± 0.05‰, 1σ; Fig. 1; Beard et al. 2003b) suggests that the continents and oceanic crust formed with a δ^{56}Fe value of zero, providing a baseline with which to compare Fe isotope variations in low-temperature environments. Aeolian deposits, such as loess and modern aerosol particles, and fluvial sediments, such as modern oceanic turbidites and the suspended load from rivers, have homogenous Fe isotope compositions, similar to that of igneous rocks (Fig. 16). Low organic-carbon (C_{org}) grey shales from the Proterozoic and Phanerozoic are slightly more variable in Fe isotope composition where δ^{56}Fe = −0.36 to +0.39‰, but all of these clastic sedimentary rocks have a normal distribution of Fe isotope compositions whose average is the same as that of terrestrial igneous rocks. (Fig. 16). In contrast to the homogenous Fe isotope composition of these organic-poor clastic sediments, black shales (C_{org} > 0.5 wt%) define a wide range of Fe isotope compositions, (δ^{56}Fe = −2.28 to +0.64‰; Fig. 16). In addition to the significant Fe isotope variations recognized in C_{org}-rich shales, there are even larger variations in chemically precipitated sediments that formed at low temperatures. Pliocene to Recent Fe-Mn crusts from the North Atlantic Ocean and the Pacific Ocean typically have Fe isotope compositions that are shifted to lower δ^{56}Fe values as compared to igneous rocks (Fig. 16). Similarly, hydrothermal vent fluids from Mid-Ocean Ridges (MOR) have δ^{56}Fe values that are shifted to lower values as compared to terrestrial igneous rocks. Late Archean to Early Proterozoic Banded Iron Formations (BIFs) define a significant range of Fe isotope compositions from −2.5 to +1.1‰. The large Fe isotope variations recorded in BIFs are generally correlated with the mineralogy of individual BIF layers (Johnson et al. 2003a), where δ^{56}Fe values increase in the order pyrite-Fe carbonate-oxides (Fig. 16). These isotopic differences may be related to equilibrium Fe isotope fractionations that are intrinsic to specific minerals, reflecting precipitation or diagenetic reactions that occurred in relation to biotic processing of Fe.

Fe isotope variations in clastic sediments

The very restricted range in Fe isotope compositions of low-C_{org} clastic sedimentary rocks is quite remarkable considering the fact that these samples have been extensively processed through the sedimentary rock cycle, including mechanical and chemical weathering of pre-existing rock, sedimentary transport by wind and water, and deposition, diagenesis, and lithification. Low-temperature surficial processes that involve weathering and transport, diagenesis, and lithification of sediments produce some of the largest variations in light stable isotope compositions (e.g., H, C, and O). The large range in H, C, and O isotope compositions of such materials is partly a function of the fact that isotopic fractionations are large at low temperatures, but also reflects the complex interactions that take place between fluids and rocks, where significant reservoirs of H and O, for example, may exist in both fluids and minerals. The δ^{18}O values of sediments are much more variable (+10 to +30‰; Arthur et al. 1983) than those of fresh igneous rocks (e.g., Eiler 2001); the majority of oxygen isotope variation in igneous rocks is attributed to variable amounts of mixing between juvenile igneous sources that have a homogenous oxygen isotope composition with crustal material that has experienced surface alteration processes (see Eiler 2001 for a review). The roles that low temperature surficial processes have in controlling Fe isotope variations are largely unknown, but based on the similar range of δ^{56}Fe values that clastic sediments and igneous rocks share, it would appear that surficial processes that occur under at least oxic conditions do not significantly change the Fe isotope composition of bulk clastic sedimentary rocks.

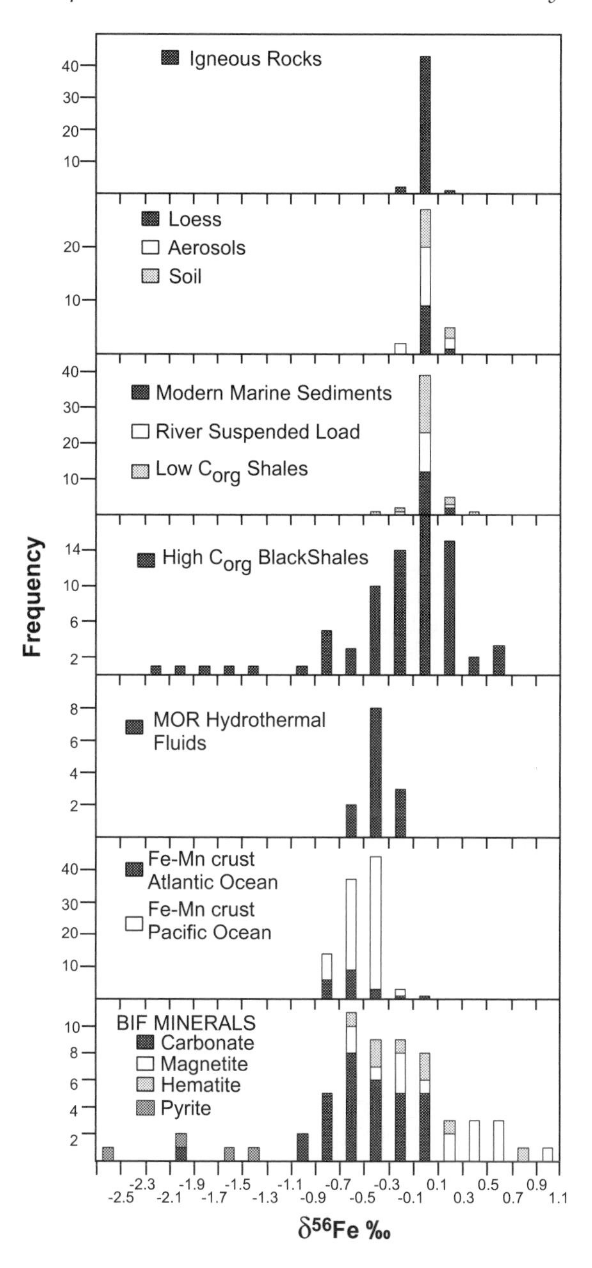

Figure 16. Frequency diagram of Fe isotope compositions measured in igneous and sedimentary rocks and MOR fluids. Inter-laboratory bias in Fe isotope measurements has been corrected by normalizing to a δ^{56}Fe value of −0.09‰ for the IRMM-014 standard, which places Fe isotope measurements relative to the average of terrestrial igneous rocks. Data sources are: terrestrial igneous rocks from Beard et al. (2003b); aeolian sediments and rocks from Beard et al. (2003a); fluvial low-C$_{org}$ sediments and rocks from Beard et al. (2003a), high-C$_{org}$ black shales from Yamaguchi et al. (2003) and Matthews et al. (2004); hydrothermal fluids from Sharma et al. (2001) and Beard et al. (2003a); Atlantic Ocean Fe-Mn crust data from Zhu et al. (2000), Pacific Ocean Fe-Mn crust data from Chu et al. (2003); Banded Iron Formations from Johnson et al. (2003a).

The significant difference in the relative behavior of Fe and O isotopes for clastic sedimentary rocks in part reflects the redox behavior of Fe. Iron solubility is highly dependent on redox state, where for example, in neutral pH oxic environments, the solubility of Fe is so low that significant transport of aqueous Fe is precluded. If Fe is not partitioned into different reservoirs, no isotopic variability will be produced in bulk samples. However for O isotopes, in fluid-rock systems there is typically sub-equal amounts of O in the two components, so that even at moderate fluid to rock ratios the bulk composition of the rock can be shifted toward the isotopic composition of the fluid, and vice versa. In contrast the quantity of Fe that is dissolved in oxic fluids is very low, and therefore it will be relatively difficult to shift the Fe isotope composition of a rock through fluid-rock interactions in an oxygenated environment. Moreover, Fe typically behaves as a conservative element during weathering (e.g., Canfield 1997) in oxic environments where, for example, nearly all Fe transported by rivers is contained in the suspended and bed loads and very little occurs in the dissolved load. Therefore, even if there are isotope fractionations associated with weathering of silicates and the production of Fe oxides and oxyhydroxides from aqueous Fe solutions, it is anticipated that there will be minimal net isotope shift in the bulk sample because Fe is not easily separated into different pools.

The δ^{56}Fe values of igneous rocks and suspended river loads are compared to the modal fraction of ferric oxide/hydroxide in suspended river loads in Figure 17. Despite markedly higher ferric oxide/hydroxide contents produced during weathering, the range in δ^{56}Fe values of bulk suspended river loads overlaps that of igneous rocks. Moreover, because there does not appear to be differences in the isotopic compositions of oxides and various silicate minerals in igneous rocks, suggesting that differential weathering is unlikely to produce significant Fe isotope variability. It is possible, however, that weathering in the presence of organic ligands will produce Fe isotope fractionations (Brantley et al. 2001). It remains unclear if Fe in poorly crystalline or labile components in suspended river loads is isotopically variable, or if the dissolved load of Fe is isotopically distinct from bulk crustal sources, and this is an important future avenue of research. Iron isotope variations have been measured in some extreme chemical weathering environments, as shown by analysis of highly altered oceanic crust where 80% of the Fe has been mobilized from the basaltic crust through interaction with mid-ocean ridge fluids, followed by production of ferric iron-rich clays like celadonite,

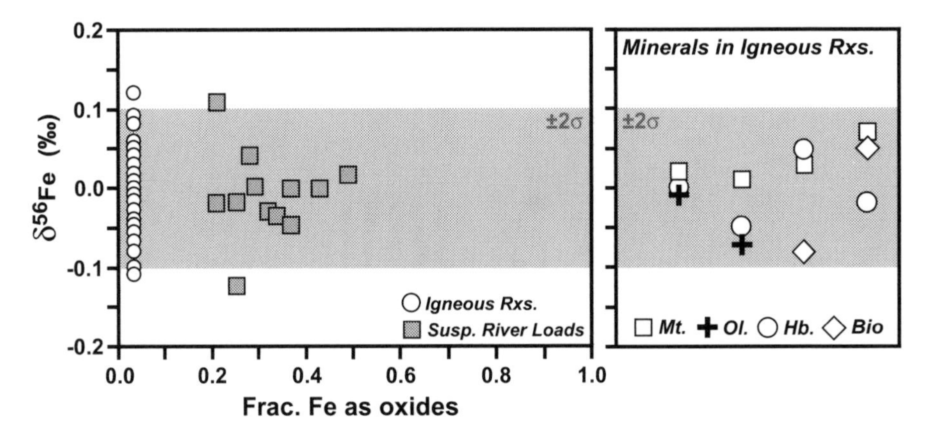

Figure 17. Comparison of δ^{56}Fe values for bulk igneous rocks (○), suspended river loads (■) and individual minerals from volcanic rocks (right panel). Left panel shows the fraction of Fe that exists as oxides (data for igneous rocks arbitrarily plotted as 0.02). Right panel shows δ^{56}Fe values for coexisting magnetite (□), olivine (✚), hornblende (○), and biotite (◇) in four volcanic rocks. Data from Canfield (1997), Beard et al. (2003a), and Beard and Johnson (2004).

leaving behind an altered basaltic crust that has high δ^{56}Fe values relative to unaltered basalt (e.g., Rouxel et al. 2003; see below).

The variability of Fe isotope compositions in black shales seems likely to reflect redox cycling of iron in anoxic environments. Reductive processes may include bacterial iron reduction or interaction with H_2S or other reducing agents, and if Fe is separated into isotopically distinct reservoirs, Fe isotope variations may be recorded in bulk samples. Bacterial iron reduction produces $Fe(II)_{aq}$ that has low δ^{56}Fe values (Johnson et al. 2004), and $Fe(II)_{aq}$ in equilibrium with $Fe(III)_{aq}$ will also have low δ^{56}Fe values (Johnson et al. 2002; Welch et al. 2003). Preliminary Fe isotope analyses of pore fluids from C_{org}-rich sediments indicate that the $Fe(II)_{aq}$ has a low δ^{56}Fe value, as low as −5‰ (Severmann et al. 2002), consistent with the expectation that reduced species should have relatively low δ^{56}Fe values. An additional pathway that may produce Fe isotope variability is precipitation of diagenetic siderite and iron sulfides. Initial experimental studies suggest that siderite and Fe monosulfide should have δ^{56}Fe values that are low relative to $Fe(II)_{aq}$ (Table 2). Empirical studies agree with these experiments. For example, an Fe isotope investigation of the Kimmeridge Clay Formation (a C_{org}-rich sequence of shales and mudstones, and carbonate material) concluded that mudstones that contain pyrite and siderite tend to have δ^{56}Fe values that are up to 0.6‰ lower than dolostones from the same formation (Matthews et al. 2004). Moreover, Matthews et al. (2004) demonstrated that pyrite from sedimentary sequences tends to have δ^{56}Fe that are ~0.4‰ less than the Fe isotope composition of most lithologic Fe, based on analysis of pyrite nodules and pyritized ammonites. Similarly low δ^{56}Fe values for pyrite were found in black shales that are interbedded with banded iron formation (Johnson et al. 2003). Another study of Archean-age black shales highlighted that magnetite-rich samples have very low δ^{56}Fe values (as low as −2.3‰), and may be explained by dissimilatory Fe(III) reducing bacteria, providing evidence for the ancient origin of Fe metabolism (Yamaguchi et al. 2003). The Fe isotope fractionations that may be produced during biological processing, and possible examples from the rock record, are discussed in the next chapter (Johnson et al. 2004).

Fe isotope variations in MOR hydrothermal fluids

Mid-Ocean Ridge (MOR) hydrothermal fluids have distinctly lower δ^{56}Fe values as compared to terrestrial igneous rocks and low-C_{org} clastic sedimentary rocks; thirteen analyses of vent fluids from the Atlantic and Pacific oceans define a range in δ^{56}Fe from −0.69 to −0.21‰ with an average of −0.39 ± 0.13‰ (Sharma et al. 2001; Beard et al. 2003a). Sharma et al. (2001) speculated that the origin of the low δ^{56}Fe values in MOR hydrothermal vent fluids is a product of 1) complex leaching reactions of basalts by fluids, 2) precipitation of sulfides from vent fluids, and 3) phase separation between vapor and brine solutions. At the time, it was not possible to evaluate all of these ideas because of sparse data, and poor precision for some samples (±0.04–0.68‰). More recent studies that include high-precision data on 1) MOR hydrothermal fluids that include analyses of brine and vapor from the same vent (Beard et al. 2003a), 2) sulfide deposits from a MOR spreading center (Rouxel et al. 2004), and 3) altered oceanic crust (Rouxel et al. 2003) allows some of these ideas to be evaluated.

Vapor and brine from the Brandon vent of the East Pacific Rise have identical Fe isotope compositions, implying that phase separation does not produce an isotopic fractionation (Beard et al. 2003a). The role that sulfide precipitation plays in controlling the Fe isotope composition of the fluid remains unknown. The precision of the two sulfide analyses reported by Sharma et al. (2001) was not sufficient to resolve if sulfide precipitation would produce Fe isotope fractionation in the vent fluid. In a detailed study of sulfides from the Lucky Strike hydrothermal field from the mid Atlantic Ridge, however, Rouxel et al. (2004) found that sulfides span a range in δ^{56}Fe values from −2.0 to +0.2‰, and that pyrite/marcasite has lower δ^{56}Fe values (~1‰) as compared to chalcopyrite. The variations in mineralogy and isotope composition are inferred to represent open-system equilibrium fractionation of Fe whereby chalcopyrite (a high-temperature sulfide) precipitated from fluids, producing a residual fluid that had low δ^{56}Fe values. Pyrite that precipitated from lower temperature fluids are thought to be the explanation for the relatively low δ^{56}Fe values measured for pyrite. So far, the evidence seems to support sulfide precipitation from hydrothermal fluids as a means to produce solutions that have low δ^{56}Fe values. Additional data on vent fluid-sulfide chimney pairs indicate that the sulfides are 0.1–0.3‰ higher in ^{56}Fe/^{54}Fe ratios as compared to the matching vent fluid (S. Severmann, et al., unpublished data). These empirically measured sulfide-aqueous Fe fractionation factors are consistent with the direction of Fe isotope fractionation inferred by Rouxel et al (2004), but they are smaller by a factor of 2 or more.

Fluid-rock interactions during hydrothermal alteration are also inferred to produce fluids that have low δ^{56}Fe values. In a survey of the Fe isotope composition of oceanic crust from ODP site 801C Rouxel et al. (2003) found that the basaltic crust is heterogeneous in its Fe isotope composition, and that δ^{56}Fe values in altered crust increase with increasing degrees of Fe leaching. Rouxel et al. (2003) inferred these variations to reflect kinetic isotopic fractionations that were associated with Fe leaching, or alternatively formation of high δ^{56}Fe minerals during clay formation. Notably, Rouxel et al. (2003) found one celadonite vein that had a δ^{56}Fe value of +1.1‰, consistent with the predictions of Polyakov and Mineev (2000). Based on these additional data it seems that the most likely origins of the low δ^{56}Fe values measured in MOR hydrothermal vent fluids are a result of sulfide precipitation as well as basalt-seawater interaction.

The fate of dissolved Fe from MOR vents has been investigated at the Rainbow plume from the mid-Atlantic Ridge. Plume particles were sampled from the buoyant part of the Rainbow plume, proximal to the vent, as well as particles from neutrally buoyant portions of the plume that were more distal from the vent (Severmann et al. 2003). Particles from the buoyant part of the plume have positive δ^{56}Fe values (up to +1.2‰), whereas in the neutrally buoyant sections of the plume, the particles have a near-constant δ^{56}Fe value of −0.2‰ that matches the Fe isotope composition of the vent fluid. The high δ^{56}Fe values of plume particles that were proximal to the vent probably reflect oxidation processes. In the neutrally-buoyant plume, all aqueous Fe(II) had been oxidized and it appears that there was no net loss in Fe because the neutrally buoyant plume particles have the same isotopic composition as the vent fluid. Moreover, metalliferous sediments sampled below the plume match the isotopic composition of the plume particles. The implication of these data is that for at least one plume, the Fe isotope composition of the vent fluid matches that of the plume particles. The Rainbow vent fluid, however, has an unusually high Fe to S ratio and hence it is uncertain if these results can be extrapolated to other plumes that originated from vent fluids that had lower Fe to S ratios.

Iron isotope variations in Fe-Mn nodules

Well-dated layers of an Fe-Mn crust from the Atlantic Ocean provide a potential record of the Fe isotope compositions of this part of the north Atlantic over the past 6 m.y. (Zhu et al. 2000). Layers that span an age from 6 to 1.7 Ma have a constant Fe isotope composition (δ^{56}Fe = −0.69 ± 0.10‰), but layers that are less than 1.7 Ma have increasing δ^{56}Fe values with decreasing age, up to a value of +0.04‰ at 0.15 Ma (Fig. 18). ^{206}Pb/^{204}Pb ratios follow temporal trends that are similar to those of δ^{56}Fe values (Fig. 18). Zhu et al. (2000) concluded that the correlation of Pb and Fe isotope compositions reflect changes in the isotopic composition of lithologic material that was delivered to the north Atlantic Ocean. Based on homogenous Fe isotope composition of clastic sedimentary rocks, modern aerosol particles, and the suspended load of rivers (Beard et al. 2003a) the interpretation of Zhu et al. (2000) cannot be correct. Beard et al (2003a) instead suggested that the origin of the Fe isotope variability for Fe-Mn crusts may be changes in the relative fluxes of Fe from isotopically distinct sources that was delivered to the oceans. Figure 19 shows the fluxes of Fe that are delivered to the modern oceans and the Fe isotope compositions for fluxes that are known. If we assume that Fe

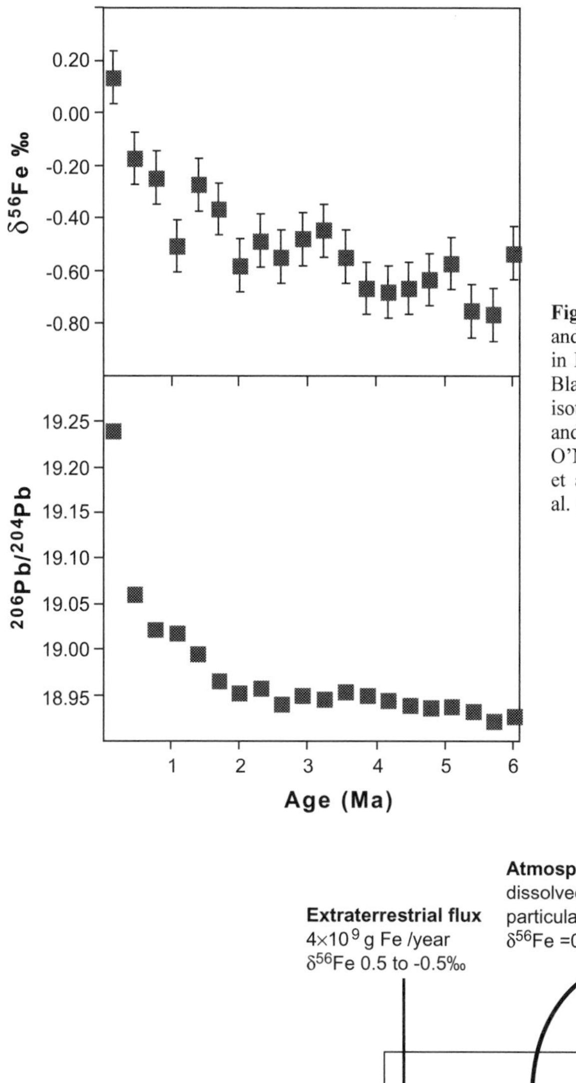

Figure 18. Temporal variations in Fe and Pb isotope composition for layers in Fe-Mn crust BM1969.05 from the Blake Plateau, Atlantic Ocean. Iron isotope data from Zhu et al. (2000) and Pb isotope data and ages from O'Nions et al. (1998) and Reynolds et al. (1999). Modified from Zhu et al. (2000).

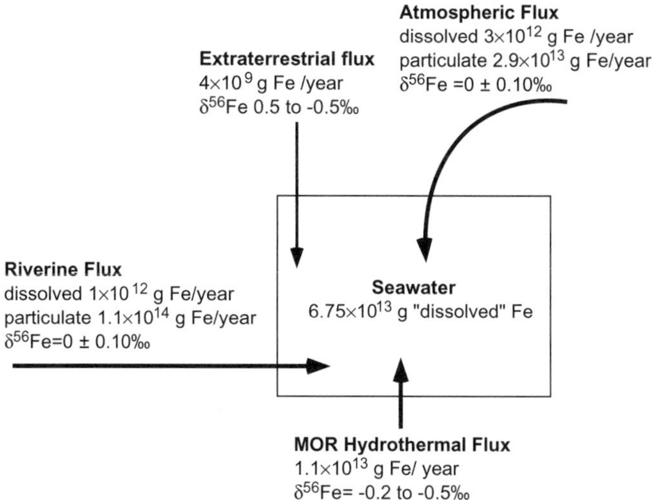

Figure 19. Fluxes of Fe delivered to the oceans and their measured or inferred Fe isotope compositions, modified from Beard et al. (2003a).

delivered to the open oceans results solely from the atmospheric particulate flux and the MOR hydrothermal flux, the $\delta^{56}Fe$ values for specific ocean masses may be predicted through simple mixing lines (Fig. 20). Based on this simple model, the calculated Fe isotope composition of the North Atlantic Ocean ($\delta^{56}Fe = -0.1‰$) agrees reasonably well with that measured in the youngest layer of the crust analyzed by Zhu et al. (2000; $\delta^{56}Fe = +0.13‰$). The outer-most layer from a hydrogenous crust from the south-central Pacific Ocean has a $\delta^{56}Fe$ value of −0.47‰ (Chu et al. 2003), which also agrees reasonably well with that predicted for the South Pacific Ocean ($\delta^{56}Fe = -0.32‰$; Fig. 20). Superimposed on these global variations are local variations that may reflect differences in the relative contributions of hydrothermal Fe, where Chu et al. (2003) noted distinct temporal trends in $\delta^{56}Fe$ values for Fe-Mn crusts from a back-arc basin for localities that were 80 km apart.

This simple two component model for the Fe isotope composition of seawater does not consider the effects of the Fe isotope composition of dissolved Fe from rivers or from rain. Although the dissolved Fe fluxes are small (Fig. 19) the dissolved fluxes may have an important control on the overall Fe isotope composition of the oceans if they represent an Fe source that is preferentially added to the hydrogenous Fe budget that is ultimately sequestered into Fe-Mn nodules. In particular riverine components may be very important in the Pacific Ocean where a significant amount of Fe to the oceans can be delivered from rivers that drain oceanic islands (Sholkovitz et al. 1999). An additional uncertainty lies in how Fe from particulate matter is utilized in seawater. For example, does the solubilization of Fe from aerosol particles result in a significant Fe isotope fractionation, and does Fe speciation lead to Fe isotope fractionation?

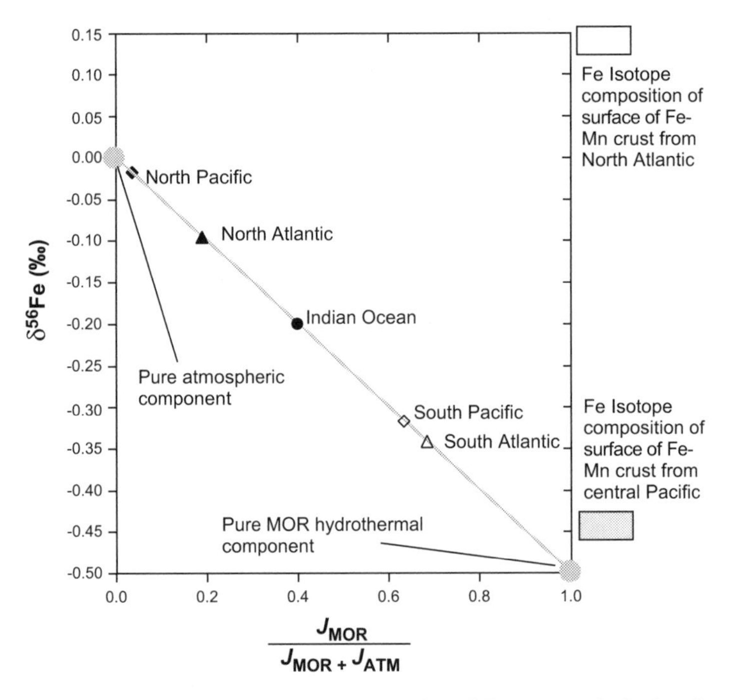

Figure 20. Calculated Fe isotope composition of seawater from different ocean basins based on a simple two-component mixing between Fe from aerosol particles and Fe from mid-oceanic-ridge (MOR) hydrothermal solutions. Atmospheric Fe fluxes (J_{ATM}) for different ocean basins from Duce and Tindale (1991); MOR hydrothermal Fe flux (J_{MOR}) to different ocean basins were proportioned relative to ridge-axis length. Modified from Beard et al. (2003a).

Crucial to unraveling these uncertainties are Fe isotope analyses of the dissolved load from rivers and rainwater, as well as direct analysis of seawater. These analyses will be analytically challenging because of low Fe concentrations. For example, in seawater, it will require 3 to 20 liters of seawater to obtain ~100 ng of Fe which is enough Fe to make a single Fe isotope analysis following the methods of Beard et al. (2003b). In contrast, other methods that rely on using microgram quantities of Fe to achieve large enough Fe ion signals that are not impacted by Ar isobars (e.g., Belshaw et al. 2000) will require significantly more seawater.

One of the key issues for future studies is to determine if the 0.5‰ difference in δ^{56}Fe values between aerosols and hydrothermal fluids is preserved as these materials are dispersed and changed by interaction with seawater. A complete knowledge of the Fe isotope composition of these Fe pools may allow a link to be established between Fe seeding in the open oceans by increased delivery of Fe from aerosols and its possible link to global climate change. For example, Beard et al. (2003a) proposed that the increase in δ^{56}Fe values for the Atlantic Ocean Fe-Mn crust analyzed by Zhu et al. (2000) may have been due to initiation of Northern Hemisphere glaciation at 2.6 Ma, thus increasing detrital loads to the open oceans by ice rafting or by an increase in atmospheric fluxes due to increased mechanical weathering at the expense of chemical weathering. Increases in atmospheric Fe flux delivered to seawater from aerosol particles has been inferred to have occurred during the last glacial maximum (Kumar et al. 1995; Edwards et al. 1998), perhaps indicating that there is a link to glaciation and the content of Fe delivered to the open oceans (Martin 1990). Based on the simple two-component mixing model in Figure 20, such a phenomenon should manifest itself as an increase in δ^{56}Fe values in ocean water as the quantity of atmospheric Fe delivered to the oceans increases. If this model is correct we would expect significant Fe isotope excursions to have occurred in the oceans during the Paleozoic Gondwanaland glaciations (Hambrey and Harland 1981) as well as the large-scale Neoproterozoic glaciations; indeed in the Neoproterozoic (the "snowball Earth" model; Kirschvink 1992), we would expect very large deviations in δ^{56}Fe values for ocean water, possibly shifting to entirely MOR sources.

Banded iron formations

Individual layers in Late Archean to Early Proterozoic BIFs have variable Fe isotope compositions from −2.5 to +1.1‰. Iron isotope variations are correlated with mineralogy, where the δ^{56}Fe values increase in the order: pyrite - Fe carbonates - hematite - magnetite (Johnson et al. 2003a). The δ^{56}Fe values for adjacent magnetite and siderite layers in the Kuruman Iron Formation (South Africa) and the Dales Gorge member of the Brockman Iron Formation (Western Australia) are generally correlated (Fig. 21). The magnetite-siderite fractionations from adjacent layers vary from zero to approximately +1‰, over a 2‰ range in δ^{56}Fe values for siderite. Although it is unknown if magnetite and siderite formed in isotopic equilibrium in many of these samples, it seems unlikely that the relatively large magnetite-siderite fractionations predicted by Polyakov and Mineev (2000) are accurate (Fig. 21). These observations highlight the utility of using natural mineral assemblages to constrain isotopic fractionation factors. Although the magnetite-siderite fractionation factor proposed by Polyakov and Mineev (2000) appears to be correct in sign, it appears to be too large by several per mil at low temperatures. In the case of Fe carbonates, there may be additional complexity due to cation substitution, where ankerite or impure siderite is predicted to have lower δ^{56}Fe values relative to pure siderite (Polyakov and Mineev 2000). Because natural siderites generally contain other cations such as Ca, Mg, and Mn (e.g., Klein and Beukes 1989), the measured $\Delta^{56}Fe_{Magnetite\text{-}Siderite}$ fractionations in Figure 21 are likely to be maximums, and therefore the discrepancy between the measured and predicted fractionations is probably even larger than is illustrated.

Johnson et al. (2003a) presented a preferred set of fluid-mineral fractionation factors based in part on data from BIFs, and proposed that the equilibrium $\Delta^{56}Fe_{Magnetite\text{-}Siderite}$ fractionation is

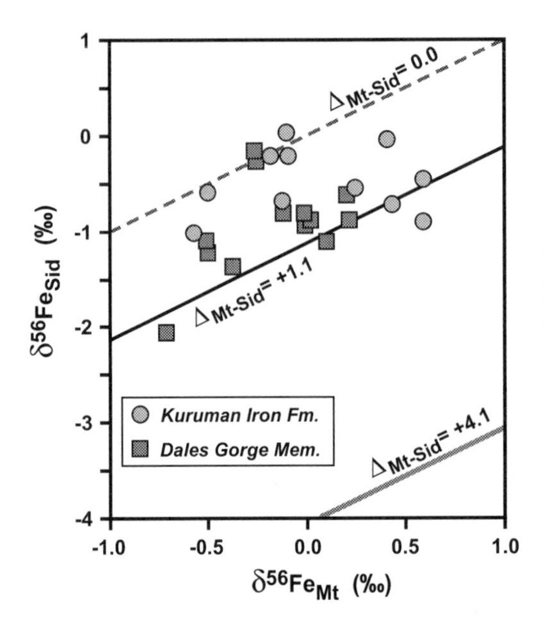

Figure 21. Plot of the measured $\delta^{56}Fe$ value of pairs of magnetite and siderite layers from Banded Iron Formations. Data from Johnson et al. (2003a) and C. Johnson and B. Beard unpublished.

+1.1‰ at 100°C, based on magnetite-siderite pairs where magnetite had positive $\delta^{56}Fe$ values, and many data scatter near this line (Fig. 21). Magnetite that had low $\delta^{56}Fe$ values were thought to have precipitated from unusually low $\delta^{56}Fe$ fluids, and were not considered in deriving the preferred set of fractionation factors. As discussed in the next chapter (Johnson et al. 2004), magnetite that has low $\delta^{56}Fe$ values is now thought to reflect precipitation from aqueous Fe(II) that was produced by bacterial Fe(III) reduction, whereas magnetite that has higher $\delta^{56}Fe$ values seems likely to have precipitated in equilibrium with aqueous Fe(II) that was derived from MOR fluids. These conclusions are possible given the new fluid-mineral fractionation factors that are available (Table 2; Johnson et al. 2004), which did not exist at the time Johnson et al. (2003a) published their study. $\delta^{56}Fe$ values for primary hematite in the Kuruman Iron Formation overlap those of magnetite, and Johnson et al. (2003a) speculated that oxides which have high $\delta^{56}Fe$ values might reflect incomplete oxidation of aqueous Fe(II) by photosynthetic Fe(II) oxidation. It is also possible, however, that abiotic oxidation may produce similar $\delta^{56}Fe$ values (Croal et al. 2004). As highlighted earlier in this chapter, interpretation of the $\delta^{56}Fe$ values of natural ferric oxide/hydroxides likely depends on an understanding of the relative contributions of equilibrium and kinetic effects that may have variable contributions in specific settings.

CONCLUSIONS

The field of Fe isotope geochemistry is rapidly becoming established. Numerous laboratories are making Fe isotope measurements and the emerging picture is that redox processes by both abiologic and biologic means (Johnson et al. 2004) produce some of the most significant Fe isotope variations in nature. The first-order variations in the rock record are now relatively well established by hundreds of analyses of natural samples including many types of low- and high-temperature rocks. Terrestrial igneous rocks and low-C_{org} clastic sedimentary rocks have a narrow range in $\delta^{56}Fe$ values, reflecting the homogenizing process of magmatism and conservative behavior of Fe during weathering under oxygenated surface conditions. Organic-C rich sedimentary rocks and chemically precipitated sediments have variable $\delta^{56}Fe$ values, reflecting partitioning between minerals and Fe(II)-rich fluids. There are also some significant variations in the Fe isotope composition of extraterrestrial samples where vaporization and condensation of Fe can produce significant Fe isotope variability, particularly in individual chondrules (Kehm et al. 2003), as well as the lunar regolith (Wiesli et al. 2003a). Moreover, there may be important mass-dependent Fe isotope differences between planetary bodies associated with Fe vaporization during planetary accretion (Poitrasson et al. 2002). In some mantle-derived rocks, inter-mineral differences in Fe isotope compositions appear to reflect open-system behavior for olivine, orthopyroxene, and clinopyroxene from spinel peridotites which has lead to significant Fe isotope variability (~0.5‰ in $^{56}Fe/^{54}Fe$). The best case for equilibrium fractionation of Fe isotopes at high temperatures appears to be between garnet and clinopyroxene. Isotopic variability in high-temperature environments, however, appears to be rarely expressed in crustal rocks apparently reflecting homogenization during magma generation, as recorded in the constant Fe isotope composition of a wide variety of igneous rocks ($\delta^{56}Fe = 0.00 \pm 0.05$‰, 1$\sigma$; Beard et al. 2003b), as well as most low-C_{org} clastic sedimentary rocks.

The ability to understand measured Fe isotope variations in nature, however, is limited because the database of well constrained Fe isotope fractionation factors is small. Determination of equilibrium fractionation factors provides the broadest application to different environments, although kinetic effects are also important, particularly in cases where reactions in nature may be rapid, such as oxidation of aqueous Fe(II). It is critical that experimental investigations take a mechanistic approach to determining isotopic fractionation, given the fact that fluid-mineral fractionations at low temperature must rely on synthesis methods. Additionally, as a check on these laboratory investigations, it is important to conduct studies on modern natural samples where the Fe isotope compositions of the fluids and minerals may be determined. The analysis of fluids that have low Fe contents, but overall high ionic strengths (such as pore fluids or seawater) will be analytically challenging, requiring superb chemical purification techniques and mass spectrometry methods that only require nanogram quantities of Fe for a mass analysis.

ACKNOWLEDGMENTS

We thank Ashley Hubbard, Rebecca Poulson, Silke Severman, Joe Skulan, Sue Welch, René Wiesli, and Kosei Yamaguchi for their assistance in the lab and numerous helpful discussions. We thank three anonymous reviewers and Francis Albarède for their helpful comments. We acknowledge support from NSF and NASA for our Fe isotope research.

REFERENCES

Albarède F, Beard BL (2004) Analytical methods for non-traditional isotopes. Rev Mineral Geochem 55: 113-151

Alexander CMO'D, Wang J (2001) Iron isotope in chondrules: implications for the role of evaporation during chondrule formation. Meteoritics Planet Sci 36:419-428

Anbar AD (2004) Iron stable isotopes: beyond biosignatures. Earth Planet Sci Lett 217:223-236

Anbar AD, Roe JE, Barling J, Nealson KH (2000) Nonbiological fractionation of iron isotopes. Science 288: 126-128

Anbar AD, Jarzecki AA, Spiro TG (2004) Theoretical investigation of iron isotope fractionation between $Fe(H_2O)_6^{3+}$ and $Fe(H_2O)_6^{2+}$: implications for iron stable isotope geochemistry. Geochim Cosmochim Acta, in press

Archer C, Vance DA (2002) Large fractionations in Fe, Cu, and Zn isotopes associated with Archean microbially mediated sulphides. Geochim Cosmochim Acta 66:A26

Arthur MA, Anderson TF, Kaplan IR (1983) Stable isotopes in sedimentary geology. SEPM short course, 10, 432 pp

Beard BL, Johnson CM (1999) High Precision iron isotope measurements of terrestrial and lunar material. Geochim Cosmochim Acta 63:1653-1660
Beard BL, Johnson CM (2004) Inter-mineral Fe isotope variations in mantle derived rocks and implications for the Fe geochemical cycle. Geochim Cosmochim Acta, in press
Beard BL, Johnson CM, Cox L, Sun H, Nealson KH, Aguiliar C (1999) Iron isotope biosignatures. Science 285:1889-1892
Beard BL, Johnson CM, Von Damm KL, Poulson RL (2003a) Iron isotope constraints on Fe cycling and mass balance in the oxygenated Earth oceans. Geology 31:629-632
Beard BL, Johnson CM, Skulan JL, Nealson KH, Cox L, Sun H. (2003b) Application of Fe isotopes to tracing the geochemical and biological cycling of Fe. Chem Geol 195:87-117
Belshaw NS, Zhu XK, Guo Y, O'Nions RK (2000) High precision measurement of iron isotopes by plasma source mass spectrometry. Int J Mass Spectrom 197:191-195
Berger A, von Blanckenburg F (2001) High-temperature fractionation of Fe isotopes. Eos Trans. Am Geophys Union 82:V21A-0958
Bergquist B, Boyle E (2002) Iron isotopic composition of Amazon River. EOS Trans. Am Geophys Union 83: OS12C-0290
Bernatowicz TJ, Nichols RH, Hohenberg CM (1994) Origin of amorphous rims on lunar soil grains. Lunar Planet Sci Conf XXV, 105-106
Beukes NJ, Klein C (1992) Models for Iron-formation deposition. In: The Proterozoic Biosphere: A multidisciplinary study. Schopf JW, Klein C (eds) Cambridge University Press, New York, p 147-151
Birck JL (2004) An overview of isotopic anomalies in extraterrestrial materials and their nucleosynthetic heritage. Rev Mineral Geochem 55:25-64
Brantley SL, Liermann L, Bullen TD (2001) Fractionation of Fe isotopes by soil microbes and organic acids. Geology 29: 535-538
Bruland KW, Odonat JR, Hutchins DA (1991) Interactive influences of bioactive trace metals on biological production in oceanic waters. Limnol Oceanography 36:1555-1577
Bullen TD, White AF, Childs CW, Vivet DV, Schultz MS (2001) Demonstration of significant abiotic iron isotope fractionation in nature. Geology 29:699-702
Bullen TD, White AF, Childs CW (2003) Comment on "Isotopic fractionation between Fe(III) and Fe(II) in aqueous solutions" by Clark Johnson et al., [Earth Planet Sci Lett 195 (2002) 141-153]. Earth Planet Sci Lett 206:229-232
Butler IP, Archer C, Rickard D, Vance D, Oldroyd A (2003) Fe isotope fractionation during Fe(II) monosulfide precipitation from aqueous Fe solutions at pH 8 and ambient temperature. Geochim Cosmochim Acta 67:A51
Canfield DE (1997) The geochemistry of river particulates from the continental USA: major elements. Geochim Cosmochim Acta 61:3349-3365
Chu N-C, Johnson CM, Beard BL, German CR, Nesbitt RW, Usui A (2003) Secular Fe isotope variations in the central Pacific Ocean. Geochim Cosmochim Acta 67:A66
Clayton DD (1999) Radiogenic iron. Meteor Planet Sci 34:A145-A160
Clayton RN (1993) Oxygen isotopes in meteorites. Ann Rev Earth Planet Sci 21:115-149
Clayton RN, Mayeda TK, Hurd JM (1974) Loss of oxygen, silicon, sulfur, and potassium from the lunar regolith. Proc Lunar Sci Conf 5:1801-1809
Criss RE (1999) Principles of stable isotope distribution. Oxford Univ Press, New York
Croal LR, Johnson CM, Beard BL, Newman DK (2004) Iron isotope fractionation by anoxygenic Fe(II)-phototrophic bacteria. Geochim Cosmochim Acta: in press
Duce RA, Tindale NW (1991) Atmospheric transport of iron and its deposition in the ocean. Limnol Oceanography 36:1715-1726
Edwards R, Sedwick PN, Morgan V, Boutron CF, Hong S (1998) Iron in ice cores from Law Dome, east Antarctica: implications for past deposition of aerosol iron. Ann Glaciol 27:365-370
Eiler JM (2001) Oxygen isotope variations of basaltic lavas and upper mantle rocks. Rev Mineral Geochem 43:319-364
Epstein S, Taylor HP (1971) O^{18}/O^{16}, Si^{30}/Si^{28}, D/H, and C^{13}/C^{12} ratios in lunar samples. Proc Lunar Sci Conf 2:1421-1441
Esat TM, Taylor SR (1992) Magnesium isotope fractionation in lunar soils. Geochim Cosmochim Acta 56: 1025-1031
Ewers WE (1983) Chemical factors in the deposition and diagenesis of banded iron-formation. In: Iron Formations: Facts and Problems. Trendall AF, Morris RC (eds) Elsevier, Amsterdam, p 491-512
Fantle MS, DePaolo DJ (2002) The isotopic composition of continental iron and implications for the global Fe cycle. EOS Trans Am Geophys Union 83:V22B-1234

Grandstaff DE (1980) Origin of uraniferous conglomerates at Elliot Lake, Canada and Witwatersrand, South Africa, implications for oxygen in the Precambrian atmosphere. Precambrian Research 13:1-26

Gregory RT, Criss RE (1986) Isotopic exchange in open and closed systems. Rev Mineral 16:91-128

Hambrey MJ, Harland WB (1981) Earth's Pre-Pleistocene Glacial Record, Cambridge Univ Press, Cambridge

Hapke B (2001) Space weathering from Mercury to the asteroid belt. J Geophys Res 106:10039-10073

Hapke B, Cassidy W, Wells E (1994) Vapor deposits in the lunar regolith: Technical comment. Science 264: 1779

Herzog GF, Xue S, Hall GS, Nyquist LE, Shih C-Y, Wiesmann H, Brownlee DE (1999) Isotopic and elemental composition of iron, nickel, and chromium, in type I deep-sea spherules: Implications for origin and composition of the parent micrometeoroids. Geochim Cosmochim Acta 63:1443-1457

Holland HD (1994) Early Proterozoic atmospheric change. In: Early Life on Earth, Noble Symposium No 84. Bengston S (ed), Columbia University Press, p. 237-244

Humayun M, Clayton RN (1995) Precise determination of the isotopic composition of potassium: application to terrestrial rocks and lunar soils. Geochim Cosmochim Acta 59:2115-3130

Johnson CM, Beard BL (1999) Correction of instrumentally produced mass fractionation during isotopic analysis of Fe by thermal ionization mass spectrometry. Int J Mass Spectrom 193:87-99

Johnson CM, Skulan JL, Beard BL, Sun H, Nealson KH, Braterman PS (2002) Isotopic fractionation between Fe(III) and Fe(II) in aqueous solutions. Earth Planet Sci Lett 195:141-153

Johnson CM, Beard BL, Beukes NJ, Klein C, O'Leary JM (2003a) Ancient geochemical cycling in the Earth as inferred from Fe isotope studies of banded iron formations from the Transvaal Craton. Contrib Mineral Petrol 144:523-547

Johnson CM, Beard BL, Braterman PS, Welch SA (2003b) Reply to comment on 'Isotopic fractionation between Fe(III) and Fe(II) in aqueous solutions' by Thomas D. Bullen, Arthur F. White and Cyril W. Childs. Earth Planet Sci Lett 206:233-236

Johnson TM, Bullen TD (2004) Mass-dependent fractionation of selenium and chromim isotopes in low-temperature environments. Rev Mineral Geochem 55:289-317

Johnson KS, Gordon RM, Coale KH (1997) What controls dissolved iron concentrations in the world ocean? Mar Chem 57:137-161

Kasting JF, Liu SC, Honahue TM (1979) Oxygen levels in the prebiological atmosphere. J Geophys Res 84: 3097-3107

Kehm K, Hauri EH, Alexander CMO'D, Carlson RW (2003) High precision iron isotope measurements of meteoritic material by cold plasma ICP-MS. Geochim Cosmochim Acta 67:2879-2891

Keller LP, McKay DS (1993) Discovery of vapor deposits in the lunar regolith. Science 261:1305-1307

Kirschvink JL (1992) Late Proterozoic low-latitude global glaciation: the snowball Earth. In: The Proterozoic Biosphere: A multidisciplinary study. Schopf JW, Klein C (eds) Cambridge University Press, New York, p 147-151

Klein C, Beukes NJ (1989) Geochemistry and sedimentology of a facies transition from limestone to iron-formation deposition in the early Proterozoic Transvaal Supergroup, South Africa. Econ Geol 84:1733-1774

Kumar N, Anderson RF, Mortlock RA, Froelich PN, Kubik P, Dittrich-Hannen B, Suter M (1995) Increased biological productivity and export production in the glacial southern ocean. Nature 378:675-680

Malinovsky D, Stenberg A, Rodushkin I, Andren H, Ingri J, Öhlander B, Baxter DC (2003) Performance of high resolution MC-ICP-MS for Fe isotope ratio measurements in sedimentary geologic materials. J Anal At Spectrom 18:687-695

Mandernack KW, Bazylinski DA, Shanks WC, III, Bullen TD (1999) Oxygen and iron isotope studies of magnetite produced by magnetotactic bacteria. Science 285:1892-1896

Martin JH (1990) Glacial-interglacial CO_2 change: the iron hypothesis. Paleooceanography 5:1-13

Martin JH (1992) Iron as a limiting factor in oceanic productivity. In: Primary productivity and biogeochemical cycles in the sea. Falkowski P, Woodhead AD (eds) Plenum, New York, p 123-127

Martin JH, Fitzwater EE (1988) Iron limits phytoplankton growth in the northeast Pacific subarctic. Nature 331:341-343

Martin JH, Gordon RM (1988) Northeast Pacific iron distributions in relation to phytoplankton productivity. Deep-Sea Res 34:267-285

Martin JH, Gordon RM, Fitzwater S, Broenkow WW (1989) VERTEX: Phytoplankton/iron studies in the Gulf of Alaska. Deep-Sea Res 36:649-680

Martin JH, Coale KH, Johnson KS, et al. (1994) Testing the iron hypothesis in ecosystems of the Equatorial Pacific Ocean. Nature 371:123-129

Matthews A, Zhu X-K, O'Nions RK (2001) Kinetic iron stable isotope fractionation between iron (-II) and (-III) complexes in solution. Earth Planet Sci Lett 192:81-92

Matthews A, Morgans-Bell HS, Emmanuel S, Jenkyns HC, Erel Y, Halicz L (2004) Controls on iron-isotope fractionation in organic-rich sediments. Geochim Cosmochim Acta, in press

Millero FJ, Yao W, Aicher J (1995) The speciation of Fe(II) and Fe(III) in natural waters. Marine Chemistry 50:21-39

Morris RV (1976) Surface exposure indices of lunar soils: a comparative FMR study. Proc Lunar Planet Sci Conf 7:315-335

Mullane E, Russel SS, Gounelle M (2002) Iron isotope fractionation within a differentiated asteroidial sample suite. 65th Annual Meteoritical Society Meeting, abstract number 5157

Mullane E, Russel SS, Gounelle M, Mason TFD (2003a) Iron isotope composition of Allende and Chainpur chondrules: effects of equilibration and thermal history. Lunar Planet Sci Conf XXXIV, abstract number 1027

Mullane E, Russell SS, Gounelle M, Mason TFD (2003b) Iron isotope composition of Allende matrix, CAIs and chondrules. 66th Annual Meteoritical Society Meeting, abstract number 5117

Mullane E, Russell SS, Gounelle M, Mason TFD, Weiss D, Coles B (2003c) Magmatic and impact processing on the HED parent body: Effects on iron isotope signatures. Geochim Cosmochim Acta 67:A311

O'Nions RK, Frank M, von Blanckenburg F, Ling H-F (1998) Secular variation of Nd and Pb isotopes in ferromanganese crusts from the Atlantic, Indian, and Pacific Oceans. Earth Planet Sci Lett 155:15-28

Poitrasson F, Halliday AN, Lee D, Levasseur S, Teutsch N (2002) Origin of the Moon unveiled by its heavy iron isotope composition. Eos Trans Am Geophys Union 83:P11A-0355

Polyakov VB, Mineev SD (2000) The use of Mössbauer spectroscopy in stable isotope geochemistry. Geochim Cosmochim Acta 64:849-865

Poulson RL, Beard BL, Johnson CM (2003) Investigating isotopic exchange between dissolved aqueous and precipitated iron species in natural and synthetic systems. Geochim Cosmochim Acta 67:A382

Reynolds BC, Frank M, O'Nions RK (1999) Nd- and Pb-isotope time series from Atlantic ferromanganese crusts: implications for changes in provenance and paleocirculation over the last 8 Myr. Earth Planet Sci Lett 173:381-396

Roe JE, Anbar AD, Barling J (2003) Nonbiological fractionation of Fe isotopes: evidence of an equilibrium isotope effect. Chem Geol 195:69-85

Rouxel O, Dobbek N, Ludden J, Fouquet Y (2003) Iron isotope fractionation during oceanic crust alteration. Chem Geol 202:155-182

Rouxel O, Fouquet Y, Ludden JN (2004) Subsurface processes at the Lucky Strike Hydrothermal Field, Mid-Atlantic Ridge: Evidence from sulfur, selenium, and iron isotopes. Geochim Cosmochim Acta, submitted

Russell WA, Papanastassiou DA, Tombrello TA, Epstein S (1977) Ca isotope fractionation on the moon, Proc Lunar Sci Conf 8:3791-3805

Sands DG, Rosman KJR, Laeter de JR (2001) A preliminary study of cadmium mass fractionation in lunar soils. Earth Planet Sci Lett 186:103-111

Schauble EA (2004) Applying stable isotope fractionation theory to new systems. Rev Mineral Geochem 55: 65-111

Schauble EA, Rossman GR, Taylor HP (2001) Theoretical estimates of equilibrium Fe-isotope fractionations from vibrational spectroscopy. Geochim Cosmochim Acta 65:2487-2497

Severmann S, Larsen O, Palmer MR, Nüster J (2002) The isotopic signature of Fe-mineralization during early diagenesis. Geochim Cosmochim Acta 66:A698

Severmann S, German CR, Edmonds HN, Beard BL, Johnson CM (2003) The modification of hydrothermal Fe-isotopic signature during plume-processing. Geochim Cosmochim Acta 67:A424

Sharma M, Polizzotto M, Anbar AD (2001) Iron isotopes in hot springs along the Juan de Fuca Ridge. Earth Planet Sci Lett 194:39-51

Sholkovitz ER, Elderfield H, Szymczak R, Casey K (1999) Island weathering: riverine sources of rare earth elements to the western Pacific Ocean. Marine Chem 68:39-57

Skulan JL, Beard BL, Johnson CM (2002) Kinetic and equilibrium Fe isotope fractionation between aqueous Fe(III) and hematite. Geochim Cosmochim Acta 66:2995-3015

Sumner DY (1997) Carbonate precipitation and oxygen stratification in Late Archean seawater as deduced from facies and stratigraphy of the Gamohaan and Frisco Formations, Transvaal Supergroup, South Africa. Am J Sci 297:455-487

Taylor LA, Pieters CM, Keller LP, Morris RV, McKay DS (2001) Lunar Mare Soils: Space weathering and the major effect of surface-correlated nanophase Fe. J Geophys Res 106:27985-27999

Taylor PDP, Maeck R, De Bièvre P (1992) Determination of the absolute isotopic composition and Atomic Weight of a reference sample of natural iron. Int J Mass Spectrom Ion Proc 121:111-125

Taylor PDP, Maeck R, Hendricks F, De Bièvre P (1993) The gravimetric preparation of synthetic mixtures of iron isotopes. Int J Mass Spectrom Ion Proc 128:91-97

Thode HG, Rees CE (1976) Sulphur isotopes in grain size fractions of lunar soils, Proc Lunar Sci Conf 7: 459-468

Völkening J, Papanastassiou DA (1989) Iron isotope anomalies. Astrophys J 347:L43-L46

Walczyk T, von Blanckenburg F (2002) Natural iron isotope variations in human blood. Science 295:2065-2066

Wasson JT, Choi B-G (2003) Main-group pallasites: Chemical composition, relationship to IIIAB irons, and origin. Geochim Cosmochim Acta 67:3079-3096

Welch SA, Beard BL, Johnson CM, Braterman PS (2003) Kinetic and equilibrium Fe isotope fractionation between aqueous Fe(II) and Fe(III). Geochim Cosmochim Acta 67:4231-4250

Weyer S, Schwieters JB (2003)High precision Fe isotope measurements with high mass resolution MC-ICPMS. Int J Mass Spectrom 226:355-368

Wiesli RA, Beard BL, Taylor LA, Johnson CM (2003a) Space weathering processes on airless bodies: Fe isotope fractionation in the lunar regolith. Earth Planet Sci Lett 216:457-465

Wiesli RA, Beard BL, Johnson CM (2003b) Experimental determination of Fe isotope fractionation between aq. Fe(II), "green rust", and siderite. Geochim Cosmochim Acta 67:A533

Williams H, Lee D-C, Levasseur S, Teutsch N, Poitrasson F., Halliday AN (2002) Iron isotope composition of mid-ocean ridge basalts and mantle peridotites. Geochim Cosmochim Acta 66:A838

Yamaguchi KE, Beard BL, Johnson CM, Ohkouchi N, Ohmoto H (2003) Iron isotope evidence for redox stratification of the Archean oceans. Geochim Cosmochim Acta 67:550

Zhu XK, O'Nions RK, Guo Y, Reynolds BC (2000) Secular variation of iron isotopes in north Atlantic Deep Water. Science 287:2000-2002

Zhu XK, Guo Y, O'Nions RK, Young ED, Ash RD (2001) Isotopic homogeneity of iron in the early solar nebula. Nature 412:311-313

Zhu XK, Guo Y, Williams RJP, O'Nions RK, Matthews A, Belshaw NS, Canters GW, de Waal EC, Weser U, Burgess BK, Salvato B (2002) Mass fractionation processes of transition metal isotopes. Earth Planet Sci Lett 200:47-62

Reviews in Mineralogy & Geochemistry
Vol. 55, pp. 359-408, 2004
Copyright © Mineralogical Society of America

10B

Isotopic Constraints on Biogeochemical Cycling of Fe

Clark M. Johnson and Brian L. Beard

Department of Geology and Geophysics
University of Wisconsin
Madison, Wisconsin, 53706, U.S.A.

Eric E. Roden

Department of Biological Sciences
University of Alabama
Tuscaloosa, Alabama, 35487, U.S.A.

Dianne K. Newman

Division of Geological and Planetary Sciences
California Institute of Technology
Pasadena, California, 91125, U.S.A.

Kenneth H. Nealson

Department of Earth Sciences
University of Southern California
Los Angeles, California, 90089, U.S.A.

INTRODUCTION

Cycling of redox-sensitive elements such as Fe is affected by not only ambient Eh-pH conditions, but also by a significant biomass that may derive energy through changes in redox state (e.g., Nealson 1983; Lovely et al. 1987; Myers and Nealson 1988; Ghiorse 1989). The evidence now seems overwhelming that biological processing of redox-sensitive metals is likely to be the rule in surface- and near-surface environments, rather than the exception. The Fe redox cycle of the Earth fundamentally begins with tectonic processes, where "juvenile" crust (high-temperature metamorphic and igneous rocks) that contains Fe which is largely in the divalent state is continuously exposed on the surface. If the surface is oxidizing, which is likely for the Earth over at least the last two billion years (e.g., Holland 1984), exposure of large quantities of Fe(II) at the surface represents a tremendous redox disequilibrium. Oxidation of Fe(II) early in Earth's history may have occurred through increases in ambient O_2 contents through photosynthesis (e.g., Cloud 1965, 1968), UV-photo oxidation (e.g., Braterman and Cairns-Smith 1987), or anaerobic photosynthetic Fe(II) oxidation (e.g., Hartman 1984; Widdel et al. 1993; Ehrenreich and Widdel 1994). Iron oxides produced by oxidation of Fe(II) represent an important sink for Fe released by terrestrial weathering processes, which will generally be quite reactive. In turn, dissimilatory microbial reduction of ferric oxides, coupled to oxidation of organic carbon and/or H_2, is an important process by which Fe(III) is reduced in both modern and ancient sedimentary environments (Lovley 1991; Nealson and Saffarini 1994). Recent microbiological evidence (Vargas et al.

1529-6466/04/0055-010B$05.00

1998), together with a wealth of geochemical information, suggests that microbial Fe(III) reduction may have been one of the earliest forms of respiration on Earth. It therefore seems inescapable that biological redox cycling of Fe has occurred for at least several billion years of Earth's history.

Significant Fe isotope variations in nature are generally restricted to relatively low-temperature systems, including hydrothermal fluids and chemically precipitated minerals (Beard and Johnson 1999; Beard et al. 1999; Zhu et al. 2000; Bullen et al. 2001; Sharma et al. 2001; Beard et al. 2003b; Johnson et al. 2003; Matthews et al. 2004). Experiments investigating metabolic processing of Fe have shown that measurable Fe isotope fractionations are produced during dissimilatory Fe(III) reduction by bacteria (Beard et al. 1999, 2003a; Icopini et al. 2004; Johnson et al. 2004a), as well as anaerobic photosynthetic Fe(II) oxidation (Croal et al. 2004). In addition, the role of organic ligands in promoting mineral dissolution has been investigated in experiments (Brantley et al. 2001, 2004). Iron isotopes may also be fractionated in abiologic systems, including ion-exchange chromatography (Anbar et al. 2000; Roe et al. 2003), abiotic precipitation of ferric oxides or oxyhydroxides (Bullen et al. 2001; Skulan et al. 2002), and sorption of aqueous Fe(II) to ferric hydroxides (Icopini et al. 2004). The largest abiotic fractionations in experiment have been measured between Fe(III) and Fe(II) species in solution, and both kinetic (Matthews et al. 2001) and equilibrium (Johnson et al. 2002; Welch et al. 2003) fractionations have been observed.

In this chapter we review some of the major pathways of biological Fe metabolism, and discuss experimental studies that have investigated Fe isotope fractionations in several systems. We largely focus on the results from experiments because our goal is to understand Fe isotope fractionations at a mechanistic level. We find that experimental results reflect both kinetic and equilibrium effects, sometimes in the same experiment, making it important to scale laboratory results to the kinetic realm of natural environments. An important concept that we develop is consideration of the residence time of Fe in the various reservoirs involved in biological processing of Fe relative to the timescales of isotopic exchange; this, in addition to evaluating the isotopic mass balance in a system, form the underpinnings for understanding the measured Fe isotope fractionations in experimental studies. We compare Fe isotope fractionations that are produced in abiologic and biologic systems, looking to situations where isotopic variations produced by biochemical cycling of Fe is likely to be found in nature. Throughout the chapter, we highlight some areas of future research that are critical to our understanding of Fe isotope geochemistry and its uses in tracing biogeochemical cycling of Fe.

Nomenclature

We discuss Fe isotope variations using standard δ notation, in units of per mil (parts per 1000, or ‰):

$$\delta^{56}\mathrm{Fe} = \left(\frac{^{56}\mathrm{Fe}/^{54}\mathrm{Fe}_{\mathrm{SAMPLE}}}{^{56}\mathrm{Fe}/^{54}\mathrm{Fe}_{\mathrm{BULK\ EARTH}}} - 1 \right) 10^{3} \tag{1}$$

where $^{56}Fe/^{54}Fe_{BULK\ EARTH}$ is defined by a wide variety of terrestrial and lunar igneous rocks that have $\delta^{56}Fe = 0.00 \pm 0.05$‰ (Beard et al. 2003a). Data from different laboratories may be compared using the IRMM-14 standard, which on the Bulk Earth scale defined above, has a $\delta^{56}Fe$ value of −0.09‰ (Beard et al. 2003a). Other notations are used in the literature, and the reader is referred to the previous chapter for a detailed discussion (Chapter 10A, Beard and Johnson 2004). When describing Fe isotope fractionations between coexisting phases A and B, we follow the traditional definitions for the isotope fractionation factor $\alpha_{A\text{-}B}$, which, in the case of Fe, is defined as:

$$\alpha_{\text{A-B}} = \frac{\left({}^{56}\text{Fe}/{}^{54}\text{Fe}\right)_{\text{A}}}{\left({}^{56}\text{Fe}/{}^{54}\text{Fe}\right)_{\text{B}}} \tag{2}$$

Note that $\alpha_{A\text{-}B}$ may reflect either kinetic or equilibrium isotope partitioning between phases A and B. As discussed in the previous chapter, $\alpha_{A\text{-}B}$ for $^{56}Fe/^{54}Fe$ ratios typically varies between 0.997 and 1.003 (Chapter 10A; Beard and Johnson 2004). In general, we will describe isotopic fractionations using $\Delta_{A\text{-}B}$, following standard definitions:

$$\Delta_{\text{A-B}} = \delta^{56}\text{Fe}_{\text{A}} - \delta^{56}\text{Fe}_{\text{B}} \approx 10^3 \ln \alpha_{\text{A-B}} \tag{3}$$

Given the range in Fe isotope compositions measured so far, use of the approximation $\Delta_{A\text{-}B}$ to describe isotopic fractionations introduces an error of at most 0.02‰, which is not significant relative to analytical uncertainties.

BIOLOGICAL PROCESSING OF IRON

Detailed understanding of the pathways by which biological processing of Fe occurs is required before we may identify the potential steps in which Fe isotope fractionation may be produced. Organisms process Fe in three general ways: (1) lithotrophic or phototrophic metabolism, where Fe(II) acts as an electron donor for energy generation and/or carbon fixation (e.g., Emerson 2000; Straub et al. 2001); (2) dissimilatory Fe(III) reduction, where Fe(III) acts as an electron acceptor for respiration (e.g., Nealson and Saffarini 1994); and (3) assimilatory Fe metabolism, which involves uptake and incorporation into biomolecules (e.g., Lowenstam 1981) (Table 1). For lithotrophic, phototrophic, and dissimilatory Fe metabolism, electron transfer occurs between the cell and Fe that is bound to the cell surface or incorporated in the outer membrane. Bacteria may cycle Fe through valence changes when it is energetically favorable for them to do so, and where they are able to out-compete abiotic redox reactions and other metabolic pathways that would naturally occur under specific conditions. For example, photosynthetic Fe(II) oxidation is generally restricted to anaerobic environments because high ambient O_2 contents would convert aqueous Fe(II) to ferric oxides at a rate that is substantially faster than oxidation by bacteria.

Fe(II) oxidation

Aqueous Fe(II) is common in many subaerial and submarine hot springs, and mid-ocean ridge hydrothermal activity likely provided a large source of soluble and reactive Fe(II) to the ancient oceans (e.g., Ewers 1983; Bau et al. 1997; Sumner 1997). Terrestrial weathering in the Archean, if it occurred under conditions of low atmospheric O_2 (e.g., Holland 1984), would have provided additional sources of soluble Fe(II). Photosynthesis, which may or may not involve a direct role for Fe, is likely to have been a major process for oxidizing Fe(II) over Earth's history. Molecular evidence suggests that anoxygenic photosynthesis evolved quite early in Earth's history (e.g., Xiong et al. 2000), although finding definitive morphologic or isotopic biomarkers in the rock record that would indicate the existence of photosynthesis or cycling of elements by bacteria in the Hadean or Early Archean has been challenging (e.g., Schopf 1993; Mojzsis et al. 1996; Eiler et al. 1997; Brasier et al. 2002; Fedo and Whitehouse 2002; Schopf et al. 2002).

Modern layered microbial communities provide a view into biochemical redox cycling. Oxidation of Fe(II) through high O_2 contents generated by cyanobacteria generally occurs in the top most (photic) portions of microbial mats. The upper, near-surface layers of microbial mats that are rich in cyanobacteria are commonly underlain by purple and green anoxygenic photosynthetic bacteria that thrive in the IR photic spectra (Stahl et al. 1985; Nicholson

Table 1. Biogeochemical cycles for Fe.

Metabolic pathways/organisms	*Electron donor/ acceptor*	*Oxygen levels*	*Product*
Fe(II) oxidation			
Photosynthesis			
Cyanobacteria	H_2O (donor)	oxygenic	Ferric oxide and hydroxide precipitation due to high ambient O_2.
Phototrophic Fe(II) oxidation			
"Purple bacteria" (e.g., *Rhodomicrobium*) "Green bacteria" (e.g., *Chlorobium*)	Fe(II) (donor)	anaerobic	Ferric oxide and hydroxide precipitation under anoxic conditions.
Chemolithotrophs			
Acidophiles (e.g., *Thiobacillus*)		oxygenic (low pH)	Ferric oxide and hydroxide precipitation.
Neutrophiles (e.g. *Gallionella*, ES1/ES2/PV1, TW2)		oxygenic (neutral pH)	Aqueous Fe(III).
Nitrate reduction	NO_3^- (acceptor)	anaerobic (in dark)	Aqueous Fe(III); Ferric oxide and hydroxide precipitation.
Dissimilatory Fe(III) reduction			
Shewanella; Geobacter; many others	Fe(III) (acceptor)		Aqueous Fe(II), $FeCO_3$, $CaFe(CO_3)_2$, Fe_3O_4, FeOOH.
Assimilatory Fe metabolism			Fe biomolecules (siderophores, ferritin); magnetite (magnetotactic bacteria).

et al. 1987); if atmospheric O_2 contents were low early in Earth's history, anoxygenic photosynthetic bacteria likely thrived on the surface of layered microbial communities. Although ferric oxide precipitates formed by cyanobacterial activity would not be directly related to metabolic processing of Fe, anoxygenic photosynthetic bacteria would derive energy from Fe(II) oxidation. It is expected that geochemical cycling of Fe in the upper layers of layered microbial communities will be dominated by Fe(II) oxidation, both in the presence and absence of oxygen, where Fe(II) may be supplied by the sediment substrate through dissolution of silicates, carbonates, or sulfides. In the case of mats associated with thermal springs, or in shallow marine settings, dissolved Fe(II) may be supplied by thermal or upwelling marine waters. The magnitude of visible radiation, oxygen gradients, and iron supply would be among some of the factors that govern the rate of ferric oxide deposition in layered microbial communities (e.g., Pierson et al. 1999; Pierson and Parenteau 2000). The primary products of Fe(II) oxidation are ferric oxides or oxyhydroxides, and this process may provide a source of oxidants that may be exploited by anaerobic Fe(III)-reducing bacteria.

The importance of anaerobic photosynthetic Fe(II)-oxidation by bacteria was highlighted by Widdel et al. (1993), and these organisms may have played a major role in formation of ferric iron deposits in the absence of an oxygenated atmosphere. Experimental studies have used a variety of Fe(II) species as the electron donor, including $FeCO_3$, FeS, and $FeSO_4$, suggesting that these organisms may utilize a wide variety of Fe(II) sources. Some purple (non-sulfur) bacteria, such as *Rhodomicrobium vannielli*, become encased in ferric oxides, which ultimately limits their growth, and this has been taken as evidence that Fe(II) oxidation is a "side activity" in some cases (Heising and Schink 1998). In contrast, Fe(II)-oxidizing green bacteria such as *Chlorobium ferrooxidans* appear to precipitate ferric oxides that fall away from the cell and form ferric oxide deposits outside the cell region (Heising et al. 1999). Some Fe(III) may remain in solution, possibly bound by "iron solubilizing compounds" that are postulated to be excreted by the cells (Straub et al. 2001), although this hypothesis has not been confirmed in recent experimental studies (Kappler and Newman 2004).

Several other pathways for bacterially-related Fe(II) oxidation exist. In both circumneutral and low-pH hydrothermal systems, chemolithotrophic iron-oxidizing microorganisms may thrive under aerobic conditions (e.g., Emerson 2000; Ehrlich 1996). In addition, Fe(II)-oxidizing metabolism that involves NO_3^- as the terminal electron acceptor may occur in the absence of light (e.g., Straub et al. 2001). Nitrate-reducing bacteria may grow with or without Fe(II) as the electron donor where other compounds such as acetate may supply electrons (e.g., Straub et al. 1996). Although it is possible that anaerobic nitrate-reducing bacteria may have been a globally important means by which Fe(II) oxidation occurred in dark conditions in the absence of photosynthesis (e.g., Benz et al. 1998; Straub et al. 1996, 2001), it is not clear where the abundant nitrate needed for this process would come from.

Fe(III) reduction

Dissimilatory reduction of ferric oxide/hydroxide minerals such as hematite, goethite, and ferrihydrite occurs by a number of pathways (e.g., Lovley 1987; Nealson and Myers 1990). The occurrence of reductase components such as *c*-type cytochromes or proteins in the outer cell membranes (e.g., Gaspard et al. 1998; Magnuson et al. 2000; Myers and Myers 1993, 2000; Beliaev et al. 2001; DiCristina et al. 2002), although debated for some species (Seeliger et al. 1998; Lloyd et al. 1999), has been taken as evidence that direct contact between microorganisms and ferric oxide substrates is required for Fe(III) reduction. Experiments in which ferric substrates were physically isolated from the cells support the apparent requirement of direct contact for the strict anaerobes *Geobacter sulfurreducens* and *Geobacter metallireducens* (e.g., Lloyd et al. 1999; Nevin and Lovley 2000, 2002a,b). A number of workers, however, have proposed that Fe(III) may be solubilized prior to reduction (by Fe(III) chelating compounds) or reduced by redox-active electron shuttle compounds such as excreted quinones, phenazines, or natural humic substances (e.g., Lovley et al. 1996; Newman and Kolter 2000; Hernandez and Newman 2001; Shyu et al. 2002; Hernandez et al. 2004), raising the possibility that direct contact between microorganism and ferric iron substrate may not be required in all cases. Although the energy required to secrete large quantities of chelators or electron shuttle compounds may be an issue, many natural groundwater systems that are rich in organic carbon are also rich in humic substances that may act as shuttle compounds (e.g., Nevin and Lovley 2002b). Experiments in which the ferric substrate was physically isolated from the cells (Nevin and Lovley 2002a,b), as well as those which used adhesion-deficient strains of *Shewanella algae*, where physical attachment of cells to ferric iron substrate was significantly reduced as compared to other strains (Caccavo et al. 1997; Caccavo and Das 2002), have shown that ferric oxyhydroxide may be reduced by some species of bacteria without physical contact. Using genetics to knock out flagellar biosynthesis, Dubiel et al. (2002) also showed that attachment does not appreciably affect Fe(III)-reduction rates. Although ferric oxyhydroxides have been found inside the cell, it remains unclear how common this may be and what role such intracellular minerals may play in dissimilatory Fe(III) reduction (Glasauer et al. 2002).

In addition to production of aqueous Fe(II) under anaerobic conditions, the end-products of dissimilatory Fe(III)-reduction may include Fe carbonates (siderite and ankerite) and magnetite (e.g., Lovley et al 1987; Roden and Zachara 1996; Fredrickson et al. 1998; Roden et al. 2002; Zachara et al. 2002). Goethite and lepidocrocite may also be produced where aqueous Fe(II) catalyzes phase transformation of poorly crystalline ferric hydroxides such as ferrihydrite (Hansel et al. 2003; Glasauer et al. 2003). Banded iron formations (BIFs) have been proposed as repositories for magnetite and Fe carbonate that may have formed through dissimilatory Fe(III) reduction (Nealson and Myers 1990). In addition, magnetite produced by magnetotactic bacteria has been proposed to occur in rocks spanning ages from the Proterozoic (Gunflint Formation) to modern sediments (e.g., Frankel et al. 1979, 1981; Chang and Kirschvink 1985; Chang et al. 1989). The rate of magnetite production on a per cell basis by dissimilatory Fe(III)-reducing bacteria is ~5,000 times greater than that at which magnetotactic bacteria produce magnetite (Konhauser 1998). Nevertheless, despite the vastly more rapid rate of magnetite production by dissimilatory Fe(III)-reducing bacteria, with a few notable exceptions (Karlin et. al. 1987; Eggar-Gibbs et al. 1999), production of magnetite by these organisms is relatively rare in modern environments.

Iron carbonate is a common early diagenetic phase in sedimentary rocks and occurs as either disseminated fine-grained authigenic material in sandstones and mud-rocks or as large (cm-sized or greater) concretions in mudrocks (Mozley 1989; Mozley and Wersin 1992; Mozley and Burns 1993; Macquaker et al. 1997; Uysal et al. 2000; Raiswell and Fisher 2000). In general, Fe carbonate is formed in anoxic diagenetic environments where the rate of Fe(III) reduction is greater than that of bacterial sulfate reduction (Pye et al. 1990; Coleman 1993). The growth of siderite apparently requires biotic mediation to provide the source of both Fe(II) and carbonate through oxidation of organic carbon coupled to dissimilatory Fe(III) reduction (Coleman 1993; Coleman et al. 1993). Authigenic Fe carbonates typically have much higher Mg and Ca contents than can be accommodated if they formed in thermodynamic equilibrium at low temperatures (e.g., Pearson 1974; Curtis et al. 1986; Mozley and Carothers 1992; Laverne 1993; Baker et al. 1995; Hendry 2002). These metastable compositions are significantly different as compared to Fe(II)-carbonates in, for example, BIFs, which tend to have compositions that are similar to those expected for formation under equilibrium conditions (Machamer 1968; Butler 1969; Floran and Papike 1975; Klein 1974, 1978; Lesher 1978; Klein and Gole 1981; Klein and Beukes 1989; Beukes and Klein 1990). The apparently non-equilibrium major-element compositions of many authigenic carbonates suggests that such compositions may identify biologically-mediated carbonate precipitation, and may represent a "biomarker" in and of themselves.

Implications for Fe isotope fractionations

Phase transformations that occur through biological processing of Fe may produce isotopic fractionations that are distinct from those in "equivalent" abiologic systems if organic ligands produce distinct isotopic effects, or if biology produces kinetic or equilibrium conditions or Fe products that are not commonly found in abiologic systems. For example, of the four pathways for Fe(II) oxidation illustrated in Figure 1, those which involve Fe(II) as the electron donor (e.g., anoxic photosynthetic Fe(II) oxidation) are most likely to be intimately bound to the cell so that electron transfer is facilitated, raising the possibility that Fe isotope fractionations produced by these pathways may be distinct from those where oxidation occurs indirectly, such as through oxygenic photosynthesis. The large number of different pathways and related reactions

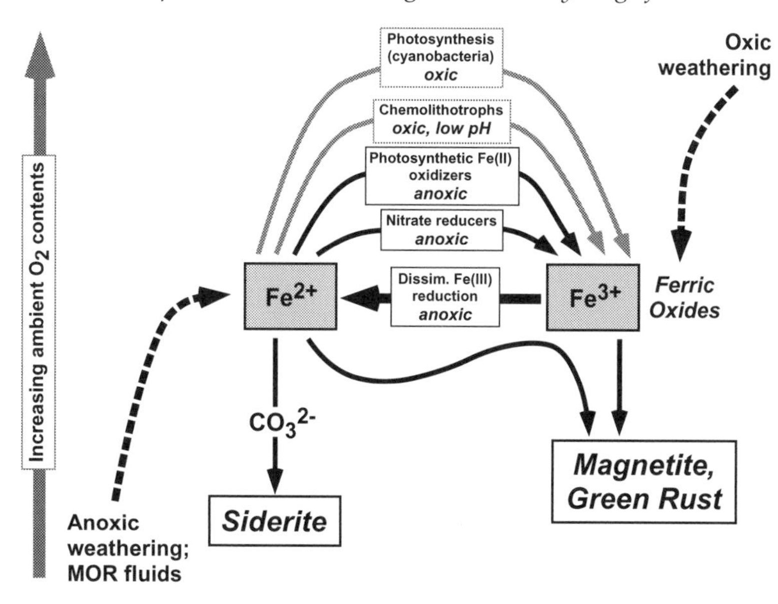

Figure 1. Schematic diagram of Fe redox cycling through biological processes. A large number of pathways are involved in dissimilatory Fe(III) reduction, as listed in Table 2. Processes that occur under oxic conditions are placed near the upper part of the diagram, and those that occur under anoxic conditions are placed in the lower part of the diagram. Major lithologic sources of Fe are noted for high and low oxygen environments.

involved in reduction of ferric hydroxides listed in Table 2 highlights the many opportunities to produce Fe isotope fractionations among the various reduction products, including aqueous Fe(II), sorbed Fe, and a wide variety of solid products. If some of these components are isotopically distinct and unlikely to be present in the absence of biology, it is possible to produce isotopic variations that may be confidently ascribed to biochemical cycling. Finally, the diverse pathways involved in Fe biochemical cycling may record distinct fractionations under equilibrium and kinetic conditions, requiring consideration of the rates of biochemical processing of Fe in laboratory experiments as compared to those expected in nature.

ISOTOPIC FRACTIONATION DURING DISSOLUTION AND PRECIPITATION OF MINERALS

Dissolution of minerals, such as may occur during dissimilatory Fe(III) reduction, or precipitation of new biominerals during reductive or oxidative processing of Fe, represent important steps in which Fe isotope fractionation may occur. We briefly review several experiments that have investigated the isotopic effects during mineral dissolution, as well as calculated and measured isotopic fractionations among aqueous Fe species and in fluid-mineral systems. In some studies, the speciation of aqueous Fe is unknown, and we will simply denote such cases as $Fe(III)_{aq}$ or $Fe(II)_{aq}$.

Congruent dissolution

The term "congruent dissolution" often refers to a process by which a mineral is dissolved in stoichiometric proportions into solution without formation of a new solid phase, and this usage is convenient for isotopic studies because it constrains the phases or components that may

Table 2. Biogenic Fe(III) reduction pathways.

Process	*Example Reactions*
HFO reductive dissolution[1]	$8Fe(OH)_3(s) + CH_3COO^- + 15H^+ = 8Fe^{2+} + 2HCO_3^- + 20H_2O$ (acetate oxidation)
	$4Fe(OH)_3(s) + CH_3CHOHCOO^- + 7H^+ = 4Fe^{2+} + CH_3COO^- + HCO_3^- + 10H_2O$ (lactate oxidation)
Ferric hydroxide phase conversion[2]	$Fe(OH)_3 + Fe^{2+} = \alpha\text{-}FeO{\cdot}OH + H_2O + Fe^{2+}$
	$Fe(OH)_3 + Fe^{2+} = \gamma\text{-}FeO{\cdot}OH + H_2O + Fe^{2+}$
Fe carbonate precipitation[3]	$Fe^{2+} + HCO_3^- = FeCO_3(s) + H^+$
	$Ca^{2+} + Fe^{2+} + 2HCO_3^- = CaFe(CO_3)_2(s) + 2H^+$
Ferrous hydroxide precipitation[4]	$Fe^{2+} + 2H_2O = Fe(OH)_2(s) + 2H^+$
Magnetite formation[5]	$2Fe(OH)_3 + Fe^{2+} = Fe_3O_4 + 2H_2O + 2H^+$
Mixed Fe(II)-Fe(III) hydroxide ("green rust") formation[6]	$2Fe(OH)_3 + 4Fe^{2+} + A^{2-}(OH^-, CO_3^{2-}, SO_4^{2-}, Cl^-) + 9H_2O = [Fe^{II}_4Fe^{III}_2(OH)_{12}][A^{2-}{\bullet}3H_2O] + 6H^+$
Fe(II) surface complexation (to HFO, magnetite, or "green rust" surfaces)[7]	$\equiv Fe^{III}OH + Fe^{2+} + H_2O = \equiv Fe^{III}OFe^{II}OH_2^+ + H^+$
$Fe(OH)_2$ surface precipitation (on HFO, magnetite, or "green rust" surfaces)[8]	$\equiv Fe^{III}OH + nFe^{2+} + (2n-1)H_2O = \equiv Fe^{III}O(Fe(OH)_{2(s)})_{n-1}{=}FeOH_2^+ + (2n-1)H^+$

Reduction pathways listed are those that occur for HFO as the terminal electron acceptor.

References:
[1] Fredrickson et al. (1998); Lovley and Phillips (1988).
[2] Zachara et al. (2002); Glasauer et al. (2003); Hansel et al. (2003).
[3] Fredrickson et al. (1998); Lovley and Phillips (1988); Postma (1977; 1981; 1982); Roden and Lovley (1993); Roden et al. (2002); Zachara et al. (2002).
[4] Mandal (1961).
[5] Ardizzone and Formaro (1983); Lovley et al. (1987); Lovley and Phillips (1988); Mann et al. (1989); Roden and Lovley (1993); Fredrickson et al. (1998); Zachara et al. (2002).
[6] Fredrickson et al. (1998); Roden et al. (2002); Zachara et al. (2002).
[7] Farley et al. (1985); Dzombak and Morel (1990); Fredrickson et al. (1998); Appelo et al. (2002); Roden and Urrutia (2002).
[8] Farley et al. (1985); Dzombak and Morel (1990); Haderlein and Pecher (1999); Roden and Urrutia (2002).

develop isotopically distinct compositions. Congruent dissolution may occur through complex interactions with heterogeneous surfaces and boundary layers (e.g., Jeschke and Dreybrodt 2002), but may be satisfactorily described as a progressive stripping of layers or sections from the mineral surface. If the mineral is isotopically homogeneous, no isotopic contrast should exist between solid and dissolved components. Although preferential extraction of one isotope might be envisioned to occur within surface monolayers or other surface sections, such a process would create an excess of the excluded isotope as successive surface sections are removed, balancing the overall isotope composition of the dissolved material to be equal to that of the bulk solid. Partial dissolution of μm- to 100 nm-size hematite crystals in HCl confirms that no measurable Fe isotope fractionation occurs during congruent partial dissolution (Fig. 2). The low pH of these experiments ensured that $Fe(III)_{aq}$ does not precipitate and form an additional phase that might have a different isotopic composition than Fe in solution. In the experiments illustrated in Figure 2, partial dissolution of hematite in HCl likely dissolves complete surface sections of hematite crystals, and is distinct from a leaching process that selectively removes a specific element from a portion of the crystal in non-stoichiometric proportions; the later process would be considered incongruent dissolution, and for the purposes of characterizing isotopically distinct reservoirs, this may be considered a new phase. In addition to the studies of hematite dissolution, Brantley et al. (2004) investigated abiotic dissolution of goethite in the presence of the siderophore desferrioxamine mesylate (DFAM), and found that dissolution occurred congruently, where the isotopic composition of Fe in solution was identical to that of the ferric hydroxide starting material.

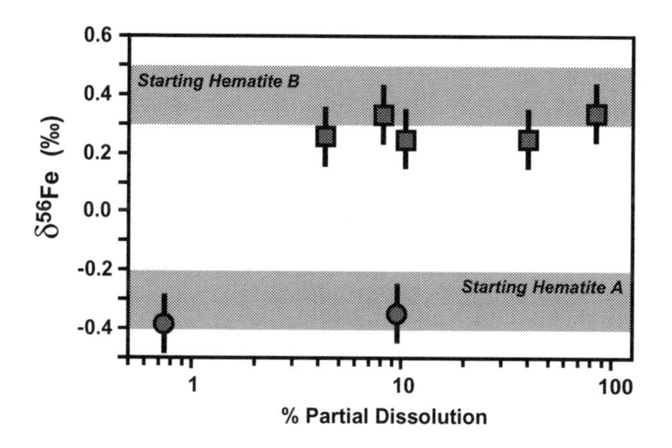

Figure 2. Isotopic effects of congruent partial dissolution of hematite. Within the 2σ error of the analyses, there is no significant Fe isotope fractionation over wide ranges of percent dissolution. Gray bars denote bulk composition (2σ) of the two hematite grains. Data from Skulan et al. (2002) and Beard et al. (2003a).

Incongruent dissolution

Significant Fe isotope fractionations may be produced during incongruent dissolution in the presence of organic ligands (Brantley et al. 2001, 2004). Dissolution of hornblende by siderophore-producing bacteria was shown to be enhanced relative to abiologic dissolution, and was interpreted to reflect preferential extraction of Fe from the mineral, accompanied by creation of an Fe-depleted layer in at least the upper 100 Å of the mineral surface (Kalinowski et al. 2000; Liermann et al. 2000). Because preferential mobilization of elements occurred, as shown by changing Fe/Si and Fe/Al ratios, the Fe-depleted layer represents a new phase. The extent of Fe leaching and apparent isotopic shift for aqueous Fe increased with increasing ligand association constant for abiotic experiments, where the largest effects (~0.3‰) were observed for the siderophore DFAM (Fig. 3). Larger isotopic contrasts between aqueous Fe and the initial hornblende were observed for leaching in bacterial cultures, on the order of 0.5 to 0.8‰ (Fig. 3).

Brantley et al. (2001) interpret Fe isotope fractionation to occur during hydrolysis of surface complexes, and to reflect a kinetic isotope fractionation that is dependent upon the strength of the organic ligand. The isotopic fractionations measured in the experiments of Brantley et al. (2001) reflect those produced during very small extents of dissolution or leaching, where at least 0.037% dissolution/leaching occurred based on estimation of the Fe content of the hornblende and reported Si/Al and Fe/Si ratios. Assuming spherical grains of an average diameter of 340 μm (Brantley et al. 2001), the minimum thickness of the leached layer may be calculated at ~210 Å. Based on isotopic mass-balance, the δ^{56}Fe value of the Fe-depleted layer would be relatively high, where the highest values would be associated with a relatively thin leached layer, and rapidly decrease if the thickness of the leach layer was greater. If the leached layer was 250 Å thick, its δ^{56}Fe value, based on mass balance, should be almost 5‰ higher that that of the Fe in solution; at 500 Å thickness, the leached layer should be ~1.5‰ higher than Fe in solution.

In a more detailed study, Brantley et al. (2004) observed non-stoichiometric dissolution of hornblende in abiotic experiments using acetate, oxalate, citrate, and siderophore (DFAM), where the δ^{56}Fe values of the aqueous Fe that was released became more negative in this order, relative to the starting material. The isotopic fractionations were modeled as a kinetic process during transport of Fe through the surface leached layer. This model predicts that significant changes in the δ^{56}Fe value for aqueous Fe will occur before the system reaches steady-state conditions during dissolution.

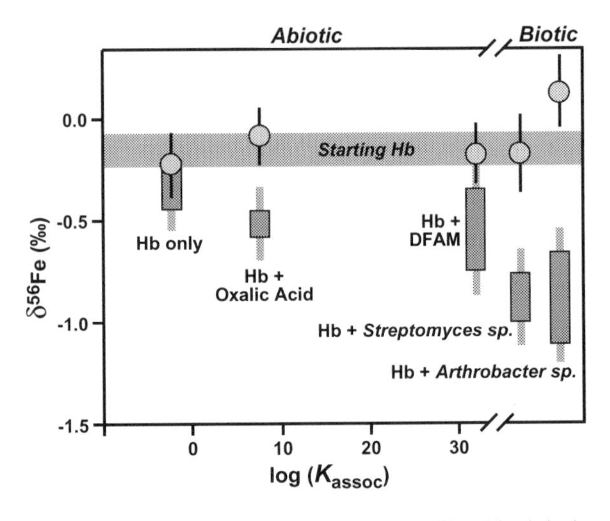

Figure 3. Effects on Fe isotope compositions of partial dissolution of hornblende in the presence of various organic ligands, as well as *Streptomyces* and *Arthrobacter* bacteria (identification of *Arthrobacter sp.* has been revised to *Bacillus sp.*; S. Brantley, pers. commun. 2004). For abiotic experiments, Fe isotope data are plotted relative to the association constant (K_{assoc}) for the indicated ligands. Fe isotope composition of starting hornblende shown in gray bar. Gray circles reflect Fe isotope composition of residual hornblende (error bars noted), and gray boxes denote range in δ^{56}Fe values measured for aqueous Fe (error bars shown) after partial dissolution. Partial dissolution occurs incongruently in these experiments, accompanied by formation of an Fe-poor leached layer in the hornblende. Data from Brantley et al. (2001).

The processes by which Fe isotope fractionations are associated with mineral dissolution and leaching by organic molecules, including those produced by bacteria, remains little explored, but, as Brantley et al. (2001, 2004) point out, are important components to understanding Fe isotope variations in natural weathering systems. At very small extents of dissolution or leaching (<0.1%; Brantley et al. 2001), it is difficult to constrain the reservoirs involved in producing isotopic fractionations because the Fe reservoirs that are complementary to the isotopically light aqueous Fe are difficult to identify, let alone analyze. Partial extractions of labile Fe components from soils show that these components may have δ^{56}Fe values that are significantly lower than the bulk solid, which is interpreted to reflect the effects of incongruent, non-stoichiometric dissolution of silicate minerals by Fe-organic complexes (Brantley et al. 2001, 2004). These results indicate that soils may contain isotopically variable components when soil formation involves organic substances that dissolve minerals incongruently. In terms of large-scale Fe cycling, it seems most likely that the isotopic variations of these labile components will be seen in the dissolved load of Fe in hydrologic systems; they are not apparently recorded in the bulk clastic detritus of eolian and fluvial systems, given the isotopic homogeneity of bulk clastic material (Beard et al. 2003b).

Isotopic fractionations among aqueous species and minerals in abiologic systems

Significant isotopic fractionations are observed among several Fe-bearing aqueous species and solid phases in abiologic systems that are pertinent to biochemical processing of Fe (Tables 1 and 2). Some of the largest Fe isotope fractionations occur between oxidized and reduced species (Polyakov and Mineev 2000; Schauble et al. 2001), where, for example, the experimentally determined equilibrium isotope fractionation between $[Fe^{III}(H_2O)_6]^{3+}$ and $[Fe^{II}(H_2O)_6]^{2+}$ ($\Delta_{Fe(III)aq-Fe(II)aq}$) at room temperature is +2.9‰ (Johnson et al. 2002; Welch et al. 2003). The equilibrium isotope fractionation between $[Fe^{III}(H_2O)_6]^{3+}$ and Fe_2O_3 (hematite) is estimated to be ~ −0.1‰ at room temperature (Skulan et al. 2002). Experimental determination of the equilibrium isotope fractionation between $[Fe^{II}(H_2O)_6]^{2+}$ and $FeCO_3$ (siderite) is estimated to be +0.5‰ at room temperature (Wiesli et al. 2004). It remains unknown if the large isotopic effects that are predicted for Ca, Mg, and Mn substitution into Fe carbonates (Polyakov and Mineev 2000), up to 1.5‰ at room temperature, will be confirmed by experimental studies in abiologic systems. Isotopic fractionation between $[Fe^{II}(H_2O)_6]^{2+}$ and Fe_3O_4 (magnetite) is predicted to be −4.2‰ at room temperature under equilibrium conditions, but this has yet to be confirmed experimentally; experimental confirmation of predicted fractionations is important, because in many cases the predicted Fe isotope fractionations based on theory have been shown to be significantly different from those determined by experiment. The reader is referred to the previous chapter (Chapter 10A; Beard and Johnson 2004) for a more detailed discussion of predicted and measured Fe isotope fractionation factors in abiologic systems. We will return to the fractionations in abiologic systems in later sections.

ISOTOPIC MASS BALANCE

In complex systems that involve multiple Fe-bearing species and phases, such as those that are typical of biologic systems (Tables 1 and 2), it is often difficult or impossible to identify and separate all components for isotopic analysis. Commonly only the initial starting materials and one or more products may be analyzed for practical reasons, and this approach may not provide isotope fractionation factors between intermediate components but only assess a net overall isotopic effect. In the discussions that follow on biologic reduction and oxidation, we will conclude that significant isotopic fractionations are likely to occur among intermediate components.

We illustrate some examples of the differences between apparent (measured) Fe isotope fractionations between starting materials and a single product and those that are postulated to occur among intermediate products in Figure 4. In calculating the isotopic mass balance during mineral leaching (Brantley et al. 2001), a leached layer thickness of 500 Å is assumed in calculating the δ^{56}Fe value for the leach layer, which results in a calculated fractionation between the leached layer and $Fe(II)_{aq}$ of +1.5‰ (Fig. 4). The correlation of increasing percent leaching with increasing ligand association constant (K_{assoc}) observed by Brantley et al. (2001, 2004) indicates that the isotopic mass balance between the leached layer and aqueous Fe was likely to have changed. Brantley et al. (2004) interpret their data to reflect kinetic fractionation during transport through a leached layer. It is, however, possible that the observed correlation between δ^{56}Fe values for $Fe(II)_{aq}$ and K_{assoc} (Fig. 3) in part reflects changes in the isotopic mass balance of the leached layer and aqueous Fe rather than a kinetic isotope fractionation that is dependent upon K_{assoc}, and demonstrates the importance of assessing isotopic mass balance in experiments.

In the sorption experiments of Icopini et al. (2004), the measured isotopic contrast between $Fe(II)_{aq}$ and the goethite starting material was −0.8‰ after Fe(II) had sorbed to the surface over 24 hours; in this case, the isotopic fractionation between sorbed Fe(II) and $Fe(II)_{aq}$ is not the 0.8‰ *measured* difference, but is approximately +2.1‰ based on an inferred δ^{56}Fe value for the sorbed component as calculated from Fe mass balance (Fig. 4), as was noted in that study. Measured differences in Fe isotope compositions between ferric oxide/hydroxide and $Fe(II)_{aq}$ during dissimilatory Fe(III) reduction and photosynthetic Fe(II) oxidation have been proposed to reflect fractionation between soluble Fe(II) and Fe(III) species, where the soluble Fe(III) component is postulated to be bound to the cell and is not directly measured (Beard et al. 2003a; Croal et al. 2004). In the case of dissimilatory Fe(III) reduction, assuming a static model simply for purposes of illustration, if 50% of the Fe in a pool that is open to

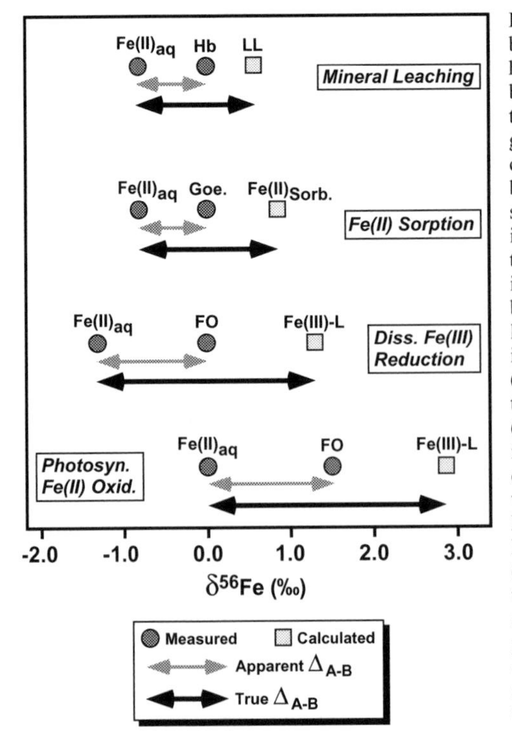

Figure 4. Examples of Fe isotope mass balance in experimental systems that have studied Fe isotope fractionation in biological systems or those that are pertinent to biogeochemical processing of Fe. Dark gray circles reflect measured Fe isotope compositions (where one component has been normalized to δ^{56}Fe=0.0) Light gray squares reflect calculated or inferred Fe isotope compositions for a component that is inferred from isotopic mass balance in the experiments, but which has not been analyzed directly. Data for mineral leaching from Brantley et al. (2001), and inferred δ^{56}Fe value for leached Fe layer ("LL") based on calculations described in the text. Data for Fe(II) sorption to goethite ("Goe.") from Icopini et al. (2004), and inferred δ^{56}Fe value for sorbed Fe based on calculations presented in that study. δ^{56}Fe value for ligand-bound Fe(III) ("Fe(III)-L") based on discussion of dissimilatory Fe(III) reduction of ferric oxide/hydroxide ("FO") in Beard et al. (2003a) and in the text, assuming a 50:50 mixture of $Fe(II)_{aq}$ and Fe(III)-L. δ^{56}Fe values for anaerobic photosynthetic Fe(II) oxidation based on inferred soluble Fe(III) component ("Fe(III)-L") as discussed in text and in Croal et al. (2004).

isotopic exchange occurs as ligand-bound Fe(III) (not measured) and the remaining 50% exists as $Fe(II)_{aq}$ (measured), then the 1.3‰ *measured* differences in isotopic compositions between the ferric oxide/hydroxide starting material and $Fe(II)_{aq}$ may actually reflect a +2.6‰ fractionation between the intermediate species of ligand-bound Fe(III) and the final $Fe(II)_{aq}$ product (Fig. 4). Similarly, the differences in isotopic compositions between $Fe(II)_{aq}$ starting material and ferric hydroxide precipitate formed during anaerobic photosynthetic Fe(II) oxidation may in fact reflect the combined effect of a large fractionation between ligand-bound Fe(III) and $Fe(II)_{aq}$, followed by a second fractionation between Fe(III) and the ferric hydroxide precipitate (Croal et al. 2004); the $\delta^{56}Fe$ value of the ligand-bound Fe(III) in such a model would be quite high (Fig. 4).

ISOTOPIC FRACTIONATION PRODUCED DURING DISSIMILATORY Fe(III) REDUCTION

Of the variety of ways in which Fe may be biologically processed (Table 1; Fig. 1), experimental investigation of dissimilatory Fe(III) reduction, which we will hereafter refer to as DIR, has received by far the most attention. Experimental studies of DIR have highlighted the importance of intermediate phases or species such as Fe(III) and Fe(II) surface complexes or poorly-crystalline solids (Table 2) to producing a range of end products. Ferrihydrite has been most commonly used as the terminal electron acceptor, which is often referred to as hydrous ferric oxide (HFO) in the biological literature, and we will follow this convention. More crystalline sources of electron acceptors such as hematite (Fe_2O_3), goethite (α-FeOOH), or lepidocrosite (γ-FeOOH) have been used as well. Reduction of Fe(III)-bearing clay minerals such as smectite is now recognized as an important component to DIR in natural systems (e.g., Kostka et al. 1996, 2002; Kim et al. 2004). Production of magnetite and Fe carbonates is readily identifiable in XRD spectra, but several non-aqueous, Fe(II)-bearing components that may be produced during DIR are more difficult to identify. Production of $Fe(OH)_2$(s) has been identified, as well as mixed Fe(II)-Fe(III) hydroxide, or "green rust" (Table 2), which may also contain carbonate ion in high-carbonate systems. Surface complexation or sorption of Fe(II) on oxide minerals, either ferric oxide/hydroxide starting materials or a product such as magnetite may represent a significant repository of labile Fe(II) during DIR (Table 2). Fredrickson et al. (1998) demonstrated that intermediate Fe(II) and Fe(III) phases may be identified through extraction from partially reacted HFO or reduced products in 0.5 M HCl, followed by careful assays of Fe(III) and Fe(II) contents.

A key component to understanding Fe isotope fractionations produced by DIR seems likely to be the fate of Fe(III) following dissolution of the ferric oxide/hydroxide starting material, given the large isotopic fractionations that occur between oxidized and reduced phases (Polyakov and Mineev 2000; Schauble et al. 2001; Johnson et al. 2002; Welch et al. 2003; Anbar et al. 2004). Experiments in which the ferric substrate was isolated from contact with bacterial cells using *Geothrix fermentans* (a strict anaerobe) or *Shewanella algae* (a facultative anaerobe) did not prevent Fe(III) reduction (Nevin and Lovley 2002a,b), and significant quantities of soluble Fe(III) were measured in these experiments. In both the *G. fermentans* and *S. algae* experiments, ~10% of the soluble Fe in solution existed as $Fe(III)_{aq}$ over the course of the 15–20 day experiments. The $Fe(III)_{aq}/Fe(II)_{aq}$ ratio was relatively constant over time, including time periods when total aqueous Fe was increasing, as well as decreasing at later time periods when precipitation of Fe(II)-bearing phases may have occurred (Nevin and Lovley 2002a,b). The significant quantity of $Fe(III)_{aq}$ in these experiments stands in contrast to parallel experiments using *Geobacter metallireducens*, which did not produce measurable quantities of $Fe(III)_{aq}$ when *G. metallireducens* was isolated from physical contact with the ferric substrate (Nevin and Lovley 2002b). It has been hypothesized that the $Fe(III)_{aq}$ produced in the *G. fermentans* and *S. algae* experiments reflects release of electron shuttling compounds and/or Fe(III) chelators (Nevin and Lovley 2002a,b). In experiments using *S. algae* where the cells were allowed free contact with the ferric substrate, however, no detectable $Fe(III)_{aq}$ existed (Johnson et al. 2004a). Nevertheless, the evidence seems to support a model where Fe(III) is first solubilized by an organic ligand, followed by reduction and release to the ambient aqueous solution and/or reacted to form a solid phase, although the quantity of "soluble" Fe(III) may be quite variable depending upon the organism and experimental conditions.

The process of DIR may be conceptualized as a series of reactions that occur at various rates and produce changing pools of Fe that are able to undergo isotopic exchange to variable degrees. Isotopic fractionations generally occur through uni-directional processes such as precipitation, or through exchange between pools of an element that are open to isotopic exchange over the timescales of the process. Based in part on the evidence that at least some species of bacteria solubilize Fe(III) during DIR, Beard et al. (2003a) hypothesized that the −1.3‰ fractionation in $^{56}Fe/^{54}Fe$ ratios measured between $Fe(II)_{aq}$ and ferric oxide/hydroxide substrate (the electron acceptor) may primarily reflect isotopic fractionation between pools of ligand-bound Fe(III) and Fe(II) that are open to isotopic exchange (Fig. 5). For the moment, we will ignore the effects of other reduced products such as Fe carbonate and magnetite, as well as the effects of Fe(II) that may be sorbed to solid phases, although these issues will be discussed in detail later. The first stage in which Fe isotope fractionation may occur during DIR is upon dissolution (Δ_1, Fig. 5). If dissolution occurs congruently, however, no Fe isotope fractionation is expected, as discussed above. The second step in which isotopic fractionation may occur is during reduction (Δ_2, Fig. 5), between ligand-bound or ambient Fe(II) and soluble Fe(III) that is delivered to the cell via an electron shuttle or extracellular protein. It is also possible that isotopic fractionation may occur between ligand-bound Fe(III) and the ferric substrate (Δ_3, Fig. 5), although isotopic exchange with solids at low temperatures will be quite slow. Finally, isotopic fractionation may occur between $Fe(II)_{aq}$ and Fe(II) that is sorbed to the ferric substrate (Δ_4, Fig. 5).

When viewed in a temporal context, the model illustrated in Figure 5 implicitly predicts changes in the proportions of ferric substrate, ligand-bound Fe(III), and the Fe(II) product over

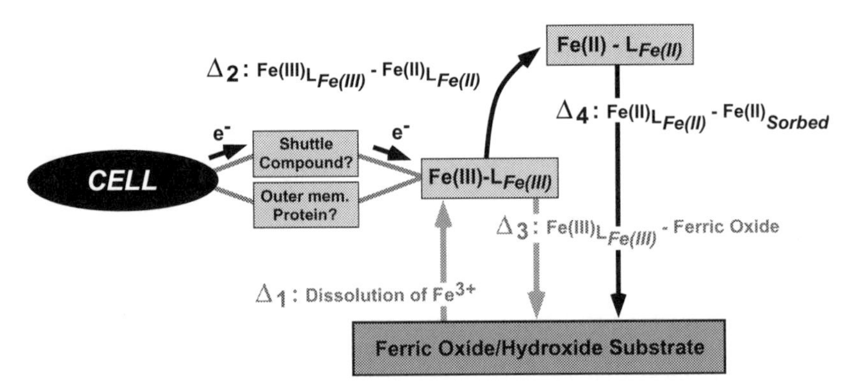

Figure 5. Possible pathways by which Fe isotopes may be fractionated during dissimilatory Fe(III) reduction (DIR). Dissolution, if it occurs congruently, is unlikely to produce isotopic fractionation (Δ_1). If Fe(II) is well complexed in solution and conditions are anaerobic, precipitation of new ferric oxides (Δ_3) is unlikely to occur. Significant isotopic fractionation is expected during the reduction step (Δ_2), possibly reflecting isotopic fractionation between soluble pools of Fe(III) and Fe(II). The soluble Fe(III) component is expected to interact with the cell through an electron shuttle compound and/or an outer membrane protein, and is not part of the ambient pool of aqueous Fe. Sorption of aqueous or soluble Fe(II) to the ferric oxide/hydroxide substrate (Δ_4) is another step in which isotopic fractionation may occur. Modified from Beard et al. (2003a).

time as reduction proceeds. For completeness, we hypothesize that two Fe(II) components exist, one bound to the cell immediately after reduction, and one that accumulates in the ambient pool of $Fe(II)_{aq}$. We therefore assume that the reduction process, where the initial ferric substrate is hematite (Fe_2O_3) or HFO/ferrihydrite (denoted as $Fe(OH)_3$ for simplicity), may be described as:

$$Fe_2O_3 \text{ or } Fe(OH)_3 \rightarrow \text{Fe(III)-L}_{\text{Fe(III)}} \rightarrow \text{Fe(II)-L}_{\text{Fe(II)}} \rightarrow \text{Fe(II)}_{\text{aq}} \quad (4)$$

where the ligand-bound Fe(III) is the terminal electron acceptor ("Fe(III)-$L_{Fe(III)}$"; Fig. 5), and Fe(II) is the immediate reduced product, possibly bound to organic ligands ("Fe(II)-$L_{Fe(II)}$"), and eventually released to a larger pool of Fe(II), including $Fe(II)_{aq}$. In terms of quantities that may be measured for their Fe isotope compositions, these would include the ferric substrate, $Fe(II)_{aq}$, and likely Fe(II)-$L_{Fe(II)}$ (although not a discrete phase) if this exists in the aqueous solution component. We assume that the Fe(III)-$L_{Fe(III)}$ component is *not* represented in a sample of the ambient aqueous solution, but instead is closely bound to or associated with the cells.

Fe(II) production during experiments investigating Fe isotope fractionation coupled to DIR generally followed a first-order rate law in terms of the rate of production of total Fe(II) (Beard et al. 1999, 2003a; Johnson et al. 2004a). Exponential regressions of the total Fe(II) contents (liquid + solid) versus time for these experiments produce high R^2 values (>0.95 for all but one case), and show that the initial reduction rates are a function of substrate, cell densities, and growth media, varying over a factor of 300 for these studies. The proportions of possible phases involved in DIR may be calculated using first-order kinetics, assuming a closed system and no initial product phases. The pertinent rate equations are:

$$\frac{d[Fe_2O_3]}{dt} = -k_1[Fe_2O_3] \quad (5)$$

$$\frac{d\left[\text{Fe(III)-L}_{\text{Fe(III)}}\right]}{dt} = k_1[Fe_2O_3] - k_2\left[\text{Fe(III)-L}_{\text{Fe(III)}}\right] \quad (6)$$

$$\frac{d\left[\text{Fe(II)-L}_{\text{Fe(II)}}\right]}{dt} = k_2\left[\text{Fe(III)-L}_{\text{Fe(III)}}\right] - k_3\left[\text{Fe(II)-L}_{\text{Fe(II)}}\right] \quad (7)$$

$$\frac{d[\text{Fe(II)}]}{dt} = k_3\left[\text{Fe(II)-L}_{\text{Fe(II)}}\right] \quad (8)$$

where we have represented the substrate as hematite and where k_1, k_2, and k_3 are first-order rate constants.

In this model, Fe isotope exchange is envisioned to most likely occur between "soluble" Fe(III) and Fe(II) components, such as Fe(III)-$L_{Fe(III)}$ and Fe(II)-$L_{Fe(II)}$ (Beard et al. 2003a), although only the $Fe(II)_{aq}$ component is measured. If this model is valid, a critical issue is whether isotopic equilibrium may be attained between these soluble pools of Fe, despite the changing reservoir sizes and fluxes that occur through Equations (5)–(8). As noted in the previous chapter (Chapter 10A; Beard and Johnson 2004), attainment of isotopic equilibrium will depend upon the elemental residence time in an Fe pool relative to the time required for isotopic exchange. We can define the residence time for Fe(III)-L, for example, using standard definitions, as:

$$\tau_{\text{Fe(III)-L}} = \frac{M_{\text{Fe(III)-L}}}{J_{\text{Fe(III)-L}}} \quad (9)$$

where $M_{\text{Fe(III)-L}}$ is the total moles of Fe in the Fe(III)-$L_{\text{Fe(III)}}$ reservoir, and $J_{\text{Fe(III)-L}}$ is the flux of Fe through the Fe(III)-$L_{\text{Fe(III)}}$ reservoir. Substituting the rate equations into Equation (9), we obtain:

$$\tau_{\text{Fe(III)}_{\text{aq}}} = \frac{\left[\text{Fe(III)-L}_{\text{Fe(III)}}\right]}{k_1\left[\text{Fe}_2\text{O}_3\right] - k_2\left[\text{Fe(III)-L}_{\text{Fe(III)}}\right]} \tag{10}$$

If we define the time to reach isotopic equilibrium through isotopic exchange as t_{ex}, then we can expect isotopic equilibrium to be attained between [Fe(III)-$L_{\text{Fe(III)}}$] and [Fe(II)-$L_{\text{Fe(II)}}$] when:

$$\tau_{\text{Fe(III)-L}} \gg t_{\text{ex}} \tag{11}$$

It is important to note that, as discussed in the previous chapter (Chapter 10A; Beard and Johnson 2004), attainment of isotopic equilibrium is not related to the time required to reach steady-state in terms of the concentration or molar ratios of Fe in various species, such as the [Fe_2O_3]/[Fe(III)-$L_{\text{Fe(III)}}$] ratio. The time required to reach isotopic equilibrium (t_{ex}) has been determined for Fe(III)-Fe(II) exchange in dilute aqueous solutions of varying Cl^- contents, and isotopic equilibrium is essentially complete within ~10 seconds at room temperature (Welch et al. 2003). It is possible that t_{ex} is significantly longer if Fe(III) and/or Fe(II) are bound to strong organic ligands, but it seems unlikely that t_{ex} could exceed timescales of hours; this inference, however, needs to be confirmed by experiments.

Assuming that $k_2 > k_1$ and $k_3 > k_1$, the system will eventually reach steady-state in terms of the concentration ratios of [Fe_2O_3]/[Fe(III)-$L_{\text{Fe(III)}}$] and [Fe_2O_3]/[Fe(II)-$L_{\text{Fe(II)}}$]. The time to reach steady-state in, for example, the [Fe_2O_3]/[Fe(III)-$L_{\text{Fe(III)}}$] ratio, may be defined as:

$$t_{\text{Steady State}} = 1/k_2 \tag{12}$$

(Lasaga 1981), although more conservative formulations tend to be used in the short-lived radionuclide literature that are related to the half-life, such as:

$$t_{\text{Steady State}} = n\left(\ln 2/k_2\right) \tag{13}$$

We will use Equation (13) to define the time to reach steady-state, setting $n = 7$ because this provides a close match to the time required to reach steady-state in Figure 6.

Calculation of Fe isotope compositions

Once temporal changes in the relative proportions of the species Fe_2O_3, Fe(III)-$L_{\text{Fe(III)}}$, Fe(II)-$L_{\text{Fe(II)}}$, and $\text{Fe(II)}_{\text{aq}}$ are computed, the δ^{56}Fe values for the various components may be defined by a simple mass-balance equation:

$$\delta_{\text{Sys}} = \delta_{\text{Hem}} X_{\text{Hem}} + \delta_{\text{Fe(III)-L}} X_{\text{Fe(III)-L}} + \delta_{\text{Fe(II)-L}} X_{\text{Fe(II)-L}} + \delta_{\text{Fe(II)aq}} X_{\text{Fe(II)aq}} \tag{14}$$

where X is the mole fraction of various components, "Sys" refers to the total system, "Hem" refers to hematite, and "Fe(III)-L", "Fe(II)-L", and "Fe(II)aq" are defined above. Equation (14) assumes that the final product of DIR is $\text{Fe(II)}_{\text{aq}}$. We cannot use Equation (14) where the final products are magnetite and carbonate, or where there is significant quantities of surface-sorbed Fe if these components have distinct isotopic compositions (see discussion below). We may, however, simplify Equation (14) because $\delta_{\text{Sys}} = \delta_{\text{Hem}}$, and implicit in this relation is that there is no Fe isotope fractionation during the dissolution step or isotopic exchange between hematite and aqueous Fe, as discussed above. Additionally, we define the following Fe isotope fractionation factors:

$$\Delta_{\text{Fe(III)L}-\text{Fe(II)L}} = \delta_{\text{Fe(III)-L}} - \delta_{\text{Fe(II)-L}} \tag{15}$$

$$\Delta_{\text{Fe(II)}_{\text{aq}}-\text{Fe(II)L}} = \delta_{\text{Fe(II)}_{\text{aq}}} - \delta_{\text{Fe(II)-L}} \tag{16}$$

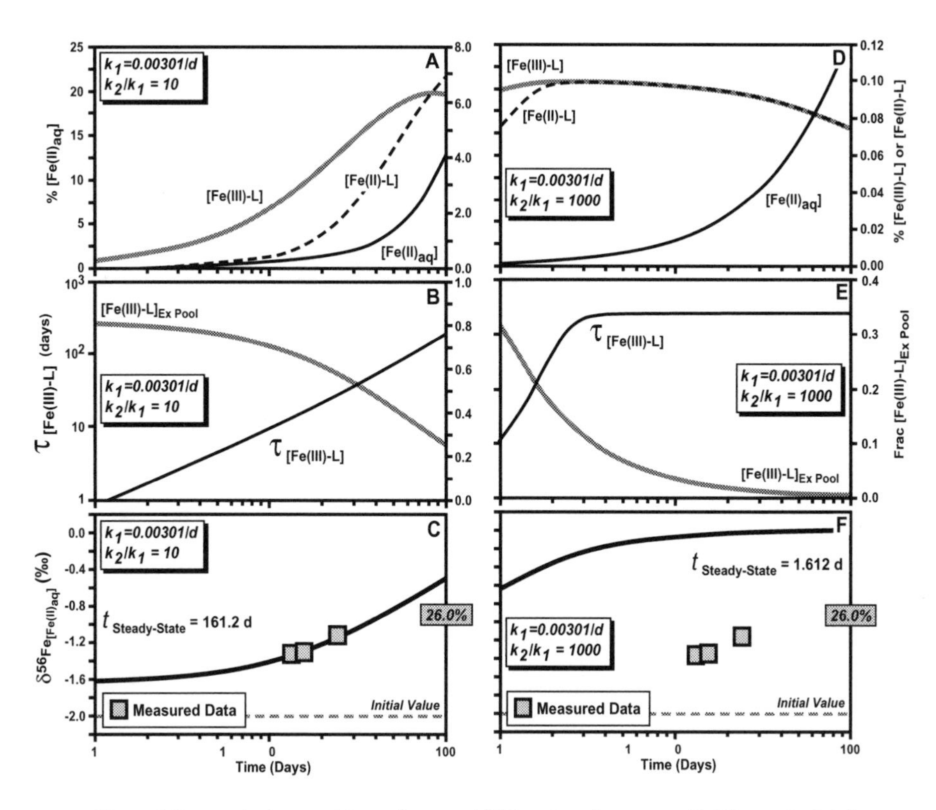

Figure 6. Reservoir sizes, residence times, and δ^{56}Fe values for aqueous Fe(II), as calculated for DIR assuming first-order rate laws. Timescale arbitrarily set to 100 days. Calculations based on rate constant k_1 determined for a 23 day DIR experiment involving hydrous ferric oxide (HFO) by *S. algae* (Beard et al. 1999). The percent total reduction at 100 days is shown in the grey box on the lower right side of the lower diagrams, based on the value of k_1. Parts A-C assume a k_2/k_1 ratio of 10, whereas parts D-F assume a k_2/k_1 ratio of 1000. As constrained by first-order rate laws, the proportion of the intermediate products Fe(III)-L, followed by Fe(II)-L, increase before substantial accumulation of the final $Fe(II)_{aq}$ product (Parts A and D). The fraction of Fe(III)-L in the exchangeable pool of Fe (Fe(III)-L + Fe(II)-L + $Fe(II)_{aq}$) decreases with time, primarily due to accumulation of the $Fe(II)_{aq}$ end product, where the rate of change is a function of the k_2/k_1 ratio.

Left panels (A-C): Where the k_2/k_1 ratio is 10, the system will not reach steady-state until 161.2 days, producing large changes in the residence time (τ) of Fe(III)-L; however, for virtually all time periods, the residence time of Fe(III)-L is several orders of magnitude longer than the time required for isotopic exchange, indicating that isotopic equilibrium will be maintained among the exchangeable pool of Fe, despite the fact that steady-state conditions are not attained when the k_2/k_1 ratio is 10. The difference in predicted δ^{56}Fe values for ferrihydrite substrate and aqueous Fe(II) are similar to those measured by Beard et al. (1999) if the Fe(III)-L - Fe(II)-L fractionation is +2.0‰.

Right panels (D-F): For a k_2/k_1 ratio of 1000, the system reaches steady-state conditions in 1.6 days. The proportion of Fe(III)-L in the exchangeable pool is exceedingly small under steady-state conditions and at high k_2/k_1 ratios, resulting in a shift in the isotopic mass balance such that the predicted δ^{56}Fe values for ferrihydrite substrate and aqueous Fe(II) are far from the inferred Fe(III)-L - Fe(II)-L fractionation.

Note that Equation (15) is the parameter that is of primary interest in this discussion, and the ultimate goal is determining the relation between the measured δ^{56}Fe value for $Fe(II)_{aq}$ and the true $\Delta_{Fe(III)L\text{-}Fe(II)L}$ fractionation, which is not measured directly. Substitution of the fractionation factors into Equation (14) produces:

$$\delta_{Fe(II)aq} = \delta_{Hem} - \frac{X_{Fe(III)\text{-}L}\,\Delta_{Fe(III)L\text{-}Fe(II)L} - \Delta_{Fe(II)aq\text{-}Fe(II)L}\left(X_{Fe(III)L} + X_{Fe(II)L}\right)}{X_{Fe(III)L} + X_{Fe(II)L} + X_{Fe(II)aq}} \tag{17}$$

We will further assume that $\Delta_{Fe(II)aq\text{-}Fe(II)L}$ is zero, that is, there is no Fe isotope fractionation between Fe(II)-$L_{Fe(II)}$ and $Fe(II)_{aq}$. This is primarily for convenience, but is equivalent to letting Fe(II)-$L_{Fe(II)}$ and $Fe(II)_{aq}$ be the same species. Setting $\Delta_{Fe(II)aq\text{-}Fe(II)L}$ to zero, and re-arrangement into measured quantities (δ_{Hem} and $\delta_{Fe(II)aq}$) on the left side, produces:

$$\delta_{Hem} - \delta_{Fe(II)aq} = \frac{X_{Fe(III)\text{-}L}\Delta_{Fe(III)L\text{-}Fe(II)L}}{X_{Fe(III)L} + X_{Fe(II)L} + X_{Fe(II)aq}} \tag{18}$$

Equation (18) illustrates that the measured δ^{56}Fe value for $Fe(II)_{aq}$ is dependent not only on $\Delta_{Fe(III)L\text{-}Fe(II)L}$, but on the proportion of Fe(III)-$L_{Fe(III)}$ in the components that are open to isotopic exchange, which additionally includes Fe(II)-$L_{Fe(II)}$ and $Fe(II)_{aq}$; we will refer to these three components as the "exchangeable pool" of Fe in the system. We stress that the isotopic mass balance described by Equation (18) assumes that the ligand-bound Fe(III) component is not sampled in the aqueous phase component, but instead exists as a component that is bound to the cells.

Temporal changes in reservoir sizes and isotopic compositions

Figure 6 illustrates the proportions of Fe(III)-$L_{Fe(III)}$, Fe(II)-$L_{Fe(II)}$, and $Fe(II)_{aq}$ using the rate constants (k_1) determined for the experiments of Beard et al. (1999). Because k_2 is unknown, we illustrate two examples, arbitrarily setting the ratio k_2/k_1 to 10 and 1000. The system will not reach steady-state in terms of the concentration ratio [HFO]/[Fe(III)-$L_{Fe(III)}$] until 161.2 days (when k_2/k_1 = 10) using the criterion of Equation (13), well past the length of the 23 day experiment (Fig. 6). However, during the 23-day run, the residence time for Fe(III)-$L_{Fe(III)}$ varies from ~1 day to ~100 days if k_2/k_1 = 10, which exceeds the time required for isotopic exchange (t_{ex}) by several orders of magnitude, even if we assume that t_{ex} is relatively long if strong organic ligands are involved. In the case of k_2/k_1 = 1000, the system will reach steady-state in 1.6 days, which is reflected in establishment of a constant residence time early in the experiment (Fig. 6), commensurate with the invariant $M_{Fe(III)\text{-}L}/J_{Fe(III)\text{-}L}$ ratio (Eqn. 9) under steady-state conditions.

Based on Equation (18), we expect the calculated variations in the δ^{56}Fe value for $Fe(II)_{aq}$ to be related to the fraction of Fe(III)-$L_{Fe(III)}$ in the exchangeable pool, where the difference between the measured δ^{56}Fe value for the ferric substrate and $Fe(II)_{aq}$ will most closely reflect the isotopic fractionation between Fe(III)-$L_{Fe(III)}$ and Fe(II)-$L_{Fe(II)}$ as the fraction of Fe(III)-$L_{Fe(III)}$ in the exchangeable pool (Fe(III)-$L_{Fe(III)}$ + Fe(II)-$L_{Fe(II)}$ + $Fe(II)_{aq}$) approaches unity; this condition will exist at the start of an experiment. For example, in the case of the HFO reduction experiment of Beard et al. (1999), the calculated fraction of Fe(III)-$L_{Fe(III)}$ in the exchangeable pool varies from ~0.8 at 1 day to ~0.25 at 100 days, which corresponds to changes in the δ^{56}Fe value for $Fe(II)_{aq}$ from −1.6‰ at 1 day to −0.5‰ at 100 days (Fig. 6). These calculations assume that the $\Delta_{Fe(III)L\ -\ Fe(II)L}$ fractionation was +2.0‰, which produces δ^{56}Fe values for the measured $Fe(II)_{aq}$ component that reasonably describe the observed data (Fig. 6). The variations in the relative proportions of species in the soluble Fe pool indicate that the measured quantity $\delta_{HFO/Hem}$-$\delta_{Fe(II)aq}$ may deviate significantly from $\Delta_{Fe(III)L\text{-}Fe(II)L}$, preventing a precise estimate of the Fe isotope fractionation between Fe(III)-$L_{Fe(III)}$ and Fe(II)-$L_{Fe(II)}$.

Two important observations may be made from the calculations illustrated in Figure 6. First, isotopic equilibrium should be maintained between Fe(III)-$L_{Fe(III)}$ and Fe(II)-$L_{Fe(II)}$ despite the generally large changes that occur in their relative proportions prior to reaching steady-state conditions because the residence times of Fe in the exchangeable pools are many orders of magnitude longer than the expected time required for isotopic exchange. The second observation is that the measured quantity $\delta_{HFO/Hem}$-$\delta_{Fe(II)aq}$ only lies close to the isotopic fractionation between Fe(III)-$L_{Fe(III)}$ and Fe(II)-$L_{Fe(II)}$ at low total % reaction, when the system is actually very far from steady-state conditions, and where the proportion of Fe(III)-$L_{Fe(III)}$ in the exchangeable pool is high. If the time required for isotopic exchange is short, as expected for soluble components, then isotopic equilibrium will be maintained among the Fe(III)-$L_{Fe(III)}$, Fe(II)-$L_{Fe(II)}$, and $Fe(II)_{aq}$ components even early in the experiment. The true $\Delta_{Fe(III)L\text{-}Fe(II)L}$ fractionation may be significantly larger, however, than the measured differences in the $\delta^{56}Fe$ values for the ferric substrate and $Fe(II)_{aq}$.

Sorption effects

Sorption of $Fe(II)_{aq}$ to the ferric oxide/hydroxide starting material (substrate) may represent a significant proportion of total Fe(II) that is produced during DIR, particularly in the early stages of Fe(III) reduction (e.g., Burgos et al. 2002; Zachara et al. 2002). In addition, sorption of $Fe(II)_{aq}$ may occur on the surfaces of bacteria (e.g., Urrutia et al. 1998; Liu et al. 2001; Roden and Urrutia 2002). Icopini et al. (2004) investigated Fe isotope fractionation upon abiotic sorption of $Fe(II)_{aq}$ to goethite under anaerobic conditions. Sorption occurred rapidly, where after 24 hours, ~39% of the initial $Fe(II)_{aq}$ was sorbed to goethite. Based on isotopic analyses of the initial $FeCl_2$ reagent and $Fe(II)_{aq}$ after sorption, the $Fe(II)_{sorbed}$-$Fe(II)_{aq}$ fractionation calculated from mass balance was estimated at +2.1‰. These results are important because they suggest that sorption may play an important role in producing low $\delta^{56}Fe$ values for $Fe(II)_{aq}$ that is produced during DIR. Icopini et al. (2004) concluded that a large component of the fractionation observed between $Fe(II)_{aq}$ and ferric oxide/hydroxide substrates in DIR experiments may occur by entirely abiotic means, which stands in contrast to the model discussed above where the major fractionation step is inferred to occur between ligand-bound pools of Fe(III) and Fe(II).

A different view on the role of Fe(II) sorption during DIR was presented by Johnson et al. (2004a), who suggested that early rapid sorption of $Fe(II)_{aq}$ during the initial stages of DIR is accompanied by significant kinetic isotope fractionation, followed by approach toward isotopic equilibrium over longer timescales, when the $Fe(II)_{sorbed}$-$Fe(II)_{aq}$ fractionation converges toward zero. Because the $\delta^{56}Fe$ values of $Fe(II)_{aq}$ was correlated with Fe reduction rates (Fig. 7), where the lowest values occurred when the major repository of Fe(II) existed as sorbed Fe, Johnson et al. (2004a) estimated that the initial kinetic fractionation between $Fe(II)_{sorbed}$ and $Fe(II)_{aq}$ during sorption was ~2‰ or larger, which is similar to that observed by Icopini et al. (2004). HFO was progressively converted to lepidocrocite over time in an experiment using *Geobacter sulfurreducens* in the study of Johnson et al. (2004a), such that the major proportion of HFO, if not all, was converted within ~16 days. The solid phase conversion was likely accompanied by a large reduction in surface area, which released isotopically heavy Fe that had been sorbed early in the experiment. At relatively low Fe reduction rates, the isotopic contrast between $Fe(II)_{aq}$ and ferric oxyhydroxide substrate approached that observed in longer-term experiments (Fig. 7), which Johnson et al. (2004a) inferred to reflect the equilibrium fractionation between ligand-bound Fe(III) and Fe(II) and a zero fractionation between sorbed Fe(II) and $Fe(II)_{aq}$. Johnson et al. (2004a) note that if abiotic sorption is the primary explanation for the low $\delta^{56}Fe$ values observed for $Fe(II)_{aq}$ during DIR, as argued by Icopini et al. (2004), this suggests that the fractionation between any ligand-bound Fe(III) and Fe(II) components ("Fe(III)-$L_{Fe(III)}$" and "Fe(II)-$L_{Fe(II)}$" in Fig. 5) must be near zero, which seems unlikely given the major effect of redox state on Fe isotope fractionations.

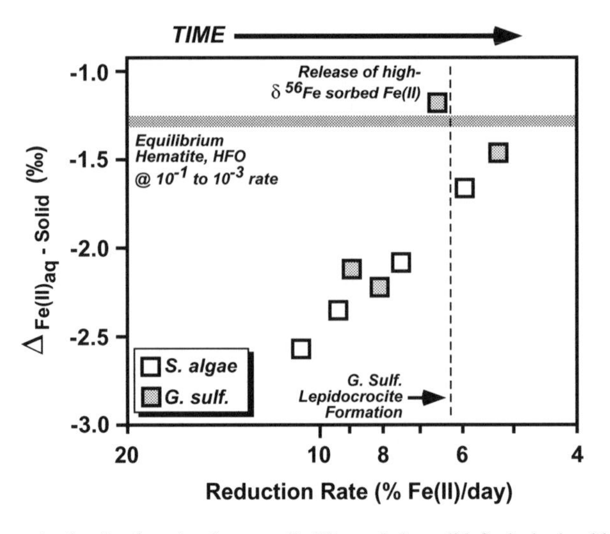

Figure 7. Changes in the fractionation between $Fe(II)_{aq}$ and the solid ferric hydroxide substrate for a comparative study of DIR by *G. sulfurreducens* and *S. algae*, as a function of Fe reduction rates. Reduction rate plotted as decreasing to the right, which reflects increasing time. The initial ferric hydroxide substrate is HFO, which undergoes phase conversion to lepidocrocite in the *G. sulfurreducens* experiment by the end of the third time sample. The decrease in $\Delta_{Fe(II)aq\text{-}Solid}$ fractionations with decreasing Fe reduction rates are interpreted by Johnson et al. (2004a) to reflect an early, large-magnitude fractionation due to kinetic effects upon rapid sorption of Fe(II), followed by an approach to isotopic equilibrium with time. The large decrease in sorption capacity during conversion of HFO to lepidocrocite that occurs in the *G. sulfurreducens* experiment is accompanied by an increase in the $\delta^{56}Fe$ value for $Fe(II)_{aq}$, producing a decrease in the magnitude of the $\Delta_{Fe(II)aq\text{-}Solid}$ fractionation. The equilibrium fractionation between $Fe(II)_{aq}$ and ferric substrate is taken to be that measured in the experiments of Beard et al. (1999, 2003a), which involved much slower Fe reduction rates (10^{-1} to 10^{-3} % Fe(II)/day). Modified from Johnson et al. (2004a).

The isotopic effects of sorption may also be explored using mass-balance relations between sorbed and aqueous Fe. If, for example, the isotopic fractionation observed by Icopini et al. (2004) reflects closed-system equilibrium exchange, the $\delta^{56}Fe$ value of $Fe(II)_{aq}$ should be a function of the proportion of Fe sorbed (Fig. 8). The proportion of sorbed Fe(II) may be calculated for the DIR experiments of Beard et al. (1999, 2003a) using surface areas and saturation capacities from the literature (e.g., Roden and Zachara 1996, and references within), and these, in addition to the *measured* sorption from Johnson et al. (2004a), are plotted in Figure 8, relative to mass-balance sorption lines based on the single determination by Icopini et al. (2004). Although there is some uncertainty in the calculated sorption capacities for HFO and hematite in the experiments of Beard et al. (1999, 2003a), it seems likely that the proportion of sorbed Fe(II) must have been quite different among these studies, and yet all produced similar $\delta^{56}Fe$ values for $Fe(II)_{aq}$ (Fig. 8). If the isotopic fractionations determined by Beard et al. (1999, 2003a) and Johnson et al. (2004a) reflected only sorption of $Fe(II)_{aq}$ and a +2.1‰ $Fe(II)_{Sorbed}$-$Fe(II)_{aq}$ fractionation (Icopini et al. 2004), the data should lie along the diagonal mass-balance lines if closed-system equilibrium fractionation describes the process; instead, they have a relatively constant $\delta^{56}Fe$ values for $Fe(II)_{aq}$ over a wide range of % sorption.

The isotopic effects of sorption remain relatively unknown, with disparate interpretations of its significance during DIR. An important avenue of future research will be to quantify the rates of isotopic exchange between sorbed Fe(II) and $Fe(II)_{aq}$, and to investigate a range of ferric oxide/hydroxide substrates. Because traditional sorption experiments generally

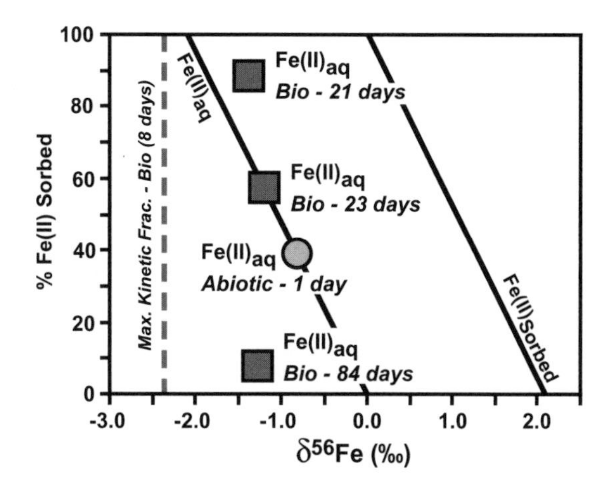

Figure 8. Isotopic mass-balance relations between sorbed Fe(II) and fraction of Fe sorbed for abiotic sorption experiment of Icopini et al. (2004) and those calculated for long-term DIR experiments of Beard et al. (1999, 2003a) and Johnson et al. (2004a). The lowest δ^{56}Fe value for $Fe(II)_{aq}$ measured by Johnson et al. (2004a) is shown by the vertical dashed line, which is interpreted to be a kinetic effect due to rapid sorption. Data normalized to a system δ^{56}Fe value of zero to aid in comparison. Measured data for $Fe(II)_{aq}$ after sorption shown in light gray circle for study of Icopini et al. (2004), and solid diagonal lines indicate calculated isotopic mass balance from that study, assuming a closed-system equilibrium model. Measured data for $Fe(II)_{aq}$ from DIR studies of Beard et al. (1999, 2003a) shown in dark gray squares. Percent Fe(II) sorbed for experiment by Johnson et al. (2004a) is calculated from the measured total and $Fe(II)_{aq}$ contents; if calculated using measured $Fe(II)_{aq}$ contents and an assumed surface area of 600 m^2/g and capacity of 3×10^{-6} mol/g, the % sorbed Fe(II) is 83.8% instead of the 89.0% based on the measured data. Percent Fe(II) sorbed for the experiments of Beard et al. (1999, 2003a) calculated using measured $Fe(II)_{aq}$ contents and assuming surface areas for HFO and hematite of 600 m^2/g and 5 m^2/g, respectively, and capacity of 3×10^{-6} mol/g. A factor of 2 change in surface area correlates with a 6% (hematite) and 17% (HFO) change in the % Fe(II) sorbed.

involve very rapid uptake from high-Fe(II) solutions (e.g., Burgos et al. 2002), which are conditions that are far from the very slow Fe(II) production rates that were associated with the DIR experiments of Beard et al. (1999, 2003a), as well as DIR in nature (e.g., Glausauer et al. 2003), constraining the kinetic effects of sorption will be important. If the equilibrium isotope fractionation between sorbed and aqueous Fe(II) is near zero as suggested by Johnson et al. (2004a), then a sorption term does not need to be introduced into Equations (5)–(8). If, however, early kinetic isotope fractionation occurs during rapid DIR, then the equations would need to be modified, although they may still be applicable to slow rates of DIR where isotopic equilibrium may be approached.

Solid products of DIR

A wide variety of solid products may form during DIR (Table 2). Magnetite readily forms through reaction of soluble Fe(II) that is produced during DIR with the ferric substrate (e.g., Lovley et al. 1987; Roden and Lovely 1993; Fredrickson et al. 1998; Zachara et al. 2002). *G. sulfurreducens*, for example, produces magnetite and siderite as primary end-products of synthetic HFO reduction when a bicarbonate-buffered medium is used (e.g., Caccavo et al. 1994; Lloyd et al. 2000). Recent studies have demonstrated that production of siderite occurs during HFO reduction by *Shewanella putrefaciens* strain CN32, without formation of magnetite (Fredrickson et al. 1998; Parmar et al. 2000; Roden et al. 2002). We first focus on the results of experimental studies where a single phase was produced because of the additional complexities in interpreting Fe isotope fractionations where multiple solid phases are formed.

Production of solid phases during DIR may be highly path dependent, determined by bacterial species and experimental conditions such as growth media and ferric substrate. Because formation of the solid phases during DIR is analogous to mineral synthesis experiments, the same uncertainties exists in interpreting measured fluid-mineral isotope fractionations in terms of kinetic or equilibrium effects as are present in abiologic synthesis experiments (e.g., Chapter 1; Johnson et al. 2004b). For example, large crystals that formed during early rapid precipitation may record highly variable Fe isotope compositions because they may not have been in isotopic equilibrium with aqueous Fe when they formed, and may not fully respond to changes in the Fe isotope composition of the aqueous phase with time due to limited solid-state equilibration or dissolution/re-precipitation at low temperatures. Fine-grained crystals, or the outer portions of large crystals, however, may be closer to isotopic equilibrium with the aqueous phase if they are actively undergoing dissolution and re-precipitation, provided that such processes occurred sufficiently slowly so as to maintain isotopic equilibrium.

Calculated and measured Fe isotope fractionation factors between $Fe(II)_{aq}$ and magnetite vary between −4 and 0‰ (Fig. 9), which is a significant range given the few per mil variations seen in chemically precipitated sediments to date (Beard and Johnson 1999; Zhu et al. 2000; Johnson et al. 2003; Matthews et al. 2004). The largest magnitude $\Delta_{Fe(II)\text{-}Magnetite}$ fractionation, −4.2‰ (Fig. 9), is calculated from spectroscopic data, using the reduced partition function ratios (β factors) from Polyakov and Mineev (2000) and Schauble et al. (2001). A more

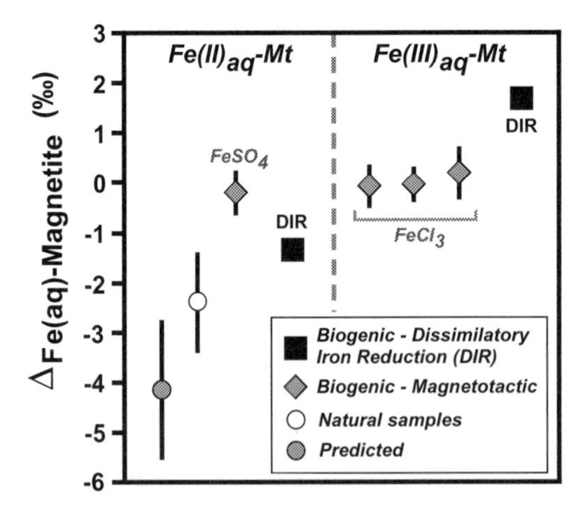

Figure 9. Comparison of isotopic fractionations between Fe in solution ($Fe(II)_{aq}$ and $Fe(III)_{aq}$) and magnetite from predictions based on spectroscopic data (Polyakov and Mineev 2000; Schauble et al. 2001), natural samples (Johnson et al. 2003), and experimental studies of biogenic magnetite formation (Mandernack et al. 1999; Johnson et al. 2004a). Error bars shown reflect estimated uncertainties from specific studies; analytical errors for data reported for DIR by Johnson et al. (2004a) are smaller than the size of the symbol. $Fe(II)_{aq}$–magnetite fractionations shown on left side of figure, and $Fe(III)_{aq}$ - magnetite fractionations shown on right side of figure. The measured $Fe(II)_{aq}$–magnetite fractionation in the DIR study also has been converted to $Fe(III)_{aq}$–magnetite using the $Fe(III)_{aq}$–$Fe(II)_{aq}$ fractionation reported by Johnson et al. (2002) and Welch et al. (2003), so that these results may be compared to the $Fe(III)_{aq}$–magnetite fractionations measured for magnetotactic bacteria by Mandernack et al. (1999). The results for DIR contrast strongly with those calculated from spectroscopic data, but just overlap those predicted using natural data, within the estimated uncertainties. Isotopic fractionations measured by Johnson et al. (2004a) are significantly different, however, from those measured by Mandernack et al. (1999) for magnetotactic bacteria; the later also show unexplained inconsistencies between $Fe(II)_{aq}$–magnetite and $Fe(III)_{aq}$–magnetite fractionations. Adapted from Johnson et al. (2004a).

moderate $\Delta_{Fe(II)\text{-}Magnetite}$ fractionation of approximately −2.4‰ (Fig. 9) is calculated using natural samples (Johnson et al. 2003) and the $Fe(II)_{aq}$ β factor from Schauble et al. (2001), and this fractionation more closely matches that estimated to be produced during DIR (Johnson et al. 2004a). The $Fe(II)_{aq}$ - magnetite fractionation factor measured by Johnson et al. (2004a) during DIR is interpreted to reflect an equilibrium isotope fractionation based on experimental runs that produced simple solid phase assemblages and isotopically homogeneous crystals as determined by partial dissolutions. The Fe isotope fractionations measured in the study of magnetite formation by DIR are, however, significantly different from those measured for magnetotactic bacteria (Fig. 9; Mandernack et al. 1999).

It is unclear if the $\Delta_{Fe(II)aq\text{-}Magnetite}$ fractionations that were measured for magnetite formation by magnetotactic bacteria (Mandernack et al. 1999) reflect equilibrium conditions because of inconsistencies in the $Fe(II)_{aq}$–magnetite and $Fe(III)_{aq}$–magnetite fractionations (Fig. 9). For example, Mandernack et al. (1999) report similar $Fe(II)_{aq}$–magnetite and $Fe(III)_{aq}$–magnetite fractionations, which is not at all expected if the experimental conditions reflected equilibrium conditions, given the +2.9‰ fractionation between $[Fe^{III}(H_2O)_6]^{3+}$ and $[Fe^{II}(H_2O)_6]^{2+}$ in solution (Johnson et al. 2002; Welch et al. 2003). Using the experimentally determined $Fe(III)_{aq}$–$Fe(II)_{aq}$ fractionation, recalculation of the estimated equilibrium $Fe(II)_{aq}$–magnetite fractionation measured in the study of Johnson et al. (2004a) to $Fe(III)_{aq}$–magnetite produces $\Delta_{Fe(III)\text{-}Magnetite} = +1.7$‰, which stands in marked contrast to that measured for three experiments using $Fe(III)Cl_3$ by Mandernack et al. (1999) (Fig. 9). Calculation of $\Delta_{Fe(III)\text{-}Magnetite}$ by this method assumes isotopic equilibrium exists between $Fe(III)_{aq}$ and magnetite, independent of the pathways in which isotopic equilibrium may be attained in a system that contains mixed valance states of Fe. Most of the magnetotactic experiments were done at 28°C, which produced very rapid formation of magnetite, and a possible explanation is that the measured isotope fractionations reflect kinetic isotope effects.

The equilibrium $\Delta_{Fe(II)aq\text{-}Fe\ Carbonate}$ fractionation for pure siderite during DIR is estimated to be near zero, and that for Ca-bearing Fe carbonates is ≥1‰ (Fig. 10) (Johnson et al. 2004a). In contrast, kinetic $\Delta_{Fe(II)aq\text{-}Fe\ Carbonate}$ fractionations produced during DIR are ~1‰ higher than those estimated for equilibrium fractionations, for both pure siderite and Ca-bearing siderite (Fig. 10). A wide range in $Fe(II)_{aq}$–Fe-carbonate fractionations are predicted from spectroscopic and natural data, spanning values from −0.7 to +3.5‰ (Fig. 10; Polyakov and Mineev 2000; Schauble et al. 2001; Johnson et al. 2003). The equilibrium $Fe(II)_{aq}$–siderite fractionation in abiotic systems is estimated at +0.5‰ (Fig. 10; Wiesli et al. 2004), which is somewhat larger than that estimated from DIR experiments. The $Fe(II)_{aq}$–Fe-carbonate fractionations inferred from natural mineral assemblages and those predicted from theory suggest that the $\Delta_{Fe(II)aq\text{-}Fe\ carbonate}$ fractionations should increase with decreasing mole fraction of Fe, from "siderite" to "ankerite" (Fig. 10). The effect of increasing $\Delta_{Fe(II)aq\text{-}Fe\ carbonate}$ upon Ca substitution was observed by Johnson et al. (2004a), suggesting that the bonding changes and distortions in the crystal lattice that accompany even small amounts of Ca substitution into siderite produce significant Fe isotope effects. Because natural Fe carbonates commonly contain significant Ca, Mn, and Mg, it seems likely that carbonate stoichiometry may exert a substantial control on Fe isotope fractionations between $Fe(II)_{aq}$ and carbonate, and it is anticipated that this will be seen in both biologic and abiotic systems.

In addition to producing biogenic magnetite and siderite during DIR, formation of poorly crystalline Fe(II)-bearing solid phases commonly occurs (Table 2), and these may be difficult to detect using XRD spectra. Partial dissolution of solids using weak acids has been used to isolate non-magnetic, non-carbonate Fe(II) solids (e.g., Fredrickson et al. 1998; Roden et al. 2002), and such approaches may also be used to isolate such components for isotopic analysis (Johnson et al. 2004a). Treatments with weak acids may also, however, dissolve unreduced ferric oxide/hydroxide substrates or sorbed Fe components, and determination of Fe(III) and

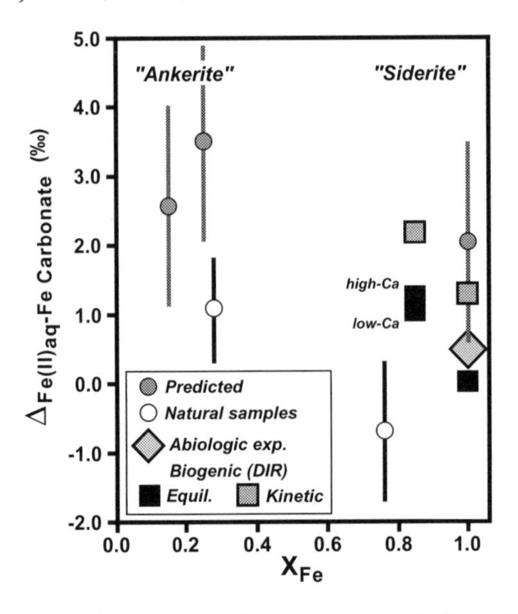

Figure 10. Comparison of isotopic fractionations determined between $Fe(II)_{aq}$ and Fe carbonates relative to mole fraction of Fe from predictions based on spectroscopic data (Polyakov and Mineev 2000; Schauble et al. 2001), natural samples (Johnson et al. 2003), DIR (Johnson et al. 2004a), and abiotic formation of siderite under equilibrium conditions (Wiesli et al. 2004). $Fe(II)_{aq}$ exists as the hexaquo complex in the study of Wiesli et al. (2004); hexaquo Fe(II) is assumed for the other studies. Total cations normalized to unity, so that end-member siderite is plotted at $X_{Fe} = 1.0$. Error bars shown reflect reported uncertainties; analytical errors for data reported by Johnson et al. (2004a) and Wiesli et al. (2004) are smaller than the size of the symbol. Fractionations measured on bulk carbonate produced by DIR are interpreted to reflect kinetic isotope fractionations, whereas those estimated from partial dissolutions are interpreted to lie closer to those of equilibrium values because they reflect the outer layers of the crystals. Also shown are data for a Ca-bearing DIR experiment, where the bulk solid has a composition of approximately $Ca_{0.15}Fe_{0.85}CO_3$, "high-Ca" and "low-Ca" refer to the range measured during partial dissolution studies (Johnson et al. 2004a). Adapted from Johnson et al. (2004a).

Fe(II) contents of the material that has been partially dissolved is required to confidently identify the solid components that have been sampled. In the DIR experiments of Johnson et al. (2004a), significant quantities of a non-magnetic, non-carbonate Fe(II)(s) phase, which we will refer to as NMNC Fe(II)(s), were produced under some conditions, and this phase may exert a strong influence on the Fe isotope compositions of $Fe(II)_{aq}$. In cases where rapid formation of NMNC Fe(II)(s) occurs early, such material appears to have very low $\delta^{56}Fe$ values relative to those of $Fe(II)_{aq}$, driving the remaining $Fe(II)_{aq}$ to very high $\delta^{56}Fe$ values (Johnson et al. 2004a). In contrast, slow production of small quantities of NMNC Fe(II)(s) appears to have $\delta^{56}Fe$ values that are similar to those of $Fe(II)_{aq}$, suggesting that the equilibrium fractionation between NMNC Fe(II)(s) and $Fe(II)_{aq}$ is near zero (Johnson et al. 2004a).

The isotopic effects of DIR using Fe(III)-bearing clay minerals are unknown, but represent an important avenue of future research. Reduction of smectite by DIR bacteria produces significant changes in clay structures (e.g., Kostka et al. 1999a,b). In addition, DIR may catalyze phase transformations in clay minerals, such as conversion of smectite to illite (Kim et al. 2004). The very high surface areas of clay minerals, as well as their high sorption capacity, suggest that there may be significant Fe isotope fractionations during DIR involving Fe(III)-bearing clay minerals as the terminal electron acceptor.

Summary of Fe isotope fractionations produced during DIR

The Fe isotope fractionations that have been observed to date in experimental studies of DIR are summarized in Figure 11. The isotopic compositions are calculated using the fractionation factors listed in Table 3, setting the δ^{56}Fe value of the ferric substrate starting material (HFO, goethite, hematite, etc.) or magnetite solid product to zero, consistent with the general observation that Fe mass balance is dominated by these phases in DIR experiments. During the initial stages of reduction and over short time periods, the δ^{56}Fe values for $Fe(II)_{aq}$, NMNC Fe(II)(s), and sorbed Fe(II) are interpreted to largely reflect kinetic isotope fractionations. Note, however, that although any ligand-bound Fe(III) components might be in isotopic equilibrium with $Fe(II)_{aq}$, based on the long residence times that are calculated (e.g., Fig. 6), the δ^{56}Fe values for $Fe(II)_{aq}$ are likely to be controlled by rapid formation of other components at early stages. These components may include sorbed Fe(II) and NMNC Fe(II)(s). If early, rapid sorption of $Fe(II)_{aq}$ produces a kinetic isotope fractionation that results in anomalously low δ^{56}Fe values for $Fe(II)_{aq}$, this is opposite the effect that is produced by rapid precipitation of NMNC Fe(II)(s). The net changes in δ^{56}Fe values for $Fe(II)_{aq}$ during the initial stages of rapid DIR are therefore likely to be strongly dependent on the relative rates of Fe(II) sorption and formation of NMNC Fe(II)(s).

The general progression in importance of sorbed Fe(II) to NMNC Fe(II)(s) illustrated in Figure 11 follows that observed in experimental studies. The changes in isotopic fractionations between $Fe(II)_{aq}$ and NMNC Fe(II)(s) are well documented by partial dissolutions of the solid phases produced during DIR (Johnson et al. 2004a). The inferred changes in $Fe(II)_{aq}$–$Fe(II)_{sorbed}$ fractionations, however, are less well constrained and this component of DIR needs further study, as noted above. For simplicity, we have omitted Fe carbonate as a solid product in Figure 11, but note that similar changes in isotopic fractionation between $Fe(II)_{aq}$ and solid are inferred to occur, recording initial kinetic conditions that existed during rapid reduction, as well as movement toward equilibrium with time (Johnson et al. 2004a). Finally, we note that besides magnetite, many of the components illustrated in Figure 11 have negative δ^{56}Fe values, suggesting that there is a missing, relatively high-δ^{56}Fe component that has not been analyzed in these experiments. Following the model developed above (Fig. 6), we assume this high-δ^{56}Fe component is a reservoir of Fe(III), which we termed "Fe(III)-$L_{Fe(III)}$" to refer to a ligand- or cell-bound Fe(III) component that would not be present in samples of $Fe(II)_{aq}$ or other Fe-bearing solids that are analyzed in a typical experiment.

ANAEROBIC PHOTOSYNTHETIC Fe(II) OXIDATION

Investigation of Fe isotope fractionation produced by anaerobic photosynthetic Fe(II) oxidation is much less extensive than it is for DIR. Possible pathways where Fe isotopes may be fractionated during anaerobic photosynthetic Fe(II) oxidation, which we will refer to as APIO, are illustrated in Figure 12. There are four possible steps in which Fe isotope fractionation may occur (Croal et al. 2004). Isotopic fractionation may occur during binding of Fe from the ambient aqueous $Fe(II)Cl_2$ starting material to a redox-active site or ligand. Because there is no change in oxidation state at this stage, however, it is anticipated that Fe isotope fractionation, if any, will be relatively small for this step. Iron oxidation occurs in the next step (Fig. 12), and Croal et al. (2004) postulate that a significant step in which Fe isotope fractionation may occur is between ligand-bound Fe(III) and Fe(II) (Δ_1, Fig. 12), given the fact that relatively large Fe isotope fractionations occur between Fe(III) and Fe(II) phases. Isotopic fractionation at this step in the model of Figure 12 is similar to that illustrated for DIR in Figure 5, although the process is occurring in the opposite direction, and, as will be discussed below, likely involves different pools of exchangeable Fe. Formation of the final ferric hydroxide end products may also be accompanied by Fe isotope fractionation, which is essentially an isotopic fractionation upon precipitation (Δ_2, Fig. 12). Finally, it is possible that

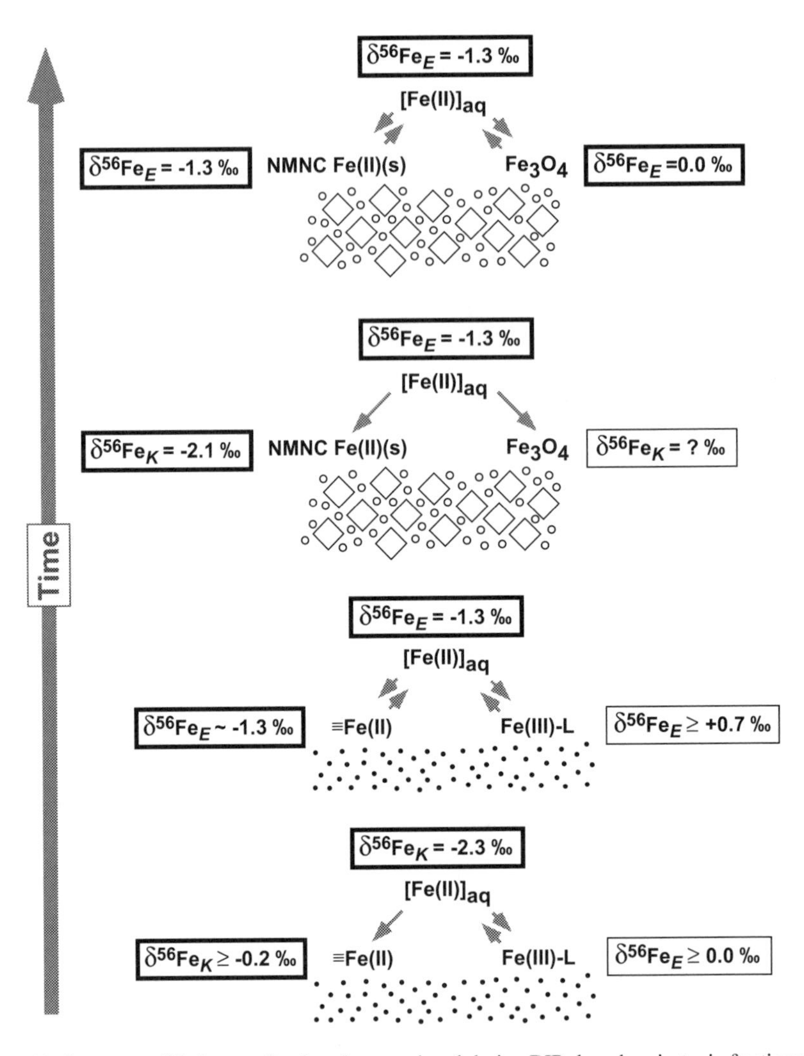

Figure 11. Summary of Fe isotope fractionations produced during DIR, based on isotopic fractionations factors in Table 3. δ^{56}Fe values for measured species shown by boxes with heavy lines, and those that are inferred for species not directly measured shown by boxes with thin lines. Isotopic compositions calculated assuming the δ^{56}Fe values for ferric oxide/hydroxide substrate and magnetite equal zero. During the initial stages of DIR (bottom figure), when Fe reduction rates are highest, rapid formation of $Fe(II)_{aq}$ results in rapid sorption to the ferric oxide/hydroxide substrate (small solid dots), producing very low δ^{56}Fe values that are interpreted to reflect kinetic isotope fractionation. At later stages of DIR, when Fe reduction rates are slower, but prior to formation of significant quantities of reduced solid products (second from bottom figure), the δ^{56}Fe values of $Fe(II)_{aq}$ are interpreted to reflect equilibrium fractionation between ligand-bound Fe(III) and $Fe(II)_{aq}$, and the δ^{56}Fe value of sorbed Fe(II) is inferred to be similar to that of $Fe(II)_{aq}$. Formation of significant quantities of Fe(II)-bearing solid products is illustrated in the third from the bottom figure, including magnetite (Fe_3O_4) in diamonds and non-magnetic, non-carbonate Fe(II)(s) [NMNC Fe(II)(s)] in small open circles; the later phase, if formed rapidly, is interpreted to have very low δ^{56}Fe values that reflect kinetic isotope fractionation. Over the long term, when the system more closely reflects equilibrium conditions (top figure), the isotopic effects of NMNC Fe(II)(s) are minimal, and the δ^{56}Fe values of $Fe(II)_{aq}$ are interpreted to reflect equilibrium fractionation with magnetite.

Table 3. Summary of Fe isotope fractionations during biogenic mineral formation.

Species	*Kinetic* $\Delta_{A\text{-}B}$	*Equilibrium* $\Delta_{A\text{-}B}$	*Ref.*
Dissimilatory Fe(III) reduction:			
$Fe(II)_{aq}$ - $FeCO_3$	+1.2‰	+0.0‰	1
$Fe(II)_{aq}$ - $Ca_{0.15}Fe_{0.85}CO_3$	+2.2‰	+0.9‰	1
$Fe(II)_{aq}$ - Fe_3O_4		−1.3‰	1
$Fe(II)_{aq}$ - Ferric oxide/hydroxide substrate (DIR)	−2.3‰	−1.3‰	2
$Fe(II)_{aq}$ - NMNC Fe(II)(s)	+0.8‰	~0‰	1
$Fe(II)_{aq}$ - ≡Fe(II) [HFO]	−2.1‰	~0‰	3
Fe_3O_4 - $FeCO_3$	+3.5‰	+1.3‰	1
Fe_3O_4 - $Ca_{0.15}Fe_{0.85}CO_3$	+4.5‰	+2.2‰	1
Anaerobic photosynthetic Fe(II) oxidation:			
$Fe(II)_{aq}$ - Ferric oxide/hydroxide precipitate (APIO)	−1.5‰		4

References:
1 - Johnson et al. (2004a).
2 - Beard et al. (1999; 2003a); Johnson et al. (2004a).
3 - Icopini et al. (2004); Johnson et al. (2004a).
4 - Croal et al. (2004).

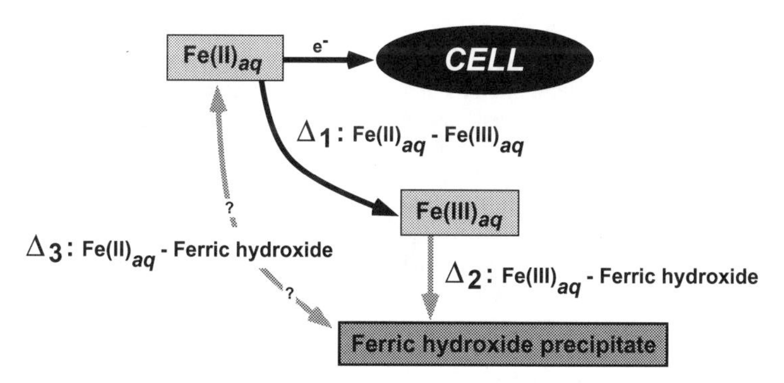

Figure 12. Possible isotope fractionation steps during anaerobic photosynthetic Fe(II) oxidation (APIO). It is assumed that the process of oxidation proceeds through an oxidation step, where $Fe(II)_{aq}$ is converted to soluble Fe(III) in close proximity to the cell, followed by precipitation as ferric oxides/hydroxides. As in DIR (Fig. 5), the most likely step in which the measured Fe isotope fractionations are envisioned to occur is during oxidation, where isotopic exchange is postulated to occur between pools of Fe(II) and Fe(III) (Δ_1). As discussed in the text and in Croal et al. (2004), however, it is also possible that significant Fe isotope fractionation occurs between $Fe(III)_{aq}$ and the ferrihydrite precipitate (Δ_2); in this case the overall isotopic fractionation measured between $Fe(II)_{aq}$ and the ferrihydrite precipitate would reflect the sum of Δ_1 and Δ_2, assuming the proportion of Fe(III) is small (see text for discussion). Isotopic exchange may also occur between $Fe(II)_{aq}$ and the ferric hydroxide precipitate (Δ_3), although this is considered unlikely.

isotopic exchange may occur between the ferric hydroxide precipitates and $Fe(II)_{aq}$ (Δ_3, Fig. 12); such exchange is most likely to occur if the precipitate is nm-size ferrihydrite, but seems unlikely if ferrihydrite consists of larger diameter crystals or is converted to more crystalline forms such as goethite, given the very low rate of isotopic exchange between aqueous Fe and oxides at low temperatures (Poulson et al. 2003).

Although there may be several steps that produce isotopic fractionation during APIO, the process is simpler in some ways than DIR in that a single solid phase is produced, and sorption effects on the isotopic composition of $Fe(II)_{aq}$ are insignificant during the early stages of oxidation when the proportion of $Fe(II)_{aq}$ to ferrihydrite precipitate is very large (Croal et al. 2004). We can describe the APIO process as:

$$Fe(II)_{aq} \rightarrow Fe(III)_{aq} \rightarrow Fe(OH)_3(s) \tag{19}$$

which is identical to that used to describe abiotic oxidation of $Fe(II)_{aq}$ in the previous chapter (Chapter 10A; Beard and Johnson 2004). Equation (19) explicitly assumes that precipitation of ferric hydroxide is preceded by formation of $Fe(III)_{aq}$. Although the overall process of APIO may appear similar to abiotic oxidation of $Fe(II)_{aq}$, the former occurs in the absence of O_2 and involves binding to biologic sites, whereas the later involves inorganic oxidation of Fe(II) by oxygen. In the case of APIO, it may be that one of more pools of soluble Fe are bound to biological ligands, where, for example, $Fe(III)_{aq}$ would be better represented as $Fe(III)\text{-}L_{Fe(III)}$. The quantities of ligand-bound Fe(III) present during APIO, however, is likely to be very small, given the very low levels of Fe(III)-specific organic ligands that have been detected (Kappler and Newman 2004). Nevertheless, the binding environments may be sufficiently distinct for soluble Fe(II) and Fe(III) during oxidation in the outer cell membrane or at the surface, as compared to simple aqueous solutions of Fe(II) and Fe(III), raising the possibility that Fe isotope fractionations produced by APIO may be distinct from those measured between $Fe(III)_{aq}$ and $Fe(II)_{aq}$ in abiotic solutions.

Proportions of reacting species

If the process of APIO is properly described by Equation (19), which infers the presence of a soluble Fe(III) intermediate species, it will be difficult to analyze this species directly, given the low levels that are expected. We must therefore develop mathematical approaches to estimating the isotopic composition of this component, as was done for DIR. The equations used in the previous chapter (Chapter 10A; Beard and Johnson 2004) to describe abiotic Fe(II) oxidation are useful for illustrating possible isotopic fractionations that may occur during APIO. We will assume that the overall oxidation process occurs through a series of first-order rate equations, where relatively slow oxidation of $Fe(II)_{aq}$ to a soluble Fe(III) component occurs, which we will denote as $Fe(III)_{aq}$ for simplicity. The oxidation step is followed by precipitation of $Fe(III)_{aq}$ to ferrihydrite at a much faster rate, which maintains a relatively low level of $Fe(III)_{aq}$ relative to $Fe(II)_{aq}$. The assumption of first-order kinetics is not strictly valid for the experiments reported in Croal et al. (2004), where decreasing $Fe(II)_{aq}$ contents with time do not closely follow either zeroth-, first-, or second-order rate laws. However, use of a first-order rate law allows us to directly compare calculations here with those that are appropriate for abiologic Fe(II) oxidation, where experimental data are well fit to a first-order rate law (Chapter 10A; Beard and Johnson 2004).

Croal et al. (2004) investigated APIO using an enrichment culture of Fe(II)-oxidizing photoautotrophs, as well as a pure culture of the genus *Thiodictyon*, which was cultured at three different light intensities (40, 80, and 120 cm distances from a 40W light source) to investigate possible kinetic effects on isotopic fractionations. Using the three different rates (k_1) that fit the extent of reactions at the end of the 20 day experiments at the three different light intensities, we calculate the relative proportions of $Fe(II)_{aq}$, $Fe(III)_{aq}$, and ferrihydrite precipitate (noted as $Fe(OH)_3(s)$ for simplicity) at various ratios of k_2/k_1 (Fig. 13). We illustrate

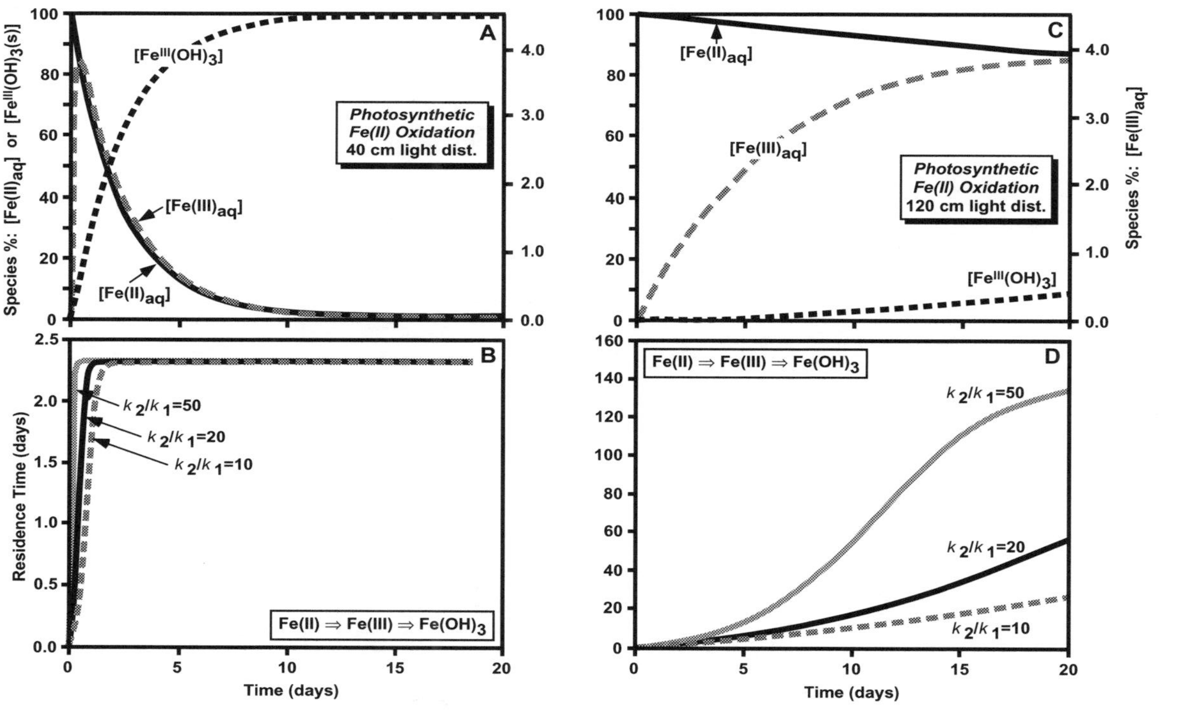

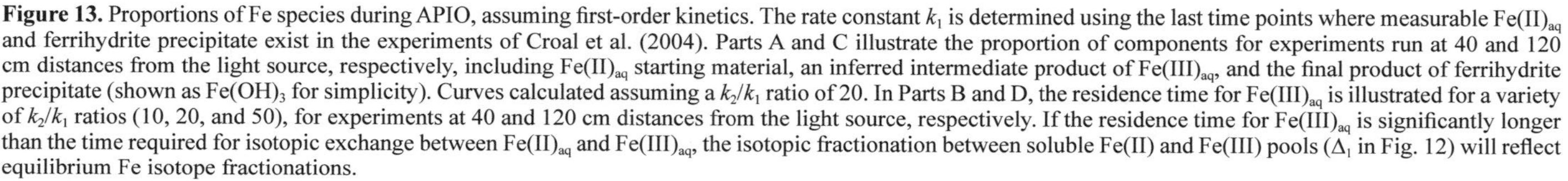

Figure 13. Proportions of Fe species during APIO, assuming first-order kinetics. The rate constant k_1 is determined using the last time points where measurable $Fe(II)_{aq}$ and ferrihydrite precipitate exist in the experiments of Croal et al. (2004). Parts A and C illustrate the proportion of components for experiments run at 40 and 120 cm distances from the light source, respectively, including $Fe(II)_{aq}$ starting material, an inferred intermediate product of $Fe(III)_{aq}$, and the final product of ferrihydrite precipitate (shown as $Fe(OH)_3$ for simplicity). Curves calculated assuming a k_2/k_1 ratio of 20. In Parts B and D, the residence time for $Fe(III)_{aq}$ is illustrated for a variety of k_2/k_1 ratios (10, 20, and 50), for experiments at 40 and 120 cm distances from the light source, respectively. If the residence time for $Fe(III)_{aq}$ is significantly longer than the time required for isotopic exchange between $Fe(II)_{aq}$ and $Fe(III)_{aq}$, the isotopic fractionation between soluble Fe(II) and Fe(III) pools (Δ_1 in Fig. 12) will reflect equilibrium Fe isotope fractionations.

calculations for several k_2/k_1 ratios (10, 20, and 50), all of which would produce small quantities of $Fe(III)_{aq}$. Oxidation rates were fastest for the experiment that involved the highest light intensity (40 cm distance), and, for a k_2/k_1 ratio of 20, steady-state conditions in terms of $Fe(II)_{aq}/Fe(III)_{aq}$ ratios would be attained in 0.56 days (Fig. 13). Under such conditions, the residence time of $Fe(III)_{aq}$ quickly reaches a maximum value of 2.3 days early in the experimental run. Such long residence times, relative to the timescales of isotopic exchange, ensures that isotopic equilibrium would have been maintained between $Fe(II)_{aq}$ and $Fe(III)_{aq}$ during the 20 day experiment. The very slow k_1 rate constant for the experiment at 120 cm light distance would prevent this experiment from attaining steady-state conditions over its 20-day timescale (Fig. 13), and therefore the residence time for $Fe(III)_{aq}$ would not reach its maximum prior to termination of the experiment. We calculate, however, that during virtually the entire experiment the residence time of $Fe(III)_{aq}$ would be on the order of days for the experiments run at distances of 80 and 120 cm from the light source, indicating that isotopic equilibrium should have been maintained between $Fe(II)_{aq}$ and $Fe(III)_{aq}$.

Constraints on isotopic variations

Using simple mass-balance relations, we may calculate the δ^{56}Fe values for $Fe(II)_{aq}$, $Fe(III)_{aq}$, and $Fe(OH)_3(s)$, as was done in the section on DIR above, as well as the approach outlined in the previous chapter for abiotic oxidation of $Fe(II)_{aq}$ (Chapter 10A; Beard and Johnson 2004). For example, if isotopic equilibrium is maintained between $Fe(II)_{aq}$ and $Fe(III)_{aq}$, the δ^{56}Fe value for the $Fe(III)_{aq}$ component is given by:

$$\delta_{\text{Fe(III)aq}} = \delta_{\text{Fe(aq)-Total}} + \Delta_{\text{Fe(III) - Fe(II)}}\left(\frac{X_{\text{Fe(II)}}}{X_{\text{Fe(II)}} + X_{\text{Fe(III)}}}\right) \tag{20}$$

where $\delta_{\text{Fe(aq)-Total}}$ is the δ^{56}Fe value of the total aqueous Fe pool ($Fe(II)_{aq}$ and $Fe(III)_{aq}$), and $X_{\text{Fe(II)}}$ and $X_{\text{Fe(III)}}$ are the mole fractions of $Fe(II)_{aq}$ and $Fe(III)_{aq}$ in the total system, respectively. Note that if $X_{\text{Fe(III)}}$ is very small, which is the case for large k_2/k_1 ratios (Fig. 13), then $\delta_{\text{Fe(aq)-Total}}$ is essentially equal to the δ^{56}Fe value of $Fe(II)_{aq}$. Note also that the ratio $X_{\text{Fe(II)}}/(X_{\text{Fe(II)}} + X_{\text{Fe(III)}})$ will be unity at the beginning of the experiment (prior to oxidation), and will move to a constant value under steady-state conditions that is determined by the k_2/k_1 ratio.

At high k_2/k_1 ratios, the isotopic mass balance among the soluble (exchangeable) pools of Fe ($Fe(II)_{aq}$ and $Fe(III)_{aq}$) is essentially invariant with time. When $\delta_{\text{Fe(aq)-Total}} \approx \delta_{\text{Fe(II)aq}}$ and $X_{\text{Fe(II)}}/(X_{\text{Fe(II)}} + X_{\text{Fe(III)}}) \approx 1$, we may simplify Equation (20) to:

$$\delta_{\text{Fe(III)aq}} \approx \delta_{\text{Fe(II)aq}} + \Delta_{\text{Fe(III)-Fe(II)}} \tag{21}$$

Equation (21) is an excellent approximation to Equation (20) for moderate to high k_2/k_1 ratios (~10 and higher) for processes that occur by first-order kinetics. It is important to note, however, that a specific rate law does not appear anywhere in Equations (20) and (21), and they are equally valid for any reaction process where $X_{\text{Fe(III)}}$ is small. Equation (21) illustrates that oxidation of $Fe(II)_{aq}$ to $Fe(III)_{aq}$ produces a markedly different isotopic mass balance than that associated with DIR. In cases where the product of DIR is $Fe(II)_{aq}$, the concentration of this component is continually increasing, changing the relative mass balance among the exchangeable pools of Fe over time.

Next we explore using the δ^{56}Fe value of the ferric oxide/oxyhydroxide precipitate as a proxy for $\delta_{\text{Fe(III)aq}}$, which allows Equation (21) to be used to calculate the $\Delta_{\text{Fe(III)-Fe(II)}}$ fractionation from the measured δ^{56}Fe values for the ferric precipitate and $Fe(II)_{aq}$. This approach is valid when the molar proportion of $Fe(III)_{aq}$ is very small. However, if there is a significant Fe isotope fractionation between $Fe(III)_{aq}$ and ferric hydroxide precipitate, this must be taken into account. As discussed in the previous chapter (Chapter 10A; Beard and Johnson 2004), at low

$Fe(III)_{aq}/Fe(II)_{aq}$ ratios, and where isotopic equilibrium is maintained between $Fe(III)_{aq}$ and $Fe(II)_{aq}$, the effect of Fe isotope fractionation upon precipitation may be incorporated into the form of Equation (21) as:

$$\delta_{Fe(II)aq} - \delta_{FH} \approx \Delta_{Fe(III)\text{-}FH} - \Delta_{Fe(III)\text{-}Fe(II)} \tag{22}$$

where the $\delta^{56}Fe$ values for $Fe(II)_{aq}$ and the ferrihydrite precipitate are defined as $\delta_{Fe(II)aq}$ and δ_{FH}, respectively, and the Fe isotope fractionation between $Fe(III)_{aq}$ and ferrihydrite precipitate, and $Fe(III)_{aq}$ and $Fe(II)_{aq}$, is given by $\Delta_{Fe(III)\text{-}FH}$ and $\Delta_{Fe(III)\text{-}Fe(II)}$, respectively.

Croal et al. (2004) measured an approximate −1.5‰ fractionation between $Fe(II)_{aq}$ and ferrihydrite precipitate that formed during APIO. This fractionation is independent of the overall rate of oxidation, particularly if data early in the experiment are considered, when the measured isotopic differences between phases is least dependent upon a specific mechanistic model, such as closed-system equilibration or Rayleigh fractionation (Fig. 14). Precipitates that are removed early in the experiment are least likely to "back exchange" with the ambient aqueous Fe unless exchange rates are very rapid because they are not in contact with aqueous Fe for extended periods (Δ_2, Fig. 12). In interpreting these data, however, we are immediately faced with the ambiguity posed by Equation (22), which shows that in the absence of independent knowledge of the $\Delta_{Fe(III)\text{-}FH}$ fractionation, we cannot infer the Fe isotope fractionation produced during the oxidation step ($\Delta_{Fe(III)\text{-}Fe(II)}$). The Fe isotope fractionations measured by Croal et al. (2004), therefore, have several possible interpretations. First, the data may be interpreted to reflect Fe isotope fractionation between $Fe(II)_{aq}$ and $Fe(III)_{aq}$ of ~ −1.5‰, if $\Delta_{Fe(III)\text{ - }FH}$ is zero. In this case, the long residence time that is expected for $Fe(III)_{aq}$ (Fig. 13), relative to the time required to reach isotopic equilibrium between $Fe(II)_{aq}$ and $Fe(III)_{aq}$, suggests that the −1.5‰ $Fe(II)_{aq}$-$Fe(III)_{aq}$ fractionation would reflect an equilibrium

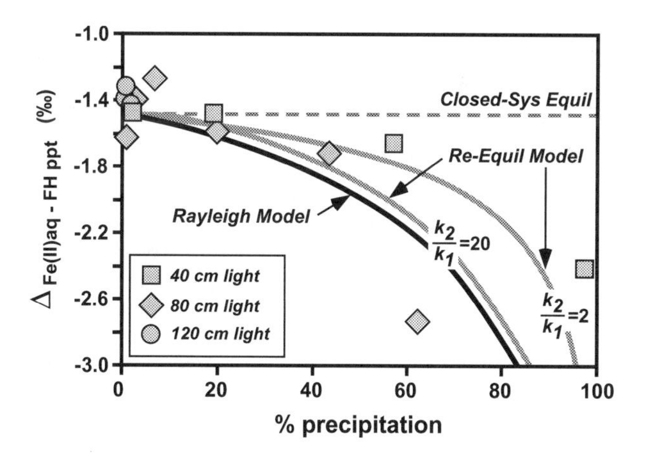

Figure 14. Measured Fe isotope fractionations between $Fe(II)_{aq}$ and ferrihydrite precipitates for the APIO experiments by Croal et al. (2004), relative to % total reaction (precipitation). A simple Rayleigh fractionation model is illustrated in the solid black curve, using a $\Delta_{Fe(II)aq\text{-}FH}$ fractionation of −1.5‰. Solid grey curves illustrate a two-step model, where the overall $Fe(II)_{aq}$–ferrihydrite fractionation proceeds first through a −2.9‰ equilibrium fractionation between $Fe(II)_{aq}$ and $Fe(III)_{aq}$, followed by a +1.4‰ between $Fe(III)_{aq}$ and ferrihydrite. Because the k_2/k_1 ratio in the first-order rate model affects the relative proportions of $Fe(II)_{aq}$ and $Fe(III)_{aq}$, the two-step model is a function of k_2/k_1 ratio; two examples are illustrated, one where the proportion of $Fe(III)_{aq}$ is relatively high (k_2/k_1 = 2), and one where the proportion of $Fe(III)_{aq}$ is relatively low (k_2/k_1 = 20). As the k_2/k_1 ratio increases beyond ~20 (very low proportion of $Fe(III)_{aq}$), the two-step model is well approximated by a single −1.5‰ fractionation between $Fe(II)_{aq}$ and ferrihydrite.

fractionation factor. This would be approximately half the −2.9‰ fractionation measured between Fe(II) and Fe(III) in dilute aqueous solutions (Johnson et al. 2002; Welch et al. 2003), and might suggest unique bonding environments for at least one Fe species during APIO as compared to $Fe(II)_{aq}$ and $Fe(III)_{aq}$ in abiologic systems.

The Fe(III)aq–ferrihydrite fractionation factor

The data of Croal et al. (2004) may also be interpreted to reflect a two-step process, where a −2.9‰ fractionation occurs between $Fe(II)_{aq}$ and $Fe(III)_{aq}$, accompanied by a +1.4‰ fractionation between $Fe(III)_{aq}$ and ferrihydrite upon precipitation, produces a net fractionation of −1.5‰. When cast in terms of common mechanistic models for separation of solid and liquid phases such as Rayleigh fractionation, it becomes clear that the two-step model produces essentially the same fractionation trend as a single −1.5‰ fractionation step between $Fe(II)_{aq}$ and ferrihydrite if the $Fe(III)_{aq}/Fe(II)_{aq}$ ratio is low (Fig. 14). As the $Fe(III)_{aq}/Fe(II)_{aq}$ ratio increases, however, the calculated net $Fe(II)_{aq}$-ferrihydrite fractionation in the two-step model deviates from that of simple Rayleigh fractionation (Fig. 14). Unfortunately, the scatter in the data reported by Croal et al. (2004), which likely reflects minor contamination of $Fe(II)_{aq}$ in the ferrihydrite precipitate, prevents distinguishing between these various models without consideration of additional factors.

A non-zero $\Delta_{Fe(III)-FH}$ fractionation might reflect kinetic or equilibrium isotope partitioning, and in the absence of independent measurements of the $\Delta_{Fe(III)-FH}$ fractionation factor under equilibrium conditions, as well as at different precipitation rates that are far from equilibrium, we must approach the problem indirectly. Equation (22) illustrates that the uncertainty in inferring the $\Delta_{Fe(III)-Fe(II)}$ fractionation is directly related to the uncertainty posed by the $\Delta_{Fe(III)-FH}$ fractionation factor when the proportion of $Fe(II)_{aq}$ is very small. The potential range of $\Delta_{Fe(III)-FH}$ under equilibrium and kinetic conditions may be explored through the experiments of Skulan et al. (2002), who investigated kinetic and equilibrium $[Fe^{III}(H_2O)_6]^{3+}$–hematite fractionations. There is some difficulty in comparing the experiments of Skulan et al. (2002) with those of Croal et al. (2004), because the ferric precipitates and temperatures were different. Skulan et al. (2002) observed that the kinetic $[Fe^{III}(H_2O)_6]^{3+}$–hematite fractionation, at 98°C, varied linearly with precipitation rate when precipitation rates were relatively low, on the order of 10^{-3} F/hour, where F is the fraction of the total precipitate. The largest kinetic $[Fe^{III}(H_2O)_6]^{3+}$–hematite isotope fractionation observed by Skulan et al. (2002) was +1.3‰, where near-complete precipitation occurred over ~12 hours, equivalent to a relatively high rate of ~10^{-1} F/hour. In contrast, Skulan et al. (2002) estimated that the equilibrium $[Fe^{III}(H_2O)_6]^{3+}$–hematite fractionation was near zero at 98°C, and it is possible, though unknown, that the equilibrium $[Fe^{III}(H_2O)_6]^{3+}$–ferrihydrite fractionation is similarly small at room temperature. If we assume that similar effects occurred during precipitation of ferrihydrite at the room temperature (~22°C) of the experiments by Croal et al. (2004), we would infer that moderately rapid ferrihydrite precipitation would produce significant kinetic $Fe(III)_{aq}$–ferrihydrite fractionations, but that slow precipitation of ferrihydrite may be associated with a $\Delta_{Fe(III)-FH}$ fractionation that is closer to zero.

We consider the relations between precipitation rate and the fractionation between $Fe(III)_{aq}$ and ferric oxide/hydroxide precipitates in Figure 15. The overall rates of Fe(II) oxidation and precipitation of ferrihydrite in the experiments by Croal et al. (2004) were several orders of magnitude lower than those studied by Skulan et al. (2002) where significant kinetic $[Fe^{III}(H_2O)_6]^{3+}$–hematite fractionations were observed, and this may suggest that the $\Delta_{Fe(III)-FH}$ fractionation was low in the experiments of Croal et al. (2004). Through variations in light intensity, the Fe(II) oxidation rates during APIO varied by nearly an order of magnitude (Fig. 15), and it is striking that the measured $Fe(II)_{aq}$–ferrihydrite fractionations were relatively constant, as determined for $Fe(II)_{aq}$–ferrihydrite pairs early in the experiment. This might suggest that kinetic isotope effects during precipitation were not important, provided the rate-

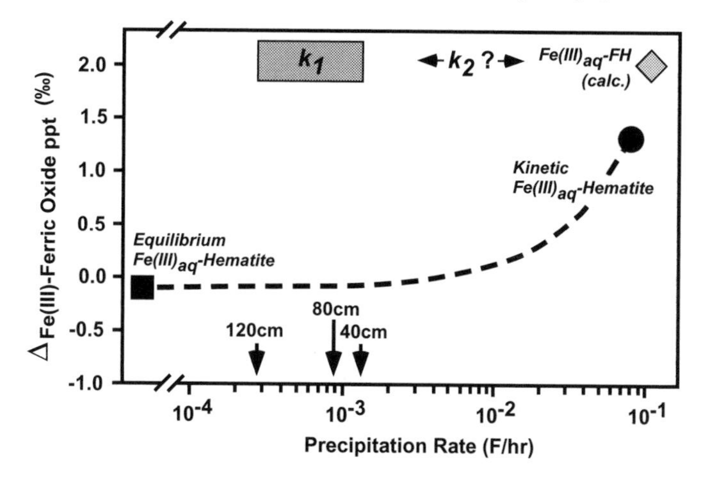

Figure 15. Illustration of possible variations in isotopic fractionation between $Fe(III)_{aq}$ and ferric oxide/hydroxide precipitate ($\Delta_{Fe(III)aq\text{-}Ferric\ ppt}$) and precipitation rate. Skulan et al. (2002) noted that the kinetic $\Delta_{Fe(III)aq\text{-}Ferric\ ppt}$ fractionation produced during precipitation of hematite from $Fe(III)_{aq}$ was linearly related to precipitation rate, which is shown in the dashed curve (precipitation rate plotted on log scale). The most rapid precipitation rate measured by Skulan et al. (2002) is shown in the black circle. The equilibrium $Fe(III)_{aq}$-hematite fractionation is near zero at 98°C, and this is plotted (black square) to the left of the break in scale for precipitation rate. Also shown for comparison is the calculated $Fe(III)_{aq}$-ferrihydrite fractionation from the experiments of Bullen et al. (2001) (grey diamond), as discussed in the previous chapter (Chapter 10A; Beard and Johnson 2004). The average oxidation-precipitation rates for the APIO experiments of Croal et al. (2004) are also noted, where the overall process is limited by the rate constant k_1. As discussed in the text, if the proportion of $Fe(III)_{aq}$ is small relative to total aqueous Fe, the rate constant for the precipitation of ferrihydrite from $Fe(III)_{aq}$ (k_2) will be higher, assuming first-order rate laws, although its value is unknown.

limiting step was varied in the experiments during changes in the overall oxidation rate, but this remains unknown in the experiments of Croal et al. (2004). The rate constant for precipitation of $Fe(III)_{aq}$ to ferrihydrite (k_2) is unknown, but is assumed to be significantly larger than k_1 in our first-order rate model. If the variations in light intensity varied only the rate constant for the oxidation step, k_1, and not, for example, the precipitation rate, k_2, then the range in overall oxidation rates place *no constraint* on inferring the value of the $\Delta_{Fe(III)\text{-}FH}$ fractionation based on the experimental observations of Skulan et al. (2002) (Fig. 15). Paradoxically, if k_2 is extremely high, as might be the case if the proportion of $Fe(III)_{aq}$ is quite low, very rapid precipitation of $Fe(III)_{aq}$ to ferrihydrite is unlikely to produce a significant $\Delta_{Fe(III)\text{-}FH}$ fractionation, as has been observed for very rapid (~1 s) oxidation of $Fe(II)_{aq}$ (Johnson et al. 2002).

Comparison with abiologic Fe(II) oxidation

The fractionation between $Fe(II)_{aq}$ and ferrihydrite of −1.5‰ measured during APIO by Croal et al. (2004) is somewhat similar to the −0.9‰ fractionation measured by Bullen et al. (2001) during abiotic oxidation of $Fe(II)_{aq}$ to ferrihydrite through reaction with an oxygenated solution. Bullen et al. (2001) interpreted this fractionation to reflect isotopic equilibrium between aqueous Fe(II) species, such as $[Fe^{II}(OH)(H_2O)_5]^+$ and $[Fe^{II}(H_2O)_6]^{2+}$, with the implicit assumption that there was no fractionation between aqueous Fe(II) and Fe(III), or upon precipitation of $Fe(III)_{aq}$ to ferrihydrite. As discussed in the previous chapter (Chapter 10A; Beard and Johnson 2004), the Bullen et al. (2001) model is unlikely, given the −2.9‰ equilibrium fractionation between $[Fe^{II}(H_2O)_6]^{2+}$ and $[Fe^{III}(H_2O)_6]^{3+}$ (Johnson et al. 2002; Welch et al. 2003), and the fact that isotopic exchange is so rapid between these species that isotopic equilibrium will be maintained over the oxidation rates of the Bullen et al. (2001) experiment. Instead, these experiments likely reflect the combination of a −2.9‰ equilibrium fractionation between $[Fe^{II}(H_2O)_6]^{2+}$ and $[Fe^{III}(H_2O)_6]^{3+}$, and a +2.0‰ kinetic fractionation between $[Fe^{III}(H_2O)_6]^{3+}$ and ferrihydrite upon precipitation, although this later fractionation is inferred and has yet to be measured experimentally. The issue of the $\Delta_{Fe(III)-FH}$ fractionation factor, under both equilibrium and kinetic conditions, once again becomes important. If the experiments by Bullen et al. (2001) in part reflect a +2.0‰ kinetic fractionation between $Fe(III)_{aq}$ and ferrihydrite, this is comparable to the kinetic $[Fe^{III}(H_2O)_6]^{3+}$–hematite fractionation measured by Skulan et al. (2002) when considered in terms of precipitation rates (Fig. 15), and the fact that the Bullen et al. (2001) experiments were run at lower temperatures.

If the rate constant for precipitation of $Fe(III)_{aq}$ to ferrihydrite (k_2) was $\geq 10^2$ times greater than the overall oxidation rate (k_1) during APIO, it is possible, based on the relations shown in Figure 15, that the $\Delta_{Fe(III)-FH}$ fractionation factor was +1 to +2‰, as noted in the "two-step" scenario above. If, however, k_2 was $\sim \leq 10$ times greater than k_1, it is possible that the $\Delta_{Fe(III)-FH}$ fractionation factor was close to zero (Fig. 15). If the former case is correct, then we would infer that the isotopic fractionation between $Fe(II)_{aq}$ and $Fe(III)_{aq}$ during APIO is similar to that in dilute aqueous solutions. However, if the latter case is correct, the isotopic fractionation between soluble pools of $Fe(II)_{aq}$ and $Fe(III)_{aq}$ in the biologic system is unique relative to that in dilute aqueous solutions, perhaps indicating unique bonding environments caused by binding to biological ligands. If the $\Delta_{Fe(III)-FH}$ fractionation is zero in the biologic experiment, the relatively similar fractionations between $Fe(II)_{aq}$ and ferrihydrite in the experiments of Croal et al. (2004) and Bullen et al. (2001) are merely coincidental, reflecting the effects of kinetic fractionations during precipitation in the later study. Key resolutions to these issues include determining the $Fe(III)_{aq}$–ferrihydrite fractionation, as well as developing strategies to analyze the isotopic composition of the $Fe(III)_{aq}$ component directly, and these are important avenues for future research.

We wish to stress that comparison of the isotopic effects in biologic and abiologic systems will be most meaningful if experimental conditions are identical, where the only difference is the presence or absence of bacteria. The wide variety of buffers, growth media, and others conditions that are involved in biological experiments raise the possibility that spurious results may be obtained if these factors are not carefully controlled. Because speciation may exert a strong control on Fe isotope fractionations (Schauble et al. 2001), even small differences across experimental studies may be significant.

ISOTOPIC VARIATIONS PRODUCED DURING BIOGEOCHEMICAL CYCLING OF IRON

The kinetic and equilibrium Fe isotope fractionations associated with DIR and anaerobic APIO are summarized in Table 3. In this section, we use the Fe isotope fractionation factors determined for biologic systems to investigate the isotopic variations that may be produced during biogeochemical cycling of Fe. We discuss below two settings in which Fe cycling may occur. The first is redox cycling of Fe in a low-carbonate surface environment that involves oxidation of Fe(II) and reduction of Fe(III), such as might be found in a hydrothermal or hot spring pool. The second involves diagenetic reactions, with or without super-saturation of carbonate ion, where reductive dissolution of ferric oxide/hydroxide produces magnetite ± Fe carbonate.

Redox cycling of Fe by bacteria

Modern iron-depositing hot springs that are fed by Fe(II)-rich waters have been invoked as analogs to environments where active metal cycling most likely occurred in the

Precambrian (e.g., Pierson et al. 1999; Pierson and Parenteau 2000), which eventually may have been preserved as iron deposits (e.g., Wade et al. 1999). Oxidation of Fe(II) in the upper layers of microbial communities may occur abiotically through interaction of Fe(II) with an oxygenated atmosphere, or through biologic activity. For example, Fe(II) may be oxidized through locally increased oxygen contents generated by cyanobacteria in the upper photic region of a microbial community. Alternatively, phototrophic Fe(II) oxidation may occur in the absence of oxygen by purple and green bacteria, and this is illustrated in Figure 16. Finally, reductive dissolution of ferric oxide/hydroxide precipitates may occur at the sediment-water interface at the bottom of hot springs by DIR (Fig. 16), completing the Fe redox cycle (e.g., Nealson and Stahl 1997). To illustrate the isotopic variations that may be produced in such a model, we define the following fluxes:

$$J_{\text{Fe(II)-Ext}} = J_{2\text{IE}} \tag{23}$$

the external flux of Fe(II) into the system;

$$J_{\text{Fe(III)-ppt}} = J_{3\text{O}} \tag{24}$$

the flux of Fe precipitated from the Fe(II) pool as ferric oxyhydroxides, produced by Fe(II)-oxidizing bacteria;

$$J_{\text{Fe(II)-Bio}} = J_{2\text{IB}} \tag{25}$$

the flux of Fe returned to the main pool as Fe(II), generated by Fe(III)-reducing bacteria.

We further constrain the system to have no Fe loss. In addition, we define the following:

$$\Delta_{32} = \delta^{56}\text{Fe}_{\text{Ferric Oxide ppt}} - \delta^{56}\text{Fe}_{\text{Fe(II)}} = +1.5‰ \tag{26}$$

which is the Fe isotope fractionation factor between ferric oxide/hydroxide precipitate and Fe(II) in the pool, produced by Fe(II)-oxidizing bacteria.

We define the δ^{56}Fe of the Fe(II) that is input into the pool at any time as:

$$\delta^{56}\text{Fe}_{2\text{I}} = \delta^{56}\text{Fe}_{2\text{IE}}\left(\frac{J_{2\text{IE}}}{J_{2\text{IE}} + J_{2\text{IB}}}\right) + \delta^{56}\text{Fe}_{2\text{IB}}\left(\frac{J_{2\text{IB}}}{J_{2\text{IE}} + J_{2\text{IB}}}\right) \tag{27}$$

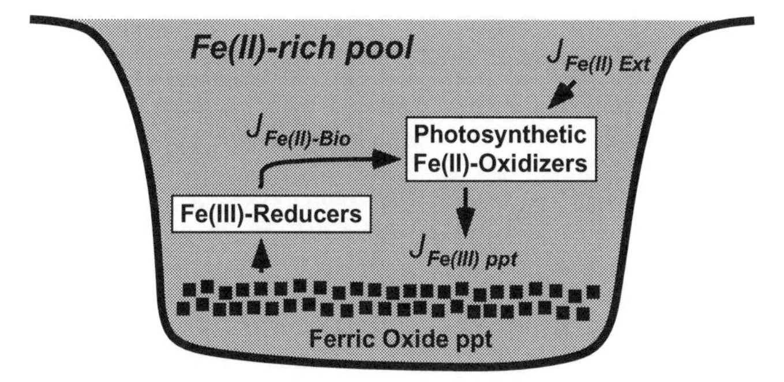

Figure 16. Conceptual model for biological redox cycling in a hot spring environment. Influx of external aqueous Fe(II) [$J_{\text{Fe(II)-Ext}}$] may reflect hydrothermal fluids or other sources of Fe(II)_{aq}. Oxidation of Fe(II)_{aq} is envisioned to occur by Fe(II)-oxidizing phototrophs in anaerobic conditions, but could also occur through interaction of Fe(II)_{aq} with an oxygen-rich atmosphere. Oxidation of Fe(II) produces a flux of ferric oxide/hydroxide precipitates [$J_{\text{Fe(III) ppt}}$] that settle to the lower, anaerobic sections of the pool. These ferric precipitates are in turn partially reduced by DIR bacteria, returning a flux of Fe(II)_{aq} to the pool [$J_{\text{Fe(II)-Bio}}$].

Following the mass-flux equations developed for open-system magma chambers by DePaolo (1981), we will define a flux ratio of Fe(II) into the pool relative to Fe(III) out of the pool as:

$$r = \frac{J_{2I}}{J_{3O}} = \frac{J_{2IE} + J_{2IB}}{J_{3O}} \tag{28}$$

We will restrict r to values less than unity, which is necessary to accumulate an Fe deposit. Solution of the time derivatives, expressed as F (fraction of liquid remaining in the pool), following the approach of DePaolo (1981), but cast in terms of fluxes, produces:

$$\delta^{56}\text{Fe}_{\text{POOL}} = \delta^{56}\text{Fe}^{\text{I}}_{\text{POOL}} + \left(1 - F^{[-r/(r-1)]}\right)\left(\delta^{56}\text{Fe}_{2\text{I}} - \frac{\Delta_{32} + r\delta^{56}\text{Fe}^{\text{I}}_{\text{POOL}}}{r}\right) \tag{29}$$

where $\delta^{56}\text{Fe}^{\text{I}}_{\text{POOL}}$ is the initial δ^{56}Fe value of the pool (set equal to $\delta^{56}\text{Fe}_{2\text{IE}}$).

An end-member case would be precipitation of ferric oxyhydroxide by photosynthetic Fe(II)-oxidizing bacteria through simple Rayleigh fractionation, with no external Fe(II) flux or return of Fe(II) to the pool from Fe(III)-reducing bacterial activity, which will produce extreme Fe isotope compositions, but only in the latest fluids and precipitates (Fig. 17). In

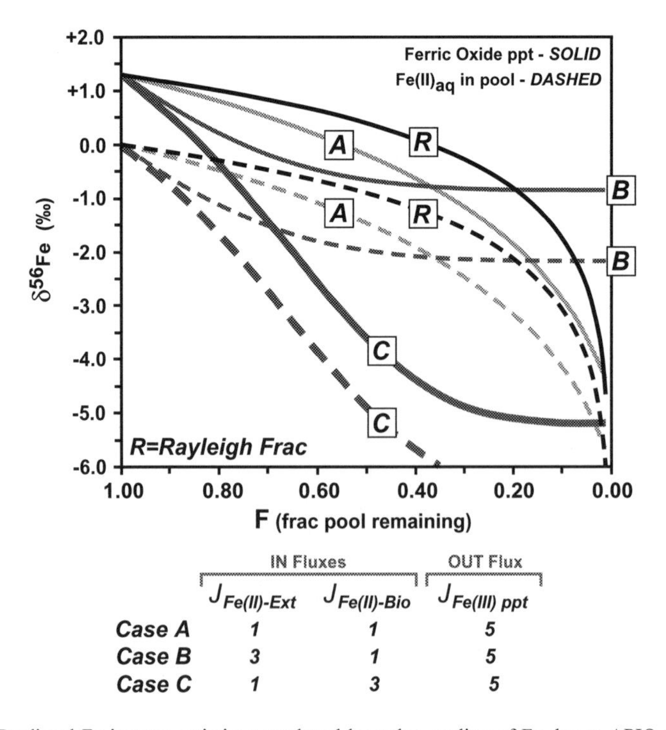

Figure 17. Predicted Fe isotope variations produced by redox cycling of Fe due to APIO and DIR by bacteria, as envisioned from Fig. 16. The initial δ^{56}Fe of the pool, as well as that of the influx of $Fe(II)_{aq}$ [$J_{Fe(II)\text{-}Ext}$] is assumed to be zero. Solid lines show the δ^{56}Fe values for ferric oxide/oxyhydroxide deposits as a function of solidification of the pool as an iron deposit is formed, and dashed lines show the δ^{56}Fe values for $Fe(II)_{aq}$. Depending upon the relative fluxes of external $Fe(II)_{aq}$ [$J_{Fe(II)\text{-}Ext}$], return of Fe(II) to the pool by DIR [$J_{Fe(II)\text{-}Bio}$], and precipitation of ferric oxide/oxyhydroxides by APIO [$J_{Fe(III)\ ppt}$], a wide range of δ^{56}Fe values can be produced.

the case of a modest return of Fe(II) to the pool from Fe(III)-reducing bacteria, as well as a continued modest influx of external Fe(II) (J_{2E}:J_{2B}:J_{3O} of 1:1:5), the δ^{56}Fe values of the pool and ferric oxide/hydroxide precipitates decrease slightly more rapidly than in the Rayleigh model. In the case where there is a very high flux of external Fe(II) into the pool (J_{2E}:J_{2B}:J_{3O} of 3:1:5), the δ^{56}Fe values of the pool and the resulting ferric oxide deposit initially drop, but then level off to a pseudo steady-state condition after about 40% of the pool has been solidified.

The most dramatic shifts in Fe isotope compositions are predicted for cases where there is a large return of Fe(II) from DIR bacterial activity, but small external Fe(II) flux (i.e., J_{2E}:J_{2B}:J_{3O} of 1:3:5), where, after ~40–50% solidification of the pool, δ^{56}Fe values of the ferric oxyhydroxide precipitate will drop 4–5‰. The calculations in Figure 17 illustrate that a wide range in Fe isotope compositions may be expected in the relatively simple system $Fe(II)_{aq}$-$Fe(III)_{aq}$-ferric oxide/hydroxide if extensive re-processing of Fe occurs, despite the fact that the Fe isotope fractionation factor used in the model is very similar for APIO and DIR (Table 3). Experimental verification of the predicted variations in δ^{56}Fe values during biological redox cycling of Fe in a system of Fe(II)-oxidizing and Fe(III)-reducing bacteria would be useful in predicting Fe isotope fractionations expected for modern or ancient microbial communities.

The large range in δ^{56}Fe values that are predicted for a ferric oxide/hydroxide deposit from the above model reflects incomplete reduction of ferric oxide/hydroxide precipitates by DIR bacteria. Although experimental studies of DIR using poorly crystalline HFO and rich growth media commonly run to completion, such conditions may be unrepresentative of natural conditions (e.g., Glasauer et al. 2003). Natural environments will generally involve more crystalline ferric oxide/hydroxide minerals and be poorer in nutrients, resulting in incomplete dissolution by DIR bacteria. In contrast, natural Fe(III)-bearing clay minerals may have very high surface areas, and, in fact, may be equally reactive as poorly crystalline ferric hydroxides (e.g., Kostka et al. 1999a). An important, yet largely unexplored, component to preservation of the large range in δ^{56}Fe values predicted in Figure 17 is the degree of isotopic exchange between oxide/hydroxide precipitates and aqueous Fe, which will be an important factor in determining the extent to which the predicted Fe isotope variations will be preserved in the rock record. Initial results investigating isotopic exchange between $Fe(III)_{aq}$ and ferrihydrite suggest that exchange is limited to surface Fe atoms (Poulson et al. 2003); for all but the smallest ferric hydroxide particles, these results suggest that little "back exchange" occurs between aqueous Fe and solid ferric hydroxide.

Isotopic variations in marine settings

The occurrence of isotopic and morphologic evidence for bacterial activity in marine sedimentary rocks makes such sequences a logical target for Fe isotope studies as a tracer for bacterial metabolism. One of the largest repositories of Fe that was sequestered from the oceans lies in Archean and Proterozoic Banded Iron Formations (BIFs). The ultimate source of Fe in BIFs is generally thought to be Mid-Ocean Ridge (MOR) hydrothermal fluids, based largely on REE and Nd isotope data and the assumption that REE and Fe sources would be similar (e.g., Klein and Beukes 1989; Beukes and Klein 1990; Bau and Dulski 1996). Possible relations between BIF formation and biologic activity have been discussed for many decades (e.g., Harder 1919; Barghoorn 1981; Baur et al. 1985; LaBerge et al. 1987; Nealson and Myers 1990; Schopf 1992; Skinner 1993; Aisen 1994; Brown et al. 1995; Konhauser et al. 2002). Iron isotope fractionations produced during APIO and DIR may be used to evaluate the role bacteria may have played in BIF genesis.

Many exposed BIF sequences have been subjected to deep weathering and metamorphism, where hematite and goethite are, for the most part, secondary alteration products (e.g., Beukes and Klein 1992). In a few cases, however, primary hematite is found as finely disseminated grains in carbonate- and chert-facies BIFs from fresh drill cores (e.g., Beukes et al. 1990). The $\delta^{56}Fe$ values of primary hematite from the Kuruman Iron Formation in South Africa varies from −0.7 to +0.8‰ (Johnson et al. 2003), which is significantly more variable than the near-zero values expected for ferric oxide/hydroxides derived from terrestrial weathering processes (Chapter 10A; Beard and Johnson 2004). The $\delta^{56}Fe$ values measured for primary hematite in BIFs may be explained through complete oxidation of MOR $Fe(II)_{aq}$ hydrothermal sources ($\delta^{56}Fe < 0$), incomplete oxidation of $Fe(II)_{aq}$ during anaerobic photosynthesis ($\delta^{56}Fe > 0$), or might reflect incomplete abiotic oxidation if atmospheric oxygen contents were high ($\delta^{56}Fe > 0$) (Fig. 18) (Johnson et al. 2003; Croal et al. 2004). As noted by Croal et al. (2004), distinction between abiotic oxidation and anaerobic photosynthetic oxidation may not be possible based on Fe isotope compositions alone and likely requires independent knowledge of ambient atmospheric oxygen contents.

Although the majority of attention in discussions on the origins of BIFs has been on the oxide facies, siderite facies rocks are equally important in many BIF sequences. Reaction of $Fe(II)_{aq}$ and dissolved carbonate with hematite to form siderite and magnetite has been hypothesized to be an important diagenetic process in marine basins during formation of some BIFs if sulfate contents were low (e.g., Klein and Beukes 1989; Beukes et al. 1990; Kaufman 1996; Sumner 1997). In Figure 18 we assume that $Fe(II)_{aq}$ was derived either from MOR sources or DIR, or a combination of the two, which reacted with ferric oxide precipitates to form magnetite or dissolved carbonate to produce siderite.

Under oxic conditions, primary ferric oxides derived from weathering will have $\delta^{56}Fe$ values near zero (Chapter 10A; Beard and Johnson 2004), and we will assume this isotope composition for simplicity. If the majority of $Fe(II)_{aq}$ was derived through DIR, the $\delta^{56}Fe$ value of $Fe(II)_{aq}$ is taken as ~ −1.3‰ (Beard et al. 1999, 2003a), assuming equilibrium conditions; lower $\delta^{56}Fe$ values would be expected if kinetic conditions prevailed (Johnson et al. 2004a). In the case where the $Fe(II)_{aq}$ reservoir is largely produced by DIR, and when there is excess $Fe(II)_{aq}$, the $\delta^{56}Fe$ values for magnetite, siderite, and "ankerite" would be approximately 0.0, −1.3, and −2.2‰, respectively, using the equilibrium Fe isotope fractionation factors summarized in Table 3 (Fig. 18). If $Fe(II)_{aq}$ sources were dominated by hydrothermal fluids that were associated with MOR activity, and we assume this source had a $\delta^{56}Fe$ value of ~ −0.5‰ as measured today (Sharma et al. 2001; Beard et al. 2003b), the predicted $\delta^{56}Fe$ values for magnetite, siderite, and "ankerite" would be slightly higher, at ~ +0.8, −0.5, and −1.4‰, respectively (Fig. 18), again assuming excess $Fe(II)_{aq}$. These ranges in $\delta^{56}Fe$ values for magnetite, siderite, and ankerite indeed span those measured in BIFs (Johnson et al. 2003). If ambient $Fe(II)_{aq}$ contents were low, and diagenetic reactions ran to completion, the final $\delta^{56}Fe$ values for magnetite and Fe carbonates will reflect the relative proportions of Fe sources, and will likely lie much closer to the $\delta^{56}Fe$ values of primary ferric oxide/hydroxides. That the $\delta^{56}Fe$ values for the Fe(II)-bearing minerals in BIFs do not lie near zero suggests that they formed in the presence of substantial quantities of $Fe(II)_{aq}$, and do not simply reflect "closed-system" redox processing that ran to completion.

The wide range in $\delta^{56}Fe$ values for magnetite relative to those of siderite that are observed in BIFs may reflect mixing between two different sources of Fe (Johnson et al. 2004a). Based on the isotopic fractionations listed in Table 3, magnetite that has the highest $\delta^{56}Fe$ values, and accompanying large $\Delta_{Magnetite\text{-}Siderite}$ fractionations between adjacent bands, appears to be well explained by $Fe(II)_{aq}$ sources derived from MOR hydrothermal sources (Fig. 19), assuming a $\Delta_{Fe(II)\text{-}Magnetite}$ fractionation of −1.3‰. Assuming that the $\delta^{56}Fe$ value for MOR sources of $Fe(II)_{aq}$ was the same during formation of BIFs in the late Archean to early Proterozoic as it is today, magnetite that has moderate to low $\delta^{56}Fe$ values, from near zero to negative values, seems to require different sources for $Fe(II)_{aq}$. Johnson et al. (2004a) argued that the source for very low $\delta^{56}Fe$ values was mostly likely $Fe(II)_{aq}$ which was produced by DIR of ferric oxides (Fig. 19).

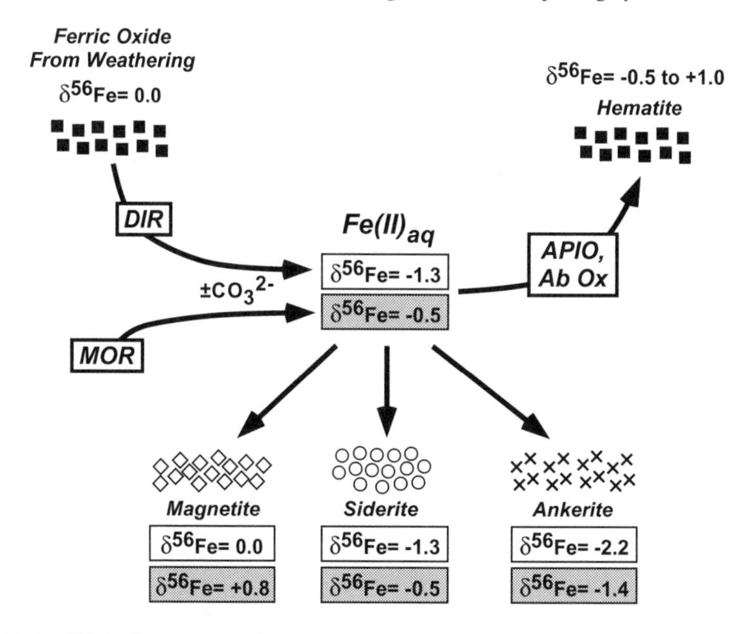

Figure 18. Possible Fe isotope compositions that may be produced by diagenetic reactions associated with DIR of ferric oxide/oxyhydroxide sediments, where $Fe(II)_{aq}$ reacts with hematite/goethite/ferrihydrite to produce magnetite, or, in the presence of carbonate ion ± Ca, siderite and ankerite. Such reactions may be envisioned to be characteristic of those that occur in relatively anoxic marine basins. $Fe(II)_{aq}$-magnetite, $Fe(II)_{aq}$-siderite, and $Fe(II)_{aq}$-ankerite isotope fractionation factors pertinent to biologic systems are from Table 3. δ^{56}Fe values in the non-shaded boxes assume that $Fe(II)_{aq}$ is dominated by DIR, where the ferric oxide (electron acceptor) has a δ^{56}Fe value of zero. δ^{56}Fe values in the shaded boxes assume that the isotopic composition of $Fe(II)_{aq}$ is dominated by Mid-Ocean Ridge (MOR) hydrothermal sources (Sharma et al. 2001; Beard et al. 2003b); this does not rule out a biologic input, but emphasizes an "external" (MOR) control of the δ^{56}Fe value for $Fe(II)_{aq}$. Oxidation of $Fe(II)_{aq}$ generated by DIR or MOR sources produces ferric oxides that have slightly negative to positive δ^{56}Fe values (upper left), dependent upon the sources of Fe(II). Oxidation may occur by APIO or by abiotic processes (Ab Ox); fractionation factors from Table 3.

Magnetite that has δ^{56}Fe values near zero might reflect formation during slow (equilibrium) reduction rates, whereas the lowest δ^{56}Fe values might reflect rapid Fe(III) reduction and kinetic isotope fractionation (Fig. 19), or multiple reduction cycles. The correlation between $\Delta_{\text{Magnetite-Siderite}}$ and δ^{56}Fe of magnetite observed in BIFs therefore probably reflects changes in the δ^{56}Fe values of aqueous Fe(II) involved in magnetite formation, reflecting a spectrum between abiotic MOR sources and biologic sources (Fig. 19). Based on the data at hand, magnetite that has negative δ^{56}Fe values appears to be an Fe isotope fingerprint for biological processing of Fe that is recorded in the ancient rock record.

The contrast in δ^{56}Fe values for Fe carbonates and magnetite from adjacent bands in BIFs can be explained through formation from a common fluid in cases where the δ^{56}Fe value of magnetite is positive; in these cases, the major source for $Fe(II)_{aq}$ for both siderite and magnetite appears to be MOR hydrothermal sources. For magnetite that has δ^{56}Fe values ≤ 0‰, however, layers of adjacent magnetite and siderite would not to be in isotopic equilibrium, where magnetite appears to be related to DIR. The relatively constant δ^{56}Fe values for siderite from BIFs studied to date suggests that most precipitated from MOR sources of $Fe(II)_{aq}$. Although DIR may produce siderite, the Fe isotope data obtained on siderite-facies BIFs so far suggests that DIR bacteria were not involved. A possible explanation would be that

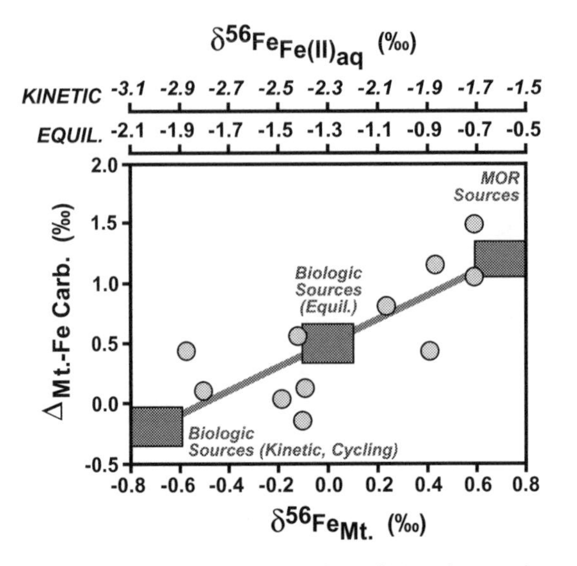

Figure 19. Iron isotope fractionations between magnetite and Fe carbonates from adjacent bands in Banded Iron Formations (Johnson et al. 2004a), compared to $Fe(II)_{aq}$ sources and Fe pathways inferred from the Fe isotope fractionations given in Table 3. $\delta^{56}Fe_{Mt}$ values are as measured, and the scales for $\delta^{56}Fe_{Fe(II)aq}$ values (top) are based on the equilibrium $\Delta_{Fe(II)aq\text{-}Magnetite}$ fractionation of −1.3‰, or −2.3‰ for kinetic fractionations (Table 3). These results suggest that magnetite which has moderately positive $\delta^{56}Fe$ values probably formed in equilibrium with MOR hydrothermal sources for $Fe(II)_{aq}$. In contrast, the lower inferred $\delta^{56}Fe$ values for $Fe(II)_{aq}$ for magnetite that has $\delta^{56}Fe \leq 0$‰ would apparently require production of $Fe(II)_{aq}$ through DIR, using the fractionation factors determined by Johnson et al. (2004a), where the lowest values would be expected to have the largest contribution from kinetic isotope fractionation, or reflect biologic cycling of Fe. The solid line is a mixing line based on changing $\delta^{56}Fe$ values for $Fe(II)_{aq}$, a constant $\Delta_{Fe(II)aq\text{-}Magnetite}$ fractionation of −1.3‰, and a constant $\delta^{56}Fe$ value for siderite of −0.5‰ (the average of that in BIFs; Johnson et al. 2003). The model assumes that isotopic equilibrium is maintained between $Fe(II)_{aq}$ and magnetite, but not between adjacent siderite and magnetite bands. Regression of the measured $\Delta_{Magnetite\text{-}Siderite}$–$\delta^{56}Fe_{Mt}$ relations produces a slope that is identical to that of the illustrated mixing line, suggesting that the measured variations reflects precipitation of magnetite from variable sources of $Fe(II)_{aq}$, scattering between an MOR (high $\delta^{56}Fe$) and a biogenic (low $\delta^{56}Fe$) component. Adapted from Johnson et al. (2004a).

precipitation of ferric oxides/hydroxides in the upper water column was minimal during times of siderite formation, which would depress DIR activity.

Identification of significant masses of magnetite-rich BIF layers that formed in equilibrium with a low-$\delta^{56}Fe$, Fe(II)-rich fluid would have important implications for the sizes of various aqueous Fe reservoirs in the oceans if they were relatively anoxic. The long residence time of Fe in an anoxic water mass would make it resistant to changes in $\delta^{56}Fe$ values (Johnson et al. 2003) from, for example, a MOR hydrothermal source ($\delta^{56}Fe \geq -0.5$‰) to one dominated by DIR ($\delta^{56}Fe \leq -1.3$‰). If the low $\delta^{56}Fe$ values for $Fe(II)_{aq}$ inferred from the low-$\delta^{56}Fe$ magnetite in BIFs (Fig. 19) reflect those of the open ocean, the biomass required to overwhelm the MOR sources must have been tremendous. Such a scenario, however, is inconsistent with the relative homogeneity in $\delta^{56}Fe$ values for interbedded siderite-rich layers in BIFs. Large changes in $\delta^{56}Fe$ values for ambient $Fe(II)_{aq}$ over the timescales involved in deposition of alternating cm-thick magnetite- and siderite-rich layers is very unlikely if the residence time of Fe was long. Instead, the most plausible explanation may be that the $\delta^{56}Fe$ values for siderite-rich layers do indeed reflect those of the open ocean water masses, dominated by a

MOR hydrothermal component, but that the very low $\delta^{56}Fe$ values inferred for $Fe(II)_{aq}$ from low $\delta^{56}Fe$ magnetite reflect interstitial pore waters and/or bottom waters that were closely associated with DIR bacteria, and not those of the open oceans. A substantial biomass of DIR bacteria is still required to process the very large inventory of Fe that is sequestered in BIFs as low-$\delta^{56}Fe$ magnetite, although not so extensive as that which would be required to lower the $\delta^{56}Fe$ values of the open oceans if the oceans were rich in $Fe(II)_{aq}$.

CONCLUSIONS

A significant proportion of the Fe isotope literature has focused on use of Fe isotopes as a "biosignature" for life, and the mechanisms by which Fe isotope variations may be produced in biologic systems has been the focus of this chapter. At first glance, an Fe isotope biosignature would seem to require isotopic fractionations that can only be produced by biology, which a substantial body of data clearly show is not the case. Although there is no question that metabolic processing of Fe produces isotopic fractionation (Beard et al. 1999, 2003a; Croal et al. 2004; Johnson et al. 2004a), it is also true that similar Fe isotope fractionations may be produced by abiologic processes (Anbar et al. 2000; Bullen et al. 2001; Johnson et al. 2002; Roe et al. 2003; Welch et al. 2003; Icopini et al. 2004). A rapidly growing set of experimental studies in abiologic and biologic systems allows us to place some constraints on the issue, and identify future directions of research that are needed.

Experimental studies of abiologic Fe isotope fractionation that seem most applicable to natural systems include fractionations between $Fe(III)_{aq}$ and $Fe(II)_{aq}$, $[Fe^{III}(H_2O)_6]^{3+}$–hematite, $[Fe^{II}(H_2O)_6]^{2+}$–siderite, oxidation of $Fe(II)_{aq}$ to ferrihydrite, and sorption of Fe(II) to ferric hydroxides (Table 3). Additional studies such as those involving concentrated HCl solutions and ion-exchange chromatography (e.g., Anbar et al. 2000; Matthews et al. 2001; Roe et al. 2003), while providing insight into the effects of different bonding environments, have less applicability to natural systems. For example, the octahedral Fe(III) chloro complexes are expected to be lower in $^{56}Fe/^{54}Fe$ ratios by 5 to 10‰ as compared to the hexaquo Fe(III) complex at room temperature (Schauble et al. 2001; Anbar et al. 2004), highlighting the fact that the isotopic compositions of very acidic, high-Cl^- fluids which essentially only exist in the laboratory will be markedly different than those found in virtually all natural environments that may support life.

The equilibrium isotopic fractionation between $[Fe^{III}(H_2O)_6]^{3+}$ and $[Fe^{II}(H_2O)_6]^{2+}$ of +2.9‰ at room temperature (Table 3), which appears to be constant over a range of Cl^- contents, provides a benchmark with which to compare isotopic fractionations during redox cycling of Fe. Although the measured isotopic contrast between ferric oxide/hydroxide substrate and $Fe(II)_{aq}$ during DIR is approximately half that measured between $Fe(III)_{aq}$ and $Fe(II)_{aq}$ (Table 3), the uncertainty in calculating the isotopic compositions of the soluble Fe(III) component based on mass-balance modeling discussed in this chapter makes it difficult at present to compare the isotopic fractionations observed during DIR with an "equivalent" abiologic system. There is little doubt that many DIR bacteria are capable of producing soluble reservoirs of Fe(III), but this component has yet to be analyzed directly so that it may be compared to the isotopic compositions of $Fe(II)_{aq}$ that is produced. An important goal for future work will be isolation of the soluble Fe(III) component, identification of the ligands that are bound to it, and direct isotopic analysis.

The observation that DIR produces $Fe(II)_{aq}$ that has low $\delta^{56}Fe$ values might be taken as a biosignature for this type of Fe metabolism. Complicating factors, however, include the effects of intermediate Fe(II) species, including sorbed Fe(II) and poorly defined NMNC Fe(II) solids. Although the evidence at hand suggests that the effects of these intermediate species in determining the $\delta^{56}Fe$ values for $Fe(II)_{aq}$ are significant during rapid Fe(III) reduction rates when kinetic effects are most likely, these issues require further study before we may fully understand their role in determining the Fe isotope fractionations that are produced by DIR in natural systems. Because the rates of Fe(III) reduction are expected to be generally far slower in nature than in most experiments, due, for example, to the greater crystallinity of ferric oxide/hydroxides and limited nutrients in natural systems, it seems likely that Fe isotope fractionations will tend to reflect equilibrium conditions in nature. As such, the low $\delta^{56}Fe$ values for $Fe(II)_{aq}$ that have been determined in experiments that involve very slow rates of DIR using crystalline substrates seem to be the most analogous to those expected in natural systems. An important, though completely unexplored avenue of research, is DIR involving Fe(III)-bearing clay minerals.

For mineral end products of DIR that are most likely to be preserved in the rock record, such as siderite and magnetite, if their $\delta^{56}Fe$ values are low, this may reflect precipitation from Fe(II)-bearing fluids that have the low-$\delta^{56}Fe$ fingerprint of DIR. It is not yet clear if the small (~0.5‰) difference in the $Fe(II)_{aq}$-siderite fractionation factor in abiologic and biologic experiments (Table 3) is significant, but their gross similarity suggests that siderite which has low $\delta^{56}Fe$ values probably precipitated from Fe(II)-bearing fluids of equal or slightly higher $^{56}Fe/^{54}Fe$ ratios; such low $\delta^{56}Fe$ values would appear to be best explained by DIR. A critical issue, however, remains the isotopic effect of cation substitution, which commonly occurs in natural Fe carbonates and may substantially affect the fluid-mineral Fe isotope fractionation factors. The case for a DIR fingerprint in low-$\delta^{56}Fe$ magnetite appears stronger, where the $^{56}Fe/^{54}Fe$ ratios for $Fe(II)_{aq}$ in equilibrium with magnetite would be ~ 1.3‰ lower, again consistent with $Fe(II)_{aq}$ that was produced by DIR.

The confidence with which $Fe(II)_{aq}$ that has $\delta^{56}Fe$ values ≤ -1.3‰ is a biosignature for DIR may be evaluated through abiotic reductive dissolution experiments of ferric oxide/hydroxides (e.g., Cornell and Schwertmann 1996; Larsen and Postma 2001), which has not yet been pursued in terms of possible Fe isotope fractionations. Could it be that the mixed Fe(III)-Fe(II) surface complexes that are likely to be present during abiotic reduction of ferrihydrite, goethite, or hematite (e.g., Hering and Stumm 1990) may produce an apparent Fe isotope fractionation between $Fe(II)_{aq}$ and ferric substrate that is similar to that observed during DIR? Although rapid sorption of Fe(II) to ferric oxide/hydroxide substrates produces low $\delta^{56}Fe$ values for the remaining aqueous Fe(II), the isotopic effects under equilibrium conditions are inferred to be less extreme. It is possible, however, that sorbed Fe(III) may have relatively high $\delta^{56}Fe$ values, even under equilibrium conditions, and this is a potential means to produce $Fe(II)_{aq}$ that has low $\delta^{56}Fe$ values. As demonstrated by the mass-balance calculations discussed in this chapter, a sorbed ≡[Fe(II), Fe(III)] phase would have to form a significant proportion of the exchangeable pool of Fe for it to influence the Fe isotope compositions of $Fe(II)_{aq}$. Such detailed experimental studies, including assessment of isotopic mass balance and exchange kinetics, have only just begun.

APIO produces ferric hydroxide precipitates that have relatively high $\delta^{56}Fe$ values, but similar effects may be observed during moderately rapid abiotic oxidation of $Fe(II)_{aq}$. Although the pathways involved in APIO are less complicated than those associated with DIR, direct measurement of intermediate species such as soluble or $Fe(III)_{aq}$ has not been done, raising the possibility that isotopic fractionation between $Fe(III)_{aq}$ and ferric hydroxide may be a significant contribution to the overall measured fractionation. The problem may be addressed through experiments aimed at determining the $Fe(III)_{aq}$-ferric hydroxide fractionation factor under equilibrium and kinetic conditions, including a range of precipitation rates. It is therefore unclear at present if ferric oxides that have positive $\delta^{56}Fe$ values in the rock record reflect APIO or abiotic oxidation of $Fe(II)_{aq}$ by high oxygen contents, or, perhaps, UV photo-oxidation. In addition, only one of the pathways in which Fe(II) is oxidized by biological processing has been explored in experiments, and experimental studies of other oxidative pathways such as chemolithotrophic oxidation or nitrate reduction would be valuable.

A useful thought experiment might be to pose the question that if abiotic reductive dissolution of ferric oxides/hydroxides and abiotic oxidation of Fe(II) to ferric oxide/hydroxide precipitates are eventually shown to produce similar overall Fe isotope fractionations as DIR and APIO, respectively, would Fe isotopes be useful as a biosignature at all? We do not know if the preceding hypothesis is correct, of course, but the implications of such a question bear on future directions Fe isotope geochemistry may take. If redox cycling of Fe involves ferric oxides/hydroxides, a mechanism for oxidation of the primary "lithologic" sources of Fe(II) is required. In the absence of UV photo-oxidation as a major process for oxidizing Fe, we are left with APIO, or increases in ambient O_2 contents due to photosynthesis where H_2O is the electron donor; either of these later cases would indicate that Fe isotopes either directly or indirectly indicate the presence of life, although the isotopic compositions may not distinguish between possible metabolic processes. Assuming a mechanism for producing ferric Fe oxides/hydroxides, reduction in the absence of biology would seem to require a redox-stratified environment, where, for example, $Fe(II)_{aq}$ in anaerobic environments reacted with ferric oxide/hydroxides that formed under oxidizing conditions. One possibility for production of reducing environments is through sulfide emanation from MOR vents, but the very low solubility of sulfide minerals suggests that large quantities of $Fe(II)_{aq}$ are unlikely to exist in sulfide-rich environments (e.g., Canfield et al. 1992). It may be that formation of redox stratifications on a planetary body, which should produce significant Fe isotope fractionations, are most likely to be produced by life. In summary, evaluation of the usefulness of Fe isotopes in tracing biological processing requires detailed consideration of the processes and environments in which biological cycling of Fe is likely to occur. Simple comparisons of the range in Fe isotope compositions produced by biological and abiological processes (e.g., Rouxel et al. 2003, 2004) may have limited usefulness in this regard.

A clear avenue of future research is to explore the S-Fe redox couple in biologic systems. Bacterial sulfate reduction and DIR may be spatially decoupled, dependent upon the distribution of poorly crystalline ferric hydroxides and sulfate (e.g., Canfield et al. 1993; Thamdrup and Canfield 1996), or may be closely associated in low-sulfate environments. Production of H_2S from bacterial sulfate reduction may quickly react with $Fe(II)_{aq}$ to form iron sulfides (e.g., Sørensen and Jeørgensen 1987; Thamdrup et al. 1994). In addition to these reactions, Fe(III) reduction may be coupled to oxidation of reduced S (e.g., Thamdrup and Canfield 1996), where the net result is that S and Fe may be cycled extensively before they find themselves in the inventory of sedimentary rocks (e.g., Canfield et al. 1993). Investigation of both S and Fe isotope fractionations produced during biochemical cycling of these elements will be an important future avenue of research that will bear on our understanding of the isotopic variations of these elements in both modern and ancient environments.

ACKNOWLEDGMENTS

Reviews by Francis Albarède, Ariel Anbar, and Susan Glasauer are appreciated. Sue Brantley is thanked for sharing several preprints and for additional comments on the paper. We also thank the Fe isotope group at UW Madison for their discussions and comments on drafts of the manuscript. Financial support for the research embodied here was provided by NASA, NSF, the Packard Foundation, and the University of Wisconsin. In particular, the NASA Astrobiology Institute supported a large portion of our work on Fe isotope fractionations in biologic systems. Collaborations with Carmen Agular, Nic Beukes, Paul Braterman, Lea Cox, Laura Croal, Andreas Kappler, Kase Klein, Hiroshi Ohmoto, Rebecca Poulson, Silke Severmann, Joseph Skulan, Henry Sun, Sue Welch, Rene Wiesli, and Kosei Yamaguchi have added greatly to our understanding of Fe isotope geochemistry in experimental and natural systems.

REFERENCES

Aisen P (1994) Iron metabolism: an evolutionary perspective. *In*: Iron Metabolism in Health and Disease. Brock JH, Halliday JW, Pippard MJ, Powell LW (eds) WB Saunders, London, p 1-30

Anbar AD, Roe JE, Barling J, Nealson KH (2000) Nonbiological fractionation of iron isotopes. Science 288: 126-128

Anbar AD, Jarzecki AA, Spiro TG (2004) Theoretical investigation of iron isotope fractionation between $Fe(H_2O)_6^{3+}$ and $Fe(H_2O)_6^{2+}$: implications for iron stable isotope geochemistry. Geochim Cosmochim Acta, in press

Appelo CAJ, VanDerWeiden MJJ, Tournassat C, Charlet L (2002) Surface complexation of ferrous iron and carbonate on ferrihydrite and the mobilization of arsenic. Environ Sci Techno 36:3096-3103

Ardizzone S, Formaro L (1983) Temperature induced phase transformation of metastable $Fe(OH)_3$ in the presence of ferrous ions. Mat Chem Phys 8:125-133

Baker JC, Kassan J, Hamilton PJ (1995) Early diagenetic siderite as an indicator of depositional environment in the Triassic Rewan Group, southern Bowen Basin, eastern Australia. Sedimentology 43:77-88

Barghoorn ES (1981) Aspects of Precambrian paleobiology: the early Precambrian. Paleobot Paleoecol Evolution 1:1-16

Bau M, Dulski P (1996) Distribution of yttrium and rare-earth elements in the Penge and Kuruman iron-formations, Transvaal Supergroup, South Africa. Precam Res 79:37-55

Bau M, Hohndorf A, Dulski P, Beukes NJ (1997) Sources of rare-earth elements and iron in Paleoproterozoic iron-formations from the Transvaal Supergroup, South Africa: evidence from neodymium isotopes. J Geol 105:121-129

Baur ME, Hayes JM, Studley SA, Walter MR (1985) Millimeter-scale variations of stable isotope abundances in carbonates from banded iron-formations in the Hamersley Group of Western Australia. Econ Geol 80: 270-282

Beard BL, Johnson CM (1999) High precision iron isotope measurements of terrestrial and lunar materials. Geochim Cosmochim Acta 63:1653-1660

Beard BL, Johnson CM (2004) Fe isotope variations in the modern and ancient earth and other planetary bodies. Rev Mineral Geochem 55:319-357

Beard BL, Johnson CM, Cox L, Sun H, Nealson KH, Aguilar C (1999) Iron isotope biosignatures. Science 285:1889-1892

Beard BL, Johnson CM, Skulan JL, Nealson KH, Cox L, Sun H (2003a) Application of Fe isotopes to tracing the geochemical and biological cycling of Fe. Chem Geol 195:87-117

Beard BL, Johnson CM, Von Damm KL, Poulson RL (2003b) Iron isotope constraints on Fe cycling and mass balance in oxygenated Earth oceans. Geology 31:629-632

Beliaev AS, Saffarini DA, McLaughlin JL, Hunicutt D (2001) MtrC, an outer membrane decahaem *c* cytochrome required for metal reduction in *Shewanella putrefaciens* MR-1. Mol Microbio 39:722-730

Benz M, Brune A, Schink B (1998) Anaerobic and aerobic oxidation of ferrous iron neutral pH by chemoheterotrophic nitrate-reducing bacteria. Arch Microbiol 169:159-165

Beukes NJ, Klein C (1990) Geochemistry and sedimentology of a facies transition - from microbanded to granular iron-formation - in the early Proterozoic Transvaal Supergroup, South Africa. Precam Res 47: 99-139

Beukes NJ, Klein C, Kaufman AJ, Hayes JM (1990) Carbonate petrography, kerogen distribution, and carbon and oxygen isotope variations in an Early Proterozoic transition from limestone to iron-formation deposition, Transvaal Supergroup, South Africa. Econ Geol 85:663-690

Brantley SL, Liermann L, Bullen TD (2001) Fractionation of Fe isotopes by soil microbes and organic acids. Geology 29:535-538

Brantley SL, Liermann LJ, Guynn RL, Anbar A, Icopini GA, Barling J (2004) Fe isotopic fractionation during mineral dissolution with and without bacteria. Geochim Cosmochim Acta, in press

Brasier MD, Green OR, Jephcoat AP, Kleppe AK, Van Kranendonk MJ, Lindsay JF, Steele A, Grassineau NV (2002) Questioning the evidence for Earth's oldest fossils. Nature 416:76-81

Braterman PS, Cairns-Smith AG (1987) Photoprecipitation and the banded iron-formations. In: The Precambrian Iron-Formations. Appel PWU, LaBerge GL (eds) Theophrastus Pub, Athens, p 215-245

Brown DA, Gross GA, Sawicki JA (1995) A review of the microbial geochemistry of banded iron-formations. Canad Mineral 33:1321-1333

Bullen TD, White AF, Childs CW, Vivit DV, Schultz MS (2001) Demonstration of significant abiotic iron isotope fractionation in nature. Geology 29:699-702

Burgos WD, Royer RA, Fang Y, Yeh GT, Fisher AS, Jeon BH, Dempsey BA (2002) Theoretical and experimental considerations related to reaction-based modeling: a case study using Iron(III) oxide bioreduction. Geomicro J 19:253-287

Butler P (1969) Mineral compositions and equilibria in the metamorphosed iron-formation of the Gagnon region, Quebec, Canada. J Petrol 10:56-101

Caccavo F, Das A (2002) Adhesion of dissimilatory Fe(III)-reducing bacteria to Fe(III) minerals. Geomicro J 19:161-177

Caccavo F, Lonergan DJ, Lovley DR, Davis M, Stolz JF, McInerney MJ (1994) *Geobacter sulfurreducens sp. nov.*, a hydrogen- and acetate-oxidizing dissimilatory metal-reducing microorganism. App Environ Microbio 60:3752-3759

Caccavo F, Schamberger PC, Keiding K, Nielsen PH (1997) Role of hydrophobicity in adhesion of the dissimilatory Fe(III)-reducing bacterium *Shewanella alga* to amorphous Fe(III) oxide. App Environ Microbio 63:3837-3843

Canfield DE, Raiswell R, Bottrell S (1992) The reactivity of sedimentary iron minerals toward sulfide. Am J Sci 292:659-683

Canfield DE, Jørgensen BB, Fossing H, Glud R, Gundersen J, Ramsing NB, Thamdrup B, Hansen JW, Nielsen LP, Hall POJ (1993) Pathways of organic carbon dioxide oxidation in three continental margin sediments. Marine Geol 113:27-40

Chang SBR, Kirschvink JL (1985) Possible biogenic magnetite fossils from the Late Miocene Potamida clays of Crete. *In*: Magnetite Biomineralization and Magnetoreception in Organisms. Kirschvink JL, Jones DS, MacFadden BJ (eds) Plenum Press, New York, p 647-669

Chang SBR, Stolz JF, Kirschvink JL, Awramik SM (1989) Biogenic magnetite in stromatolites II. Occurrences in ancient sedimentary environments. Precam Res 43:305-315

Cloud PE (1965) Significance of the Gunflint (Precambrian) microflora. Science 148:27-45

Cloud P (1968) Atmospheric and hydrospheric evolution on the primitive earth. Science 160:729-736

Coleman ML (1993) Microbial processes - controls on the shape and composition of carbonate concretions. Marine Geol 113:127-140

Coleman ML, Hedrick DB, Lovely DR, White DC, Pye K (1993) Reduction of Fe(III) in sediments by sulphate-reducing bacteria. Nature 361:436-438

Cornell RM, Schwertmann U (1996) The iron oxides: structure, properties, reaction, occurrence and uses. VCH, Weinheim, Germany

Croal LR, Johnson CM, Beard BL, Newman DK (2004) Iron isotope fractionation by anoxygenic Fe(II)-phototrophic bacteria. Geochim Cosmochim Acta 68:1227-1242

Curtis CD, Coleman ML, Love LG (1986) Pore water evolution during sediment burial from isotopic and mineral chemistry of calcite, dolomite and siderite concretions. Geochim Cosmochim Acta 50:2321-2334

DePaolo DJ (1981) Trace element and isotopic effects of combined wallrock assimilation and fractional crystallization. Earth Planet Sci Lett 53:189-201

DiChristina TJ, Moore CM, Haller CA (2002) Dissimilatory Fe(III) and Mn(IV) reduction by *Shewanella putrefaciens* requires *ferE*, a homolog of the *pulE* (*gspE*) Type II protein secretion gene. J Bacteriol 185: 142-151

Dubiel M, Hsu CH, Chien CC, Mansfeld F, Newman DK (2002) Microbial iron respiration can protect steel from corrosion. App Environ Microbio 68:1440-1445

Dzombak DA, Morel FMM (1990) Surface Complexation Modeling: Hydrous Ferric Oxide. John Wiley and Sons

Eggar-Gibbs ZG, Jude B, Dominik J, Loizeau JL, Oldfield F (1999) Possible evidence for dissimilatory bacterial magnetite dominating the magnetic properties of recent lake sediments. Earth Planet Sci Lett 168:1-6

Ehrenreich A, Widdel F (1994) Anaerobic oxidation of ferrous iron by purple bacteria, a new type of phototrophic metabolism. App Environ Microbio 60:4517-4526

Ehrlich HL (1996) Geomicrobiology. 3rd Edn. Marcel Dekker, New York, NY

Eiler JM, Mojzsis SJ, Arrhenius G (1997) Carbon isotope evidence for early life; discussion. Nature 386:665

Emerson D (2000) Microbial oxidation of Fe(II) and Mn(II) at circumneutral pH. *In*: Environmental metal-microbe interactions. Lovley DR (ed) ASM Press, Washington DC, p 31-52

Ewers WE (1983) Chemical factors in the deposition and diagenesis of banded iron-formation. *In*: Iron formations: facts and problems. Trendall AF, Morris RC (eds) Elsevier, Amsterdam, p 491-512

Farley KJ, Dzombak DA, Morel FMM (1985) A surface precipitation model for the sorption of cations on metal oxides. J Colloid Interface Sci 106:226-242

Fedo CM, Whitehouse MJ (2002) Metasomatic origin of quartz-pyroxene rock, Akilia, Greenland, and implications for Earth's earliest life. Science 296:1448-1452

Floran RJ, Papike JJ (1975) Petrology of the low-grade rocks of the Gunflint Iron Formation, Ontario-Minnesota. Geol Soc Amer Bull 86:1169-1190

Frankel RB, Blakemore RP, Wolfe RS (1979) Magnetite in freshwater magnetotactic bacteria. Science 203: 1355-1356

Frankel RB, Blakemore RP, Torres de Araujo FF, Esquivel DMS, Danon J (1981) Magnetotactic bacteria at the geomagnetic equator. Science 212:1269-1270

Fredrickson JK, Zachara JM, Kennedy DW, Dong H, Onstott TC, Hinman NW, Li S (1998) Biogenic iron mineralization accompanying the dissimilatory reduction of hydrous ferric oxide by a groundwater bacterium. Geochim Cosmochim Acta 62:3239-3257

Gaspard S, Vazques F, Holliger C (1998) Localization and solubilization of the Fe(II) reductase of *Geobacter sulfurreducens*. App Environ Microbio 64:3188-3194

Ghiorse WC (1989) Manganese and iron as physiological electron donors and acceptors in aerobic-anaerobic transition zones. *In:* Microbial mats. Cohen Y, Rosenberg E (eds) ASM Press, Washington DC, p 163-179

Glasauer S, Langley S, Beveridge TJ (2002) Intracellular iron minerals in a dissimilatory iron-reducing bacterium. Science 295:117-119

Glasauer S, Weidler PG, Langley S, Beveridge TJ (2003) Controls on Fe reduction and mineral formation by a subsurface bacterium. Geochim Cosmochim Acta 67:1277-1288

Haderlein SB, Pecher K (1999) Pollutant reduction in heterogeneous Fe(II)-Fe(III) systems. *In*: Mineral-water interfacial reactions. Sparks DL, Grundl TJ (eds) American Chemical Society, Washington, DC, p 342-347

Hansel CM, Benner SG, Neiss J, Dohnalkova A, Kukkadapu RK, Fendorf S (2003) Secondary mineralization pathways induced by dissimilatory iron reduction of ferrihydrite under advective flow. Geochim Cosmochim Acta 67:2977-2992

Harder EC (1919) Iron-depositing bacteria and their geologic relations. US Geol Surv Prof Pap 113

Hartman H (1984) The evolution of photosynthesis and microbial mats: a speculation on banded iron formations. *In:* Microbial Mats: Stromatolites. Cohen Y, Castenholz RW, Halvorson HO (eds) Alan Liss Pub, New York, p 451–453

Heising S, Schink B (1998) Phototrophic oxidation of ferrous iron by *Rhodomicrobium vanniellii* strain. Microbiology 144:2263-2269

Heising S, Richter L, Ludwig W, Schink B (1999) *Chlorobium ferrooxidans sp. nov.*, a phototrophic green sulfur bacterium that oxidizes ferrous iron in coculture with a *"Geospirillum" sp.* strain. Arch Microbiol 172:116-124

Hendry JP (2002) Geochemical trends and paleohydrological significance of shallow burial calcite and ankerite cements in Middle Jurassic strata on the East Midlands Shelf (onshore UK). Sed Geol 151:149-176

Hering JG, Stumm W (1990) Oxidative and reductive dissolution of minerals. Rev Mineral 23:427-465

Hernandez ME, Newman DK (2001) Review: Extracellular electron transfer. Cell Mol Life Sci 58:1562-1571

Hernandez ME, Kappler A, Newman DK (2004) Phenazines and other redox-active antibiotics promote microbial mineral reduction. App Environ Microbio 70:921-928

Holland HD (1984) The Chemical Evolution of the Atmosphere and Oceans. Princeton Univ Press, Princeton

Icopini GA, Anbar AD, Ruebush SS, Tien M, Brantley SL (2004) Iron isotope fractionation during microbial reduction of iron: the importance of adsorption. Geology 32:205-208

Jeschke AA, Dreybrodt W (2002) Dissolution rates of minerals and their relation to surface morphology. Geochim Cosmochim Acta 66:3055-3062

Johnson CM, Skulan JL, Beard BL, Sun H, Nealson KH, Braterman PS (2002) Isotopic fractionation between Fe(III) and Fe(II) in aqueous solutions. Earth Planet Sci Lett 195:141-153

Johnson CM, Beard BL, Beukes NJ, Klein C, O'Leary JM (2003) Ancient geochemical cycling in the Earth as inferred from Fe isotope studies of banded iron formations from the Transvaal Craton. Contrib Mineral Petrol 144:523-547

Johnson CM, Roden EE, Welch SA, Beard BL (2004a) Experimental constraints on Fe isotope fractionation during magnetite and Fe carbonate formation coupled to dissimilatory hydrous ferric oxide reduction. Geochim Cosmochim Acta, in press

Johnson CM, Beard BL, Albarède F (2004b) Overview and general concepts. Rev Mineral Geochem 55:1-24

Kalinowski BE, Liermann LJ, Brantley SL, Barnes A, Pantano CG (2000) X-ray photoelectron evidence for bacteria-enhanced dissolution of hornblende. Geochim Cosmochim Acta 64:1331–1343

Kappler A, Newman DK (2004) Formation of Iron(III)-minerals by Iron(II)-oxidizing photoautotrophic bacteria. Geochim Cosmochim Acta 68:1217-1226

Karlin R, Lyle M, Heath CR (1987) Authigenic magnetite formation in suboxic marine sediments. Nature 326: 490-493

Kaufman AJ (1996) Geochemical and mineralogic effects of contact metamorphism on banded iron-formation: an example from the Transvaal Basin, South Africa. Precam Res 79:171-194

Kim J, Dong H, Seabaugh J, Newell SW, Eberl DD (2004) Role of microbes in the smectite-to-illite reaction. Science 203:830-832

Klein C (1974) Greenalite, stilpnomelane, minnesotaite, crocidolite, and carbonates in a very low-grade metamorphic Precambrian iron-formation. Can Min 12:475-498

Klein C (1978) Regional metamorphism of Proterozoic iron-formation, Labrador Trough, Canada. Am Min 63:898-912

Klein C, Gole MJ (1981) Mineralogy and petrology of parts of the Marra Mamba Iron Formation, Hamersley Basin, western Australia. Am Min 66:507-525

Klein C, Beukes NJ (1989) Geochemistry and sedimentology of a facies transition from limestone to iron-formation deposition in the early Proterozoic Transvaal Supergroup, South Africa. Econ Geol 84:1733-1774

Konhauser K (1998) Diversity of bacterial iron mineralization. Earth Sci Rev 43:91-121

Konhauser KO, Hamade T, Raiswell R, Morrice RC, Ferris FG, Southam G, Canfield DE (2002) Could bacteria have formed the Precambrian banded iron formations? Geology 30:1079-1082

Kostka JE, Stucki JW, Nealson KH, Wu J (1996) Reduction of structural Fe(III) in smectite by a pure culture of *Shewanella putrefaciens* strain MR-1. Clays Clay Min 44:522-529

Kostka JE, Haefele E, Viehweger R, Stucki JW (1999a) Respiration and dissolution of Iron(III)-containing clay minerals by bacteria. Environ Sci Tech 33:3127-3133

Kostka JE, Wu J, Nealson KH, Stucki JW (1999b) The impact of structural Fe(III) reduction by bacteria on the surface chemistry of smectite clay minerals. Geochem Cosmochim Acta 63:3705-3713

Kostka JE, Dalton DD, Skelton H, Dollhopf S, Stucki JW (2002) Growth of Iron(III)-reducing bacteria on clay minerals as the sole electron acceptor and comparison of growth yields on a variety of oxidized iron forms. App Environ Microbio 68:6256-6262

LaBerge GL, Robbins EI, Han TM (1987) A model for the biological precipitation of Precambrian iron-formations geological evidence. In: The Precambrian Iron-Formations. Appel PWU, LaBerge GL (eds) Theophrastus Pub, Athens, p 69-96

Larson O, Postma D (2001) Kinetics of reductive bulk dissolution of lepidocrocite, ferrihydrite, and goethite. Geochim Cosmochim Acta 65:1367-1379

Lasaga AC (1981) Rate laws of chemical reactions. Rev Mineral 8:1-68

Laverne C (1993) Occurrence of siderite and ankerite in young basalts from the Galápagos Spreading Center (DSDP Holes 506G and 507B). Chem Geol 106:27-46

Lesher CM (1978) Mineralogy and petrology of the Sokoman Iron Formation near Ardua Lake, Quebec. Can J Earth Sci 15:480-500

Liermann LJ, Kalinowski BE, Brantley SL, Ferry JG (2000) Role of bacterial siderophores in dissolution of hornblende. Geochim Cosmochim Acta 64:587–602

Liu C, Zachara JM, Gorby YA, Szecsody JE, Brown CF (2001) Microbial reduction of Fe(III) and sorption/precipitation of Fe(II) on *Shewanella putrefaciens* strain CN32. Environ Sci Technol 35:1385-1393

Lloyd JR, Blunt-Harris EL, Lovley DR (1999) The periplasmic 9.6 Kilodalton *c*-type cytochrome of *Geobacter sulfurreducens* is not an electron shuttle to Fe(III). J Bact 181:7647-7649

Lloyd JR, Sole VA, van Praagh CVG, Lovley DR (2000) Direct and Fe(II)-mediated reduction of technetium by Fe(III)-reducing bacteria. App Environ Microbio 66:3743-3749

Lovley DR (1987) Organic matter mineralization with the reduction of ferric iron: a review. Geomicro J 5: 375-399

Lovley DR (1991) Dissimilatory Fe and Mn reduction. Microbio Rev 55:259-287

Lovley DR, Phillips EJP (1988) Novel mode of microbial energy metabolism: organic carbon oxidation coupled to dissimilatory reduction of iron or manganese. App Environ Microbio 54:1472-1480

Lovley DR, Stolz JF, Nord Jr GL, Phillips EJP (1987) Anaerobic production of magnetite by a dissimilatory iron-reducing microorganism. Nature 330:252-254

Lovley DR, Coates JD, Blunt-Harris EL, Phillips EJP, Woodward JC (1996) Humic substances as electron acceptors for microbial respiration. Nature 382:445-448

Lowenstam HA (1981) Minerals formed by organisms. Science 211:1126-1131

Machamer JF (1968) Geology and origin of the iron ore deposits of the Zenith mine, Vermilion district, Minnesota. Minn Geol Surv Spec Pub Series 2:1-56

Macquaker JHS, Curtis CD, Coleman ML (1997) The role of iron in mudstone diagenesis: Comparison of Kimmeridge Clay Formation mudstones from onshore and offshore (UKCS) localities. J Sed Res 67: 871-878

Magnuson TS, Hodges-Myerson AL, Lovley DR (2000) Characterization of a membrane-bound NADH-dependent Fe^{3+} reductase from the dissimilatory Fe^{3+}-reducing bacterium *Geobacter sulfurreducens*. FEMS Microbio Lett 185:205-211

Mandal LN (1961) Transformation of iron and manganese in water-logged rice soils. Soil Sci 121-126

Mandernack KW, Bazylinski DA, Shanks III WC, Bullen TD (1999) Oxygen and iron isotope studies of magnetite produced by magnetotactic bacteria. Science 285:1892-189

Mann S, Sparks NHC, Couling SB, Larcombe MC, Franke RB (1989) Crystallochemical characterization of magnetite spinels prepared from aqueous solution. J Chem Soc Far Trans 85:3033-3044

Matthews A, Zhu XK, O'Nions K (2001) Kinetic iron stable isotope fractionation between iron (-II) and (-III) complexes in solution. Earth Planet Sci Lett 192:81-92

Matthews A, Morgans-Bell HS, Emmanuel S, Jenkyns HC, Erel Y, Halicz L (2004) Controls on iron-isotope fractionation in organic-rich sediments (Kimmeridge Clay, Upper Jurassic, southern England). Geochem Cosmochim Acta, in press

Mojzsis SJ, Arrhenius G, McKeegan KD, Harrison TM, Nutman AP, Friend CRL (1996) Evidence for life on Earth before 3,800 million years ago. Nature 384:55-59

Mozely PS (1989) Relationship between depositional environment and the elemental composition of early diagenetic siderite. Geology 17:704-706

Mozley PS, Carothers WW (1992) Elemental and isotopic compositions of siderite in the Kuparuk formation, Alaska: effect of microbial activity and water/sediment interaction on early pore-water chemistry. J Sed Pet 62:681-692

Mozley PS, Wersin P (1992) Isotopic composition of siderite as an indicator of depositional environment. Geology 20:817-820

Mozley PS, Burns SJ (1993) Oxygen and carbon isotopic composition of marine carbonate concretions: an overview. J Sed Pet 63:73-83

Myers CR, Myers JM (1993) Ferric reductase is associated with the membranes of anaerobically grown *Shewanella putrefaciens* MR-1. FEMS Microbio Lett 108:15-22

Myers CR, Nealson KH (1988) Bacterial manganese reduction and growth with manganese oxide as the sole electron acceptor. Science 240:1319-1321

Myers JM, Myers CR (2000) Role of the tetraheme cytochrome *CymA* in anaerobic electron transport in cells of *Shewanella putrefaciens* MR-1 with normal levels of menaquinone. Am Soc Microbio J Bact 183: 67-75

Nealson KH (1983) The microbial iron cycle. *In:* Microbial geochemistry. Krumbein W (ed) Blackwell Sci, Boston, p 159-190

Nealson KH, Myers CR (1990) Iron reduction by bacteria: a potential role in the genesis of banded iron formations. Amer Jour Sci 290A:35-45

Nealson KH, Saffarini D (1994) Iron and manganese in anaerobic respiration: environmental significance, phylogeny, and regulation. Ann Rev Microbio 48:311-343

Nealson KH, Stahl DA (1997) Microorganisms and biogeochemical cycles: what can we learn from layered microbial communities? Rev Mineral 35:5-34

Nevin KP, Lovley DR (2000) Lack of production of electron-shuttling compounds or solubilization of Fe(III) during reduction of insoluble Fe(III) oxide by *Geobacter metallireducens*. App Environ Microbio 66: 2248-2251

Nevin KP, Lovley DR (2002a) Mechanisms for Fe(III) oxide reduction in sedimentary environments. Geomicrobio J 19:141-159

Nevin KP, Lovley DR (2002b) Mechanisms for accessing insoluble Fe(III) oxide during dissimilatory Fe(III) reduction by *Geothrix fermentans*. App Environ Microbio 68:2294-2299

Newman DK, Kolter R (2000) A role for excreted quinones in extracellular electron transfer. Nature 405: 94-97

Nicholson AM, Stolz JF, Pierson BK (1987) Structure of a microbial mat at Great Sippewissett Marsh, Cape Cod, Massachusetts. FEMS Microb Ecol 45:343-363

Parmar N, Warren LA, Roden EE, Ferris FG (2000) Solid phase capture of strontium by the iron reducing bacteria *Shewanella alga* strain BrY. Chem Geol 169:281-288

Pearson MJ (1974) Sideritic concretions from the Westphalian of Yorkshire: a chemical investigation of the carbonate phase. Min Mag 39:696-699

Pierson BK, Parenteau MN (2000) Phototrophs in high iron microbial mats: microstructure of mats in iron-depositing hot springs. FEMS Microb Ecol 32:181-196

Pierson BK, Parenteau MN, Griffin BM (1999) Phototrophs in high-iron concentration microbial mats: physiological ecology of phototrophs in an iron-depositing hot spring. App Environ Microbio 65: 5474-5483

Polyakov VB, Mineev SD (2000) The use of Mössbauer spectroscopy in stable isotope geochemistry. Geochim Cosmochim Acta 64:849-865

Postma D (1977) The occurrence and chemical composition of recent Fe-rich mixed carbonates in a river bog. J Sed Pet 47:1089-1098

Postma D (1981) Formation of siderite and vivianite and the pore-water composition of a recent bog sediment in Denmark. Chem Geol 31:225-244

Postma D (1982) Pyrite and siderite formation in brackish and freshwater swamp sediments. Amer J Sci 282: 1151-1183

Poulson RL, Beard BL, Johnson CM (2003) Investigating isotope exchange between dissolved aqueous and precipitated amorphous iron species in natural and synthetic systems. 13th Ann Goldschmidt Meeting, abstract A382

Pye K, Dickson JAD, Schiavon N, Coleman ML, Cox M (1990) Formation of siderite-Mg-calcite-iron sulphide concretions in intertidal marsh and sandflat sediments, north Norfolk, England. Sedimentology 37:325-343

Raiswell R, Fisher QJ (2000) Mudrock-hosted carbonate concretions: a review of growth mechanisms and their influence on chemical and isotopic composition. J Geol Soc 157: 239-251

Roden EE, Lovley DR (1993) Dissimilatory Fe (III) reduction by the marine microorganism *Desulfuromonas acetoxidans*. App Environ Microbio 59:734-742

Roden EE, Zachara JM (1996) Microbial reduction of crystalline iron(III) oxides: influence of oxide surface area and potential for cell growth. Environ Sci Technol 30:1618-1628

Roden EE, Urrutia MM (2002) Influence of biogenic Fe(II) on bacterial reduction of crystalline Fe(III) oxides. Geomicrobio J 19:209-251

Roden EE, Leonardo MR, Ferris FG (2002) Immobilization of strontium during iron biomineralization coupled to dissimilatory hydrous ferric oxide reduction. Geochim Cosmochim Acta 66:2823-2839

Roe JE, Anbar AD, Barling J (2003) Nonbiological fractionation of Fe isotopes: evidence of an equilibrium isotope effect. Chem Geol 195: 69-85

Rouxel O, Dobbek N, Ludden J, Fouquet Y (2003) Iron isotope fractionation during oceanic crust alteration. Chem Geol 202:155-182

Rouxel O, Fouquet Y, Ludden JN (2004) Subsurface processes at the Lucky Strike hydrothermal field, Mid-Atlantic Ridge: evidence from sulfur, selenium and iron isotopes. Geochim Cosmochim Acta, in press

Schauble EA, Rossman GR, Taylor HP (2001) Theoretical estimates of equilibrium Fe-isotope fractionations from vibrational spectroscopy. Geochim Cosmochim Acta 65:2487-2497

Schopf JW (1992) Paleobiology of the Archaen. *In*: The Proterozoic Biosphere: A Multidisciplinary Study. Schopf JW, Klein C (eds) p 25-39

Schopf JW (1993) Microfossils of the early Archean Apex Chert: new evidence of the antiquity of life. Science 260:640-646

Schopf JW, Kudryavtsev AB, Agresti DG, Wdowiak TJ, Czaja AD (2002) Laser-Raman imagery of Earth's earliest fossils. Nature 416:73-76

Seeliger S, Corn-Ruwisch R, Schink B (1998) A periplasmic and extracellular c-type cytochrome of Geobacter sulfurreducens acts as a ferric iron reductase and as an electron carrier to other acceptors or to partner bacteria. J Bacteriol 180:3686-3691

Sharma M, Polizzotto M, Anbar AD (2001) Iron isotopes in hot springs along the Juan de Fuca Ridge. Earth Planet Sci Lett 194:39-51

Shyu JBH, Lies DP, Newman DK (2002) Protective role of tolC in efflux of the electron shuttle anthraquinone-2,6-disulfonate. J Bacteriol 184:1806-1810

Skinner HCW (1993) A review of apatites, iron and manganese minerals and their roles as indicators of biological activity in black shales. Precam Res 61:209-229

Skulan JL, Beard BL, Johnson CM (2002) Kinetic and equilibrium Fe isotope fractionation between aqueous Fe(III) and hematite. Geochim Cosmochim Acta 66:2995-3015

Sørensen J, Jeørgensen BB (1987) Early diagenesis in sediments from Danish coastal waters: microbial activity and Mn-Se-S geochemistry. Geochim Cosmochim Acta 51:1583-1590

Stahl LJ, Van Gemerden H, Krumbein WE (1985) Structure and development of a benthic microbial mat. FEMS Microbio Ecol 31:111-125

Straub KL, Benz M, Schink B, Widdel F (1996) Anaerobic, nitrate-dependent microbial oxidation of ferrous iron. App Environ Microbio 62:1458-1460

Straub KL, Benz M, Schink B (2001) Iron metabolism in anoxic environments at near neutral pH. FEMS Microbio Ecol 34:181-186

Sumner DY (1997) Carbonate precipitation and oxygen stratification in Late Archean seawater as deduced from facies and stratigraphy of the Gamohaan and Frisco Formations, Transvaal Supergroup, South Africa. Am J Sci 297:455-487

Thamdrup B, Canfield DE (1996) Pathways of carbon oxidation in continental margin sediments off central Chile. Limnol Oceanogr 41:1629-1650

Thamdrup B, Fossing H, Jeørgensen BB (1994) Manganese, iron and sulfur cycling in a coastal marine sediment, Aarhus Bay, Denmark. Geochim Cosmochim Acta 58:5115-5129

Urrutia MM, Roden EE, Fredrickson JK, Zachara JM (1998) Microbial and geochemical controls on synthetic Fe(III) oxide reduction by *Shewanella alga* strain BrY. Geomicrobio J 15:269-291

Uysal IT, Golding SD, Glikson M (2000) Petrographic and isotope constraints on the origin and evolution of authigenic carbonate minerals and the associated fluid evolution in Late Permian coal measures, Bowen Basin (Queensland), Australia. Sed Geol 136:189-206

Vargas M, Kashefi K, Blunt-Harris EL, Lovley DR (1998) Microbiological evidence for Fe(III) reduction of early Earth. Nature 395:65-67

Wade ML, Agresti DG, Wdowiak TJ, Armendarez LP, Farmer JD (1999) A Mössbauer investigation of iron-rich terrestrial hydrothermal vent systems: lessons for Mars exploration. J Geophys Res 104:8489-8507

Welch SA, Beard BL, Johnson CM, Braterman PS (2003) Kinetic and equilibrium Fe isotope fractionation between aqueous Fe(II) and Fe(III). Geochim Cosmochim Acta 67:4231-4250

Widdel F, Schnell S, Heising S, Ehrenreich A, Assmus B, Schink B (1993) Ferrous iron oxidation by anoxygenic phototrophic bacteria. Nature 362:834-836

Wiesli R, Beard BL, Johnson CM (2004) Experimental determination of Fe isotope fractionation between aqueous Fe(II), siderite and "green rust" in abiotic systems. Chem Geol, submitted

Xiong J, Fischer WM, Inoue K, Nakahara M, Bauer CE (2000) Molecular evidence for the early evolution of photosynthesis. Science 289:1724-1730

Zachara JM, Kukkadapu RK, Fredrickson JK, Gorby YA, Smith SC (2002) Biomineralization of poorly crystalline Fe(III) oxides by dissimilatory metal reducing bacteria (DMRB). Geomicrobio J 19:179-206

Zhu XK, O'Nions RK, Guo YL, Reynolds BC (2000) Secular variation of iron isotopes in North Atlantic deep water. Science 287:2000-2002

Reviews in Mineralogy & Geochemistry
Vol. 55, pp. 409-427, 2004
Copyright © Mineralogical Society of America

11

The Stable Isotope Geochemistry of Copper and Zinc

Francis Albarède

Ecole Normale Supérieure de Lyon
46 Allée d'Italie
69007 Lyon, France

INTRODUCTION

Copper and zinc are the last two elements of the first row of transition metals (d-block). Interest in these elements arises because they are both strongly chalcophile and, thanks to a rich coordination chemistry, participate in a large number of important biological compounds and reactions. Copper has two isotopes, ^{63}Cu and ^{65}Cu with respective abundances of 69.174% and 30.826% in the reference metal SRM-NIST 976 (Shields et al. 1964). Zn has five stable isotopes ^{64}Zn, ^{66}Zn, ^{67}Zn, ^{68}Zn, and ^{70}Zn with average natural abundances of 48.63, 27.90, 4.10, 18.75, and 0.62%, respectively (Rosman and Taylor 1998). The first attempts at describing the stable isotope geochemistry of copper go back to Walker et al. (1958) and Shields et al. (1965) who identified isotopic variations in the range of several per mil but the analytical difficulties were then such that these pioneering investigations remained isolated. With the exception of a paper by Rosman (1972) who determined the isotopic abundances of Zn isotopes and concluded that there is no noticeable isotopic fractionation in terrestrial samples, the stable isotope geochemistry of this element remained essentially unexplored. It was not until the advent of inductively-coupled plasma mass spectrometry (ICP-MS) instruments equipped with a magnetic sector and multiple collection that precise isotopic measurements became possible and that the isotope geochemistry of these two elements took off. Maréchal et al. (1999) published the first measurements of Cu and Zn isotope compositions in a variety of minerals and biological materials. Maréchal et al. (2000) and Pichat et al. (2003) demonstrated the variability of Zn isotopes in ferromanganese nodules, sediment trap material, and marine carbonates. Maréchal et al. (1999) and Zhu et al. (2000) confirmed the broad range of isotopic variations in copper ores observed by the earlier workers. The same authors and Zhu et al. (2002) observed that biological material has isotopic compositions which vary and differ from the solution in which organisms grow. Maréchal et al. (1999) and Zhu et al. (2000) found the first evidence of isotope fractionation of transition element isotopes on chromatographic columns and this important observation was extended to Fe isotopes by Anbar et al. (2000) and to Zn isotopes by Maréchal and Albarède (2002). So far, very little is known about isotopic fractionation of Cu and Zn isotopes between coexisting phases in controlled environments and most of the existing work is still scattered in abstracts and PhD memoirs (Rouxel 2002; Maréchal and Sheppard 2002; Young 2003; Ehrlich et al., submitted). There is no doubt anymore that the broad variability of Cu and Zn isotopic compositions in the mineral and biological worlds is an enormous potential resource awaiting more experimental and theoretical work.

1529-6466/04/0055-0011$05.00

THE CHEMICAL PROPERTIES OF Cu AND Zn

In their atomic state, Cu and Zn have electronic configurations of $[Ar]3d^{10}4s^1$ and $[Ar]3d^{10}4s^2$, respectively, where [Ar] represents the filled argon electronic shells. Copper has only two important oxidation states Cu^+ ($[Ar]3d^{10}4s^0$) and Cu^{2+} ($[Ar]3d^9 4s^0$), while Zn has only one Zn^{2+} ($[Ar]3d^{10}4s^0$). Thanks to an unpaired electron, Cu^{2+}, the most widespread state of Cu in many natural conditions, gives many copper salts their paramagnetic properties and allows d–d transitions observable through vivid colors. Native copper in the form of Cu^0 is an occasional low-temperature ore. The cation Zn^{2+} is an ubiquitous acceptor of electron pairs (which is the character of a Lewis acid), which gives this element its outstanding importance in biological processes. The energies of ionization of Cu and Zn are 7.73 and 9.39 eV for the first and 20.29 and 17.96 eV for the second ionization, respectively. The ionic radii of Cu^{2+} and Zn^{2+} in octahedral coordination are very similar, 0.73 and 0.74 Å, respectively. At room temperature, the pe^0 ($-\ln[e^-]$ in standard conditions) of the reaction $Cu^{2+} + e^- \leftrightarrow Cu^+$ is 2.6, which shows that the O_2-H_2O (20.75), Fe^{3+}-Fe^{2+} (13.0), and even the SO_4^--H_2S (5.13) pairs normally oxidize Cu into its cupric form (Morel and Hering 1993).

A particularly important feature of the coordination chemistry of transition elements is the effect of the crystal field exerted by the first neighboring ions (e.g., Cl^- or OH^-). When inserted in octahedral or tetrahedral sites with surrounding ligands, the five degenerate d orbitals of the transition elements are split into two groups and form with the ligands three t_{2g} and two e_g molecular orbitals separated by the mean field-splitting energy Δ. In an octahedral environment, the metal ion occupies the center of a ligand octahedron. The lobes of the two e_g orbitals point towards the ligands while the three t_{2g} point in between. It therefore requires more energy to counterbalance electronic repulsion and fill the e_g orbitals than the t_{2g} orbitals. The situation is opposite in a tetrahedral environment, in which the three t_{2g} orbitals experience the strongest repulsion by the corner ligands. This is particularly important for Cu^{2+}, since only three electrons occupy the two e_g orbitals with two electrons on the vertical axis (d_{z^2}) and one on the equatorial plane ($d_{x^2-y^2}$) of the octahedron. The vertical bond will therefore be harder than the equatorial bond. This asymmetry is known as Jahn-Teller effect and produces heavily distorted octahedra with a complex distribution of bond energies.

Both Zn^0 and Cu^0 alloy with metallic Fe. In silicates, Cu^{2+} and Zn^{2+} substitute for Fe^{2+} and Mg^{2+} in octahedral sites but the octahedral preference energy is small for Cu^{2+} and zero for Zn^{2+}, which, beyond their trace abundance in nature, explains why these two elements are not particularly abundant in any major FeMg-silicates. Cu and Zn form a very large number of sulfides (often in association with Fe) with Cu^{2+} and Cu^+ occupying either octahedral or tetrahedral sites. They also form scores of carbonates and hydroxides. Cu in chalcopyrite and Zn in sphalerite are tetrahedrally coordinated with S but other sulfides may contain trigonally coordinated metals. Cu in malachite and Zn in smithsonite are octahedrally coordinated with O.

In aqueous solutions, the stable complexes depend on the dominant ligands (Cotton and Wilkinson 1999). At increasing chlorine concentrations, the water molecules of the octahedrally coordinated aqua ion $[Zn(H_2O)_6)]^{2+}$ are replaced by Cl^- ions to give $[ZnCl(H_2O)_5)]^+$, $[ZnCl_2(H_2O)_4)]^0$, and $[ZnCl_4(H_2O)_2^{2-})]$. The tetrahedral complex $[ZnCl_4]^{2-}$ is also known. Although XAFS studies show that Cu^{2+} in pure water also forms an octahedrally coordinated aqua ion (D'Angelo et al. 1997; Fulton et al. 2000) (Fig. 1), in the presence of chlorine ions, H_2O is replaced by Cl^- ions. The presence of tetrahedral $[CuCl_4]^{2-}$ has also been recognized (Collings et al. 2000b; Brugger et al. 2001). It has recently been suggested that the octahedral complex may be destabilized by the Jahn-Teller effect and that Cu^{2+} is actually in fivefold coordination (Brugger et al. 2001; Pasquarello et al. 2001). Most importantly, $[Cu_2Cl_6]^{2-}$ ions are formed by two tetrahedra sharing an edge (Cotton and Wilkinson 1999; Brugger et al. 2001). Maréchal and Albarède (2002) suggested that such a multinuclear configuration accounts for the particularly strong fractionation of Cu isotopes in Cl-rich

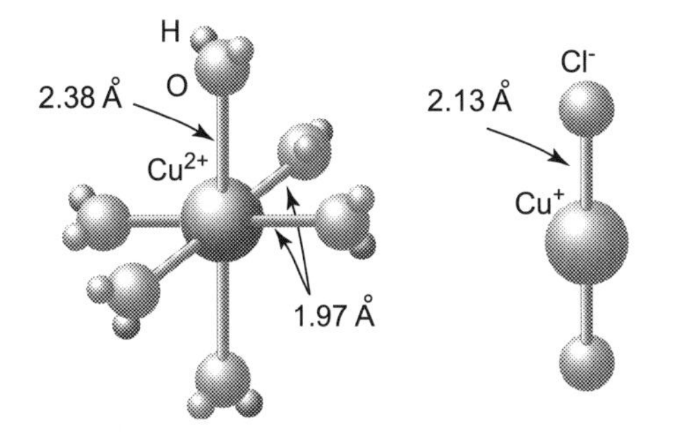

Figure 1. *Left:* octahedral coordination of Cu^{2+} in hydrous environment as an aqua $[Cu(H_2O)_6)]^{2+}$ ion. As a result of Jahn-Teller distortion, the lengths of equatorial and vertical bonds are different. *Right:* the linear $CuCl^-$ complex of Cu^+ in chlorine-rich solutions. After Fulton et al. (2000b).

solutions. Likewise, the existence of multinuclear Cu sulfide complexes (Helz et al. 1993) suggests a similar possibility in sulfide-rich aqueous systems. In contrast, Cu^+ in reduced Cl- and Br-rich solutions form the linear $CuCl^-$ and $CuBr^-$ ions (Fulton et al. 2000a,b) (Fig. 1).

THE GEOCHEMICAL PROPERTIES OF Cu AND Zn

The temperature of 50% condensation of a given element in the Solar Nebula defined by Wasson (1985) is 1037 K for Cu and 660 K for Zn. The much more volatile character of Zn with respect to Cu conditions the relative abundances of the two elements among the different classes of chondrites. Copper concentrations vary from 80 to 120 ppm in carbonaceous and ordinary chondrites (Newsom 1995). In contrast, Zn concentrations decrease from 310 ppm in the volatile-rich CI to 100 ppm in CO and CV, and to 50 ppm in ordinary chondrites. McDonough and Sun (1995) estimate the Cu and Zn content of the Bulk Silicate Earth to be 30 and 55 ppm, respectively.

In mafic and ultramafic rocks, Zn likely is concentrated in oxides, notably spinel and magnetite, while Cu is mostly hosted in sulfides. On average, a fresh MORB glass contains 72 ppm Cu and 87 ppm Zn (data compiled from the Lamont-Doherty PETDB database). The geochemical behavior of Cu and Zn in MORB has been reviewed by Doe (1994). Zn correlates inversely with mg # but is only a mildly incompatible element. Mauna Kea basalts (Albarède 1996; Rhodes 1996) and Réunion basalts (Albarède et al. 1997a) have almost equal concentrations of Cu and Zn ($\approx$ 100 ppm). Copper in basalts generally does not correlate well with other elements. In basalts with mg # > 70, Cu has a tendency to behave incompatibly (Doe 1994), while separation of sulfides, in which Cu is markedly enriched, is ubiquitous in differentiated magmas. The solubility of Cu in sulfur-saturated basaltic melt has been investigated by Holzheid and Lodders (2001) and Ripley et al. (2002) and found to be dependent on the oxygen fugacity. The relationship between subduction zones and porphyry copper deposits has long been noticed (Sillitoe 1972). Stolper and Newman (1994) suggested that the fluids expelled by slab dehydration may be rich in Cu and Zn. De Hoog et al. (2001) actually found that melt inclusions from subduction zone basalts contain more Cu than inclusions from MORB and concluded that this enrichment was already present in the source.

The aqua Zn ion is dominant in organic matter-free freshwater while the free Zn^{2+} ion and chloride complexes dominate in seawater (Stanley and Byrne 1990; Millero 1996). The free Cu^{2+} ion is dominant in freshwater, while the carbonate complexes $CuCO_3$ and $[Cu(CO_3)_2]^{2-}$ are preponderant in seawater. Speciation and solubility of Zn in Cl-rich hydrothermal solutions has been investigated by Wesolowski et al. (1998). Speciation and solubility of Cu have been investigated by Mountain and Seward (1999) for hydrothermal solutions dominated by sulfides and by Xiao et al. (1998), Liu et al. (2002), and Archibald et al. (2002) for solutions dominated by chlorides.

In the ocean, Zn behaves as a refractory nutrient associated with the hard part of the organisms such as foraminifer carbonate shells (Boyle 1981) and diatom frustules (Collier and Edmond 1984). Its depletion is nearly complete in surface water and a deep maximum indicates a deep regeneration cycle (Bruland 1980, 1983; Bruland and Frank 1983; Bruland et al. 1991). It has been considered as one of the micronutrients for which extreme depletion may limit phytoplankton productivity in surface water (Price and Morel 1990). Zinc has been suggested as a proxy for the CO_3^{2-} of the ancient oceans (Marchitto et al. 2000). Copper also shows a downward concentration increase, which signals both its uptake as a biologically active element, but also its scavenging by settling particles (Boyle et al. 1977; Bruland 1983; Bruland and Franks 1983). Copper speciation in the surface ocean is particularly complex due to the presence of many organic ligands (Moffett 1995; Moffett and Brand 1996).

Cu and Zn enter sedimentary material in substantial proportions, both in the structure of minerals (carbonates, clays) and adsorbed on surfaces. Boyle (1981) showed that foraminiferal tests may contain Zn in excess of a few ppm. Partitioning of Cu and Zn between water and carbonates has been investigated by Rimstidt et al. (1998). The crystal chemistry of Cu and Zn in goethite has been investigated by EXAFS by Manceau et al. (2000). Typical Zn and Cu concentrations in FeMn nodules and encrustations are 500–1000 ppm and 800–6000 ppm, respectively (e.g., Albarède et al. 1997b).

Both Cu and Zn are base metals. Copper is largely mined from chalcopyrite ($CuFeS_2$), chalcocite (Cu_2S), cuprite (Cu_2O), and malachite ($Cu_2CO_3(OH)_2$). Zinc is essentially produced from sphalerite (ZnS) and smithsonite ($ZnCO_3$).

THE BIOCHEMICAL USE OF Cu AND Zn

Both Cu and Zn are extensively used in biological processes (Lippard and Berg 1994; Frausto da Silva and Williams 2001). Zn concentration varies from a few ppt to a few tens of ppm in the cytoplasm of most cells (Outten and O'Halloran 2001) and in some vesicles. Copper is normally one to two orders of magnitude less concentrated. Copper and zinc play a fundamental role in the functioning of a number of enzymes. For example, hemocyanin is a complex copper protein analogous to hemoglobin and is involved in the oxidative transport of di-oxygen in the blood stream of some mollusks and arthropods. Cytochrom oxidase is a Fe-Cu protein (enzyme) with three copper sites and is found pinned in the inner mitochondrial membrane. Its role is to shuttle electrons and through the simplified respiratory reaction $O_2 + 4H^+ + 4e^- \leftrightarrow 2H_2O$ helps maintain the proton gradient across the membrane required for bidirectional ATP-ADP conversion. A site with complex coordination accommodates both Cu^{2+} and Cu^+ with little deformation in blue proteins such as plastocyanin (photosynthesis) and azurin, which allows the cell to shuttle electrons between different sites at low energy cost. Superoxide dismutase, a cytoplasmic Cu-Zn enzyme, disposes of the mutagenic O_2^-, which is transformed into molecular oxygen and hydrogen peroxide. Metallothionein is a Cu-Zn (Cd) enzyme which transports toxic metals through the cytoplasm. Carbonic anhydrase is another cytoplasmic Zn protein, which catalyzes the transformation of H_2CO_3 into HCO_3^-. Copper turnover in gastropod proteins was investigated by Mason and Borja (2002). An important group of Zn proteins are the Zn fingers, only found in eukaryotes and which play a structural role during transcription: the Zn complexes only attach themselves to specific segments of DNA. Newly discovered Cu and Zn proteins in cells have been recently reviewed by Finney and O'Halloran (2003). Morel and Price (2003) discuss the role of complexing agents in the uptake of transition elements from seawater by plankton and in the detoxification of its environment.

ANALYTICAL TECHNIQUES

The earliest attempts at measuring Cu (Shields et al. 1964; 1965) and Zn isotope compositions used thermal ionization mass spectrometry (Rosman 1972). Because Cu has only two isotopes, the mass bias could not be easily controlled. In spite of errors of several per mil, Shields et al. (1965) were able to document substantial isotopic variations in ores and other minerals. Double-spike techniques (for a description, see the chapter by Albarède and Beard in this volume) allowed Rosman (1972) to obtain TIMS isotopic data with a typical 1-2‰ error, but this study could not identify any terrestrial isotopic variation. The analytical breakthrough came with the advent of multiple-collector inductively-coupled plasma mass spectrometers (MC-ICPMS) equipped with a magnetic mass analyzer. Maréchal et al. (1999) demonstrated that a combination of standard bracketing and normalization with an external element (typically Zn for Cu and Cu for Zn) are efficient at correcting the mass bias (instrumental mass fractionation) and produced data on a variety of ores, rocks, and biological materials with a precision of 40–50 ppm (0.04–0.05 delta units). Studies by Zhu et al. (2000, 2002) used standard bracketing only and collected further data on the isotopic compositions of Cu and Zn of different materials, which terrestrial and meteoritic rocks and minerals but also included enzymes. Tanimizu et al. (2002) conducted a very careful determination of the isotopic compositions of reference materials using double spike and pointed out that the ultra-pure SRM-NIST 682 Zn standard metal is heavily fractionated with respect to other reference materials.

Although the principles of stable isotope analysis by MC-ICPMS are reviewed in this volume by Albarède and Beard, I will here summarize the principal features of the method relevant to Cu and Zn. It will first be assumed that each of these elements can be isolated from their host rock or biological matrix without isotopic fractionation, which essentially amounts to achieving full extraction yields. Maréchal et al. (1999) demonstrated that failure to achieve 100% recovery from ion exchange columns results in severe isotopic biases (see below). Mason et al. (2004a) provide a comprehensive account of the isobaric mass interferences in the mass range of Cu and Zn. The argon plasma of the ICPMS source is hot enough to ionize Cu and Zn efficiently: the low ionization energy of Cu accounts for a better sensitivity of this element. For Zn isotope analyses, a preliminary step is to ensure that the interference of ^{64}Ni with ^{64}Zn is either negligible or adequately corrected by measuring ^{62}Ni. The mass bias is normally expressed by the exponential law:

$$r_i \approx R_i \left(\frac{M_i}{M_k} \right)^{\beta} \tag{1}$$

where r and R refer to measured and true isotopic ratios, M to the atomic mass of the isotope, i and k to the numerator and denominator (reference) isotope, respectively, and β to the element-dependent fractionation coefficient. The standard bracketing method (e.g., Zhu et al. 2000) consists in running the unknown sample solution in between two standard solutions with a known isotopic ratio R_i and to interpolate the value of β of the sample between its values for the bracketing standard runs. The underlying assumption of this method is that β varies smoothly from one run to the next and only holds true for matrix-free solutions. The external normalization method consists of mixing a reference solution of another element with the sample solution (e.g., introducing standard Zn into a Cu sample and vice-versa). Let us illustrate this method by writing the previous equation for the ^{65}Cu/^{63}Cu and ^{68}Zn/^{64}Zn results as:

$$\frac{\ln\left({}^{65}\mathrm{Cu}/{}^{63}\mathrm{Cu}\right)_2 - \ln\left({}^{65}\mathrm{Cu}/{}^{63}\mathrm{Cu}\right)_1}{\ln\left({}^{68}\mathrm{Zn}/{}^{64}\mathrm{Zn}\right)_2 - \ln\left({}^{68}\mathrm{Zn}/{}^{64}\mathrm{Zn}\right)_1} = \frac{\beta_{\mathrm{Cu}}}{\beta_{\mathrm{Zn}}} \frac{\ln M_{^{65}\mathrm{Cu}}/M_{^{63}\mathrm{Cu}}}{\ln M_{^{68}\mathrm{Zn}}/M_{^{64}\mathrm{Zn}}} \tag{2}$$

which, for a constant β_{Cu}/β_{Zn} ratio, is the equation of a straight-line in a ln(^{65}Cu/^{63}Cu) vs. ln(^{68}Zn/^{64}Zn) plot. The original technique called for the condition $\beta_{Cu} = \beta_{Zn}$, which Maréchal et al. (1999) showed to be both incorrect and unnecessary. The ratio of beta values for two elements can be determined from the slope of the alignment made by mixtures of any Cu-Zn standard solutions (Fig. 2). This technique is less sensitive to matrix effects than the simple standard bracketing method used in Figure 2. When both methods are used in conjunction, unprecedented precision of 0.04 delta units is obtained (Maréchal et al. 1999). A recent discussion of the comparison between the two methods is provided by Mason et al. (2004b).

The double-spike technique of Rosman (1972) has been revived by Tanimizu et al. (2002), who used a ^{64}Zn-^{68}Zn spike and obtained precisions in the range of a fraction of a per mil. Jackson and Günther (2003) describe a laser-ablation technique of isotopic measurement, which provides a precision comparable to the standard solution nebulization methods.

As with other elements for which the stable isotope proportions are sought, extraction of Cu and Zn from rock and biological samples must be quantitative to prevent mass fractionation at this stage. After digestion in HF (rock) or HNO_3 (biological sample) media, these two elements are usually extracted on anion-exchange columns. Some advantage may be found in the similar extraction chemistry of Zn and Pb. Using the original observation by van der Walt et al. (1985) that macroporous anion-exchange resins have higher partition coefficients than regular beads, Maréchal et al. (1999) designed a chemical separation procedure, which allowed the separation of Cu, Fe, and Zn in a single pass.

The results can be reported in the conventional delta notation of stable isotopes, which is the relative deviation in parts per one thousand of a given isotopic ratio in a sample with respect to the same ratio in a standard sample. Reporting results in the epsilon notation (in parts per 10,000 as in Zhu et al. 2000, 2002) may seem legitimate, but so far stable isotope data have overwhelmingly been reported in per mil and competing notations are a source of confusion.

A natural standard for Cu is the NIST-SRM 976. Since the original work of Maréchal et al. most existing results have been reported with respect to JMC 3-0749C solution. It is not known whether the JMC metal analyzed by Tanimizu et al. (2002) was produced out of the same metal batch as the JMC 3-0749C solution. We suggest that future values should be reported with respect to the NIST-SRM 683 Zn metal, which is only 0.03‰ heavier per mass

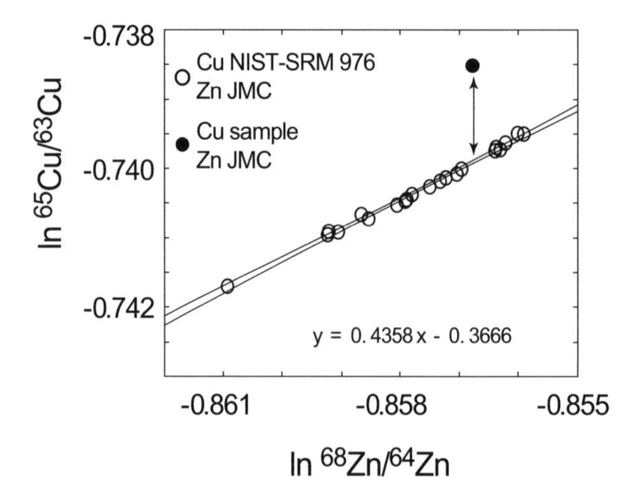

Figure 2. Correlation between the ^{65}Cu/^{63}Cu ratio of the NIST-SRM 976 and the ^{68}Zn/^{64}Zn ratio of the JMC 3-0749C standard solution in a log-log plot during 24 hours (open circles). The elevation of the sample (full circle) above the correlation line represents the deviation of its ^{65}Cu/^{63}Cu with respect to the NIST-SRM 976 value. Note the strong deviation of the slope from 0.5, the value expected if Cu and Zn isotopes were being discriminated to the same extent.

unit than the JMC metal. The NIST-SRM 682 Zn metal standard has higher purity than 683 but has an unusual fractionation of −1.2‰ per mass unit, probably inherited from the purification process. We suggest retaining the usual practice of using the most abundant isotope for reference and defining fractionation as:

$$\delta^{65}\text{Cu}(\text{‰}) = \left[\frac{\left({}^{65}\text{Cu}/{}^{63}\text{Cu}\right)_{\text{sample}}}{\left({}^{65}\text{Cu}/{}^{63}\text{Cu}\right)_{\text{NIST976}}} - 1 \right] \times 1000 \quad (3)$$

and

$$\delta^{66}\text{Zn}(\text{‰}) = \left[\frac{\left({}^{68}\text{Zn}/{}^{64}\text{Zn}\right)_{\text{sample}}}{\left({}^{68}\text{Zn}/{}^{64}\text{Zn}\right)_{\text{NIST683}}} - 1 \right] \times 1000 \quad (4)$$

For Zn, the mass dependence of isotopic fractionation should be verified, i.e., the relationship δ^{66}Zn = 2/3 δ^{67}Zn = 1/2 δ^{68}Zn (Fig. 3). Since, however, the isotopes 64 and 66 are the most abundant, δ^{66}Zn should be known with the best precision.

NATURAL ISOTOPIC VARIATIONS

It must be emphasized that we are at a very early stage of Cu and Zn isotope geochemistry and therefore that the data on which this review is based are relatively few and often have appeared in abstract form or PhD memoirs only.

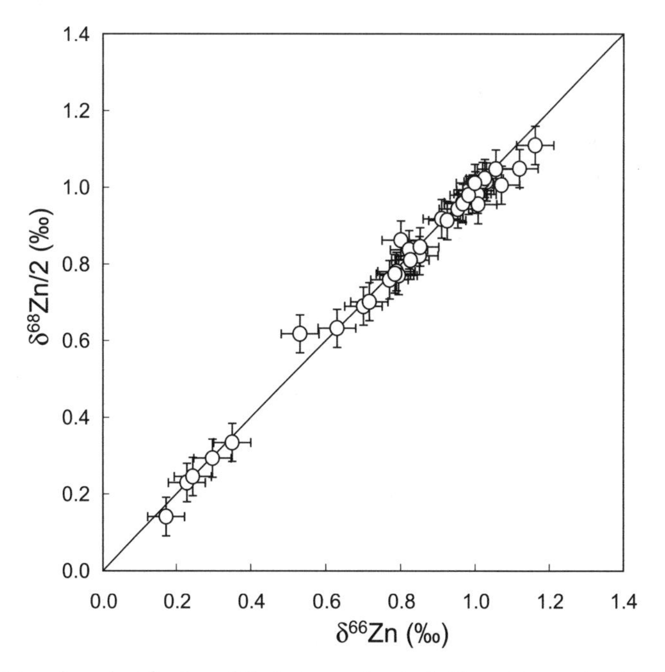

Figure 3. Mass-dependent fraction of Zn isotopes. For the FeMn-nodules analyzed by Maréchal et al. (2000), the mass dependence of isotopic fractionation holds with δ^{66}Zn = δ^{68}Zn/2.

Meteorites and igneous rocks

Copper in meteorites is depleted in the heavier 65 isotope with respect to the Earth (Luck et al. 2003; Russell et al. 2003). Luck et al.'s (2003) study of the four main groups of carbonaceous chondrites CI-CM-CO-CV showed that ^{65}Cu depletion is maximum (−1.5‰) for the CV chondrites (e.g., Allende) for which the depletion of volatile elements is strongest, which indicates that volatilization does not account for the observed isotopic heterogeneity (Fig. 4). Luck et al. (2003) found that δ^{65}Cu in CI-CM-CO classes correlates with ^{16}O excess, but this does not seems to be the case for CV (Luck et al. 2003) nor for the CR, CB, and the particularly ^{65}Cu-depleted CH-like classes (Russell et al. 2003). In contrast, chondritic Zn is relatively heavy with δ^{66}Zn up to 1‰ (Luck et al. 2001). The rather high δ^{66}Zn values of iron meteorites (up to ≈4‰)is reminiscent of a similar fractionation of Fe isotopes between metal and silicates (Zhu et al. 2002).

Very few data have been published so far on basaltic rocks. Maréchal (1998) and Maréchal et al. (2000) report values of δ^{65}Cu = −0.03‰ and δ^{66}Zn = 0.25‰ on a basalt from the active Piton de la Fournaise volcano (Réunion Island, Indian Ocean). This result is consistent with Ben Othman et al.'s (2003) findings, which show that the δ^{65}Cu values of MORB and OIB scatter around zero, while their δ^{66}Zn fall in the range 0.2–0.3‰.

With the exception of metal-silicate equilibrium, (which is relevant to mantle-core segregation), igneous processes seem to be inefficient at fractionating Cu and Zn isotopes.

Sedimentary rocks: zinc

At this point, nothing is known about Cu and Zn isotopic variability in seawater. Zn is very depleted in surface waters because it behaves as a nutrient. A substantial amount of Zn isotope data is available for sediments. Maréchal et al. (2000) found that the δ^{66}Zn values of clay minerals from different environments (Paleozoic shales, including a black shale, Mediterranean sapropels, Pacific and Atlantic sediments, a eolian dust) fall within a narrow range (0.17–0.35‰) centered around the magmatic values and therefore reflect the Zn isotope composition

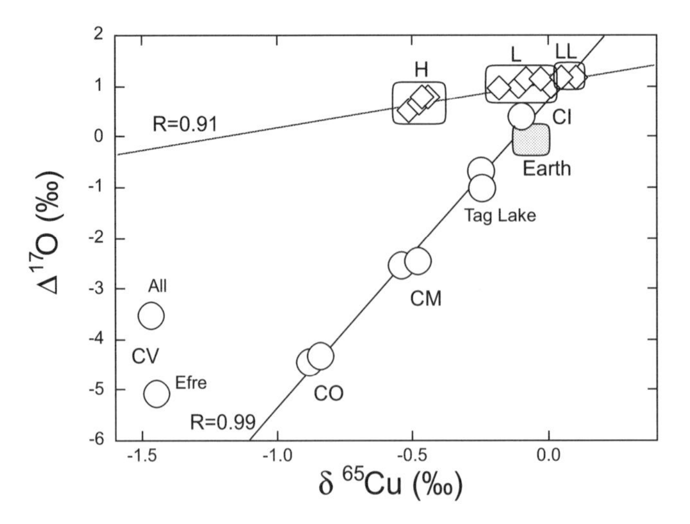

Figure 4. Correlation between the excesses of ^{16}O of the different classes of meteorites, as measured by their Δ^{17}O and their δ^{65}Cu values. The volatile-depleted carbonaceous chondrites (CV) show a deficit in the heavier isotope ^{65}Cu, which led Luck et al. (2003) to suggest that the observed correlations result from a nucleosynthetic excess of ^{63}Cu.

of their precursors. Ben Othman et al. (2001) report a range of 0.2–0.3‰ for loesses from different provenances. A downcore profile in the Central Pacific (Fig. 5) show constant values, suggesting that early diagenesis does not alter Zn isotope compositions significantly.

In contrast, limestones from the Equatorial Pacific show much heavier Zn. The surface sediment of the carbonate-dominated core TT013 19MC in the Equatorial Pacific (1° 52′ S, 139° 43′ W, z = 4500 m) has a δ^{66}Zn value of 0.79‰, while Pichat et al. (2003) analyzed the carbonate fraction of the core ODP 849 (0° 11.59′ N, 110° 31.18′ W, z = 3851 m) over the last 175 ka and found a downcore variation from 0.6 ± 0.2 to 1.2 ± 0.2‰ (Fig. 6). The reproducibility of the measurements is consistent with the internal errors. Small scale fluctuations along the core are dominated by two significant 35.2 and 21.2 ka periods. The former is difficult to assign to a known astronomical feature but the latter corresponds to a forcing by the precession of the Earth's axis of rotation.

The Zn isotope composition of sediment trap material collected over more than one year at the site of the 1991-1992 EUMELI experiment (18° 28′ N, 21° 03′ W, z = 3851 m)(0° 11.59′ N, 110° 31.18′ W, z = 3100 m) near the upwelling off the coast of Mauritania, central Atlantic (Maréchal et al. 2000), is on average similar to that of clay minerals (0.24 ± 0.14‰), which is consistent with the composition of the settling material. A subtle increase of δ^{66}Zn values of ≈0.20‰ over Spring and Summer in the fractions collected at depths of 1000 and 2500 m suggests, however, that surface productivity preferentially removes the light isotopes from surface water.

The surface layers of 41 ferromanganese nodules from different oceans also show elevated values of δ^{66}Zn (Maréchal et al. 2000) with a mean value of 0.90 ± 0.28‰ (Fig. 7). Nodules from the Southern Ocean are remarkably homogeneous with δ^{66}Zn values in the range 0.95−1.15‰, while those of lower latitudes tend to be significantly lighter (0.5−1.0‰). Since the presence of the equatorial upwelling does not affect this pattern, the zonality cannot be ascribed to productivity. Maréchal et al. (2000) suggest instead that Zn depletion above the seasonal thermocline at high latitude and preferential uptake of the lighter Zn isotopes

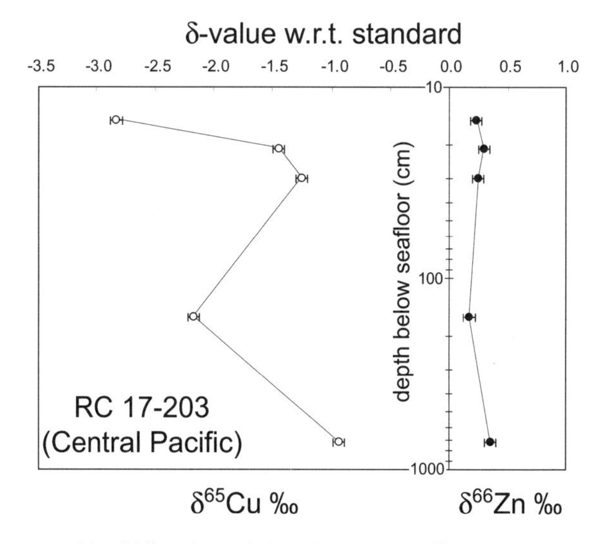

Figure 5. Downcore profile of δ^{66}Zn (Maréchal et al. 2000) and δ^{65}Cu values (unpublished) in Central Pacific core RC 17-203 (21° 50′ S, 132° 53′ W, z = 3900 m). The water-sediment interface is located below the carbonate compensation depth and deep-sea clays dominate the mineralogy of the samples.

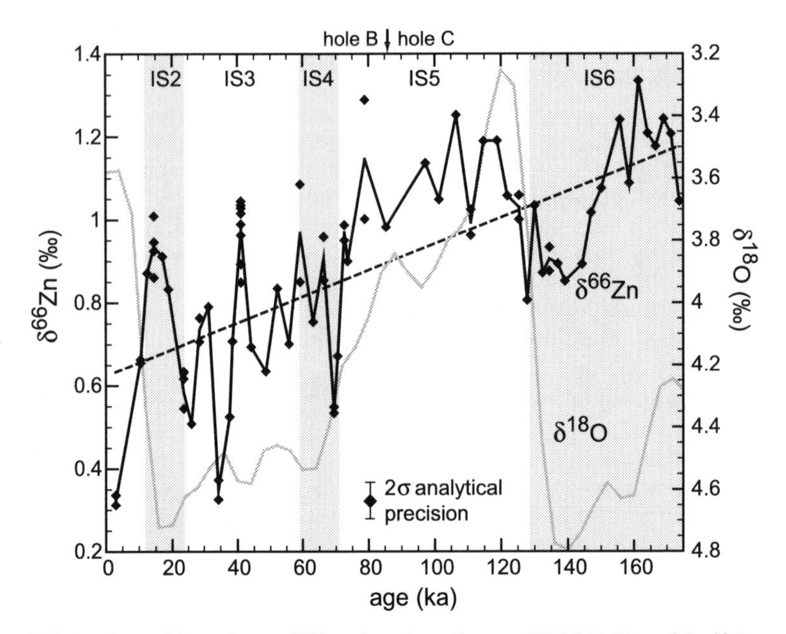

Figure 6. Fluctuations of the carbonate δ^{66}Zn values down the core ODP 849 (Central Pacific) compared with the SPECMAP scale of δ^{18}O. The Zn isotope fluctuations are modulated by two significant periods of 35.2 and 21.2 ka, the latter being ascribed to precession. Pichat et al. (2003) assigned this modulation to changes in the stability of the thermocline.

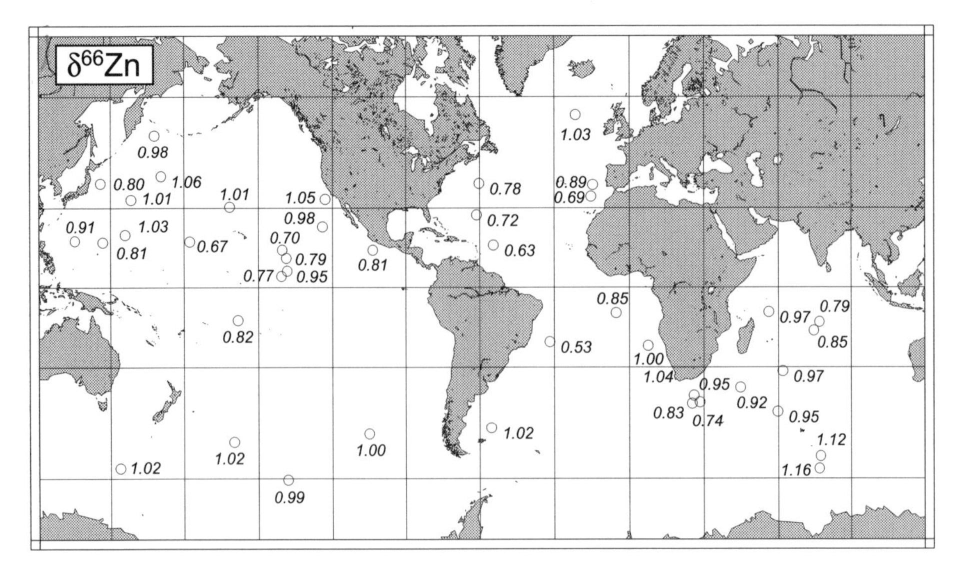

Figure 7. Map of δ^{66}Zn (‰) in the surface layer of FeMn-nodules. High-latitude samples have isotopically heavier Zn than low-latitude samples. This feature was interpreted by Maréchal et al. (2000) as reflecting the presence of a Zn-depleted seasonal thermocline at high latitude. Map drawn using the GMT software package (Wessel and Smith 1991).

increases δ^{66}Zn in the shallow ocean during the warm season, a property which is passed on to the settling sedimentary Fe-Mn hydroxides. Pichat et al. (2003) also interpreted the presence of the precession signal in the ODP 849 core as reflecting a variable amplitude of the trade winds, which, in turn, affects to a variable extent the overturn of the thermocline. In this respect, Zn isotopes behave similarly to nitrogen isotopes (Sigman et al. 1999), but the potential advantage of the former is that their fossil record will survive much longer. Whether the isotopically heavy Zn from marine carbonates and Fe-Mn nodules is compensated for by a lighter complement of possible biological affinity or is subordinate enough that it only represents a small fraction of the element flux through the ocean is not known. Archer and Vance (2002) report substantially heavier values (up to +1.1‰) for δ^{66}Zn in 2.7 Ga old black shales and enclosed sulfides from Belingwe, Zimbabwe. They interpret these values as a biological effect, but Zn from diagenetically reduced ferromanganese deposits would also provide an adequate source.

Sedimentary rocks: copper

Only scant data exist for Cu isotopes in sedimentary material. In the Central Pacific core RC 17-203, the δ^{65}Cu values of the material dominated by deep-sea clay vary from −0.94 to −2.83‰ over the first 760 cm (Fig. 5). This relatively large dispersion is probably indicative of preferential reductive removal of isotopically light Cu during early diagenesis, which may correspond to the decay of Cu-bearing organic material. The δ^{65}Cu of the surface layers of 31 ferro-manganese nodules analyzed by Maréchal et al. (2000) are shown in Figure 8. The mean value of 0.31 ± 0.23‰ (2σ) shows little deviation from the basaltic values and very small dispersion with no apparent geographic trend. This isotopic homogeneity probably reflects the fact that Cu is in much less demand by biological activity than Zn.

Maréchal (1998) reports a relatively large seasonal fluctuation of 0.2‰ in sediment trap material at the EUMELI site (see above) at depths of 1000 and 2500 m with maximum depletion of ^{63}Cu during Spring and Summer. In contrast to Zn, Archer and Vance (2002) found that Cu (and Fe) in Belingwe black shales is significantly lighter with δ^{65}Cu values down to −1.0‰.

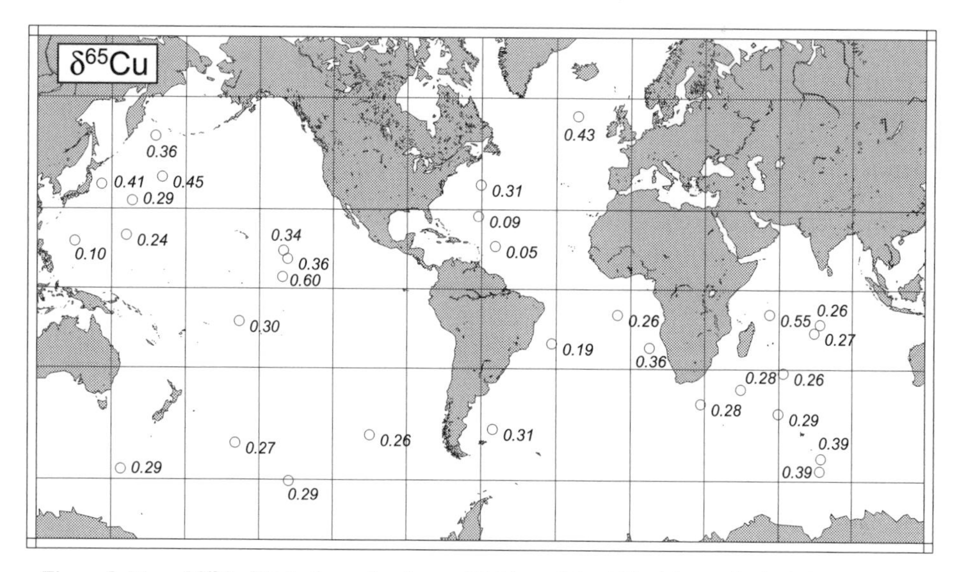

Figure 8. Map of δ^{65}Cu (‰) in the surface layer of FeMn-nodules (Albarède unpub. data). In contrast to Zn (Fig. 7), Cu in nodules is isotopically homogeneous.

Biological products

Very little is known about the isotopic composition of Cu and Zn in biological material. Maréchal et al. (1999) report δ^{65}Cu values (0.30‰) for human blood, while Zhu et al. (2002) found that octopus hemocyanin δ^{65}Cu is significantly higher (0.62‰). Maréchal et al. (1999) also report, however, near-zero δ^{65}Cu values for mussel tissue. Evidence of isotopic fractionation of the bulk biological material with respect to the ambient medium therefore remains ambiguous. However, substantial isotopic fractionation is observed between enzymes and metalloproteins on the one hand and the solution used as a feeding medium by cells on the other hand. More significant ^{65}Cu/^{63}Cu fractionation of −0.5 to −1.5‰ with the source material were observed in controlled cultures for azurin, metallothionein, and superoxide dismutase (Zhu et al. 2003), which confirms that these specific complexes preferentially incorporate the lighter Cu isotopes. This does not seem to be the case for Zn in superoxide dismutase which is preferentially enriched in ^{66}Zn over ^{64}Zn by one per mil. Maréchal et al. (2000) report δ^{66}Zn values in the range of marine carbonates and ferromanganese nodules for lobster liver and mussel tissue (0.51 and 0.82‰, respectively). Zinc in phytoplankton (0.16‰), however, seem to be distinctively lighter, which is consistent with the seasonal variations observed on EUMELI sediment trap sediments.

Cu and Zn ores

Relatively abundant data exist on the isotope composition of Cu ores from both continental deposits and black smokers (Maréchal et al. 1999; Zhu et al. 2000; Larson et al. 2003; Rouxel et al. in press; Young et al. in press; Graham et al. in press). Larson et al. (2003) (Fig. 9) summarized most of the recent data, while Rouxel et al. (in press) focuses on black smokers from the Mid-Atlantic Ridge. The δ^{65}Cu values of most chalcopyrite samples from mafic intrusions cluster tightly between −0.10 and −0.20‰. A relative isotopic homogeneity (0−1‰) is observed by Graham et al. (in press) in chalcopyrite associated with the Grasberg porphyry copper deposit (Irian Jaya, West Papua). Chalcopyrite, bornite, covellite, and atacamite from black smokers define a much broader range between ≈−1.0 and 4.0‰. Larson et al. (2003) observed isotopic variations of nearly one per mil over distances on the order of one meter. Larson et al. (2003) and Graham et al. (in press) also suggest that Cu is lighter in bornite by up to 0.4‰ with respect to coexisting chalcopyrite, which is in agreement with the black smoker data of Rouxel et al. (in press). Maréchal (1998), Larson et al. (2003), and Rouxel et al. (in press) concur that low-temperature deposits, such as chrysocolle, azurite, malachite, cuprite, and native copper show a broader range of Cu isotopic compositions (from −3.0‰ in some minerals from Ray, Arizona up to +5.6‰ for three concordant minerals from Morenci, Arizona), which strongly suggests that the isotopic variability may be ascribed to the alteration of primary high-temperature deposits. Rouxel et al. (in press) also found that the hydrous copper chloride atacamite often forming on the outer rim of black smoker conduits may be up to 2.5‰ heavier than the chalcopyrite forming the conduit walls.

Very few Zn isotope compositions have been produced on ores. Maréchal (1998) (see also Maréchal and Albarède 2000) analyzed a variety of Zn carbonates (smithsonite) and sulfides (sphalerite) from different localities in Europe. The maximum range of δ^{66}Zn values is from −0.06 to +0.69‰ with little apparent sulfide/carbonate fractionation (smithsonite may be up to 0.3‰ heavier than associated sphalerite). Unpublished data (Table 1) from the Lyon group on sphalerite from the Cévennes, Southeastern France fall within this range and show that the δ^{66}Zn values from a single mine such as Les Malines cluster within ≈0.2‰ of each other.

EXPERIMENTAL PARTITIONING

The data described above demonstrate a substantial fractionation of isotopes for both Cu and Zn. Zinc isotopic variability is particularly noticeable in the marine environment: the δ^{66}Zn

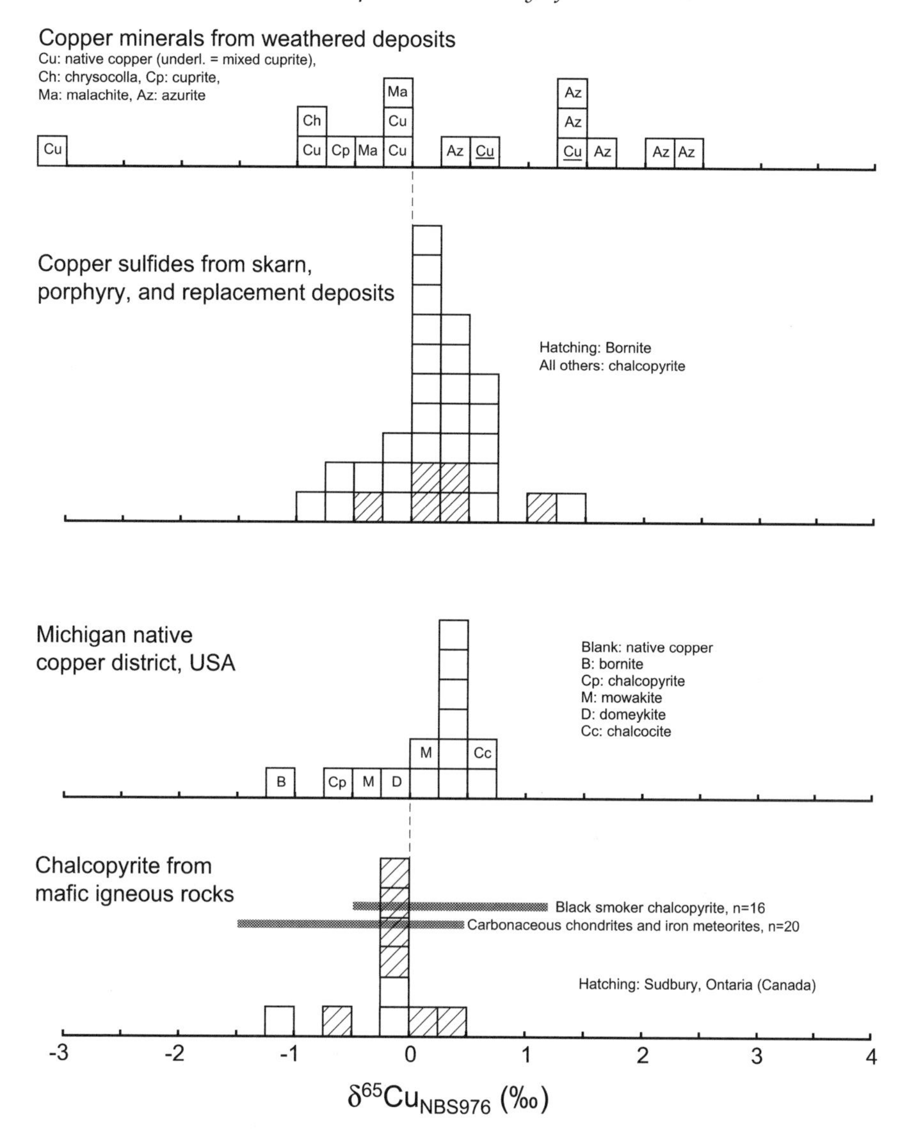

Figure 9. Compilation of δ^{65}Cu (‰) values in various copper ores (redrawn from Larson et al. 2003). Source of data: Maréchal et al. (1999), Zhu et al. (2000), Larsen et al. (in press) for ores, Maréchal et al. (1999) and Zhu et al. (2000) for black smokers, Luck et al. (2003) for meteorites.

values of carbonates and ferromanganese nodules deviate from a relatively homogeneous stock of igneous rocks, sediments, and ore deposits. In contrast, maximum variability of Cu isotope compositions has been observed in low-temperature ores. For both elements, the perspective of isotopic fractionation in proteins with respect to their environment is real (Maréchal et al. 2000; Zhu et al. 2002).

Very few experiments so far have been conducted to document isotopic fractionation of Cu and Zn under well controlled conditions. Both Zhu et al. (2002) and Rouxel (2002)

Table 1. $\delta^{66}Zn$ (‰) values of sphalerite from the Cévennes area, Southeastern France.

District	*Shaft*	*Sample #*	*$\delta^{66}Zn$ ‰*
Malines	Q Espérance	91-E-2-7	0.18
		dupl.	0.18
	Q Petralba	OMP-4	0.08
	Q Vieille Mine	OMO 3	−0.02
	Q Vieille Mine	OMA 460	−0.03
		dupl.	0.00
	Sanguinède Panneau 483	89-S-7	0.06
		dupl.	0.07
	Sanguinède Panneau 100	90-S-67	0.12
	Montdardier P41-P10	89-M-38	0.11
Largentières	Roubreau	OLA R1	−0.06
Trèves	Amas T8	OTR-T8	0.40
Bois Madame		OMB 1	0.42
		dupl.	0.47
Cévennes 1		RR-116	0.38
Cévennes 2		BELO1	0.34

investigated Cu isotope fractionation in the reductive precipitation of cuprous iodide:

$$2Cu^{2+} + 4I^{-} \leftrightarrow 2CuI + I_2 \tag{5}$$

and found that the dissolved Cu^{2+}ion is about 4‰ heavier than the Cu^{+} ion from the iodide salt. They both assign this difference to the reduction step ($Cu^{2+} + e^{-} \leftrightarrow Cu^{+}$) rather than to the precipitation step ($Cu^{+} + I^{-} \leftrightarrow CuI$).

Ehrlich et al. (submitted) measured Cu isotopic fractionation between aqueous Cu(II) and covellite between 2 and 40°C (Fig. 10). The temperature-dependent isotope fraction is fairly large (≈ 3‰) and hints at a redox control of Cu isotopic variability in abiotic systems. Maréchal and Sheppard (2002) conducted experiments at 30 and 50°C between malachite and a chloride solution for Cu isotope fractionation and between smithsonite and a nitrate solution for Zn. They found that, in this temperature range, Cu in malachite is 0.2 to 0.4‰ lighter than in the chloride solution. Replacing the chloride by nitrate ion reduces fractionation which indicates that the coordination of the Cu ion dictates isotopic fractionation. In contrast, Zn isotope fractionation between smithsonite and fluid is extremely small (<0.1‰).

Indirect evidence of isotopic fractionation among different complexes was obtained by Maréchal et al. (1999) and Maréchal and Albarède (2002) who observed different elution rates of ^{63}Cu and ^{65}Cu on anion-exchange columns (Fig. 11). These experiments were confirmed by Zhu et al. (2002) and Rouxel (2002) with similar results on fractionation coefficients. Figure 11 shows that, in HCl medium, the heavier isotope 65 is less well retained on the column than the lighter isotope 63. Maréchal and Albarède (2002) used an error function approximation to the elution curve to derive the ratio of fractionation coefficients for the 63 and 65 isotopes between the resin and the eluent. From the relationship between the elution volume (position

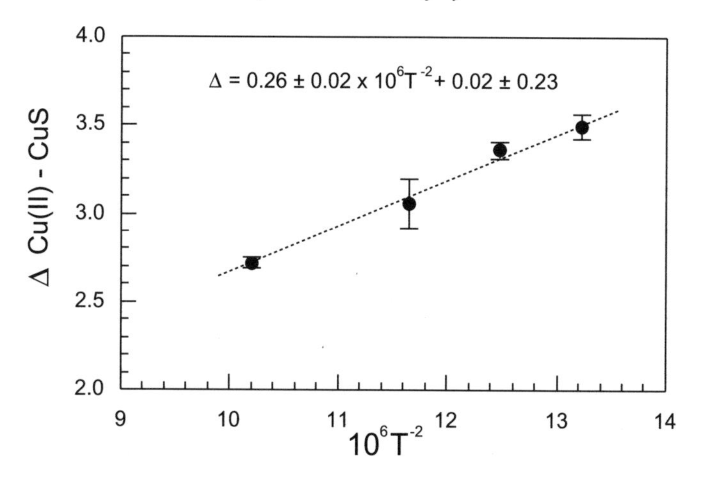

Figure 10. Experimental partitioning of Cu isotopes between hydrous cupric Cu and cuprous Cu in covellite (CuS) from 2 to 40°C (Ehrlich et al. submitted). The symbol Δ denotes the difference of $\delta^{65}Cu$ values between the two species.

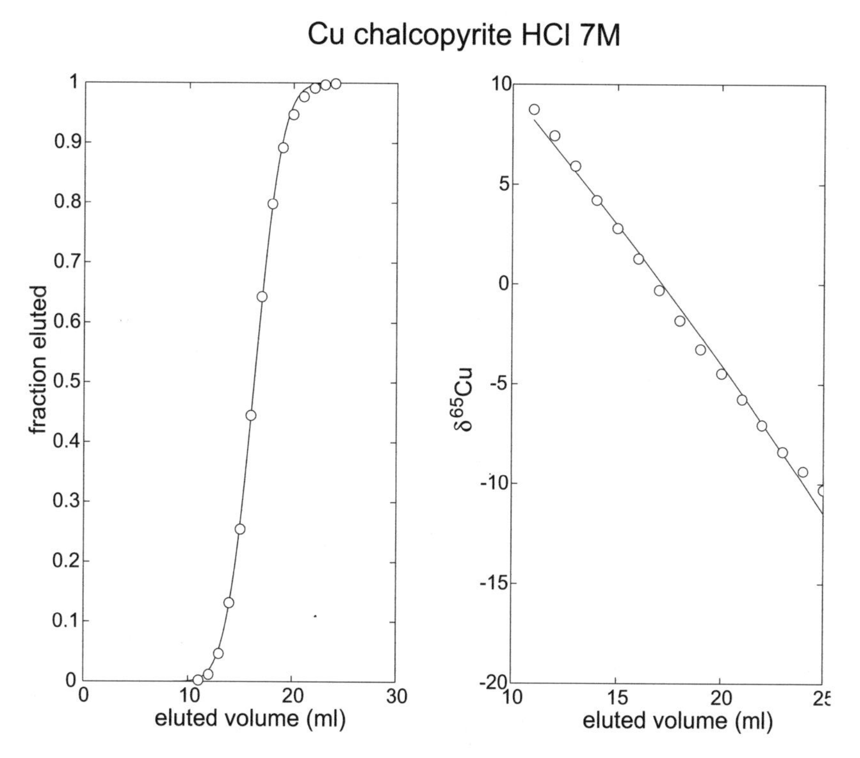

Figure 11. *Left:* elution curve of Cu from a chalcopyrite sample in HCl 7 M on anion exchange resin AG-MP1 (Maréchal and Albarède 2002). The points are fitted by an error function. The small misfit of the curve from the curve for small and large fractions eluted reflects that this function is only an approximation. *Right:* evolution of the $\delta^{65}Cu$ values in each fraction. The curve represents the isotopic values derived from the error function model.

of the peak) v and fractionation coefficient D:

$$v = V_i\,(1 + D) \tag{6}$$

in which V_i is the interstitial volume, they derived the isotope fractionation coefficient ${}^{63}_{65}\alpha = D({}^{63}\mathrm{Cu})/D({}^{65}\mathrm{Cu})$ as

$${}^{63}_{65}\alpha = 1 + \frac{{}^{63}v - {}^{65}v}{{}^{63}v - V_i} \tag{7}$$

They obtained in this way ${}^{63}_{65}\alpha$ values for different HCl normalities indicating a preferential uptake of ^{63}Cu by the resin varying between 0.5 and 0.7‰, values which compare reasonably well with batch experiments of 0.4‰. Again, the same experiments conducted for Zn produced very little fractionation (<0.1‰). These results indicate that the ability of Cu to engage in a variety of complexes, notably hydrates and halides, with very different coordination, including polynuclear components, is the source of rather strong isotopic variation. An in-depth evaluation of the isotopic effects of coordination chemistry in solution has been provided for iron by Schauble et al. (2001). Copper and zinc isotope geochemistry is awaiting a similar effort. The results would not solve all the difficulties, however, since Cu speciation and coordination in natural solutions are not fully elucidated, but they would nevertheless represent a major step towards a quantitative use of Cu isotopes as a geochemical tool.

Young et al. (in press) conducted related experiments. They examined the Cu isotope compositions of the solutions produced during acid sulfate leaching experiments aimed at extracting Cu from ore minerals (chalcopyrite, chalcocite, djurleite, bornite). The leaching experiment for chalcopyrite reads:

$$CuFeS_2 + 2Fe^{3+} \leftrightarrow Cu^{2+} + 3Fe^{2+} + 2S^0 \tag{8}$$

These authors observed that the leach solutions of chalcocite become more and more depleted in ^{63}Cu and that this depletion is accompanied by a decrease of the Cu/S ratios of the solution from 2 to 1, which these authors ascribe to fractionation between diversely coordinated Cu in the different minerals. In contrast to chalcocite, chalcopyrite leaching produces no isotope fractionation. These authors also conclude from a comparison between columns seeded with bacteria and sterile columns that bacterial mediation had little if any influence on Cu isotopic fractionation in this specific experiment, which simply reflects that bacteria do not store significant amounts of metal.

ACKNOWLEDGMENTS

I thank in particular Philippe Télouk for his exceptional talents and dedication in setting up the Lyon Plasma 54 for successful runs and Chantal Douchet for help with the analytical work. In the course of their PhDs, Chloé Maréchal and Sylvain Pichat developed outstanding expertise with Cu and Zn isotope geochemistry and became, together with Simon Sheppard, indispensable companions for scientific interaction. Steady and visionary support by the Institut des Sciences de l'Univers and by the Ministère de l'Education Nationale, in particular Philippe Vidal, was instrumental in developing these new techniques. Financial support by the Bureau de Recherches Géologiques et Minières through the Geofrance 3D program at an early stage of this work was greatly appreciated. Preprints and memoirs were made available by Peter Larson, Alan Matthews, Olivier Rouxel, and Steve Young. Comments on a first draft of the manuscript by Janne Blichert-Toft and Bruno Reynard are gratefully acknowledged. Reviews by Peter Larson, Jean-Marc Luck, and Eta Mullane helped considerably.

REFERENCES

Albarède F (1996) High resolution geochemical stratigraphy of Mauna Kea flows from the Hawaii Scientific Drilling Project core. J Geophys Res 101:11841-11853

Albarède F, Luais B, Fitton G, Semet M, Kaminski E, Upton BGJ, Bachelery P, Cheminee JL (1997a) The geochemical regimes of Piton de la Fournaise volcano (R union) during the last 530.000 years. J Petrol 38:171-201

Albarède F, Goldstein SL, Dautel D (1997b) The Nd isotopic composition of Mn-nodules from the Southern and Indian Oceans, the global oceanic Nd budget, and their bearing on deep ocean circulation. Geochim Cosmochim Acta 61:1277-1291

Anbar AD, Roe JE, Barling J, Nealson KH (2000) Nonbiological fractionation of iron isotopes. Science 288: 126-128

Archer C, Vance D (2002) Large fractionation in Fe, Cu and Zn isotopes associated with Archean microbially-mediated sulphides (abstr.). Geochim Cosmochim Acta (suppl.) 66:A26

Archibald SM, Migdisov AA, Williams-Jones AE (2002) An experimental study of the stability of copper chloride complexes in water vapor at elevated temperatures and pressures. Geochim Cosmochim Acta 66:1611-1619

Ben Othman D, Luck JM, Grousset F, Didier-Rousseau D, Albarède F (2001) Cu, Zn (and Pb) isotopes in aerosols and loesses (abstr.). Strasbourg:Europ Union Geosci XI:688

Ben Othman D, Luck JM, Tchalikian A, Albarède F (2003) Cu-Zn isotope systematics in terrestrial basalts (abstr.). Nice:Geophys Res Abstr (Europ Geophys Soc) 5:9669

Boyle EA, Sclater FR, Edmond JM (1977) The distribution of dissolved copper in the Pacific. Earth Planet Sci Lett 37:38-54

Boyle EA (1981) Cadmium, zinc, copper, and barium in foraminifera tests. Earth Planet Sci Lett 53:11-35

Brugger J, McPhail DC, Black J, Spiccia L (2001) Complexation of metal ions in brines:application of electronic spectroscopy in the study of the Cu(II)-LiCl-H_2O system between 25 and 90°C. Geochim Cosmochim Acta 65:2691-2708

Bruland KW (1980) Oceanographic distributions of cadmium, zinc, nickel, and copper in the North Pacific. Earth Planet Sci Lett 47:176-198

Bruland KW (1983) Trace elements in seawater. In: Chemical Oceanography. Riley JP, Chester R (eds) Acad. Press, London, 8:157-220

Bruland KW, Franks RP (1983) Mn, Ni, Cu, Zn, and Cd in the western North Atlantic. In: Trace Metals in Seawater, NATA Conf Ser. 4 Mar. Chem Geol. Wong CS, Boyle EA, Bruland KW, Burton JD (eds) Plenum, New York, 9:395-414

Bruland KW, Orians KJ, Cowen JP (1994) Reactive trace metals in the stratified central North Pacific. Geochim Cosmochim Acta 58:3171-3182

Collier RW, Edmond JM (1984) The trace element geochemistry of marine biogenic particulate matter. Prog Oceanogr 13:113-199

Collings MD, Sherman DM, Ragnarsdottir KV (2000) Complexation of Cu^{2+} in oxidized NaCl brines from 25°C to 175°C: results from in situ EXAFS spectroscopy. Chem Geol 167:65-73

Cotton FA, Wilkinson G (1999) Advanced Inorganic Chemistry. Wiley, Chichester

D'Angelo P, Bottari E, Festa MR, Nolting HF, Pavel NV (1997) Structural investigation of copper(II) chloride solutions using X-ray absorption spectroscopy. J Chem. Phys. 107:2807-2812

de Hoog JCM, Mason PRD, van Bergen MJ (2001) Sulfur and chalcophile elements in subduction zones: constraints from a laser ablation ICP-MS study of melt inclusions from Galunggung Volcano, Indonesia. Geochim Cosmochim Acta 65:3147-3164

Doe BR (1994) Zinc, copper, and lead in mid-ocean ridge basalts and the source rock control on Zn/Pb in ocean-ridge hydrothermal deposits. Geochim Cosmochim Acta 58:2215-2223

Ehrlich S, Butler I, Halicz L, Rickard D, Oldroyd A, Matthews A (submitted) Experimental study of copper isotope fractionation between aqueous Cu(II) and covellite, CuS. Chem Geol

Finney LA, O'Halloran TV (2003) Transition metal speciation in the cell: insights from the chemistry of metal ion receptors. Science 300:931-936

Frausto da Silva JJR, Williams RJP (2001) The Biological Chemistry of the Elements. The Inorganic Chemistry of Life. Oxford:Univ. Press

Fulton JL, Hoffmann MM, Darab JG (2000a) An X-ray absorption fine structure study of copper(I) chloride coordination structure in water up to 325°C. Chem Phys Lett 330:300-308

Fulton JL, Hoffmann MM, Darab JG, Palmer BJ, Stern EA (2000b) Copper(I) and Copper(II) Coordination structure under hydrothermal conditions at 325°C: An X-ray Absorption Fine Structure and Molecular Dynamics Study. J Phys Chem A104:11651-11663

Graham S, Pearson N, Jackson S, Griffin W, O'Reilly SY (in press) Tracing Cu and Fe from source to porphyry: in situ determination of Cu and Fe isotope ratios in sulfides from the Grasberg Cu-Au deposit. Chem Geol
Helz GR, Charnock JM, Vaughan DJ, Garner CD (1993) Multinuclearity of aqueous copper and zinc bisulfide complexes: An EXAFS investigation. Geochim Cosmochim Acta 57:15-25
Holzheid A, Lodders K (2001) Solubility of copper in silicate melts as function of oxygen and sulfur fugacities, temperature, and silicate composition. Geochim Cosmochim Acta 65:1933-1951
Jackson SE, Günther D (2003) The nature and sources of laser induced isotopic fractionation in laser ablation-multicollector-inductively coupled plasma-mass spectrometry. J Anal At Spectrom 18:205-212
Larson PB, Maher K, Ramos FC, Chang Z, Gaspar M, Meinert LD (2003) Copper isotope ratios in magmatic and hydrothermal ore forming environments. Chem Geol 201:337-350
Lippard SJ, Berg JM (1994) Principles of Bioinorganic Chemistry. Mill Valley:University Science Books
Liu W, Brugger J, McPhail DC, Spiccia L (2002) A spectrophotometric study of aqueous copper(I)-chloride complexes in LiCl solutions between 100°C and 250°C. Geochim Cosmochim Acta 66:3615-3633
Luck JM, Ben Othman D, Albarède F, Barrat JA (2001) Cu and Zn Isotopes in carbonaceous chondrites and iron meteorites (abstr.). 11th Goldschmidt Conf., Hot Springs:3638.
Luck JM, Ben Othman D, Barrat JA, Albarède F (2003) Coupled ^{63}Cu and ^{16}O excesses in chondrites. Geochim Cosmochim Acta 67:143-151
Manceau A, Schlegel ML, Musso M, Sole VA, Gauthier C, Petit PE, Trolard F (2000) Crystal chemistry of trace elements in natural and synthetic goethite. Geochim Cosmochim Acta 64:3643-3661
Marchitto Jr. TM, Curry WB, Oppo DW (2000) Zinc concentrations in benthic foraminifera reflect seawater chemistry. Paleoceanogr 15:299-306.
Maréchal C (1998) Géochimie des isotopes du Cuivre et du Zinc. Méthode, variabilités naturelles, et application océanographique, Lyon: Thesis Ecole Normale Supérieure, 253 p
Maréchal C, Albarède F (2002) Ion-exchange fractionation of copper and zinc isotopes. Geochim Cosmochim Acta 66:1499-1509
Maréchal C, Télouk P, Albarède F (1999) Precise analysis of copper and zinc isotopic compositions by plasma-source mass spectrometry. Chem Geol 156:251-273
Maréchal CN, Douchet C, Nicolas E, Albarède F (2000) The abundance of zinc isotopes as a marine biogeochemical tracer. Geochem Geophys Geosyst 1:1999GC-000029
Maréchal CN, Sheppard SMF (2002) Isotopic fractionation of Cu and Zn between chloride and nitrate solutions and malachite or smithsonite at 30° and 50°C (abstr.). Geochim Cosmochim Acta (suppl.) 66:A84.
Mason AZ, Borja MR (2002) A study of Cu turnover in proteins of the visceral complex of Littorina littorea by stable isotopic analysis using coupled HPLC-ICP-MS. Mar Envir Res 54:351-355
Mason TFD, Weiss DJ, Horstwood M, Parrish RR, Russell SS, Mullane E, Coles BJ (2004a) High-precision Cu and Zn isotope analysis by plasma source mass spectrometry. Part 1. Spectral interferences and their correction. J Anal At Spectrom 19:209-217
Mason TFD, Weiss DJ, Horstwood M, Parrish RR, Russell SS, Mullane E, Coles BJ (2004b) High-precision Cu and Zn isotope analysis by plasma source mass spectrometry. Part 2. Correcting for mass discrimination effects. J Anal At Spectrom 19:218-226
McDonough WF, Sun S-s (1995) The composition of the Earth. Chem Geol 120:223-253
Millero FJ (1996) Chemical Oceanography. Boca Raton:CRC Press
Moffett JW (1995) Temporal and spatial variability of copper complexation by strong chelators in the Sargasso Sea. Deep Sea Res I 42:1273-1295.
Moffett JW, Brand LE (1996) The production of strong, extracellular Cu chelators by marine cyanobacteria in response to Cu stress. Limnol Oceanogr 41:288-293.
Morel FMM, Hering JG (1993) Principles and Applications of Aquatic Chemistry. New York:John Wiley, New York, 588 p.
Morel FMM, Price NM (2003) The biogeochemical cycles of trace metals in the oceans. Science 300:944-947
Mountain BW, Seward TM (1999) The hydrosulphide/sulphide complexes of copper(I):experimental determination of stoichiometry and stability at 22°C and reassessment of high temperature data. Geochim Cosmochim Acta 63:11-29
Newsom HE (1995) Composition of the solar system, planets, meteorites, and major terrestrial reservoirs. In: Global Earth Physics. A handbook of physical constants. Ahrens TJ (ed) American Geophysical Union, Washington, 159-189
Outten CE, O'Halloran TV (2001) Femtomolar sensitivity of metalloregulatory proteins controlling zinc homeostasis. Science 292:2488-2492

Pasquarello A, Petri I, Salmon PS, Parisel O, Car R, Toth E, Powell DH, Fischer HE, Helm L, Merbach AE (2001) First solvation shell of the Cu(II) aqua ion: evidence for fivefold coordination. Science 291: 856-859

Pichat S, Douchet C, Albarède F (2003) Zinc isotope variations in deep-sea carbonates from the eastern equatorial Pacific over the last 175 ka. Earth Planet Sci Lett 210:167-178

Price NM, Morel FMM (1990) Cadmium and cobalt substitution for zinc in a zinc-deficient marine diatom. Nature 344:658-660

Rhodes JM (1996) Geochemical stratigraphy of lava flows sampled by the Hawaii Scientific Drilling Project. J Geophys Res 101:11,729-11,746

Rimstidt JD, Balog A, Webb J (1998) Distribution of trace elements between carbonate minerals and aqueous solutions. Geochim Cosmochim Acta 62:1851-1863

Ripley EM, Brophy JG, Li C (2002) Copper solubility in a basaltic melt and sulfide liq-uid/silicate melt partition coefficients of Cu and Fe. Geochim Cosmochim Acta 66:2791-2800

Rosman KJR (1972) A survey of the isotopic and elemental abundances of zinc. Geochim Cosmochim Acta 36: 801-819.

Rosman KJR, Taylor PN (1998) Isotopic compositions of the elements 1997. Pure Appl Chem 70:217-235

Rouxel OD (2002) Géochimie isotopique des métaux (Fe, Cu, Pb) et des metalloïdes (S, Se) dans la croûte océanique. Thesis Inst. Nat. Polytechn. Lorraine, Nancy

Rouxel OD, Fouquet Y, Ludden JN (in press) Copper isotope systematics of the Lucky Strike, Rainbow, and Logatchev hydrothermal fields on the Mid-Atlantic Ridge. Econ Geol

Russel SS, Zhu X, Guo Y, Belshaw N, Gounelle M, and Mullane E (2003) Copper isotope systematics in CR, CH-like, and CB meteorites:a preliminary study (abstr). Meteorit Planet Chem Geol, 38

Schauble EA, Rossman GR, Taylor JHP (2001) Theoretical estimates of equilibrium Fe-isotope fractionations from vibrational spectroscopy. Geochim Cosmochim Acta 65:2487-2497

Shields WR, Goldich SS, Garner EL, Murphy TJ (1965) Natural variations in the abundance ratio and the atomic weight of copper. J Geophys Res:479-491

Shields WR, Murphy TJ, Garner EL (1964) Absolute isotopic abundance ratio and the atomic weight of a reference sample of copper. J Res NBS 68A:589-592

Sigman DM, Altabet MA, McCorkle DC, Francois R, Fisher G (1999) The $\delta^{15}N$ of nitrate in the Southern Ocean: Consumption of nitrate in surface waters. Global Geochem Cycles 13:1149-1166

Sillitoe RH (1972) A plate tectonic model for the origin of porphyry copper deposits. Econ Geol 67:184-197

Stanley JK Jr, Byrne RH (1990) Inorganic complexation of Zinc (II) in seawater. Geochim Cosmochim Acta 54:753-760

Stolper E, Newman S (1994) The role of water in the petrogenesis of Mariana trough magmas.Earth Planet Sci Lett 121:293-325

Tanimizu M, Asada Y, Hirata T (2002) Absolute isotopic composition and atomic weight of commercial zinc using inductively coupled plasma mass spectrometry. Anal Chem 74:5814-5819

Van der Walt TN, Strelow FWE, Verheij R (1985) The influence of crosslinkage on the distribution coefficients and anion exchange behavior of some elements in hydrochloric acid. Solvent Extract Ion Exchange 3: 723-740

Walker EC, Cuttitta F, Senftle FE (1958) Some natural variations in the relative abundance of copper isotopes. Geochim Cosmochim Acta 15:183-194

Wasson JT (1985) Meteorites: Their Record of Early Solar System History. New York: Freeman

Wesolowski DJ, Benezeth P, Palmer DA (1998) ZnO solubility and Zn^{2+} complexation by chloride and sulfate in acidic solutions to 290°C with *in-situ* pH measurement. Geochim Cosmochim Acta 62:971-984

Wessel P, Smith WHF (1991) Free software helps map and display data. EOS Trans AGU 72:445-446

Xiao Z, Gammons CH, Williams-Jones AE (1998) Experimental study of copper(I) chloride complexing in hydrothermal solutions at 40 to 300°C and saturated water vapor pressure. Geochim Cosmochim Acta 62:2949-2964

Young SE (2003) Crystal chemical control on intra-structural Cu isotope fractionation in natural Cu-(Fe)-S minerals. PhD Mem. Univ. Arizona

Young SE, Ruiz J, Mathur R, Hiskey JB, Uhrie JL (in press) Inorganic copper fractionation as a function of average copper coordination in natural copper sulfide minerals. Chem Geol

Zhu XK, O'Nions RK, Guo Y, Belshaw NS, Rickard D (2000) Determination of Cu-isotope variation by plasma souce mass spectrometry: implications for use as geochemical tracers. Chem Geol 163:139-149

Zhu XK, Guo Y, Williams RJP, O'Nions RK, Matthews A, Belshaw NS, Canters GW, de Waal EC, Weser U, Burgess BK, Salvato B (2002) Mass fractionation processes of transition metal isotopes. Earth Planet Sci Lett 200:47-62

Reviews in Mineralogy & Geochemistry
Vol. 55, pp. 429-454, 2004
Copyright © Mineralogical Society of America

Molybdenum Stable Isotopes: Observations, Interpretations and Directions

Ariel D. Anbar

Dept. of Earth and Environmental Sciences
and Dept. of Chemistry
University of Rochester
Rochester, New York, 14627, U.S.A.

INTRODUCTION

The unusual chemistry of molybdenum (Mo) makes this trace element interesting to both geochemists and biochemists. Geochemically, Mo is relatively unreactive in oxygenated, aqueous solutions, and hence is a nominally conservative element in the oceans. In fact, Mo is removed so slowly from seawater that it is the most abundant transition metal in the oceans despite being a ppm-level constituent of the crust. In contrast, Mo is readily removed from solution in anoxic-sulfidic ("euxinic") settings, so that Mo enrichments in sediments are considered diagnostic of reducing depositional conditions. Few elements possess such bimodal redox behavior at the Earth's surface.

Biochemically, Mo draws attention because it is an essential enzyme cofactor in nearly all organisms, with particular importance for nitrogen fixation, nitrate reduction and sulfite oxidation. Such biochemical ubiquity is surprising in view of the general scarcity of Mo at the Earth's surface.

Isotopically, Mo initially catches the eye because it has seven stable isotopes of 10–25% abundance, covering a mass range of ~8% (Fig. 1). Thus, from an analyst's perspective, Mo offers both an unusually large mass spread and a number of options for isotope ratio

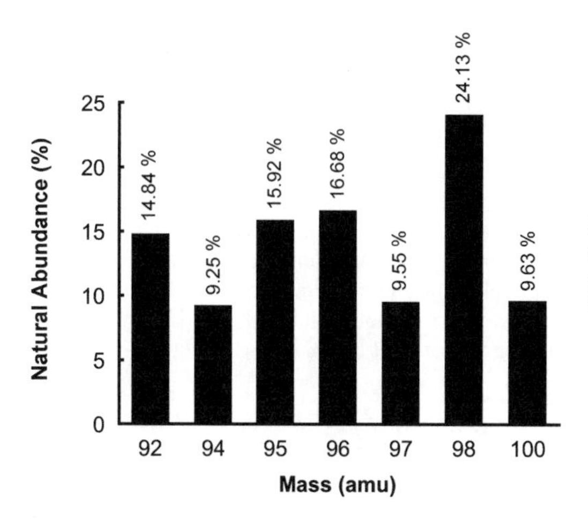

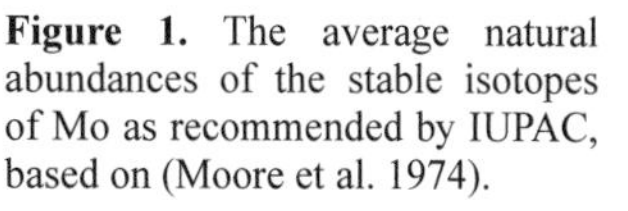
Figure 1. The average natural abundances of the stable isotopes of Mo as recommended by IUPAC, based on (Moore et al. 1974).

1529-6466/04/0055-0012$05.00

determination. Combined with rich redox chemistry and covalent-type bonding, both of which tend to drive isotope fractionation, these factors make the Mo isotope system a particularly promising target for stable isotope investigation.

In the environment, Mo isotope research began in earnest with the application of multiple-collector inductively coupled plasma mass spectrometry. While much work remains to be done, this early research points to promising applications in paleoceanography, and beyond.

This review is intended to provide an overview of this emerging stable isotope system in the context of Mo environmental biogeochemistry. Special attention is given to Mo in marine systems because this is where most initial work has focused. We begin by reviewing the environmental biogeochemistry of Mo, emphasizing the ocean budget and the role of this element in biology, and drawing comparisons and contrasts with other elements. Subsequent sections discuss analytical methods, summarize Mo isotope observations in natural samples and laboratory experiments, provide a first-order interpretation of Mo isotope systematics in the oceans, and consider logical directions for future investigations.

MOLYBDENUM IN THE ENVIRONMENT AND IN BIOLOGY: SETTING THE STAGE

Molybdenum environmental geochemistry

The distribution of Mo at the Earth's surface is unique among the transition metals. Mo is a trace constituent of the upper crust, with an average abundance of 1–2 ppm (Taylor and McLennan 1985). Yet, Mo is the most abundant transition metal in the oceans, with a concentration of ~105 nmol kg^{-1} (Morris 1975; Bruland 1983; Collier 1985). In seeking to understand this distribution, we gain insight into fundamental aspects of Mo geochemistry.

As for all elements, the distribution of Mo in the environment depends critically on chemical speciation, including oxidation state (Bertine and Turekian 1973; Morford and Emerson 1999). However, Mo is somewhat unusual in both respects. In terms of ligand coordination, Mo is one of a small number of transition metals that commonly form oxyanions and coordinate only weakly with other environmentally common ligands such as Cl^- or OH^-. Other such metals include Cr and W, which sit above and below Mo, respectively, in Group VI of the Periodic Table, as well as Tc, Re, Os and U. Hence, Mo chemistry has some analogies with these metals, as well as with nonmetals such as S, Se, P and As which also form oxyanions.

However, in terms of redox chemistry, Mo differs significantly from Cr, Se and many other elements in that it is so easily oxidized that its most highly oxidized form, Mo(VI), is prevalent when Eh > 0 V in solutions with pH > 5 (Fig. 2). As a result, Mo speciation in most oxygenated waters is believed dominated by the tetrahedrally coordinated oxyanion MoO_4^{2-}. Protonated species such as $HMoO_4^-$ or H_2MoO_4 ("molybdic acid') are quantitatively important only at pH < 5 (Fig. 2). Polynuclear Mo species such as $Mo_7O_{24}^{6-}$ or $Mo_8O_{26}^{4-}$ are observed when pH < 6, but only at Mo concentrations too high to be generally relevant in the environment (e.g., >10^{-3} M). Mo environmental chemistry in oxygenated systems is therefore loosely analogous to S, for which the dominant species is SO_4^{2-}. In particular, both elements are readily transferred to solution during oxidative weathering, are highly mobile in oxidizing fluids, and relatively non-reactive in the oceans. In fact, the chemical similarities between MoO_4^{2-} and SO_4^{2-} (e.g., charge; coordination; ionic radii) have led to the suggestion that biological assimilation of Mo may be inhibited by SO_4^{2-} (e.g., Marino et al. 2003).

The geochemical character of Mo changes sharply in H_2S-bearing systems, so much so that it has been likened to a "geochemical switch" (Helz et al. 1996; Erickson and Helz 2000). This change is seen in the concentration depth profiles of these elements in the Black Sea

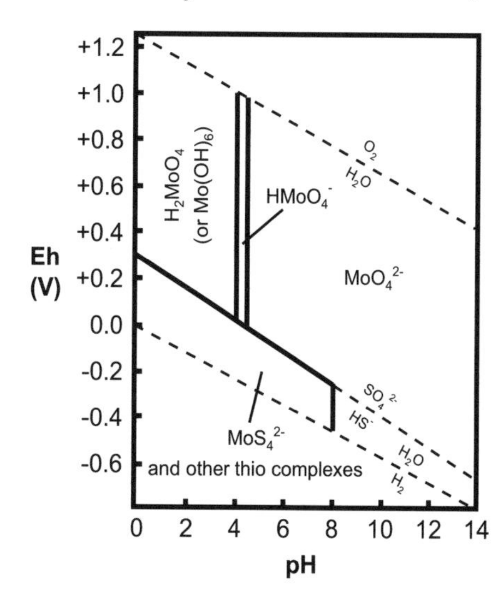

Figure 2. Eh-pH diagram of dissolved Mo speciation in the system Mo-H_2O-S. $\Sigma Mo = 10^{-6}$ M; $\Sigma S = 10^{-4}$ M. Modified after Manheim and Landergren (1974), using molybdate protonation constants from Baes and Mesmer (1986). H_2MoO_4 is related to $Mo(OH)_6$ by addition of two water molecules (see text). MoO^{2+}, included in earlier Eh-pH diagrams, is omitted because this and other Mo(V) species are typically unstable except as dimers (e.g., $Mo_2O_4^{2+}$) at higher ΣMo than common in nature. Speciation at Eh below the SO_4^{2-} - H_2S transition is not well characterized and is commonly out of equilibrium. The boundary between MoS_4^{2-} and MoO_4^{2-} is based on Erickson and Helz (2000); intermediate oxythiomolybdates are metastable and hence not indicted.

(Fig. 3) (Emerson and Huested 1991). Here, oxygenated surface waters give way to anoxia and $[H_2S] > 100$ μmol kg^{-1} below ~100 m. The total Mo concentration across this redox transition declines from ~40 nmol kg^{-1} at the surface to ~3 nmol kg^{-1} below the chemocline. Investigations of Mo in marine sediments and porewaters indicate that Mo is also removed from solution under less intensely reducing conditions; in general, authigenic accumulation of Mo occurs in sediments overlain by waters in which $O_2 < 5$ μmol kg^{-1} (Emerson and Huested 1991; Crusius et al. 1996; Dean et al. 1997; Zheng et al. 2000a; Nameroff et al. 2002).

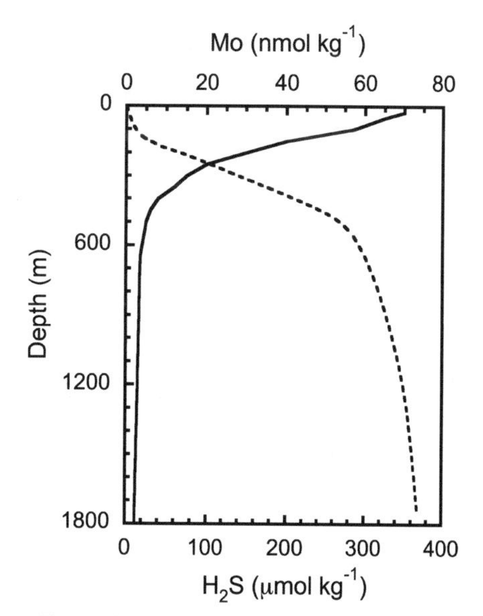

Figure 3. Depth profiles of Mo (solid line) and H_2S (dashed line) in the Black Sea (Mo: Emerson and Huested 1991; H_2S: Neretin et al. 2001). The concentration of Mo falls sharply in the euxinic zone.

The Mo removal mechanisms in reducing settings are not well understood, in part because Mo speciation in such settings is complicated by slow reaction kinetics. The reduction of Mo(VI) to form insoluble MoS_2 appears to play less of a role than might be intuitively expected, as a result of kinetic hindrances (Helz et al. 1996). Instead, interactions with HS^- are probably important; Mo may be removed as particle-reactive oxythiomolybdates ($MoO_{4-x}S_x^{2-}$) which form when $[H_2S]$ exceeds a "switch point" of ~10 μmol kg^{-1} (Helz et al. 1996; Erickson and Helz 2000). Additionally, a general association of Mo with C_{org} is commonly observed which could reflect a direct connection between Mo and organic compounds in reducing environments (e.g., Brumsack and Gieskes 1983). Interestingly, most Mo removal apparently occurs below the sediment-water interface rather than in the water column even when the water column is strongly reducing (Francois 1987; Emerson and Huested 1991; Crusius et al. 1996), a phenomenon that may result from acid-catalysis of oxythiomolybdate formation in low-pH sediment pore waters (Erickson and Helz 2000).

Regardless of mechanistic uncertainties, sedimentation under low-O_2 waters is a critical part of the global Mo ocean budget. Despite their scarcity in the oceans, such sedimentary settings account for removal of 10–50% of the Mo entering the oceans via rivers each year (Bertine and Turekian 1973; Morford and Emerson 1999). Most of this removal probably occurs in sulfidic ("euxinic") basins like the Black Sea or the Cariaco Basin. Hence, relative to its average crustal abundance, Mo is strongly enriched in black shales, saprolites and other carbonaceous sediments, with concentrations invariably >10 ppm and often exceeding 100 ppm.

Some 10–20% of Mo removal may also occur in so-called "suboxic" settings of intermediate redox potential, where bottom water O_2 is <5 μmol kg^{-1} and H_2S << 10 μmol kg^{-1} (Morford and Emerson 1999; Emerson personal communication).

Surprisingly in view of the stability of MoO_4^{2-} in solution, Mo enrichment to concentrations of 100's–1000's of ppm, correlated with Mn content, is seen in ferromanganese oxide sediments, especially crusts, nodules and some oxic pelagic sediments (Bertine and Turekian 1973; Calvert and Price 1977; Cronan 1980; Calvert and Piper 1984; Shimmield and Price 1986). Such enrichment most likely reflects authigenic accumulation of Mo by adsorption to and/or coprecipitation with Mn oxide phases. This phenomenon is observed in the laboratory (Chan and Riley 1966; Barling and Anbar 2004). Therefore, the bulk of Mo not removed to euxinic sediments is generally thought to be removed in association with sedimentary Mn oxides (Bertine and Turekian 1973; Morford and Emerson 1999). Ferromanganese crusts and nodules are probably not themselves dominant sinks because these sediments accumulate very slowly; Mo associated with Mn oxides in widely disseminated pelagic sediments may be quantitatively more important although Mo enrichments in pelagic sediments are relatively small.

The net result of these considerations is that Mo is thought to have a comparatively straightforward ocean budget (Fig. 4), entering largely dissolved in river waters and leaving primarily in association with authigenic Mn-oxides and anoxic sediments. The accumulation of Mo in the oceans relative to other transition metals therefore reflects the strongly bimodal character of Mo geochemistry with respect to redox conditions. Essentially, Mo is readily transferred from crust to oceans during oxidative weathering but, because settings in which bottom water O_2 < 5 μmol kg^{-1} cover only ~0.3% of the modern seafloor, Mo is only slowly removed from the oceans. Hence, the Mo ocean residence time is ~10^5–10^6 years (Morford and Emerson 1999), and the largest variations in Mo concentration in oxygenated seawater are only ~5% on a salinity-normalized basis (Tuit and Ravizza 2003). Re and U have similar budgets, reflecting similar geochemical considerations (Klinkhammer and Palmer 1991; Anbar et al. 1992; Colodner et al. 1993).

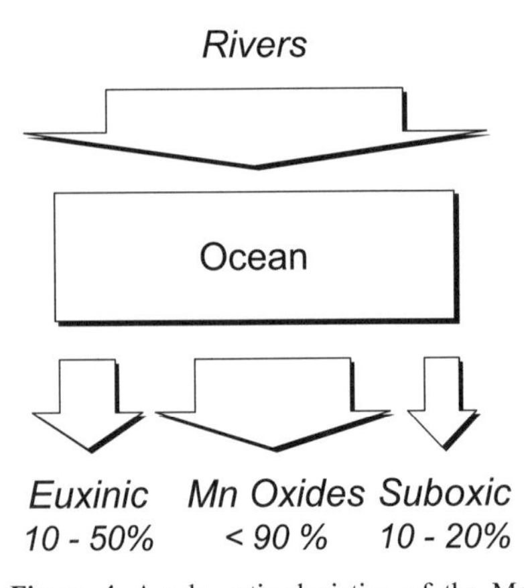

Figure 4. A schematic depiction of the Mo elemental budget in the modern oceans. Rough estimates of the percent removal to each sediment type, drawn from the literature, are indicated (see text).

A consequence of this geochemical behavior is that the abundance of Mo in sediments (as well as the abundances of U and Re) can be used as an indicator of local bottom water redox conditions (e.g., Crusius et al. 1996; Dean et al. 1997; Dean et al. 1999; Yarincik et al. 2000; Zheng et al. 2000b; Calvert et al. 2001; Werne et al. 2002; Algeo and Maynard 2003; Lyons et al. 2003; Sageman et al. 2003). Broadly speaking, Mo enrichments of 10s to 100s of ppm in sediments (typically normalized to tracers of detrital input, such as Al) may indicate water column Eh below the "switch point" during sediment deposition.

Finally, because the Mo ocean budget should be strongly sensitive to bottom water redox conditions, particularly perturbations associated with expanded deep sea anoxia, determination of Mo concentrations or their variations in ancient seawater, via sedimentary proxies, has been proposed as an indicator of global ocean paleoredox change (e.g., Emerson and Huested 1991).

Molybdenum in biology

Biologically, Mo, like Fe, has a prominent role in biochemical pathways that impact global biogeochemical cycles, particularly the nitrogen and sulfur cycles (Stiefel 1997; Frausto da Silva and Williams 2001). However, unlike Fe, or Mn, Mo is not widely used as a terminal electron receptor or donor in microbial respiration. This is probably the result of the low reduction potentials of the Mo oxidation states compared to other metals (Frausto da Silva and Williams 2001) (Fig. 5); the fact that multiple Mo oxidation states can be accessed over a narrow range of voltages makes Mo relatively "redox labile" at low environmental Eh, but also means that the energy gain from Mo redox transformations is small compared to many other elements.

At the same time, this redox lability makes Mo well suited as a cofactor in enzymes that catalyze redox reactions. An example is the prominence of Mo in nitrogen fixation. This prokaryotic metabolism, the dominant pathway for conversion of atmospheric N_2 to biologically-useful NH_3, utilizes Mo (along with Fe) in the active site of the nitrogenase enzyme that catalyzes N_2 reduction. Alternative nitrogenases that do not incorporate Mo have been identified, but are markedly less efficient (Miller and Eady 1988; Eady 1996).

The redox properties of Mo also make it useful in enzymes that catalyze reactions involving two-electron or oxygen-atom transfer (Frausto da Silva and Williams 2001). Such enzymes include nitrate reductase, sulfite oxidase, formate dehydrogenase and aldehyde oxidase (Hille 1996; Stiefel 1997; Kroneck and Abt 2002). Hence, while Mo is rarely a terminal electron

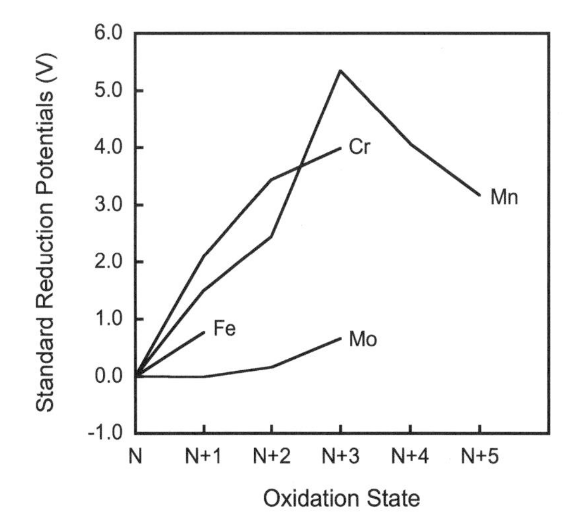

Figure 5. An oxidation state diagram for Mo, Cr, Fe and Mn. For Mo and Cr, N = III; for Fe and Mn, N = II. Potentials are given at standard states in acid solution relative to the hydrogen electrode. On such a diagram, the slope between any two points equals the redox potential. In contrast to most other metals, multiple Mo oxidation states are accessible over a small range of potentials. Note also that Mo is oxidized to Mo(VI) at relatively low potential (similar to Fe(III). Figure modified after Frausto da Silva and Williams (2001).

acceptor or donor, it does have a role in respiratory metabolism by microorganisms that reduce nitrate or oxidize reduced forms of sulfur to sulfate.

Mo use is not confined to prokaryotes; similar enzymes are also found in eukaryotes (including humans). Mo deficiency is rare, as are disorders of Mo metabolism, but symptoms may be induced if diets are rich in Cu or W, which are Mo antagonists. The direct toxicity of Mo is low, but Mo is an antagonist for Cu in cattle, where thiomolybdates formed in the rumen act as high-affinity ligands for Cu. Hence intake of foods high in Mo can induce the disease molybdenosis (Mills and Davis 1987).

It is intriguing that Mo is an important element in biology despite its scarcity at the Earth's surface—indeed, no element of similar scarcity is so biochemically ubiquitous. The reason for this paradoxical combination is unknown. Possible explanations include the unique chemical character of this element, evolutionary adaptation to the abundance of Mo in oxygenated oceans relative to other transition metals, a legacy of prebiotic chemical evolution in Mo-rich environments (e.g., in association with sulfide minerals), or some combination of these factors. Regardless, this paradox has inspired creative hypotheses about the importance of Mo in evolution (Crick and Orgel 1973; Anbar and Knoll 2002).

Isotopic expectations

From such chemical and biological considerations it is apparent *a priori* that Mo stable isotope studies could provide novel biogeochemical insights if natural Mo isotope variations are measurable. Measurable isotope variations are seen for Se and Cr (reviewed in Johnson et al. 2004), other elements with Mo-like oxyanion-dominated environmental chemistry. Mo isotope variations might provide information on ancient ocean redox conditions (Barling et al. 2001; Siebert et al. 2003), or on changes in the availability of this bioessential element (particularly in response to changing redox conditions (e.g., Anbar and Knoll 2002)). Mo isotopes could also provide insights into Mo biochemistry (Williams and Frausto da Silva 2002), an issue of non-trivial importance.

As discussed at length in this review, Mo isotope variations are now known to be pervasive. This finding was not unexpected. First, the Mo isotope system spans a mass range of ~8% (Fig. 1), which compares favorably with many other stable isotope systems. Second, the dynamic redox chemistry of Mo, and the fact that Mo-ligand interactions have strong covalent character, lends itself to chemical isotope effects. Finally, with seven stable isotopes of relative abundance ~10–25%, high-precision Mo isotope analyses are analytically tractable.

Less expected, initial Mo isotope variations in ocean sediments (and seawater) are proving to be quite systematic and, at least to a first approximation, interpretable in terms of a limited number of fractionation processes. This situation presumably reflects the relatively straightforward ocean geochemical cycle of Mo and its long ocean residence time. In addition, it is probably important that Mo is fundamentally a catalyst in respiration rather than a reductant or oxidant, so that Mo isotope systematics are not strongly complicated by isotope effects associated with microbial respiration. Regardless, early observations already point to important geoscience applications for this emerging non-traditional isotope system.

ANALYTICAL CONSIDERATIONS

Data reporting conventions

The stable isotopes of Mo include ^{92}Mo, ^{94}Mo, ^{95}Mo, ^{96}Mo, ^{97}Mo, ^{98}Mo and ^{100}Mo (Fig. 1) (with a half life of 10^{19} yr, ^{100}Mo is effectively stable). Because of this large number of stable isotopes, the Mo isotope analyst is presented with a number of options in both analysis and data reporting. It is common to analyze the relative abundances of as many isotopes as possible to confirm that observed variations are, indeed, mass dependent. Such confirmation is particularly important because of potential polyatomic interferences at masses 95 and 96 from $^{40}Ar^{55}Mn^+$ and $^{40}Ar^{56}Fe^+$ in ICP-MS if residual Mn or Fe are present after Mo preconcentration (an occasional problem given the high abundances of Mn and Fe relative to Mo in many geological samples). Isobaric interferences from Zr are also a potential problem in many geological samples, and particularly when using Zr as an "element spike" (see below). Interferences from Ru are less problematic in natural samples because of low abundance. However, these interferences can be corrected to high degree of accuracy (Anbar et al. 2001).

Reporting conventions vary. Some workers report variations as $\delta^{97/95}$Mo (= $[(^{97}Mo/^{95}Mo)_{sample}/(^{97}Mo/^{95}Mo)_{standard} - 1] \times 1000$) because ^{97}Mo and ^{95}Mo are the only Mo isotopes completely free of isobaric interferences from other elements (Anbar et al. 2001; Barling et al. 2001). Because ^{97}Mo is used by Siebert et al. (2001) in their isotope double spike procedure (see below), this group reports $\delta^{98/95}$Mo (= $[(^{98}Mo/^{95}Mo)_{sample}/(^{98}Mo/^{95}Mo)_{standard} - 1] \times 1000$) in their publications (Siebert et al. 2001; McManus et al. 2002; Siebert et al. 2003). These conventions are related by $\delta^{97/95}$Mo ~2/3 × $\delta^{98/95}$Mo (although this relationship is not exact for all fractionation laws, e.g., Criss 1999, Young et al. 2001). Data are presented here as $\delta^{97/95}$Mo.

There is no internationally-certified Mo isotope standard. Various groups have used different working laboratory standards. Data are reported here relative to the Rochester JMC standard (Barling et al. 2001), a laboratory standard which seems similar to average crustal Mo and therefore is a convenient reference point. Rochester JMC Mo is also similar to the JMC Mo standard used by Siebert et al. 2003 in reporting their data, based on cross-comparison of $\delta^{97/95}$Mo of similar samples measured by both groups. As proposed by Siebert et al. (2003), data from different groups can also be compared with reference to the mean Mo isotope composition of seawater ("mean ocean Mo," a.k.a. MOMO; see below for discussion of uniformity of MOMO). However, MOMO is not a practical standard because the need for significant preconcentration of Mo from seawater creates a risk of contamination during sample handling or chromatographic isotope fractionation, because $\delta^{97/95}$Mo of seawater is isotopically heavier than most other materials (see below), and because it is possible that measurable $\delta^{97/95}$Mo variations in the oceans remain to be discovered. Certification of a international reference standard for inter-lab comparison should be a high priority for the Mo stable isotope community.

Historical perspective

Efforts to measure Mo isotope variations, using thermal ionization mass spectrometry (TIMS), can be traced back to the early 1960s. Early studies were provocative: Murthy (1962; 1963) reported that the Mo isotope composition in some iron meteorites deviated from that of others, and from terrestrial Mo, by ~10 ‰/amu, while Crouch and Tuplin (1964) reported mass dependent variations and mass independent isotopic anomalies among terrestrial molybdenites. As discussed below, Wetherill (1964) subsequently demonstrated isotopic uniformity among iron meteorites to better than 1 ‰/amu. It is likely that earlier reports were affected by analytical artifacts due to inadequate correction of Mo isotope fractionation during analysis.

Although TIMS-based studies of Mo isotopic composition continued (Turnlund et al. 1993; Lu and Masuda 1994; Dauphas et al. 2002a,b; Yin et al. 2002), little attention was given to the possibility of mass-dependent variations until the advent of multiple-collector inductively coupled plasma mass spectrometery (MC-ICP-MS) in the mid 1990s. MC-ICP-MS is generally preferred over TIMS for Mo isotope studies because of the high ionization efficiency of Mo in the ICP source (permitting sample sizes <1 μg), the elimination of blank contamination from Mo in TIMS filaments, and more rapid sample throughput (Halliday et al. 1995; Lee and Halliday 1995). As discussed below, MC-ICP-MS also offers more than one approach to correct for analytical mass fractionation to extremely high precision.

Chemical separations

Preconcentration of Mo is necessary from most natural samples, in which typical Mo abundances ranges from 1 ppm (e.g., average crust) to ~100 ppm (e.g., black shales and marine ferromanganese crusts and nodules). The major exception is the Mo-bearing mineral molybdenite (MoS_2) which can be found in hydrothermal deposits and igneous rocks. Purification is also essential for most natural samples (again, molybdenite is the major exception). Removal of Fe and Mn is particularly critical to minimize formation of their argides which potentially generate significant interferences at masses 94–97.

Both preconcentration and purification are accomplished with traditional ion exchange chromatographic methods. More than one procedure has been used—the interested reader is referred to relevant publications for analytical details (Barling et al. 2001, Siebert et al. 2001).

A key finding is that Mo isotopes can be fractionated during elution in at least some ion exchange systems (Anbar et al. 2001; Siebert et al. 2001), similar to observations of Ca, Cu and Fe isotope effects (Russell and Papanastassiou 1978; Maréchal et al. 1999; Anbar et al. 2000). Therefore, it is often necessary to ensure that yields through these chemical processes are essentially quantitative.

Analytical mass fractionation

The major analytical complication in Mo isotope analysis is precise correction for isotope fractionation during Mo purification and mass spectrometric analysis. This subject is reviewed in general by Albarède and Beard (2004), and is discussed here in particular reference to Mo. It is important to recognize that this challenge is fundamentally different in mass dependent stable isotope studies as compared to investigations of mass-independent Mo isotope variations produced by nucleosynthesis. The latter have received attention in recent years for high-precision determination of Mo isotope composition (e.g., Dauphas et al. 2002a,b; Yin et al. 2002), but are not relevant here.

Sample-standard bracketing. In the earliest Mo isotope studies (Murthy 1962; Murthy 1963; Crouch and Tulpin 1964), analytical mass fractionation was assumed to be identical for samples and standards even when analyzed separately. However, in TIMS this assumption is risky because the isotopic composition of an analyte changes continuously during analysis, a result of isotope enrichment during thermal evaporation and ionization (a problem acknowledged by Murthy). The time-course of this fractionation is highly sensitive to sample purity and operating conditions.

Sample-standard comparison is more applicable in MC-ICP-MS, in which instrument mass fractionation is fundamentally a "steady state" phenomenon (Maréchal et al. 1999). This method has been used successfully for some non-traditional stable isotopes, particularly involving Fe, in which analyses of samples are bracketed by standards to cope with systematic instrumental drift (e.g., Zhu et al. 2002; Beard et al. 2003). However, other methods have been used for Mo stable isotope work published to date because of concerns about non-systematic changes in instrument mass fractionation, particularly arising from differences in matrices, between samples and standards. Such concerns are more acute for Mo than for Fe and many other elements because Mo is a trace constituent of most samples, increasing the challenge of rigorous, high-yield sample purification.

Element spike. An additional benefit of MC-ICP-MS is the ability to use an "element spike" approach. In this method, the isotopic composition of a different element, added to the sample, is to monitor variations in instrument mass fractionation (Longerich et al. 1987; Maréchal et al. 1999). This method is possible with MC-ICP-MS because the relative fractionation behavior of different elements in ICP sources is more-or-less constant, although it is critical to note that the data reduction must account for the fact that the absolute fractionation behavior of each element may be different (Maréchal et al. 1999; Anbar et al. 2001).

The first modern observations of Mo isotope fractionations in geological materials were made using this method (Barling et al. 2001), employing Zr and Ru spikes with on-line correction for isobaric interferences from these spikes (Anbar et al. 2001). Precision of ±0.25‰ was reported in measurement of $\delta^{97/95}$Mo, with comparable results for other ratios.

Isotopic double spike. The most rigorous approach is to use an isotopic "double spike", in which samples are doped with a known quantity of "spike" Mo which consists of two isotopes in a known ratio (Wetherill 1964; Siebert et al. 2001). These spike isotopes serve as an internal standard to monitor mass fractionation of the sample subsequent to spiking. The fundamental advantage over the element spike is that the spike isotopes follow exactly the same fractionation behavior as the isotopes of interest. A disadvantage is the need to carefully prepare and calibrate the double spike.

Mo is particularly suitable for double spike analysis because it has a large number of stable isotopes. It is not surprising, therefore, that Wetherill (1964) used this approach to demonstrate isotopic homogeneity between terrestrial and meteorite Mo samples, employing a ^{95}Mo-^{98}Mo spike and TIMS. This study was one of the earliest applications of the double spike method.

In a more recent TIMS Mo isotope double spike study using a multiple collector system, Wieser and de Laeter (Wieser and DeLaeter 2003) reported precision of better than ±0.1 ‰/amu mass difference (±2σ) using a ^{94}Mo-^{100}Mo spike.

The utility of the double spike is not limited to TIMS. The method has also been applied very successfully in MC-ICP-MS (Siebert et al. 2001). Using a ^{97}Mo-^{100}Mo spike, precision of ±0.1‰ (±2σ) was reported in measurement of $\delta^{98/95}$Mo, and has been applied to a range of natural materials (McManus et al. 2002; Siebert et al. 2003).

Methods compared. Looking to the future, when working with geological samples the double spike has an important practical advantage over both the sample-standard bracketing and the element spike methods. It can obviate the need for high yield chemistry if the spike isotopes are added prior to Mo purification and extraction, thereby providing an internal reference point for isotope fractionation produced by the instrument as well as by the separation chemistry. This advantage is important because it is difficult in practice to design Mo chemical separation schemes that achieve both high yield and high purity for a wide range of sample types.

In terms of the existing database, the element spike and isotope double spike methods have both been used successfully for Mo isotope surveys in nature (Barling et al. 2001; Siebert et al. 2003). The reason for the lower precision of the element spike data (±2σ of ±0.25‰ vs. ±0.1‰) is unclear. This difference may reflect inherent limitations of the element spike technique, performance differences among MC-ICP-MS instruments of differing designs, or matrix complications from sample impurities residual from the higher yield chemistry used in initial Mo isotope element spike research. However, within these uncertainties, $\delta^{97/95}$Mo (or $\delta^{98/95}$Mo) offsets observed between similar sample types using both methods are statistically indistinguishable (Barling et al. 2001; Siebert et al. 2003; see next section).

OBSERVATIONS

Examinations of Mo isotope variations in nature include the initial reconnaissance of Barling et al. (2001), the more comprehensive survey of Siebert et al. (2003), and some more focused studies of specific sample types (McManus et al. 2002; Wieser and DeLaeter 2003; Arnold et al. 2004). These studies emphasized reservoirs, sources and sinks important to the Mo ocean geochemical cycle, including igneous rocks, ferromanganese oxide sediments, euxinic sediments, suboxic sediments and seawater. Laboratory studies have emphasized associated fractionation mechanisms.

The findings are summarized below, and in Figures 6, 7 and 8. Complete references are provided in the text and in the figure captions. Implications of these data for understanding Mo geochemical cycling, and for application of this understanding to paleoredox questions, are discussed in the subsequent section.

Igneous rocks

Granites and basalts. Mo isotope data from igneous rocks are sparse but revealing (Siebert et al. 2003). A number of basalts and granites have $\delta^{97/95}$Mo close to 0‰ (relative to the Rochester JMC standard). Although the basalts are all subduction volcanics from Kamchatka, and hence from a single, geographically coherent suite, the granite samples are

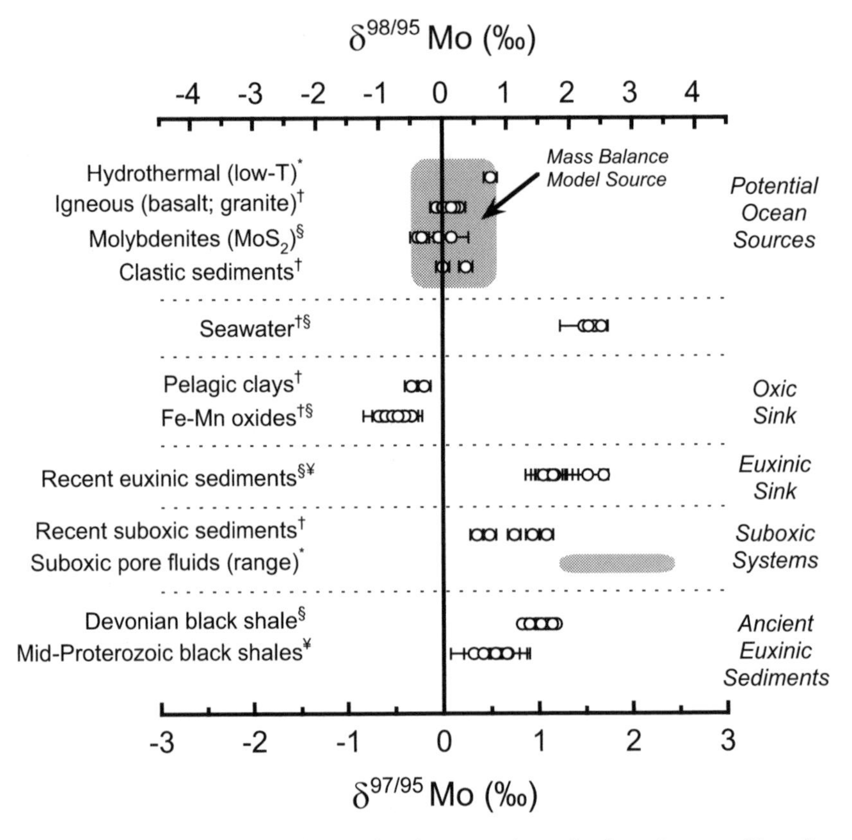

Figure 6. Summary of existing Mo isotope data from natural samples. Isotopic composition of ocean source is based on mass balance (see text). Data are presented as $\delta^{97/95}$Mo and $\delta^{98/95}$Mo relative to the Rochester JMC standard ($\delta^{97/95}$Mo ~2/3 × $\delta^{98/95}$Mo). References: (*) McManus et al. 2002; (†) Siebert et al. 2003; (§) Barling et al. 2001; (¥) Arnold et al. 2004. Molybdenite values of Wieser and DeLaeter (2003) are omitted because of standard normalization problems (see text). Data obtained by different research groups using different standards are cross-calibrated by comparing seawater δ values.

Archean and Himalayan (Tertiary). The coherence of these samples suggests that Mo isotope fractionation during igneous processing is small compared to current analytical limits (Siebert et al. 2003).

Molybdenites. Detectable Mo isotope variation is seen in molybdenites from hydrothermal ore deposits (Barling et al. 2001; Wieser and DeLaeter 2003). Unfortunately, the results of Wieser and DeLaeter (2003) cannot be directly compared to those of Barling et al. (2001) because no common materials were analyzed. However, if it is assumed that the JMC Mo standard of Wieser and DeLaeter (2003) is identical to other JMC Mo standards (Barling et al. 2001; Siebert et al. 2003), then the total range of $\delta^{97/95}$Mo is nearly 1‰. It is plausible that the variations among these samples are somewhat larger than among granites and basalts because formation temperatures for hydrothermal molybdenites are 350–600°C, substantially cooler than silicate melting temperatures (Wieser and DeLaeter 2003). Additionally, hydrothermal molybdenites commonly occur in veins, raising the possibility of Rayleigh effects during progressive precipitation of MoS_2 from migrating fluids. Such effects could amplify variations relative to expectations from single-stage fractionation factors. Consistent with this scenario, the molybdenite $\delta^{97/95}$Mo overlap those of granites and basalts.

Marine sediments

Ferromanganese oxides. An extensive Mo isotope dataset is now available for ferromanganese crusts and nodules, which are enriched in Mo and hence are logical targets for early isotopic investigations (Barling et al. 2001; Siebert et al. 2003). These include nodules from the Pacific and Atlantic Oceans, and crusts from the Pacific, Atlantic and Indian Oceans. Nodule data are of bulk samples, but Pacific and Atlantic crust data are time-resolved at ~1–3 million year intervals from the present back to 60 million years ago.

This dataset has three striking features. First, the Mo isotope composition of all these samples is remarkably uniform. Overall, Mo in these sediments varies within a range of $\pm<0.25$‰. This uniformity includes the 60 million year records in the two crusts. Non-systematic variations in Mo concentration through these crusts and comparison with U diffusion models suggest that this uniformity is not the result of diffusional homogenization of an originally variable isotope signal (Siebert et al. 2003). Second, there is no significant offset between the isotopic compositions of crusts and nodules despite the greatly different accumulation rates of these sediment types. Third, the mean isotopic composition of these samples is offset from $\delta^{97/95}$Mo of igneous rocks by approximately −0.5‰.

The only significant $\delta^{97/95}$Mo variation in these samples is a small offset of ~0.15‰ seen between Pacific and Atlantic crusts, just outside the $\pm 2\sigma$ uncertainties of Siebert et al. (2003). A similar offset is seen in nodule data, although well within the analytical uncertainties of Barling et al. (2001). Such offsets seem unlikely to reflect variations in $\delta^{97/95}$Mo in seawater in view of the long ocean residence time and relatively invariant seawater concentrations of Mo. Systematic differences in sediment composition may play a role (Barling and Anbar 2004).

Euxinic sediments. The Black Sea and the Cariaco Basin are both euxinic basins in which the sediment-water interface is overlain by anoxic, H_2S-bearing waters. The Black Sea is the classic "type euxinic" basin: At depths exceeding ~100 m, [H_2S] exceeds 100 μM. Mo concentrations fall sharply at these depths (Emerson and Huested 1991), and underlying sediments are enriched in Mo by 10–100 fold over average crust.

The most significant Mo isotope feature in these settings is their $\delta^{97/95}$Mo enrichment (Barling et al. 2001; Arnold et al. 2004): in Holocene, organic-rich sediments from both basins, $\delta^{97/95}$Mo is >1‰ heavier than in igneous rocks, molybdenites or ferromanganese sediments. Black Sea Unit I samples are more enriched in heavy isotopes than any other rocks analyzed to date.

Uniformity among these euxinic sediments is not as tight as in the case of ferromanganese crusts and nodules. For example. some variability is seen between Unit I and Unit II of the Black Sea, with Unit I enriched in $^{97}Mo/^{95}Mo$ by ~0.4‰. $\delta^{97/95}Mo$ in ~14,000 year old sediments from the Cariaco basin, which is second only to the Black Sea in its extent of modern anoxic-sulfidic water-column conditions, are comparable to Black Sea Unit II.

Suboxic sediments. Siebert et al. (2003) analyzed Mo isotopes in a limited number of samples from marine sediments accumulating under a range of water column conditions best described as "suboxic" or "dysoxic" (e.g., $O_2 < 5$ μmol kg^{-1}). Mo concentrations in these sediments are enriched over average crust by a factor of ~3–100, in some cases exceeding enrichments seen in Black Sea sediments (e.g., Chile Margin sediments). Isotopically, $\delta^{97/95}Mo$ in these sediments covers a large range, essentially filling the gap between igneous rocks and euxinic sediments. A rough correlation is suggested between Mo concentration and $\delta^{97/95}Mo$.

Clastic and pelagic sediments. Only a handful of clastic and pelagic samples have been analyzed for $\delta^{97/95}Mo$ (Siebert et al. 2003). Clastic sediments (two samples) are indistinguishable from igneous rocks in their Mo isotope compositions. Pelagic clays (two samples) show enrichment in Mo (3.4 and 185 ppm), and also a shift toward lighter $\delta^{97/95}Mo$ similar to ferromanganese sediments.

Waters

Seawater. A half dozen analyses of Mo isotopes in the Pacific, Atlantic and Indian oceans, covering depths to ~3000 m (Barling et al. 2001, Siebert et al. 2003), reveal two important facts. First, there is no detectable $\delta^{97/95}Mo$ variation in the oceans with location or depth. Second, $\delta^{97/95}Mo$ in the oceans is similar to the heaviest of euxinic sediments, and is heavier than in igneous rocks or ferromanganese sediments by ~1.5‰ and ~2‰, respectively. A uniform isotopic composition in the oceans is consistent with the 10^5–10^6 year ocean residence time. The explanation for the heavy isotopic composition is discussed further below.

Low temperature hydrothermal fluids. Warm fluids emerging from hydrothermal systems on the flanks of mid ocean ridges have recently emerged as a net source of Mo and other elements to the oceans (Wheat and Mottl 2000). Mo isotopes have recently been investigated (McManus et al. 2002). These fluids are found to be isotopically lighter than seawater, but heavier than igneous rocks. Mixing relationships point to an endmember hydrothermal fluid with $\delta^{97/95}Mo$ ~0.5‰.

Suboxic porewaters. As described above, $\delta^{97/95}Mo$ in suboxic sediments is intermediate between igneous rocks and euxinic sediments (and seawater). McManus et al. (2002) examined $\delta^{97/95}Mo$ in sediment porewaters from the Santa Monica basin (Fig. 7), a well-studied suboxic setting ($O_2 < 5$ μmol kg^{-1}). The upper ~25 cm were analyzed. Fe and Mn remobilization is seen in these fluids in association with a down-core redox gradient. $\delta^{97/95}Mo$ immediately below the sediment water interface is ~0.25‰ lighter than seawater, potentially reflecting release of light Mo associated with Mn oxides; elevated Mn and Mo concentrations are seen in fluids at these depths. The concentration of Mo is progressively depleted at depths >5 cm. In conjunction with this depletion, $\delta^{97/95}Mo$ is progressively ^{97}Mo-enriched, ultimately reaching a value heavier than seawater by nearly 1‰—the heaviest Mo isotopic composition measured to date.

Laboratory experiments

Adsorption to Mn oxides. The only thorough experimental study of Mo isotope fractionation to date focused on adsorption to Mn oxides (Barling and Anbar 2004). The intent of this study was to examine the possibility that Mo isotopes are fractionated during this process. In these experiments, Mo dissolved in aqueous solution of defined ionic strength at room temperature was exposed to synthetic δ-MnO_2. pH (6.5–8.5) and solution-solid contact

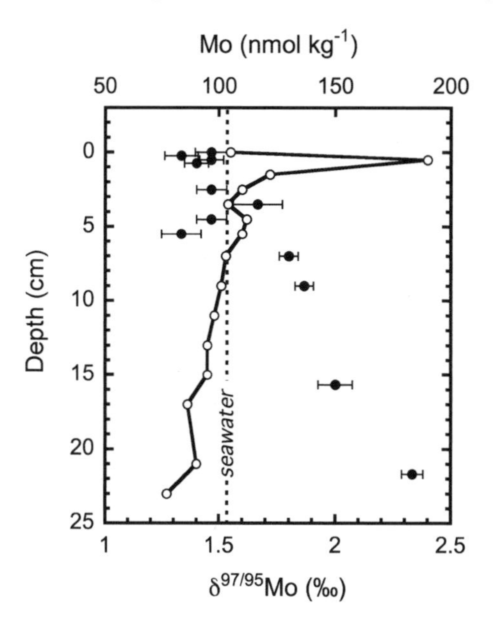

Figure 7. Mo concentrations (○) and isotopic compositions (●) from reducing pore fluids in Santa Monica Basin (McManus et al. 2002). Dotted line indicates seawater values for both variables. The data can be fit by a 1-D reaction-diffusion model with a fractionation factor of ~1.005. The effective fractionation factor for Mo removal across the sediment-water interface is smaller, <1.0025 (see text).

time (2–96 hr) were varied so that the extent of total Mo adsorbed ranged from ~40% to >95% in the various experiments. Mo was extracted both from solutions and from Mn oxides for $\delta^{97/95}$Mo determination.

The major isotopic findings are summarized in Figure 8. A mean $\delta^{97/95}$Mo offset of ~1.8‰ was observed between dissolved and oxide-bound Mo in all experiments. The data do not fit the expectations of a Rayleigh model, as might be expected from irreversible adsorption. Instead, the data closely approximate closed system exchange between adsorbed and dissolved Mo, suggesting that Mo isotope fractionation in this system is an equilibrium process.

Leaching. Mo isotope fractionation has also been investigated during mild HCl dissolution of one granite sample (Siebert et al. 2003). The goal of this experiment was to provide a crude indication of Mo isotope fractionation during rock weathering. After transfer of >50% of Mo to solution, no significant fractionation was detected between solution and the granite starting material.

INTERPRETATIONS

Some of the above observations are particularly striking and provide a focus for efforts to interpret the Mo ocean isotope system. These observations include:

a. The $\delta^{97/95}$Mo uniformity of the oceans

b. The ~2‰ $\delta^{97/95}$Mo offset between seawater and ferromanganese oxide sediments, and the uniformity of $\delta^{97/95}$Mo in these sediments

c. The 1.5–2.0‰ offset between $\delta^{97/95}$Mo in ferromanganese oxides and in euxinic sediments (and hence the similarity of $\delta^{97/95}$Mo in euxinic sediments and seawater).

d. The apparent ~1.5‰ $\delta^{97/95}$Mo offset between seawater and igneous rocks.

Interpretations of these observations are discussed below.

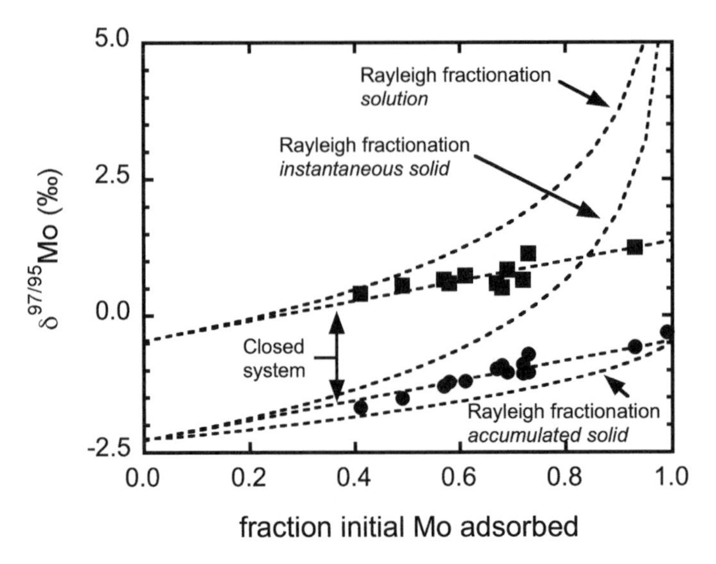

Figure 8. Results of Mo adsorption experiments of Barling and Anbar (2004). Mo-bearing solutions were exposed to synthetic Mn oxides (δ-MnO_2) for 2–96 hours at pH 6.5–8.5. Residual Mo in solution (■) was measured for all experiments. Mo adsorbed to oxide particle surfaces (●) was either measured or inferred from mass balance. Dissolved Mo was systematically heavier than adsorbed Mo with a fractionation factor of 1.0018 ± 0.0005. The data are consistent with closed system equilibrium, in which isotopes exchange continuously between surface and solution, but incompatible with an irreversible, Rayleigh-type process. Figure modified after Barling and Anbar (2004).

Mo isotope fractionation in the oceans

The uniform isotopic composition of dissolved Mo in the oceans is readily understood in terms of the 10^5–10^6 year Mo ocean residence time. For example, given an ocean mixing time of ~10^3 years, and assuming that the $\delta^{97/95}$Mo contrast of different Mo isotope sources is <5‰ (consistent with the range of values measured to date), Mo isotope variations between ocean basins or with depth should be <5‰/($10^5/10^3$) = 0.05‰. Such variations are smaller than present analytical uncertainties.

But why is $\delta^{97/95}$Mo of seawater ~1.5‰ heavier than igneous rocks and ~2‰ heavier than Mo in ferromanganese sediments? A number of ideas related to Mo inputs can be quickly deemed unlikely, if not altogether ruled out. An obvious possibility is Mo isotope fractionation during weathering, because Mo probably enters the oceans largely via rivers. However, evidence from leaching experiments suggests weathering imparts a relatively minor isotopic effect (Siebert et al. 2003), and it seems likely that, because of the efficient transfer of Mo from minerals to solution during oxic weathering, mass balance would minimize the influence of any such effect on riverine Mo. Input of heavy Mo from hydrothermal sources cannot be completely eliminated as an alternative hypothesis, given the limited available data, but there is no indication that such sources are of sufficient volumetric importance, or sufficiently heavy in terms of ^{97}Mo/^{95}Mo, to be the answer. Ancient euxinic sediments (e.g., Paleozoic black shales) clearly constitute a reservoir with appropriate isotopic composition, and the weathering of such sediments when they are exhumed on land or exposed to oxidizing conditions on the seafloor is probably not a trivial source of Mo to the oceans. However, Mo in such sediments is, itself, derived from ancient ocean Mo. Hence, such an Mo input cannot be the *primary* cause of heavy Mo isotope enrichment in the oceans.

It seems much more likely that Mo isotopes are fractionated within the ocean system during removal from seawater, as first proposed by Barling et al. (2001). This possibility is considered below with respect to Mo removal to ferromanganese oxides, euxinic sediments and suboxic sediments.

Ferromanganese oxides. The ~2‰ $\delta^{97/95}$Mo offset between Mo associated with ferromanganese sediments and Mo in seawater is by far the largest Mo isotope variation observed to date. This offset is remarkably systematic, as documented above (Barling et al. 2001; Siebert et al. 2003). Combined with knowledge that Mo and Mn concentrations correlate in such sediments, these observations strongly suggest that there is a large, systematic fractionation of Mo isotopes, favoring removal of light isotopes from solution, when Mo is authigenically enriched in association with Mn oxides (Barling et al. 2001; Siebert et al. 2003).

The simplest hypothesis consistent with such observations is an equilibrium Mo isotope fractionation during adsorption of Mo onto the surfaces of Mn oxide particles (Barling et al. 2001; Siebert et al. 2003). Although the enrichment of light isotopes in Mn oxides is suggestive of a kinetic isotope effect (Barling et al. 2001), the similarity of $\delta^{97/95}$Mo in crusts, nodules and pelagic clays from different ocean basins indicates that this fractionation is largely independent of sedimentation rate. While not ruling out a kinetic effect, an equilibrium effect seems more likely. Recent examination of this fractionation in the laboratory supports this hypothesis (Barling and Anbar 2004).

Based on the $\delta^{97/95}$Mo offset between seawater and such sediments, a fractionation factor, $\alpha_{sw\text{-}MnO_x}$, of ~1.002 is inferred ($\delta^{97/95}Mo_a - \delta^{97/95}Mo_b = \Delta^{97/95}Mo_{a\text{-}b} \sim 1000 \times \ln(\alpha_{a-b})$). This value is indistinguishable from $\alpha_{dissolved\text{-}adsorbed} = 1.0018 \pm 0.0005$ determined experimentally (Barling and Anbar 2004). Therefore, isotope fractionation during Mo adsorption is fully capable of creating the offset in $\delta^{97/95}$Mo between ferromanganese sediments and seawater.

The fractionation mechanism is not entirely clear. MoO_4^{2-} may adsorb directly to Mn-oxide surfaces by an inner-sphere mechanism in which Mn–O–Mo bonds are formed (Barling and Anbar 2004):

$$\equiv MnOH + MoO_4^{2-} + 2H^+ \leftrightarrow \equiv MnHMoO_4 + H_2O$$

If so, equilibrium isotope exchange of the following sort is possible between dissolved and adsorbed species:

$$^{97}MoO_4^{2-} + \equiv MnH^{95}MoO_4 \leftrightarrow {}^{95}MoO_4^{2-} + \equiv MnH^{97}MoO_4$$

Alternatively, Siebert et al. (2003) proposed that adsorption may proceed via a minor species such as $Mo(OH)_6$. If so, then fractionation could be driven by the following exchange, in solution:

$$^{97}MoO_4^{2-} + {}^{95}Mo(OH)_6 \leftrightarrow {}^{95}MoO_4^{2-} + {}^{97}Mo(OH)_6$$

The coordination geometries of MoO_4^{2-} and $Mo(OH)_6$ are tetrahedral and octahedral, respectively. Such differences may drive isotope fractionation (Anbar et al. 2000; Roe et al. 2003). While seawater speciation of Mo is dominated by MoO_4^{2-}, trace intermediates like $Mo(OH)_6$ may still play a role in isotope fractionation if they adsorb because, at equilibrium, the overall fractionation factor for $MoO_4^{2-} \leftrightarrow$ Mo-surface necessarily equals the product of the fractionation factors for $MoO_4^{2-} \leftrightarrow Mo(OH)_6$ and $Mo(OH)_6 \leftrightarrow$ Mo-surface regardless of the concentrations of the three species.

Mo adsorption chemistry offers few clues as to which mechanism is operative. Mo adsorption is found to increase at decreasing pH (Chan and Riley 1966; Barling and Anbar 2004). As pH decreases, oxide surfaces become positively charged, facilitating adsorption of anions like MoO_4^{2-} (Dzombak and Morel 1990; Stumm and Morgan 1996). However, $Mo(OH)_6$ becomes increasingly important at pH<5 due to protonation of oxygen atoms and a change from tetrahedral to octahedral coordination via the following reactions (Aveston et al. 1964; Busey and Keller 1964; Cruywagen and Rohwer 1975; Mitchell 1999):

$$MoO_4^{2-} + H^+ \leftrightarrow HMoO_4^-$$

$$HMoO_4^- + H^+ + 2H_2O \leftrightarrow Mo(OH)_6$$

Therefore, the observed pH dependence of Mo adsorption is consistent with either mechanism.

Regardless of mechanism, it is likely that this isotope effect contributes substantially to the $\delta^{97/95}$Mo offset between igneous rocks and seawater because of the importance of ferromanganese sediments for the ocean Mo budget. This possibility is discussed further below. This hypothesis is supported by the similarity of $\delta^{97/95}$Mo in Mo-rich pelagic clays and ferromanganese sediments. It is likely that the Mo enrichment in these clays results from adsorption to authigenic Mn oxides. Further study of Mo isotopes in pelagic sediments is needed to validate this hypothesis.

Euxinic sediments. The similarity of $\delta^{97/95}$Mo in sediments accumulating in euxinic basins and seawater, to within <0.5‰, suggests that net fractionation when Mo is removed from the water column in strongly reducing depositional settings is small in comparison to the ~2‰ effect during uptake by Mn oxides. This observation may be a consequence of mass balance: Mo removal from aqueous systems when $[H_2S] > 100$ μM is nearly quantitative (Fig. 3), and a very large fraction is removed even when $[H_2S]$ is a tenth this value (Helz et al. 1996; Erickson and Helz 2000; Zheng et al. 2000a).

Evidence for this hypothesis can be found in the rough correlation between $\delta^{97/95}$Mo and [Mo] in suboxic sediments (Siebert et al. 2003): Higher [Mo] is associated with $\delta^{97/95}$Mo approaching the seawater value, as expected from mass balance in a closed reservoir (the reservoir is the diffusive zone beneath the sediment-water interface in suboxic settings; see following section).

Importantly, this hypothesis does not exclude the possibility of large equilibrium Mo isotope effects between equilibrating species in reducing systems or large kinetic isotope effects during unidirectional or incomplete reactions. The controlling factor on the magnitude of fractionation between sediments and dissolved Mo is *not* the existence or absence of such specific isotope effects, but rather the mass balance constraints in the water column or sediment pore waters. Indeed, specific isotope effects seem likely because speciation and redox transformations of the sort that occur in these settings commonly drive isotope fractionation. For example, equilibrium or kinetic isotope effects might well occur between MoO_4^{2-} and $MoO_{4-x}S_x^{2-}$ because of the substantial differences in Mo bonding environments between such complexes.

The small offsets between sediments from Black Sea Unit I and from Black Sea Unit II and the Cariaco Basin and seawater may provide a hint that removal is not necessarily quantitative even in H_2S-rich systems, or that $[H_2S]$ was not as high as generally believed during deposition of these sediments. Alternatively, these offsets could reflect regional or basin-scale reservoir effects linked to riverine Mo fluxes (Arnold et al. 2004). For example, during the deposition of the analyzed Cariaco interval, glacial melt water influx to the Atlantic ocean was approximately three times greater than today's, affecting the seawater chemistry of the Gulf of Mexico (Bertine and Keene 1974; Fairbanks 1989). Similarly, during the deposition of Black Sea Unit II, sea level was lower than today and flow through the Bosphorus was more restricted (Hay 1988). Consequently, the relative importance of local riverine input was presumably greater, and $\delta^{97/95}$Mo of the overlying water may have deviated more from open ocean values relative to more recent intervals.

However, the most important observation regarding euxinic sediments is that $\delta^{97/95}$Mo in euxinic sediments is *not* strongly shifted from seawater, and therefore that $\delta^{97/95}$Mo in sediments accumulating at the bottom of euxinic basins is similar to that of the waters entering such basins even if there are large isotope effects during Mo removal to these sediments. Consequently, the contribution of fractionation during the removal of Mo to euxinic sediments to global-scale Mo isotope fractionation is likely to be small in comparison to the effect during removal to ferromanganese sediments (Barling et al. 2001; Siebert et al. 2003; Arnold et al. 2004).

Suboxic systems. The observations of Siebert et al. (2003) and McManus et al. (2002) in suboxic sediments and porewaters, respectively, are complementary. Siebert et al. (2003) found that $\delta^{97/95}$Mo in such sediments is lighter than in seawater (Fig. 6), while McManus et al. (2002) found that $\delta^{97/95}$Mo in porewaters were heavier than in seawater (Fig. 7). These data demonstrate that light isotopes are preferentially removed from porewaters to sediments in such settings. The isotope effect is not small: using a 1-D reaction-diffusion model, McManus et al. (2002) determined $\alpha \sim 1.005$ (reported here on a ^{97}Mo/^{95}Mo basis), and hence is larger than that occurring during adsorption to Mn oxides.

The mechanism of Mo removal in suboxic systems is unclear, and so the fundamental nature of this fractionation requires further study. However, the effect may be understood in terms of a two layer diffusion-reaction model in which a reaction zone in the sediment (where Mo is chemically removed) is separated from seawater by a purely diffusive zone in which there is no chemical reaction (Brandes and Devol 1997). The presence of a diffusive zone is likely because Mo removal presumably occurs in sulfidic porewaters that lie a finite distance L below the sediment-water interface (Wang and van Cappellen 1996; Zheng et al. 2000a). If H_2S is present in the reactive zone such that Mo is removed below this depth, then Mo isotope fractionation in the diffusive zone may be driven by isotope effects in the reactive zone.

An important consequence of such a model is that the effect of such sedimentary systems on the ocean Mo isotope budget is *not* represented by α, but rather by the relative fluxes of the isotopes across the sediment-water interface. This "effective" fractionation factor, α_{eff}, is likely to be smaller than α (Bender 1990; Brandes and Devol 1997) because the diffusive zone acts as a barrier to isotope exchange with overlying waters, approximating a closed system.

These considerations can be illustrated by adapting the mathematical treatment of Brandes and Devol (1997) for O_2 and N_2 to the case of Mo (Emerson and Breckel personal communication). Following this treatment, at steady state:

$$0 = \phi D \frac{d^2[\mathrm{Mo}]}{dz^2} \quad \text{for } z < L$$

$$0 = \phi D \frac{d^2[\mathrm{Mo}]}{dz^2} - \phi k[\mathrm{Mo}] \quad \text{for } z > L$$

where: [Mo] is the concentration of Mo in porewaters; ϕ is the porosity; D is the molecular diffusion coefficient for Mo (presumably diffusing as MoO_4^{2-}); and k is the reaction rate constant for Mo removal (a first order reaction is assumed). For present purposes, advection is discounted. These equations can be solved subject to the boundary conditions $[\mathrm{Mo}]_{z=0} = [\mathrm{Mo}]_{sw}$ and $[\mathrm{Mo}]_{z=\infty} = 0$ ("sw" denotes seawater). It is also assumed that where the layers meet, at $z = L$, [Mo] and d[Mo]/dz are identical in both layers. Since we are interested in isotopic fluxes (F) across the sediment-water interface, we apply Fick's first law and solve for d[Mo]/dz at $z = 0$:

$$F_{z=0} = -\phi D\left(\frac{\mathrm{d}[\mathrm{Mo}]}{\mathrm{d}z}\right)_{z=0} = \phi D[\mathrm{Mo}]_{\mathrm{sw}}\left(\frac{\sqrt{\frac{k}{D}}}{1+L\sqrt{\frac{k}{D}}}\right)$$

Analogous expressions can be derived for ^{95}Mo and ^{97}Mo in terms of ^{95}k and ^{97}k, where $^{97}k = \alpha^{95}k$. D is assumed identical for all isotopes. The relative isotope fluxes are therefore:

$$\frac{^{97}F}{^{95}F} = \frac{[^{97}\mathrm{Mo}]_{\mathrm{sw}}}{[^{95}\mathrm{Mo}]_{\mathrm{sw}}}\sqrt{\alpha}\left(\frac{1+L\sqrt{\frac{^{95}k}{D}}}{1+L\sqrt{\frac{\alpha^{95}k}{D}}}\right) = R_{\mathrm{sw}}\alpha_{eff}$$

where $\alpha = {}^{97}k/{}^{95}k$ and $R_{\mathrm{sw}} = [^{97}\mathrm{Mo}]_{\mathrm{sw}}/[^{95}\mathrm{Mo}]_{\mathrm{sw}}$. It is apparent that if $L = 0$, $\alpha_{eff} = \sqrt{(^{97}k/^{95}k)} = \sqrt{\alpha}$ (Bender 1990). If L is >0 then the effective fractionation between the sediments and the bottom water is less than $\sqrt{\alpha}$. Therefore, $\alpha_{eff} < \sqrt{(1.005)} < 1.0025$; this magnitude is comparable to the fractionation during Mo adsorption by Mn oxides ($\alpha_{\mathrm{sw\text{-}MnO_x}}$).

Based on these considerations, the importance of fractionation in suboxic sedimentary systems for the global ocean Mo isotope budget is probably modest. This conclusion follows from the magnitude of α_{eff} relative to $\alpha_{\mathrm{sw\text{-}MnO_x}}$ and the relatively modest importance of suboxic sediments as compared to Mn oxides for the global ocean Mo elemental budget (Morford and Emerson 1999; Emerson and Breckel personal communication). However, this prediction must be assessed by further studies in open ocean reducing sediments.

Mo isotope mass balance in the oceans

The preceding considerations point to a first order model in which Mo entering the oceans primarily from continental sources undergoes fractionation during removal to Mn oxides, resulting in an isotopically heavy reservoir of Mo dissolved in seawater. This model is somewhat analogous to the Ca isotope system (DePaolo 2004), wherein $\delta^{44/40}$Ca of seawater is offset approximately +1.5‰ relative to riverine sources because of a fractionation during carbonate deposition. Within such a framework for Mo, euxinic and suboxic sediments are seen as having only a minor impact on the isotope budget, and Mn oxides and euxinic sediments would be seen to have complementary isotopic compositions—one light and the other heavy—relative to the input (Fig. 9).

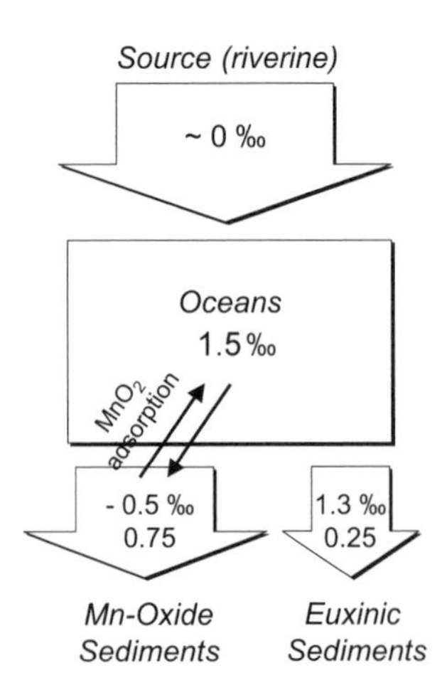

Figure 9. A schematic depiction of the apparent Mo isotope budget in the modern oceans, neglecting suboxic sediments. Mean isotopic values for inputs and outputs are indicated, along with the fractional importance of each sediment type for Mo removal. See text for details.

The model depicted in Figure 9 can be tested by a simple mass balance calculation (Barling et al. 2001, Siebert et al. 2003):

$$\delta^{97/95}\mathrm{Mo}_{\mathrm{input}} = f_{\mathrm{ox}} \times \delta^{97/95}\mathrm{Mo}_{\mathrm{ox}} + f_{\mathrm{eux}} \times \delta^{97/95}\mathrm{Mo}_{\mathrm{eux}}$$

where the subscripts *input*, *ox* and *eux* denote the riverine input and oxic and euxinic sediments, respectively, and f_{ox} and f_{eux} denote the fraction of total Mo removed to each sediment type ($f_{\mathrm{ox}} + f_{\mathrm{eux}} = 1$). If the isotopic compositions presented above are representative of the major Mo sources and sinks, then to a first approximation the global ocean Mo isotope budget can be expressed in terms of an input of ~0‰ and outputs to oxic and euxinic sinks with isotopic compositions of approximately −0.5‰ and 1.3‰, respectively. Using these mean values and the range of literature estimates of f_{ox} and f_{eux} in the modern oceans (Bertine and Turekian 1973; Emerson and Huested 1991; Morford and Emerson 1999), the predicted range of $\delta^{97/95}Mo_{input}$ matches reasonably well with measurements of Mo in igneous rocks, used here as a proxy for riverine input (Fig. 6). This agreement implies that the oxic and euxinic observations to date provide a first-order representation of Mo sources to, and sinks from, the oceans, and that the simple model in Figure 9 is a reasonable first-order description of the Mo isotope system in the oceans. Similar conclusions were reached by both Barling et al. (2001) and Siebert et al. (2003).

Conversely (Siebert et al. 2003; Arnold et al. 2004), mean $\delta^{97/95}Mo_{input}$, $\delta^{97/95}Mo_{ox}$ and $\delta^{97/95}Mo_{eux}$ data can be used to derive $f_{ox} \sim 0.75$ and $f_{eux} \sim 0.25$ (Fig. 9). These may be the best constraints to date on the relative magnitudes of these sedimentary sinks.

Nevertheless, it is important to stress that uncertainties persist in our understanding of the Mo ocean budget, particularly as regards Mo removal to suboxic sediments. The Mo isotope system may ultimately prove useful in reducing such uncertainties.

POTENTIAL APPLICATIONS

Although understanding of Mo stable isotope systematics is still evolving, some paleoceanographic applications are already coming into focus. It is not too early to explore these applications, nor to speculate on other possibilities.

Mo stable isotopes and global ocean paleoredox

The Mo isotope system is being developed in particular for investigations of global ocean paleoredox because of the strong redox sensitivities of the Mo elemental and isotopic budgets (Barling et al. 2001; Siebert et al. 2003; Arnold et al. 2004). Large variations in ocean oxygenation may have occurred repeatedly during Earth history. During the Pleistocene, the coupling of pCO_2, the ocean "carbon pump" and oxygen concentrations could have caused variations in ocean redox and the rate of organic carbon burial on glacial-interglacial timescales (e.g., Sarmiento and Toggweiler 1984). Globally-extensive deposits of carbonaceous shales are indicative of so-called "Ocean Anoxic Events" in the Cretaceous (e.g., Bralower et al. 1994), and similar deposits occur repeatedly in the Paleozoic (e.g., Wignall and Twitchett 1996). In the mid-Proterozoic, sulfur isotopes and other data are taken to indicate widespread anoxia or euxinia under the low pO_2 atmosphere of the period (Canfield 1998). This condition may have persisted for hundreds of millions of years, with evolutionary consequences (Anbar and Knoll 2002). The Archean oceans are generally believed to have been essentially anoxic, albeit not euxinic (Holland 1984).

There are many established or exploratory sedimentary paleoredox proxies, such as sedimentary iron chemistry (Canfield et al. 1996; Raiswell and Canfield 1998), S isotopes (e.g., Lyons 1997) and various elemental ratios (e.g., Algeo and Maynard 2003; Calvert et al. 2001; Crusius et al. 1996; Dean et al. 1997; Dean et al. 1999; Lyons et al. 2003; Sageman et al. 2003; Werne et al. 2002; Yarincik et al. 2000; Zheng et al. 2000b). However, these proxies provide information only about oxygenation of the *local* overlying water column. As outlined below, the Mo stable isotope system could provide a means of assessing the extent of ocean anoxia on *regional* or *global* scale (Barling et al. 2001; Siebert et al. 2003; Arnold et al. 2004). Hence, this non-traditional isotope system is likely to attract wide interest.

Conceptual basis. The foundation of this type of application can be illustrated straightforwardly by modification of the mass balance equation above. If it is assumed that

Mo isotope fractionation in the oceans results from an equilibrium isotope effect between dissolved and Mn oxide-associated Mo (Barling et al. 2001; Siebert et al. 2003; Arnold et al. 2004; Barling and Anbar 2004), then $\delta^{97/95}Mo_{sw} - \delta^{97/95}Mo_{ox} \sim 1000 \times \ln(\alpha_{sw\text{-}MnO_X})$. Additionally, it seems reasonable that $\delta^{97/95}Mo_{eux}$ approximates $\delta^{97/95}Mo_{sw}$, at least in settings similar to the Black Sea and Cariaco Basin. Using these relationships and assuming $f_{ox} + f_{eux} = 1$ (i.e., neglecting suboxic sediments), we obtain:

$$\delta^{97/95}Mo_{eux} = \delta^{97/95}Mo_{input} + 1000\ln(\alpha_{sw\text{-}MnO_X}) \times f_{ox}$$

An analogous expression can be derived for $\delta^{97/95}Mo_{ox}$ in terms of $\alpha_{sw\text{-}MnO_X}$, $\delta^{97/95}Mo_{input}$ and f_{ox} (or f_{eux}). Such a model illustrates that changes in ocean redox, which affect the relative proportions of f_{eux} and f_{ox}, should result in changes in the isotopic composition of hydrogenous Mo in both euxinic and Mn-oxide sediments. Hence, measurements of $\delta^{97/95}Mo_{eux}$ or $\delta^{97/95}Mo_{ox}$ could yield information about changes in ocean redox through time.

It must be recognized that this model is highly idealized, particularly in the neglect of suboxic sedimentary environments and in the underlying assumption that $\delta^{97/95}Mo_{eux} \sim \delta^{97/95}Mo_{sw}$, and hence should not be applied with too much quantitative rigor. Nevertheless, based on present knowledge it seems inescapable that an expansion of euxinic environments should lead to a decrease in the extent of Mo isotope fractionation in the oceans, and vice-versa and that this should be reflected in sedimentary Mo. This conclusion follows from the observation that Mo isotope fractionation is much smaller during removal to euxinic sediments as compared to either Mn oxides or suboxic sediments.

The Mo isotope record should be an especially effective global redox proxy because of the long Mo ocean residence time relative to the ~1000 year timescale for ocean mixing. The residence time would decrease along with Mo concentrations during periods of greatly decreased oxygenation, but $\delta^{97/95}Mo_{sw}$ could be reasonably well-mixed unless ocean mixing times were substantially slower than today, or Mo concentrations decreased by more than a factor of ~50. For this reason, the Mo isotope system has the potential to provide paleoredox information on a global-scale.

A disadvantage of the long Mo ocean residence time is that it hinders the application of this system to ocean redox variations on timescales shorter than $\sim 10^6$ years—timescales of particular paleoclimatological interest. Substantial improvements in analytical precision will be required to detect the small $\delta^{97/95}Mo$ variations that may exist on such scales.

Initial efforts to apply the Mo isotope system have targeted sediments deposited under oxic conditions (marine ferromanganese crusts) as well as sediments deposited under reducing conditions (black shales). Results of these studies are summarized below.

$\delta^{97/95}Mo_{ox}$ in Cenozoic ferromanganese crusts. Siebert et al. (2003) applied the ocean Mo isotope mass balance concept to the interpretation of $\delta^{97/95}Mo$ in 60 million year records from Atlantic and Pacific ferromanganese crusts. The isotopic uniformity of this record, summarized above and in Figure 6, is taken to reflect invariance of $\delta^{97/95}Mo_{sw}$ at the sampled resolution of 1–3 million years. This interpretation assumes that $\alpha_{sw\text{-}MnO_X}$ is relatively insensitive to temperature, pH or the details of Mn-oxide mineralogy, which seems reasonable based on available data.

When analytical uncertainties are propagated through a mass balance model, $f_{ox} : f_{eux}$ is inferred to have varied less than 10% from modern values (~0.75 : 0.25, based on the mean $\delta^{97/95}Mo$ of each sediment type) (Siebert et al. 2003). This conclusion is not unreasonable. Although larger ocean redox perturbations are possible during this time in association with glacial pCO_2 drawdown, their duration does not approach the $\sim 10^6$ year timescale of crust sampling and the Mo residence.

$\delta^{97/95}Mo_{eux}$ in mid-Proterozoic black shales. Arnold et al. (2004) investigated $\delta^{97/95}Mo_{eux}$ in black shales from ~1.6 and ~1.3 billion years ago in the McArthur and Tawallah Basins of Northern Australia to assess the anoxic ocean hypothesis of Canfield (1998). It is critical to note that the application of $\delta^{97/95}Mo$ in black shales for this purpose requires independent knowledge of local water column redox conditions because such sediments are not produced exclusively under euxinic water column conditions (Sageman et al. 2003). In the case of the McArthur and Tawallah basin samples, recent examination of sedimentary iron chemistry and sulfur isotopes indicates a water column that was at least locally euxinic (Shen et al. 2002, 2003).

As seen in Figure 6, mean $\delta^{97/95}Mo$ in these ancient sediments is significantly offset from that of modern seawater and from mean $\delta^{97/95}Mo$ of recent euxinic sediments. The offset is in the direction of less fractionation of Mo isotopes in the oceans (closer to the likely $\delta^{97/95}Mo_{input}$), as would be expected if removal of Mo to euxinic sediments expanded at the expense of removal to oxic (and suboxic) sediments. The result is consistent with the hypothesis of expanded ocean anoxia, and consequent increase in the areal extent of euxinic depositional settings, during this time. A similar, albeit smaller, offset is seen in repeated measurements of a Devonian black shale sample (USGS SDO-1; Fig. 6). This shift could be interpreted as consistent with somewhat expanded euxinic deposition during this time—an interpretation consistent with other evidence of expanded Paleozoic ocean anoxia in inland basins. While further work is needed, these initial results are promising.

Other potential applications

Although current efforts are focused on global ocean paleoredox, the Mo stable isotope system should also prove useful in other oceanographic and environmental applications and in geobiological research. Such possibilities, while as yet unexplored, are previewed briefly below.

Local water column paleoredox. The offset between recent euxinic and suboxic sediments (Fig. 6) suggests that such data could provide an additional tool to differentiate between euxinic and suboxic conditions in ancient sediments. Such application requires independent constraint on $\delta^{97/95}Mo_{sw}$, as might be obtained from analyses of coeval Mn oxide sediments.

Paleoweathering intensity. Large variations in weathering rates may have occurred at various times in Earth history in association with changes in climate or terrestrial vegetation (e.g., Walker et al. 1981; Raymo 1991; Algeo and Scheckler 1998; Kump et al. 2000; Ravizza et al. 2001). Radiogenic isotope systems have been widely used to investigate such changes, especially $^{87}Sr/^{86}Sr$ and $^{187}Os/^{188}Os$, because of contrasts between the isotopic compositions in seawater and crustal (weathering) sources. The same logic might apply to $^{97}Mo/^{95}Mo$. Changes in the flux of Mo to the oceans, or in the isotopic composition of riverine Mo due to changing weathering sources, should impact the ocean Mo isotope budget. In the case of flux changes the perturbation to $\delta^{97/95}Mo_{sw}$ will be transient (Siebert et al. 2003), recovering on the ~10^6 year timescale of the Mo ocean residence time. Changes in riverine $\delta^{97/95}Mo$ could result in longer lived shifts in $\delta^{97/95}Mo_{sw}$.

Mechanisms of Mo environmental chemistry. Many of the chemical reactions that shape the distribution of Mo in the environment are still not well understood, as seen above in discussions of Mo adsorption to Mn oxides and removal in reducing systems. It is likely that different reaction pathways will impart different isotope effects. Hence, coupling of laboratory and well-constrained field studies should provide new insights into Mo environmental chemistry. Theoretical modeling of Mo isotope effects would also be useful in this context, as is proving true for Fe isotopes (Schauble et al. 2001).

Mo biological uptake and processing. As an essential trace nutrient, Mo uptake and incorporation into enzymes is tightly controlled at the cellular level via selective "pumps" and "chaperone" proteins. There are considerable uncertainties about the mechanisms involved, and their response to changes in Mo availability and other environmental variables. Mo isotope fractionation is likely during key steps in this biochemistry, which involve potentially significant changes in Mo coordination. For example, while Mo is present in solution in tetrahedral coordination as MoO_4^{2-}, it is bound to high-affinity uptake proteins in octahedral coordination. Such coordination differences and associated bond cleavage can drive isotope fractionation, as has been suggested experimentally and theoretically for Fe (Matthews et al. 2001; Schauble et al. 2001; Roe et al. 2003). Hence, the Mo isotope system could provide a useful new tool to examine how organisms extract this element from the environment and manipulate it internally (Williams and Frausto da Silva 2002).

CONCLUSIONS

This chapter summarizes the results of the first wave of examinations of Mo stable isotopes in nature. These efforts reveal large, systematic variations in Mo isotope compositions in the ocean system that are apparently associated with the redox-sensitive marine geochemistry of this element. Laboratory studies of fractionation mechanisms, while limited in number, converge with predictions from natural observations. Combined with the relative simplicity of the Mo ocean budget and the likely preservation of oceanographically-interpretable Mo isotope signatures in both black shales and ferromanganese oxides, applications of this isotope system for paleoredox research are already apparent. Further work is needed to validate such applications, and to develop a framework to quantitatively relate $\delta^{97/95}$Mo variations in sedimentary records to changes in $\delta^{97/95}Mo_{sw}$ and hence in the relative importance of oxic and anoxic sedimentary reservoirs through time. Better constraints on the modern Mo element budget, which would permit more robust "calibration" of the Mo isotope budget, would be extremely helpful in this regard. However, even in advance of such refinements, the available data make clear that the Mo isotope system is poised to be among the more immediately useful of the emerging "non-traditional" stable isotope systems.

ACKNOWLEDGMENTS

This manuscript benefited substantially from thoughtful, constructive reviews by Francis Albarède, Clark Johnson, Thomas Nägler and an anonymous reviewer. The author is grateful to Steve Emerson and Erin Breckel for sharing their unpublished calculations on Mo isotope fractionation in reducing sedimentary systems, and to Gail Arnold, Jane Barling and Tim Lyons for many valuable discussions. George Helz is thanked for providing valuable insight into Mo speciation in aqueous solutions. Preparation of this manuscript was supported by NSF (EAR 0230183) and NASA (EXB02-0000-0073 and the NASA Astrobiology Institute).

REFERENCES

Albarède F, Beard BL (2004) Analytical methods for non-traditional isotopes. Rev Mineral Geochem 55: 113-151

Algeo TJ, Maynard JB (2003) Trace element behavior and redox facies analysis of core shales of upper Pennsylvanian Kansas-type cyclothems. Geochim Cosmochim Acta 67(18):A12-A12 Suppl. 1

Algeo TJ, Scheckler SE (1998) Terrestrial-marine teleconnections in the Devonian: links between the evolution of land plants, weathering processes, and marine anoxia. Phil Trans Royal Soc B 353:113-128

Anbar AD, Creaser RA, Papanastassiou DA, Wasserburg GJ (1992) Rhenium in seawater: confirmation of generally conservative behavior. Geochim Cosmochim Acta 56:4099-4103

Anbar AD, Knab KA, Barling J (2001) Precise determination of mass-dependent variations in the isotopic composition of molybdenum using MC-ICPMS. Anal Chem 73:1425-1431

Anbar AD, Knoll AH (2002) Proterozoic ocean chemistry and evolution: A bioinorganic bridge? Science 297: 1137-1142

Anbar AD, Roe JE, Barling J, Nealson KH (2000) Nonbiological fractionation of iron isotopes. Science 288: 126-128

Arnold GL, Anbar AD, Barling J and Lyons TW (2004) Molybdenum isotope evidence for widespread anoxia in mid-Proterozoic oceans. Science, in press

Aveston J, Anacker EW, Johnson JS (1964) Hydrolysis of molybdenum (VI). Ultracentrifugation, acidity measurements, and Raman spectra of polymolybdates. Inorg Chem 3:735-746

Baes Jr. CF, Mesmer RE (1986) The Hydrolysis of Cations. Krieger Publishing Company, Melbourne, FL

Barling J, Anbar AD (2004) Molybdenum isotope fractionation during adsorption by manganese oxides. Earth Planet Sci Lett, in press

Barling J, Arnold GL, Anbar AD (2001) Natural mass-dependent variations in the isotopic composition of molybdenum. Earth Planet Sci Lett 193:447-457

Beard BL, Johnson CM, Cox L, Sun H, Nealson KH, Aguilar C (1999) Iron isotope biosignatures. Science 285:1889-1892

Beard BL, Johnson CM, Skulan JL, Nealson KH, Cox L, Sun H (2003) Application of Fe isotopes to tracing the geochemical and biological cycling of Fe. Chem Geol 195:87-117

Bender ML (1990) The delta-O-18 of dissolved O_2 in seawater - a unique tracer of circulation and respiration in the deep-sea. J Geophys Res 95:22243-22252

Bertine KK, Keene JB (1975) Laurentide ice sheet meltwater recorded in Gulf of Mexico deep-sea cores. Science 188:147-152

Bertine KK, Turekian KK (1973) Molybdenum in marine deposits. Geochim Cosmochim Acta 37:1415-1434

Bralower TJ, Arthur MA, Leckie RM, Sliter WV, Allard DJ, Schlanger SO (1994) Timing and paleoceanography of oceanic dysoxia and anoxia in the late Barremian to early Aptian (early Cretaceous). Palois 9:335-369

Brandes JA, Devol AH (1997) Isotopic fractionation of oxygen and nitrogen in coastal marine sediments. Geochim Cosmochim Acta 61:1793-1801

Bruland KW (1983) Trace elements in seawater. *In:* Chemical Oceanography. Volume 8. Riley JP, Chester R (eds) London, Academic Press. p 157-220

Brumsack HJ, Gieskes JM (1983) Interstitial water trace metal chemistry of laminated sediments from the Gulf of California, Mexico. Mar Chem 14:89-106

Bullen TD, White AF, Childs CW, Vivit DV, Schulz MS (2001) Demonstration of significant abiotic iron isotope fractionation in nature. Geology 29:699-702

Busey RH, Keller OL (1964) Structure of the aqueous pertechnetate ion by Raman and infrared spectroscopy. Raman and infrared spectra of crystalline $KTcO_4$, $KReO_4$, Na_2MoO_4, Na_2WO_4, $Na_2MoO_4{\bullet}2H_2O$, and $Na_2WO_4{\bullet}2H_2O$. J Chem Phys 41:215-225

Calvert SE, Pedersen TF, Karlin RE (2001) Geochemical and isotopic evidence for post-glacial paleoceanographic changes in Saanich Inlet, British Columbia. Mar Geol 174:287-305

Calvert SE, Piper DZ (1984) Geochemistry of ferromanganese nodules from DOMES site A, Northern Equatorial Pacific: Multiple diagenetic metal sources in the deep sea. Geochim Cosmochim Acta 48: 1913-1928

Calvert SE, Price NB (1977) Geochemical variation in ferromanganese nodules and associated sediments from the Pacific Ocean. Mar Geochem 5:43-47

Canfield DE (1998) A new model for Proterozoic ocean chemistry. Nature 396:450-453

Canfield DE, Lyons TW, Raiswell R (1996) A model for iron deposition to euxinic Black Sea sediments. Am J Sci 296:818-834

Chan KM, Riley JP (1966) The determination of molybdenum in natural waters, silicates and biological materials. Anal Chim Acta 36:220-229

Collier RW (1985) Molybdenum in the northeast Pacific-ocean. Limnol Oceanogr 30:1351-1354

Colodner DC, Sachs J, Ravizza G, Turekian K, Edmond J, Boyle E (1993) The geochemical cycle of rhenium: a reconnaissance. Earth Planet Sci Lett 117:205-221

Crick FHC, Orgel LE (1973) Directed panspermia. Icarus 19:341-346

Criss RE (1999) Principles of Stable Isotope Distribution. Oxford University Press, Inc., New York

Cronan DS (1980) Underwater Minerals. Academic Press, New York

Crouch AC, Tulpin TA (1964) Isotopic composition and the atomic weight of naturally occurring molybdenum; a possible reflexion of the creation process. Nature 202:1282-1284

Crusius J, Calvert S, Pedersen T, Sage D (1996) Rhenium and molybdenum enrichments in sediments as indicators of oxic, suboxic and sulfidic conditions of deposition. Earth Planet Sci Lett 145:65-78

Cruywagen JJ, Rohwer EFCH (1975) Coordination number of molybdenum(VI) in monomeric molybdic acid. Inorg Chem 14:3136-3137

Dauphas N, Marty B, Reisberg L (2002a) Molybdenum nucleosynthetic dichotomy revealed in primitive meteorites. Astrophys J 569:L139-L142
Dauphas N, Marty B, Reisberg L (2002b) Molybdenum evidence for inherited planetary scale isotope heterogeneity of the protosolar nebula. Astrophys J 565:640-644.
Dean WE, Gardner JV, Piper DZ (1997) Inorganic geochemical indicators of glacial-interglacial changes in productivity and anoxia on the California continental margin. Geochim Cosmochim Acta 61:4507-4518
Dean WE, Piper DZ, Peterson LC (1999) Molybdenum accumulation in Cariaco basin sediment over the past 24 ky: a record of water-column anoxia and climate. Geology 27:507-510
DePaolo DJ (2004) Calcium isotopic variations produced by biological, kinetic, radiogenic and nucleosynthetic processes. Rev Mineral Geochem 55:255-288
Dzombak DA, Morel FMM (1990) Surface complexation modeling: hydrous ferric oxide. John Wiley and Sons, New York
Eady RR (1996) Structure-function relationships of alternative nitrogenases. Chem Rev 96:3013-3030
Emerson SR, Huested SS (1991) Ocean anoxia and the concentrations of molybdenum and vanadium in seawater. Mar Chem 34:177-196
Erickson BE, Helz GR (2000) Molybdenum (VI) speciation in sulfidic waters: Stability and lability of thiomolybdates. Geochim Cosmochim Acta 64:1149-1158
Fairbanks RG (1989) A 17,000-year glacio-eustatic sea-level record - influence of glacial melting rates on the Yonger Dryas event and deep-ocean circulation. Nature 342:637-642
Francois R (1987) A study of sulphur enrichment in the humic fraction of marine sediments during early diagenesis. Geochim Cosmochim Acta 51:17-27
Frausto da Silva JJR, Williams RJP (2001) The Biological Chemistry of the Elements: The Inorganic Chemistry of Life. Clarendon Press, Oxford
Halliday AN, Lee D-C, Christensen JN, Walder AJ, Freedman PA, Jones CE, Hall CM, Yi W, Teagle D (1995) Recent developments in inductively coupled plasma magnetic sector multiple collector mass spectrometry. Int J Mass Spectr Ion Proc 146:21-33
Hay BJ (1988) Sediment accumulation in the central western Black Sea over the past 5100 years. Paleoceanography 3:491-508
Helz GR, Miller CV, Charnock JM, Mosselmans JFW, Pattrick RAD, Garner CD, Vaughn DJ (1996) Mechanism of molybdenum removal from the sea and its concentration in black shales: EXAFS evidence. Geochim Cosmochim Acta 60:3631-3642
Hille R (1996) The mononuclear molybdenum enzymes. Chem Rev 96:2757-2816
Holland HD (1984) The Chemical Evolution of the Atmosphere and Oceans. Princeton University Press, Princeton
Johnson CM, Skulan JL, Beard BL, Sun H, Nealson KH, Braterman PS (2002) Isotopic fractionation between Fe(III) and Fe(II) in aqueous solutions. Earth Planet Sci Lett 195:141-153
Johnson TM, Bullen TD (2004) Mass-dependent fractionation of selenium and chromim isotopes in low-temperature environments. Rev Mineral Geochem 55:289-317
Klinkhammer GP, Palmer MR (1991) Uranium in the oceans: Where it goes and why. Geochim Cosmochim Acta 55:1799-1806
Kroneck PMH, Abt DJ (2002) Molybdenum in nitrate reductase and nitrite oxidoreductase. *In:* Molybdenum and Tungsten- Their Roles in Biological Processes. Sigel A, Sigel H (eds) Marcel Dekker, Inc., New York, 369-403
Kump LR, Brantley SL, Arthur MA (2000) Chemical weathering, atmospheric CO_2 and climate. Ann Rev Earth Planet Sci 28:611-667
Lee DC, Halliday AN (1995) Precise determinations of the isotopic compositions and atomic weights of molybdenum, tellurium, tin and tungsten using ICP magnetic-sector multiple collector mass-spectrometry. Int J Mass Spectr Ion Proc 146:35-46
Lewis BL, Landing WM (1991) The biogeochemistry of manganese and iron in the Black Sea. Deep-Sea Res 38:S773-S803
Longerich HP, Fryer BJ, Strong DF (1987) Determination of lead isotope ratios by inductively coupled plasma-mass spectrometry (ICP-MS). Spectrochim Acta B 42:39-48
Lyons TW (1997) Sulfur isotopic trends and pathways of iron sulfide formation in upper Holocene sediments of the anoxic Black Sea. Geochim Cosmochim Acta 61:3367-3382
Lyons TW, Werne JP, Hollander DJ, Murray RW (2003) Contrasting sulfur geochemistry and Fe/Al and Mo/Al ratios across the last oxic-to-anoxic transition in the Cariaco Basin, Venezuela. Chem Geol 195:131-157
Manheim FT, Landergren S (1974) Molybdenum. *In:* Handbook of Geochemistry. Wedepohl (ed) Springer-Verlag, New York, p 42-H-1–42-H-5
Maréchal CN, Telouk P, Albarede F (1999) Precise analysis of copper and zinc isotopic composition by plasma-source mass spectrometry. Chem Geol 156:251-273

Marino R, Howarth RW, Chan F, Cole JJ, Likens GE (2003) Sulfate inhibition of molybdenum-dependent nitrogen fixation by planktonic cyanobacteria under seawater conditions: a non-reversible effect. Hydrobiol 500:277-293

Matthews A, Zhu X-K, O'Nions K (2001) Kinetic iron stable isotope fractionation between iron (−II) and (−III) complexes in solution. Earth Planet Sci Lett 192:81-92

McManus J, Nagler TF, Siebert C, Wheat CG, Hammond DE (2002) Oceanic molybdenum isotope fractionation: Diagenesis and hydrothermal ridge-flank alteration. Geochem Geophys Geosys 3: 2002GC000356

Miller RW, Eady RR (1988) Molybdenum and vanadium nitrogenases of Azotobacter chroococcum. Biochem J 256:429-432

Mills CF, Davis GK (1987) Molybdenum. *In*: Trace Elements in Human and Animal Nutrition, Volume 1. Mertz W (ed) Academic Press, New York, p 429-457

Mitchell PCH (1999) Molybdenum compounds. *In*: Ullmann's Encyclopedia of Industrial Chemistry. Volume A16. Arpe H-J (ed) John Wiley and Sons, New York, p 675-82

Moore LJ, Machlan LA, Shields WR, Garner EL (1974) Internal normalization techniques for high accurate isotope dilution analyses. Application to molybdenum and nickel in standard reference materials. Anal Chem 46:1082-1089

Morford JL, Emerson S (1999) The geochemistry of redox sensitive trace metals in sediments. Geochim Cosmochim Acta 63:1735-1750

Morris AW (1975) Dissolved molybdenum and vanadium in the northeast Atlantic ocean. Deep-Sea Res 22: 49-54

Murthy VR (1962) Isotopic anomalies of molybdenum in some iron meteorites. J Geophys Res 67:905-907

Murthy VR. 1963. Elemental and isotopic abundances of molybdenum in some meteorites. Geochim. Cosmochim. Acta 27:1171-1178

Nameroff TJ, Balistrieri LS, Murray JW (2002) Suboxic trace metal geochemistry in the eastern tropical North Pacific. Geochim Cosmochim Acta 66:1139-1158

Neretin L.N., Volkov II, Böttcher ME, Grinenko VA (2001) A sulfur budget for the Black Sea anoxic zone. Deep-Sea Res I 48:2569-2593

Lu Q, Masuda A (1994) The isotopic composition and atomic-weight or molybdenum. Int J Mass Spectr Ion Proc 130:65-72

Raiswell R, Canfield DE (1998) Sources of iron for pyrite formation in marine sediments. Am J Sci 298:219-245

Ravizza G, Norris RN, Blusztajn J, Aubry MP (2001) An osmium isotope excursion associated with the late Paleocene thermal maximum: Evidence of intensified chemical weathering. Paleoceanogr 16:155-163

Raymo ME (1991) Geochemical evidence supporting Chamberlain, T. C., theory of glaciation. Geology 19: 344-347

Roe JE, Anbar AD, Barling J (2003) Nonbiological fractionation of Fe isotopes: evidence of an equilibrium isotope effect. Chem Geol 195:69-85

Russell WA, Papanastassiou DA (1978) Calcium isotope fractionation in ion-exchange chromatography. Anal Chem 50:1151-1154

Sageman BB, Murphy AE, Werne JP, Straeten CAV, Hollander DJ, Lyons TW (2003) A tale of shales: the relative roles of production, decomposition, and dilution in the accumulation of organic-rich strata, Middle-Upper Devonian, Appalachian basin. Chem Geol 195:229-273

Sarmiento JL, Toggweiler JR (1984) A new model for the role of the oceans in determining atmospheric pCO_2. Nature 308:621-624

Schauble EA, Rossman GR, Taylor HP (2001) Theoretical estimates of equilibrium Fe-isotope fractionations from vibrational spectroscopy. Geochim Cosmochim Acta 65:2487-2597

Shen Y, Canfield DE, Knoll AH (2002) Middle Proterozoic ocean chemistry: Evidence from the McArthur Basin, northern Australia. Am J Sci 302:81-109

Shen Y, Knoll AH, Walter MR (2003) Evidence for low sulphate and anoxia in a mid-Proterozoic marine basin. Nature 423:632-635

Shimmield GB, Price NB (1986) The behavior of molybdenum and manganese during early sediment diagenesis—offshore Baja California, Mexico. Mar Chem 19:261-280

Siebert C, Nagler TF, Kramers JD (2001) Determination of molybdenum isotope fractionation by double-spike multicollector inductively coupled plasma mass spectrometry. Geochem Geophys Geosys 2: 2000GC000124

Siebert C, Nagler TF, von Blanckenburg F, Kramers JD (2003) Molybdenum isotope records as a potential new proxy for paleoceanography. Earth Planet Sci Lett 211:159-171

Stiefel EI (1997) Chemical keys to molybdenum enzymes. J Chem Soc Dalton Trans 3915-3923

Stumm W, Morgan JJ (1996) Aquatic Chemistry. New York: John Wiley and Sons

Taylor SR, McLennan SM (1985) The Continental Crust: Its Composition and Evolution. Blackwell, Boston

Tuit CB, Ravizza G (2003) The marine distribution of molybdenum. Geochim Cosmochim Acta 67:A4950

Turnlund JR, Keyes WR, Peiffer GL (1993) Isotope ratios of molybdenum determined by thermal ionization mass spectrometry for stable isotope studies of molybdenum metabolism in humans. Anal Chem 65: 1717-1722

Walker JCG, Hays PB, Kasting JF (1981) A negative feedback mechanism for the long-term stabilization of Earth's surface temperature. J Geophys Res 86:9776-9782

Wang Y, van Cappellen P (1996) A multicomponent reactive transport model of early diagenesis: Application to redox cycling in coastal marine sediment. Geochim Cosmochim Acta 60:2993-3014

Werne JP, Sageman BB, Lyons TW, Hollander DJ (2002) An integrated assessment of a "type euxinic" deposit: evidence for multiple controls on black shale deposition in the Middle Devonian Oatka Creek Formation. Am J Sci 302:110-143

Wetherill GW (1964) Isotopic composition and concentration of molybdenum in iron meteorites. J Geophys Res 69:4403-4408

Wheat CG, Mottl MJ (2000) Composition of pore and spring waters from Baby Bare: Global implications of geochemical fluxes from a ridge flank hydrothermal system. Geochim Cosmochim Acta 64:629-642

Wieser ME, DeLaeter JR (2003) A preliminary study of isotope fractionation in molybdenites. Int. J Mass Spectr Ion Proc 225:177-183

Wignall PB, Twitchett RJ (1996) Oceanic anoxia and the end Permian mass extinction. Science 272:1155-1158

Williams RJP, Frausto da Silva JJR (2002) The involvement of molybdenum in life. Biochem Biophys Res Comm 292:293-299

Yarincik KM, Murray RW, Lyons TW, Peterson LC, Haug GH (2000) Oxygenation history of bottom waters in the Cariaco Basin, Venezuela, over the past 578,000 years: Results from redox-sensitive metals (Mo, V, Mn, and Fe). Paleoceanogr 15:593-604

Yin QZ, Jacobsen SB, Yamashita K (2002) Diverse supernova sources of pre-solar material inferred from molybdenum isotopes in meteorites. Nature 415:881-883

Young ED, Galy A, Nagahara H (2001) Kinetic and equilibrium mass-dependent isotope fractionation laws in nature and their geochemical and cosmochemical significance. Geochim Cosmochim Acta 66:1095-1104

Zheng Y, Anderson RF, van Geen A, Kuwabara J (2000a) Authigenic molybdenum formation in marine sediments: A link to pore water sulfide in the Santa Barbara Basin. Geochim Cosmochim Acta 64:4165-4178

Zheng Y, van Geen A, Anderson RF, Gardner JV, Dean WE (2000b) Intensification of the northeast Pacific oxygen minimum zone during the Bolling-Allerod warm period. Paleoceanogr 15:528-536

Zhu XK, Guo Y, Williams RJP, O'Nions RK, Matthews A, Belshaw NS, Canters GW, de Waal EC, Weser U, Burgess BK, Salvato B (2002) Mass fractionation processes of transition metal isotopes. Earth Planet Sci Lett 200:47-62